Universitext

Universitext

Universitext is a series of textbooks that presents material from a wide variety of mathematical disciplines at master's level and beyond. The books, often well class-tested by their author, may have an informal, personal even experimental approach to their subject matter. Some of the most successful and established books in the series have evolved through several editions, always following the evolution of teaching curricula, to very polished texts.

Thus as research topics trickle down into graduate-level teaching, first textbooks written for new, cutting-edge courses may make their way into *Universitext*.

For further volumes:
www.springer.com/series/223

Vladimir A. Zorich

Mathematical Analysis II

Second Edition

 Springer

Vladimir A. Zorich
Department of Mathematics
Moscow State University
Moscow, Russia

Translators:
Roger Cooke (first English edition translated from the 4th Russian edition)
Burlington, Vermont, USA
and
Octavio Paniagua T. (Appendices A–E and new problems of the 6th Russian edition)
Berlin, Germany

Original Russian edition: Matematicheskij Analïz (Part II, 6th corrected edition, Moscow, 2012) MCCME (Moscow Center for Continuous Mathematical Education Publ.)

ISSN 0172-5939   ISSN 2191-6675 (electronic)
Universitext
ISBN 978-3-662-48991-8   ISBN 978-3-662-48993-2 (eBook)
DOI 10.1007/978-3-662-48993-2

Library of Congress Control Number: 2016931909

Mathematics Subject Classification (2010): 26-01, 26Axx, 26Bxx, 42-01

Springer Heidelberg New York Dordrecht London
© Springer-Verlag Berlin Heidelberg 2004, 2016

Printed on acid-free paper

Springer is part of Springer Science+Business Media (www.springer.com)

Prefaces

Preface to the Second English Edition

Science has not stood still in the years since the first English edition of this book was published. For example, Fermat's last theorem has been proved, the Poincaré conjecture is now a theorem, and the Higgs boson has been discovered. Other events in science, while not directly related to the contents of a textbook in classical mathematical analysis, have indirectly led the author to learn something new, to think over something familiar, or to extend his knowledge and understanding. All of this additional knowledge and understanding end up being useful even when one speaks about something apparently completely unrelated.[1]

In addition to the original Russian edition, the book has been published in English, German, and Chinese. Various attentive multilingual readers have detected many errors in the text. Luckily, these are local errors, mostly misprints. They have assuredly all been corrected in this new edition.

But the main difference between the second and first English editions is the addition of a series of appendices to each volume. There are six of them in the first and five of them in the second. So as not to disturb the original text, they are placed at the end of each volume. The subjects of the appendices are diverse. They are meant to be useful to students (in mathematics and physics) as well as to teachers, who may be motivated by different goals. Some of the appendices are surveys, both prospective and retrospective. The final survey contains the most important conceptual achievements of the whole course, which establish connections between analysis and other parts of mathematics as a whole.

---

[1] There is a story about Erdős, who, like Hadamard, lived a very long mathematical and human life. When he was quite old, a journalist who was interviewing him asked him about his age. Erdős replied, after deliberating a bit, "I remember that when I was very young, scientists established that the Earth was two billion years old. Now scientists assert that the Earth is four and a half billion years old. So, I am approximately two and a half billion years old."

I was happy to learn that this book has proven to be useful, to some extent, not only to mathematicians, but also to physicists, and even to engineers from technical schools that promote a deeper study of mathematics.

It is a real pleasure to see a new generation that thinks bigger, understands more deeply, and is able to do more than the generation on whose shoulders it grew.

Moscow, Russia   V. Zorich
2015

Preface to the First English Edition

An entire generation of mathematicians has grown up during the time between the appearance of the first edition of this textbook and the publication of the fourth edition, a translation of which is before you. The book is familiar to many people, who either attended the lectures on which it is based or studied out of it, and who now teach others in universities all over the world. I am glad that it has become accessible to English-speaking readers.

This textbook consists of two parts. It is aimed primarily at university students and teachers specializing in mathematics and natural sciences, and at all those who wish to see both the rigorous mathematical theory and examples of its effective use in the solution of real problems of natural science.

The textbook exposes classical analysis as it is today, as an integral part of Mathematics in its interrelations with other modern mathematical courses such as algebra, differential geometry, differential equations, complex and functional analysis.

The two chapters with which this second book begins, summarize and explain in a general form essentially all most important results of the first volume concerning continuous and differentiable functions, as well as differential calculus. The presence of these two chapters makes the second book formally independent of the first one. This assumes, however, that the reader is sufficiently well prepared to get by without introductory considerations of the first part, which preceded the resulting formalism discussed here. This second book, containing both the differential calculus in its generalized form and integral calculus of functions of several variables, developed up to the general formula of Newton–Leibniz–Stokes, thus acquires a certain unity and becomes more self-contained.

More complete information on the textbook and some recommendations for its use in teaching can be found in the translations of the prefaces to the first and second Russian editions.

Moscow, Russia   V. Zorich
2003

Preface to the Sixth Russian Edition

On my own behalf and on behalf of future readers, I thank all those, living in different countries, who had the possibility to inform the publisher or me personally about errors (typos, errors, omissions), found in Russian, English, German and Chinese editions of this textbook.

As it turned out, the book has been also very useful to physicists; I am very happy about that. In any case, I really seek to accompany the formal theory with meaningful examples of its application both in mathematics and outside of it.

The sixth edition contains a series of appendices that may be useful to students and lecturers. Firstly, some of the material is actually real lectures (for example, the transcription of two introductory survey lectures for students of first and third semesters), and, secondly, this is some mathematical information (sometimes of current interest, such as the relation between multidimensional geometry and the theory of probability), lying close to the main subject of the textbook.

Moscow, Russia   V. Zorich
2011

Prefaces to the Fifth, Fourth, Third and Second Russian Editions

In the fifth edition all misprints of the fourth edition have been corrected.

Moscow, Russia   V. Zorich
2006

In the fourth edition all misprints that the author is aware of have been corrected.

Moscow, Russia   V. Zorich
2002

The third edition differs from the second only in local corrections (although in one case it also involves the correction of a proof) and in the addition of some problems that seem to me to be useful.

Moscow, Russia   V. Zorich
2001

In addition to the correction of all the misprints in the first edition of which the author is aware, the differences between the second edition and the first edition of this book are mainly the following. Certain sections on individual topics – for example, Fourier series and the Fourier transform – have been recast (for the better, I hope). We have included several new examples of applications and new substantive problems relating to various parts of the theory and sometimes significantly extending it. Test questions are given, as well as questions and problems from the midterm examinations. The list of further readings has been expanded.

Further information on the material and some characteristics of this second part of the course are given below in the preface to the first edition.

Moscow, Russia   V. Zorich
1998

Preface to the First Russian Edition

The preface to the first part contained a rather detailed characterization of the course as a whole, and hence I confine myself here to some remarks on the content of the second part only.

The basic material of the present volume consists on the one hand of multiple integrals and line and surface integrals, leading to the generalized Stokes' formula and some examples of its application, and on the other hand the machinery of series and integrals depending on a parameter, including Fourier series, the Fourier transform, and the presentation of asymptotic expansions.

Thus, this Part 2 basically conforms to the curriculum of the second year of study in the mathematics departments of universities.

So as not to impose rigid restrictions on the order of presentation of these two major topics during the two semesters, I have discussed them practically independently of each other.

Chapters 9 and 10, with which this book begins, reproduce in compressed and generalized form, essentially all of the most important results that were obtained in the first part concerning continuous and differentiable functions. These chapters are starred and written as an appendix to Part 1. This appendix contains, however, many concepts that play a role in any exposition of analysis to mathematicians. The presence of these two chapters makes the second book formally independent of the first, provided the reader is sufficiently well prepared to get by without the numerous examples and introductory considerations that, in the first part, preceded the formalism discussed here.

The main new material in the book, which is devoted to the integral calculus of several variables, begins in Chap. 11. One who has completed the first part may begin the second part of the course at this point without any loss of continuity in the ideas.

The language of differential forms is explained and used in the discussion of the theory of line and surface integrals. All the basic geometric concepts and analytic constructions that later form a scale of abstract definitions leading to the generalized Stokes' formula are first introduced by using elementary material.

Chapter 15 is devoted to a similar summary exposition of the integration of differential forms on manifolds. I regard this chapter as a very desirable and systematizing supplement to what was expounded and explained using specific objects in the mandatory Chaps. 11–14.

The section on series and integrals depending on a parameter gives, along with the traditional material, some elementary information on asymptotic series and asymptotics of integrals (Chap. 19), since, due to its effectiveness, the latter is an unquestionably useful piece of analytic machinery.

For convenience in orientation, ancillary material or sections that may be omitted on a first reading, are starred.

The numbering of the chapters and figures in this book continues the numbering of the first part.

Biographical information is given here only for those scholars not mentioned in the first part.

As before, for the convenience of the reader, and to shorten the text, the end of a proof is denoted by $\square$. Where convenient, definitions are introduced by the special symbols $:=$ or $=:$ (equality by definition), in which the colon stands on the side of the object being defined.

Continuing the tradition of Part 1, a great deal of attention has been paid to both the lucidity and logical clarity of the mathematical constructions themselves and the demonstration of substantive applications in natural science for the theory developed.

Moscow, Russia   V. Zorich
1982

Contents

Chapter 9
*Continuous Mappings (General Theory)

In this chapter we shall generalize the properties of continuous mappings established earlier for numerical-valued functions and mappings of the type $f : \mathbb{R}^m \to \mathbb{R}^n$ and discuss them from a unified point of view. In the process we shall introduce a number of simple, yet important concepts that are used everywhere in mathematics.

9.1 Metric Spaces

9.1.1 Definition and Examples

Definition 1 A set X is said to be endowed with a *metric* or *a metric space structure* or to be a *metric space* if a function

$$d : X \times X \to \mathbb{R} \tag{9.1}$$

is exhibited satisfying the following conditions:

 a) $d(x_1, x_2) = 0 \Leftrightarrow x_1 = x_2$,
 b) $d(x_1, x_2) = d(x_2, x_1)$ (symmetry),
 c) $d(x_1, x_3) \leq d(x_1, x_2) + d(x_2, x_3)$ (the triangle inequality),

where x_1, x_2, x_3 are arbitrary elements of X.

In that case, the function (9.1) is called a *metric* or *distance* on X.

Thus a *metric space* is a pair (X, d) consisting of a set X and a metric defined on it.

In accordance with geometric terminology the elements of X are called *points*.

We remark that if we set $x_3 = x_1$ in the triangle inequality and take account of conditions a) and b) in the definition of a metric, we find that

$$0 \leq d(x_1, x_2),$$

that is, a distance satisfying axioms a), b), and c) is nonnegative.

© Springer-Verlag Berlin Heidelberg 2016
V.A. Zorich, *Mathematical Analysis II*, Universitext,
DOI 10.1007/978-3-662-48993-2_1

Let us now consider some examples.

Example 1 The set $\mathbb{R}$ of real numbers becomes a metric space if we set $d(x_1, x_2) = |x_2 - x_1|$ for any two numbers x_1 and x_2, as we have always done.

Example 2 Other metrics can also be introduced on $\mathbb{R}$. A trivial metric, for example, is the discrete metric in which the distance between any two distinct points is 1.

The following metric on $\mathbb{R}$ is much more substantive. Let $x \mapsto f(x)$ be a nonnegative function defined for $x \geq 0$ and vanishing for $x = 0$. If this function is strictly convex upward, then, setting

$$d(x_1, x_2) = f\big(|x_1 - x_2|\big) \tag{9.2}$$

for points $x_1, x_2 \in \mathbb{R}$, we obtain a metric on $\mathbb{R}$.

Axioms a) and b) obviously hold here, and the triangle inequality follows from the easily verified fact that f is strictly monotonic and satisfies the following inequalities for $0 < a < b$:

$$f(a + b) - f(b) < f(a) - f(0) = f(a).$$

In particular, one could set $d(x_1, x_2) = \sqrt{|x_1 - x_2|}$ or $d(x_1, x_2) = \frac{|x_1 - x_2|}{1 + |x_1 - x_2|}$. In the latter case the distance between any two points of the line is less than 1.

Example 3 Besides the traditional distance

$$d(x_1, x_2) = \sqrt{\sum_{i=1}^{n} |x_1^i - x_2^i|^2} \tag{9.3}$$

between points $x_1 = (x_1^1, \ldots, x_1^n)$ and $x_2 = (x_2^1, \ldots, x_2^n)$ in $\mathbb{R}^n$, one can also introduce the distance

$$d_p(x_1, x_2) = \left(\sum_{i=1}^{n} |x_1^i - x_2^i|^p\right)^{1/p}, \tag{9.4}$$

where $p \geq 1$. The validity of the triangle inequality for the function (9.4) follows from Minkowski's inequality (see Sect. 5.4.2).

Example 4 When we encounter a word with incorrect letters while reading a text, we can reconstruct the word without too much trouble by correcting the errors, provided the number of errors is not too large. However, correcting the error and obtaining the word is an operation that is sometimes ambiguous. For that reason, other conditions being equal, one must give preference to the interpretation of the incorrect text that requires the fewest corrections. Accordingly, in coding theory the metric (9.4) with $p = 1$ is used on the set of all finite sequences of length n consisting of zeros and ones.

Geometrically the set of such sequences can be interpreted as the set of vertices of the unit cube $I = \{x \in \mathbb{R}^n \mid 0 \le x^i \le 1, \ i = 1, \ldots, n\}$ in $\mathbb{R}^n$. The distance between two vertices is the number of interchanges of zeros and ones needed to obtain the coordinates of one vertex from the other. Each such interchange represents a passage along one edge of the cube. Thus this distance is the shortest path along the edges of the cube from one of the vertices under consideration to the other.

Example 5 In comparing the results of two series of n measurements of the same quantity the metric most commonly used is (9.4) with $p = 2$. The distance between points in this metric is usually called their *mean-square deviation*.

Example 6 As one can easily see, if we pass to the limit in (9.4) as $p \to +\infty$, we obtain the following metric in $\mathbb{R}^n$:

$$d(x_1, x_2) = \max_{1 \le i \le n} \left| x_1^i - x_2^i \right|. \tag{9.5}$$

Example 7 The set $C[a, b]$ of functions that are continuous on a closed interval becomes a metric space if we define the distance between two functions f and g to be

$$d(f, g) = \max_{a \le x \le b} \left| f(x) - g(x) \right|. \tag{9.6}$$

Axioms a) and b) for a metric obviously hold, and the triangle inequality follows from the relations

$$\left| f(x) - h(x) \right| \le \left| f(x) - g(x) \right| + \left| g(x) - h(x) \right| \le d(f, g) + d(g, h),$$

that is,

$$d(f, h) = \max_{a \le x \le b} \left| f(x) - h(x) \right| \le d(f, g) + d(g, h).$$

The metric (9.6) – the so-called *uniform* or *Chebyshev* metric in $C[a, b]$ – is used when we wish to replace one function by another (for example, a polynomial) from which it is possible to compute the values of the first with a required degree of precision at any point $x \in [a, b]$. The quantity $d(f, g)$ is precisely a characterization of the precision of such an approximate computation.

The metric (9.6) closely resembles the metric (9.5) in $\mathbb{R}^n$.

Example 8 Like the metric (9.4), for $p \ge 1$ we can introduce in $C[a, b]$ the metric

$$d_p(f, g) = \left(\int_a^b |f - g|^p (x) \, dx \right)^{1/p}. \tag{9.7}$$

It follows from Minkowski's inequality for integrals, which can be obtained from Minkowski's inequality for the Riemann sums by passing to the limit, that this is indeed a metric for $p \geq 1$.

The following special cases of the metric (9.7) are especially important: $p = 1$, which is the integral metric; $p = 2$, the metric of mean-square deviation; and $p = +\infty$, the uniform metric.

The space $C[a, b]$ endowed with the metric (9.7) is often denoted $C_p[a, b]$. One can verify that $C_\infty[a, b]$ is the space $C[a, b]$ endowed with the metric (9.6).

Example 9 The metric (9.7) could also have been used on the set $\mathcal{R}[a, b]$ of Riemann-integrable functions on the closed interval $[a, b]$. However, since the integral of the absolute value of the difference of two functions may vanish even when the two functions are not identically equal, axiom a) will not hold in this case. Nevertheless, we know that the integral of a nonnegative function $\varphi \in \mathcal{R}[a, b]$ equals zero if and only if $\varphi(x) = 0$ at almost all points of the closed interval $[a, b]$.

Therefore, if we partition $\mathcal{R}[a, b]$ into equivalence classes of functions, regarding two functions in $\mathcal{R}[a, b]$ as equivalent if they differ on at most a set of measure zero, then the relation (9.7) really does define a metric on the set $\widetilde{\mathcal{R}}[a, b]$ of such equivalence classes. The set $\widetilde{\mathcal{R}}[a, b]$ endowed with this metric will be denoted $\widetilde{\mathcal{R}}_p[a, b]$ and sometimes simply by $\mathcal{R}_p[a, b]$.

Example 10 In the set $C^{(k)}[a, b]$ of functions defined on $[a, b]$ and having continuous derivatives up to order k inclusive one can define the following metric:

$$d(f, g) = \max\{M_0, \ldots, M_k\}, \tag{9.8}$$

where

$$M_i = \max_{a \leq x \leq b} \left| f^{(i)}(x) - g^{(i)}(x) \right|, \quad i = 0, 1, \ldots, k.$$

Using the fact that (9.6) is a metric, one can easily verify that (9.8) is also a metric.

Assume for example that f is the coordinate of a moving point considered as a function of time. If a restriction is placed on the allowable region where the point can be during the time interval $[a, b]$ and the particle is not allowed to exceed a certain speed, and, in addition, we wish to have some assurance that the accelerations cannot exceed a certain level, it is natural to consider the set $\{\max_{a \leq x \leq b} |f(x)|, \max_{a \leq x \leq b} |f'(x)|, \max_{a \leq x \leq b} |f''(x)|\}$ for a function $f \in C^{(2)}[a, b]$ and using these characteristics, to regard two motions f and g as close together if the quantity (9.8) for them is small.

These examples show that a given set can be metrized in various ways. The choice of the metric to be introduced is usually controlled by the statement of the problem. At present we shall be interested in the most general properties of metric spaces, the properties that are inherent in all of them.

9.1.2 Open and Closed Subsets of a Metric Space

Let (X, d) be a metric space. In the general case, as was done for the case $X = \mathbb{R}^n$ in Sect. 7.1, one can also introduce the concept of a ball with center at a given point, open set, closed set, neighborhood of a point, limit point of a set, and so forth.

Let us now recall these concepts, which are basic for what is to follow.

Definition 2 For $\delta > 0$ and $a \in X$ the set

$$B(a, \delta) = \{x \in X \mid d(a, x) < \delta\}$$

is called the *ball with center $a \in X$ of radius δ* or the *δ-neighborhood of the point a*.

This name is a convenient one in a general metric space, but it must not be identified with the traditional geometric image we are familiar with in $\mathbb{R}^3$.

Example 11 The unit ball in $C[a, b]$ with center at the function that is identically 0 on $[a, b]$ consists of the functions that are continuous on the closed interval $[a, b]$ and whose absolute values are less than 1 on that interval.

Example 12 Let X be the unit square in $\mathbb{R}^2$ for which the distance between two points is defined to be the distance between those same points in $\mathbb{R}^2$. Then X is a metric space, while the square X considered as a metric space in its own right can be regarded as the ball of any radius $\rho \geq \sqrt{2}/2$ about its center.

It is clear that in this way one could construct balls of very peculiar shape. Hence the term ball should not be understood too literally.

Definition 3 A set $G \subset X$ is *open in the metric space* (X, d) if for each point $x \in G$ there exists a ball $B(x, \delta)$ such that $B(x, \delta) \subset G$.

It obviously follows from this definition that X itself is an open set in (X, d). The empty set $\varnothing$ is also open. By the same reasoning as in the case of $\mathbb{R}^n$ one can prove that a ball $B(a, r)$ and its exterior $\{x \in X : d(a, x) > r\}$ are open sets. (See Examples 3 and 4 of Sect. 7.1.)

Definition 4 A set $F \subset X$ is *closed in* (X, d) if its complement $X \backslash F$ is open in (X, d).

In particular, we conclude from this definition that the *closed ball*

$$\widetilde{B}(a, r) := \{x \in X \mid d(a, x) \leq r\}$$

is a closed set in a metric space (X, d).

The following proposition holds for open and closed sets in a metric space (X, d).

Proposition 1 a) *The union* $\bigcup_{\alpha \in A} G_\alpha$ *of the sets in any system* $\{G_\alpha, \alpha \in A\}$ *of sets* G_α *that are open in* X *is an open set in* X.

b) *The intersection* $\bigcap_{i=1}^{n} G_i$ *of any finite number of sets that are open in* X *is an open set in* X.

a') *The intersection* $\bigcap_{\alpha \in A} F_\alpha$ *of the sets in any system* $\{F_\alpha, \alpha \in A\}$ *of sets* F_α *that are closed in* X *is a closed set in* X.

b') *The union* $\bigcup_{i=1}^{n} F_i$ *of any finite number of sets that are closed in* X *is a closed set in* X.

The proof of Proposition 1 is a verbatim repetition of the proof of the corresponding proposition for open and closed sets in $\mathbb{R}^n$, and we omit it. (See Proposition 1 in Sect. 7.1.)

Definition 5 An open set in X containing the point $x \in X$ is called a *neighborhood* of the point x in X.

Definition 6 Relative to a set $E \subset X$, a point $x \in X$ is called

an interior point of E if some neighborhood of it is contained in X,

an exterior point of E if some neighborhood of it is contained in the complement of E in X,

a boundary point of E if it is neither interior nor exterior to E (that is, every neighborhood of the point contains both a point belonging to E and a point not belonging to E).

Example 13 All points of a ball $B(a, r)$ are interior to it, and the set $C_X \widetilde{B}(a, r) = X \setminus \widetilde{B}(a, r)$ consists of the points exterior to the ball $B(a, r)$.

In the case of $\mathbb{R}^n$ with the standard metric d the *sphere* $S(a, r) := \{x \in \mathbb{R}^n \mid d(a, x) = r \geq 0\}$ is the set of boundary points of the ball $B(a, r)$.[1]

Definition 7 A point $a \in X$ is a *limit point* of the set $E \subset X$ if the set $E \cap O(a)$ is infinite for every neighborhood $O(a)$ of the point.

Definition 8 The union of the set E and the set of all its limit points is called the *closure* of the set E in X.

As before, the closure of a set $E \subset X$ will be denoted $\overline{E}$.

Proposition 2 *A set* $F \subset X$ *is closed in* X *if and only if it contains all its limit points.*

Thus

$$(F \text{ is closed in } X) \Longleftrightarrow (F = \overline{F} \text{ in } X).$$

---

[1]In connection with Example 13 see also Problem 2 at the end of this section.

We omit the proof, since it repeats the proof of the analogous proposition for the case $X = \mathbb{R}^n$ discussed in Sect. 7.1.

9.1.3 Subspaces of a Metric Space

If (X, d) is a metric space and E is a subset of X, then, setting the distance between two points x_1 and x_2 of E equal to $d(x_1, x_2)$, that is, the distance between them in X, we obtain the metric space (E, d), which is customarily called a *subspace* of the original space (X, d).

Thus we adopt the following definition.

Definition 9 A metric space (X_1, d_1) is a *subspace of the metric space* (X, d) if $X_1 \subset X$ and the equality $d_1(a, b) = d(a, b)$ holds for any pair of points a, b in X_1.

Since the ball $B_1(a, r) = \{x \in X_1 \mid d_1(a, x) < r\}$ in a subspace (X_1, d_1) of the metric space (X, d) is obviously the intersection

$$B_1(a, r) = X_1 \cap B(a, r)$$

of the set $X_1 \subset X$ with the ball $B(a, r)$ in X, it follows that every open set in X_1 has the form

$$G_1 = X_1 \cap G,$$

where G is an open set in X, and every closed set F_1 in X_1 has the form

$$F_1 = X_1 \cap F,$$

where F is a closed set in X.

It follows from what has just been said that the properties of a set in a metric space of being open or closed are relative properties and depend on the ambient space.

Example 14 The open interval $|x| < 1$, $y = 0$ of the x-axis in the plane $\mathbb{R}^2$ with the standard metric in $\mathbb{R}^2$ is a metric space (X_1, d_1), which, like any metric space, is closed as a subset of itself, since it contains all its limit points in X_1. At the same time, it is obviously not closed in $\mathbb{R}^2 = X$.

This same example shows that openness is also a relative concept.

Example 15 The set $C[a, b]$ of continuous functions on the closed interval $[a, b]$ with the metric (9.7) is a subspace of the metric space $\mathcal{R}_p[a, b]$. However, if we consider the metric (9.6) on $C[a, b]$ rather than (9.7), this is no longer true.

9.1.4 The Direct Product of Metric Spaces

If (X_1, d_1) and (X_2, d_2) are two metric spaces, one can introduce a metric d on the direct product $X_1 \times X_2$. The commonest methods of introducing a metric in $X_1 \times X_2$ are the following. If $(x_1, x_2) \in X_1 \times X_2$ and $(x_1', x_2') \in X_1 \times X_2$, one may set

$$d\big((x_1, x_2), (x_1', x_2')\big) = \sqrt{d_1^2(x_1, x_1') + d_2^2(x_2, x_2')},$$

or

$$d\big((x_1, x_2), (x_1', x_2')\big) = d_1(x_1, x_1') + d_2(x_2, x_2'),$$

or

$$d\big((x_1, x_2), (x_1', x_2')\big) = \max\{d_1(x_1, x_2'), d_2(x_2, x_2')\}.$$

It is easy to see that we obtain a metric on $X_1 \times X_2$ in all of these cases.

Definition 10 if (X_1, d_1) and (X_2, d_2) are two metric spaces, the space $(X_1 \times X_2, d)$, where d is a metric on $X_1 \times X_2$ introduced by any of the methods just indicated, will be called the *direct product* of the original metric spaces.

Example 16 The space $\mathbb{R}^2$ can be regarded as the direct product of two copies of the metric space $\mathbb{R}$ with its standard metric, and $\mathbb{R}^3$ is the direct product $\mathbb{R}^2 \times \mathbb{R}^1$ of the spaces $\mathbb{R}^2$ and $\mathbb{R}^1 = \mathbb{R}$.

9.1.5 Problems and Exercises

1. a) Extending Example 2, show that if $f : \mathbb{R}_+ \to \mathbb{R}_+$ is a continuous function that is strictly convex upward and satisfies $f(0) = 0$, while (X, d) is a metric space, then one can introduce a new metric d_f on X by setting $d_f(x_1, x_2) = f(d(x_1, x_2))$.

b) Show that on any metric space (X, d) one can introduce a metric $d'(x_1, x_2) = \frac{d(x_1, x_2)}{1 + d(x_1, x_2)}$ in which the distance between the points will be less than 1.

2. Let (X, d) be a metric space with the trivial (*discrete*) metric shown in Example 2, and let $a \in X$. For this case, what are the sets $B(a, 1/2)$, $B(a, 1)$, $\overline{B}(a, 1)$, $\widetilde{B}(a, 1)$, $B(a, 3/2)$, and what are the sets $\{x \in X \mid d(a, x) = 1/2\}$, $\{x \in X \mid d(a, x) = 1\}$, $\overline{B}(a, 1)\backslash B(a, 1)$, $\widetilde{B}(a, 1)\backslash B(a, 1)$?

3. a) Is it true that the union of any family of closed sets is a closed set?

b) Is every boundary point of a set a limit point of that set?

c) Is it true that in any neighborhood of a boundary point of a set there are points in both the interior and exterior of that set?

d) Show that the set of boundary points of any set is a closed set.

4. a) Prove that if (Y, d_Y) is a subspace of the metric space (X, d_X), then for any open (resp. closed) set G_Y (resp. F_Y) in Y there is an open (resp. closed) set G_X (resp. F_X) in X such that $G_Y = Y \cap G_X$, (resp. $F_Y = Y \cap F_X$).

b) Verify that if the open sets G'_Y and G''_Y in Y do not intersect, then the corresponding sets G'_X and G''_X in X can be chosen so that they also have no points in common.

5. Having a metric d on a set X, one may attempt to define the distance $\overline{d}(A, B)$ between sets $A \subset X$ and $B \subset X$ as follows:

$$\overline{d}(A, B) = \inf_{a \in A, b \in B} d(a, b).$$

a) Give an example of a metric space and two nonintersecting subsets of it A and B for which $\overline{d}(A, B) = 0$.

b) Show, following Hausdorff, that on the set of closed sets of a metric space (X, d) one can introduce the *Hausdorff metric* D by assuming that for $A \subset X$ and $B \subset X$

$$D(A, B) := \max\left\{ \sup_{a \in A} \overline{d}(a, B),\ \sup_{b \in B} \overline{d}(A, b) \right\}.$$

9.2 Topological Spaces

For questions connected with the concept of the limit of a function or a mapping, what is essential in many cases is not the presence of any particular metric on the space, but rather the possibility of saying what a neighborhood of a point is. To convince oneself of that it suffices to recall that the very definition of a limit or the definition of continuity can be stated in terms of neighborhoods. Topological spaces are the mathematical objects on which the operation of passage to the limit and the concept of continuity can be studied in maximum generality.

9.2.1 Basic Definitions

Definition 1 A set X is said to be endowed with the structure of a *topological space* or a *topology* or is said to be a *topological space* if a system τ of subsets of X is exhibited (called *open sets in X*) possessing the following properties:

a) $\varnothing \in \tau$; $X \in \tau$.
b) $(\forall \alpha \in A(\tau_\alpha \in \tau)) \Longrightarrow \bigcup_{\alpha \in A} \tau_\alpha \in \tau$.
c) $(\tau_i \in \tau; i = 1, \ldots, n) \Longrightarrow \bigcap_{i=1}^{n} \tau_i \in \tau$.

Thus, a topological space is a pair (X, τ) consisting of a set X and a system τ of distinguished subsets of the set having the properties that τ contains the empty set and the whole set X, the union of any number of sets of τ is a set of τ, and the intersection of any finite number of sets of τ is a set of τ.

As one can see, in the axiom system a), b), c) for a topological space we have postulated precisely the properties of open sets that we already proved in the case of a metric space. Thus any metric space with the definition of open sets given above is a topological space.

Thus *defining a topology* on X means exhibiting a system τ of subsets of X satisfying the axioms a), b), and c) for a topological space.

Defining a metric in X, as we have seen, automatically defines the topology on X induced by that metric. It should be remarked, however, that different metrics on X may generate the same topology on that set.

Example 1 Let $X = \mathbb{R}^n$ $(n > 1)$. Consider the metric $d_1(x_1, x_2)$ defined by relation (9.5) in Sect. 9.1, and the metric $d_2(x_1, x_2)$ defined by formula (9.3) in Sect. 9.1.

The inequalities

$$d_1(x_1, x_2) \leq d_2(x_1, x_2) \leq \sqrt{n} d_1(x_1, x_2),$$

obviously imply that every ball $B(a, r)$ with center at an arbitrary point $a \in X$, interpreted in the sense of one of these two metrics, contains a ball with the same center, interpreted in the sense of the other metric. Hence by definition of an open subset of a metric space, it follows that the two metrics induce the same topology on X.

Nearly all the topological spaces that we shall make active use of in this course are metric spaces. One should not think, however, that every topological space can be metrized, that is, endowed with a metric whose open sets will be the same as the open sets in the system τ that defines the topology on X. The conditions under which this can be done form the content of the so-called *metrization theorems*.

Definition 2 If (X, τ) is a topological space, the sets of the system τ are called the *open sets*, and their complements in X are called the *closed sets* of the topological space (X, τ).

A topology τ on a set X is seldom defined by enumerating all the sets in the system τ. More often the system τ is defined by exhibiting only a certain set of subsets of X from which one can obtain any set in the system τ through union and intersection. The following definition is therefore very important.

Definition 3 A *base of the topological space* (X, τ) (an *open base* or *base for the topology*) is a family $\mathfrak{B}$ of open subsets of X such that every open set $G \in \tau$ is the union of some collection of elements of the family $\mathfrak{B}$.

Example 2 If (X, d) is a metric space and (x, τ) the topological space corresponding to it, the set $\mathfrak{B} = \{B(a, r)\}$ of all balls, where $a \in X$ and $r > 0$, is obviously a base of the topology τ. Moreover, if we take the system $\mathfrak{B}$ of all balls with positive rational radii r, this system is also a base for the topology.

Thus a topology can be defined by describing only a base of that topology. As one can see from Example 2, a topological space may have many different bases for the topology.

Definition 4 The minimal cardinality among all bases of a topological space is called its *weight*.

As a rule, we shall be dealing with topological spaces whose topologies admit a countable base (see, however, Problems 4 and 6).

Example 3 If we take the system $\mathfrak{B}$ of balls in $\mathbb{R}^k$ of all possible rational radii $r = \frac{m}{n} > 0$ with centers at all possible rational points $(\frac{m_1}{n_1}, \ldots, \frac{m_k}{n_k}) \in \mathbb{R}^k$, we obviously obtain a countable base for the standard topology of $\mathbb{R}^k$. It is not difficult to verify that it is impossible to define the standard topology in $\mathbb{R}^k$ by exhibiting a finite system of open sets. Thus the standard topological space $\mathbb{R}^k$ has countable weight.

Definition 5 A *neighborhood* of a point of a topological space (X, τ) is an open set containing the point.

It is clear that if a topology τ is defined on X, then for each point the system of its neighborhoods is defined.

It is also clear that the system of all neighborhoods of all possible points of topological space can serve as a base for the topology of that space. Thus a topology can be introduced on X by describing the neighborhoods of the points of X. This is the way of defining the topology in X that was originally used in the definition of a topological space.[2] Notice, for example, that we have introduced the topology in a metric space itself essentially by saying what a δ-neighborhood of a point is. Let us give one more example.

Example 4 Consider the set $C(\mathbb{R}, \mathbb{R})$ of real-valued continuous functions defined on the entire real line. Using this set as foundation, we shall construct a new set – the set of germs of continuous functions. We shall regard two functions $f, g \in C(\mathbb{R}, \mathbb{R})$ as equivalent at the point $a \in \mathbb{R}$ if there is a neighborhood $U(a)$ of that point such that $\forall x \in U(a)$ $(f(x) = g(x))$. The relation just introduced really is an equivalence relation (it is reflexive, symmetric, and transitive). An equivalence class of continuous functions at the point $a \in \mathbb{R}$ is called *germ of continuous function* at that point. If f is one of the functions generating the germ at the point a, we shall denote the germ itself by the symbol f_a. Now let us define a neighborhood of a germ. Let $U(a)$ be a neighborhood of the point a and f a function defined on $U(a)$ generating the germ f_a at a. This same function generates its germ f_x at any point $x \in U(a)$. The set $\{f_x\}$ of all germs corresponding to the points $x \in U(a)$ will be called a *neighborhood of the germ f_a*. Taking the set of such neighborhoods of all germs as the base of a topology, we turn the set of germs of continuous functions into a topological space. It is worthwhile to note that in the resulting topological

---

[2]The concepts of a metric space and a topological space were explicitly stated early in the twentieth century. In 1906 the French mathematician M. Fréchet (1878–1973) introduced the concept of a metric space, and in 1914 the German mathematician F. Hausdorff (1868–1942) defined a topological space.

Fig. 9.1

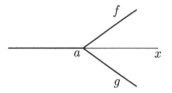

space two different points (germs) f_a and g_a may not have disjoint neighborhoods (see Fig. 9.1).

Definition 6 A topological space is *Hausdorff* if the *Hausdorff axiom* holds in it: *any two distinct points of the space have nonintersecting neighborhoods.*

Example 5 Any metric space is obviously Hausdorff, since for any two points $a, b \in X$ such that $d(a, b) > 0$ their spherical neighborhoods $B(a, \frac{1}{2}d(a, b))$, $B(b, \frac{1}{2}d(a, b))$ have no points in common.

At the same time, as Example 4 shows, there do exist non-Hausdorff topological spaces. Perhaps the simplest example here is the topological space (X, τ) with the trivial topology $\tau = \{\varnothing, X\}$. If X contains at least two distinct points, then (X, τ) is obviously not Hausdorff. Moreover, the complement $X \setminus x$ of a point in this space is not an open set.

We shall be working exclusively with Hausdorff spaces.

Definition 7 A set $E \subset X$ is *(everywhere) dense* in a topological space (X, τ) if for any point $x \in X$ and any neighborhood $U(x)$ of it the intersection $E \cap U(X)$ is nonempty.

Example 6 If we consider the standard topology in $\mathbb{R}$, the set $\mathbb{Q}$ of rational numbers is everywhere dense in $\mathbb{R}$. Similarly the set $\mathbb{Q}^n$ of rational points in $\mathbb{R}^n$ is dense in $\mathbb{R}^n$.

One can show that in every topological space there is an everywhere dense set whose cardinality does not exceed the weight of the topological space.

Definition 8 A metric space having a countable dense set is called a *separable* space.

Example 7 The metric space $(\mathbb{R}^n, d)$ in any of the standard metrics is a separable space, since $\mathbb{Q}^n$ is dense in it.

Example 8 The metric space $(C([0, 1], \mathbb{R}), d)$ with the metric defined by (9.6) is also separable. For, as follows from the uniform continuity of the functions $f \in C([0, 1], \mathbb{R})$, the graph of any such function can be approximated as closely as desired by a broken line consisting of a finite number of segments whose nodes have rational coordinates. The set of such broken lines is countable.

We shall be dealing mainly with separable spaces.

We now remark that, since the definition of a neighborhood of a point in a topological space is verbally the same as the definition of a neighborhood of a point in a metric space, the concepts of *interior point, exterior point, boundary point*, and *limit point* of a set, and the concept of the *closure* of a set, all of which use only the concept of a neighborhood, can be carried over without any changes to the case of an arbitrary topological space.

Moreover, as can be seen from the proof of Proposition 2 in Sect. 7.1, it is also true that a set in a Hausdorff space is closed if and only if it contains all its limit points.

9.2.2 Subspaces of a Topological Space

Let (X, τ_X) be a topological space and Y a subset of X. The topology τ_X makes it possible to define the following topology τ_Y in Y, called the *induced* or *relative* topology on $Y \subset X$.

We define an *open set in* Y to be any set G_Y of the form $G_Y = Y \cap G_X$, where G_X is an open set in X.

It is not difficult to verify that the system τ_Y of subsets of Y that arises in this way satisfies the axioms for open sets in a topological space.

As one can see, the definition of open sets G_Y in Y agrees with the one we obtained in Sect. 9.1.3 for the case when Y is a subspace of a metric space X.

Definition 9 A subset $Y \subset X$ of a topological space (X, τ) with the topology τ_Y induced on Y is called a *subspace of the topological space* X.

It is clear that a set that is open in (Y, τ_Y) is not necessarily open in (X, τ_X).

9.2.3 The Direct Product of Topological Spaces

If (X_1, τ_1) and (X_2, τ_2) are two topological spaces with systems of open sets $\tau_1 = \{G_1\}$ and $\tau_2 = \{G_2\}$, we can introduce a topology on $X_1 \times X_2$ by taking as the base the sets of the form $G_1 \times G_2$.

Definition 10 The topological space $(X_1 \times X_2, \tau_1 \times \tau_2)$ whose topology has the base consisting of sets of the form $G_1 \times G_2$, where G_i is an open set in the topological space (X_i, τ_i), $i = 1, 2$, is called the *direct product* of the topological spaces (X_1, τ_1) and (X_2, τ_2).

Example 9 If $\mathbb{R} = \mathbb{R}^1$ and $\mathbb{R}^2$ are considered with their standard topologies, then, as one can see, $\mathbb{R}^2$ is the direct product $\mathbb{R}^1 \times \mathbb{R}^1$. For every open set in $\mathbb{R}^2$ can be obtained, for example, as the union of "square" neighborhoods of all its points. And squares (with sides parallel to the axes) are the products of open intervals, which are open sets in $\mathbb{R}$.

It should be noted that the sets $G_1 \times G_2$, where $G_1 \in \tau_1$ and $G_2 \in \tau_2$, constitute only a base for the topology, not all the open sets in the direct product of topological spaces.

9.2.4 Problems and Exercises

1. Verify that if (X, d) is a metric space, then $(X, \frac{d}{1+d})$ is also a metric space, and the metrics d and $\frac{d}{1+d}$ induce the same topology on X. (See also Problem 1 of the preceding section.)

2. a) In the set $\mathbb{N}$ of natural numbers we define a neighborhood of the number $n \in \mathbb{N}$ to be an arithmetic progression with difference d relatively prime to n. Is the resulting topological space Hausdorff?

b) What is the topology of $\mathbb{N}$, regarded as a subset of the set $\mathbb{R}$ of real numbers with the standard topology?

c) Describe all open subsets of $\mathbb{R}$.

3. If two topologies τ_1 and τ_2 are defined on the same set, we say that τ_2 is *stronger* than τ_1 if $\tau_1 \subset \tau_2$, that is τ_2 contains all the sets in τ_1 and some additional open sets not in τ_1.

a) Are the two topologies on $\mathbb{N}$ considered in the preceding problem comparable?

b) If we introduce a metric on the set $C[0, 1]$ of continuous real-valued functions defined on the closed interval $[0, 1]$ first by relation (9.6) of Sect. 9.1, and then by relation (9.7) of the same section, two topologies generally arise on $C[a, b]$. Are they comparable?

4. a) Prove in detail that the space of germs of continuous functions defined in Example 4 is not Hausdorff.

b) Explain why this topological space is not metrizable.

c) What is the weight of this space?

5. a) State the axioms for a topological space in the language of closed sets.

b) Verify that the closure of the closure of a set equals the closure of the set.

c) Verify that the boundary of any set is a closed set.

d) Show that if F is closed and G is open in (X, τ), then the set $G \backslash F$ is open in (X, τ).

e) If (Y, τ_Y) is a subspace of the topological space (X, τ), and the set E is such that $E \subset Y \subset X$ and $E \in \tau_X$, then $E \in \tau_Y$.

6. A topological space (X, τ) in which every point is a closed set is called a *topological space in the strong sense* or a τ_1-space. Verify the following statements.

a) Every Hausdorff space is a τ_1-space (partly for this reason, Hausdorff spaces are sometimes called τ_2-spaces).

b) Not every τ_1-space is a τ_2-space. (See Example 4.)

c) The two-point space $X = \{a, b\}$ with the open sets $\{\varnothing, X\}$ is not a τ_1-space.

d) In a τ_1-space a set F is closed if and only if it contains all its limit points.

7. a) Prove that in any topological space there is an everywhere dense set whose cardinality does not exceed the weight of the space.

b) Verify that the following metric spaces are separable: $C[a, b]$, $C^{(k)}[a, b]$, $\mathcal{R}_1[a, b]$, $\mathcal{R}_p[a, b]$ (for the formulas giving the respective metrics see Sect. 9.1).

c) Verify that if max is replaced by sup in relation (9.6) of Sect. 9.1 and regarded as a metric on the set of all bounded real-valued functions defined on a closed interval $[a, b]$, we obtain a nonseparable metric space.

9.3 Compact Sets

9.3.1 Definition and General Properties of Compact Sets

Definition 1 A set K in a topological space (X, τ) is *compact* (or *bicompact*[3]) if from every covering of K by sets that are open in X one can select a finite number of sets that cover K.

Example 1 An interval $[a, b]$ of the set $\mathbb{R}$ of real numbers in the standard topology is a compact set, as follows immediately from the lemma of Sect. 2.1.3 asserting that one can select a finite covering from any covering of a closed interval by open intervals.

In general an m-dimensional closed interval $I^m = \{x \in \mathbb{R}^m \mid a^i \leq x^i \leq b^i, i = 1, \ldots, m\}$ in $\mathbb{R}^m$ is a compact set, as was established in Sect. 7.1.3.

It was also proved in Sect. 7.1.3 that a subset of $\mathbb{R}^m$ is compact if and only if it is closed and bounded.

In contrast to the relative properties of being open and closed, the property of compactness is absolute, in the sense that it is independent of the ambient space. More precisely, the following proposition holds.

Proposition 1 *A subset K of a topological space (X, τ) is a compact subset of X if and only if K is compact as a subset of itself with the topology induced from (X, τ).*

Proof This proposition follows from the definition of compactness and the fact that every set G_K that is open in K can be obtained as the intersection of K with some set G_X that is open in X. □

---

[3]The concept of compactness introduced by Definition 1 is sometimes called *bicompactness* in topology.

Thus, if (X, τ_X) and (Y, τ_Y) are two topological spaces that induce the same topology on $K \subset X \cap Y$, then K is simultaneously compact or not compact in both X and Y.

Example 2 Let d be the standard metric on $\mathbb{R}$ and $I = \{x \in \mathbb{R} \mid 0 < x < 1\}$ the unit interval in $\mathbb{R}$. The metric space (I, d) is closed (in itself) and bounded, but is not a compact set, since for example, it is not a compact subset of $\mathbb{R}$.

We now establish the most important properties of compact sets.

Lemma 1 (Compact sets are closed) *If K is a compact set in a Hausdorff space (X, τ), then K is a closed subset of X.*

Proof By the criterion for a set to be closed, it suffices to verify that every limit point of K, $x_0 \in X$, belongs to K.

Suppose $x_0 \notin K$. For each point $x \in K$ we construct an open neighborhood $G(x)$ such that x_0 has a neighborhood disjoint from $G(x)$. The set $G(x)$, $x \in K$, of all such neighborhoods forms an open covering of K, from which one can select a finite covering $G(x_1), \ldots, G(x_n)$. Now if $O_i(x_0)$ is a neighborhood of x_0 such that $G(x_i) \cap O_i(x_0) = \varnothing$, the set $O(x) = \bigcap_{i=1}^{n} O_i(x_0)$ is also a neighborhood of x_0, and $G(x_i) \cap O(x_0) = \varnothing$ for all $i = 1, \ldots, n$. But this means that $K \cap O(x_0) = \varnothing$, and then x_0 cannot be a limit point for K.   □

Lemma 2 (Nested compact sets) *If $K_1 \supset K_2 \supset \cdots \supset K_n \supset \cdots$ is a nested sequence of nonempty compact sets, then the intersection $\bigcap_{i=1}^{\infty} K_i$ is nonempty.*

Proof By Lemma 1 the sets $G_i = K_1 \backslash K_i$, $i = 1, \ldots, n, \ldots$ are open in K_1. If the intersection $\bigcap_{i=1}^{\infty} K_i$ is empty, then the sequence $G_1 \subset G_2 \subset \cdots \subset G_n \subset \cdots$ forms a covering of K_1. Extracting a finite covering from it, we find that some element G_m of the sequence forms a covering of K_1. But by hypothesis $K_m = K_1 \backslash G_m \neq \varnothing$. This contradiction completes the proof of Lemma 2.   □

Lemma 3 (Closed subsets of compact sets) *A closed subset F of a compact set K is itself compact.*

Proof Let $\{G_\alpha, \alpha \in A\}$ be an open covering of F. Adjoining to this collection the open set $G = K \backslash F$, we obtain an open covering of the entire compact set K. From this covering we can extract a finite covering of K. Since $G \cap F = \varnothing$, it follows that the set $\{G_\alpha, \alpha \in A\}$ contains a finite covering of F.   □

9.3.2 Metric Compact Sets

We shall establish below some properties of metric compact sets, that is, metric spaces that are compact sets with respect to the topology induced by the metric.

Definition 2 The set $E \subset X$ is called an ε-*grid* in the metric space (X, d) if for every point $x \in X$ there is a point $e \in E$ such that $d(e, x) < \varepsilon$.

Lemma 4 (Finite ε-grids) *If a metric space (K, d) is compact, then for every $\varepsilon > 0$ there exists a finite ε-grids in X.*

Proof For each point $x \in K$ we choose an open ball $B(x, \varepsilon)$. From the open covering of K by these balls we select a finite covering $B(x_1, \varepsilon), \ldots, B(x_n, \varepsilon)$. The points $x_1, \ldots, x_n$ obviously form the required ε-grid.   □

In analysis, besides arguments that involve the extraction of a finite covering, one often encounters arguments in which a convergent subsequence is extracted from an arbitrary sequence. As it happens, the following proposition holds.

Proposition 2 (Criterion for compactness in a metric space) *A metric space (K, d) is compact if and only if from each sequence of its points one can extract a subsequence that converges to a point of K.*

The convergence of the sequence $\{x_n\}$ to some point $a \in K$, as before, means that for every neighborhood $U(a)$ of the point $a \in K$ there exists an index $N \in \mathbb{N}$ such that $x_n \in U(a)$ for $n > N$.

We shall discuss the concept of limit in more detail below in Sect. 9.6.

We preface the proof of Proposition 2 with two lemmas.

Lemma 5 *If a metric space (K, d) is such that from each sequence of its points one can select a subsequence that converges in K, then for every $\varepsilon > 0$ there exists a finite ε-grid.*

Proof If there were no finite ε_0-grid for some $\varepsilon_0 > 0$, one could construct a sequence $\{x_n\}$ of points in K such that $d(x_n, x_i) > \varepsilon_0$ for all $n \in \mathbb{N}$ and all $i \in \{1, \ldots, n - 1\}$. Obviously it is impossible to extract a convergent subsequence of this sequence. □

Lemma 6 *If the metric space (K, d) is such that from each sequence of its points one can select a subsequence that converges in K, then every nested sequence of nonempty closed subsets of the space has a nonempty intersection.*

Proof If $F_1 \supset \cdots \supset F_n \supset \cdots$ is the sequence of closed sets, then choosing one point of each, we obtain a sequence $x_1, \ldots, x_n, \ldots$, from which we extract a convergent subsequence $\{x_{n_i}\}$. The limit $a \in K$ of this sequence, by construction, necessarily belongs to each of the closed sets F_i, $i \in \mathbb{N}$.   □

We can now prove Proposition 2.

Proof We first verify that if (K, d) is compact and $\{x_n\}$ a sequence of points in it, one can extract a subsequence that converges to some point of K. If the sequence $\{x_n\}$ has only a finite number of different values, the assertion is obvious.

Therefore we may assume that the sequence $\{x_n\}$ has infinitely many different values. For $\varepsilon_1 = 1/1$, we construct a finite 1-grid and take a closed ball $\widetilde{B}(a_1, 1)$ that contains an infinite number of terms of the sequence. By Lemma 3 the ball $\widetilde{B}(a_1, 1)$ is itself a compact set, in which there exists a finite $\varepsilon_2 = 1/2$-grid and a ball $\widetilde{B}(a_2, 1/2)$ containing infinitely many elements of the sequence. In this way a nested sequence of compact sets $\widetilde{B}(a_1, 1) \supset \widetilde{B}(a_2, 1/2) \supset \cdots \supset \widetilde{B}(a_n, 1/n) \supset \cdots$ arises, and by Lemma 2 has a common point $a \in K$. Choosing a point x_{n_1} of the sequence $\{x_n\}$ in the ball $\widetilde{B}(a_1, 1)$, then a point x_{n_2} in $\widetilde{B}(a_2, 1/2)$ with $n_2 > n_1$, and so on, we obtain a subsequence $\{x_{n_i}\}$ that converges to a by construction.

We now prove the converse, that is, we verify that if from every sequence $\{x_n\}$ of points of the metric space (K, d) one can select a subsequence that converges in K, then (K, d) is compact.

In fact, if there is some open covering $\{G_\alpha, \alpha \in A\}$ of the space (K, d) from which one cannot select a finite covering, then using Lemma 5 to construct a finite 1-grid in K, we find a closed ball $\widetilde{B}(a_1, 1)$, that also cannot be covered by a finite collection of sets of the system $\{B_\alpha, \alpha \in A\}$.

The ball $\widetilde{B}(a_1, 1)$ can now be regarded as the initial set, and, constructing a finite 1/2-grid in it, we find in it a ball $\widetilde{B}(a_2, 1/2)$ that does not admit covering by a finite number of sets in the system $\{G_\alpha, \alpha \in A\}$.

The resulting nested sequence of closed sets $\widetilde{B}(a_1, 1) \supset \widetilde{B}(a_2, 1/2) \supset \cdots \supset \widetilde{B}(a_n, 1/n) \supset \cdots$ has a common point $a \in K$ by Lemma 6, and the construction shows that there is only one such point. This point is covered by some set G_{α_0} of the system; and since G_{α_0} is open, all the sets $\widetilde{B}(a_n, 1/n)$ must be contained in G_{α_0} for sufficiently large values of n. This contradiction completes the proof of the proposition.   □

9.3.3 Problems and Exercises

1. A subset of a metric space is *totally bounded* if for every $\varepsilon > 0$ it has a finite ε-grid.

a) Verify that total boundedness of a set is unaffected, whether one forms the grid from points of the set itself or from points of the ambient space.

b) Show that a subset of a complete metric space is compact if and only if it is totally bounded and closed.

c) Show by example that a closed bounded subset of a metric space is not always totally bounded, and hence not always compact.

2. A subset of a topological space is *relatively* (*or conditionally*) *compact* if its closure is compact.

Give examples of relatively compact subsets of $\mathbb{R}^n$.

3. A topological space is *locally compact* if each point of the space has a relatively compact neighborhood.

Give examples of locally compact topological spaces that are not compact.

4. Show that for every locally compact, but not compact topological space (X, τ_X) there is a compact topological space (Y, τ_Y) such that $X \subset Y$, $Y \backslash X$ consists of a single point, and the space $(X, \tau x)$ is a subspace of the space (Y, τ_Y).

9.4 Connected Topological Spaces

Definition 1 A topological space (X, τ) is *connected* if it contains no open-closed sets[4] except X itself and the empty set.

This definition will become more transparent to intuition if we recast it in the following form.

A topological space is connected if and only if it cannot be represented as the union of two disjoint nonempty closed sets (or two disjoint nonempty open sets).

Definition 2 A set E in a topological space (X, τ) is *connected* if it is connected as a topological subspace of (X, τ) (with the induced topology).

It follows from this definition and Definition 1 that the property of a set of being connected is independent of the ambient space. More precisely, if (X, τ_X) and (Y, τ_Y) are topological spaces containing E and inducing the same topology on E, then E is connected or not connected simultaneously in both X and Y.

Example 1 Let $E = \{x \in \mathbb{R} \mid x \neq 0\}$. The set $E_- = \{x \in E \mid x < 0\}$ is nonempty, not equal to E, and at the same time open-closed in E (as is $E_+ = \{x \in \mathbb{R} \mid x > 0\}$), if E is regarded as a topological space with the topology induced by the standard topology of $\mathbb{R}$. Thus, as our intuition suggests, E is not connected.

Proposition (Connected subsets of $\mathbb{R}$) *A nonempty set $E \subset \mathbb{R}$ is connected if and only if for any x and z belonging to E, the inequalities $x < y < z$ imply that $y \in E$.*

Thus, the only connected subsets of the line are intervals (finite or infinite): open, half-open, and closed.

Proof Necessity. Let E be a connected subset of $\mathbb{R}$, and let the triple of points a, b, c be such that $a \in E$, $b \in E$, but $c \notin E$, even though $a < c < b$. Setting $A = \{x \in E \mid x < c\}$, $B = \{x \in E \mid x > c\}$, we see that $a \in A$, $b \in B$, that is, $A \neq \varnothing$, $B \neq \varnothing$, and $A \cap B = \varnothing$. Moreover $E = A \cup B$, and both sets A and B are open in E. This contradicts the connectedness of E.

Sufficiency. Let E be a subspace of $\mathbb{R}$ having the property that together with any pair of points a and b belonging to it, every point between them in the closed interval $[a, b]$ also belongs to E. We shall show that E is connected.

---

[4]That is, sets that are simultaneously open and closed.

Suppose that A is an open-closed subset of E with $A \neq \varnothing$ and $B = E \backslash A \neq \varnothing$. Let $a \in A$ and $b \in B$. For definiteness we shall assume that $a < b$. (We certainly have $a \neq b$, since $A \cap B = \varnothing$.) Consider the point $c_1 = \sup\{A \cap [a, b]\}$. Since $A \ni a \leq c_1 \leq b \in B$, we have $c_1 \in E$. Since A is closed in E, we conclude that $c_1 \in A$.

Considering now the point $c_2 = \inf\{B \cap [c_1, b]\}$ we conclude similarly, since B is closed, that $c_2 \in B$. Thus $a \leq c_1 < c_2 \leq b$, since $c_1 \in A$, $c_2 \in B$, and $A \cap B = \varnothing$. But it now follows from the definition of c_1 and c_2 and the relation $E = A \cup B$ that no point of the open interval $]c_1, c_2[$ can belong to E. This contradicts the original property of E. Thus the set E cannot have a subset A with these properties, and that proves that E is connected. $\qquad\qquad\qquad\qquad\qquad\qquad\qquad\qquad\qquad\qquad\qquad\qquad$ $\square$

9.4.1 Problems and Exercises

1. a) Verify that if A is an open-closed subset of (X, τ), then $B = X \backslash A$ is also such a set.

b) Show that in terms of the ambient space the property of connectedness of a set can be expressed as follows: *A subset E of a topological space (X, τ) is connected if and only if there is no pair of open (or closed) subsets G'_X, G''_X that are disjoint and such that $E \cap G'_X \neq \varnothing$, $E \cap G''_X \neq \varnothing$, and $E \subset G'_X \cup G''_X$.*

2. Show the following:

a) The union of connected subspaces having a common point is connected.
b) The intersection of connected subspaces is not always connected.
c) The closure of a connected subspace is connected.

3. One can regard the group $GL(n)$ of nonsingular $n \times n$ matrices with real entries as an open subset in the product space $\mathbb{R}^{n^2}$, if each element of the matrix is associated with a copy of the set $\mathbb{R}$ of real numbers. Is the space $GL(n)$ connected?

4. A topological space is *locally connected* if each of its points has a connected neighborhood.

a) Show that a locally connected space may fail to be connected.

b) The set E in $\mathbb{R}^2$ consists of the graph of the function $x \mapsto \sin \frac{1}{x}$ (for $x \neq 0$) plus the closed interval $\{(x, y) \in \mathbb{R}^2 \mid x = 0 \wedge |y| \leq 1\}$ on the y-axis. The set E is endowed with the topology induced from $\mathbb{R}^2$. Show that the resulting topological space is connected but not locally connected.

5. In Sect. 7.2.2 we defined a connected subset of $\mathbb{R}^n$ as a set $E \subset \mathbb{R}^n$ any two of whose points can be joined by a path whose support lies in E. In contrast to the definition of topological connectedness introduced in the present section, the concept we considered in Chap. 7 is usually called *path connectedness* or *arcwise connectedness*. Verify the following:

a) A path-connected subset of $\mathbb{R}^n$ is connected.

b) Not every connected subset of $\mathbb{R}^n$ with $n > 1$ is path connected. (See Problem 4.)

c) Every connected open subset of $\mathbb{R}^n$ is path connected.

9.5 Complete Metric Spaces

In this section we shall be discussing only metric spaces, more precisely, a class of such spaces that plays an important role in various areas of analysis.

9.5.1 Basic Definitions and Examples

By analogy with the concepts that we already know from our study of the space $\mathbb{R}^n$, we introduce the concepts of fundamental (Cauchy) sequences and convergent sequences of points of an arbitrary metric space.

Definition 1 A sequence $\{x_n; n \in \mathbb{N}\}$ of points of a metric space (X, d) is a *fundamental* or *Cauchy* sequence if for every $\varepsilon > 0$ there exists $N \in \mathbb{N}$ such that $d(x_m, x_n) < \varepsilon$ for any indices $m, n \in \mathbb{N}$ larger than N.

Definition 2 A sequence $\{x_n; n \in \mathbb{N}\}$ of points of a metric space (X, d) *converges to the point* $a \in X$ and a is its *limit* if $\lim_{n \to \infty} d(a, x_n) = 0$.

A sequence that has a limit will be called *convergent*, as before.
We now give the basic definition.

Definition 3 A metric space (X, d) is *complete* if every Cauchy sequence of its points is convergent.

Example 1 The set $\mathbb{R}$ of real numbers with the standard metric is a complete metric space, as follows from the Cauchy criterion for convergence of a numerical sequence.

We remark that, since every convergent sequence of points in a metric space is obviously a Cauchy sequence, the definition of a complete metric space essentially amounts to simply postulating the Cauchy convergence criterion for it.

Example 2 If the number 0, for example, is removed from the set $\mathbb{R}$, the remaining set $\mathbb{R}\backslash 0$ will not be a complete space in the standard metric. Indeed, the sequence $x_n = 1/n$, $n \in \mathbb{N}$, is a Cauchy sequence of points of this set, but has no limit in $\mathbb{R}\backslash 0$.

Example 3 The space $\mathbb{R}^n$ with any of its standard metrics is complete, as was explained in Sect. 7.2.1.

Example 4 Consider the set $C[a, b]$ of real-valued continuous functions on a closed interval $[a, b] \subset \mathbb{R}$, with the metric

$$d(f, g) = \max_{a \leq x \leq b} |f(x) - g(x)| \tag{9.9}$$

(see Sect. 9.1, Example 7). We shall show that the metric space $C[a, b]$ is complete.

Proof Let $\{f_n(x): n \in \mathbb{N}\}$ be a Cauchy sequence of functions in $C[a, b]$, that is

$$\forall \varepsilon > 0 \, \exists N \in \mathbb{N} \, \forall m \in \mathbb{N} \, \forall n \in \mathbb{N} \, \big((m > N \wedge n > N) \Longrightarrow$$

$$\Longrightarrow \forall x \in [a, b] \, \big(|f_m(x) - f_n(x)| < \varepsilon\big)\big). \tag{9.10}$$

For each fixed value of $x \in [a, b]$, as one can see from (9.10), the numerical sequence $\{f_n(x); n \in \mathbb{N}\}$ is a Cauchy sequence and hence has a limit $f(x)$ by the Cauchy convergence criterion.

Thus

$$f(x) := \lim_{n \to \infty} f_n(x), \quad x \in [a, b]. \tag{9.11}$$

We shall verify that the function $f(x)$ is continuous on $[a, b]$, that is, $f \in C[a, b]$. It follows from (9.10) and (9.11) that the inequality

$$|f(x) - f_n(x)| \leq \varepsilon \quad \forall x \in [a, b] \tag{9.12}$$

holds for $n > N$.

We fix the point $x \in [a, b]$ and verify that the function f is continuous at this point. Suppose the increment h is such that $(x + h) \in [a, b]$. The identity

$$f(x + h) - f(x) = f(x + h) - f_n(x + h) + f_n(x + h) - f_n(x) + f_n(x) - f(x)$$

implies the inequality

$$|f(x + h) - f(x)| \leq |f(x + h) - f_n(x + h)| +$$

$$+ |f_n(x + h) - f_n(x)| + |f_n(x) - f(x)|. \tag{9.13}$$

By virtue of (9.12) the first and last terms on the right-hand side of this last inequality do not exceed ε if $n > N$. Fixing $n > N$, we obtain a function $f_n \in C[a, b]$, and then choosing $\delta = \delta(\varepsilon)$ such that $|f_n(x + h) - f_n(x)| < \varepsilon$ for $|h| < \delta$, we find that $|f(x + h) - f(x)| < 3\varepsilon$ if $|h| < \delta$. But this means that the function f is continuous at the point x. Since x was an arbitrary point of the closed interval $[a, b]$, we have shown that $f \in C[a, b]$. $\qquad \square$

Thus the space $C[a, b]$ with the metric (9.9) is a complete metric space. This is a very important fact, one that is widely used in analysis.

Fig. 9.2

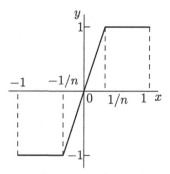

Example 5 If instead of the metric (9.9) we consider the integral metric

$$d(f, g) = \int_a^b |f - g|(x)\,dx \tag{9.14}$$

on the same set $C[a, b]$, the resulting metric space is no longer complete.

Proof For the sake of notational simplicity, we shall assume $[a, b] = [-1, 1]$ and consider, for example, the sequence $\{f_n \in C[-1, 1]; n \in \mathbb{N}\}$ of functions defined as follows:

$$f_n(x) = \begin{cases} -1, & \text{if } -1 \leq x \leq -1/n, \\ nx, & \text{if } -1/n < x < 1/n, \\ 1, & \text{if } 1/n \leq x \leq 1. \end{cases}$$

(See Fig. 9.2.)

It follows immediately from properties of the integral that this sequence is a Cauchy sequence in the sense of the metric (9.14) in $C[-1, 1]$. At the same time, it has no limit in $C[-1, 1]$. For if a continuous function $f \in C[-1, 1]$ were the limit of this sequence in the sense of metric (9.14), then f would have to be constant on the interval $-1 \leq x < 0$ and equal to -1 while at the same time it would have to be constant and equal to 1 on the interval $0 < x \leq 1$, which is incompatible with the continuity of f at the point $x = 0$. □

Example 6 It is slightly more difficult to show that even the set $\mathcal{R}[a, b]$ of real-valued Riemann-integrable functions defined on the closed interval $[a, b]$ is not complete in the sense of the metric (9.14).[5] We shall show this, using the Lebesgue criterion for Riemann integrability of a function.

Proof We take $[a, b]$ to be the closed interval $[0, 1]$, and we shall construct a Cantor set on it that is not a set of measure zero. Let $\Delta \in]0, 1/3[$. We remove from the interval $[0, 1]$ the middle piece of it of length Δ. More precisely, we remove the

---

[5]In regard to the metric (9.14) on $\mathcal{R}[a, b]$ see the remark to Example 9 in Sect. 9.1.

$\Delta/2$-neighborhood of the midpoint of the closed interval $[0, 1]$. On each of the two remaining intervals, we remove the middle piece of length $\Delta \cdot 1/3$. On each of the four remaining closed intervals we remove the middle piece of length $\Delta \cdot 1/3^2$, and so forth. The length of the intervals removed in this process is $\Delta + \Delta \cdot 2/3 + \Delta \cdot 4/3^2 + \cdots + \Delta \cdot (2/3)^n + \cdots = 3\Delta$. Since $0 < \Delta < 1/3$ we have $1 - 3\Delta > 0$, and, as one can verify, it follows from this that the (Cantor) set K remaining on the closed interval $[0, 1]$ does not have measure zero in the sense of Lebesgue.

Now consider the following sequence: $\{f_n \in \mathcal{R}[0, 1]; n \in \mathbb{N}\}$. Let f_n be a function equal to 1 everywhere on $[0, 1]$ except at the points of the intervals removed at the first n steps, where it is set equal to zero. It is easy to verify that this sequence is a Cauchy sequence in the sense of the metric (9.14). If some function $f \in \mathcal{R}[0, 1]$ were the limit of this sequence, then f would have to be equal to the characteristic function of the set K at almost every point of the interval $[0, 1]$. Then f would have discontinuities at all points of the set K. But, since K does not have measure 0, one could conclude from the Lebesgue criterion that $f \notin \mathcal{R}[0, 1]$. Hence $\mathcal{R}[a, b]$ with the metric (9.14) is not a complete metric space.   □

9.5.2 The Completion of a Metric Space

Example 7 Let us return again to the real line and consider the set $\mathbb{Q}$ of rational numbers with the metric induced by the standard metric on $\mathbb{R}$.

It is clear that a sequence of rational numbers converging to $\sqrt{2}$ in $\mathbb{R}$ is a Cauchy sequence, but does not have a limit in $\mathbb{Q}$, that is, $\mathbb{Q}$ is not a complete space with this metric. However, $\mathbb{Q}$ happens to be a subspace of the complete metric space $\mathbb{R}$, which it is natural to regard as the completion of $\mathbb{Q}$. Note that the set $\mathbb{Q} \subset \mathbb{R}$ could also be regarded as a subset of the complete metric space $\mathbb{R}^2$, but it does not seem reasonable to call $\mathbb{R}^2$ the completion of $\mathbb{Q}$.

Definition 4 The smallest complete metric space containing a given metric space (X, d) is the *completion* of (X, d).

This intuitively acceptable definition requires at least two clarifications: what is meant by the "smallest" space, and does it exist?

We shall soon be able to answer both of these questions; in the meantime we adopt the following more formal definition.

Definition 5 If a metric space (X, d) is a subspace of a complete metric space (Y, d) and the set $X \subset Y$ is everywhere dense in Y, the space (Y, d) is called a *completion* of the metric space (X, d).

Definition 6 We say that the metric space (X_1, d_1) is *isometric* to the metric space (X_2, d_2) if there exists a bijective mapping $f : X_1 \to X_2$ such that $d_2(f(a), f(b)) = d_1(a, b)$ for any points a and b in X_1. (The mapping $f : X_1 \to X_2$ is called an *isometry* in that case.)

It is clear that this relation is reflexive, symmetric, and transitive, that is, it is an equivalence relation between metric spaces. In studying the properties of metric spaces we study not the individual space, but the properties of all spaces isometric to it. For that reason one may regard isometric spaces as identical.

Example 8 Two congruent figures in the plane are isometric as metric spaces, so that in studying the metric properties of figures we abstract completely, for example, from the location of a figure in the plane, identifying all congruent figures.

By adopting the convention of identifying isometric spaces, one can show that if the completion of a metric space exists at all, it is unique.

As a preliminary, we verify the following statement.

Lemma *The following inequality holds for any quadruple of points a, b, u, v of the metric space (X, d):*

$$\left| d(a, b) - d(u, v) \right| \leq d(a, u) + d(b, v). \tag{9.15}$$

Proof By the triangle inequality

$$d(a, b) \leq d(a, u) + d(u, v) + d(b, v).$$

By the symmetry of the points, this relation implies (9.15). □

We now prove uniqueness of the completion.

Proposition 1 *If the metric spaces (Y_1, d_1) and (Y_2, d_2) are completions of the same space (X, d), then they are isometric.*

Proof We construct an isometry $f : Y_1 \to Y_2$ as follows. For $x \in X$ we set $f(x) = x$. Then $d_2(f(x_1), f(x_2)) = d(f(x_1), f(x_2)) = d(x_1, x_2) = d_1(x_1, x_2)$ for $x_1, x_2 \in X$. If $y_1 \in Y_1 \backslash X$, then y_1 is a limit point for X, since X is everywhere dense in Y_1. Let $\{x_n; n \in \mathbb{N}\}$ be a sequence of points of X converging to y_1 in the sense of the metric d_1. This sequence is a Cauchy sequence in the sense of the metric d_1. But since the metrics d_1 and d_2 are both equal to d on X, this sequence is also a Cauchy sequence in (Y_2, d_2). The latter space is complete, and hence this sequence has a limit $y_2 \in Y_2$. It can be verified in the standard manner that this limit is unique. We now set $f(y_1) = y_2$. Since any point $y_2 \in Y_2 \backslash X$, just like any point $y_1 \in Y_1 \backslash X$, is the limit of a Cauchy sequence of points in X, the mapping $f : Y_1 \to Y_2$ so constructed is surjective.

We now verify that

$$d_2\left(f\left(y_1'\right), f\left(y_1''\right)\right) = d_1\left(y_1', y_1''\right) \tag{9.16}$$

for any pair of points y_1', y_1'' of Y_1.

If y_1' and y_1'' belong to X, this equality is obvious. In the general case we take two sequences $\{x_n'; n \in \mathbb{N}\}$ and $\{x_n''; n \in \mathbb{N}\}$ converging to y_1' and y_1'' respectively. It follows from inequality (9.15) that

$$d_1\left(y_1', y_1''\right) = \lim_{n \to \infty} d_1\left(x_n', x_n''\right),$$

or, what is the same,

$$d_1\left(y_1', y_1''\right) = \lim_{n \to \infty} d\left(x_n', x_n''\right). \tag{9.17}$$

By construction these same sequences converge to $y_2' = f(y_1')$ and $y_2'' = f(y_2'')$ respectively in the space (Y_2, d_2). Hence

$$d_2\left(y_2', y_2''\right) = \lim_{n \to \infty} d\left(x_n', x_n''\right). \tag{9.18}$$

Comparing relations (9.17) and (9.18), we obtain Eq. (9.16). This equality then simultaneously establishes that the mapping $f : Y_1 \to Y_2$ is injective and hence completes the proof that f is an isometry. $\square$

In Definition 5 of the completion (Y, d) of a metric space (X, d) we required that (X, d) be a subspace of (Y, d) that is everywhere dense in (Y, d). Under the identification of isometric spaces one could now broaden the idea of a completion and adopt the following definition.

Definition 5' A complete metric space (Y, d_Y) is a *completion* of the metric space (X, d_X) if there is a dense subspace of (Y, d_Y) isometric to (X, d_X).

We now prove the existence of a completion.

Proposition 2 *Every metric space has a completion.*

Proof If the initial space itself is complete, then it is its own completion.

We have already essentially demonstrated the idea for constructing the completion of an incomplete metric space (X, d_X) when we proved Proposition 1.

Consider the set of Cauchy sequences in the space (X, d_X). Two such sequences $\{x_n'; n \in \mathbb{N}\}$ and $\{x_n''; n \in \mathbb{N}\}$ are called *equivalent* or *confinal* if $d_X(x_n', x_n'') \to 0$ as $n \to \infty$. It is easy to see that confinality really is an equivalence relation. We shall denote the set of equivalence classes of Cauchy sequences by S. We introduce a metric in S by the following rule. If s' and s'' are elements of S, and $\{x_n'; n \in \mathbb{N}\}$ and $\{x_n''; n \in \mathbb{N}\}$ are sequences from the classes s' and s'' respectively, we set

$$d\left(s', s''\right) = \lim_{n \to \infty} d_X\left(x_n', x_n''\right). \tag{9.19}$$

It follows from inequality (9.15) that this definition is unambiguous: the limit written on the right exists (by the Cauchy criterion for a numerical sequence) and is independent of the choice of the individual sequences $\{x_n'; n \in \mathbb{N}\}$ and $\{x_n''; n \in \mathbb{N}\}$ from the classes s' and s''.

The function $d(s', s'')$ satisfies all the axioms of a metric. The resulting metric space (S, d) is the required completion of the space (X, d_X). Indeed, (X, d_X) is isometric to the subspace (S_X, d) of the space (S, d) consisting of the equivalence classes of fundamental sequences that contain constant sequences $\{x_n = x \in X;$ $n \in \mathbb{N}\}$. It is natural to identify such a class $s \in S$ with the point $x \in X$. The mapping $f : (X, d_X) \to (S_X, d)$ is obviously an isometry.

It remains to be verified that (S_X, d) is everywhere dense in (S, d) and that (S, d) is a complete metric space.

We first verify that (S_X, d) is dense in (S, d). Let s be an arbitrary element of S and $\{x_n; n \in \mathbb{N}\}$ a Cauchy sequence in (X, d_X) belonging to the class $s \in S$. Taking $\xi_n = f(x_n)$, $n \in \mathbb{N}$, we obtain a sequence $\{\xi_n; n \in \mathbb{N}\}$ of points of (S_X, d) that has precisely the element $s \in S$ as its limit, as one can see from (9.19).

We now prove that the space (S, d) is complete. Let $\{s_n; n \in \mathbb{N}\}$ be an arbitrary Cauchy sequence in the space (S, d). For each $n \in \mathbb{N}$ we choose an element ξ_n in (S_X, d) such that $d(s_n, \xi_n) < 1/n$. Then the sequence $\{\xi_n; n \in \mathbb{N}\}$, like the sequence $\{s_n; n \in \mathbb{N}\}$, is a Cauchy sequence. But in that case the sequence $\{x_n = f^{-1}(\xi_n);$ $n \in \mathbb{N}\}$ will also be a Cauchy sequence. The sequence $\{x_n; n \in \mathbb{N}\}$ defines an element $s \in S$, to which the given sequence $\{s_n; n \in \mathbb{N}\}$ converges by virtue of relation (9.19). $\qquad\square$

Remark 1 Now that Propositions 1 and 2 have been proved, it becomes understandable that the completion of a metric space in the sense of Definition 5' is indeed the smallest complete space containing (up to isometry) the given metric space. In this way we have justified the original Definition 4 and made it precise.

Remark 2 The construction of the set $\mathbb{R}$ of real numbers, starting from the set $\mathbb{Q}$ of rational numbers could have been carried out exactly as in the construction of the completion of a metric space, which was done in full generality above. That is exactly how the transition from $\mathbb{Q}$ to $\mathbb{R}$ was carried out by Cantor.

Remark 3 In Example 6 we showed that the space $\mathcal{R}[a, b]$ of Riemann-integrable functions is not complete in the natural integral metric. Its completion is the important space $\mathcal{L}[a, b]$ of Lebesgue-integrable functions.

9.5.3 Problems and Exercises

1. a) Prove the following nested ball lemma. *Let (X, d) be a metric space and $\widetilde{B}(x_1, r_1) \supset \cdots \supset \widetilde{B}(x_n, r_n) \supset \cdots$ a nested sequence of closed balls in X whose radii tend to zero. The space (X, d) is complete if and only if for every such sequence there exists a unique point belonging to all the balls of the sequence.*

b) Show that if the condition $r_n \to 0$ as $n \to \infty$ is omitted from the lemma stated above, the intersection of a nested sequence of balls may be empty, even in a complete space.

2. a) A set $E \subset X$ of a metric space (X, d) is *nowhere dense* in X if it is not dense in any ball, that is, if for every ball $B(x, r)$ there is a second ball $B(x_1, r_1) \subset B(x, r)$ containing no points of the set E.

A set E is *of first category* in X if it can be represented as a countable union of nowhere dense sets.

A set that is not of first category is *of second category* in X.

Show that a complete metric space is a set of second category (in itself).

b) Show that if a function $f \in C^{(\infty)}[a, b]$ is such that $\forall x \in [a, b] \exists n \in \mathbb{N} \forall m > n$ $(f^{(m)}(x) = 0)$, then the function f is a polynomial.

9.6 Continuous Mappings of Topological Spaces

From the point of view of analysis, the present section and the one following contain the most important results in the present chapter.

The basic concepts and propositions discussed here form a natural, some-times verbatim extension to the case of mappings of arbitrary topological or metric spaces, of concepts and propositions that are already well known to us in. In the process, not only the statement but also the proofs of many facts turn out to be identical with those already considered; in such cases the proofs are naturally omitted with a reference to the corresponding propositions that were discussed in detail earlier.

9.6.1 The Limit of a Mapping

a. The Basic Definition and Special Cases of It

Definition 1 Let $f : X \to Y$ be a mapping of the set X with a fixed base $\mathcal{B} = \{B\}$ in X into a topological space Y. The point $A \in Y$ is the *limit of the mapping* $f :$ $X \to Y$ *over the base* $\mathcal{B}$, and we write $\lim_{\mathcal{B}} f(x) = A$, if for every neighborhood $V(A)$ of A in Y there exists an element $B \in \mathcal{B}$ of the base $\mathcal{B}$ whose image under the mapping f is contained in $V(A)$.

In logical symbols Definition 1 has the form

$$\lim_{\mathcal{B}} f(x) = A := \forall V(A) \subset Y \ \exists B \in \mathcal{B} \ \big(f(B) \subset V(A)\big).$$

We shall most often encounter the case in which X, like Y, is a topological space and $\mathcal{B}$ is the base of neighborhoods or deleted neighborhoods of some point $a \in X$. Retaining our earlier notation $x \to a$ for the base of deleted neighborhoods $\{\mathring{U}(a)\}$ of the point a, we can specialize Definition 1 for this base:

$$\lim_{x \to a} f(x) = A := \forall V(A) \subset Y \ \exists \mathring{U}(a) \subset X \ \big(f(\mathring{U}(a)) \subset V(A)\big).$$

If (X, d_X) and (Y, d_Y) are metric spaces, this last definition can be restated in ε–δ language:

$$\lim_{x \to a} f(x) = A := \forall \varepsilon > 0 \; \exists \delta > 0 \; \forall x \in X$$

$$\left(0 < d_X(a, x) < \delta \Longrightarrow d_Y\big(A, f(x)\big) < \varepsilon\right).$$

In other words,

$$\lim_{x \to a} f(x) = A \Longleftrightarrow \lim_{x \to a} d_Y\big(A, f(x)\big) = 0.$$

Thus we see that, having the concept of a neighborhood, one can define the concept of the limit of a mapping $f : X \to Y$ into a topological or metric space Y just as was done in the case $Y = \mathbb{R}$ or, more generally, $Y = \mathbb{R}^n$.

b. Properties of the Limit of a Mapping

We now make some remarks on the general properties of the limit.

We first note that the uniqueness of the limit obtained earlier no longer holds when Y is not a Hausdorff space. But if Y is a Hausdorff space, then the limit is unique and the proof does not differ at all from the one given in the special cases $Y = \mathbb{R}$ or $Y = \mathbb{R}^n$.

Next, if $f : X \to Y$ is a mapping into a metric space, it makes sense to speak of the *boundedness* of the mapping (meaning the boundedness of the set $f(X)$ in Y), and of *ultimate boundedness* of a mapping with respect to the base $\mathcal{B}$ in X (meaning that there exists an element B of $\mathcal{B}$ on which f is bounded).

It follows from the definition of a limit that if a mapping $f : X \to Y$ of a set X with base $\mathcal{B}$ into a metric space Y has a limit over the base $\mathcal{B}$, then it is ultimately bounded over that base.

c. Questions Involving the Existence of the Limit of a Mapping

Proposition 1 (Limit of a composition of mappings) *Let Y be a set with base $\mathcal{B}_Y$ and $g : Y \to Z$ a mapping of Y into a topological space Z having a limit over the base $\mathcal{B}_Y$.*

Let X be a set with base $\mathcal{B}_X$ and $f : X \to Y$ a mapping of X into Y such that for every element $B_Y \in \mathcal{B}_Y$ there exists an element $B_X \in \mathcal{B}_X$ whose image is contained in B_Y, that is, $f(B_X) \subset B_Y$.

Under these hypotheses the composition $g \circ f : X \to Z$ of the mappings f and g is defined and has a limit over the base $\mathcal{B}_X$, and

$$\lim_{\mathcal{B}_X} g \circ f(x) = \lim_{\mathcal{B}_Y} g(y).$$

For the proof see Theorem 5 of Sect. 3.2.

Another important proposition on the existence of the limit is the Cauchy criterion, to which we now turn. This time we will be discussing a mapping $f : X \to Y$ into a metric space, and in fact a complete metric space.

In the case of a mapping $f : X \to Y$ of the set X into a metric space (Y, d) it is natural to adopt the following definition.

Definition 2 The *oscillation* of the mapping $f : X \to Y$ on a set $E \subset X$ is the quantity

$$\omega(f, E) = \sup_{x_1, x_2 \in E} d\big(f(x_1), f(x_2)\big).$$

The following proposition holds.

Proposition 2 (Cauchy criterion for existence of the limit of a mapping) *Let X be a set with a base $\mathcal{B}$, and let $f : X \to Y$ be a mapping of X into a complete metric space (Y, d).*

A necessary and sufficient condition for the mapping f to have a limit over the base $\mathcal{B}$ is that for every $\varepsilon > 0$ there exists an element B in $\mathcal{B}$ on which the oscillation of the mapping is less than ε.

More briefly:

$$\exists \lim_{\mathcal{B}} f(x) \iff \forall \varepsilon > 0 \, \exists B \in \mathcal{B} \, \big(\omega(f, B) < \varepsilon\big).$$

For the proof see Theorem 4 of Sect. 3.2.

It is useful to remark that the completeness of the space Y is needed only in the implication from the right-hand side to the left-hand side. Moreover, if Y is not a complete space, it is usually this implication that breaks down.

9.6.2 Continuous Mappings

a. Basic Definitions

Definition 3 A mapping $f : X \to Y$ of a topological space (X, τ_X) into a topological space (Y, τ_Y) is *continuous at a point* $a \in X$ if for every neighborhood $V(f(a)) \subset Y$ of the point $f(a) \in Y$ there exists a neighborhood $U(a) \subset X$ of the point $a \in X$ whose image $f(U(a))$ is contained in $V(f(a))$.

Thus,

$$f : X \to Y \text{ is continuous at } a \in X :=$$
$$= \forall V\big(f(a)\big) \, \exists U(a) \, \big(f\big(U(a)\big) \subset V\big(f(a)\big)\big).$$

In the case when X and Y are metric spaces (X, d_X) and (Y, d_Y), Definition 3 can of course be stated in ε–δ language:

$f : X \to Y$ is continuous at $a \in X :=$

$$= \forall \varepsilon > 0 \; \exists \delta > 0 \; \forall x \in X \; \big(d_X(a, x) < \delta \Longrightarrow d_Y\big(f(a), f(x)\big) < \varepsilon\big).$$

Definition 4 The mapping $f : X \to Y$ is *continuous* if it is continuous at each point $x \in X$.

The set of continuous mappings from X into Y will be denoted $C(X, Y)$.

Theorem 1 (Criterion for continuity) *A mapping $f : X \to Y$ of a topological space (X, τ_X) into a topological space (Y, τ_Y) is continuous if and only if the pre-image of every open (resp. closed) subset of y is open (resp. closed) in X.*

Proof Since the pre-image of a complement is the complement of the pre-image, it suffices to prove the assertions for open sets.

We first show that if $f \in C(X, Y)$ and $G_Y \in \tau_Y$, then $G_X = f^{-1}(G_Y)$ belongs to τ_X. If $G_X = \varnothing$, it is immediate that the pre-image is open. If $G_X \neq \varnothing$ and $a \in G_X$, then by definition of continuity of the mapping f at the point a, for the neighborhood G_Y of the point $f(a)$ there exists a neighborhood $U_X(a)$ of $a \in X$ such that $f(U_X(a)) \subset G_Y$. Hence $U_X(a) \subset G_X = f^{-1}(G_Y)$. Since $G_X = \bigcup_{a \in G_X} U_X(a)$, we conclude that G_X is open, that is, $G_X \in \tau_X$.

We now prove that if the pre-image of every open set in Y is open in X, then $f \in C(X, Y)$. But, taking any point $a \in X$ and any neighborhood $V_Y(f(a))$ of its image $f(a)$ in Y, we discover that the set $U_X(a) = f^{-1}(V_Y(f(a)))$ is an open neighborhood of $a \in X$, whose image is contained in $V_Y(f(a))$. Consequently we have verified the definition of continuity of the mapping $f : X \to Y$ at an arbitrary point $a \in X$. $\square$

Definition 5 A bijective mapping $f : X \to Y$ of one topological space (X, τ_X) onto another (Y, τ_Y) is a *homeomorphism* if both the mapping itself and the inverse mapping $f^{-1} : Y \to X$ are continuous.

Definition 6 Topological spaces that admit homeomorphisms onto one another are said to be *homeomorphic*.

As Theorem 1 shows, under a homeomorphism $f : X \to Y$ of the topological space (X, τ_X) onto (Y, τ_Y) the systems of open sets τ_X and τ_Y correspond to each other in the sense that $G_X \in \tau_X \Leftrightarrow f(G_X) = G_Y \in \tau_Y$.

Thus, from the point of view of their topological properties homeomorphic spaces are absolutely identical. Consequently, homeomorphism is the same kind of equivalence relation in the set of all topological spaces as, for example, isometry is in the set of metric spaces.

b. Local Properties of Continuous Mappings

We now exhibit the local properties of continuous mappings. They follow immediately from the corresponding properties of the limit.

Proposition 3 (Continuity of a composition of continuous mappings) *Let* (X, τ_X), (Y, τ_Y) *and* (Z, τ_Z) *be topological spaces. If the mapping* $g : Y \to Z$ *is continuous at a point* $b \in Y$ *and the mapping* $f : X \to Y$ *is continuous at a point* $a \in X$ *for which* $f(a) = b$, *then the composition of these mappings* $g \circ f : X \to Z$ *is continuous at* $a \in X$.

This follows from the definition of continuity of a mapping and Proposition 1.

Proposition 4 (Boundedness of a mapping in a neighborhood of a point of continuity) *If a mapping* $f : X \to Y$ *of a topological space* (X, τ) *into a metric space* (Y, d) *is continuous at a point* $a \in X$, *then it is bounded in some neighborhood of that point.*

This proposition follows from the ultimate boundedness (over a base) of a mapping that has a limit.

Before stating the next proposition on properties of continuous mappings, we recall that for mappings into $\mathbb{R}$ or $\mathbb{R}^n$ we defined the quantity

$$\omega(f; a) := \lim_{r \to 0} \omega\big(f, B(a, r)\big)$$

to be the *oscillation of* f *at the point* a. Since both the concept of the oscillation of a mapping on a set and the concept of a ball $B(a, r)$ make sense in any metric space, the definition of the oscillation $\omega(f, a)$ of the mapping f at the point a also makes sense for a mapping $f : X \to Y$ of a metric space (X, d_X) into a metric space (Y, d_Y).

Proposition 5 *A mapping* $f : X \to Y$ *of a metric space* (X, d_X) *into a metric space* (Y, d_Y) *is continuous at the point* $a \in X$ *if and only if* $\omega(f, a) = 0$.

This proposition follows immediately from the definition of continuity of a mapping at a point.

c. Global Properties of Continuous Mappings

We now discuss some of the important global properties of continuous mappings.

Theorem 2 *The image of a compact set under a continuous mapping is compact.*

Proof Let $f : K \to Y$ be a continuous mapping of the compact space (K, τ_K) into a topological space (Y, τ_Y), and let $\{G_Y^\alpha, \alpha \in A\}$ be a covering of $f(K)$ by sets that are open in Y. By Theorem 1, the sets $\{G_X^\alpha = f^{-1}(G_Y^\alpha), \alpha \in A\}$ form an open covering of K. Extracting a finite covering $G_X^{\alpha_1}, \ldots, G_X^{\alpha_n}$, we find a finite covering $G_Y^{\alpha_1}, \ldots, G_Y^{\alpha_n}$ of $f(K) \subset Y$. Thus $f(K)$ is compact in Y. $\square$

Corollary *A continuous real-valued function* $f : K \to \mathbb{R}$ *on a compact set assumes its maximal value at some point of the compact set (and also its minimal value, at some point).*

Proof Indeed, $f(K)$ is a compact set in $\mathbb{R}$, that is, it is closed and bounded. This means that $\inf f(K) \in f(K)$ and $\sup f(K) \in f(K)$. $\square$

In particular, if K is a closed interval $[a, b] \subset \mathbb{R}$, we again obtain the classical theorem of Weierstrass.

Cantor's theorem on uniform continuity carries over verbatim to mappings that are continuous on compact sets. Before stating it, we must give a necessary definition.

Definition 7 A mapping $f : X \to Y$ of a metric space (X, d_X) into a metric space (Y, d_Y) is *uniformly continuous* if for every $\varepsilon > 0$ there exists $\delta > 0$ such that the oscillation $\omega(f, E)$ of f on each set $E \subset X$ of diameter less than δ is less than ε.

Theorem 3 (Uniform continuity) *A continuous mapping* $f : K \to Y$ *of a compact metric space K into a metric space (Y, d_Y) is uniformly continuous.*

In particular, if K is a closed interval in $\mathbb{R}$ and $Y = \mathbb{R}$, we again have the classical theorem of Cantor, the proof of which given in Sect. 4.2.2 carries over with almost no changes to this general case.

Let us now consider continuous mappings of connected spaces.

Theorem 4 *The image of a connected topological space under a continuous mapping is connected.*

Proof Let $f : X \to Y$ be a continuous mapping of a connected topological space (X, τ_X) onto a topological space (Y, τ_Y). Let E_Y be an open-closed subset of Y. By Theorem 1, the pre-image $E_X = f^{-1}(E_Y)$ of the set E_Y is open-closed in X. By the connectedness of X, either $E_X = \varnothing$ or $E_X = X$. But this means that either $E_Y = \varnothing$ or $E_Y = Y = f(X)$. $\square$

Corollary *If a function* $f : X \to \mathbb{R}$ *is continuous on a connected topological space (X, τ) and assumes values $f(a) = A \in \mathbb{R}$ and $f(b) = B \in \mathbb{R}$, then for any number C between A and B there exists a point $c \in X$ at which $f(c) = C$.*

Proof Indeed, by Theorem 4 $f(X)$ is a connected set in $\mathbb{R}$. But the only connected subsets of $\mathbb{R}$ are intervals (see the Proposition in Sect. 9.4). Thus the point C belongs to $f(X)$ along with A and B. $\square$

In particular, if X is a closed interval, we again have the classical intermediate-value theorem for a continuous real-valued function.

9.6.3 Problems and Exercises

1. a) If the mapping $f : X \to Y$ is continuous, will the images of open (or closed) sets in X be open (or closed) in Y?

b) If the image, as well as the inverse image, of an open set under the mapping $f : X \to Y$ is open, does it necessarily follow that f is a homeomorphism?

c) If the mapping $f : X \to Y$ is continuous and bijective, is it necessarily a homeomorphism?

d) Is a mapping satisfying b) and c) simultaneously a homeomorphism?

2. Show the following.

a) Every continuous bijective mapping of a compact space into a Hausdorff space is a homeomorphism.

b) Without the requirement that the range be a Hausdorff space, the preceding statement is in general not true.

3. Determine whether the following subsets of $\mathbb{R}^n$ are (pairwise) homeomorphic as topological spaces: a line, an open interval on the line, a closed interval on the line; a sphere; a torus.

4. A topological space (X, τ) is *arcwise connected* or *path connected* if any two of its points can be joined by a path lying in X. More precisely, this means that for any points A and B in X there exists a continuous mapping $f : I \to X$ of a closed interval $[a, b] \subset \mathbb{R}$ into X such that $f(a) = A$ and $f(b) = B$.

a) Show that every path connected space is connected.

b) Show that every convex set in $\mathbb{R}^n$ is path connected.

c) Verify that every connected open subset of $\mathbb{R}^n$ is path connected.

d) Show that a sphere $S(a, r)$ is path connected in $\mathbb{R}^n$, but that it may fail to be connected in another metric space, endowed with a completely different topology.

e) Verify that in a topological space it is impossible to join an interior point of a set to an exterior point without intersecting the boundary of the set.

9.7 The Contraction Mapping Principle

Here we shall establish a principle that, despite its simplicity, turns out to be an effective way of proving many existence theorems.

Definition 1 A point $a \in X$ is a *fixed point* of a mapping $f : X \to X$ if $f(a) = a$.

Definition 2 A mapping $f : X \to X$ of a metric space (X, d) into itself is called a *contraction* if there exists a number q, $0 < q < 1$, such that the inequality

$$d\big(f(x_1), f(x_2)\big) \le q d(x_1, x_2) \tag{9.20}$$

holds for any points x_1 and x_2 in X.

Theorem (Picard[6]–Banach[7] fixed-point principle) *A contraction mapping $f : X \to X$ of a complete metric space (X, d) into itself has a unique fixed point a.*

Moreover, for any point $x_0 \in X$ the recursively defined sequence $x_0, x_1 = f(x_0), \ldots, x_{n+1} = f(x_n), \ldots$ converges to a. The rate of convergence is given by the estimate

$$d(a, x_n) \le \frac{q^n}{1-q} d(x_1, x_0). \tag{9.21}$$

Proof We shall take an arbitrary point $x_0 \in X$ and show that the sequence $x_0, x_1 = f(x_0), \ldots, x_{n+1} = f(x_n), \ldots$ is a Cauchy sequence. The mapping f is a contraction, so that by Eq. (9.20)

$$d(x_{n+1}, x_n) \le q d(x_n, x_{n-1}) \le \cdots \le q^n d(x_1, x_0)$$

and

$$d(x_{n+k}, x_n) \le d(x_n, x_{n+1}) + \cdots + d(x_{n+k-1}, x_{n+k}) \le$$

$$\le \big(q^n + q^{n+1} + \cdots + q^{n+k-1}\big) d(x_1, x_0) \le \frac{q^n}{1-q} d(x_1, x_0).$$

From this one can see that the sequence $x_0, x_1, \ldots, x_n, \ldots$ is indeed a Cauchy sequence.

The space (X, d) is complete, so that this sequence has a limit $\lim_{n \to \infty} x_n = a \in X$.

It is clear from the definition of a contraction mapping that a contraction is always continuous, and therefore

$$a = \lim_{n \to \infty} x_{n+1} = \lim_{n \to \infty} f(x_n) = f\left(\lim_{n \to \infty} x_n\right) = f(a).$$

Thus a is a fixed point of the mapping f.

---

[6]Ch.É. Picard (1856–1941) – French mathematician who obtained many important results in the theory of differential equations and analytic function theory.

[7]S. Banach (1892–1945) – Polish mathematician, one of the founders of functional analysis.

The mapping f cannot have a second fixed point, since the relations $a_i = f(a_i)$, $i = 1, 2$, imply, when we take account of (9.20), that

$$0 \le d(a_1, a_2) = d\big(f(a_1), f(a_2)\big) \le q d(a_1, a_2),$$

which is possible only if $d(a_1, a_2) = 0$, that is, $a_1 = a_2$.

Next, by passing to the limit as $k \to \infty$ in the relation

$$d(x_{n+k}, x_n) \le \frac{q^n}{1-q} d(x_1, x_0),$$

we find that

$$d(a, x_n) \le \frac{q^n}{1-q} d(x_1, x_0). \qquad \square$$

The following proposition supplements this theorem.

Proposition (Stability of the fixed point) *Let (X, d) be a complete metric space and (Ω, τ) a topological space that will play the role of a parameter space in what follows.*

Suppose to each value of the parameter $t \in \Omega$ there corresponds a contraction mapping $f_t : X \to X$ of the space X into itself and that the following conditions hold.

a) *The family $\{f_t; t \in \Omega\}$ is uniformly contracting, that is, there exists q, $0 < q < 1$, such that each mapping f_t is a q-contraction.*

b) *For each $x \in X$ the mapping $f_t(x) : \Omega \to X$ is continuous as a function of t at some point $t_0 \in \Omega$, that is $\lim_{t \to t_0} f_t(x) = f_{t_0}(x)$.*

Then the solution $a(t) \in X$ of the equation $x = f_t(x)$ depends continuously on t at the point t_0, that is, $\lim_{t \to t_0} a(t) = a(t_0)$.

Proof As was shown in the proof of the theorem, the solution $a(t)$ of the equation $x = f_t(x)$ can be obtained as the limit of the sequence $\{x_{n+1} = f_t(x_n); n = 0, 1, \ldots\}$ starting from any point $x_0 \in X$. Let $x_0 = a(t_0) = f_{t_0}(a(t_0))$.

Taking account of the estimate (9.21) and condition a), we obtain

$$d\big(a(t), a(t_0)\big) = d\big(a(t), x_0\big) \le$$

$$\le \frac{1}{1-q} d(x_1, x_0) = \frac{1}{1-q} d\big(f_t\big(a(t_0)\big), f_{t_0}\big(a(t_0)\big)\big).$$

By condition b), the last term in this relation tends to zero as $t \to t_0$. Thus it has been proved that

$$\lim_{t \to t_0} d\big(a(t), a(t_0)\big) = 0, \quad \text{that is} \quad \lim_{t \to t_0} a(t) = a(t_0). \qquad \square$$

Example 1 As an important example of the application of the contraction mapping principle we shall prove, following Picard, an existence theorem for the solution of the differential equation $y'(x) = f(x, y(x))$ satisfying an initial condition $y(x_0) = y_0$.

If the function $f \in C(\mathbb{R}^2, \mathbb{R})$ is such that

$$\left| f(u, v_1) - f(u, v_2) \right| \leq M |v_1 - v_2|,$$

where M is a constant, then, for any initial condition

$$y(x_0) = y_0, \tag{9.22}$$

there exists a neighborhood $U(x_0)$ of $x_0 \in \mathbb{R}$ and a unique function $y = y(x)$ defined in $U(x_0)$ satisfying the equation

$$y' = f(x, y) \tag{9.23}$$

and the initial condition (9.22).

Proof Equation (9.23) and the condition (9.22) can be jointly written as a single relation

$$y(x) = y_0 + \int_{x_0}^{x} f(t, y(t)) \, dt. \tag{9.24}$$

Denoting the right-hand side of this equality by $A(y)$, we find that $A : C(V(x_0), \mathbb{R}) \to C(V(x_0), \mathbb{R})$ is a mapping of the set of continuous functions defined on a neighborhood $V(x_0)$ of x_0 into itself. Regarding $C(V(x_0), \mathbb{R})$ as a metric space with the uniform metric (see formula (9.6) from Sect. 9.1), we find that

$$d(Ay_1, Ay_2) = \max_{x \in \bar{V}(x_0)} \left| \int_{x_0}^{x} f(t, y_1(t)) \, dt - \int_{x_0}^{x} f(t, y_2(t)) \, dt \right| \leq$$

$$\leq \max_{x \in \bar{V}(x_0)} \left| \int_{x_0}^{x} M |y_1(t) - y_2(t)| \, dt \right| \leq M |x - x_0| d(y_1, y_2).$$

If we assume that $|x - x_0| \leq \frac{1}{2M}$, then the inequality

$$d(Ay_1, Ay_2) \leq \frac{1}{2} d(y_1, y_2)$$

is fulfilled on the corresponding closed interval I, where $d(y_1, y_2) = \max_{x \in I} |y_1(x) - y_2(x)|$. Thus we have a contraction mapping

$$A : C(I, \mathbb{R}) \to C(I, \mathbb{R})$$

of the complete metric space $(C(I, \mathbb{R}), d)$ (see Example 4 of Sect. 9.5) into itself, which by the contraction mapping principle must have a unique fixed point $y = Ay$. But this means that the function in $C(I, \mathbb{R})$ just found is the unique function defined on $I \ni x_0$ and satisfying Eq. (9.24).   □

Example 2 As an illustration of what was just said, we shall seek a solution of the familiar equation

$$y' = y$$

with the initial condition (9.22) on the basis of the contraction mapping principle.

In this case

$$Ay = y_0 + \int_{x_0}^{x} y(t)\, dt,$$

and the principle is applicable at least for $|x - x_0| \leq q < 1$.

Starting from the initial approximation $y(x) \equiv 0$, we construct successively the sequence $0, y_1 = A(0), \dots, y_{n+1}(t) = A(y_n(t)), \dots$ of approximations

$$y_1(t) = y_0,$$

$$y_2(t) = y_0\big(1 + (x - x_0)\big),$$

$$y_3(t) = y_0\left(1 + (x - x_0) + \frac{1}{2}(x - x_0)^2\right),$$

$$\vdots$$

$$y_{n+1}(t) = y_0\left(1 + (x - x_0) + \frac{1}{2!}(x - x_0)^2 + \cdots + \frac{1}{n!}(x - x_0)^n\right),$$

$$\vdots$$

from which it is already clear that

$$y(x) = y_0 e^{x - x_0}.$$

The fixed-point principle stated in the theorem above also goes by the name of the *contraction mapping principle*. It arose as a generalization of Picard's proof of the existence theorem for a solution of the differential equation (9.23), which was discussed in Example 1. The contraction mapping principle was stated in full generality by Banach.

Example 3 (Newton's method of finding a root of the equation $f(x) = 0$) Suppose a real-valued function that is convex and has a positive derivative on a closed interval $[\alpha, \beta]$ assumes values of opposite signs at the endpoints of the interval. Then there is a unique point a in the interval at which $f(a) = 0$. In addition to the elementary method of finding the point a by successive bisection of the interval, there also exist more sophisticated and rapid methods of finding it, using the properties of the function f. Thus, in the present case, one may use the following method, proposed by Newton and called *Newton's method* or the *method of tangents*. Take an arbitrary point $x_0 \in [\alpha, \beta]$ and write the equation $y = f(x_0) + f'(x_0)(x - x_0)$ of the tangent to the graph of the function at the point $(x_0, f(x_0))$. We then find the point $x_1 = x_0 - [f'(x_0)]^{-1} \cdot f(x_0)$ where the tangent intersects the x-axis (Fig. 9.3). We take x_1 as the first approximation of the root a and repeat this operation, replacing x_0

Fig. 9.3

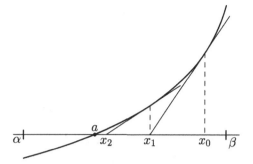

by x_1. In this way we obtain a sequence

$$x_{n+1} = x_n - \left[f'(x_n) \right]^{-1} \cdot f(x_n) \tag{9.25}$$

of points that, as one can verify, will tend monotonically to a in the present case.

In particular, if $f(x) = x^k - a$, that is, when we are seeking $\sqrt[k]{a}$, where $a > 0$, the recurrence relation (9.25) has the form

$$x_{n+1} = x_n - \frac{x_n^k - a}{k x_n^{k-1}},$$

which for $k = 2$ becomes the familiar expression

$$x_{n+1} = \frac{1}{2} \left(x_n + \frac{a}{x_n} \right).$$

The method (9.25) for forming the sequence $\{x_n\}$ is called *Newton's method*.

If instead of the sequence (9.25) we consider the sequence obtained by the recurrence relation

$$x_{n+1} = x_n - \left[f'(x_0) \right]^{-1} \cdot f(x_n), \tag{9.26}$$

we speak of the *modified Newton's method*.[8] The modification amounts to computing the derivative once and for all at the point x_0.

Consider the mapping

$$x \mapsto A(x) = x - \left[f'(x_0) \right]^{-1} \cdot f(x). \tag{9.27}$$

By Lagrange's theorem

$$\left| A(x_2) - A(x_1) \right| = \left| \left[f'(x_0) \right]^{-1} \cdot f'(\xi) \right| \cdot |x_2 - x_1|,$$

where ξ is a point lying between x_1 and x_2.

---

[8] In functional analysis it has numerous applications and is called the *Newton–Kantorovich method*. L.V. Kantorovich (1912–1986) – eminent Soviet mathematician, whose research in mathematical economics earned him the Nobel Prize.

Thus, if the conditions

$$A(I) \subset I \tag{9.28}$$

and

$$\left|\left[f'(x_0)\right]^{-1} \cdot f'(x)\right| \le q < 1, \tag{9.29}$$

hold on some closed interval $I \subset \mathbb{R}$, then the mapping $A : I \to I$ defined by relation (9.27) is a contraction of this closed interval. Then by the general principle it has a unique fixed point on the interval. But, as can be seem from (9.27), the condition $A(a) = a$ is equivalent to $f(a) = 0$.

Hence, when conditions (9.28) and (9.29) hold for a function f, the modified Newton's method (9.26) leads to the required solution $x = a$ of the equation $f(x) = 0$ by the contraction mapping principle.

9.7.1 Problems and Exercises

1. Show that condition (9.20) in the contraction mapping principle cannot be replaced by the weaker condition

$$d\big(f(x_1), f(x_2)\big) < d(x_1, x_2).$$

2. a) Prove that if a mapping $f : X \to X$ of a complete metric space (X, d) into itself is such that some iteration of it $f^n : X \to X$ is a contraction, then f has a unique fixed point.

b) Verify that the mapping $A : C(I, \mathbb{R}) \to C(I, \mathbb{R})$ in Example 2 is such that for any closed interval $I \subset \mathbb{R}$ some iteration A^n of the mapping A is a contraction.

c) Deduce from b) that the local solution $y = y_0 e^{x-x_0}$ found in Example 2 is actually a solution of the original equation on the entire real line.

3. a) Show that in the case of a function on $[\alpha, \beta]$ that is convex and has a positive derivative and assumes values of opposite signs at the endpoints, Newton's method really does give a sequence $\{x_n\}$ that converges to the point $a \in [\alpha, \beta]$ at which $f(a) = 0$.

b) Estimate the rate of convergence of the sequence (9.25) to the point a.

Chapter 10
*Differential Calculus from a More General Point of View

10.1 Normed Vector Spaces

Differentiation is the process of finding the best local linear approximation of a function. For that reason any reasonably general theory of differentiation must be based on elementary ideas connected with linear functions. From the course in algebra the reader is well acquainted with the concept of a *vector space*, as well as linear dependence and independence of systems of vectors, bases and dimension of a vector space, vector subspaces, and so forth. In the present section we shall present vector spaces with a norm, or as they are described, *normed vector spaces*, which are widely used in analysis. We begin, however, with some examples of vector spaces.

10.1.1 Some Examples of Vector Spaces in Analysis

Example 1 The real vector space $\mathbb{R}^n$ and the complex vector space $\mathbb{C}^n$ are classical examples of vector spaces of dimension n over the fields of real and complex numbers respectively.

Example 2 In analysis, besides the spaces $\mathbb{R}^n$ and $\mathbb{C}^n$ exhibited in Example 1, we encounter the space closest to them, which is the space ℓ of sequences $x = (x^1, \ldots, x^n, \ldots)$ of real or complex numbers. The vector-space operations in ℓ, as in $\mathbb{R}^n$ and $\mathbb{C}^n$, are carried out coordinatewise. One peculiarity of this space, when compared with $\mathbb{R}^n$ or $\mathbb{C}^n$ is that any finite subsystem of the countable system of vectors $\{x_i = (0, \ldots, 0, x^i = 1, 0, \ldots), i \in \mathbb{N}\}$ is linearly independent, that is, ℓ is an infinite-dimensional vector space (of countable dimension in the present case).

The set of finite sequences (all of whose terms are zero from some point on) is a vector subspace $\underset{0}{\ell}$ of the space ℓ, also infinite-dimensional.

Example 3 Let $F[a, b]$ be the set of numerical-valued (real- or complex-valued) functions defined on the closed interval $[a, b]$. This set is a vector space over the

© Springer-Verlag Berlin Heidelberg 2016
V.A. Zorich, *Mathematical Analysis II*, Universitext,
DOI 10.1007/978-3-662-48993-2_2

corresponding number field with respect to the operations of addition of functions and multiplication of a function by a number.

The set of functions of the form

$$e_\tau(x) = \begin{cases} 0, & \text{if } x \in [a, b] \text{ and } x \neq \tau, \\ 1, & \text{if } x \in [a, b] \text{ and } x = \tau \end{cases}$$

is a continuously indexed system of linearly independent vectors in $F[a, b]$.

The set $C[a, b]$ of continuous functions is obviously a subspace of the space $F[a, b]$ just constructed.

Example 4 If X_1 and X_2 are two vector spaces over the same field, there is a natural way of introducing a vector-space structure into their direct product $X_1 \times X_2$, namely by carrying out the vector-space operations on elements $x = (x_1, x_2) \in X_1 \times X_2$ coordinatewise.

Similarly one can introduce a vector-space structure into the direct product $X_1 \times \cdots \times X_n$ of any finite set of vector spaces. This is completely analogous to the cases of $\mathbb{R}^n$ and $\mathbb{C}^n$.

10.1.2 Norms in Vector Spaces

We begin with the basic definition.

Definition 1 Let X be a vector space over the field of real or complex numbers.

A function $\| \ \| : X \to \mathbb{R}$ assigning to each vector $x \in X$ a real number $\|x\|$ is called a *norm* in the vector space X if it satisfies the following three conditions:

a) $\|x\| = 0 \Leftrightarrow x = 0$ (nondegeneracy);
b) $\|\lambda x\| = |\lambda| \|x\|$ (homogeneity);
c) $\|x_1 + x_2\| \leq \|x_1\| + \|x_2\|$ (the triangle inequality).

Definition 2 A vector space with a norm defined on it is called a *normed vector space*.

Definition 3 The value of the norm at a vector is called the *norm of that vector*.

The norm of a vector is always nonnegative and, as can be seen by a), equals zero only for the zero vector.

Proof Indeed, by c), taking account of a) and b), we obtain for every $x \in X$,

$$0 = \|0\| = \|x + (-x)\| \leq \|x\| + \|-x\| = \|x\| + |-1| \|x\| = 2\|x\|. \qquad \square$$

By induction, condition c) implies the following general inequality.

$$\|x_1 + \cdots + x_n\| \leq \|x_1\| + \cdots + \|x_n\|, \tag{10.1}$$

and taking account of b), one can easily deduce from c) the following useful inequality.

$$\left| \|x_1\| - \|x_2\| \right| \leq \|x_1 - x_2\|. \tag{10.2}$$

Every normed vector space has a natural metric

$$d(x_1, x_2) = \|x_1 - x_2\|. \tag{10.3}$$

The fact that the function $d(x_1, x_2)$ just defined satisfies the axioms for a metric follows immediately from the properties of the norm. Because of the vector-space structure in X the metric d in X has two additional special properties:

$$d(x_1 + x, x_2 + x) = \left\|(x_1 + x) - (x_2 + x)\right\| = \|x_1 - x_2\| = d(x_1, x_2),$$

that is, the metric is translation-invariant, and

$$d(\lambda x_1, \lambda x_2) = \|\lambda x_1 - \lambda x_2\| = \left\|\lambda(x_1 - x_2)\right\| = |\lambda| \|x_1 - x_2\| = |\lambda| d(x_1, x_2),$$

that is, it is homogeneous.

Definition 4 If a normed vector space is complete as a metric space with the natural metric (10.3), it is called a *complete normed vector space* or *Banach space*.

Example 5 If for $p \geq 1$ we set

$$\|x\|_p := \left(\sum_{i=1}^{n} |x^i|^p \right)^{\frac{1}{p}} \tag{10.4}$$

for $x = (x^1, \ldots, x^n) \in \mathbb{R}^n$, it follows from Minkowski's inequality that we obtain a norm on $\mathbb{R}^n$. The space $\mathbb{R}^n$ endowed with this norm will be denoted $\mathbb{R}^n_p$.

One can verify that

$$\|x\|_{p_2} \leq \|x\|_{p_1}, \quad \text{if } 1 \leq p_1 \leq p_2, \tag{10.5}$$

and that

$$\|x\|_p \rightarrow \max\{|x^1|, \ldots, |x^n|\} \tag{10.6}$$

as $p \rightarrow +\infty$. Thus, it is natural to set

$$\|x\|_\infty := \max\{|x^1|, \ldots, |x^n|\}. \tag{10.7}$$

It then follows from (10.4) and (10.5) that

$$\|x\|_\infty \leq \|x\|_p \leq \|x\|_1 \leq n\|x\|_\infty \quad \text{for } p \geq 1. \tag{10.8}$$

It is clear from this inequality, as in fact it is from the very definition of the norm $\|x\|_p$ in Eq. (10.4), that $\mathbb{R}^n_p$ is a complete normed vector space.

Example 6 The preceding example can be usefully generalized as follows. If $X = X_1 \times \cdots \times X_n$ is the direct product of normed vector spaces, one can introduce the norm of a vector $x = (x_1, \ldots, x_n)$ in the direct product by setting

$$\|x\|_p := \left(\sum_{i=1}^n \|x_i\|^p \right)^{\frac{1}{p}}, \quad p \geq 1, \tag{10.9}$$

where $\|x_i\|$ is the norm of the vector $x_i \in X_i$.

Naturally, inequalities (10.8) remain valid in this case as well.

From now on, when the direct product of normed spaces is considered, unless the contrary is explicitly stated, it is assumed that the norm is defined in accordance with formula (10.9) (including the case $p = +\infty$).

Example 7 Let $p \geq 1$. We denote by ℓ_p the set of sequences $x = (x^1, \ldots, x^n, \ldots)$ of real or complex numbers such that the series $\sum_{n=1}^\infty |x^n|^p$ converges, and for $x \in \ell_p$ we set

$$\|x\|_p := \left(\sum_{n=1}^\infty |x^n|^p \right)^{\frac{1}{p}}. \tag{10.10}$$

Using Minkowski's inequality, one can easily see that ℓ_p is a normed vector space with respect to the standard vector-space operations and the norm (10.10). This is an infinite-dimensional space with respect to which $\mathbb{R}^n_p$ is a vector subspace of finite dimension.

All the inequalities (10.8) except the last are valid for the norm (10.10). It is not difficult to verify that ℓ_p is a Banach space.

Example 8 In the vector space $C[a, b]$ of numerical-valued functions that are continuous on the closed interval $[a, b]$, one usually considers the following norm:

$$\|f\| := \max_{x \in [a,b]} |f(x)|. \tag{10.11}$$

We leave the verification of the norm axioms to the reader. We remark that this norm generates a metric on $C[a, b]$ that is already familiar to us (see Sect. 9.5), and we know that the metric space that thereby arises is complete. Thus the vector space $C[a, b]$ with the norm (10.11) is a Banach space.

Example 9 One can also introduce another norm in $C[a, b]$

$$\|f\|_p := \left(\int_a^b |f|^p(x)\, \mathrm{d}x \right)^{\frac{1}{p}}, \quad p \geq 1, \tag{10.12}$$

which becomes (10.11) as $p \to +\infty$.

It is easy to see (for example, Sect. 9.5) that the space $C[a, b]$ with the norm (10.12) is not complete for $1 \le p < +\infty$.

10.1.3 Inner Products in Vector Spaces

An important class of normed spaces is formed by the spaces with an inner product. They are a direct generalization of Euclidean spaces.

We recall their definition.

Definition 5 We say that a *Hermitian form* is defined in a vector space X (over the field of complex numbers) if there exists a mapping $\langle, \rangle : X \times X \to \mathbb{C}$ having the following properties:

a) $\langle x_1, x_2 \rangle = \overline{\langle x_2, x_1 \rangle}$,
b) $\langle \lambda x_1, x_2 \rangle = \lambda \langle x_1, x_2 \rangle$,
c) $\langle x_1 + x_2, x_3 \rangle = \langle x_1, x_3 \rangle + \langle x_2, x_3 \rangle$,

where x_1, x_2, x_3 are vectors in X and $\lambda \in \mathbb{C}$.

It follows from a), b), and c), for example, that

$$\langle x_1, \lambda x_2 \rangle = \overline{\langle \lambda x_2, x_1 \rangle} = \overline{\lambda \langle x_2, x_1 \rangle} = \overline{\lambda}\,\overline{\langle x_2, x_1 \rangle} = \overline{\lambda} \langle x_1, x_2 \rangle;$$
$$\langle x_1, x_2 + x_3 \rangle = \overline{\langle x_2 + x_3, x_1 \rangle} = \overline{\langle x_2, x_1 \rangle} + \overline{\langle x_3, x_1 \rangle} = \langle x_1, x_2 \rangle + \langle x_1, x_3 \rangle;$$
$$\langle x, x \rangle = \overline{\langle x, x \rangle}, \quad \text{that is,} \quad \langle x, x \rangle \text{ is a real number.}$$

A Hermitian form is called *nonnegative* if

d) $\langle x, x \rangle \ge 0$

and *nondegenerate* if

e) $\langle x, x \rangle = 0 \Leftrightarrow x = 0$.

If X is a vector space over the field of real numbers, one must of course consider a real-valued form $\langle x_1, x_2 \rangle$. In this case a) can be replaced by $\langle x_1, x_2 \rangle = \langle x_2, x_1 \rangle$, which means that the form is symmetric with respect to its vector arguments x_1 and x_2.

An example of such a form is the dot product familiar from analytic geometry for vectors in three-dimensional Euclidean space. In connection with this analogy we make the following definition.

Definition 6 A nondegenerate nonnegative Hermitian form in a vector space is called an *inner product* in the space.

Example 10 An inner product of vectors $x = (x^1, \ldots, x^n)$ and $y = (y^1, \ldots, y^n)$ in $\mathbb{R}^n$ can be defined by setting

$$\langle x, y \rangle := \sum_{i=1}^{n} x^i y^i, \tag{10.13}$$

and in $\mathbb{C}^n$ by setting

$$\langle x, y \rangle := \sum_{i=1}^{n} x^i \overline{y^i}. \tag{10.14}$$

Example 11 In ℓ_2 the inner product of the vectors x and y can be defined as

$$\langle x, y \rangle := \sum_{i=1}^{\infty} x^i \overline{y^i}.$$

The series in this expression converges absolutely since

$$2 \sum_{i=1}^{\infty} |x^i \overline{y^i}| \le \sum_{i=1}^{\infty} |x^i|^2 + \sum_{i=1}^{\infty} |y^i|^2.$$

Example 12 An inner product can be defined in $C[a, b]$ by the formula

$$\langle f, g \rangle := \int_a^b (f \cdot \bar{g})(x) \, dx. \tag{10.15}$$

It follows easily from properties of the integral that all the requirements for an inner product are satisfied in this case.

The following important inequality, known as the *Cauchy–Bunyakovskii inequality*, holds for the inner product:

$$\left| \langle x, y \rangle \right|^2 \le \langle x, x \rangle \cdot \langle y, y \rangle, \tag{10.16}$$

where equality holds if and only if the vectors x and y are collinear.

Proof Indeed, let $a = \langle x, x \rangle$, $b = \langle x, y \rangle$, and $c = \langle y, y \rangle$. By hypothesis $a \ge 0$ and $c \ge 0$. If $c > 0$, the inequalities

$$0 \le \langle x + \lambda y, x + \lambda y \rangle = a + \bar{b}\lambda + b\bar{\lambda} + c\lambda\bar{\lambda}$$

with $\lambda = -\frac{b}{c}$ imply

$$0 \le a - \frac{\bar{b}b}{c} - \frac{b\bar{b}}{c} + \frac{b\bar{b}}{c}$$

or

$$0 \le ac - b\bar{b} = ac - |b|^2, \qquad (10.17)$$

which is the same as (10.16).

The case $a > 0$ can be handled similarly.

If $a = c = 0$, then, setting $\lambda = -b$ in (10.17), we find $0 \le -\bar{b}b - b\bar{b} = -2|b|^2$, that is, $b = 0$, and (10.16) is again true.

If x and y are not collinear, then $0 < \langle x + \lambda y, x + \lambda y \rangle$ and consequently inequality (10.16) is a strict inequality in this case. But if x and y are collinear, it becomes equality as one can easily verify. $\square$

A vector space with an inner product has a natural norm:

$$\|x\| := \sqrt{\langle x, x \rangle} \qquad (10.18)$$

and metric

$$d(x, y) := \|x - y\|.$$

Using the Cauchy–Bunyakovskii inequality, we verify that if $\langle x, y \rangle$ is a nondegenerate nonnegative Hermitian form, then formula (10.18) does indeed define a norm.

Proof In fact,

$$\|x\| = \sqrt{\langle x, x \rangle} = 0 \Leftrightarrow x = 0,$$

since the form $\langle x, y \rangle$ is nondegenerate.

Next,

$$\|\lambda x\| = \sqrt{\langle \lambda x, \lambda x \rangle} = \sqrt{\lambda \bar{\lambda} \langle x, x \rangle} = |\lambda| \sqrt{\langle x, x \rangle} = |\lambda| \|x\|.$$

We verify finally that the triangle inequality holds:

$$\|x + y\| \le \|x\| + \|y\|.$$

Thus, we need to show that

$$\sqrt{\langle x + y, x + y \rangle} \le \sqrt{\langle x, x \rangle} + \sqrt{\langle y, y \rangle},$$

or, after we square and cancel, that

$$\langle x, y \rangle + \langle y, x \rangle \le 2\sqrt{\langle x, x \rangle \cdot \langle y, y \rangle}.$$

But

$$\langle x, y \rangle + \langle y, x \rangle = \langle x, y \rangle + \overline{\langle x, y \rangle} = 2\,\mathrm{Re}\langle x, y \rangle \le 2|\langle x, y \rangle|,$$

and the inequality to be proved now follows immediately from the Cauchy–
Bunyakovskii inequality (10.16).   □

In conclusion we note that finite-dimensional vector spaces with an inner product
are usually called *Euclidean* or *Hermitian* (*unitary*) spaces according as the field of
scalars is $\mathbb{R}$ or $\mathbb{C}$ respectively. If a normed vector space is infinite-dimensional, it
is called a *Hilbert space* if it is complete in the metric induced by the natural norm
and a *pre-Hilbert space* otherwise.

10.1.4 Problems and Exercises

1. a) Show that if a translation-invariant homogeneous metric $d(x_1, x_2)$ is defined
in a vector space X, then X can be normed by setting $\|x\| = d(0, x)$.

b) Verify that the norm in a vector space X is a continuous function with respect
to the topology induced by the natural metric (10.3).

c) Prove that if X is a finite-dimensional vector space and $\|x\|$ and $\|x\|'$ are two
norms on X, then one can find positive numbers M, N such that

$$M\|x\| \le \|x\|' \le N\|x\| \tag{10.19}$$

for any vector $x \in X$.

d) Using the example of the norms $\|x\|_1$ and $\|x\|_\infty$ in the space ℓ, verify that
the preceding inequality generally does not hold in infinite-dimensional spaces.

2. a) Prove inequality (10.5).

b) Verify relation (10.6).

c) Show that as $p \to +\infty$ the quantity $\|f\|_p$ defined by formula (10.12) tends
to the quantity $\|f\|$ given by formula (10.11).

3. a) Verify that the normed space ℓ_p considered in Example 7 is complete.

b) Show that the subspace of ℓ_p consisting of finite sequences (ending in zeros)
is not a Banach space.

4. a) Verify that relations (10.11) and (10.12) define a norm in the space $C[a, b]$
and convince yourself that a complete normed space is obtained in one of these
cases but not in the other.

b) Does formula (10.12) define a norm in the space $\mathcal{R}[a, b]$ of Riemann-
integrable functions?

c) What factorization (identification) must one make in $\mathcal{R}[a, b]$ so that the quantity defined by (10.12) will be a norm in the resulting vector space?

5. a) Verify that formulas (10.13)–(10.15) do indeed define an inner product in the
corresponding vector spaces.

b) Is the form defined by formula (10.15) an inner product in the space $\mathcal{R}[a, b]$
of Riemann-integrable functions?

c) Which functions in $\mathcal{R}[a, b]$ must be identified so that the answer to part b) will be positive in the quotient space of equivalence classes?

6. Using the Cauchy–Bunyakovskii inequality, find the greatest lower bound of the values of the product $(\int_a^b f(x)\, dx)(\int_a^b (1/f)(x)\, dx)$ on the set of continuous real-valued functions that do not vanish on the closed interval $[a, b]$.

10.2 Linear and Multilinear Transformations

10.2.1 Definitions and Examples

We begin by recalling the basic definition.

Definition 1 If X and Y are vector spaces over the same field (in our case, either $\mathbb{R}$ or $\mathbb{C}$), a mapping $A : X \to Y$ is *linear* if the equalities

$$A(x_1 + x_2) = A(x_1) + A(x_2),$$

$$A(\lambda x) = \lambda A(x)$$

hold for any vectors x, x_1, x_2 in X and any number λ in the field of scalars.

For a linear transformation $A : X \to Y$ we often write Ax instead of $A(x)$.

Definition 2 A mapping $A : X_1 \times \cdots \times X_n \to Y$ of the direct product of the vector spaces $X_1, \ldots, X_n$ into the vector space Y is *multilinear* (*n-linear*) if the mapping $y = A(x_1, \ldots, x_n)$ is linear with respect to each variable for all fixed values of the other variables.

The set of n-linear mappings $A : X_1 \times \cdots \times X_n \to Y$ will be denoted $\mathcal{L}(X_1, \ldots, X_n; Y)$.

In particular for $n = 1$ we obtain the set $\mathcal{L}(X; Y)$ of linear mappings from $X_1 = X$ into Y.

For $n = 2$ a multilinear mapping is called *bilinear*, for $n = 3$, *trilinear*, and so forth.

One should not confuse an n-linear mapping $A \in \mathcal{L}(X_1, \ldots, X_n; Y)$ with a linear mapping $A \in \mathcal{L}(X; Y)$ of the vector space $X = X_1 \times \cdots \times X_n$ (in this connection see Examples 9–11 below).

If $Y = \mathbb{R}$ or $Y = \mathbb{C}$, linear and multilinear mappings are usually called linear or multilinear *functionals*. When Y is an arbitrary vector space, a linear mapping $A : X \to Y$ is usually called a *linear transformation* from X into Y, and a *linear operator* in the special case when $X = Y$.

Let us consider some examples of linear mappings.

Example 1 Let $\ell\atop 0$ be the vector space of finite numerical sequences. We define a transformation $A : \ell\atop 0$ $\to$ $\ell\atop 0$ as follows:

$$A\big((x_1, x_2, \ldots, x_n, 0, \ldots)\big) := (1x_1, 2x_2, \ldots, nx_n, 0, \ldots).$$

Example 2 We define the functional $A : C[a, b] \to \mathbb{R}$ by the relation

$$A(f) := f(x_0),$$

where $f \in C([a, b], \mathbb{R})$ and x_0 is a fixed point of the closed interval $[a, b]$.

Example 3 We define the functional $A : C([a, b], \mathbb{R}) \to \mathbb{R}$ by the relation

$$A(f) := \int_a^b f(x)\,\mathrm{d}x.$$

Example 4 We define the transformation $A : C([a, b], \mathbb{R}) \to C([a, b], \mathbb{R})$ by the formula

$$A(f) := \int_a^x f(t)\,\mathrm{d}t,$$

where x is a point ranging over the closed interval $[a, b]$.

All of these transformations are obviously linear.

Let us now consider some familiar examples of multilinear mappings.

Example 5 The usual product $(x_1, \ldots, x_n) \mapsto x_1 \cdot \ldots \cdot x_n$ of n real numbers is a typical example of an n-linear functional $A \in \mathcal{L}(\underbrace{\mathbb{R}, \ldots, \mathbb{R}}_{n}; \mathbb{R})$.

Example 6 The inner product $(x_1, x_2) \overset{A}{\longmapsto} \langle x_1, x_2\rangle$ in a Euclidean vector space over the field $\mathbb{R}$ is a bilinear function.

Example 7 The cross product $(x_1, x_2) \overset{A}{\longmapsto} [x_1, x_2]$ of vectors in three-dimensional Euclidean space E^3 is a bilinear transformation, that is, $A \in \mathcal{L}(E^3, E^3; E^3)$.

Example 8 If X is a finite-dimensional vector space over the field $\mathbb{R}$, $\{e_1, \ldots, e_n\}$ is a basis in X, and $x = x^i e_i$ is the coordinate representation of the vector $x \in X$, then, setting

$$A(x_1, \ldots, x_n) = \det \begin{pmatrix} x_1^1 & \cdots & x_1^n \\ \vdots & \ddots & \vdots \\ x_n^1 & \cdots & x_n^n \end{pmatrix},$$

we obtain an n-linear function $A : X^n \to \mathbb{R}$.

As a useful supplement to the examples just given, we investigate in addition the structure of the linear mappings of a product of vector spaces into a product of vector spaces.

Example 9 Let $X = X_1 \times \cdots \times X_m$ be the vector space that is the direct product of the spaces $X_1, \ldots, X_m$, and let $A : X \to Y$ be a linear mapping of X into a vector space Y. Representing every vector $x = (x_1, \ldots, x_m) \in X$ in the form

$$x = (x_1, \ldots, x_m) =$$
$$= (x_1, 0, \ldots, 0) + (0, x_2, 0, \ldots, 0) + \cdots + (0, \ldots, 0, x_m) \qquad (10.20)$$

and setting

$$A_i(x_i) := A\big((0, \ldots, 0, x_i, 0, \ldots, 0)\big) \qquad (10.21)$$

for $x_i \in X_i$, $i = \{1, \ldots, m\}$, we observe that the mappings $A_i : X_i \to Y$ are linear and that

$$A(x) = A_1(x_1) + \cdots + A_m(x_m). \qquad (10.22)$$

Since the mapping $A : X = X_1 \times \cdots \times X_m \to Y$ is obviously linear for any linear mappings $A_i : X_i \to Y$, we have shown that formula (10.22) gives the general form of any linear mapping $A \in \mathcal{L}(X = X_1 \times \cdots \times X_m; Y)$.

Example 10 Starting from the definition of the direct product $Y = Y_1 \times \cdots \times Y_n$ of the vector spaces $Y_1, \ldots, Y_n$ and the definition of a linear mapping $A : X \to Y$, one can easily see that any linear mapping

$$A : X \to Y = Y_1 \times \cdots \times Y_n$$

has the form $x \mapsto Ax = (A_1 x, \ldots, A_n x) = (y_1, \ldots, y_n) = y \in Y$, where $A_i : X \to Y_i$ are linear mappings.

Example 11 Combining Examples 9 and 10, we conclude that any linear mapping

$$A : X_1 \times \cdots \times X_m = X \to Y = Y_1 \times \cdots \times Y_n$$

of the direct product $X = X_1 \times \cdots \times X_m$ of vector spaces into another direct product $Y = Y_1 \times \cdots \times Y_n$ has the form

$$y = \begin{pmatrix} y_1 \\ \vdots \\ y_n \end{pmatrix} = \begin{pmatrix} A_{11} & \cdots & A_{1m} \\ \vdots & \ddots & \vdots \\ A_{n1} & \cdots & A_{nm} \end{pmatrix} \begin{pmatrix} x_1 \\ \vdots \\ x_m \end{pmatrix} = Ax, \qquad (10.23)$$

where $A_{ij} : X_j \to Y_i$ are linear mappings.

In particular, if $X_1 = X_2 = \cdots = X_m = \mathbb{R}$ and $Y_1 = Y_2 = \cdots = Y_n = \mathbb{R}$, then $A_{ij} : X_j \to Y_i$ are the linear mappings $\mathbb{R} \ni x \mapsto a_{ij} x \in \mathbb{R}$, each of which is given by a single number a_{ij}. Thus in this case relation (10.23) becomes the familiar numerical notation for a linear mapping $A : \mathbb{R}^m \to \mathbb{R}^n$.

10.2.2 The Norm of a Transformation

Definition 3 Let $A : X_1 \times \cdots \times X_n \to Y$ be a multilinear transformation mapping the direct product of the normed vector spaces $X_1, \ldots, X_n$ into a normed space Y.

The quantity

$$\|A\| := \sup_{\substack{x_1,\ldots,x_n \\ x_i \neq 0}} \frac{|A(x_1,\ldots,x_n)|_Y}{|x_1|_{X_1} \times \cdots \times |x_n|_{X_n}}, \tag{10.24}$$

where the supremum is taken over all sets $x_1, \ldots, x_n$ of nonzero vectors in the spaces $X_1, \ldots, X_n$, is called the *norm* of the multilinear transformation A.

On the right-hand side of Eq. (10.24) we have denoted the norm of a vector x by the symbol $|\cdot|$ subscripted by the symbol for the normed vector space to which the vector belongs, rather than the usual symbol $\|\cdot\|$ for the norm of a vector. From now on we shall adhere to this notation for the norm of a vector; and, where no confusion can arise, we shall omit the symbol for the vector space, taking for granted that the norm (absolute value) of a vector is always computed in the space to which it belongs. In this way we hope to introduce for the time being some distinction in the notation for the norm of a vector and the norm of a linear or multilinear transformation acting on a normed vector space.

Using the properties of the norm of a vector and the properties of a multilinear transformation, one can rewrite formula (10.24) as follows:

$$\|A\| = \sup_{\substack{x_1,\ldots,x_n \\ x_i \neq 0}} \left| A\left(\frac{x_1}{|x_1|}, \ldots, \frac{x_n}{|x_n|} \right) \right| = \sup_{e_1,\ldots,e_n} \left| A(e_1,\ldots,e_n) \right|, \tag{10.25}$$

where the last supremum extends over all sets $e_1, \ldots, e_n$ of unit vectors in the spaces $X_1, \ldots, X_n$ respectively (that is, $|e_i| = 1, i = 1, \ldots, n$).

In particular, for a linear transformation $A : X \to Y$, from (10.24) and (10.25) we obtain

$$\|A\| = \sup_{x \neq 0} \frac{|Ax|}{|x|} = \sup_{|e|=1} |Ae|. \tag{10.26}$$

It follows from Definition 3 for the norm of a multilinear transformation A that if $\|A\| < \infty$, then the inequality

$$\left| A(x_1,\ldots,x_n) \right| \leq \|A\| |x_1| \times \cdots \times |x_n| \tag{10.27}$$

holds for any vectors $x_i \in X_i, i = 1, \ldots, n$.

In particular, for a linear transformation we obtain

$$|Ax| \leq \|A\| |x|. \tag{10.28}$$

In addition, it follows from Definition 3 that if the norm of a multilinear transformation is finite, it is the greatest lower bound of all numbers M for which the inequality

$$\left|A(x_1, \ldots, x_n)\right| \le M|x_1| \times \cdots \times |x_n| \tag{10.29}$$

holds for all values of $x_i \in X_i$, $i = 1, \ldots, n$.

Definition 4 A multilinear transformation $A : X_1 \times \cdots \times X_n \to Y$ is *bounded* if there exists $M \in \mathbb{R}$ such that inequality (10.29) holds for all values of $x_1, \ldots, x_n$ in the spaces $X_1, \ldots, X_n$ respectively.

Thus the bounded transformations are precisely those that have a finite norm.

On the basis of relation (10.26) one can easily understand the geometric meaning of the norm of a linear transformation in the familiar case $A : \mathbb{R}^m \to \mathbb{R}^n$. In this case the unit sphere in $\mathbb{R}^m$ maps under the transformation A into some ellipsoid in $\mathbb{R}^n$ whose center is at the origin. Hence the norm of A in this case is simply the largest of the semiaxes of the ellipsoid.

On the other hand, one can also interpret the norm of a linear transformation as the least upper bound of the coefficients of dilation of vectors under the mapping, as can be seen from the first equality in (10.26).

It is not difficult to prove that for mappings of finite-dimensional spaces the norm of a multilinear transformation is always finite, and hence in particular the norm of a linear transformation is always finite. This is no longer true in the case of infinite-dimensional spaces, as can be seen from the first of the following examples.

Let us compute the norms of the transformations considered in Examples 1–8.

Example 1′ If we regard ℓ_0 as a subspace of the normed space ℓ_p, in which the vector $e_n = (0, \ldots, 0, \underbrace{1, 0 \ldots}_{n-1})$ has unit norm, then, since $Ae_n = ne_n$, it is clear that $\|A\| = \infty$.

Example 2′ If $|f| = \max_{a \le x \le b} |f(x)| \le 1$, then $|Af| = |f(x_0)| \le 1$, and $|Af| = 1$ if $f(x_0) = 1$, so that $\|A\| = 1$.

We remark that if we introduce, for example, the integral norm

$$|f| = \int_a^b |f|(x)\,dx$$

on the same vector space $C([a, b], \mathbb{R})$, the result of computing $\|A\|$ may change considerably. Indeed, set $[a, b] = [0, 1]$ and $x_0 = 1$. The integral norm of the function $f_n = x^n$ on $[0, 1]$ is obviously $\frac{1}{n+1}$, while $Af_n = Ax^n = x^n|_{x=1} = 1$. It follows that $\|A\| = \infty$ in this case.

Throughout what follows, unless the contrary is explicitly stated, the space $C([a, b], \mathbb{R})$ is assumed to have the norm defined by the maximum of the absolute value of the function on the closed interval $[a, b]$.

Example 3′ If $|f| = \max_{a \leq x \leq b} |f(x)| \leq 1$, then

$$|Af| = \left| \int_a^b f(x)\,dx \right| \leq \int_a^b |f|(x)\,dx \leq \int_a^b 1\,dx = b - a.$$

But for $f(x) \equiv 1$, we obtain $|A1| = b - a$, and therefore $\|A\| = b - a$.

Example 4′ If $|f| = \max_{a \leq x \leq b} |f(x)| \leq 1$, then

$$\max_{a \leq x \leq b} \left| \int_a^x f(t)\,dt \right| \leq \max_{a \leq x \leq b} \int_a^x |f|(t)\,dt \leq \max_{a \leq x \leq b} (x - a) = b - a.$$

But for $|f|(t) \equiv 1$, we obtain

$$\max_{a \leq x \leq b} \int_a^x 1\,dt = b - a,$$

and therefore in this example $\|A\| = b - a$.

Example 5′ We obtain immediately from Definition 3 that $\|A\| = 1$ in this case.

Example 6′ By the Cauchy–Bunyakovskii inequality

$$\left| \langle x_1, x_2 \rangle \right| \leq |x_1| \cdot |x_2|,$$

and if $x_1 = x_2$, this inequality becomes equality. Hence $\|A\| = 1$.

Example 7′ We know that

$$\left| [x_1, x_2] \right| = |x_1||x_2| \sin \varphi,$$

where φ is the angle between the vectors x_1 and x_2, and therefore $\|A\| \leq 1$. At the same time, if the vectors x_1 and x_2 are orthogonal, then $\sin \varphi = 1$. Thus $\|A\| = 1$.

Example 8′ If we assume that the vectors lie in a Euclidean space of dimension n, we note that $A(x_1, \ldots, x_n) = \det(x_1, \ldots, x_n)$ is the volume of the parallelepiped spanned by the vectors $x_1, \ldots, x_n$, and this volume is maximal if the vectors $x_1, \ldots, x_n$ are made pairwise orthogonal while keeping their lengths constant.
     Thus,

$$\left| \det(x_1, \ldots, x_n) \right| \leq |x_1| \cdot \ldots \cdot |x_n|,$$

equality holding for orthogonal vectors. Hence in this case $\|A\| = 1$.
     Let us now estimate the norms of the operators studied in Examples 9–11. We shall assume that in the direct product $X = X_1 \times \cdots \times X_m$ of the normed spaces $X_1, \ldots, X_m$ the norm of the vector $x = (x_1, \ldots, x_m)$ is introduced in accordance with the convention in Sect. 10.1 (Example 6).

Example 9′ Defining a linear transformation

$$A : X_1 \times \cdots \times X_m = X \to Y,$$

as has been shown, is equivalent to defining the m linear transformations $A_i : X_i \to Y$ given by the relations $A_i x_i = A((0, \ldots, 0, x_i, 0, \ldots, 0))$, $i = 1, \ldots, m$. When this is done, formula (10.22) holds, by virtue of which

$$|Ax|_Y \le \sum_{i=1}^{m} |A_i x_i|_Y \le \sum_{i=1}^{m} \|A_i\| |x_i|_{X_i} \le \left(\sum_{i=1}^{m} \|A_i\| \right) |x|_X.$$

Thus we have shown that

$$\|A\| \le \sum_{i=1}^{m} \|A_i\|.$$

On the other hand, since

$$|A_i x_i| = \left| A\big((0, \ldots, 0, x_i, 0, \ldots, 0)\big) \right| \le$$

$$\le \|A\| \big|(0, \ldots, 0, x_i, 0, \ldots, 0)\big|_X = \|A\| |x_i|_{X_i},$$

we can conclude that the estimate

$$\|A_i\| \le \|A\|$$

also holds for all $i = 1, \ldots, m$.

Example 10′ Taking account of the norm introduce in $Y = Y_1 \times \cdots \times Y_n$, in this case we immediately obtain the two-sided estimates

$$\|A_i\| \le \|A\| \le \sum_{i=1}^{n} \|A_i\|.$$

Example 11′ Taking account of the results of Examples 9 and 10, one can conclude that

$$\|A_{ij}\| \le \|A\| \le \sum_{i=1}^{m} \sum_{j=1}^{n} \|A_{ij}\|.$$

10.2.3 The Space of Continuous Transformations

From now on we shall not be interested in all linear or multilinear transformations, only continuous ones. In this connection it is useful to keep in mind the following proposition.

Proposition 1 *For a multilinear transformation $A : X_1 \times \cdots \times X_n \to Y$ mapping a product of normed spaces $X_1, \ldots, X_n$ into a normed space Y the following conditions are equivalent:*

a) *A has a finite norm,*
b) *A is a bounded transformation,*
c) *A is a continuous transformation,*
d) *A is continuous at the point $(0, \ldots, 0) \in X_1 \times \cdots \times X_n$.*

Proof We prove a closed chain of implications a) $\Rightarrow$ b) $\Rightarrow$ c) $\Rightarrow$ d) $\Rightarrow$ a).

It is obvious from relation (10.27) that a) $\Rightarrow$ b).

Let us verify that b) $\Rightarrow$ c), that is, that (10.29) implies that the operator A is continuous. Indeed, taking account of the multilinearity of A, we can write that

$$A(x_1 + h_1, x_2 + h_2, \ldots, x_n + h_n) - A(x_1, x_2, \ldots, x_n) =$$
$$= A(h_1, x_2, \ldots, x_n) + \cdots + A(x_1, x_2, \ldots, x_{n-1}, h_n) =$$
$$+ A(h_1, h_2, x_3, \ldots, x_n) + \cdots + A(x_1, \ldots, x_{n-2}, h_{n-1}, h_n) +$$
$$+ \cdots + A(h_1, \ldots, h_n).$$

From (10.29) we now obtain the estimate

$$\left| A(x_1 + h_1, x_2 + h_2, \ldots, x_n + h_n) - A(x_1, x_2, \ldots, x_n) \right| \le$$
$$\le M \big(|h_1| \cdot |x_2| \cdot \ldots \cdot |x_n| + \cdots + |x_1| \cdot |x_2| \cdot \ldots \cdot |x_{n-1}| \cdot |h_n| +$$
$$+ \cdots + |h_1| \cdot \ldots \cdot |h_n| \big),$$

from which it follows that A is continuous at each point $(x_1, \ldots, x_n) \in X_1 \times \cdots \times X_n$.

In particular, if $(x_1, \ldots, x_n) = (0, \ldots, 0)$ we obtain d) from c).

It remains to be shown that d) $\Rightarrow$ a).

Given $\varepsilon > 0$ we find $\delta = \delta(\varepsilon) > 0$ such that $|A(x_1, \ldots, x_n)| < \varepsilon$ when $\max\{|x_1|, \ldots, |x_n|\} < \delta$. Then for any set $e_1, \ldots, e_n$ of unit vectors we obtain

$$\left| A(e_1, \ldots, e_n) \right| = \frac{1}{\delta^n} \left| A(\delta e_1, \ldots, \delta e_n) \right| < \frac{\varepsilon}{\delta^n},$$

that is, $\|A\| < \frac{\varepsilon}{\delta^n} < \infty$. $\square$

We have seen above (Example 1) that not every linear transformation has a finite norm, that is, a linear transformation is not always continuous. We have also pointed out that continuity can fail for a linear transformation only when the transformation is defined on an infinite-dimensional space.

From here on $\mathcal{L}(X_1, \ldots, X_n; Y)$ will denote the set of `continuous` multilinear transformations mapping the direct product of the normed vector spaces $X_1, \ldots, X_n$ into the normed vector space Y.

In particular, $\mathcal{L}(X; Y)$ is the set of continuous linear transformations from X into Y.

In the set $\mathcal{L}(X_1, \ldots, X_n; Y)$ we introduce a natural vector-space structure:

$$(A + B)(x_1, \ldots, x_n) := A(x_1, \ldots, x_n) + B(x_1, \ldots, x_n)$$

and

$$(\lambda A)(x_1, \ldots, x_n) := \lambda A(x_1, \ldots, x_n).$$

It is obvious that if $A, B \in \mathcal{L}(X_1, \ldots, X_n; Y)$, then $(A + B) \in \mathcal{L}(X_1, \ldots, X_n; Y)$ and $(\lambda A) \in \mathcal{L}(X_1, \ldots, X_n; Y)$.

Thus $\mathcal{L}(X_1, \ldots, X_n; Y)$ can be regarded as a vector space.

Proposition 2 *The norm of a multilinear transformation is a norm in the vector space $\mathcal{L}(X_1, \ldots, X_n; Y)$ of continuous multilinear transformations.*

Proof We observe first of all that by Proposition 1 the nonnegative number $\|A\| < \infty$ is defined for every transformation $A \in \mathcal{L}(X_1, \ldots, X_n; Y)$.

Inequality (10.27) shows that

$$\|A\| = 0 \Leftrightarrow A = 0.$$

Next, by definition of the norm of a multilinear transformation

$$\|\lambda A\| = \sup_{\substack{x_1, \ldots, x_n \\ x_i \neq 0}} \frac{(\lambda A)(x_1, \ldots, x_n)|}{|x_1| \cdot \ldots \cdot |x_n|} =$$

$$= \sup_{\substack{x_1, \ldots, x_n \\ x_i \neq 0}} \frac{|\lambda| |A(x_1, \ldots, x_n)|}{|x_1| \cdot \ldots \cdot |x_n|} = |\lambda| \|A\|.$$

Finally, if A and B are elements of the space $\mathcal{L}(X_1, \ldots, X_n; Y)$, then

$$\|A + B\| = \sup_{\substack{x_1, \ldots, x_n \\ x_i \neq 0}} \frac{|(A + B)(x_1, \ldots, x_n)|}{|x_1| \cdot \ldots \cdot |x_n|} =$$

$$= \sup_{\substack{x_1, \ldots, x_n \\ x_i \neq 0}} \frac{|A(x_1, \ldots, x_n) + B(x_1, \ldots, x_n)|}{|x_1| \cdot \ldots \cdot |x_n|} \leq$$

$$\leq \sup_{\substack{x_1, \ldots, x_n \\ x_i \neq 0}} \frac{|A(x_1, \ldots, x_n)|}{|x_1| \cdot \ldots \cdot |x_n|} + \sup_{\substack{x_1, \ldots, x_n \\ x_i \neq 0}} \frac{|B(x_1, \ldots, x_n)|}{|x_1| \cdot \ldots \cdot |x_n|} = \|A\| + \|B\|. \qquad \square$$

From now on when we use the symbol $\mathcal{L}(X_1, \ldots, X_n; Y)$ we shall have in mind the vector space of *continuous n-linear transformations* normed by this *transformation norm*. In particular $\mathcal{L}(X, Y)$ is the normed space of continuous linear transformations from X into Y.

We now prove the following useful supplement to Proposition 2.

Supplement *If* X, Y, *and* Z *are normed spaces and* $A \in \mathcal{L}(X; Y)$ *and* $B \in \mathcal{L}(Y; Z)$, *then*

$$\|B \circ A\| \leq \|B\| \cdot \|A\|.$$

Proof Indeed,

$$\|B \circ A\| = \sup_{x \neq 0} \frac{|(B \circ A)x|}{|x|} \leq \sup_{x \neq 0} \frac{\|B\||Ax|}{|x|} =$$

$$= \|B\| \sup_{x \neq 0} \frac{|Ax|}{|x|} = \|B\| \cdot \|A\|. \qquad \square$$

Proposition 3 *If* Y *is a complete normed space, then* $\mathcal{L}(X_1, \ldots, X_n; Y)$ *is also a complete normed space.*

Proof We shall carry out the proof for the space $\mathcal{L}(X; Y)$ of continuous linear transformations. The general case, as will be clear from the reasoning below, differs only in requiring a more cumbersome notation.

Let $A_1, A_2, \ldots, A_n, \ldots$ be a Cauchy sequence in $\mathcal{L}(X; Y)$. Since for any $x \in X$ we have

$$|A_m x - A_n x| = \big|(A_m - A_n)x\big| \leq \|A_m - A_n\||x|,$$

it is clear that for any $x \in X$ the sequence $A_1 x, A_2 x, \ldots, A_n x, \ldots$ is a Cauchy sequence in Y. Since Y is complete, it has a limit in Y, which we denote by Ax.

Thus,

$$Ax := \lim_{n \to \infty} A_n x.$$

We shall show that $A : X \to Y$ is a continuous linear transformation.

The linearity of A follows from the relations

$$\lim_{n \to \infty} A_n(\lambda_1 x_1 + \lambda_2 x_2) = \lim_{n \to \infty} (\lambda_1 A_n x_1 + \lambda_2 A_n x_2) =$$

$$= \lambda_1 \lim_{n \to \infty} A_n x_1 + \lambda_2 \lim_{n \to \infty} A_n x_2.$$

Next, for any fixed $\varepsilon > 0$ and sufficiently large values of $m, n \in \mathbb{N}$ we have $\|A_m - A_n\| < \varepsilon$, and therefore

$$|A_m x - A_n x| \leq \varepsilon |x|$$

at each vector $x \in X$. Letting m tend to infinity in this last relation and using the continuity of the norm of a vector, we obtain

$$|Ax - A_n x| \leq \varepsilon |x|.$$

Thus $\|A - A_n\| \leq \varepsilon$, and since $A = A_n + (A - A_n)$, we conclude that

$$\|A\| \leq \|A_n\| + \varepsilon.$$

Consequently, we have shown that $A \in \mathcal{L}(X; Y)$ and $\|A - A_n\| \to 0$ as $n \to \infty$, that is, $A = \lim_{n \to \infty} A_n$ in the sense of the norm of the space $\mathcal{L}(X; Y)$. $\quad\square$

In conclusion, we make one special remark relating to the space of multilinear transformations, which we shall need when studying higher-order differentials.

Proposition 4 *For each $m \in \{1, \ldots, n\}$ there is a bijection between the spaces*

$$\mathcal{L}\big(X_1, \ldots, X_m; \mathcal{L}(X_{m+1}, \ldots, X_n; Y)\big) \quad and \quad \mathcal{L}(X_1, \ldots, X_n; Y)$$

that preserves the vector-space structure and the norm.

Proof We shall exhibit this isomorphism.
Let $\mathfrak{B} \in \mathcal{L}(X_1, \ldots, X_m; \mathcal{L}(X_{m+1}, \ldots, X_n; Y))$, that is, $\mathfrak{B}(x_1, \ldots, x_m) \in \mathcal{L}(X_{m+1}, \ldots, X_n; Y)$.
We set

$$A(x_1, \ldots, x_n) := \mathfrak{B}(x_1, \ldots, x_m)(x_{m+1}, \ldots, x_n). \tag{10.30}$$

Then

$$
\|\mathfrak{B}\| = \sup_{\substack{x_1,\ldots,x_m \\ x_i \neq 0}} \frac{\|\mathfrak{B}(x_1, \ldots, x_m)\|}{|x_1| \cdot \ldots \cdot |x_m|} =
$$

$$
= \sup_{\substack{x_1,\ldots,x_m \\ x_i \neq 0}} \frac{\sup_{\substack{x_{m+1},\ldots,x_n \\ x_j \neq 0}} \frac{|\mathfrak{B}(x_1,\ldots,x_m)(x_{m+1},\ldots,x_n)|}{|x_{m+1}| \cdot \ldots \cdot |x_n|}}{|x_1| \cdot \ldots \cdot |x_m|} =
$$

$$
= \sup_{\substack{x_1,\ldots,x_n \\ x_k \neq 0}} \frac{|A(x_1, \ldots, x_n)|}{|x_1| \cdot \ldots \cdot |x_n|} = \|A\|.
$$

We leave to the reader the verification that relation (10.30) defines an isomorphism of these vector spaces. $\quad\square$

Applying Proposition 4 n times, we find that the space

$$\mathcal{L}\big(X_1; \mathcal{L}\big(X_2; \ldots; \mathcal{L}(X_n; Y)\big) \cdots \big)$$

is isomorphic to the space $\mathcal{L}(X_1, \ldots, X_n; Y)$ of n-linear transformations.

10.2.4 Problems and Exercises

1. a) Prove that if $A : X \to Y$ is a linear transformation from the normed space X into the normed space Y and X is finite-dimensional, then A is a continuous operator.

b) Prove the proposition analogous to that stated in a) for a multilinear operator.

2. Two normed vector spaces are *isomorphic* if there exists an isomorphism between them (as vector spaces) that is continuous together with its inverse transformation.

a) Show that normed vector spaces of the same finite dimension are isomorphic.

b) Show that for the infinite-dimensional case assertion a) is generally no longer true.

c) Introduce two norms in the space $C([a, b], \mathbb{R})$ in such a way that the identity mapping of $C([a, b], \mathbb{R})$ is not a continuous mapping of the two resulting normed spaces.

3. Show that if a multilinear transformation of n-dimensional Euclidean space is continuous at some point, then it is continuous everywhere.

4. Let $A : E^n \to E^n$ be a linear transformation of n-dimensional Euclidean space and $A^* : E^n \to E^n$ the adjoint to this transformation.
Show the following.

a) All the eigenvalues of the operator $A \cdot A^* : E^n \to E^n$ are nonnegative.

b) If $\lambda_1 \le \cdots \le \lambda_n$ are the eigenvalues of the operator $A \cdot A^*$, then $\|A\| = \sqrt{\lambda_n}$.

c) If the operator A has an inverse $A^{-1} : E^n \to E^n$, then $\|A^{-1}\| = \frac{1}{\sqrt{\lambda_1}}$.

d) If (a_j^i) is the matrix of the operator $A : E^n \to E^n$ in some basis, then the estimates

$$\max_{1 \le i \le n} \sqrt{\sum_{j=1}^n (a_j^i)^2} \le \|A\| \le \sqrt{\sum_{i,j=1}^n (a_j^i)^2} \le \sqrt{n} \|A\|$$

hold.

5. Let $\mathbb{P}[x]$ be the vector space of polynomials in the variable x with real coefficients. We define the norm of the vector $P \in \mathbb{P}[x]$ by the formula

$$|P| = \sqrt{\int_0^1 P^2(x) \, dx}.$$

a) Is the operator $D : \mathbb{P}[x] \to \mathbb{P}[x]$ given by differentiation $(D(P(x)) := P'(x))$ continuous in the resulting space?

b) Find the norm of the operator $F : \mathbb{P}[x] \to \mathbb{P}[x]$ of multiplication by x, which acts according to the rule $F(P(x)) = x \cdot P(x)$.

6. Using the example of projection operators in $\mathbb{R}^2$, show that the inequality $\| B \circ A\| \le \|B\| \cdot \|A\|$ may be a strict inequality.

10.3   The Differential of a Mapping

10.3.1   Mappings Differentiable at a Point

Definition 1 Let X and Y be normed spaces. A mapping $f : E \to Y$ of a set $E \subset X$ into Y is *differentiable at an interior point* $x \in E$ if there exists a continuous linear transformation $L(x) : X \to Y$ such that

$$f(x+h) - f(x) = L(x)h + \alpha(x; h), \qquad (10.31)$$

where $\alpha(x; h) = o(h)$ as $h \to 0$, $x + h \in E$.[1]

Definition 2 The function $L(x) \in \mathcal{L}(X; Y)$ that is linear with respect to h and satisfies relation (10.31) is called the *differential*, the *tangent mapping*, or the *derivative of the mapping* $f : E \to Y$ *at the point* x.

As before, we shall denote $L(x)$ by $\mathrm{d}f(x)$, $Df(x)$, or $f'(x)$.

We thus see that the general definition of differentiability of a mapping at a point is a nearly verbatim repetition of the one already familiar to us from Sect. 8.2, where it was considered in the case $X = \mathbb{R}^m$, $Y = \mathbb{R}^n$. For that reason, from now on we shall allow ourselves to use such concepts introduced there as *increment of a function, increment of the argument*, and *tangent space at a point* without repeating the explanations, preserving the corresponding notation.

We shall, however, verify the following proposition in general form.

Proposition 1 *If a mapping* $f : E \to Y$ *is differentiable at an interior point* x *of a set* $E \subset X$, *its differential* $L(x)$ *at that point is uniquely determined.*

Proof Thus we are verifying the uniqueness of the differential.

Let $L_1(x)$ and $L_2(x)$ be linear mappings satisfying relation (10.31), that is

$$f(x+h) - f(x) - L_1(x)h = \alpha_1(x; h),$$
$$f(x+h) - f(x) - L_2(x)h = \alpha_2(x; h), \qquad (10.32)$$

where $\alpha_i(x; h) = o(h)$ as $h \to 0$, $x + h \in E$, $i = 1, 2$.

---

[1] The notation "$\alpha(x; h) = o(h)$ as $h \to 0$, $x + h \in E$", of course, means that

$$\lim_{h \to 0, x+h \in E} |\alpha(x; h)|_Y \cdot |h|_X^{-1} = 0.$$

Then, setting $L(x) = L_2(x) - L_1(x)$ and $\alpha(x; h) = \alpha_2(x; h) - \alpha_1(x; h)$ and subtracting the second equality in (10.32) from the first, we obtain

$$L(x)h = \alpha(x; h).$$

Here $L(x)$ is a mapping that is linear with respect to h, and $\alpha(x; h) = o(h)$ as $h \to 0$, $x + h \in E$. Taking an auxiliary numerical parameter λ, we can now write

$$\left| L(x)h \right| = \frac{|L(x)(\lambda h)|}{|\lambda|} = \frac{|\alpha(x; \lambda h)|}{|\lambda h|} |h| \to 0 \quad \text{as } \lambda \to 0.$$

Thus $L(x)h = 0$ for any $h \neq 0$ (we recall that x is an interior point of E). Since $L(x)0 = 0$, we have shown that $L_1(x)h = L_2(x)h$ for every value of h. □

If E is an open subset of X and $f : E \to Y$ is a mapping that is differentiable at each point $x \in E$, that is, *differentiable on* E, by the uniqueness of the differential of a mapping at a point, which was just proved, a function $E \ni x \mapsto f'(x) \in \mathcal{L}(X; Y)$ arises on the set E, which we denote $f' : E \to \mathcal{L}(X; Y)$. This mapping is called the *derivative of* f, or the *derivative mapping* relative to the original mapping $f : E \to Y$. The value $f'(x)$ of this function at an individual point $x \in E$ is the continuous linear transformation $f'(x) \in \mathcal{L}(X; Y)$ that is the differential or derivative of the function f at the particular point $x \in E$.

We note that by the requirement of `continuity` of the linear mapping $L(x)$ Eq. (10.31) implies that a mapping that is differentiable at a point is necessarily continuous at that point.

The converse is of course not true, as we have seen in the case of numerical functions.

We now make one more important remark.

Remark If the condition for differentiability of the mapping f at some point a is written as

$$f(x) - f(a) = L(x)(x - a) + \alpha(a; x),$$

where $\alpha(a; x) = o(x - a)$ as $x \to a$, it becomes clear that Definition 1 actually applies to a mapping $f : A \to B$ of any affine spaces (A, X) and (B, Y) whose vector spaces X and Y are normed. Such affine spaces, called *normed affine spaces*, are frequently encountered, so that it is useful to keep this remark in mind when using the differential calculus.

Everything that follows, unless specifically stated otherwise, applies equally to both normed vector spaces and normed affine spaces, and we use the notation for vector spaces only for the sake of simplicity.

10.3.2 The General Rules for Differentiation

The following general properties of the operation of differentiation follow from Definition 1. In the statements below X, Y, and Z are normed spaces and U and V open sets in X and Y respectively.

a. Linearity of Differentiation

If the mappings $f_i : U \to Y$, $i = 1, 2$, are differentiable at a point $x \in U$, a linear combination of them $(\lambda_1 f_1 + \lambda_2 f_2) : U \to Y$ is also differentiable at x, and

$$(\lambda_1 f_1 + \lambda_2 f_2)'(x) = \lambda_1 f_1'(x) + \lambda_2 f_2'(x).$$

Thus the differential of a linear combination of mappings is the corresponding linear combination of their differentials.

b. Differentiation of a Composition of Mappings (Chain Rule)

If the mapping $f : U \to V$ is differentiable at a point $x \in U \subset X$, and the mapping $g : V \to Z$ is differentiable at $f(x) = y \in V \subset Y$, then the composition $g \circ f$ of these mappings is differentiable at x, and

$$(g \circ f)'(x) = g'(f(x)) \circ f'(x).$$

Thus, the differential of a composition is the composition of the differentials.

c. Differentiation of the Inverse of a Mapping

Let $f : U \to Y$ be a mapping that is continuous at $x \in U \subset X$ and has an inverse $f^{-1} : V \to X$ that is defined in a neighborhood of $y = f(x)$ and continuous at that point.

If the mapping f is differentiable at x and its tangent mapping $f'(x) \in \mathcal{L}(X; Y)$ has a continuous inverse $[f'(x)]^{-1} \in \mathcal{L}(Y; X)$, then the mapping f^{-1} is differentiable at $y = f(x)$ and

$$\left[f^{-1} \right]'(f(x)) = \left[f'(x) \right]^{-1}.$$

Thus, the differential of an inverse mapping is the linear mapping inverse to the differential of the original mapping at the corresponding point.

We omit the proofs of a, b, and c, since they are analogous to the proofs given in Sect. 8.3 for the case $X = \mathbb{R}^m$, $Y = \mathbb{R}^n$.

10.3.3 Some Examples

Example 1 If $f : U \to Y$ is a constant mapping of a neighborhood $U = U(x) \subset X$ of the point x, that is, $f(U) = y_0 \in Y$, then $f'(x) = 0 \in \mathcal{L}(X; Y)$.

Proof Indeed, in this case it is obvious that

$$f(x + h) - f(x) - 0h = y_0 - y_0 - 0 = 0 = o(h). \qquad \square$$

Example 2 If the mapping $f : X \to Y$ is a continuous linear mapping of a normed vector space X into a normed vector space Y, then $f'(x) = f \in \mathcal{L}(X; Y)$ at any point $x \in A$.

Proof Indeed,

$$f(x + h) - f(x) - fh = fx + fh - fx - fh = 0. \qquad \square$$

We remark that strictly speaking $f'(x) \in \mathcal{L}(TX_x; TY_{f(x)})$ here and h is a vector of the tangent space TX_x. But parallel translation of a vector to any point $x \in X$ is defined in a vector space, and this allows us to identify the tangent space TX_x with the vector space X itself. (Similarly, in the case of an affine space (A, X) the space TA_a of vectors "attached" to the point $a \in A$ can be identified with the vector space X of the given affine space.) Consequently, after choosing a basis in X, we can extend it to all the tangent spaces TX_x. This means that if, for example, $X = \mathbb{R}^m$, $Y = \mathbb{R}^n$, and the mapping $f \in \mathcal{L}(\mathbb{R}^m; \mathbb{R}^n)$ is given by the matrix (a_i^j), then at every point $x \in \mathbb{R}^m$ the tangent mapping $f'(x) : T\mathbb{R}_x^m \to T\mathbb{R}_{f(x)}^n$ will be given by the same matrix.

In particular, for a linear mapping $x \xrightarrow{f} ax = y$ from $\mathbb{R}$ to $\mathbb{R}$ with $x \in \mathbb{R}$ and $h \in T\mathbb{R}_x \sim \mathbb{R}$, we obtain the corresponding mapping $T\mathbb{R}_x \ni h \xrightarrow{f'} ah \in T\mathbb{R}_{f(x)}$.

Taking account of these conventions, we can provisionally state the result of Example 2 as follows: The mapping $f' : X \to Y$ that is the derivative of a linear mapping $f : X \to Y$ of normed spaces is constant, and $f'(x) = f$ at each point $x \in X$.

Example 3 From the chain rule for differentiating a composition of mappings and the result of Example 2 one can conclude that if $f : U \to Y$ is a mapping of a neighborhood $U = U(x) \subset X$ of the point $x \in X$ and is differentiable at x, while $A \in \mathcal{L}(Y; Z)$, then

$$(A \circ f)'(x) = A \circ f'(x).$$

For numerical functions, when $Y = Z = \mathbb{R}$, this is simply the familiar possibility of moving a constant factor outside the differentiation sign.

Example 4 Suppose once again that $U = U(x)$ is a neighborhood of the point x in a normed space X, and let

$$f : U \to Y = Y_1 \times \cdots \times Y_n$$

be a mapping of U into the direct product of the normed spaces $Y_1, \ldots, Y_n$.

Defining such a mapping is equivalent to defining the n mappings $f_i : U \to Y_i$, $i = 1, \ldots, n$, connected with f by the relation

$$x \mapsto f(x) = y = (y_1, \ldots, y_n) = \big(f_1(x), \ldots, f_n(x)\big),$$

which holds at every point of U.

If we now take account of the fact that in formula (10.31) we have

$$f(x + h) - f(x) = \big(f_1(x + h) - f_1(x), \ldots, f_n(x + h) - f_n(x)\big),$$

$$L(x)h = \big(L_1(x)h, \ldots, L_n(x)h\big),$$

$$\alpha(x; h) = \big(\alpha_1(x; h), \ldots, \alpha_n(x; h)\big),$$

then, referring to the results of Example 6 of Sect. 10.1 and Example 10 of Sect. 10.2, we can conclude that the mapping f is differentiable at x if and only if all of its components $f_i : U \to Y_i$ are differentiable at x, $i = 1, \ldots, n$; and when the mapping f is differentiable, we have the equality

$$f'(x) = \big(f_1'(x), \ldots, f_n'(x)\big).$$

Example 5 Now let $A \in \mathcal{L}(X_1, \ldots, X_n; Y)$, that is, A is a continuous n-linear transformation from the product $X_1 \times \cdots \times X_n$ of the normed vector spaces $X_1, \ldots, X_n$ into the normed vector space Y.

We shall prove that the mapping

$$A : X_1 \times \cdots \times X_n = X \to Y$$

is differentiable and find its differential.

Proof Using the multilinearity of A, we find that

$$A(x + h) - A(x) = A(x_1 + h_1, \ldots, x_n + h_n) - A(x_1, \ldots, x_n) =$$

$$= A(x_1, \ldots, x_n) + A(h_1, x_2, \ldots, x_n) +$$

$$+ \cdots + A(x_1, \ldots, x_{n-1}, h_n) + A(h_1, h_2, x_3, \ldots, x_n) +$$

$$+ \cdots + A(x_1, \ldots, x_{n-2}, h_{n-1}, h_n) +$$

$$+ \cdots + A(h_1, \ldots, h_n) - A(x_1, \ldots, x_n).$$

Since the norm in $X = X_1 \times \cdots \times X_n$ satisfies the inequalities

$$|x_i|_{X_i} \le |x|_X \le \sum_{i=1}^{n} |x_i|_{X_i},$$

and the norm $\|A\|$ of the transformation A is finite and satisfies

$$\left| A(\xi_1, \ldots, \xi_n) \right| \leq \|A\| |\xi_1| \times \cdots \times |\xi_n|,$$

we can conclude that

$$A(x+h) - A(x) = A(x_1 + h_1, \ldots, x_n + h_n) - A(x_1, \ldots, x_n) =$$
$$= A(h_1, x_2, \ldots, x_n) + \cdots + A(x_1, \ldots, x_{n-1}, h_n) + \alpha(x; h),$$

where $\alpha(x; h) = o(h)$ as $h \to 0$.

But the transformation

$$L(x)h = A(h_1, x_2, \ldots, x_n) + \cdots + A(x_1, \ldots, x_{n-1}, h_n)$$

is a continuous transformation (because A is continuous) that is linear in $h = (h_1, \ldots, h_n)$.

Thus we have established that

$$A'(x)h = A'(x_1, \ldots, x_n)(h_1, \ldots, h_n) =$$
$$= A(h_1, x_2, \ldots, x_n) + \cdots + A(x_1, \ldots, x_{n-1}, h_n),$$

or, more briefly,

$$dA(x_1, \ldots, x_n) = A(dx_1, x_2, \ldots, x_n) + \cdots + A(x_1, \ldots, x_{n-1}, dx_n). \qquad \square$$

In particular, if:

a) $x_1 \cdot \ldots \cdot x_n$ is the product of n numerical variables, then

$$d(x_1 \cdot \ldots \cdot x_n) = dx_1 \cdot x_2 \cdot \ldots \cdot x_n + \cdots + x_1 \cdot \ldots \cdot x_{n-1} \cdot dx_n;$$

b) $\langle x_1, x_2 \rangle$ is the inner product in E^3, then

$$d\langle x_1, x_2 \rangle = \langle dx_1, x_2 \rangle + \langle x_1, dx_2 \rangle;$$

c) $[x_1, x_2]$ is the vector cross product in E^3, then

$$d[x_1, x_2] = [dx_1, x_2] + [x_1, dx_2];$$

d) (x_1, x_2, x_3) is the scalar triple product in E^3, then

$$d(x_1, x_2, x_3) = (dx_1, x_2, x_3) + (x_2, dx_2, x_3) + (x_2, x_2, dx_3);$$

e) $\det(x_1, \ldots, x_n)$ is the determinant of the matrix formed from the coordinates of n vectors $x_1, \ldots, x_n$ in an n-dimensional vector space X with a fixed basis, then

$$d\big(\det(x_1, \ldots, x_n)\big) = \det(dx_1, x_2, \ldots, x_n) + \cdots + \det(x_1, \ldots, x_{n-1}, dx_n).$$

Example 6 Let U be the subset of $\mathcal{L}(X; Y)$ consisting of the continuous linear transformations $A : X \to Y$ having continuous inverse transformations $A^{-1} : Y \to X$ (belonging to $\mathcal{L}(Y; X)$). Consider the mapping

$$U \ni A \mapsto A^{-1} \in \mathcal{L}(Y; X),$$

which assigns to each transformation $A \in U$ its inverse $A^{-1} \in \mathcal{L}(Y; X)$.

Proposition 2 proved below makes it possible to determine whether this mapping is differentiable.

Proposition 2 *If X is a complete space and $A \in U$, then for any $h \in \mathcal{L}(X; Y)$ such that $\|h\| < \|A^{-1}\|^{-1}$, the transformation $A + h$ also belongs to U and the following relation holds:*

$$(A + h)^{-1} = A^{-1} - A^{-1}hA^{-1} + o(h) \quad as\ h \to 0. \tag{10.33}$$

Proof Since

$$(A + h)^{-1} = \left(A\left(E + A^{-1}h\right)\right)^{-1} = \left(E + A^{-1}h\right)^{-1} A^{-1}, \tag{10.34}$$

it suffices to find the operator $(E + A^{-1}h)^{-1}$ inverse to $(E + A^{-1}h) \in \mathcal{L}(X; X)$, where E is the identity mapping e_X of X into itself.

Let $\Delta := -A^{-1}h$. Taking account of the supplement to Proposition 2 of Sect. 10.2, we can observe that $\|\Delta\| \leq \|A^{-1}\| \cdot \|h\|$, so that by the assumptions made with respect to the operator h we may assume that $\|\Delta\| \leq q < 1$.

We now verify that

$$(E - \Delta)^{-1} = E + \Delta + \Delta^2 + \cdots + \Delta^n + \cdots, \tag{10.35}$$

where the series on the right-hand side is formed from the linear operators $\Delta^n = (\Delta \circ \cdots \circ \Delta) \in \mathcal{L}(X; X)$.

Since X is a complete normed vector space, it follows from Proposition 3 of Sect. 10.2 that the space $\mathcal{L}(X; X)$ is also complete. It then follows immediately from the relation $\|\Delta^n\| \leq \|\Delta\|^n \leq q^n$ and the convergence of the series $\sum_{n=0}^{\infty} q^n$ for $|q| < 1$ that the series (10.35) formed from the vectors in that space converges.

The direct verification that

$$\left(E + \Delta + \Delta^2 + \cdots\right)(E - \Delta) =$$
$$= \left(E + \Delta + \Delta^2 + \cdots\right) - \left(\Delta + \Delta^2 + \Delta^3 + \cdots\right) = E$$

and

$$(E - \Delta)\left(E + \Delta + \Delta^2 + \cdots\right) =$$
$$= \left(E + \Delta + \Delta^2 + \cdots\right) - \left(\Delta + \Delta^2 + \Delta^3 + \cdots\right) = E$$

shows that we have indeed found $(E - \Delta)^{-1}$.

It is worth remarking that the freedom in carrying out arithmetic operations on series (rearranging the terms!) in this case is guaranteed by the absolute convergence (convergence in norm) of the series under consideration.

Comparing relations (10.34) and (10.35), we conclude that

$$(A + h)^{-1} = A^{-1} - A^{-1}hA^{-1} + \left(A^{-1}h\right)^2 A^{-1} -$$
$$- \cdots + (-1)^n \left(A^{-1}h\right)^n A^{-1} + \cdots \qquad (10.36)$$

for $\|h\| \le \|A^{-1}\|^{-1}$.

Since

$$\left\| \sum_{n=2}^{\infty} \left(-A^{-1}h\right)^n A^{-1} \right\| \le \sum_{n=2}^{\infty} \left\| A^{-1}h \right\|^n \left\| A^{-1} \right\| \le$$

$$\le \left\| A^{-1} \right\|^3 \|h\|^2 \sum_{m=0}^{\infty} q^m = \frac{\left\| A^{-1} \right\|^3}{1-q} \|h\|^2,$$

Eq. (10.33) follows in particular from (10.36). □

Returning now to Example 6, we can say that when the space X is complete the mapping $A \xmapsto{f} A^{-1}$ under consideration is necessarily differentiable, and

$$\mathrm{d}f(A)h = \mathrm{d}\left(A^{-1}\right)h = -A^{-1}hA^{-1}.$$

In particular, this means that if A is a nonsingular square matrix and A^{-1} is its inverse, then under a perturbation of the matrix A by a matrix h whose elements are close to zero, we can write the inverse matrix $(A + h)^{-1}$ in first approximation in the following form:

$$(A + h)^{-1} \approx A^{-1} - A^{-1}hA^{-1}.$$

More precise formulas can obviously be obtained starting from Eq. (10.36).

Example 7 Let X be a complete normed vector space. The important mapping

$$\exp : \mathcal{L}(X; X) \to \mathcal{L}(X; X)$$

is defined as follows:

$$\exp A := E + \frac{1}{1!}A + \frac{1}{2!}A^2 + \cdots + \frac{1}{n!}A^n + \cdots, \qquad (10.37)$$

if $A \in \mathcal{L}(X; X)$.

The series in (10.37) converges, since $\mathcal{L}(X; X)$ is a complete space and $\left\| \frac{1}{n!}A^n \right\| \le \frac{\|A\|^n}{n!}$, while the numerical series $\sum_{n=0}^{\infty} \frac{\|A\|^n}{n!}$ converges.

It is not difficult to verify that

$$\exp(A + h) = \exp A + L(A)h + o(h) \quad \text{as } h \to \infty, \tag{10.38}$$

where

$$L(A)h = h + \frac{1}{2!}(Ah + hA) + \frac{1}{3!}\left(A^2h + AhA + hA^2\right) +$$

$$+ \cdots + \frac{1}{n!}\left(A^{n-1}h + A^{n-2}hA + \cdots + AhA^{n-2} + hA^{n-1}\right) + \cdots$$

and $\|L(A)\| \le \exp\|A\| = e^{\|A\|}$, that is, $L(A) \in \mathcal{L}(\mathcal{L}(X; X), \mathcal{L}(X; X))$.

Thus, the mapping $\mathcal{L}(X; X) \ni A \mapsto \exp A \in \mathcal{L}(X; X)$ is differentiable at every value of A.

We remark that if the operators A and h commute, that is, $Ah = hA$, then, as one can see from the expression for $L(A)h$, in this case we have $L(A)h = (\exp A)h$. In particular, for $X = \mathbb{R}$ or $X = \mathbb{C}$, instead of (10.38) we again obtain

$$\exp(A + h) = \exp A + (\exp A)h + o(h) \quad \text{as } h \to 0. \tag{10.39}$$

Example 8 We shall attempt to give a mathematical description of the instantaneous angular velocity of a rigid body with a fixed point o (a top). Consider an orthonormal frame $\{\mathbf{e}_1, \mathbf{e}_2, \mathbf{e}_3\}$ at the point o rigidly attached to the body. It is clear that the position of the body is completely characterized by the position of this orthoframe, and the triple $\{\dot{\mathbf{e}}_1, \dot{\mathbf{e}}_2, \dot{\mathbf{e}}_3\}$ of instantaneous velocities of the vectors of the frame obviously give a complete characterization of the instantaneous angular velocity of the body. The position of the frame itself $\{\mathbf{e}_1, \mathbf{e}_2, \mathbf{e}_3\}$ at time t can be given by an orthogonal matrix (α_i^j), $i, j = 1, 2, 3$ composed of the coordinates of the vectors $\mathbf{e}_1, \mathbf{e}_2, \mathbf{e}_3$ with respect to some fixed orthonormal frame in space. Thus, the motion of the top corresponds to a mapping $t \mapsto O(t)$ from $\mathbb{R}$ (the time axis) into the group $SO(3)$ of special orthogonal 3×3 matrices. Consequently, the angular velocity of the body, which we have agreed to describe by the triple $\{\dot{\mathbf{e}}_1, \dot{\mathbf{e}}_2, \dot{\mathbf{e}}_3\}$, is the matrix $\dot{O}(t) =: (\omega_i^j)(t) = (\dot{\alpha}_i^j)(t)$, which is the derivative of the matrix $O(t) = (\alpha_i^j)(t)$ with respect to time.

Since $O(t)$ is an orthogonal matrix, the relation

$$O(t)O^*(t) = E \tag{10.40}$$

holds at any time t, where $O^*(t)$ is the transpose of $O(t)$ and E is the identity matrix.

We remark that the product $A \cdot B$ of matrices is a bilinear function of A and B, and the derivative of the transposed matrix is obviously the transpose of the derivative of the original matrix. Differentiating (10.40) and taking account of these things, we find that

$$\dot{O}(t)O^*(t) + O(t)\dot{O}^*(t) = 0$$

or

$$\dot{O}(t) = -O(t)\dot{O}^*(t)O(t), \tag{10.41}$$

since $O^*(t)O(t) = E$.

In particular, if we assume that the frame $\{e_1, e_2, e_3\}$ coincides with the spatial frame of reference at time t, then $O(t) = E$, and it follows from (10.41) that

$$\dot{O}(t) = -\dot{O}^*(t), \tag{10.42}$$

that is, the matrix $\dot{O}(t) =: \Omega(t) = (\omega_i^j)$ of coordinates of the vectors $\{\dot{e}_1, \dot{e}_2, \dot{e}_3\}$ in the basis $\{e_1, e_2, e_3\}$ turns out to be skew-symmetric:

$$\Omega(t) = \begin{pmatrix} \omega_1^1 & \omega_1^2 & \omega_1^3 \\ \omega_2^1 & \omega_2^2 & \omega_2^3 \\ \omega_3^1 & \omega_3^2 & \omega_3^3 \end{pmatrix} = \begin{pmatrix} 0 & -\omega^3 & \omega^2 \\ \omega^3 & 0 & -\omega^1 \\ -\omega^2 & \omega^1 & 0 \end{pmatrix}.$$

Thus the instantaneous angular velocity of a top is actually characterized by three independent parameters, as follows in our line of reasoning from relation (10.40) and is natural from the physical point of view, since the position of the frame $\{e_1, e_2, e_3\}$, and hence the position of the body itself, can be described by three independent parameters (in mechanics these parameters may be, for example, the Euler angles).

If we associate with each vector $\omega = \omega^1 e_1 + \omega^2 e_2 + \omega^3 e_3$ in the tangent space at the point o a right-handed rotation of space with angular velocity $|\omega|$ about the axis defined by this vector, it is not difficult to conclude from these results that at each instant of time t the body has an instantaneous angular velocity and that the velocity at that time can be adequately described by the instantaneous angular velocity vector $\omega(t)$ (see Problem 5 below).

10.3.4 The Partial Derivatives of a Mapping

Let $U = U(a)$ be a neighborhood of the point $a \in X = X_1 \times \cdots \times X_m$ in the direct product of the normed spaces $X_1, \ldots, X_m$, and let $f : U \to Y$ be a mapping of U into the normed space V. In this case

$$y = f(x) = f(x_1, \ldots, x_m), \tag{10.43}$$

and hence, if we fix all the variables but x_i in (10.43) by setting $x_k = a_k$ for $k \in \{1, \ldots, m\}\setminus i$, we obtain a function

$$f(a_1, \ldots, a_{i-1}, x_i, a_{i+1}, \ldots, a_m) =: \varphi_i(x_i), \tag{10.44}$$

defined in some neighborhood U_i of a_i in X.

Definition 3 Relative to the original mapping (10.43) the mapping $\varphi_i : U_i \to Y$ is called the *partial mapping with respect to the variable x_i at $a \in X$.*

Definition 4 If the mapping (10.44) is differentiable at $x_i = a_i$, its derivative at that point is called the *partial derivative* or *partial differential of f at a with respect to the variable x_i.*

We usually denote this partial derivative by one of the symbols

$$\partial_i f(a), \quad D_i f(a), \quad \frac{\partial f}{\partial x_i}(a), \quad f'_{x_i}(a).$$

In accordance with these definitions $D_i f(a) \in \mathcal{L}(X_i; Y)$. More precisely, $D_i f(a) \in \mathcal{L}(T X_i(a_i); T Y(f(a)))$.

The differential $df(a)$ of the mapping (10.43) at the point a (if f is differentiable at that point) is often called the *total differential* in this situation in order to distinguish it from the partial differentials with respect to the individual variables.

We have already encountered all these concepts in the case of real-valued functions of m real variables, so that we shall not give a detailed discussion of them. We remark only that by repeating our earlier reasoning, taking account of Example 9 in Sect. 9.2, one can prove easily that the following proposition holds in general.

Proposition 3 *If the mapping (10.43) is differentiable at the point $a = (a_1, \ldots, a_m)$ $\in X_1 \times \cdots \times X_m = X$, it has partial derivatives with respect to each variable at that point, and the total differential and the partial differentials are related by the equation*

$$df(a)h = \partial_1 f(a)h_1 + \cdots + \partial_m f(a)h_m, \tag{10.45}$$

where $h = (h_1, \ldots, h_m) \in T X_1(a_1) \times \cdots \times T X_m(a_m) = T X(a)$.

We have already shown by the example of numerical functions that the existence of partial derivatives does not in general guarantee the differentiability of the function (10.43).

10.3.5 Problems and Exercises

1. a) Let $A \in \mathcal{L}(X; X)$ be a *nilpotent operator,* that is, there exists $k \in \mathbb{N}$ such that $A^k = 0$. Show that the operator $(E - A)$ has an inverse in this case and that $(E - A)^{-1} = E + A + \cdots + A^{k-1}$.

b) Let $D : \mathbb{P}[x] \to \mathbb{P}[x]$ be the operator of differentiation on the vector space $\mathbb{P}[x]$ of polynomials. Remarking that D is a nilpotent operator, write the operator $\exp(aD)$, where $a \in \mathbb{R}$, and show that $\exp(aD)(P(x)) = P(x + a) =: T_a(P(x))$.

c) Write the matrices of the operators $D : \mathbb{P}_n[x] \to \mathbb{P}_n[x]$ and $T_a : \mathbb{P}_n[x] \to \mathbb{P}_n[x]$ from part b) in the basis $e_i = \frac{x^{n-i}}{(n-i)!}$, $1 \le i \le n$, in the space $\mathbb{P}_n[x]$ of real polynomials of degree n in one variable.

2. a) If $A, B \in \mathcal{L}(X; X)$ and $\exists B^{-1} \in \mathcal{L}(X; X)$, then $\exp(B^{-1}AB) = B^{-1}(\exp A)B$.

b) If $AB = BA$, then $\exp(A + B) = \exp A \cdot \exp B$.

c) Verify that $\exp 0 = E$ and that $\exp A$ always has an inverse, namely $(\exp A)^{-1} = \exp(-A)$.

3. Let $A \in \mathcal{L}(X; X)$. Consider the mapping $\varphi_A : \mathbb{R} \to \mathcal{L}(X; X)$ defined by the correspondence $\mathbb{R} \ni t \mapsto \exp(tA) \in \mathcal{L}(X; X)$. Show the following.

a) The mapping φ_A is continuous.

b) φ_A is a homomorphism of $\mathbb{R}$ as an additive group into the multiplicative group of invertible operators in $\mathcal{L}(X; X)$.

4. Verify the following.

a) If $\lambda_1, \ldots, \lambda_n$ are the eigenvalues of the operator $A \in \mathcal{L}(\mathbb{C}^n; \mathbb{C}^n)$, then $\exp \lambda_1, \ldots, \exp \lambda_n$ are the eigenvalues of $\exp A$.

b) $\det(\exp A) = \exp(\operatorname{tr} A)$, where $\operatorname{tr} A$ is the trace of the operator $A \in \mathcal{L}(\mathbb{C}^n, \mathbb{C}^n)$.

c) If $A \in \mathcal{L}(\mathbb{R}^n, \mathbb{R}^n)$, then $\det(\exp A) > 0$.

d) If A^* is the transpose of the matrix $A \in \mathcal{L}(\mathbb{C}^n, \mathbb{C}^n)$ and $\bar{A}$ is the matrix whose elements are the complex conjugates of those of A, then $(\exp A)^* = \exp A^*$ and $\overline{\exp A} = \exp \bar{A}$.

e) The matrix $\begin{pmatrix} -1 & 0 \\ 1 & -1 \end{pmatrix}$ is not of the form $\exp A$ for any 2×2 matrix A.

5. We recall that a set endowed with both a group structure and a topology is called a *topological group* or *continuous group* if the group operation is continuous. If there is a sense in which the group operation is even analytic, the topological group is called a *Lie group*.[2]

A *Lie algebra* is a vector space X with an anticommutative bilinear operation $[,] : X \times X \to X$ satisfying the *Jacobi identity*: $[[a, b], c] + [[b, c], a] + [[c, a], b] = 0$ for any vectors $a, b, c \in X$. Lie groups and algebras are closely connected with each other, and the mapping exp plays an important role in establishing this connection (see Problem 1 above).

An example of a Lie algebra is the oriented Euclidean space E^3 with the operation of the vector cross product. For the time being we shall denote this Lie algebra by LA_1.

a) Show that the real 3×3 skew-symmetric matrices form a Lie algebra (which we denote LA_2) if the product of the matrices A and B is defined as $[A, B] = AB - BA$.

---

[2]For the precise definition of a Lie group and the corresponding reference see Problem 8 in Sect. 15.2.

b) Show that the correspondence

$$\Omega = \begin{pmatrix} 0 & -\omega^3 & \omega^2 \\ \omega^3 & 0 & -\omega^1 \\ -\omega^2 & \omega^1 & 0 \end{pmatrix} \leftrightarrow (\omega_1, \omega_2, \omega_3) = \omega$$

is an isomorphism of the Lie algebras LA_2 and LA_1.

c) Verify that if the skew-symmetric matrix Ω and the vector ω correspond to each other as shown in b), then the equality $\Omega \mathbf{r} = [\omega, \mathbf{r}]$ holds for any vector $\mathbf{r} \in E^3$, and the relation $P \Omega P^{-1} \leftrightarrow P\omega$ holds for any matrix $P \in SO(3)$.

d) Verify that if $\mathbb{R} \ni t \mapsto O(t) \in SO(3)$ is a smooth mapping, then the matrix $\Omega(t) = O^{-1}(t)\dot{O}(t)$ is skew-symmetric.

e) Show that if $\mathbf{r}(t)$ is the radius vector of a point of a rotating top and $\Omega(t)$ is the matrix $(O^{-1}\dot{O})(t)$ found in d), then $\dot{\mathbf{r}}(t) = (\Omega\mathbf{r})(t)$.

f) Let $\mathbf{r}$ and ω be two vectors attached at the origin of E^3. Suppose a right-handed frame has been chosen in E^3, and that the space undergoes a right-handed rotation with angular velocity $|\omega|$ about the axis defined by ω. Show that $\dot{\mathbf{r}}(t) = [\omega, \mathbf{r}(t)]$ in this case.

g) Summarize the results of d), e), and f) and exhibit the instantaneous angular velocity of the rotating top discussed in Example 8.

h) Using the result of c), verify that the velocity vector ω is independent of the choice of the fixed orthoframe in E^3, that is, it is independent of the coordinate system.

6. Let $\mathbf{r} = \mathbf{r}(s) = (x^1(s), x^2(s), x^3(s))$ be the parametric equations of a smooth curve in E^3, the parameter being arc length along the curve (the *natural parametrization of the curve*).

a) Show that the vector $\mathbf{e}_1(s) = \frac{d\mathbf{r}}{ds}(s)$ tangent to the curve has unit length.

b) The vector $\frac{d\mathbf{e}_1}{ds}(s) = \frac{d^2\mathbf{r}}{ds^2}(s)$ is orthogonal to $\mathbf{e}_1$. Let $\mathbf{e}_2(s)$ be the unit vector formed from $\frac{d\mathbf{e}_1}{ds}(s)$. The coefficient $k(s)$ in the equality $\frac{d\mathbf{e}_1}{ds}(s) = k(s)\mathbf{e}_2(s)$ is called the *curvature* of the curve at the corresponding point.

c) By constructing the vector $\mathbf{e}_3(s) = [\mathbf{e}_1(s), \mathbf{e}_2(s)]$ we obtain a frame $\{\mathbf{e}_1, \mathbf{e}_2, \mathbf{e}_3\}$ at each point, called the *Frenet frame*[3] or *companion trihedral* of the curve. Verify the following Frenet formulas:

$$\frac{d\mathbf{e}_1}{ds}(s) = k(s)\mathbf{e}_2(s),$$

$$\frac{d\mathbf{e}_2}{ds}(s) = -k(s)\mathbf{e}_1(s) \qquad \varkappa(s)\mathbf{e}_3(s),$$

$$\frac{d\mathbf{e}_3}{ds}(s) = -\varkappa(s)\mathbf{e}_2(s).$$

---

[3] J.F. Frenet (1816–1900) – French mathematician.

Explain the geometric meaning of the coefficient $\varkappa(s)$ called the *torsion* of the curve at the corresponding point.

10.4 The Finite-Increment Theorem and Some Examples of Its Use

10.4.1 The Finite-Increment Theorem

In our study of numerical functions of one variable in Sect. 5.3.2 we proved the finite-increment theorem for them and discussed in detail various aspects of this important theorem of analysis. In the present section the finite-increment theorem will be proved in its general form. So that its meaning will be fully obvious, we advise the reader to recall the discussion in that subsection and also to pay attention to the geometric meaning of the norm of a linear operator (see Sect. 10.2.2).

Theorem 1 (The finite-increment theorem) *Let $f : U \to Y$ be a continuous mapping of an open set U of a normed space X into a normed space Y.*

If the closed interval $[x, x+h] = \{\xi \in X \mid \xi = x + \theta h, 0 \le \theta \le 1\}$ is contained in U and the mapping f is differentiable at all points of the open interval $]x, x+h[= \{\xi \in X \mid \xi = x + \theta h, 0 < \theta < 1\}$, then the following estimate holds:

$$\left| f(x+h) - f(x) \right|_Y \le \sup_{\xi \in]x,x+h[} \left\| f'(\xi) \right\|_{\mathcal{L}(X,Y)} |h|_X. \qquad (10.46)$$

Proof We remark first of all that if we could prove the inequality

$$\left| f(x'') - f(x') \right| \le \sup_{\xi \in [x',x'']} \left\| f'(\xi) \right\| \left| x'' - x' \right| \qquad (10.47)$$

in which the supremum extends over the whole interval $[x', x'']$, for every closed interval $[x', x''] \subset]x, x+h[$, then, using the continuity of f and the norm together with the fact that

$$\sup_{\xi \in [x',x'']} \left\| f'(\xi) \right\| \le \sup_{\xi \in]x,x+h[} \left\| f'(\xi) \right\|,$$

we would obtain inequality (10.46) in the limit as $x' \to x$ and $x'' \to x + h$.

Thus, it suffices to prove that

$$\left| f(x+h) - f(x) \right| \le M|h|, \qquad (10.48)$$

where $M = \sup_{0 \le \theta \le 1} \| f'(x + \theta h) \|$ and the function f is assumed differentiable on the entire closed interval $[x, x+h]$.

The very simple computation

$$\left| f(x_3) - f(x_1) \right| \le \left| f(x_3) - f(x_2) \right| + \left| f(x_2) - f(x_1) \right| \le$$
$$\le M|x_3 - x_2| + M|x_2 - x_1| = M \left(|x_3 - x_2| + |x_2 - x_1| \right) =$$
$$= M|x_3 - x_1|,$$

which uses only the triangle inequality and the properties of a closed interval, shows that if an inequality of the form (10.48) holds on the portions $[x_1, x_2]$ and $[x_2, x_3]$ of the closed interval $[x_1, x_3]$, then it also holds on $[x_1, x_3]$.

Hence, if estimate (10.48) fails for the closed interval $[x, x + h]$, then by successive bisections, one can obtain a sequence of closed intervals $[a_k, b_k] \subset \,]x, x + h[$ contracting to some point $x_0 \in [x, x + h]$ such that (10.48) fails on each interval $[a_k, b_k]$. Since $x_0 \in [a_k, b_k]$, consideration of the closed intervals $[a_k, x_0]$ and $[x_0, b_k]$ enables us to assume that we have found a sequence of closed intervals of the form $[x_0, x_0 + h_k] \subset [x, x + h]$, where $h_k \to 0$ as $k \to \infty$ on which

$$\left| f(x_0 + h_k) - f(x_0) \right| > M|h_k|. \tag{10.49}$$

If we prove (10.48) with M replaced by $M + \varepsilon$, where ε is any positive number, we will still obtain (10.48) as $\varepsilon \to 0$, and hence we can also replace (10.49) by

$$\left| f(x_0 + h_k) - f(x_0) \right| > (M + \varepsilon)|h_k| \tag{10.49'}$$

and we can now show that this is incompatible with the assumption that f is differentiable at x_0.

Indeed, by the assumption that f is differentiable,

$$\left| f(x_0 + h_k) - f(x_0) \right| = \left| f'(x_0)h_k + o(h_k) \right| \le$$
$$\le \left\| f'(x_0) \right\| |h_k| + o(|h_k|) \le (M + \varepsilon)|h_k|$$

as $h_k \to 0$. □

The finite-increment theorem has the following useful, purely technical corollary.

Corollary *If $A \in \mathcal{L}(X; Y)$, that is, A is a continuous linear mapping of the normed space X into the normed space Y and $f : U \to Y$ is a mapping satisfying the hypotheses of the finite-increment theorem, then*

$$\left| f(x + h) - f(x) - Ah \right| \le \sup_{\xi \in]x, x+h[} \left\| f'(\xi) - A \right\| |h|.$$

Proof For the proof it suffices to apply the finite-increment theorem to the mapping

$$t \mapsto F(t) = f(x + th) - Ath$$

of the unit interval $[0, 1] \subset \mathbb{R}$ into Y, since

$$F(1) - F(0) = f(x + h) - f(x) - Ah,$$
$$F'(\theta) = f'(x + \theta h)h - Ah \quad \text{for } 0 < \theta < 1,$$
$$\|F'(\theta)\| \leq \|f'(x + \theta h) - A\| |h|,$$
$$\sup_{0 < \theta < 1} \|F'(\theta)\| \leq \sup_{\xi \in]x, x+h[} \|f'(\xi) - A\| |h|.$$

$\square$

Remark As can be seen from the proof of Theorem 1, in its hypotheses there is no need to require that f be differentiable as a mapping $f : U \to Y$; it suffices that its restriction to the closed interval $[x, x + h]$ be a continuous mapping of that interval and differentiable at the points of the open interval $]x, x + h[$.

This remark applies equally to the corollary of the finite-increment theorem just proved.

10.4.2 Some Applications of the Finite-Increment Theorem

a. Continuously Differentiable Mappings

Let

$$f : U \to Y \tag{10.50}$$

be a mapping of an open subset U of a normed vector space X into a normed space Y. If f is differentiable at each point $x \in U$, then, assigning to the point x the mapping $f'(x) \in \mathcal{L}(X; Y)$ tangent to f at that point, we obtain the derivative mapping

$$f' : U \to \mathcal{L}(X; Y). \tag{10.51}$$

Since the space $\mathcal{L}(X; Y)$ of continuous linear transformations from X into Y is, as we know, a normed space (with the transformation norm), it makes sense to speak of the continuity of the mapping (10.51).

Definition When the derivative mapping (10.51) is continuous in U, the mapping (10.50), in complete agreement with our earlier terminology, will be said to be *continuously differentiable*.

As before, the set of continuously differentiable mappings of type (10.50) will be denoted by the symbol $C^{(1)}(U, Y)$, or more briefly, $C^{(1)}(U)$, if it is clear from the context what the range of the mapping is.
Thus, by definition

$$f \in C^{(1)}(U, Y) \Leftrightarrow f' \in C\big(U, \mathcal{L}(X; Y)\big).$$

Let us see what continuous differentiability of a mapping means in different particular cases.

Example 1 Consider the familiar situation when $X = Y = \mathbb{R}$, and hence $f : U \to \mathbb{R}$ is a real-valued function of a real argument. Since any linear mapping $A \in \mathcal{L}(\mathbb{R}; \mathbb{R})$ reduces to multiplication by some number $a \in \mathbb{R}$, that is, $Ah = ah$ and obviously $\|A\| = |a|$, we find that $f'(x)h = a(x)h$, where $a(x)$ is the numerical derivative of the function f at the point x.

Next, since

$$\big(f'(x + \delta) - f'(x)\big)h = f'(x + \delta)h - f'(x)h =$$

$$= a(x + \delta)h - a(x)h = \big(a(x + \delta) - a(x)\big)h, \qquad (10.52)$$

it follows that

$$\big\| f'(x + \delta) - f'(x) \big\| = \big| a(x + \delta) - a(x) \big|$$

and hence in this case continuous differentiability of the mapping f is equivalent to the concept of a continuously differentiable numerical function (of class $C^{(1)}(U, \mathbb{R})$) studied earlier.

Example 2 This time suppose that X is the direct product $X_1 \times \cdots \times X_n$ of normed spaces. In this case the mapping (10.50) is a function $f(x) = f(x_1, \ldots, x_m)$ of m variables $x_i \in X_i$, $i = 1, \ldots, m$, with values in Y.

If the mapping f is differentiable at $x \in U$, its differential $df(x)$ at that point is an element of the space $\mathcal{L}(X_1 \times \cdots \times X_m = X; Y)$.

The action of $df(x)$ on a vector $h = (h_1, \ldots, h_m)$, by formula (10.45), can be represented as

$$df(x)h = \partial_1 f(x)h_1 + \cdots + \partial_m f(x)h_m,$$

where $\partial_i f(x) : X_i \to Y$, $i = 1, \ldots, m$, are the partial derivatives of the mapping f at the point x under consideration.

Next,

$$\big(df(x + \delta) - df(x)\big)h = \sum_{i=1}^{m} \big(\partial_i f(x + \delta) - \partial_i f(x)\big)h_i. \qquad (10.53)$$

But by the properties of the standard norm in the direct product of normed spaces (see Example 6 in Sect. 10.1.2) and the definition of the norm of a transformation, we find that

$$\big\| \partial_i f(x + \delta) - \partial_i f(x) \big\|_{\mathcal{L}(X_i; Y)} \leq \big\| df(x + \delta) - df(x) \big\|_{\mathcal{L}(X; Y)} \leq$$

$$\leq \sum_{i=1}^{m} \big\| \partial_i f(x + \delta) - \partial_i f(x) \big\|_{\mathcal{L}(X_i; Y)}. \qquad (10.54)$$

Thus in this case the differentiable mapping (10.50) is continuously differentiable in U if and only if all its partial derivatives are continuous in U.

In particular, if $X = \mathbb{R}^m$ and $Y = \mathbb{R}$, we again obtain the familiar concept of a continuously differentiable numerical function of m real variables (a function of class $C^{(1)}(U, \mathbb{R})$, where $U \subset \mathbb{R}^m$).

Remark It is worth noting that in writing (10.52) and (10.53) we have made essential use of the canonical identification $TX_x \sim X$, which makes it possible to compare or identify vectors lying in different tangent spaces.

We shall now show that continuously differentiable mappings satisfy a Lipschitz condition.

Proposition 1 *If K is a convex compact set in a normed space X and $f \in C^{(1)}(K, Y)$, where Y is also a normed space, then the mapping $f : K \to Y$ satisfies a Lipschitz condition on K, that is, there exists a constant $M > 0$ such that the inequality*

$$\left| f(x_2) - f(x_1) \right| \leq M |x_2 - x_1| \tag{10.55}$$

holds for any points $x_1, x_2 \in K$.

Proof By hypothesis $f' : K \to \mathcal{L}(X; Y)$ is a continuous mapping of the compact set K into the metric space $\mathcal{L}(X; Y)$. Since the norm is a continuous function on a normed space with its natural metric, the mapping $x \mapsto \| f'(x) \|$, being the composition of continuous functions, is itself a continuous mapping of the compact set K into $\mathbb{R}$. But such a mapping is necessarily bounded. Let M be a constant such that $\| f'(x) \| \leq M$ at any point $x \in K$. Since K is convex, for any two points $x_1 \in K$ and $x_2 \in K$ the entire interval $[x_1, x_2]$ is contained in K. Applying the finite-increment theorem to that interval, we immediately obtain relation (10.55). $\square$

Proposition 2 *Under the hypotheses of Proposition 1 there exists a non-negative function $\omega(\delta)$ tending to 0 as $\delta \to +0$ such that*

$$\left| f(x + h) - f(x) - f'(x)h \right| \leq \omega(\delta) |h| \tag{10.56}$$

at any point $x \in K$ for $|h| < \delta$ if $x + h \in K$.

Proof By the corollary to the finite-increment theorem we can write

$$\left| f(x + h) - f(x) - f'(x)h \right| \leq \sup_{0 < \theta < 1} \left\| f'(x + \theta h) - f'(x) \right\| |h|$$

and, setting

$$\omega(\delta) = \sup_{\substack{x_1, x_2 \in K \\ |x_1 - x_2| < \delta}} \left\| f'(x_2) - f'(x_1) \right\|,$$

we obtain (10.56) in view of the uniform continuity of the function $x \mapsto f'(x)$, which is continuous on the compact set K.   □

b. A Sufficient Condition for Differentiability

We shall now show that by using the general finite-increment theorem, we can obtain a general sufficient condition for differentiability of a mapping in terms of its partial derivatives.

Theorem 2 *Let U be a neighborhood of the point x in a normed space $X = X_1 \times \cdots \times X_m$, which is the direct product of the normed spaces $X_1 \times \cdots \times X_m$, and let $f : U \to Y$ be a mapping of U into a normed space Y. If the mapping f has partial derivatives with respect to all its variables in U, then it is differentiable at the point x if the partial derivatives are all continuous at that point.*

Proof To simplify the writing we carry out the proof for the case $m = 2$. We verify immediately that the mapping

$$Lh = \partial_1 f(x)h_1 + \partial_2 f(x)h_2,$$

which is linear in $h = (h_1, h_2)$, is the total differential of f at x.

Making the elementary transformations

$$f(x + h) - f(x) - Lh =$$
$$= f(x_1 + h_1, x_2 + h_2) - f(x_1, x_2) - \partial_1 f(x)h_1 - \partial_2 f(x)h_2 =$$
$$= f(x_1 + h_1, x_2 + h_2) - f(x_1, x_2 + h_2) - \partial_1 f(x_1, x_2)h_1 +$$
$$+ f(x_1, x_2 + h_2) - f(x_1, x_2) - \partial_2 f(x_1, x_2)h_2,$$

by the corollary to Theorem 1 we obtain

$$\left| f(x_1 + h_1, x_2 + h_2) - f(x_1, x_2) - \partial_1 f(x_1, x_2)h_1 - \partial_2 f(x_1, x_2)h_2 \right| \leq$$
$$\leq \sup_{0 < \theta_1 < 1} \left\| \partial_1 f(x_1 + \theta_1 h_1, x_2 + h_2) - \partial_1 f(x_1, x_2) \right\| |h_1| +$$
$$+ \sup_{0 < \theta_2 < 1} \left\| \partial_2 f(x_1, x_2 + \theta_2 h_2) - \partial_2 f(x_1, x_2) \right\| |h_2|. \tag{10.57}$$

Since $\max\{|h_1|, |h_2|\} \leq |h|$, it follows obviously from the continuity of the partial derivatives $\partial_1 f$ and $\partial_2 f$ at the point $x = (x_1, x_2)$ that the right-hand side of inequality (10.57) is $o(h)$ as $h = (h_1, h_2) \to 0$.   □

Corollary *A mapping $f : U \to Y$ of an open subset U of the normed space $X = X_1 \times \cdots \times X_m$ into a normed space Y is continuously differentiable if and only if all the partial derivatives of the mapping f are continuous.*

Proof We have shown in Example 2 that when the mapping $f : U \to Y$ is differentiable, it is continuously differentiable if and only if its partial derivatives are continuous.

We now see that if the partial derivatives are continuous, then the mapping f is automatically differentiable, and hence (by Example 2) also continuously differentiable. □

10.4.3 Problems and Exercises

1. Let $f : I \to Y$ be a continuous mapping of the closed interval $I = [0, 1] \subset \mathbb{R}$ into a normed space Y and $g : I \to \mathbb{R}$ a continuous real-valued function on I. Show that if f and g are differentiable in the open interval $]0, 1[$ and the relation $\| f'(t) \| \leq g'(t)$ holds at points of this interval, then the inequality $|f(1) - f(0)| \leq g(1) - g(0)$ also holds.

2. a) Let $f : I \to Y$ be a continuously differentiable mapping of the closed interval $I = [0, 1] \subset \mathbb{R}$ into a normed space Y. It defines a smooth path in Y. Define the length of that path.

b) Recall the geometric meaning of the norm of the tangent mapping and give an upper bound for the length of the path considered in a).

c) Give a geometric interpretation of the finite-increment theorem.

3. Let $f : U \to Y$ be a continuous mapping of a neighborhood U of the point a in a normed space X into a normed space Y. Show that if f is differentiable in $U \setminus a$ and $f'(x)$ has a limit $L \in \mathcal{L}(X; Y)$ as $x \to a$, then the mapping f is differentiable at a and $f'(a) = L$.

4. a) Let U be an open convex subset of a normed space X and $f : U \to Y$ a mapping of U into a normed space Y. Show that if $f'(x) \equiv 0$ on U, then the mapping f is constant.

b) Generalize the assertion of a) to the case of an arbitrary domain U (that is, when U is an open connected subset of X).

c) The partial derivative $\frac{\partial f}{\partial y}$ of a smooth function $f : D \to \mathbb{R}$ defined in a domain $D \subset \mathbb{R}^2$ of the xy-plane is identically zero. Is it true that f is then independent of y in this domain? For which domains D is this true?

10.5 Higher-Order Derivatives

10.5.1 Definition of the nth Differential

Let U be an open set in a normed space X and

$$f : U \to Y \tag{10.58}$$

a mapping of U into a normed space Y.

If the mapping (10.58) is differentiable in U, then the derivative of f, given by

$$f' : U \to \mathcal{L}(X; Y), \tag{10.59}$$

is defined in U.

The space $\mathcal{L}(X; Y) =: Y_1$ is a normed space relative to which the mapping (10.59) has the form (10.58), that is, $f' : U \to Y_1$, and it makes sense to speak of differentiability for it.

If the mapping (10.59) is differentiable, its derivative

$$\left(f' \right)' : U \to \mathcal{L}(X; Y_1) = \mathcal{L}\big(X; \mathcal{L}(X; Y)\big)$$

is called the *second derivative* or *second differential* of f and denoted f'' or $f^{(2)}$. In general, we adopt the following inductive definition.

Definition 1 The *derivative of order* $n \in \mathbb{N}$ or *nth differential* of the mapping (10.58) at the point $x \in U$ is the mapping tangent to the derivative of f of order $n - 1$ at that point.

If the derivative of order $k \in \mathbb{N}$ at the point $x \in U$ is denoted $f^{(k)}(x)$, Definition 1 means that

$$f^{(n)}(x) := \left(f^{(n-1)} \right)'(x). \tag{10.60}$$

Thus, if $f^{(n)}(x)$ is defined, then

$$f^{(n)}(x) \in \mathcal{L}(X; Y_n) = \mathcal{L}\big(X; \mathcal{L}(X; Y_{n-1})\big) =$$

$$= \cdots = \mathcal{L}\big(X; \mathcal{L}\big(X; \ldots; \mathcal{L}(X; Y)\big) \ldots\big).$$

Consequently, by Proposition 4 of Sect. 10.2, $f^{(n)}(x)$, the differential of order n of the mapping (10.58) at the point x can be interpreted as an element of the space $\underbrace{\mathcal{L}(X, \ldots, X; Y)}_{n \text{ factors}}$ of continuous n-linear transformations.

We note once again that the tangent mapping $f'(x) : TX_x \to TY_{f(x)}$ is a mapping of tangent spaces, each of which, because of the affine or vector-space structure of the spaces being mapped, we have identified with the corresponding vector space and said on that basis that $f'(x) \in \mathcal{L}(X; Y)$. It is this device of regarding elements $f'(x_1) \in \mathcal{L}(TX_{x_1}; TY_{f(x_1)})$ and $f'(x_2) \in \mathcal{L}(TX_{x_2}, TY_{f(x_2)})$, which lie in different spaces, as vectors in the same space $\mathcal{L}(X; Y)$ that provides the basis for defining higher-order differentials of mappings of normed vector spaces. In the case of an affine or vector space there is a natural connection between vectors in the different tangent spaces corresponding to different points of the original space. In the final analysis, it is this connection that makes it possible to speak of the continuous differentiability of both the mapping (10.58) and its higher-order differentials.

10.5.2 Derivative with Respect to a Vector and Computation of the Values of the nth Differential

When we are making the abstract Definition 1 specific, the concept of the derivative with respect to a vector may be used to advantage. This concept is introduced for the general mapping (10.58) just as was done earlier in the case $X = \mathbb{R}^m$, $Y = \mathbb{R}$.

Definition 2 If X and Y are normed vector spaces over the field $\mathbb{R}$, the *derivative of the mapping* (10.58) *with respect to the vector* $h \in TX_x \sim X$ *at the point* $x \in U$ is defined as the limit

$$D_h f(x) := \lim_{\mathbb{R} \ni t \to 0} \frac{f(x+th) - f(x)}{t},$$

provided this limit exists.

It can be verified immediately that

$$D_{\lambda h} f(x) = \lambda D_h f(x) \tag{10.61}$$

and that if the mapping f is differentiable at the point $x \in U$, it has a derivative at that point with respect to every vector; moreover

$$D_h f(x) = f'(x)h, \tag{10.62}$$

and, by the linearity of the tangent mapping,

$$D_{\lambda_1 h_1 + \lambda_2 h_2} f(x) = \lambda_1 D_{h_1} f(x) + \lambda_2 D_{h_2} f(x). \tag{10.63}$$

It can also be seen from Definition 2 that the value $D_h f(x)$ of the derivative of the mapping $f : U \to Y$ with respect to a vector is an element of the vector space $TY_{f(x)} \sim Y$, and that if L is a continuous linear transformation from Y to a normed space Z, then

$$D_h(L \circ f)(x) = L \circ D_h f(x). \tag{10.64}$$

We shall now try to give an interpretation to the value $f^{(n)}(h_1, \ldots, h_n)$ of the nth differential of the mapping f at the point x on the set $(h_1, \ldots, h_n)$ of vectors $h_i \in TX_x \sim X$, $i = 1, \ldots, n$.

We begin with $n = 1$. In this case, by formula (10.62)

$$f'(x)(h) = f'(x)h = D_h f(x).$$

We now consider the case $n = 2$. Since $f^{(2)}(x) \in \mathcal{L}(X; \mathcal{L}(X; Y))$, fixing a vector $h_1 \in X$, we assign a linear transformation $(f^{(2)}(x)h_1) \in \mathcal{L}(X; Y)$ to it by the rule

$$h_1 \mapsto f^{(2)}(x)h_1.$$

Then, after computing the value of this operator at the vector $h_2 \in X$, we obtain an element of Y:

$$f^{(2)}(x)(h_1, h_2) := \left(f^{(2)}(x)h_1\right)h_2 \in Y. \tag{10.65}$$

But

$$f^{(2)}(x)h = \left(f'\right)'(x)h = D_h f'(x),$$

and therefore

$$f^{(2)}(x)(h_1, h_2) = \left(D_{h_1} f'(x)\right)h_2. \tag{10.66}$$

If $A \in \mathcal{L}(X; Y)$ and $h \in X$, this pairing with Ah can be regarded not only as a mapping $h \mapsto Ah$ from X into Y, but as a mapping $A \mapsto Ah$ from $\mathcal{L}(X; Y)$ into Y, the latter mapping being linear, just like the former.

Comparing relations (10.62), (10.64), and (10.66), we can write

$$\left(D_{h_1} f'(x)\right)h_2 = D_{h_1}\left(f'(x)h_2\right) = D_{h_1} D_{h_2} f(x).$$

Thus we finally obtain

$$f^{(2)}(x)(h_1, h_2) = D_{h_1} D_{h_2} f(x).$$

Similarly, one can show that the relation

$$f^{(n)}(x)(h_1, \ldots, h_n) := \left(\ldots\left(f^{(n)}(x)h_1\right)\ldots h_n\right) = D_{h_1} D_{h_2} \cdots D_{h_n} f(x) \tag{10.67}$$

holds for any $n \in \mathbb{N}$, the differentiation with respect to the vectors being carried out sequentially, starting with differentiation with respect to h_n and ending with differentiation with respect to h_1.

10.5.3 Symmetry of the Higher-Order Differentials

In connection with formula (10.67), which is perfectly adequate for computation as it now stands, the question naturally arises: To what extent does the result of the computation depend on the order of differentiation?

Proposition *If the form $f^{(n)}(x)$ is defined at the point x for the mapping* (10.58), *it is symmetric with respect to any pair of its arguments.*

Proof The main element in the proof is to verify that the proposition holds in the case $n = 2$.

Let h_1 and h_2 be two arbitrary fixed vectors in the space $TX_x \sim X$. Since U is open in X, the following auxiliary function of t is defined for all values of $t \in \mathbb{R}$ sufficiently close to zero:

$$F_t(h_1, h_2) = f\left(x + t(h_1 + h_2)\right) - f(x + th_1) - f(x + th_2) + f(x).$$

We consider also the following auxiliary function:

$$g(v) = f\big(x + t(h_1 + v)\big) - f(x + tv),$$

which is certainly defined for vectors v that are collinear with the vector h_2 and such that $|v| \leq |h_2|$.

We observe that

$$F_t(h_1, h_2) = g(h_2) - g(0).$$

We further observe that, since the function $f : U \to Y$ has a second differential $f''(x)$ at the point $x \in U$, it must be differentiable at least in some neighborhood of x. We shall assume that the parameter t is sufficiently small that the arguments on the right-hand side of the equality that defines $F_t(h_1, h_2)$ lie in that neighborhood.

We now make use of these observations and the corollary of the mean-value theorem in the following computations:

$$\big| F_t(h_1, h_2) - t^2 f''(x)(h_1, h_2) \big| =$$
$$= \big| g(h_2) - g(0) - t^2 f''(x)(h_1, h_2) \big| \leq$$
$$\leq \sup_{0 < \theta_2 < 1} \big\| g'(\theta_2 h_2) - t^2 f''(x) h_1 \big\| |h_2| =$$
$$= \sup_{0 < \theta_2 < 1} \big\| \big(f'\big(x + t(h_1 + \theta_2 h_2)\big) - f'(x + t\theta_2 h_2) \big) t - t^2 f''(x) h_1 \big\| |h_2|.$$

By definition of the derivative mapping we can write that

$$f'\big(x + t(h_1 + \theta_2 h_2)\big) = f'(x) + f''(x)\big(t(h_1 + \theta_2 h_2)\big) + o(t)$$

and

$$f'(x + t\theta_2 h_2) = f'(x) + f''(x)(t\theta_2 h_2) + o(t)$$

as $t \to 0$. Taking this relation into account, one can continue the preceding computation, finding after cancellation that

$$\big| F_t(h_1, h_2) - t^2 f''(x)(h_1, h_2) \big| = o(t^2)$$

as $t \to 0$. But this equality means that

$$f''(x)(h_2, h_2) = \lim_{t \to 0} \frac{F_t(h_1, h_2)}{t^2}.$$

Since it is obvious that $F_t(h_1, h_2) = F_t(h_2, h_1)$, it follows from this relation that $f''(x)(h_1, h_2) = f''(x)(h_2, h_1)$.

One can now complete the proof of the proposition by induction, repeating verbatim what was said in the proof that the values of the mixed partial derivatives are independent of the order of differentiation.   □

Thus we have shown that the nth differential of the mapping (10.58) at the point $x \in U$ is a symmetric n-linear transformation

$$f^{(n)}(x) \in \mathcal{L}(TX_x, \ldots, TX_x; TY_{f(x)}) \sim \mathcal{L}(X, \ldots, X; Y)$$

whose value on the set $(h_1, \ldots, h_n)$ of vectors $h_i \in TX_x = X$, $i = 1, \ldots, n$, can be computed by formula (10.67).

If X is a finite-dimensional space having a basis $\{e_1, \ldots, e_k\}$ and $h_j = h^i_j e_i$ is the expansion of the vector h_j, $j = 1, \ldots, n$, with respect to that basis, then by the multilinearity of $f^{(n)}(x)$ we can write

$$f^{(n)}(x)(h_1, \ldots, h_n) = f^{(n)}(x)\big(h^{i_1}_1 e_{i_1}, \ldots, h^{i_n}_n e_{i_n}\big) =$$

$$= f^{(n)}(x)(e_{i_1}, \ldots, e_{i_n}) h^{i_1}_1 \cdot \ldots \cdot h^{i_n}_n.$$

Using our earlier notation $\partial_{i_1 \cdots i_n} f(x)$ for $D_{e_1} \cdots D_{e_n} f(x)$, we find finally that

$$f^{(n)}(x)(h_1, \ldots, h_n) = \partial_{i_1 \cdots i_n} f(x) h^{i_1}_1 \cdots h^{i_n}_n,$$

where as usual summation extends over the repeated indices on the right-hand side within their range of variation, that is, from 1 to k.

Let us agree to use the following abbreviation:

$$f^{(n)}(x)(h, \ldots, h) =: f^{(n)}(x) h^n. \tag{10.68}$$

In particular, if we are discussing a finite-dimensional space X and $h = h^i e_i$, then

$$f^{(n)}(x) h^n = \partial_{i_1 \cdots i_n} f(x) h^{i_1} \cdot \ldots \cdot h^{i_n},$$

which is already very familiar to us from the theory of numerical functions of several variables.

10.5.4 Some Remarks

In connection with the notation (10.68) consider the following example, which is quite useful and will be used in the next section.

Example Let $A \in \mathcal{L}(X_1, \ldots, X_n; Y)$, that is, $y = A(x_1, \ldots, x_n)$ is a continuous n-linear transformation from the product of the normed vector spaces $X_1, \ldots, X_n$ into the normed vector space Y.

It was shown in Example 5 of Sect. 10.3 that A is a differentiable mapping $A : X_1 \times \cdots \times X_n \to Y$ and

$$A'(x_1, \ldots, x_n)(h_1, \ldots, h_n) = A(h_1, x_2, \ldots, x_n) + \cdots + A(x_1, \ldots, x_{n-1}, h_n).$$

Thus, if $X_1 = \cdots = X_n = X$ and A is symmetric, then

$$A'(x, \ldots, x)(h, \ldots, h) = nA(\underbrace{x, \ldots, x}_{n-1}, h) =: \left(nAx^{n-1}\right)h.$$

Hence, if we consider the function $F : X \to Y$ defined by the condition

$$X \ni x \mapsto F(x) = A(x, \ldots, x) =: Ax^n,$$

it turns out to be differentiable and

$$F'(x)h = \left(nAx^{n-1}\right)h,$$

that is, in this case

$$F'(x) = nAx^{n-1},$$

where $Ax^{n-1} := A(\underbrace{x, \ldots, x}_{n-1}, \cdot)$.

In particular, if the mapping (10.58) has a differential $f^{(n)}(x)$ at a point $x \in U$, then the function $F(h) = f^{(n)}(x)h^n$ is differentiable, and

$$F'(h) = nf^{(n)}(x)h^{n-1}. \tag{10.69}$$

To conclude our discussion of the concept of an nth-order derivative, it is useful to add the remark that if the original function (10.58) is defined on a set U in a space X that is the direct product of normed spaces $X_1, \ldots, X_m$, one can speak of the first-order partial derivatives $\partial_1 f(x), \ldots, \partial_m f(x)$ of f with respect to the variables $x_i \in X_i$, $i = 1, \ldots, m$, and the higher-order partial derivatives $\partial_{i_1 \cdots i_n} f(x)$.

On the basis of Theorem 2 of Sect. 10.4, we obtain by induction in this case that *if all the partial derivatives $\partial_{i_1 \cdots i_n} f(x)$ of a mapping $f : U \to Y$ are continuous at a point $x \in X = X_1 \times \cdots \times X_m$, then the mapping f has an nth order differential $f^{(n)}(x)$ at that point.*

If we also take account of the result of Example 2 from the same section, we can conclude that *the mapping $U \ni x \mapsto f^{(n)}(x) \in \mathcal{L}(\underbrace{X, \ldots, X}_{n \text{ factors}}; Y)$ is contin-
uous if and only if all the nth-order partial derivatives $U \ni x \mapsto \partial_{i_1 \cdots i_n} f(x) \in \mathcal{L}(X_{i_1}, \ldots, X_{i_n}; Y)$ of the original mapping $f : U \to Y$ are continuous* (or, what is the same, the partial derivatives of all orders up to n inclusive are continuous).

The class of mappings (10.58) having continuous derivatives up to order n inclusive in U is denoted $C^{(n)}(U, Y)$, or, where no confusion can arise, by the briefer symbol $C^{(n)}(U)$ or even $C^{(n)}$.

In particular, if $X = X_1 \times \cdots \times X_n$, the conclusion reached above can be written in abbreviated form as

$$\left(f \in C^{(n)}\right) \iff (\partial_{i_1 \cdots i_n} f \in C, i_1, \ldots, i_n = 1, \ldots, m),$$

where C, as always, denotes the corresponding set of continuous functions.

10.5.5 Problems and Exercises

1. Carry out the proof of Eq. (10.64) in full.

2. Give the details at the end of the proof that $f^{(n)}(x)$ is symmetric.

3. a) Show that if the functions $D_{h_1} D_{h_2} f$ and $D_{h_2} D_{h_1} f$ are defined and continuous at a point $x \in U$ for a pair of vectors h_1, h_2 and the mapping (10.58) in the domain U, then the equality $D_{h_1} D_{h_2} f(x) = D_{h_2} D_{h_1} f(x)$ holds.

b) Show using the example of a numerical function $f(x, y)$ that, although the continuity of the mixed partial derivatives $\frac{\partial^2 f}{\partial x \partial y}$ and $\frac{\partial^2 f}{\partial y \partial x}$ implies by a) that they are equal at this point, it does not in general imply that the second differential of the function exists at the point.

c) Show that, although the existence of $f^{(2)}(x, y)$, guarantees that the mixed partial derivatives $\frac{\partial^2 f}{\partial x \partial y}$ and $\frac{\partial^2 f}{\partial y \partial x}$ exist and are equal, it does not in general guarantee that they are continuous at that point.

4. Let $A \in \mathcal{L}(X, \ldots, X; Y)$ where A is a symmetric n-linear transformation. Find the successive derivatives of the function $x \mapsto Ax^n := A(x, \ldots, x)$ up to order $n + 1$ inclusive.

10.6 Taylor's Formula and the Study of Extrema

10.6.1 Taylor's Formula for Mappings

Theorem 1 *If a mapping $f : U \to Y$ from a neighborhood $U = U(x)$ of a point x in a normed space X into a normed space Y has derivatives up to order $n - 1$ inclusive in U and has an nth order derivative $f^{(n)}(x)$ at the point x, then*

$$f(x + h) = f(x) + f'(x)h + \cdots + \frac{1}{n!} f^{(n)}(x)h^n + o(|h|^n) \qquad (10.70)$$

as $h \to 0$.

Equality (10.70) is one of the varieties of Taylor's formula, written here for rather general classes of mappings.

Proof We prove Taylor's formula by induction.

For $n = 1$ it is true by definition of $f'(x)$.

Assume formula (10.70) is true for some $(n - 1) \in \mathbb{N}$.

Then by the mean-value theorem, formula (10.69) of Sect. 10.5, and the induction hypothesis, we obtain

$$\left| f(x+h) - \left(f(x) + f'(x)h + \cdots + \frac{1}{n!} f^{(n)}(x)h^n \right) \right| \leq$$

$$\leq \sup_{0<\theta<1} \left\| f'(x+\theta h) - \left(f'(x) + f''(x)(\theta h) + \right. \right.$$

$$\left. \left. + \cdots + \frac{1}{(n-1)!} f^{(n)}(x)(\theta h)^{n-1} \right) \right\| |h| = o\left(|\theta h|^{n-1}\right)|h| = o\left(|h|^n\right)$$

as $h \to 0$.   □

We shall not take the time here to discuss other versions of Taylor's formula, which are sometimes quite useful. They were discussed earlier in detail for numerical functions. At this point we leave it to the reader to derive them (see, for example, Problem 1 below).

10.6.2 Methods of Studying Interior Extrema

Using Taylor's formula, we shall exhibit necessary conditions and also sufficient conditions for an interior local extremum of real-valued functions defined on an open subset of a normed space. As we shall see, these conditions are analogous to the differential conditions already known to us for an extremum of a real-valued function of a real variable.

Theorem 2 *Let $f : U \to \mathbb{R}$ be a real-valued function defined on an open set U in a normed space X and having continuous derivatives up to order $k - 1 \geq 1$ inclusive in a neighborhood of a point $x \in U$ and a derivative $f^{(k)}(x)$ of order k at the point x itself.*

If $f'(x) = 0, \ldots, f^{(k-1)}(x) = 0$ and $f^{(k)}(x) \neq 0$, then for x to be an extremum of the function f it is:

necessary that k be even and that the form $f^{(k)}(x)h^k$ be semidefinite,[4] and sufficient that the values of the form $f^{(k)}(x)h^k$ on the unit sphere $|h| = 1$ be bounded away from zero; moreover, x is a local minimum if the inequalities

$$f^{(k)}(x)h^k \geq \delta > 0$$

hold on that sphere, and a local maximum if

$$f^{(k)}(x)h^k \leq \delta < 0.$$

---

[4]This means that the form $f^{(k)}(x)h^k$ cannot take on values of opposite signs, although it may vanish for some values $h \neq 0$. The equality $f^{(i)}(x) = 0$, as usual, is understood to mean that $f^{(i)}(x)h = 0$ for every vector h.

Proof For the proof we consider the Taylor expansion (10.70) of f in a neighbor-hood of x. The assumptions enable us to write

$$f(x+h) - f(x) = \frac{1}{k!} f^{(k)}(x) h^k + \alpha(h) |h|^k,$$

where $\alpha(h)$ is a real-valued function, and $\alpha(h) \to 0$ as $h \to 0$.

We first prove the necessary conditions.

Since $f^{(k)}(x) \neq 0$, there exists a vector $h_0 \neq 0$ on which $f^{(k)}(x) h_0^k \neq 0$. Then for values of the real parameter t sufficiently close to zero,

$$f(x+th_0) - f(x) = \frac{1}{k!} f^{(k)}(x)(th_0)^k + \alpha(th_0)|th_0|^k =$$

$$= \left(\frac{1}{k!} f^{(k)}(x) h_0^k \pm \alpha(th_0)|h_0|^k \right) t^k$$

and the expression in the outer parentheses has the same sign as $f^{(k)}(x) h_0^k$.

For x to be an extremum it is necessary for the left-hand side (and hence also the right-hand side) of this last equality to be of constant sign when t changes sign. But this is possible only if k is even.

This reasoning shows that if x is an extremum, then the sign of the difference $f(x+th_0) - f(x)$ is the same as that of $f^{(k)}(x) h_0^k$ for sufficiently small t; hence in that case there cannot be two vectors h_0, h_1 at which the form $f^{(k)}(x)$ assumes values with opposite signs.

We now turn to the proof of the sufficiency conditions. For definiteness we consider the case when $f^{(k)}(x) h^k \geq \delta > 0$ for $|h| = 1$. Then

$$f(x+h) - f(x) = \frac{1}{k!} f^{(k)}(x) h^k + \alpha(h) |h|^k =$$

$$= \left(\frac{1}{k!} f^{(k)}(x) \left(\frac{h}{|h|} \right)^k + \alpha(h) \right) |h|^k \geq \left(\frac{1}{k!} \delta + \alpha(h) \right) |h|^k,$$

and, since $\alpha(h) \to 0$ as $h \to 0$, the last term in this inequality is positive for all vectors $h \neq 0$ sufficiently close to zero. Thus, for all such vectors h,

$$f(x+h) - f(x) > 0,$$

that is, x is a strict local minimum.

The sufficient condition for a strict local maximum is verified similarly. □

Remark 1 If the space X is finite-dimensional, the unit sphere $S(x, 1)$ with center at $x \in X$, being a closed bounded subset of X, is compact. Then the continuous function $f^{(k)}(x) h^k = \partial_{i_1 \ldots i_k} f(x) h^{i_1} \cdot \ldots \cdot h^{i_k}$ (a k-form) has both a maximal and a minimal value on $S(x, 1)$. If these values are of opposite sign, then f does not have an extremum at x. If they are both of the same sign, then, as was shown in Theorem 2, there is an extremum. In the latter case, a sufficient condition for an extremum can obviously be stated as the equivalent requirement that the form $f^{(k)}(x)h^k$ be either positive- or negative-definite.

It was this form of the condition that we encountered in studying real-valued functions on $\mathbb{R}^n$.

Remark 2 As we have seen in the example of functions $f : \mathbb{R}^n \to \mathbb{R}$, the semi-definiteness of the form $f^{(k)}h^k$ exhibited in the necessary conditions for an extremum is not a sufficient criterion for an extremum.

Remark 3 In practice, when studying extrema of differentiable functions one normally uses only the first or second differentials. If the uniqueness and type of extremum are obvious from the meaning of the problem being studied, one can restrict attention to the first differential when seeking an extremum, simply finding the point x where $f'(x) = 0$.

10.6.3 Some Examples

Example 1 Let $L \in C^{(1)}(\mathbb{R}^3, \mathbb{R})$ and $f \in C^{(1)}([a, b], \mathbb{R})$. In other words, $(u^1, u^2, u^3) \mapsto L(u^1, u^2, u^3)$ is a continuously differentiable real-valued function defined in $\mathbb{R}^3$ and $x \mapsto f(x)$ a smooth real-valued function defined on the closed interval $[a, b] \subset \mathbb{R}$.

Consider the function

$$F : C^{(1)}\big([a, b], \mathbb{R}\big) \to \mathbb{R} \tag{10.71}$$

defined by the relation

$$C^{(1)}\big([a, b], \mathbb{R}\big) \ni f \mapsto F(f) = \int_a^b L\big(x, f(x), f'(x)\big)\, dx \in \mathbb{R}. \tag{10.72}$$

Thus, (10.71) is a real-valued functional defined on the set of functions $f \in C(1)([a, b], \mathbb{R})$.

The basic variational principles connected with motion are known in physics and mechanics. According to these principles, the actual motions are distinguished among all the conceivable motions in that they proceed along trajectories along which certain functionals have an extremum. Questions connected with the extrema of functionals are central in optimal control theory. Thus, finding and studying the extrema of functionals is a problem of intrinsic importance, and the theory associated with it is the subject of a large area of analysis – the calculus of variations. We have already done a few things to make the transition from the analysis of the extrema of numerical functions to the problem of finding and studying extrema of functionals seem natural to the reader. However, we shall not go deeply into the special problems of variational calculus, but rather use the example of the functional (10.72) to illustrate only the general ideas of differentiation and study of local extrema considered above.

We shall show that the functional (10.72) is a differentiable mapping and find its differential.

We remark that the function (10.72) can be regarded as the composition of the mappings

$$F_1 : C^{(1)}([a,b], \mathbb{R}) \to C([a,b], \mathbb{R}) \tag{10.73}$$

defined by the formula

$$F_1(f)(x) = L(x, f(x), f'(x)) \tag{10.74}$$

followed by the mapping

$$C([a,b], \mathbb{R}) \ni g \mapsto F_2(g) = \int_a^b g(x)\,\mathrm{d}x \in \mathbb{R}. \tag{10.75}$$

By properties of the integral, the mapping F_2 is obviously linear and continuous, so that its differentiability is clear.

We shall show that the mapping F_1 is also differentiable, and that

$$F_1'(f)h(x) = \partial_2 L(x, f(x), f'(x))h(x) + \partial_3 L(x, f(x), f'(x))h'(x) \tag{10.76}$$

for $h \in C^{(1)}([a,b], \mathbb{R})$.

Indeed, by the corollary to the mean-value theorem, we can write in the present case

$$\left| L(u^1 + \Delta^1, u^2 + \Delta^2, u^3 + \Delta^3) - L(u^1, u^2, u^3) - \sum_{i=1}^{3} \partial_i L(u^1, u^2, u^3)\Delta^i \right| \le$$

$$\le \sup_{0 < \theta < 1} \left\| \left(\partial_1 L(u + \theta\Delta) - \partial_1 L(u), \partial_2 L(u + \theta\Delta) - \partial_2 L(u), \right. \right.$$

$$\left. \partial_3 L(u + \theta\Delta) - \partial_3 L(u) \right) \right\| \cdot |\Delta| \le$$

$$\le 3 \max_{\substack{0 \le \theta \le 1 \\ i=1,2,3}} \left| \partial_i L(u + \theta u) - \partial_i L(u) \right| \cdot \max_{i=1,2,3} \left| \Delta^i \right|, \tag{10.77}$$

where $u = (u^1, u^2, u^3)$ and $\Delta = (\Delta^1, \Delta^2, \Delta^3)$.

If we now recall that the norm $|f|_{C^{(1)}}$ of the function f in $C^{(1)}([a,b], \mathbb{R})$ is $\max\{|f|_C, |f'|_C\}$ (where $|f|_C$ is the maximum absolute value of the function on the closed interval $[a,b]$), then, setting $u^1 = x$, $u^2 = f(x)$, $u^3 = f'(x)$, $\Delta^1 = 0$, $\Delta^2 = h(x)$, and $\Delta^3 = h'(x)$, we obtain from inequality (10.77), taking account of the uniform continuity of the functions $\partial_i L(u^1, u^2, u^3)$, $i = 1, 2, 3$, on bounded subsets of $\mathbb{R}^3$, that

$$\max_{a \le x \le b} \left| L(x, f(x) + h(x), f'(x) + h'(x)) - L(x, f(x), f'(x)) - \right.$$

$$\left. - \partial_2 L(x, f(x), f'(x))h(x) - \partial_3 L(X, f(x)f'(x))h'(x) \right| =$$

$$= o(|h|_{C^{(1)}}) \quad \text{as } |h|_{C^{(1)}} \to 0.$$

But this means that Eq. (10.76) holds.

By the chain rule for differentiating a composite function, we now conclude that the functional (10.72) is indeed differentiable, and

$$F'(f)h = \int_a^b \left(\left(\partial_2 L \big(x, \, f(x) f'(x) \big) \right) h(x) + \partial_3 L \big(x, \, f(x), \, f'(x) \big) \right) h'(x) \right) dx.$$
$$(10.78)$$

We often consider the restriction of the functional (10.72) to the affine space consisting of the functions $f \in C^{(1)}([a, b], \mathbb{R})$ that assume fixed values $f(a) = A$, $f(b) = B$ at the endpoints of the closed interval $[a, b]$. In this case, the functions h in the tangent space $TC_f^{(1)}$ must have the value zero at the endpoints of the closed interval $[a, b]$. Taking this fact into account, we may integrate by parts in (10.78) and bring it into the form

$$F'(f)h = \int_a^b \left(\partial_2 L \big(x, \, f(x), \, f'(x) \big) - \frac{d}{dx} \partial_3 L \big(x, \, f(x) f'(x) \big) \right) h(x) \, dx, \quad (10.79)$$

of course under the assumption that L and f belong to the corresponding class $C^{(2)}$.

In particular, if f is an extremum (extremal) of such a functional, then by Theorem 2 we have $F'(f)h = 0$ for every function $h \in C^{(1)}([a, b], \mathbb{R})$ such that $h(a) = h(b) = 0$. From this and relation (10.79) one can easily conclude (see Problem 3 below) that the function f must satisfy the equation

$$\partial_2 L \big(x, \, f(x), \, f'(x) \big) - \frac{d}{dx} \partial_3 L \big(x, \, f(x), \, f'(x) \big) = 0. \qquad (10.80)$$

This is a frequently-encountered form of the equation known in the calculus of variations as the *Euler–Lagrange equation*.

Let us now consider some specific examples.

Example 2 (The shortest-path problem) Among all the curves in a plane joining two fixed points, find the curve that has minimal length.

The answer in this case is obvious, and it rather serves as a check on the formal computations we will be doing later.

We shall assume that a fixed Cartesian coordinate system has been chosen in the plane, in which the two points are, for example, $(0, 0)$ and $(1, 0)$. We confine ourselves to just the curves that are the graphs of functions $f \in C^{(1)}([0, 1], \mathbb{R})$ assuming the value zero at both ends of the closed interval $[0, 1]$. The length of such a curve

$$F(f) = \int_0^1 \sqrt{1 + (f')^2(x)} \, dx \qquad (10.81)$$

depends on the function f and is a functional of the type considered in Example 1. In this case the function L has the form

$$L\big(u^1, u^2, u^3\big) = \sqrt{1 + (u^3)^2},$$

and therefore the necessary condition (10.80) for an extremal here reduces to the equation

$$\frac{d}{dx}\left(\frac{f'(x)}{\sqrt{1+(f')^2(x)}}\right)=0,$$

from which it follows that

$$\frac{f'(x)}{\sqrt{1+(f')^2(x)}}\equiv\text{const} \tag{10.82}$$

on the closed interval $[0, 1]$.

Since the function $\frac{u}{\sqrt{1+u^2}}$ is not constant on any interval, Eq. (10.82) is possible only if $f'(x)\equiv\text{const}$ on $[a, b]$. Thus a smooth extremal of this problem must be a linear function whose graph passes through the points $(0, 0)$ and $(1, 0)$. It follows that $f(x)\equiv 0$, and we arrive at the closed interval of the line joining the two given points.

Example 3 (The brachistochrone problem) The classical brachistochrone problem, posed by Johann Bernoulli I in 1696, was to find the shape of a track along which a point mass would pass from a prescribed point P_0 to another fixed point P_1 at a lower level under the action of gravity in the shortest time.

We neglect friction, of course. In addition, we shall assume that the trivial case in which both points lie on the same vertical line is excluded.

In the vertical plane passing through the points P_0 and P_1 we introduce a rectangular coordinate system such that P_0 is at the origin, the x-axis is directed vertically downward, and the point P_1 has positive coordinates (x_1, y_1). We shall find the shape of the track among the graphs of smooth functions defined on the closed interval $[0, x_1]$ and satisfying the condition $f(0) = 0$, $f(x_1) = y_1$. At the moment we shall not take time to discuss this by no means uncontroversial assumption (see Problem 4 below).

If the particle began its descent from the point P_0 with zero velocity, the law of variation of its velocity in these coordinates can be written as

$$v = \sqrt{2gx}. \tag{10.83}$$

Recalling that the differential of the arc length is computed by the formula

$$ds = \sqrt{(dx)^2 + (dy)^2} = \sqrt{1 + (f')^2(x)}\,dx, \tag{10.84}$$

we find the time of descent

$$F(f) = \frac{1}{\sqrt{2g}}\int_0^{x_1}\sqrt{\frac{1+(f')^2(x)}{x}}\,dx \tag{10.85}$$

along the trajectory defined by the graph of the function $y = f(x)$ on the closed interval $[0, x_1]$.

For the functional (10.85)

$$L\left(u^1, u^2, u^3\right) = \sqrt{\frac{1 + (u^3)^2}{u^1}},$$

and therefore the condition (10.80) for an extremum reduces in this case to the equation

$$\frac{d}{dx}\left(\frac{f'(x)}{\sqrt{x(1 + (f')^2(x))}}\right) = 0,$$

from which it follows that

$$\frac{f'(x)}{\sqrt{1 + (f')^2(x)}} = c\sqrt{x}, \tag{10.86}$$

where c is a nonzero constant, since the points are not both on the same vertical line. Taking account of (10.84), we can rewrite (10.86) in the form

$$\frac{dy}{ds} = c\sqrt{x}. \tag{10.87}$$

However, from the geometric point of view

$$\frac{dx}{ds} = \cos\varphi, \qquad \frac{dy}{ds} = \sin\varphi, \tag{10.88}$$

where φ is the angle between the tangent to the trajectory and the positive x-axis. By comparing Eq. (10.87) with the second equation in (10.88), we find

$$x = \frac{1}{c^2}\sin^2\varphi. \tag{10.89}$$

But it follows from (10.88) and (10.89) that

$$\frac{dy}{d\varphi} = \frac{dy}{dx}\cdot\frac{dx}{d\varphi} = \tan\varphi\frac{dx}{d\varphi} = \tan\varphi\frac{d}{d\varphi}\left(\frac{\sin^2\varphi}{c^2}\right) = 2\frac{\sin^2\varphi}{c^2},$$

from which we find

$$y = \frac{2}{c^2}(2\varphi - \sin 2\varphi) + b. \tag{10.90}$$

Setting $2/c^2 =: a$ and $2\varphi =: t$, we write relations (10.89) and (10.90) as

$$x = a(1 - \cos t),$$
$$y = a(t - \sin t) + b. \tag{10.91}$$

Since $a \neq 0$, it follows that $x = 0$ only for $t = 2k\pi$, $k \in \mathbb{Z}$. It follows from the form of the function (10.91) that we may assume without loss of generality that the parameter value $t = 0$ corresponds to the point $P_0 = (0, 0)$. In this case Eq. (10.90) implies $b = 0$, and we arrive at the simpler form

$$x = a(1 - \cos t),$$
$$y = a(t - \sin t) \tag{10.92}$$

for the parametric definition of this curve.

Thus the brachistochrone is a cycloid having a cusp at the initial point P_0 where the tangent is vertical. The constant a, which is a scaling coefficient, must be chosen so that the curve (10.92) also passes through the point P_1. Such a choice, as one can see by sketching the curve (10.92), is by no means always unique, and this shows that the necessary condition (10.80) for an extremum is in general not sufficient. However, from physical considerations it is clear which of the possible values of the parameter a should be preferred (and this, of course, can be confirmed by direct computation).

10.6.4 Problems and Exercises

1. Let $f : U \to Y$ be a mapping of class $C^{(n)}(U; Y)$ from an open set U in a normed space X into a normed space Y. Suppose the closed interval $[x, x + h]$ is entirely contained in U, that f has a differential of order $(n + 1)$ at the points of the open interval $]x, x + h[$, and that $\|f^{(n+1)}(\xi)\| \leq M$ at every point $\xi \in]x, x + h[$.

a) Show that the function

$$g(t) = f(x + th) - \left(f(x) + f'(x)(th) + \cdots + \frac{1}{n!} f^{(n)}(x)(th)^n \right)$$

is defined on the closed interval $[0, 1] \subset \mathbb{R}$ and differentiable on the open interval $]0, 1[$, and that the estimate

$$\|g'(t)\| \leq \frac{1}{n!} M |th|^n |h|$$

holds for every $t \in]0, 1[$.

b) Show that $|g(1) - g(0)| \leq \frac{1}{(n+1)!} M |h|^{n+1}$.

c) Prove the following version of Taylor's formula:

$$\left| f(x + h) - \left(f(x) + f'(x)h + \cdots + \frac{1}{n!} f^{(n)}(x)h^n \right) \right| \leq \frac{M}{(n+1)!} |h|^{n+1}.$$

d) What can be said about the mapping $f : U \to Y$ if it is known that $f^{(n+1)}(x) \equiv 0$ in U?

2. a) If a symmetric n-linear operator A is such that $Ax^n = 0$ for every vector $x \in X$, then $A(x_1, \ldots, x_n) \equiv 0$, that is, A equals zero on every set $x_1, \ldots, x_n$ of vectors in X.

b) If a mapping $f : U \to Y$ has an nth-order differential $f^{(n)}(x)$ at a point $x \in U$ and satisfies the condition

$$f(x + h) = L_0 + L_1 h + \cdots + \frac{1}{n!} L_n h^n + \alpha(h)|h|^n,$$

where L_i, $i = 0, 1, \ldots, n$, are i-linear operators, and $\alpha(h) \to 0$ as $h \to 0$, then $L_i = f^{(i)}(x)$, $i = 0, 1, \ldots, n$.

c) Show that the existence of the expansion for f given in the preceding problem does not in general imply the existence of the nth order differential $f^{(n)}(x)$ (for $n > 1$) for the function at the point x.

d) Prove that the mapping $\mathcal{L}(X; Y) \ni A \mapsto A^{-1} \in \mathcal{L}(X; Y)$ is infinitely differentiable in its domain of definition, and that $(A^{-1})^{(n)}(A)(h_1, \ldots, h_n) = (-1)^n A^{-1} h_1 A^{-1} h_2 \cdot \ldots \cdot A^{-1} h_n A^{-1}$.

3. a) Let $\varphi \in C([a, b], \mathbb{R})$. Show that if the condition

$$\int_a^b \varphi(x) h(x) \, dx = 0$$

holds for every function $h \in C^{(2)}([a, b], \mathbb{R})$ such that $h(a) = h(b) = 0$, then $\varphi(x) \equiv 0$ on $[a, b]$.

b) Derive the Euler–Lagrange equation (10.80) as a necessary condition for an extremum of the functional (10.72) restricted to the set of functions $f \in C^{(2)}([a, b], \mathbb{R})$ assuming prescribed values at the endpoints of the closed interval $[a, b]$.

4. Find the shape $y = f(x)$, $a \le x \le b$, of a meridian of the surface of revolution (about the x-axis) having minimal area among all surfaces of revolution having circles of prescribed radius r_a and r_b as their sections by the planes $x = a$ and $x = b$ respectively.

5. a) The function L in the brachistochrone problem does not satisfy the conditions of Example 1, so that we cannot justify a direct application of the results of Example 1 in this case. Show by repeating the derivation of formula (10.79) with necessary modifications that this equation and Eq. (10.80) remain valid in this case.

b) Does the equation of the brachistochrone change if the particle starts from the point P_0 with a nonzero initial velocity (the motion is frictionless in a closed pipe)?

c) Show that if P is an arbitrary point of the brachistochrone corresponding to the pair of points P_0, P_1, the arc of that brachistochrone from P_0 to P is the brachistochrone of the pair P_0, P.

d) The assumption that the brachistochrone corresponding to a pair of points P_0, P_1 can be written as $y = f(x)$, is not always justified, as was revealed by the final formulas (10.92). Show by using the result of c) that the derivation of (10.92) can be carried out without any such assumption as to the global structure of the brachistochrone.

e) Locate a point P_1 such that the brachistochrone corresponding to the pair of points P_0, P_1 in the coordinate system introduced in Example 3 cannot be written in the form $y = f(x)$.

f) Locate a point P_1 such that the brachistochrone corresponding to the pair of points P_0, P_1 in the coordinate system introduced in Example 3 has the form $y = f(x)$, and $f \notin C^{(1)}([a, b], \mathbb{R})$. Thus it turns out that in this case the functional (10.85) we are interested in has a greatest lower bound on the set $C^{(1)}([a, b], \mathbb{R})$, but not a minimum.

g) Show that the brachistochrone of a pair of points P_0, P_1 of space is a smooth curve.

6. Let us measure the distance $d(P_0, P_1)$ of the point P_0 of space from the point P_1 in a homogeneous gravitational field by the time required for a point mass to move from one point to the other along the brachistochrone corresponding to the points.

a) Find the distance from the point P_0 to a fixed vertical line, measured in this sense.

b) Find the asymptotic behavior of the function $d(P_0, P_1)$ as the point P_1 is raised along a vertical line, approaching the height of the point P_0.

c) Determine whether the function $d(P_0, P_1)$ is a metric.

10.7 The General Implicit Function Theorem

In this concluding section of the chapter we shall illustrate practically all of the machinery we have developed by studying an implicitly defined function. The reader already has some idea of the content of the implicit theorem, its place in analysis, and its applications from Chap. 8. For that reason, we shall not go into detail here with preliminary explanations of the essence of the matter preceding the formalism. We note only that this time the implicitly defined function will be constructed by an entirely different method, one that relies on the contraction mapping principle. This method is often used in analysis and is quite useful because of its computational efficiency.

Theorem *Let X, Y, and Z be normed spaces (for example, $\mathbb{R}^m$, $\mathbb{R}^n$, and $\mathbb{R}^k$), Y being a complete space. Let $W = \{(x, y) \in X \times Y \mid |x - x_0| < \alpha \wedge |y - y_0| < \beta\}$ be a neighborhood of the point (x_0, y_0) in the product $X \times Y$ of the spaces X and Y.*

Suppose that the mapping $F : W \to Z$ satisfies the following conditions:

1. *$F(x_0, y_0) = 0$;*
2. *$F(x, y)$ is continuous at (x_0, y_0);*
3. *$F'(x, y)$ is defined in W and continuous at (x_0, y_0);*
4. *$F'_y(x_0, y_0)$ is an invertible[5] transformation.*

Then there exists a neighborhood $U = U(x_0)$ of $x_0 \in X$, a neighborhood $V = V(y_0)$ of $y_0 \in Y$, and a mapping $f : U \to V$ such that:

1'. *$U \times V \subset W$;*

---

[5]That is, $\exists [F'_y(x_0, y_0)]^{-1} \in \mathcal{L}(Z; Y)$.

$2'$. $(F(x, y) = 0 \text{ in } U \times V) \Leftrightarrow (y = f(x), \text{ where } x \in U \text{ and } f(x) \in V)$;

$3'$. $y_0 = f(x_0)$;

$4'$. f is continuous at x_0.

In essence, this theorem asserts that if the linear mapping F'_y is invertible at a point (hypothesis 4), then in a neighborhood of this point the relation $F(x, y) = 0$ is equivalent to the functional dependence $y = f(x)$ (conclusion $2'$).

Proof 1^0 To simplify the notation and obviously with no loss of generality, we may assume that $x_0 = 0$, $y_0 = 0$, and consequently

$$W = \left\{ (x, y) \in X \times Y \mid |x| < \alpha \wedge |y| < \beta \right\}.$$

2^0 The main role in the proof is played by the auxiliary family of functions

$$g_x(y) := y - \left(F'_y(0, 0) \right)^{-1} \cdot F(x, y), \tag{10.93}$$

which depend on the parameter $x \in S$, $|x| < \alpha$, and are defined on the set $\{ y \in Y \mid |y| < \beta \}$.

Let us discuss formula (10.93). We first determine whether the mappings g_x are unambiguously defined and where their values lie.

The mapping F is defined for $(x, y) \in W$, and its value $F(x, y)$ at the pair (x, y) lies in Z. The partial derivative $F'_y(x, y)$ at any point $(x, y) \in W$, as we know, is a continuous linear mapping from Y into Z.

By hypothesis 4 the mapping $F'_y(0, 0) : Y \rightarrow Z$ has a continuous inverse $(F'_y(0, 0))^{-1} : Z \rightarrow Y$. Hence the composition $(F'_y(0, 0))^{-1} \cdot F(x, y)$ really is defined, and its values lie in Y.

Thus, for any x in the α-neighborhood $B_X(0, \alpha) := \{ x \in X \mid |x| < \alpha \}$ of the point $0 \in X$, the function g_x is a mapping $g_x : B_Y(0, \beta) \rightarrow Y$ from the β-neighborhood $B_Y(0, \beta) := \{ y \in Y \mid |y| < \beta \}$ of the point $0 \in Y$ into Y.

The connection of the mappings (10.93) with the problem of solving the equation $F(x, y) = 0$ for y obviously consists of the following: the point y_x is a fixed point of g_x if and only if $F(x, y_x) = 0$.

Let us state this important observation firmly:

$$g_x(y_x) = y_x \Longleftrightarrow F(x, y_x) = 0. \tag{10.94}$$

Thus, finding and studying the implicitly defined function $y = y_x = f(x)$ reduces to finding the fixed points of the mappings (10.93) and studying the way in which they depend on the parameter x.

3^0 We shall show that there exists a positive number $\gamma < \min\{\alpha, \beta\}$ such that for each $x \in X$ satisfying the condition $|x| < \gamma < \alpha$, the mapping $g_x : B_Y(0, \gamma) \rightarrow Y$ of the ball $B_Y(0, \gamma) := \{ y \in Y \mid |y| < \gamma < \beta \}$ into Y is a contraction with a coefficient of contraction that does not exceed, say $1/2$. Indeed, for each fixed $x \in B_X(0, \alpha)$ the mapping $g_x : B_Y(0, \beta) \rightarrow Y$ is differentiable, as follows from hypothesis 3 and the theorem on differentiation of a composite mapping. Moreover,

$$g'_x(y) = e_Y - \left(F'_y(0,0)\right)^{-1} \cdot \left(F'_y(x,y)\right) =$$
$$= \left(F'_y(0,0)\right)^{-1}\left(F'_y(0,0) - F'_y(x,y)\right). \qquad (10.95)$$

By the continuity of $F'_y(x,y)$ at the point $(0,0)$ (hypothesis 3), there exists a neighborhood $\{(x,y) \in X \times Y \mid |x| < \gamma < \alpha \wedge |y| < \gamma < \beta\}$ of $(0,0) \in X \times Y$ in which

$$\left\|g'_x(y)\right\| \le \left\|\left(F'_y(0,0)\right)^{-1}\right\| \cdot \left\|F'_y(0,0) - F'_y(x,y)\right\| < \frac{1}{2}. \qquad (10.96)$$

Here we are using the relation

$$\left(F'_y(0,0)\right)^{-1} \in \mathcal{L}(Z;Y), \quad \text{that is,} \quad \left\|\left(F'_y(0,0)\right)^{-1}\right\| < \infty.$$

Throughout the following we shall assume that $|x| < \gamma$ and $|y| < \gamma$, so that estimate (10.96) holds.

Thus, at any $x \in B_X(0,\gamma)$ and for any $y_1, y_2 \in B_Y(0,\gamma)$, by the mean-value theorem, we indeed now find that

$$\left|g_x(y_1) - g_x(y_2)\right| \le \sup_{\xi \in]y_1,y_2[} \left\|g'(\xi)\right\| |y_1 - y_2| < \frac{1}{2}|y_1 - y_2|. \qquad (10.97)$$

4^0. In order to assert the existence of a fixed point y_x for the mapping g_x, we need a complete metric space that maps into (but not necessarily onto) itself under this mapping.

We shall verify that for any ε satisfying $0 < \varepsilon < \gamma$ there exists $\delta = \delta(\varepsilon)$ in the open interval $]0, \gamma[$ such that for any $x \in B_X(0,\delta)$ the mapping g_x maps the closed ball $\overline{B}_y(0,\varepsilon)$ into itself, that is, $g_x(\overline{B}_Y(0,\varepsilon)) \subset \overline{B}_Y(0,\varepsilon)$.

Indeed, we first choose a number $\delta \in]0, \gamma[$ depending on ε such that

$$\left|g_x(0)\right| = \left\|\left(F'_y(0,0)\right)^{-1} \cdot F(x,0)\right\| \le \left\|\left(F'_y(0,0)\right)^{-1}\right\| |F(x,0)| < \frac{1}{2}\varepsilon \qquad (10.98)$$

for $|x| < \delta$.

This can be done by hypotheses 1 and 2, which guarantee that $F(0,0) = 0$ and $F(x,y)$ is continuous at $(0,0)$.

Now if $|x| < \delta(\varepsilon) < \gamma$ and $|y| \le \varepsilon < \gamma$, we find by (10.97) and (10.98) that

$$\left|g_x(y)\right| \le \left|g_x(y) - g_x(0)\right| + \left|g_x(0)\right| < \frac{1}{2}|y| + \frac{1}{2}\varepsilon < \varepsilon,$$

and hence for $|x| < \delta(\varepsilon)$

$$g_x\left(\overline{B}_Y(0,\varepsilon)\right) \subset B_Y(0,\varepsilon). \qquad (10.99)$$

Being a closed subset of the complete metric space Y, the closed ball $\overline{B}_Y(0, \varepsilon)$ is itself a complete metric space.

5^0 Comparing relations (10.97) and (10.99), we can now assert by the fixed-point principle (Sect. 9.7) that for each $x \in B_X(0, \delta(\varepsilon)) =: U$ there exists a unique point $y = y_x =: f(x) \in B_Y(0, \varepsilon) =: V$ that is a fixed point of the mapping $g_x : \overline{B}_Y(0, \varepsilon) \to \overline{B}_Y(0, \varepsilon)$.

By the basic relation (10.94), it follows from this that the function $f : U \to V$ so constructed has property $2'$ and hence also property $3'$, since $F(0, 0) = 0$ by hypothesis 1.

Property $1'$ of the neighborhoods U and V follows from the fact that, by construction, $U \times V \subset B_X(0, \alpha) \times B_Y(0, \beta) = W$.

Finally, the continuity of the function $y = f(x)$ at $x = 0$, that is, property $4'$, follows from $2'$ and the fact that, as was shown in part 4^0 of the proof, for every $\varepsilon > 0$ ($\varepsilon < \gamma$) there exists $\delta(\varepsilon) > 0$ ($\delta(\varepsilon) < \gamma$) such that $g_x(\overline{B}_Y(0, \varepsilon)) \subset B_Y(0, \varepsilon)$ for any $x \in B_X(0, \delta(\varepsilon))$, that is, the unique fixed point $y_x = f(x)$ of the mapping $g_x : \overline{B}_Y(0, \varepsilon) \to \overline{B}_Y(0, \varepsilon)$ satisfies the condition $|f(x)| < \varepsilon$ for $|x| < \delta(\varepsilon)$. $\square$

We have now proved the existence of the implicit function. We now prove a number of extensions of these properties of the function, generated by properties of the original function F.

Extension 1 (Continuity of the implicit function) *If in addition to hypotheses 2 and 3 of the theorem it is known that the mappings $F : W \to Z$ and F'_y are continuous not only at the point (x_0, y_0) but in some neighborhood of this point, then the function $f : U \to V$ will be continuous not only at $x_0 \in U$ but in some neighborhood of this point.*

Proof By properties of the mapping $\mathcal{L}(Y; Z) \ni A \mapsto A^{-1} \in \mathcal{L}(Z; Y)$ it follows from hypotheses 3 and 4 of the theorem (see Example 6 of Sect. 10.3) that at each point (x, y) in some neighborhood of (x_0, y_0) the transformation $f'_y(x, y) \in \mathcal{L}(Y; Z)$ is invertible. Thus under the additional hypothesis that F is continuous all points $(\tilde{x}, \tilde{y})$ of the form $(x, f(x))$ in some neighborhood of (x_0, y_0) satisfy hypotheses 1–4, previously satisfied only by the point (x_0, y_0).

Repeating the construction of the implicit function in a neighborhood of these points $(\tilde{x}, \tilde{y})$, we would obtain a function $y = \tilde{f}(x)$ that is continuous at $\tilde{x}$ and by $2'$ would coincide with the function $y = f(x)$ in some neighborhood of x. But that means that f itself is continuous at $\tilde{x}$. $\square$

Extension 2 (Differentiability of the implicit function) *If in addition to the hypotheses of the theorem it is known that a partial derivative $F'_x(x, y)$ exists in some neighborhood W of (x_0, y_0) and is continuous at (x_0, y_0), then the function $y = f(x)$ is differentiable at x_0, and*

$$f'(x_0) = -\left(F'_y(x_0, y_0)\right)^{-1} \cdot \left(F'_x(x_0, y_0)\right). \tag{10.100}$$

Proof We verify immediately that the linear transformation $L \in \mathcal{L}(X; Y)$ on the right-hand side of formula (10.100) is indeed the differential of the function $y = f(x)$ at x_0.

As before, to simplify the notation, we shall assume that $x_0 = 0$ and $y_0 = 0$, so that $f(0) = 0$.

We begin with a preliminary computation.

$$
\begin{aligned}
\left| f(x) - f(0) - Lx \right| &= \\
&= \left| f(x) - Lx \right| = \\
&= \left| f(x) + \left(F_y'(0, 0) \right)^{-1} \cdot \left(F_x'(0, 0) \right) x \right| = \\
&= \left| \left(F_y'(0, 0) \right)^{-1} \left(F_x'(0, 0) x + F_y'(0, 0) f(x) \right) \right| = \\
&= \left| \left(F_y'(0, 0) \right)^{-1} \left(F\left(x, f(x) \right) - F(0, 0) - F_x'(0, 0) x - F_y'(0, 0) f(x) \right) \right| \le \\
&\le \left\| \left(F_y'(0, 0) \right)^{-1} \right\| \left| \left(F\left(x, f(x) \right) - F(0, 0) - F_x'(0, 0) x - F_y'(0, 0) f(x) \right) \right| \le \\
&\le \left\| \left(F_y'(0, 0) \right)^{-1} \right\| \cdot \alpha\left(x, f(x) \right) \left(|x| + |f(x)| \right),
\end{aligned}
$$

where $\alpha(x, y) \to 0$ as $(x, y) \to (0, 0)$.

These relations have been written taking account of the relation $F(x, f(x)) \equiv 0$ and the fact that the continuity of the partial derivatives F_x' and F_y' at $(0, 0)$ guarantees the differentiability of the function $F(x, y)$ at that point.

For convenience in writing we set $a := \|L\|$ and $b := \|(F_y'(0, 0))^{-1}\|$.

Taking account of the relations

$$
\left| f(x) \right| = \left| f(x) - Lx + Lx \right| \le \left| f(x) - Lx \right| + |Lx| \le \left| f(x) - Lx \right| + a|x|,
$$

we can extend the preliminary computation just done and obtain the relation

$$
\left| f(x) - Lx \right| \le b\alpha\left(x, f(x) \right)\left((a + 1)|x| + \left| f(x) - Lx \right| \right),
$$

or

$$
\left| f(x) - Lx \right| \le \frac{(a + 1)b}{1 - b\alpha(x, f(x))} \alpha\left(x, f(x) \right) |x|.
$$

Since f is continuous at $x = 0$ and $f(0) = 0$, we also have $f(x) \to 0$ as $x \to 0$, and therefore $\alpha(x, f(x)) \to 0$ as $x \to 0$.

It therefore follows from the last inequality that

$$
\left| f(x) - f(0) - Lx \right| = \left| f(x) - Lx \right| = o(|x|) \quad \text{as } x \to 0. \qquad \square
$$

Extension 3 (Continuous differentiability of the implicit function) *If in addition to the hypotheses of the theorem it is known that the mapping F has continuous partial* *derivatives F'_x and F'_y in some neighborhood W of (x_0, y_0), then the function $y = f(x)$ is continuously differentiable in some neighborhood of x_0, and its derivative is given by the formula*

$$f'(x) = -\left(F'_y\big(x, f(x)\big)\right)^{-1} \cdot \left(F'_x\big(x, f(x)\big)\right). \tag{10.101}$$

Proof We already know from formula (10.100) that the derivative $f'(x)$ exists and can be expressed in the form (10.101) at an individual point x at which the transformation $F'_y(x, f(x))$ is invertible.

It remains to be verified that under the present hypotheses the function $f'(x)$ is continuous in some neighborhood of $x = x_0$.

The bilinear mapping $(A, B) \mapsto A \cdot B$ – the product of linear transformations A and B – is a continuous function.

The transformation $B = -F'_x(x, f(x))$ is a continuous function of x, being the composition of the continuous functions $x \mapsto (x, f(x)) \mapsto -F'_x(x, f(x))$.

The same can be said about the linear transformation $A^{-1} = F'_y(x, f(x))$.

It remains only to recall (see Example 6 of Sect. 10.3) that the mapping $A^{-1} \mapsto A$ is also continuous in its domain of definition.

Thus the function $f'(x)$ defined by formula (10.101) is continuous in some neighborhood of $x = x_0$, being the composition of continuous functions. $\quad\square$

We can now summarize and state the following general proposition.

Proposition *If in addition to the hypotheses of the implicit function theorem it is known that the function F belongs to the class $C^{(k)}(W, Z)$, then the function $y = f(x)$ defined by the equation $F(x, y) = 0$ belongs to $C^{(k)}(U, Y)$ in some neighborhood U of x_0.*

Proof The proposition has already been proved for $k = 0$ and $k = 1$. The general case can now be obtained by induction from formula (10.101) if we observe that the mapping $\mathcal{L}(Y; Z) \ni A \mapsto A^{-1} \in \mathcal{L}(Z; Y)$ is (infinitely) differentiable and that when Eq. (10.101) is differentiated, the right-hand side always contains a derivative of f one order less than the left-hand side. Thus, successive differentiation of Eq. (10.101) can be carried out a number times equal to the order of smoothness of the function F. $\quad\square$

In particular, if

$$f'(x)h_1 = -\left(F'_y\big(x, f(x)\big)\right)^{-1} \cdot \left(F'_x\big(x, f(x)\big)\right)h_1,$$

then

$$f''(x)(h_1, h_2) = -\mathrm{d}\left(F'_y\big(x, f(x)\big)\right)^{-1}h_2 F'_x\big(x, f(x)\big)h_1 -$$
$$- \left(F'_y\big(x, f(x)\big)\right)^{-1}\mathrm{d}\left(F'_x\big(x, f(x)\big)h_1\right)h_2 =$$

$$= \left(F_y'(x, f(x))\right)^{-1} dF_y'(x, f(x)) h_2 \left(F_y'(x, f(x))\right)^{-1} \times$$
$$\times F_x'(x, f(x)) h_1 - \left(F_y'(x, f(x))\right)^{-1} \times$$
$$\times \left(\left(F_{xx}''(x, f(x)) + F_{xy}''(x, f(x)) f'(x)\right) h_1\right) h_2 =$$
$$= \left(F_y'(x, f(x))\right)^{-1} \left(\left(F_{yx}''(x, f(x)) + F_{yy}''(x, f(x)) f'(x)\right) h_2\right) \times$$
$$\times \left(F_y'(x, f(x))\right)^{-1} F_x'(x, f(x)) h_1 - \left(F_y'(x, f(x))\right)^{-1} \times$$
$$\times \left(\left(F_{xx}''(x, f(x)) + F_{xy}''(x, f(x)) f'(x)\right) h_1\right) h_2.$$

In less detailed, but more readable notation, this means that

$$f''(x)(h_1, h_2) = \left(F_y'\right)^{-1} \left[\left(\left(F_{yx}'' + F_{yy}'' f'\right) h_2\right) \left(F_y'\right)^{-1} F_x' h_1 - \left(\left(F_{xx}'' + F_{yy}'' f'\right) h_1\right) h_2\right]. \tag{10.102}$$

In this way one could theoretically obtain an expression for the derivative of an implicit function to any order; however, as can be seen even from formula (10.102), these expressions are generally too cumbersome to be conveniently used. Let us now see how these results can be made specific in the important special case when $X = \mathbb{R}^m$, $Y = \mathbb{R}^n$, and $Z = \mathbb{R}^n$.

In this case the mapping $z = F(x, y)$ has the coordinate representation

$$z^1 = F^1(x^1, \dots, x^m, y^1, \dots, y^n),$$
$$\vdots \tag{10.103}$$
$$z^n = F^n(x^1, \dots, x^m, y^1, \dots, y^n).$$

The partial derivatives $F_x' \in \mathcal{L}(\mathbb{R}^m; \mathbb{R}^n)$ and $F_y' \in \mathcal{L}(\mathbb{R}^n; \mathbb{R}^n)$ of the mapping are defined by the matrices

$$F_x' = \begin{pmatrix} \frac{\partial F^1}{\partial x^1} & \cdots & \frac{\partial F^1}{\partial x^m} \\ \vdots & \ddots & \vdots \\ \frac{\partial F^n}{\partial x^1} & \cdots & \frac{\partial F^n}{\partial x^m} \end{pmatrix}, \quad F_y' = \begin{pmatrix} \frac{\partial F^1}{\partial y^1} & \cdots & \frac{\partial F^1}{\partial y^n} \\ \vdots & \ddots & \vdots \\ \frac{\partial F^n}{\partial y^1} & \cdots & \frac{\partial F^n}{\partial y^n} \end{pmatrix},$$

computed at the corresponding point (x, y).

As we know, the condition that F_x' and F_y' be continuous is equivalent to the continuity of all the entries of these matrices.

The invertibility of the linear transformation $F_y'(x_0, y_0) \in \mathcal{L}(\mathbb{R}^n; \mathbb{R}^n)$ is equivalent to the nonsingularity of the matrix that defines this transformation.

Thus, in the present case the implicit function theorem asserts that if

1) $\begin{aligned} & F^1(x_0^1, \dots, x_0^m, y_0^1, \dots, y_0^n) = 0, \\ & \vdots \\ & F^n(x_0^1, \dots, x_0^m, y_0^1, \dots, y_0^n) = 0; \end{aligned}$

2) $F^i(x^1, \ldots, x^m, y^1, \ldots, y^n)$, $i = 1, \ldots, n$, are continuous functions at the point $(x_0^1, \ldots, x_0^m, y_0^1, \ldots, y_0^n) \in \mathbb{R}^m \times \mathbb{R}^n$;

3) all the partial derivatives $\frac{\partial F^i}{\partial y^j}(x^1, \ldots, x^m, y^1, \ldots, y^n)$, $i = 1, \ldots, n$, $j = 1, \ldots, n$, are defined in a neighborhood of $(x_0^1, \ldots, x_0^m, y_0^1, \ldots, y_0^n)$ and are continuous at this point;

4) the determinant

$$\begin{vmatrix} \frac{\partial F^1}{\partial y^1} & \cdots & \frac{\partial F^1}{\partial y^n} \\ \vdots & \ddots & \vdots \\ \frac{\partial F^n}{\partial y^1} & \cdots & \frac{\partial F^n}{\partial y^n} \end{vmatrix}$$

of the matrix F_y' is nonzero at the point $(x_0^1, \ldots, x_0^m, y_0^1, \ldots, y_0^n)$; then there exist a neighborhood U of $x_0 = (x_0^1, \ldots, x_0^m) \in \mathbb{R}^m$, a neighborhood V of $y_0 = (y_0^1, \ldots, y_0^n) \in \mathbb{R}^n$, and a mapping $f : U \to V$ having a coordinate representation

$$y^1 = f^1(x^1, \ldots, x^m),$$
$$\vdots \qquad\qquad\qquad (10.104)$$
$$y^n = f^n(x^1, \ldots, x^m),$$

such that

1') inside the neighborhood $U \times V$ of $(x_0^1, \ldots, x_0^m, y_0^1, \ldots, y_0^n) \in \mathbb{R}^m \times \mathbb{R}^n$ the system of equations

$$\begin{cases} F^1(x^1, \ldots, x^m, y^1, \ldots, y^n) = 0, \\ \vdots \\ F^n(x^1, \ldots, x^m, y^1, \ldots, y^n) = 0 \end{cases}$$

is equivalent to the functional relation $f : U \to V$ expressed by (10.104);

2')

$$y_0^1 = f^1(x_0^1, \ldots, x_0^m),$$
$$\vdots$$
$$y_0^n = f^n(x_0^1, \ldots, x_0^m);$$

3') the mapping (10.104) is continuous at $(x_0^1, \ldots, x_0^m, y_0^1, \ldots, y_0^n)$.

If in addition it is known that the mapping (10.103) belongs to the class $C^{(k)}$, then, as follows from the proposition above, the mapping (10.104) will also belong to $C^{(k)}$, of course within its own domain of definition.

In this case formula (10.101) can be made specific, becoming the matrix equality

$$
\begin{pmatrix} \frac{\partial f^1}{\partial x^1} & \cdots & \frac{\partial f^1}{\partial x^m} \\ \vdots & \ddots & \vdots \\ \frac{\partial f^n}{\partial x^1} & \cdots & \frac{\partial f^n}{\partial x^m} \end{pmatrix} = - \begin{pmatrix} \frac{\partial F^1}{\partial y^1} & \cdots & \frac{\partial F^1}{\partial y^n} \\ \vdots & \ddots & \vdots \\ \frac{\partial F^n}{\partial y^1} & \cdots & \frac{\partial F^n}{\partial y^n} \end{pmatrix}^{-1} \begin{pmatrix} \frac{\partial F^1}{\partial x^1} & \cdots & \frac{\partial F^1}{\partial x^m} \\ \vdots & \ddots & \vdots \\ \frac{\partial F^n}{\partial x^1} & \cdots & \frac{\partial F^n}{\partial x^m} \end{pmatrix},
$$

in which the left-hand side is computed at $(x^1, \ldots, x^m)$ and the right-hand side at the corresponding point $(x^1, \ldots, x^m, y^1, \ldots, y^n)$, where $y^i = f^i(x^1, \ldots, x^m)$, $i = 1, \ldots, n$.

If $n = 1$, that is, when the equation

$$
F(x^1, \ldots, x^m, y) = 0
$$

is being solved for y, the matrix F'_y consists of a single entry – the number $\frac{\partial F}{\partial y}(x^1, \ldots, x^m, y)$. In this case $y = f(x^1, \ldots, x^m)$, and

$$
\left(\frac{\partial f}{\partial x^1}, \ldots, \frac{\partial f}{\partial x^m} \right) = - \left(\frac{\partial F}{\partial y} \right)^{-1} \left(\frac{\partial F}{\partial x^1}, \ldots, \frac{\partial F}{\partial x^m} \right). \tag{10.105}
$$

In this case formula (10.102) also simplifies slightly; more precisely, it can be written in the following more symmetric form:

$$
f''(x)(h_1, h_2) = - \frac{(F''_{xx} + F''_{xy} f')h_1 F'_y h_2 - (F''_{yx} + F''_{yy} f')h_2 F'_x h_1}{(F'_y)^2}. \tag{10.106}
$$

And if $n = 1$ and $m = 1$, then $y = f(x)$ is a real-valued function of one real argument, and formulas (10.105) and (10.106) simplify to the maximum extent, becoming the numerical equalities

$$
f'(x) = - \frac{F'_x}{F'_y}(x, y),
$$

$$
f''(x) = - \frac{(F''_{xx} + F''_{xy} f')F'_y - (F''_{yx} + F''_{yy} f')F'_x}{(F'_y)^2}(x, y)
$$

for the first two derivatives of the implicit function defined by the equation $F(x, y) = 0$.

10.7.1 Problems and Exercises

1. a) Assume that, along with the function $f : U \to Y$ given by the implicit function theorem, we have a function $\tilde{f} : \tilde{U} \to Y$ defined in some neighborhood $\tilde{U}$ of x_0

and satisfying $y_0 = \tilde{f}(x_0)$ and $F(x, \tilde{f}(x)) \equiv 0$ in $\tilde{U}$. Prove that if $\tilde{f}$ is continuous at x_0, then the functions f and $\tilde{f}$ are equal on some neighborhood of x_0.

b) Show that the assertion in a) is generally not true without the assumption that $\tilde{f}$ is continuous at x_0.

2. Analyze once again the proof of the implicit function theorem and the extensions to it, and show the following.

a) If $z = F(x, y)$ is a continuously differentiable complex-valued function of the complex variables x and y, then the implicit function $y = f(x)$ defined by the equation $F(x, y) = 0$ is differentiable with respect to the complex variable x.

b) Under the hypotheses of the theorem X is not required to be a normed space, and may be any topological space.

3. a) Determine whether the form $f''(x)(h_1, h_2)$ defined by relation (10.102) is symmetric.

b) Write the forms (10.101) and (10.102) for the case of numerical functions $F(x^1, x^2, y)$ and $F(x, y^1, y^2)$ in matrix form.

c) Show that if $\mathbb{R} \ni t \mapsto A(t) \in \mathcal{L}(\mathbb{R}^n; \mathbb{R}^n)$ is family of nonsingular matrices $A(t)$ depending on the parameter t in an infinitely smooth manner, then

$$\frac{d^2 A^{-1}}{dt^2} = 2A^{-1}\left(\frac{dA}{dt}A^{-1}\right)^2 - A^{-1}\frac{d^2 A}{dt^2}A^{-1}, \quad \text{where } A^{-1} = A^{-1}(t)$$

denotes the inverse of the matrix $A = A(t)$.

4. a) Show that Extension 1 to the theorem is an immediate corollary of the stability conditions for the fixed point of the family of contraction mappings studied in Sect. 9.7.

b) Let $\{A_t : X \to X\}$ be a family of contraction mappings of a complete normed space into itself depending on the parameter t, which ranges over a domain Ω in a normed space T. Show that if $A_t(x) = \varphi(t, x)$ is a function of class $C^{(n)}(\Omega \times X, X)$, then the fixed point $x(t)$ of the mapping A_t belongs to class $C^{(n)}(\Omega, X)$ as a function of t.

5. a) Using the implicit function theorem, prove the following *inverse function theorem*.

Let $g : G \to X$ be a mapping from a neighborhood G of a point y_0 in a complete normed space Y into a normed space X. If

1^0 the mapping $x = g(y)$ is differentiable in G,
2^0 $g'(y)$ is continuous at y_0,
3^0 $g'(y_0)$ is an invertible transformation,

then there exists a neighborhood $V \subset Y$ of y_0 and a neighborhood $U \subset X$ of x_0 such that $g : V \to U$ is bijective, and its inverse mapping $f : U \to V$ is continuous in U and differentiable at x_0; moreover,

$$f'(x_0) = \left(g'(y_0)\right)^{-1}.$$

b) Show that if it is known, in addition to the hypotheses given in a), that the mapping g belongs to the class $C^{(n)}(V, U)$, then the inverse mapping f belongs to $C^{(n)}(U, V)$.

c) Let $f : \mathbb{R}^n \to \mathbb{R}^n$ be a smooth mapping for which the matrix $f'(x)$ is nonsingular at every point $x \in \mathbb{R}^n$ and satisfies the inequality $\|(f')^{-1}(x)\| < C$ with a constant C that is independent of x. Show that f is a bijective mapping.

d) Using your experience in solving c), try to give an estimate for the radius of a spherical neighborhood $U = B(x_0, r)$ centered at x_0 in which the mapping $f : U \to V$ studied in the inverse function theorem is necessarily defined.

6. a) Show that if the linear mappings $A \in \mathcal{L}(X; Y)$ and $B \in \mathcal{L}(X; \mathbb{R})$ are such that $\ker A \subset \ker B$ (here ker, as usual, denotes the kernel of a transformation), then there exists a linear mapping $\lambda \in \mathcal{L}(Y; \mathbb{R})$, such that $B = \lambda \cdot A$.

b) Let X and Y be normed spaces and $f : X \to \mathbb{R}$ and $g : X \to Y$ smooth functions on X with values in $\mathbb{R}$ and Y respectively. Let S be the smooth surface defined in X by the equation $g(x) = y_0$. Show that if $x_0 \in S$ is an extremum of the function $f|_S$, then any vector h tangent to S at X_0 simultaneously satisfies two conditions: $f'(x_0)h = 0$ and $g'(x_0)h = 0$.

c) Prove that if $x_0 \in S$ is an extremum of the function $f|_S$ then $f'(x_0) = \lambda \cdot g'(x_0)$, where $\lambda \in \mathcal{L}(Y; \mathbb{R})$.

d) Show how the classical *Lagrange necessary condition for an extremum with constraint* of a function on a smooth surface in $\mathbb{R}^n$ follows from the preceding result.

7. As is known, the equation $z^n + c_1 z^{n-1} + \cdots + c_n = 0$ with complex coefficients has in general n distinct complex roots. Show that the roots of the equation are smooth functions of the coefficients, at least where all the roots are distinct.

8. a) Following Hadamard, prove that a continuous locally invertible mapping $f : \mathbb{R}^n \to \mathbb{R}^n$ is globally invertible (i.e., it is bijective) if and only if $f(x) \to \infty$ as $x \to \infty$. Convince yourself that we can consider here any normed space instead of $\mathbb{R}^n$. How should we interpret (or reformulate) Hadamard's conditions if we now consider the image of $\mathbb{R}^n$ or a normed space under a homeomorphism?

b) Let $F : X \times Y \to Z$ be a continuous mapping defined on the direct product of the normed spaces X and Y. Show that the equation $F(x, y) = 0$ is solvable globally with respect to y (in the sense that the local continuous solution of $y = f(x)$ extends as such to the whole space X) exactly when the following two conditions are fulfilled: the equation has a continuous solution in a neighborhood of every point (x_0, y_0) satisfying $F(x_0, y_0) = 0$; and in the pair (x, y) satisfying the equation $F(x, y) = 0$, the second coordinate can tend to infinity (changing continuously) only if the first coordinate in its space also tends to infinity.

c) Following John,[6] show that if a continuous locally invertible mapping f : $B \to H$ from the unit ball B to a normed space H is such that locally (at every point of the ball), it changes the element of length no more than $k \geq 1$ times (expanding or contracting), then in the ball of radius k^{-2}, this mapping is injective. (Caution: an infinite-dimensional normed space can be isometrically embedded into itself as a proper subspace through a shift of coordinates, but this mapping is not invertible or locally invertible. It is invertible as a mapping only on its image.)

---

[6]F. John (1910–1994), German-born and later a famous American mathematician, student of R. Courant.

Chapter 11
Multiple Integrals

11.1 The Riemann Integral over an n-Dimensional Interval

11.1.1 Definition of the Integral

a. Intervals in $\mathbb{R}^n$ and Their Measure

Definition 1 The set $I = \{x \in \mathbb{R}^n \mid a^i \leq x^i \leq b^i, i = 1, \ldots, n\}$ is called an *interval* or a *coordinate parallelepiped* in $\mathbb{R}^n$.

If we wish to note that the interval is determined by the points $a = (a^1, \ldots, a^n)$ and $b = (b^1, \ldots, b^n)$, we often denote it $I_{a,b}$, or, by analogy with the one-dimensional case, we write it as $a \leq x \leq b$.

Definition 2 To the interval $I = \{x \in \mathbb{R}^n \mid a^i \leq x^i \leq b^i, i = 1, \ldots, n\}$ we assign the number $|I| := \prod_{i=1}^{n}(b^i - a^i)$, called the *volume* or *measure* of the interval.

The volume (measure) of the interval I is also denoted $\upsilon(I)$ and $\mu(I)$.

Lemma 1 *The measure of an interval in $\mathbb{R}^n$ has the following properties.*

a) *It is homogeneous, that is, if $\lambda I_{a,b} := I_{\lambda a, \lambda b}$, where $\lambda \geq 0$, then*

$$|\lambda I_{a,b}| = \lambda^n |I_{a,b}|.$$

b) *It is additive, that is, if the intervals $I, I_1, \ldots, I_k$ are such that $I = \bigcup_{j=1}^{k} I_j$ and no two of the intervals $I_1, \ldots, I_k$ have common interior points, then $|I| = \sum_{j=1}^{k} |I_j|$.*

c) *If the interval I is covered by a finite system of intervals $I_1, \ldots, I_k$, that is, $I \subset \bigcup_{j=1}^{k} I_j$, then $|I| \leq \sum_{j=1}^{k} |I_j|$.*

All these assertions follow easily from Definitions 1 and 2.

© Springer-Verlag Berlin Heidelberg 2016
V.A. Zorich, *Mathematical Analysis II*, Universitext,
DOI 10.1007/978-3-662-48993-2_3

b. Partitions of an Interval and a Base in the Set of Partitions

Suppose we are given an interval $I = \{x \in \mathbb{R}^n \mid a^i \leq x^i \leq b^i, i = 1, \ldots, n\}$. Partitions of the coordinate intervals $[a^i, b^i]$, $i = 1, \ldots, n$, induce a partition of the interval I into finer intervals obtained as the direct products of the intervals of the partitions of the coordinate intervals.

Definition 3 The representation of the interval I (as the union $I = \bigcup_{j=1}^k I_j$ of finer intervals I_j) just described will be called a *partition* of the interval I, and will be denoted by P.

Definition 4 The quantity $\lambda(P) := \max_{1 \leq j \leq k} d(I_j)$ (the maximum among the diameters of the intervals of the partition P) is called the *mesh* of the partition P.

Definition 5 If in each interval I_j of the partition P we fix a point $\xi_j \in I_j$, we say that we have a *partition with distinguished points*.

The set $\{\xi_1, \ldots, \xi_k\}$, as before, will be denoted by the single letter ξ, and the partition with distinguished points by (P, ξ).

In the set $\mathcal{P} = \{(P, \xi)\}$ of partitions with distinguished points on an interval I we introduce the base $\lambda(P) \to 0$ whose elements $B_d (d > 0)$, as in the one-dimensional case, are defined by $B_d := \{(P, \xi) \in \mathcal{P} \mid \lambda(P) < d\}$.

The fact that $\mathcal{B} = \{B_d\}$ really is a base follows from the existence of partitions of mesh arbitrarily close to zero.

c. Riemann Sums and the Integral

Let $f : I \to \mathbb{R}$ be a real-valued[1] function on the interval I and $P = \{I_1, \ldots, I_k\}$ a partition of this interval with distinguished points $\xi = \{\xi_1, \ldots, \xi_k\}$.

Definition 6 The sum

$$\sigma(f, P, \xi) := \sum_{i=1}^{k} f(\xi_i)|I_i|$$

is called the *Riemann sum* of the function f corresponding to the partition of the interval I with distinguished points (P, ξ).

Definition 7 The quantity

$$\int_I f(x) \, dx := \lim_{\lambda(P) \to 0} \sigma(f, P, \xi),$$

---

[1]Please note that in the following definitions one could assume that the values of f lie in any normed vector space. For example, it might be the space $\mathbb{C}$ of complex numbers or the spaces $\mathbb{R}^n$ and $\mathbb{C}^n$.

provided this limit exists, is called the *Riemann integral* of the function f over the interval I.

We see that this definition, and in general the whole process of constructing the integral over the interval $I \subset \mathbb{R}^n$ is a verbatim repetition of the procedure of defining the Riemann integral over a closed interval of the real line, which is already familiar to us. To highlight the resemblance we have even retained the previous notation $f(x)\, dx$ for the differential form. Equivalent, but more expanded notations for the integral are the following:

$$\int_I f(x^1, \ldots, x^n)\, dx^1 \cdot \ldots \cdot dx^n \quad \text{or} \quad \underbrace{\int \ldots \int_I}_{n} f(x^1, \ldots, x^n)\, dx^1 \cdot \ldots \cdot dx^n.$$

To emphasize that we are discussing an integral over a multidimensional domain I we say that this is a *multiple integral* (double, triple, and so forth, depending on the dimension of I).

d. A Necessary Condition for Integrability

Definition 8 If the finite limit in Definition 7 exists for a function $f : I \to \mathbb{R}$, then f is *Riemann integrable* over the interval I.

We shall denote the set of all such functions by $\mathcal{R}(I)$.
We now verify the following elementary necessary condition for integrability.

Proposition 1 $f \in \mathcal{R}(I) \Rightarrow f$ *is bounded on* I.

Proof Let P be an arbitrary partition of the interval I. If the function f is unbounded on I, then it must be unbounded on some interval I_{i_0} of the partition P. If (P, ξ') and (P, ξ'') are partitions P with distinguished points such that ξ' and ξ'' differ only in the choice of the points ξ'_{i_0} and ξ''_{i_0}, then

$$\left| \sigma(f, P, \xi') - \sigma(f, P, \xi'') \right| = \left| f(\xi'_{i_0}) - f(\xi''_{i_0}) \right| |I_{i_0}|.$$

By changing one of the points ξ'_{i_0} and ξ''_{i_0}, as a result of the unboundedness of f in I_{i_0}, we could make the right-hand side of this equality arbitrarily large. By the Cauchy criterion, it follows from this that the Riemann sums of f do not have a limit as $\lambda(P) \to 0$. $\square$

11.1.2 The Lebesgue Criterion for Riemann Integrability

When studying the Riemann integral in the one-dimensional case, we acquainted the reader (without proof) with the Lebesgue criterion for the existence of the Riemann integral. We shall now recall certain concepts and prove this criterion.

a. Sets of Measure Zero in $\mathbb{R}^n$

Definition 9 A set $E \subset \mathbb{R}^n$ has (n-dimensional) *measure zero* or is a *set of measure zero* (in the Lebesgue sense) if for every $\varepsilon > 0$ there exists a covering of E by an at most countable system $\{I_i\}$ of n-dimensional intervals for which the sum of the volumes $\sum_i |I_i|$ does not exceed ε.

Lemma 2 a) *A point and a finite set of points are sets of measure zero.*

b) *The union of a finite or countable number of sets of measure zero is a set of measure zero.*

c) *A subset of a set of measure zero is itself of measure zero.*

d) *A nondegenerate[2] interval $I_{a,b} \subset \mathbb{R}^n$ is not a set of measure zero.*

The proof of Lemma 2 does not differ from the proof of its one-dimensional version considered in Sect. 6.1.3, paragraph d. Hence we shall not give the details.

Example 1 The set of rational points in $\mathbb{R}^n$ (points all of whose coordinates are rational numbers) is countable and hence is a set of measure zero.

Example 2 Let $f : I \to \mathbb{R}$ be a continuous real-valued function defined on an $(n-1)$-dimensional interval $I \subset \mathbb{R}^{n-1}$. We shall show that its graph in $\mathbb{R}^n$ is a set of n-dimensional measure zero.

Proof Since the function f is uniformly continuous on I, for $\varepsilon > 0$ we find $\delta > 0$ such that $|f(x_1) - f(x_2)| < \varepsilon$ for any two points $x_1, x_2 \in I$ such that $|x_1 - x_2| < \delta$. If we now take a partition P of the interval I with mesh $\lambda(P) < \delta$, then on each interval I_i of this partition the oscillation of the function is less than ε. Hence, if x_i is an arbitrary fixed point of the interval I_i, the n-dimensional interval $\tilde{I}_i = I_i \times [f(x_i) - \varepsilon, f(x_i) + \varepsilon]$ obviously contains the portion of the graph of the function lying over the interval I_i, and the union $\bigcup_i \tilde{I}_i$ covers the whole graph of the function over I. But $\sum_i |\tilde{I}_i| = \sum_i |I_i| \cdot 2\varepsilon = 2\varepsilon |I|$ (here $|I_i|$ is the volume of I_i in $\mathbb{R}^{n-1}$ and $|\tilde{I}_i|$ the volume of $\tilde{I}_i$ in $\mathbb{R}^n$). Thus, by decreasing ε, we can indeed make the total volume of the covering arbitrarily small. □

---

[2]That is, an interval $I_{a,b} = \{x \in \mathbb{R}^n \mid a^i \leq x^i \leq b^i, i = 1, \ldots, n\}$ such that the strict inequality $a^i < b^i$ holds for each $i \in \{1, \ldots, n\}$.

Remark 1 Comparing assertion b) in Lemma 2 with Example 2, one can conclude that in general the graph of a continuous function $f : \mathbb{R}^{n-1} \to \mathbb{R}$ or a continuous function $f : M \to \mathbb{R}$, where $M \subset \mathbb{R}^{n-1}$, is a set of n-dimensional measure zero in $\mathbb{R}^n$.

Lemma 3 a) *The class of sets of measure zero remains the same whether the intervals covering the set E in Definition 9, that is, $E \subset \bigcup_i I_i$, are interpreted as an ordinary system of intervals $\{I_i\}$, or in a stricter sense, requiring that each point of the set be an interior point of at least one of the intervals in the covering.*[3]

b) *A compact set K in $\mathbb{R}^n$ is a set of measure zero if and only if for every $\varepsilon > 0$ there exists a finite covering of K by intervals the sum of whose volumes is less than ε.*

Proof a) If $\{I_i\}$ is a covering of E (that is, $E \subset \bigcup_i I_i$ and $\sum_i |I_i| < \varepsilon$), then, replacing each I_i by a dilation of it from its center, which we denote $\tilde{I}_i$, we obtain a system of intervals $\{\tilde{I}_i\}$ such that $\sum |\tilde{I}_i| < \lambda^n \varepsilon$, where λ is a dilation coefficient that is the same for all intervals. If $\lambda > 1$, it is obvious that the system $\{\tilde{I}_i\}$ will cover E in such a way that every point of E is interior to one of the intervals in the covering.

b) This follows from a) and the possibility of extracting a finite covering from any open covering of a compact set K. (The system $\{\tilde{I}_i \setminus \partial \tilde{I}_i\}$ consisting of open intervals obtained from the system $\{\tilde{I}_i\}$ considered in a) may serve as such a covering.) $\square$

b. A Generalization of Cantor's Theorem

We recall that the oscillation of a function $f : E \to \mathbb{R}$ on the set E has been defined as $\omega(f; E) := \sup_{x_1, x_2 \in E} |f(x_1) - f(x_2)|$, and the oscillation at the point $x \in E$ as $\omega(f; x) := \lim_{\delta \to 0} \omega(f; U_E^{\delta}(x))$, where $U_E^{\delta}(x)$ is the δ-neighborhood of x in the set E.

Lemma 4 *If the relation $\omega(f; x) \leq \omega_0$ holds at each point of a compact set K for the function $f : K \to \mathbb{R}$, then for every $\varepsilon > 0$ there exists $\delta > 0$ such that $\omega(f; U_K^{\delta}(x)) < \omega_0 + \varepsilon$ for each point $x \in K$.*

When $\omega_0 = 0$, this assertion becomes Cantor's theorem on uniform continuity of a function that is continuous on a compact set. The proof of Lemma 4 is a verbatim repetition of the proof of Cantor's theorem (Sect. 6.2.2) and therefore we do not take the time to give it here.

---

[3]In other words, it makes no difference whether we mean closed or open intervals in Definition 9.

c. Lebesgue's Criterion

As before, we shall say that a property holds *at almost all points of a set M* or *almost everywhere on M* if the subset of M where this property does not necessarily hold has measure zero.

Theorem 1 (Lebesgue's criterion) $f \in \mathcal{R}(I) \Leftrightarrow (f \text{ is bounded on } I) \wedge (f \text{ is continuous almost everywhere on } I)$.

Proof Necessity. If $f \in \mathcal{R}(I)$, then by Proposition 1 the function f is bounded on I. Suppose $|f| \le M$ on I.

We shall now verify that f is continuous at almost all points of I. To do this, we shall show that if the set E of its points of discontinuity does not have measure zero, then $f \notin \mathcal{R}(I)$.

Indeed, representing E in the form $E = \bigcup_{n=1}^{\infty} E_n$, where $E_n = \{x \in I \mid \omega(f; x) \ge 1/n\}$, we conclude from Lemma 2 that if E does not have measure zero, then there exists an index n_0 such that E_{n_0} is also not a set of measure zero. Let P be an arbitrary partition of the interval I into intervals $\{I_i\}$. We break the partition P into two groups of intervals A and B, where

$$A = \left\{ I_i \in P \mid I_i \cap E_{n_0} \ne \varnothing \wedge \omega(f; I_i) \ge \frac{1}{2n_0} \right\}, \quad \text{and} \quad B = P \backslash A.$$

The system of intervals A forms a covering of the set E_{n_0}. In fact, each point of E_{n_0} lies either in the interior of some interval $I_i \in P$, in which case obviously $I_i \in A$, or on the boundary of several intervals of the partition P. In the latter case, the oscillation of the function must be at least $\frac{1}{2n_0}$ on at least one of these intervals (because of the triangle inequality), and that interval belongs to the system A.

We shall now show that by choosing the set ξ of distinguished points in the intervals of the partition P in different ways we can change the value of the Riemann sum significantly.

To be specific, we choose the sets of points ξ' and ξ'' such that in the intervals of the system B the distinguished points are the same, while in the intervals I_i of the system A, we choose the points ξ_i' and ξ_i'' so that $f(\xi_i') - f(\xi_i'') > \frac{1}{3n_0}$. We then have

$$\left| \sigma\left(f, P, \xi'\right) - \sigma\left(f, P, \xi''\right) \right| = \left| \sum_{I_i \in A} \left(f(\xi_i') - f(\xi_i'') \right) |I_i| \right| > \frac{1}{3n_0} \sum_{I_i \in A} |I_i| > c > 0.$$

The existence of such a constant c follows from the fact that the intervals of the system A form a covering of the set E_{n_0}, which by hypothesis is not a set of measure zero.

Since P was an arbitrary partition of the interval I, we conclude from the Cauchy criterion that the Riemann sums $\sigma(f, P, \xi)$ cannot have a limit as $\lambda(P) \to 0$, that is, $f \notin \mathcal{R}(I)$.

Sufficiency. Let ε be an arbitrary positive number and $E_\varepsilon = \{x \in I \mid \omega(f;x) \geq \varepsilon\}$. By hypothesis, E_ε is a set of measure zero.

Moreover, E_ε is obviously closed in I, so that E_ε is compact. By Lemma 3 there exists a finite system $I_1, \ldots, I_k$ of intervals in $\mathbb{R}^n$ such that $E_\varepsilon \subset \bigcup_{i=1}^{k} I_i$ and $\sum_{i=1}^{k} |I_i| < \varepsilon$. Let us set $C_1 = \bigcup_{i=1}^{k} I_i$ and denote by C_2 and C_3 the unions of the intervals obtained from the intervals I_i by dilation with center at the center of I_i and scaling factors 2 and 3 respectively. It is clear that E_ε lies strictly in the interior of C_2 and that the distance d between the boundaries of the sets C_2 and C_3 is positive.

We note that the sum of the volumes of any finite system of intervals lying in C_3, no two of which have any common interior points is at most $3^n \varepsilon$, where n is the dimension of the space $\mathbb{R}^n$. This follows from the definition of the set C_3 and properties of the measure of an interval (Lemma 1).

We also note that any subset of the interval I whose diameter is less than d is either contained in C_3 or lies in the compact set $K = I \backslash (C_2 \backslash \partial C_2)$, where ∂C_2 is the boundary of C_2 (and hence $C_2 \backslash \partial C_2$ is the set of interior points of C_2).

By construction $E_\varepsilon \subset I \backslash K$, so that at every point $x \in K$ we must have $\omega(f;x) < \varepsilon$. By Lemma 4 there exists $\delta > 0$ such that $|f(x_1) - f(x_2)| < 2\varepsilon$ for every pair of points $x_1, x_2 \in K$ whose distance from each other is at most δ.

These constructions make it possible now to carry out the proof of the sufficient condition for integrability as follows. We choose any two partitions P' and P'' of the interval I with meshes $\lambda(P')$ and $\lambda(P'')$ less than $\lambda = \min\{d, \delta\}$. Let P be the partition obtained by intersecting all the intervals of the partitions P' and P'', that is, in a natural notation, $P = \{I_{ij} = I_i' \cap I_j''\}$. Let us compare the Riemann sums $\sigma(f, P, \xi)$ and $\sigma(f, P', \xi')$. Taking into account the equality $|I_i'| = \sum_j |I_{ij}|$, we can write

$$\left| \sigma\left(f, P, \xi'\right) - \sigma(f, P, \xi) \right| = \left| \sum_{ij} (f(\xi_i') - f(\xi_{ij})) |I_{ij}| \right| \leq$$

$$\leq \sum_1 |f(\xi_i') - f(\xi_{ij})| |I_{ij}| + \sum_2 |f(\xi_i') - f(\xi_{ij})| |I_{ij}|.$$

Here the first sum $\sum_1$ contains the intervals of the partition P lying in the intervals I_i' of the partition P' contained in the set C_3, and the remaining intervals of P are included in the sum $\sum_2$, that is, they are all necessarily contained in K (after all, $\lambda(P) < d$).

Since $|f| \leq M$ on I, replacing $|f(\xi_i') - f(\xi_{ij})|$ in the first sum by $2M$, we conclude that the first sum does not exceed $2M \cdot 3^n \varepsilon$.

Now, noting that $\xi_i', \xi_{ij} \in I_i' \subset K$ in the second sum and $\lambda(P') < \delta$, we conclude that $|f(\xi_i') - f(\xi_{ij})| < 2\varepsilon$, and consequently the second sum does not exceed $2\varepsilon |I|$.

Thus $|\sigma(f, P', \xi') - \sigma(f, P, \xi)| < (2M \cdot 3^n + 2|I|)\varepsilon$, from which (in view of the symmetry between P' and P''), using the triangle inequality, we find that

$$\left| \sigma\left(f, P', \xi'\right) - \sigma\left(f, P'', \xi''\right) \right| < 4(3^n M + |I|)\varepsilon$$

for any two partitions P' and P'' with sufficiently small mesh. By the Cauchy criterion we now conclude that $f \in \mathcal{R}(I)$. □

Remark 2 Since the Cauchy criterion for existence of a limit is valid in any complete metric space, the sufficiency part of the Lebesgue criterion (but not the necessity part), as the proof shows, holds for functions with values in any complete normed vector space.

11.1.3 The Darboux Criterion

Let us consider another useful criterion for Riemann integrability of a function, which is applicable only to real-valued functions.

a. Lower and Upper Darboux Sums

Let f be a real-valued function on the interval I and $P = \{I_i\}$ a partition of the interval I. We set

$$m_i = \inf_{x \in I_i} f(x), \qquad M_i = \sup_{x \in I_i} f(x).$$

Definition 10 The quantities

$$s(f, P) = \sum_i m_i |I_i| \quad \text{and} \quad S(f, P) = \sum_i M|I_i|$$

are called the *lower* and *upper Darboux sums* of the function f over the interval I corresponding to the partition P of the interval.

Lemma 5 *The following relations hold between the Darboux sums of a function* $f : I \to \mathbb{R}$:

a) $s(f, P) = \inf_\xi \sigma(f, P, \xi) \leq \sigma(f, P, \xi) \leq \sup_\xi \sigma(f, P, \xi) = S(f, P)$;

b) *if the partition* P' *of the interval* I *is obtained by refining intervals of the partition* P, *then* $s(f, P) \leq s(f, P') \leq S(f, P') \leq S(f, P)$;

c) *the inequality* $s(f, P_1) \leq S(f, P_2)$ *holds for any pair of partitions* P_1 *and* P_2 *of the interval* I.

Proof Relations a) and b) follow immediately from Definitions 6 and 10, taking account, of course, of the definition of the greatest lower bound and least upper bound of a set of numbers.

To prove c) it suffices to consider the auxiliary partition P obtained by intersecting the intervals of the partitions P_1 and P_2. The partition P can be regarded as a refinement of each of the partitions P_1 and P_2, so that b) now implies

$$s(f, P_1) \leq s(f, P) \leq S(f, P) \leq S(f, P_2).$$ □

b. Lower and Upper Integrals

Definition 11 The *lower* and *upper Darboux integrals* of the function $f : I \to \mathbb{R}$ over the interval I are respectively

$$\underline{\mathcal{J}} = \sup_{P} s(f, P), \qquad \overline{\mathcal{J}} = \inf_{P} S(f, P),$$

where the supremum and infimum are taken over all partitions P of the interval I.

It follows from this definition and the properties of Darboux sums exhibited in Lemma 3 that the inequalities

$$s(f, P) \le \underline{\mathcal{J}} \le \overline{\mathcal{J}} \le S(f, P)$$

hold for any partition P of the interval.

Theorem 2 (Darboux) *For any bounded function $f : I \to \mathbb{R}$,*

$$\left(\exists \lim_{\lambda(P)\to 0} s(f, P) \right) \wedge \left(\lim_{\lambda(P)\to 0} s(f, P) = \underline{\mathcal{J}} \right);$$

$$\left(\exists \lim_{\lambda(P)\to 0} S(f, P) \right) \wedge \left(\lim_{\lambda(P)\to 0} S(f, P) = \overline{\mathcal{J}} \right).$$

Proof If we compare these assertions with Definition 11, it becomes clear that in essence all we have to prove is that the limits exist. We shall verify this for the lower Darboux sums.

Fix $\varepsilon > 0$ and a partition P_ε of the interval I for which $s(f; P_\varepsilon) > \underline{\mathcal{J}} - \varepsilon$. Let Γ_ε be the set of points of the interval I lying on the boundary of the intervals of the partition P_ε. As follows from Example 2, Γ_ε is a set of measure zero. Because of the simple structure of Γ_ε, it is even obvious that there exists a number λ_ε such that the sum of the volumes of those intervals that intersect Γ_ε is less than ε for every partition P such that $\lambda(P) < \lambda_\varepsilon$.

Now taking any partition P with mesh $\lambda(P) < \lambda_\varepsilon$, we form an auxiliary partition P' obtained by intersecting the intervals of the partitions P and P_ε. By the choice of the partition P_ε and the properties of Darboux sums (Lemma 5), we find

$$\underline{\mathcal{J}} - \varepsilon < s(f, P_\varepsilon) < s(f, P') \le \underline{\mathcal{J}}.$$

We now remark that the sums $s(f, P')$ and $s(f, P)$ both contain all the terms that correspond to intervals of the partition P that do not meet Γ_ε. Therefore, if $|f(x)| \le M$ on I, then

$$\left| s(f, P') - s(f, P) \right| < 2M\varepsilon$$

and taking account of the preceding inequalities, we thereby find that for $\lambda(P) < \lambda_\varepsilon$ we have the relation

$$\underline{\mathcal{J}} - s(f, P) < (2M + 1)\varepsilon.$$

Comparing the relation just obtained with Definition 11, we conclude that the limit $\lim_{\lambda(P)\to 0} s(f, P)$ does indeed exist and is equal to $\underline{\mathcal{J}}$.

Similar reasoning can be carried out for the upper sums.   □

c. The Darboux Criterion for Integrability of a Real-Valued Function

Theorem 3 (The Darboux criterion) *A real-valued function $f : I \to \mathbb{R}$ defined on an interval $I \subset \mathbb{R}^n$ is integrable over that interval if and only if it is bounded on I and its upper and lower Darboux integrals are equal.*

Thus,

$$ f \in \mathcal{R}(I) \Longleftrightarrow (f \text{ is bounded on } I) \wedge (\underline{\mathcal{J}} = \overline{\mathcal{J}}). $$

Proof Necessity. If $f \in \mathcal{R}(I)$, then by Proposition 1 the function f is bounded on I. It follows from Definition 7 of the integral, Definition 11 of the quantities $\underline{\mathcal{J}}$ and $\overline{\mathcal{J}}$, and part a) of Lemma 5 that in this case $\underline{\mathcal{J}} = \overline{\mathcal{J}}$.

Sufficiency. Since $s(f, P) \leq \sigma(f, P, \xi) \leq S(f, P)$ when $\underline{\mathcal{J}} = \overline{\mathcal{J}}$, the extreme terms in these inequalities tend to the same limit by Theorem 2 as $\lambda(P) \to 0$. Therefore $\sigma(f, P, \xi)$ has the same limit as $\lambda(P) \to 0$.   □

Remark 3 It is clear from the proof of the Darboux criterion that if a function is integrable, its lower and upper Darboux integrals are equal to each other and to the integral of the function.

11.1.4 Problems and Exercises

1. a) Show that a set of measure zero has no interior points.

b) Show that not having interior points by no means guarantees that a set is of measure zero.

c) Construct a set having measure zero whose closure is the entire space $\mathbb{R}^n$.

d) A set $E \subset I$ is said to have content zero if for every $\varepsilon > 0$ it can be covered by a finite system of intervals $I_1, \ldots, I_k$ such that $\sum_{i=1}^{k} |I_i| < \varepsilon$. Is every bounded set of measure zero a set of content zero?

e) Show that if a set $E \subset \mathbb{R}^n$ is the direct product $\mathbb{R} \times e$ of the line $\mathbb{R}$ and a set $e \subset \mathbb{R}^{n-1}$ of $(n-1)$-dimensional measure zero, then E is a set of n-dimensional measure zero.

2. a) Construct the analogue of the Dirichlet function in $\mathbb{R}^n$ and show that a bounded function $f : I \to \mathbb{R}$ equal to zero at almost every point of the interval I may still fail to belong to $\mathcal{R}(I)$.

b) Show that if $f \in \mathcal{R}(I)$ and $f(x) = 0$, at almost all points of the interval I, then $\int_I f(x)\, dx = 0$.

3. There is a small difference between our earlier definition of the Riemann integral on a closed interval $I \subset \mathbb{R}$ and Definition 7 for the integral over an interval of arbitrary dimension. This difference involves the definition of a partition and the measure of an interval of the partition. Clarify this nuance for yourself and verify that

$$\int_a^b f(x)\,dx = \int_I f(x)\,dx, \quad \text{if } a < b$$

and

$$\int_a^b f(x)\,dx = -\int_I f(x)\,dx, \quad \text{if } a > b,$$

where I is the interval on the real line $\mathbb{R}$ with endpoints a and b.

4. a) Prove that a real-valued function $f : I \to \mathbb{R}$ defined on an interval $I \subset \mathbb{R}^n$ is integrable over that interval if and only if for every $\varepsilon > 0$ there exists a partition P of I such that $S(f; P) - s(f; P) < \varepsilon$.

b) Using the result of a) and assuming that we are dealing with a real-valued function $f : I \to \mathbb{R}$, one can simplify slightly the proof of the sufficiency of the Lebesgue criterion. Try to carry out this simplification by yourself.

11.2 The Integral over a Set

11.2.1 Admissible Sets

In what follows we shall be integrating functions not only over an interval, but also over other sets in $\mathbb{R}^n$ that are not too complicated.

Definition 1 A set $E \subset \mathbb{R}^n$ is *admissible* if it is bounded in $\mathbb{R}^n$ and its boundary is a set of measure zero (in the sense of Lebesgue).

Example 1 A cube, a tetrahedron, and a ball in $\mathbb{R}^3$ (or $\mathbb{R}^n$) are admissible sets.

Example 2 Suppose the functions $\varphi_i : I \to \mathbb{R}$, $i = 1, 2$, defined on an $(n-1)$-dimensional interval $I \subset \mathbb{R}^n$ are such that $\varphi_1(x) < \varphi_2(x)$ at every point $x \in I$. If these functions are continuous, Example 2 of Sect. 11.1 makes it possible to assert that the domain in $\mathbb{R}^n$ bounded by the graphs of these functions and the cylindrical lateral surface lying over the boundary ∂I of I is an admissible set in $\mathbb{R}^n$.

We recall that the boundary ∂E of a set $E \subset \mathbb{R}^n$ consists of the points x such that every neighborhood of x contains both points of E and points of the complement of E in $\mathbb{R}^n$. Hence we have the following lemma.

Lemma 1 *For any sets $E, E_1, E_2 \subset \mathbb{R}^n$, the following assertions hold:*

a) ∂E is a closed subset of $\mathbb{R}^n$;
b) $\partial(E_1 \cup E_2) \subset \partial E_1 \cup \partial E_2$;
c) $\partial(E_1 \cap E_2) \subset \partial E_1 \cup \partial E_2$;
d) $\partial(E_1 \backslash E_2) \subset \partial E_1 \cup \partial E_2$.

This lemma and Definition 1 together imply the following lemma.

Lemma 2 *The union or intersection of a finite number of admissible sets is an admissible set; the difference of admissible sets is also an admissible set.*

Remark 1 For an infinite collection of admissible sets Lemma 2 is generally not true, and the same is the case with assertions b) and c) of Lemma 1.

Remark 2 The boundary of an admissible set is not only closed, but also bounded in $\mathbb{R}^n$, that is, it is a compact subset of $\mathbb{R}^n$. Hence by Lemma 3 of Sect. 11.1, it can even be covered by a finite set of intervals whose total content (volume) is arbitrarily close to zero.

We now consider the characteristic function

$$\chi_E(x) = \begin{cases} 1, & \text{if } x \in E, \\ 0, & \text{if } x \notin E, \end{cases}$$

of an admissible set E. Like the characteristic function of any set E, the function $\chi_E(x)$ has discontinuities at the boundary points of the set E and at no other points. Hence if E is an admissible set, the function $\chi_E(x)$ is continuous at almost all points of $\mathbb{R}^n$.

11.2.2 The Integral over a Set

Let f be a function defined on a set E. We shall agree, as before, to denote the function equal to $f(x)$ for $x \in E$ and to 0 outside E by $f \chi_E(x)$ (even though f may happen to be undefined outside of E).

Definition 2 The *integral of f over E* is given by

$$\int_E f(x)\,dx := \int_{I \supset E} f \chi_E(x)\,dx,$$

where I is any interval containing E.

If the integral on the right-hand side of this equality does not exist, we say that f is (Riemann) *nonintegrable over E*. Otherwise f is (Riemann) *integrable over E*.
The set of all functions that are Riemann integrable over E will be denoted $\mathcal{R}(E)$.

Definition 2 of course requires some explanation, which is provided by the following lemma.

Lemma 3 *If I_1 and I_2 are two intervals, both containing the set E, then the integrals*

$$\int_{I_1} f\chi_E(x)\,dx \quad and \quad \int_{I_2} f\chi_E(x)\,dx$$

either both exist or both fail to exist, and in the first case their values are the same.

Proof Consider the interval $I = I_1 \cap I_2$. By hypothesis $I \supset E$. The points of discontinuity of $f\chi_E$ are either points of discontinuity of f on E, or the result of discontinuities of χ_E, in which case they lie on ∂E. In any case, all these points lie in $I = I_1 \cap I_2$. By Lebesgue's criterion (Theorem 1 of Sect. 11.1) it follows that the integrals of $f\chi_E$ over the intervals I, I_1, and I_2 either all exist or all fail to exist. If they do exist, we may choose partitions of I, I_1, and I_2 to suit ourselves. Therefore we shall choose only those partitions of I_1 and I_2 obtained as extensions of partitions of $I = I_1 \cap I_2$. Since the function is zero outside I, the Riemann sums corresponding to these partitions of I_1 and I_2 reduce to Riemann sums for the corresponding partition of I. It then results from passage to the limit that the integrals over I_1 and I_2 are equal to the integral of the function in question over I. $\qquad\square$

Lebesgue's criterion (Theorem 1 of Sect. 11.1) for the existence of the integral over an interval and Definition 2 now imply the following theorem.

Theorem 1 *A function $f : E \to \mathbb{R}$ is integrable over an admissible set if and only if it is bounded and continuous at almost all points of E.*

Proof Compared with f, the function $f\chi_E$ may have additional points of discontinuity only on the boundary ∂E of E, which by hypothesis is a set of measure zero. $\qquad\square$

11.2.3 The Measure (Volume) of an Admissible Set

Definition 3 The (Jordan) *measure* or *content* of a bounded *set $E \subset \mathbb{R}^n$* is

$$\mu(E) := \int_E 1 \cdot dx,$$

provided this Riemann integral exists.

Since

$$\int_E 1 \cdot dx = \int_{I \supset E} \chi_E(x)\,dx,$$

and the discontinuities of χ_E form the set ∂E, we find by Lebesgue's criterion that the measure just introduced is defined only for admissible sets.

Thus admissible sets, and only admissible sets, are measurable in the sense of Definition 3.

Let us now ascertain the geometric meaning of $\mu(E)$. If E is an admissible set then

$$\mu(E) = \int_{I \supset E} \chi_E(x)\,dx = \underline{\int_{I \supset E}} \chi_E(x)\,dx = \overline{\int_{I \supset E}} \chi_E(x)\,dx,$$

where the last two integrals are the upper and lower Darboux integrals respectively. By the Darboux criterion for existence of the integral (Theorem 3) the measure $\mu(E)$ of a set is defined if and only if these lower and upper integrals are equal. By the theorem of Darboux (Theorem 2 of Sect. 11.1) they are the limits of the upper and lower Darboux sums of the function χ_E corresponding to partitions P of I. But by definition of χ_E the lower Darboux sum is the sum of the volumes of the intervals of the partition P that are entirely contained in E (the volume of a polyhedron inscribed in E), while the upper sum is the sum of the volumes of the intervals of P that intersect E (the volume of a circumscribed polyhedron). Hence $\mu(E)$ is the common limit as $\lambda(P) \to 0$ of the volumes of polyhedra inscribed in and circumscribed about E, in agreement with the accepted idea of the volume of simple solids $E \subset \mathbb{R}^n$.

For $n = 1$ content is usually called *length*, and for $n = 2$ it is called *area*.

Remark 3 Let us now explain why the measure $\mu(E)$ introduced in Definition 3 is sometimes called Jordan measure.

Definition 4 A set $E \subset \mathbb{R}^n$ is a *set of measure zero in the sense of Jordan* or a *set of content zero* if for every $\varepsilon > 0$ it can be covered by a finite system of intervals $I_1, \ldots, I_k$ such that $\sum_{i=1}^k |I_i| < \varepsilon$.

Compared with measure zero in the sense of Lebesgue, a requirement that the covering be finite appears here, shrinking the class of sets of Lebesgue measure zero. For example, the set of rational points is a set of measure zero in the sense of Lebesgue, but not in the sense of Jordan.

In order for the least upper bound of the contents of polyhedra inscribed in a bounded set E to be the same as the greatest lower bound of the contents of polyhedra circumscribed about E (and to serve as the measure $\mu(E)$ or content of E), it is obviously necessary and sufficient that the boundary ∂E of E have measure 0 in the sense of Jordan. That is the motivation for the following definition.

Definition 5 A set E is *Jordan-measurable* if it is bounded and its boundary has Jordan measure zero.

As Remark 2 shows, the class of Jordan-measurable subsets is precisely the class of admissible sets introduced in Definition 1. That is the reason the measure $\mu(E)$

defined earlier can be called (and is called) the *Jordan measure* of the (Jordan-measurable) set E.

11.2.4 Problems and Exercises

1. a) Show that if a set $E \subset \mathbb{R}^n$ is such that $\mu(E) = 0$, then the relation $\mu(\overline{E}) = 0$ also holds for the closure $\overline{E}$ of the set.

b) Give an example of a bounded set E of Lebesgue measure zero whose closure $\overline{E}$ is not a set of Lebesgue measure zero.

c) Determine whether assertion b) of Lemma 3 in Sect. 11.1 should be understood as asserting that the concepts of Jordan measure zero and Lebesgue measure zero are the same for compact sets.

d) Prove that if the projection of a bounded set $E \subset \mathbb{R}^n$ onto a hyperplane $\mathbb{R}^{n-1}$ has $(n-1)$-dimensional measure zero, then the set E itself has n-dimensional measure zero.

e) Show that a Jordan-measurable set whose interior is empty has measure 0.

2. a) Is it possible for the integral of a function f over a bounded set E, as introduced in Definition 2, to exist if E is not an admissible (Jordan-measurable) set?

b) Is a constant function $f : E \to \mathbb{R}$ integrable over a bounded but Jordan-nonmeasurable set E?

c) Is it true that if a function f is integrable over E, then the restriction $f|A$ of this function to any subset $A \subset E$ is integrable over A?

d) Give necessary and sufficient conditions on a function $f : E \to \mathbb{R}$ defined on a bounded (but not necessarily Jordan-measurable) set E under which the Riemann integral of f over E exists.

3. a) Let E be a set of Lebesgue measure 0 and $f : E \to \mathbb{R}$ a bounded continuous function on E. Is f always integrable on E?

b) Answer question a) assuming that E is a set of Jordan measure zero.

c) What is the value of the integral of the function f in a) if it exists?

4. *The Brunn–Minkowski inequality.* Given two nonempty sets $A, B \subset \mathbb{R}^n$, we form their (vector) sum in the sense of Minkowski $A + B := \{a + b \mid a \in A, b \in B\}$. Let $V(E)$ denote the content of a set $E \subset \mathbb{R}^n$.

a) Verify that if A and B are standard n-dimensional intervals (parallelepipeds), then

$$V^{1/n}(A + B) \geq V^{1/n}(A) + V^{1/n}(B).$$

b) Now prove the preceding inequality (the *Brunn–Minkowski inequality*) for arbitrary measurable compact sets A and B.

c) Show that equality holds in the Brunn–Minkowski inequality only in the following three cases: when $V(A + B) = 0$, when A and B are singleton (one-point) sets, and when A and B are similar convex sets.

11.3 General Properties of the Integral

11.3.1 The Integral as a Linear Functional

Proposition 1 a) *The set $\mathcal{R}(E)$ of functions that are Riemann-integrable over a bounded set $E \subset \mathbb{R}^n$ is a vector space with respect to the standard operations of addition of functions and multiplication by constants.*

b) *The integral is a linear functional*

$$\int_E : \mathcal{R}(E) \to \mathbb{R} \quad \text{on the set } \mathcal{R}(E).$$

Proof Noting that the union of two sets of measure zero is also a set of measure zero, we see that assertion a) follows immediately from the definition of the integral and the Lebesgue criterion for existence of the integral of a function over an interval.

Taking account of the linearity of Riemann sums, we obtain the linearity of the integral by passage to the limit. $\Box$

Remark 1 If we recall that the limit of the Riemann sums as $\lambda(P) \to 0$ must be the same independently of the set of distinguished points ξ, we can conclude that

$$\left(f \in \mathcal{R}(E) \right) \bigwedge \left(f(x) = 0 \text{ almost everywhere on } E \right) \Rightarrow \left(\int_E f(x)\,dx = 0 \right).$$

Therefore, if two integrable functions are equal at almost all points of E, then their integrals over E are also equal. Hence if we pass to the quotient space of $\mathcal{R}(E)$ obtained by identifying functions that are equal at almost all points of E, we obtain a vector space $\widetilde{\mathcal{R}}(E)$ on which the integral is also a linear function.

11.3.2 Additivity of the Integral

Although we shall always be dealing with admissible sets $E \subset \mathbb{R}^n$, this assumption was dispensable in Sect. 11.3.1 (and we dispensed with it). From now on we shall be talking only of admissible sets.

Proposition 2 *Let E_1 and E_2 be admissible sets in $\mathbb{R}^n$ and f a function defined on $E_1 \cup E_2$.*

a) *The following relations hold:*

$$\left(\exists \int_{E_1 \cup E_2} f(x)\,dx \right) \Leftrightarrow \left(\exists \int_{E_1} f(x)\,dx \right) \bigwedge \left(\exists \int_{E_2} f(x)\,dx \right)$$

$$\Rightarrow \exists \int_{E_1 \cap E_2} f(x)\,dx.$$

b) *If in addition it is known that $\mu(E_1 \cap E_2) = 0$, the following equality holds when the integrals exist:*

$$\int_{E_1 \cup E_2} f(x)\,dx = \int_{E_1} f(x)\,dx + \int_{E_2} f(x)\,dx.$$

Proof Assertion a) follows from Lebesgue's criterion for existence of the Riemann integral over an admissible set (Theorem 1 of Sect. 11.2). Here it is only necessary to recall that the union and intersection of admissible sets are also admissible sets (Lemma 2 of Sect. 11.2).

To prove b) we begin by remarking that

$$\chi_{E_1 \cup E_2} = \chi_{E_1}(x) + \chi_{E_2}(x) - \chi_{E_1 \cap E_2}(x).$$

Therefore,

$$\int_{E_1 \cup E_2} f(x)\,dx = \int_{I \supset E_1 \cup E_2} f \chi_{E_1 \cup E_2}(x)\,dx =$$

$$= \int_I f \chi_{E_1}(x)\,dx + \int_I f \chi_{E_2}(x)\,dx - \int_I f \chi_{E_1 \cap E_2}(x)\,dx =$$

$$= \int_{E_1} f(x)\,dx + \int_{E_2} f(x)\,dx.$$

The essential point is that the integral

$$\int_I f \chi_{E_1 \cap E_2}(x)\,dx = \int_{E_1 \cap E_2} f(x)\,dx,$$

as we know from part a), exists; and since $\mu(E_1 \cap E_2) = 0$, it equals zero (see Remark 1). □

11.3.3 Estimates for the Integral

a. A General Estimate

We begin with a general estimate of the integral that is also valid for functions with values in any complete normed space.

Proposition 3 *If $f \in \mathcal{R}(E)$, then $|f| \in \mathcal{R}(E)$, and the inequality*

$$\left| \int_E f(x)\,dx \right| \le \int_E |f|(x)\,dx$$

holds.

Proof The relation $|f| \in \mathcal{R}(E)$ follows from the definition of the integral over a set and the Lebesgue criterion for integrability of a function over an interval.

The inequality now follows from the corresponding inequality for Riemann sums and passage to the limit. $\square$

b. The Integral of a Nonnegative Function

The following propositions apply only to real-valued functions.

Proposition 4 *The following implication holds for a function $f : E \to \mathbb{R}$:*

$$\left(f \in \mathcal{R}(E)\right) \wedge \left(\forall x \in E \; \left(f(x) \geq 0\right)\right) \Rightarrow \int_E f(x)\, dx \geq 0.$$

Proof Indeed, if $f(x) \geq 0$ on E, then $f\chi_E(x) \geq 0$ in $\mathbb{R}^n$. Then, by definition,

$$\int_E f(x)\, dx = \int_{I \supset E} f\chi_E(x)\, dx.$$

This last integral exists by hypothesis. But it is the limit of nonnegative Riemann sums and hence nonnegative. $\square$

From Proposition 4 just proved, we obtain successively the following corollaries.

Corollary 1

$$\left(f, g \in \mathcal{R}(E)\right) \wedge (f \leq g \text{ on } E) \Rightarrow \left(\int_E f(x)\, dx \leq \int_E g(x)\, dx\right).$$

Corollary 2 *If $f \in \mathcal{R}(E)$ and the inequalities $m \leq f(x) \leq M$ hold at every point of the admissible set E, then*

$$m\mu(E) \leq \int_E f(x)\, dx \leq M\mu(E).$$

Corollary 3 *If $f \in \mathcal{R}(E)$, $m = \inf_{x \in E} f(x)$, and $M = \sup_{x \in E} f(x)$, then there is a number $\theta \in [m, M]$ such that*

$$\int_E f(x)\, dx = \theta\mu(E).$$

Corollary 4 *If E is a connected admissible set and the function $f \in \mathcal{R}(E)$ is continuous, then there exists a point $\xi \in E$ such that*

$$\int_E f(x)\, dx = f(\xi)\mu(E).$$

Corollary 5 *If in addition to the hypotheses of Corollary 2 the function $g \in \mathcal{R}(E)$ is nonnegative on E, then*

$$m \int_E g(x)\,dx \le \int_E fg(x)\,dx \le M \int_E g(x)\,dx.$$

Corollary 4 is a generalization of the one-dimensional result and is usually called by the same name, that is, the *mean-value theorem for the integral*.

Proof Corollary 5 follows from the inequalities $mg(x) \le f(x)g(x) \le Mg(x)$ taking account of the linearity of the integral and Corollary 1. It can also be proved directly by passing from integrals over E to the corresponding integrals over an interval, verifying the inequalities for the Riemann sums, and then passing to the limit. Since all these arguments were carried out in detail in the one-dimensional case, we shall not give the details. We note merely that the integrability of the product $f \cdot g$ of the functions f and g obviously follows from Lebesgue's criterion. ☐

We shall now illustrate these relations in practice, using them to verify the following very useful lemma.

Lemma a) *If the integral of a nonnegative function $f : I \to \mathbb{R}$ over the interval I equals zero, then $f(x) = 0$ at almost all points of the interval I.*

b) *Assertion a) remains valid if the interval I in it is replaced by any admissible (Jordan-measurable) set E.*

Proof By Lebesgue's criterion the function $f \in \mathcal{R}(E)$ is continuous at almost all points of the interval I. For that reason the proof of a) will be achieved if we show that $f(a) = 0$ at each point of continuity $a \in I$ of the function f.

Assume that $f(a) > 0$. Then $f(x) \ge c > 0$ in some neighborhood $U_I(a)$ of a (the neighborhood may be assumed to be an interval). Then, by the properties of the integral just proved,

$$\int_I f(x)\,dx = \int_{U_I(a)} f(x)\,dx + \int_{I \setminus U_I(a)} f(x)\,dx \ge \int_{U_I(a)} f(x)\,dx \ge c\mu\big(U_I(a)\big) > 0.$$

This contradiction verifies assertion a). If we apply this assertion to the function $f\chi_E$ and take account of the relation $\mu(\partial E) = 0$, we obtain assertion b). ☐

Remark 2 It follows from the lemma just proved that if E is a Jordan-measurable set in $\mathbb{R}^n$ and $\widetilde{\mathcal{R}}(E)$ is the vector space considered in Remark 1, consisting of equivalence classes of functions that are integrable over E and differ only on sets of Lebesgue measure zero, then the quantity $\|f\| = \int_E |f|(x)\,dx$ is a norm on $\widetilde{\mathcal{R}}(E)$.

Proof Indeed, the inequality $\int_E |f|(x)\,dx = 0$ now implies that f is in the same equivalence class as the identically zero function. ☐

11.3.4 Problems and Exercises

1. Let E be a Jordan-measurable set of nonzero measure, $f : E \to \mathbb{R}$ a continuous nonnegative integrable function on E, and $M = \sup_{x \in E} f(x)$. Prove that

$$\lim_{n \to \infty} \left(\int_E f^n(x) \, dx \right)^{1/n} = M.$$

2. Prove that if $f, g \in \mathcal{R}(E)$, then the following are true.

a) *Hölder's inequality*

$$\left| \int_E (f \cdot g)(x) \, dx \right| \leq \left(\int_E |f|^p(x) \, dx \right)^{1/p} \left(\int_E |g|^q(x) \, dx \right)^{1/q},$$

where $p \geq 1$, $q \geq 1$, and $\frac{1}{p} + \frac{1}{q} = 1$;

b) *Minkowski's inequality*

$$\left(\int_E |f + g|^p \, dx \right)^{1/p} \leq \left(\int_E |f|^p(x) \, dx \right)^{1/p} + \left(\int_E |g|^p(x) \, dx \right)^{1/p},$$

if $p \geq 1$.

Show that

c) the preceding inequality reverses direction if $0 < p < 1$;

d) equality holds in Minkowski's inequality if and only if there exists $\lambda \geq 0$ such that one of the equalities $f = \lambda g$ or $g = \lambda f$ holds except on a set of measure zero in E;

e) the quantity $\| f \|_p = (\frac{1}{\mu(E)} \int_E |f|^p(x) \, dx)^{1/p}$, where $\mu(E) > 0$, is a monotone function of $p \in \mathbb{R}$ and is a norm on the space $\widetilde{\mathcal{R}}(E)$ for $p \geq 1$.

Find the conditions under which equality holds in Hölder's inequality.

3. Let E be a Jordan-measurable set in $\mathbb{R}^n$ with $\mu(E) > 0$. Verify that if $\varphi \in C(E, \mathbb{R})$ and $f : \mathbb{R} \to \mathbb{R}$ is a convex function, then

$$f \left(\frac{1}{\mu(E)} \int_E \varphi(x) \, dx \right) \leq \frac{1}{\mu(E)} \int_E (f \circ \varphi)(x) \, dx.$$

4. a) Show that if E is a Jordan-measurable set in $\mathbb{R}^n$ and the function $f : E \to \mathbb{R}$ is integrable over E and continuous at an interior point $a \in E$, then

$$\lim_{\delta \to +0} \frac{1}{\mu(U_E^\delta(a))} \int_{U_E^\delta(a)} f(x) \, dx = f(a),$$

where, as usual, $U_E^\delta(a)$ is the δ-neighborhood of the point in E.

b) Verify that the preceding relation remains valid if the condition that a is an interior point of E is replaced by the condition $\mu(U_E^\delta(a)) > 0$ for every $\delta > 0$.

11.4 Reduction of a Multiple Integral to an Iterated Integral

11.4.1 Fubini's Theorem

Up to now, we have discussed only the definition of the integral, the conditions under which it exists, and its general properties. In the present section we shall prove Fubini's theorem,[4] which, together with the formula for change of variable, is a tool for computing multiple integrals.

Theorem [5] *Let $X \times Y$ be an interval in $\mathbb{R}^{m+n}$, which is the direct product of intervals $X \subset \mathbb{R}^m$ and $Y \subset R^n$. If the function $f : X \times Y \to \mathbb{R}$ is integrable over $X \times Y$, then all three of the integrals*

$$\int_{X \times Y} f(x, y)\, dx\, dy, \quad \int_X dx \int_Y f(x, y)\, dy, \quad \int_Y dy \int_X f(x, y)\, dx$$

exist and are equal.

Before taking up the proof of this theorem, let us decode the meaning of the symbolic expressions that occur in the statement of it. The integral $\int_{X \times Y} f(x, y)\, dx\, dy$ is the integral of the function f over the set $X \times Y$, which we are familiar with, written in terms of the variables $x \in X$ and $y \in Y$. The iterated integral $\int_X dx \int_Y f(x, y)\, dy$ should be understood as follows: For each fixed $x \in X$ the integral $F(x) = \int_Y f(x, y)\, dy$ is computed, and the resulting function $F : X \to \mathbb{R}$ is then to be integrated over X. If, in the process, the integral $\int_Y f(x, y)\, dy$ does not exist for some $x \in X$, then $F(x)$ is set equal to any value between the lower and upper Darboux integrals $\underline{\mathcal{J}}(x) = \int_{\underline{Y}} f(x, y)\, dy$ and $\overline{\mathcal{J}}(x) = \int_Y^- f(x, y)\, dy$, including the upper and lower integrals $\underline{\mathcal{J}}(x)$ and $\overline{\mathcal{J}}(x)$ themselves. It will be shown that in that case $F \in \mathcal{R}(X)$. The iterated integral $\int_Y dy \int_X f(x, y)\, dx$ has a similar meaning.

It will become clear in the course of the proof that the set of values of $x \in X$ at which $\underline{\mathcal{J}}(x) \neq \overline{\mathcal{J}}(x)$ is a set of m-dimensional measure zero in X.

Similarly, the set of $y \in Y$ at which the integral $\int_X f(x, y)\, dx$ may fail to exist will turn out to be a set of n-dimensional measure zero in Y.

We remark finally that, in contrast to the integral over an $(m + n)$-dimensional interval, which we previously agreed to call a *multiple integral*, the successively

---

[4]G. Fubini (1870–1943) – Italian mathematician. His main work was in the area of the theory of functions and geometry.

[5]This theorem was proved long before the theorem known in analysis as Fubini's theorem, of which it is a special case. However, it has become the custom to refer to theorems making it possible to reduce the computation of multiple integrals to iterated integrals in lower dimensions as theorems of Fubini type, or, for brevity, Fubini's Theorem.

computed integrals of the function $f(x, y)$ over Y and then over X or over X and then over Y are customarily called *iterated integrals* of the function.

If X and Y are closed intervals on the line, the theorem stated here theoretically reduces the computation of a double integral over the interval $X \times Y$ to the successive computation of two one-dimensional integrals. It is clear that by applying this theorem several times, one can reduce the computation of an integral over a k-dimensional interval to the successive computation of k one-dimensional integrals.

The essence of the theorem we have stated is very simple and consists of the following. Consider a Riemann sum $\sum_{i,j} f(x_i, y_j)|X_i| \cdot |Y_j|$ corresponding to a partition of the interval $X \times Y$ into intervals $X_i \times Y_j$. Since the integral over the interval $X \times Y$ exists, the distinguished points ξ_{ij} can be chosen as we wish, and we choose them as the "direct product" of choices $x_i \in X_i \subset X$ and $y_j \in Y_j \in Y$. We can then write

$$\sum_{i,j} f(x_i, y_j)|X_i| \cdot |Y_j| = \sum_i |X_i| \sum_j f(x_i, y_j)|Y_j| = \sum_j |Y_j| \sum_i f(x_i, y_j)|X_j|,$$

and this is the prelimit form of theorem.

We now give the formal proof.

Proof Every partition P of the interval $X \times Y$ is induced by corresponding partitions P_X and P_Y of the intervals X and Y. Here every interval of the partition P is the direct product $X_i \times Y_j$ of certain intervals X_i and Y_j of the partitions P_X and P_Y respectively. By properties of the volume of an interval we have $|X_i \times Y_j| = |X_i| \cdot |Y_j|$, where each of these volumes is computed in the space $\mathbb{R}^{m+n}$, $\mathbb{R}^m$, or $\mathbb{R}^n$ in which the interval in question is situated.

Using the properties of the greatest lower bound and least upper bound and the definition of the lower and upper Darboux sums and integrals, we now carry out the following estimates:

$$s(f, P) = \sum_{\substack{i,j}} \inf_{\substack{x \in X_i \\ y \in Y_j}} f(x, y)|X_i \times Y_j| \le \sum_i \inf_{x \in X_i} \left(\sum_j \inf_{y \in Y_j} f(x, y)|Y_j| \right)|X_i| \le$$

$$\le \sum_i \inf_{x \in X_i} \left(\underline{\int_Y} f(x, y)\,dy \right)|X_i| \le \sum_i \inf_{x \in X_i} F(x)|X_i| \le$$

$$\le \sum_i \sup_{x \in X_i} F(x)|X_i| \le \sum_i \sup_{x \in X_i} \left(\overline{\int_Y} f(x, y)\,dy \right)|X_i| \le$$

$$\le \sum_i \sup_{x \in X_i} \left(\sum_j \sup_{y \in Y_j} F(x, y)|Y_j| \right)|X_i| \le$$

$$\le \sum_{\substack{i,j}} \sup_{\substack{x \in X_i \\ y \in Y_j}} f(x, y)|X_i \times Y_j| = S(f, P).$$

Since $f \in \mathcal{R}(X \times Y)$, both of the extreme terms in these inequalities tend to the value of the integral of the function over the interval $X \times Y$ as $\lambda(P) \to 0$. This fact enables us to conclude that $F \in \mathcal{R}(X)$ and that the following equality holds:

$$\int_{X \times Y} f(x, y) \, dx \, dy = \int_X F(x) \, dx.$$

We have carried out the proof for the case when the iterated integration is carried out first over Y, then over X. It is clear that similar reasoning can be used in the case when the integration over X is done first. $\square$

11.4.2 Some Corollaries

Corollary 1 *If $f \in \mathcal{R}(X \times Y)$, then for almost all $x \in X$ (in the sense of Lebesgue) the integral $\int_X f(x, y) \, dy$ exists, and for almost all $y \in Y$ the integral $\int_X f(x, y) \, dx$ exists.*

Proof By the theorem just proved,

$$\int_X \left(\overline{\int_Y} f(x, y) \, dy - \underline{\int_Y} f(x, y) \, dy \right) dx = 0.$$

But the difference of the upper and lower integrals in parentheses is nonnegative. We can therefore conclude by the lemma of Sect. 11.3 that this difference equals zero at almost all points $x \in X$.

Then by the Darboux criterion (Theorem 3 of Sect. 11.1) the integral $\int_Y f(x, y) \, dy$ exists for almost all values of $x \in X$.

The second half of the corollary is proved similarly. $\square$

Corollary 2 *If the interval $I \subset \mathbb{R}^n$ is the direct product of the closed intervals $I_i = [a^i, b^i]$, $i = 1, \ldots, n$, then*

$$\int_I f(x) \, dx = \int_{a^n}^{b^n} dx^n \int_{a^{n-1}}^{b^{n-1}} dx^{n-1} \ldots \int_{a^1}^{b^1} f(x^1, x^2, \ldots, x^n) \, dx^1.$$

Proof This formula obviously results from repeated application of the theorem just proved. All the inner integrals on the right-hand side are to be understood as in the theorem. For example, one can insert the upper or lower integral sign throughout. $\square$

Example 1 Let $f(x, y, z) = z \sin(x + y)$. We shall find the integral of the restriction of this function to the interval $I \subset \mathbb{R}^3$ defined by the relations $0 \le x \le \pi$, $|y| \le \pi/2$, $0 \le z \le 1$.

By Corollary 2

$$
\iiint_I f(x, y, z) \, dx \, dy \, dz = \int_0^1 dz \int_{-\pi/2}^{\pi/2} dy \int_0^\pi z \sin(x + y) \, dx =
$$

$$
= \int_0^1 dz \int_{-\pi/2}^{\pi/2} \left(-z \cos(x + y) \big|_{x=0}^\pi \right) dy =
$$

$$
= \int_0^1 dz \int_{-\pi/2}^{\pi/2} 2z \cos y \, dy =
$$

$$
= \int_0^1 \left(2z \sin y \big|_{y=-\pi/2}^{y=\pi/2} \right) dz = \int_0^1 4z \, dz = 2.
$$

The theorem can also be used to compute integrals over very general sets.

Corollary 3 *Let D be a bounded set in $\mathbb{R}^{n-1}$ and $E = \{(x, y) \in \mathbb{R}^n \mid (x \in D) \wedge (\varphi_1(x) \leq y \leq \varphi_2(x))\}$. If $f \in \mathcal{R}(E)$, then*

$$
\int_E f(x, y) \, dx \, dy = \int_D dx \int_{\varphi_1(x)}^{\varphi_2(x)} f(x, y) \, dy. \tag{11.1}
$$

Proof Let $E_x = \{y \in \mathbb{R} \mid \varphi_1(x) \leq y \leq \varphi_2(x)\}$ if $x \in D$ and $E_x = \varnothing$ if $x \notin D$. We remark that $\chi_E(x, y) = \chi_D(x) \cdot \chi_{E_x}(y)$. Recalling the definition of the integral over a set and using Fubini's theorem, we obtain

$$
\int_E f(x, y) \, dx \, dy = \int_{I \supset E} f \chi_E(x, y) \, dx \, dy =
$$

$$
= \int_{I_x \supset D} dx \int_{I_y \supset E_x} f \chi_E(x, y) \, dy =
$$

$$
= \int_{I_x} \left(\int_{I_y} f(x, y) \chi_{E_x}(y) \, dy \right) \chi_D(x) \, dx =
$$

$$
= \int_{I_x} \left(\int_{\varphi_1(x)}^{\varphi_2(x)} f(x, y) \, dy \right) \chi_D(x) \, dx =
$$

$$
= \int_D \left(\int_{\varphi_1(x)}^{\varphi_2(x)} f(x, y) \, dy \right) dx.
$$

The inner integral here may also fail to exist on a set of points in D of Lebesgue measure zero, and if so it is assigned the same meaning as in the theorem of Fubini proved above.   □

Remark If the set D in the hypotheses of Corollary 3 is Jordan-measurable and the functions $\varphi_i : D \to \mathbb{R}$, $i = 1, 2$, are continuous and bounded, then the set $E \subset \mathbb{R}^n$ is Jordan measurable.

Proof The boundary ∂E of E consists of the two graphs of the continuous functions $\varphi_i : D \to \mathbb{R}$, $i = 1, 2$, (which by Example 2 of Sect. 11.1) are sets of measure zero) and the set Z, which is a portion of the product of the boundary ∂D of $D \subset \mathbb{R}^{n-1}$ and a sufficiently large one-dimensional closed interval of length l. By hypothesis ∂D can be covered by a system of $(n - 1)$-dimensional intervals of total $(n - 1)$-dimensional volume less than ε / l. The direct product of these intervals and the given one-dimensional interval of length l gives a covering of Z by intervals whose total volume is less than ε.   □

Because of this remark one can say that the function $f : E \to 1 \in \mathbb{R}$ is integrable on a measurable set E having this structure (as it is on any measurable set E). Relying on Corollary 3 and the definition of the measure of a measurable set, one can now derive the following corollary.

Corollary 4 *If under the hypotheses of Corollary 3 the set D is Jordan-measurable and the functions $\varphi_i : D \to \mathbb{R}$, $i = 1, 2$, are continuous, then the set E is measurable and its volume can be computed according to the formula*

$$\mu(E) = \int_D \left(\varphi_2(x) - \varphi_1(x) \right) dx. \tag{11.2}$$

Example 2 For the disk $E = \{(x, y) \in \mathbb{R}^2 \mid x^2 + y^2 \leq r^2\}$ we obtain by this formula

$$\mu(E) = \int_{-r}^{r} \left(\sqrt{r^2 - y^2} - \left(-\sqrt{r^2 - y^2} \right) \right) dy = 2 \int_{-r}^{r} \sqrt{r^2 - y^2} \, dy =$$

$$= 4 \int_0^r \sqrt{r^2 - y^2} \, dy = 4 \int_0^{\pi/2} r \cos \varphi \, d(r \sin \varphi) =$$

$$= 4r \int_0^{\pi/2} r \cos^2 \varphi \, d\varphi = \pi r^2.$$

Corollary 5 *Let E be a measurable set contained in the interval $I \subset \mathbb{R}^n$. Represent I as the direct product $I = I_x \times I_y$ of the $(n - 1)$-dimensional interval I_x and the closed interval I_y. Then for almost all values $y_0 \in I_y$ the section $E_{y_0} = \{(x, y) \in E \mid y = y_0\}$ of the set E by the $(n - 1)$-dimensional hyperplane $y = y_0$ is a measurable subset of it, and*

$$\mu(E) = \int_{I_y} \mu(E_y) \, dy, \tag{11.3}$$

where $\mu(E_y)$ is the $(n-1)$-dimensional measure of the set E_y if it is measurable and equal to any number between the numbers $\underline{\int_{E_y}} 1 \cdot dx$ and $\overline{\int_{E_y}} 1 \cdot dx$ if E_y happens to be a nonmeasurable set.

Proof Corollary 5 follows immediately from the theorem and Corollary 1, if we set $f = \chi_E$ in both of them and take account of the relation $\chi_E(x, y) = \chi_{E_y}(x)$.   □

A particular consequence of this result is the following.

Corollary 6 (Cavalieri's[6] principle) *Let A and B be two solids in $\mathbb{R}^3$ having volume (that is, Jordan-measurable). Let $A_c = \{(x, y, z) \in A \mid z = c\}$ and $B_c = \{(x, y, z) \in B \mid z = c\}$ be the sections of the solids A and B by the plane $z = c$. If for every $c \in \mathbb{R}$ the sets A_c and B_c are measurable and have the same area, then the solids A and B have the same volumes.*

It is clear that Cavalieri's principle can be stated for spaces $\mathbb{R}^n$ of any dimension.

Example 3 Using formula (11.3), let us compute the volume V_n of the ball $B = \{x \in \mathbb{R}^n \mid |x| \leq r\}$ of radius r in the Euclidean space $\mathbb{R}^n$.

It is obvious that $V_1 = 2$. In Example 2 we found that $V_2 = \pi r^2$. We shall show that $V_n = c_n r^n$, where c_n is a constant (which we shall compute below). Let us choose some diameter $[-r, r]$ of the ball and for each point $x \in [-r, r]$ consider the section B_x of the ball B by a hyperplane orthogonal to the diameter. Since B_x is a ball of dimension $n - 1$, whose radius, by the Pythagorean theorem, equals $\sqrt{r^2 - x^2}$, proceeding by induction and using (11.3), we can write

$$V_n = \int_{-r}^{r} c_{n-1} \left(r^2 - x^2\right)^{\frac{n-1}{2}} \, dx = \left(c_{n-1} \int_{-\pi/2}^{\pi/2} \cos^n \varphi \, d\varphi \right) r^n.$$

(In passing to the last equality, as one can see, we made the change of variable $x = r \sin \varphi$.)

Thus we have shown that $V_n = c_n r^n$, and

$$c_n = c_{n-1} \int_{-\pi/2}^{\pi/2} \cos^n \varphi \, d\varphi. \tag{11.4}$$

We now find the constant c_n explicitly. We remark that for $m \geq 2$

$$I_m = \int_{-\pi/2}^{\pi/2} \cos^m \varphi \, d\varphi = \int_{-\pi/2}^{\pi/2} \cos^{m-2} \varphi \left(1 - \sin^2 \varphi\right) d\varphi =$$

$$= I_{m-2} + \frac{1}{m-1} \int_{-\pi/2}^{\pi/2} \sin \varphi \, d \cos^{m-1} \varphi = I_{m-2} - \frac{1}{m-1} I_m,$$

that is, the following recurrence relation holds:

$$I_m = \frac{m-1}{m} I_{m-2}. \tag{11.5}$$

---

[6]B. Cavalieri (1598–1647) – Italian mathematician, the creator of the so-called *method of indivisibles* for determining areas and volumes.

In particular, $I_2 = \pi/2$. It is clear immediately from the definition of I_m that $I_1 = 2$. Taking account of these values of I_1 and I_2 we find by the recurrence formula (11.5) that

$$I_{2k+1} = \frac{(2k)!!}{(2k+1)!!} \cdot 2, \qquad I_{2k} = \frac{(2k-1)!!}{(2k)!!}\pi. \qquad (11.6)$$

Returning to formula (11.4), we now obtain

$$c_{2k+1} = c_{2k}\frac{(2k)!!}{(2k+1)!!} \cdot 2 = c_{2k-1}\frac{(2k)!!}{(2k+1)!!} \cdot \frac{(2k-1)!!}{(2k)!!}\pi = \cdots = c_1 \cdot \frac{(2\pi)^k}{(2k+1)!!},$$

$$c_{2k} = c_{2k-1}\frac{(2k-1)!!}{(2k)!!}\pi = c_{2k-2}\frac{(2k-1)!!}{(2k)!!}\pi \cdot \frac{(2k-2)!!}{(2k-1)!!} \cdot 2 =$$

$$= \cdots = c_2\frac{(2\pi)^{k-1}}{(2k)!!} \cdot 2.$$

But, as we have seen above, $c_1 = 2$ and $c_2 = \pi$, and hence the final formulas for the required volume V_n are as follows:

$$V_{2k+1} = 2\frac{(2\pi)^k}{(2k+1)!!}r^{2k+1}, \qquad V_{2k} = \frac{(2\pi)^k}{(2k)!!}r^{2k}, \qquad (11.7)$$

where $k \in \mathbb{N}$, and the first of these formulas is also valid for $k = 0$.

11.4.3 Problems and Exercises

1. a) Construct a subset of the square $I \subset \mathbb{R}^2$ such that on the one hand its intersection with any vertical line and any horizontal line consists of at most one point, while on the other hand its closure equals I.

b) Construct a function $f : I \to \mathbb{R}$ for which both of the iterated integrals that occur in Fubini's theorem exist and are equal, yet $f \notin \mathcal{R}(I)$.

c) Show by example that if the values of the function $F(x)$ that occurs in Fubini's theorem, which in the theorem were subjected to the conditions $\underline{\mathcal{J}}(x) \leq F(x) \leq \overline{\mathcal{J}}(x)$ at all points where $\underline{\mathcal{J}}(x) < \overline{\mathcal{J}}(x)$, are simply set equal to zero at those points, the resulting function may turn out to be nonintegrable. (Consider, for example, the function $f(x, y)$ on $\mathbb{R}^2$ equal to 1 if the point (x, y) is not rational and to $1 - 1/q$ at the point $(p/q, m/n)$, both fractions being in lowest terms.)

2. a) In connection with formula (11.3), show that even if all the sections of a bounded set E by a family of parallel hyperplanes are measurable, the set E may yet be nonmeasurable.

b) Suppose that in addition to the hypotheses of part a) it is known that the function $\mu(E_y)$ in formula (11.3) is integrable over the closed interval I_y. Can we assert that in this case the set E is measurable?

3. Using Fubini's theorem and the positivity of the integral of a positive function, give a simple proof of the equality $\frac{\partial^2 f}{\partial x \partial y} = \frac{\partial^2 f}{\partial y \partial x}$ for the mixed partial derivatives, assuming that they are continuous functions.

4. Let $f : I_{a,b} \to \mathbb{R}$ be a continuous function defined on an interval $I_{a,b} = \{x \in \mathbb{R}^n \mid a^i \le x^i \le b^i, i = 1, \ldots, n\}$, and let $F : I_{a,b} \to \mathbb{R}$ be defined by the equality

$$F(x) = \int_{I_{a,x}} f(t)\,dt,$$

where $I_{a,x} \subset I_{a,b}$. Find the partial derivatives of this function with respect to the variables $x^1, \ldots, x^n$.

5. Let $f(x, y)$ be a continuous function defined on the rectangle $I = [a, b] \times [c, d] \subset \mathbb{R}^2$, which has a continuous partial derivative $\frac{\partial f}{\partial y}$ in I.

a) Let $F(y) = \int_a^b f(x, y)\,dx$. Starting from the equality $F(y) = \int_a^b (\int_c^y \frac{\partial f}{\partial y}(x, t)\,dt + f(x, c))\,dx$, verify the *Leibniz rule*, according to which $F'(y) = \int_a^b \frac{\partial f}{\partial y}(x, y)\,dx$.

b) Let $G(x, y) = \int_a^x f(t, y)\,dt$. Find $\frac{\partial G}{\partial x}$ and $\frac{\partial G}{\partial y}$.

c) Let $H(y) = \int_a^{h(y)} f(x, y)\,dx$, where $h \in C^{(1)}[a, b]$. Find $H'(y)$.

6. Consider the sequence of integrals

$$F_0(x) = \int_0^x f(y)\,dy, \qquad F_n(x) = \int_0^x \frac{(x - y)^n}{n!} f(y)\,dy, \qquad n \in \mathbb{N},$$

where $f \in C(\mathbb{R}, \mathbb{R})$.

a) Verify that $F_n'(x) = F_{n-1}(x)$, $F_n^{(k)}(0) = 0$ if $k \le n$, and $F_n^{(n+1)}(x) = f(x)$.

b) Show that

$$\int_0^x dx_1 \int_0^{x_1} dx_2 \ldots \int_0^{x_{n-1}} f(x_n)\,dx_n = \frac{1}{n!} \int_0^x (x - y)^n f(y)\,dy.$$

7. a) Let $f : E \to \mathbb{R}$ be a function that is continuous on the set $E = \{(x, y) \in \mathbb{R}^2 \mid 0 \le x \le 1 \wedge 0 \le y \le x\}$. Prove that

$$\int_0^1 dx \int_0^x f(x, y)\,dy = \int_0^1 dy \int_y^1 f(x, y)\,dx.$$

b) Use the example of the iterated integral $\int_0^{2\pi} dx \int_0^{\sin x} 1 \cdot dy$ to explain why not every iterated integral comes from a double integral via Fubini's theorem.

11.5   Change of Variable in a Multiple Integral

11.5.1   Statement of the Problem and Heuristic Derivation of the Change of Variable Formula

In our earlier study of the integral in the one-dimensional case, we obtained an important formula for change of variable in such an integral. Our problem now is to find a formula for change of variables in the general case. Let us make the question more precise.

Let D_x be a set in $\mathbb{R}^n$, f a function that is integrable over D_x, and $\varphi : D_t \to D_x$ a mapping $t \mapsto \varphi(t)$ of a set $D_t \subset \mathbb{R}^n$ onto D_x. We seek a rule according to which, knowing f and φ, we can find a function ψ in D_t such that the equality

$$\int_{D_x} f(x)\,dx = \int_{D_t} \psi(t)\,dt$$

holds, making it possible to reduce the computation of the integral over D_x to the computation of an integral over D_t.

We begin by assuming that D_t is an interval $I \subset \mathbb{R}^n$ and $\varphi : I \to D_x$ a diffeomorphism of this interval onto D_x. To every partition P of the interval I into intervals $I_1, I_2, \dots, I_k$ there corresponds a partition of D_x into the sets $\varphi(I_i)$, $i = 1, \dots, k$. If all these sets are measurable and intersect pairwise only in sets of measure zero, then by the additivity of the integral we find

$$\int_{D_x} f(x)\,dx = \sum_{i=1}^{k} \int_{\varphi(I_i)} f(x)\,dx. \tag{11.8}$$

If f is continuous on D_x, then by the mean-value theorem

$$\int_{\varphi(I_i)} f(x)\,dx = f(\xi_i)\mu\big(\varphi(I_i)\big),$$

where $\xi_i \in \varphi(I_i)$. Since $f(\xi_i) = f(\varphi(\tau_i))$, where $\tau_i = \varphi^{-1}(\xi_i)$, we need only connect $\mu(\varphi(I_i))$ with $\mu(I_i)$.

If φ were a linear transformation, then $\varphi(I_i)$ would be a parallelepiped whose volume, as is known from analytic geometry, would be $|\det \varphi'|\mu(I_i)$. But a diffeomorphism is locally a nearly linear transformation, and so, if the dimensions of the intervals I_i are sufficiently small, we may assume $\mu(\varphi(I_i)) \approx |\det \varphi'(\tau_i)||I_i|$ with small relative error (it can be shown that for some choice of the point $\tau_i \in I_i$ actual equality will result). Thus

$$\sum_{i=1}^{k} \int_{\varphi(I_i)} f(x)\,dx \approx \sum_{i=1}^{k} f\big(\varphi(\tau_i)\big)\big|\det \varphi'(\tau_i)\big||I_i|. \tag{11.9}$$

But, the right-hand side of this approximate equality contains a Riemann sum for the integral of the function $f(\varphi(t))|\det\varphi'(t)|$ over the interval I corresponding to the partition P of this interval with distinguished points τ. In the limit as $\lambda(P) \to 0$ we obtain from (11.8) and (11.9) the relation

$$\int_{D_x} f(x)\,dx = \int_{D_t} f\big(\varphi(t)\big)|\det\varphi'(t)|\,dt.$$

This is the desired formula together with an explanation of it. The route just followed in obtaining it can be traversed with complete rigor (and it is worthwhile to do so). However, in order to become acquainted with some new and useful general mathematical methods and facts and avoid purely technical work, we shall depart from this route slightly in the proof below.

We now proceed to precise statements. We recall the following definition.

Definition 1 The *support of a function* $f : D \to \mathbb{R}$ defined in a domain $D \subset \mathbb{R}^n$ is the closure in D of the set of points of $x \in D$ at which $f(x) \neq 0$.

In this section we shall study the situation when the integrand $f : D_x \to \mathbb{R}$ equals zero on the boundary of the domain D_x, more precisely, when the support of the function f (denoted supp f) is a compact set[7] K contained in D_x. The integrals of f over D_x and over K, if they exist, are equal, since the function equals zero in D_x outside of K. From the point of view of mappings the condition supp $f = K \subset D_x$ is equivalent to the statement that the change of variable $x = \varphi(t)$ is valid not only in the set K over which one is essentially integrating, but also in some neighborhood D_x of that set.

We now state what we intend to prove.

Theorem 1 *If* $\varphi : D_t \to D_x$ *is a diffeomorphism of a bounded open set* $D_t \subset \mathbb{R}^n$ *onto a set* $D_x = \varphi(D_t) \subset \mathbb{R}^n$ *of the same type,* $f \in \mathcal{R}(D_x)$, *and* supp f *is a compact subset of* D_x, *then* $f \circ \varphi|\det\varphi'| \in \mathcal{R}(D_t)$, *and the following formula holds:*

$$\boxed{\int_{D_x=\varphi(D_t)} f(x)\,dx = \int_{D_t} f \circ \varphi(t)|\det\varphi'(t)|\,dt.} \qquad (11.10)$$

11.5.2 Measurable Sets and Smooth Mappings

Lemma 1 *Let* $\varphi : D_t \to D_x$ *be a diffeomorphism of an open set* $D_t \subset \mathbb{R}^n$ *onto a set* $D_x \subset \mathbb{R}^n$ *of the same type. Then the following assertions hold.*

a) *If* $E_t \subset D_t$ *is a set of (Lebesgue) measure zero, its image* $\varphi(E_t) \subset D_x$ *is also a set of measure zero.*

---

[7]Such functions are naturally called *functions of compact support* in the domain.

b) *If a set E_t contained in D_t along with its closure $\overline{E}_t$ has Jordan measure zero, its image $\varphi(E_t) = E_x$ is contained in D_x along with its closure and also has measure zero.*

c) *If a (Jordan) measurable set E_t is contained in the domain D_t along with its closure $\overline{E}_t$, its image $E_x = \varphi(E_t)$ is Jordan measurable and $\overline{E}_x \subset D_x$.*

Proof We begin by remarking that every open subset D in $\mathbb{R}^n$ can be represented as the union of a countable number of closed intervals (no two of which have any interior points in common). To do this, for example, one can partition the coordinate axes into closed intervals of length Δ and consider the corresponding partition of $\mathbb{R}^n$ into cubes with sides of length Δ. Fixing $\Delta = 1$, take the cubes of the partition contained in D. Denote their union by F_1. Then taking $\Delta = 1/2$, adjoin to F_1 the cubes of the new partition that are contained in $D \backslash F_1$. In that way we obtain a new set F_2, and so forth. Continuing this process, we obtain a sequence $F_1 \subset \cdots \subset F_n \subset \cdots$ of sets, each of which consists of a finite or countable number of intervals having no interior points in common, and as one can see from the construction, $\bigcup F_n = D$.

Since the union of an at most countable collection of sets of measure zero is a set of measure zero, it suffices to verify assertion a) for a set E_t lying in a closed interval $I \subset D_t$. We shall now do this.

Since $\varphi \in C^{(1)}(I)$ (that is, $\varphi' \in C(I)$), there exists a constant M such that $\|\varphi'(t)\| \le M$ on I. By the finite-increment theorem the relation $|x_2 - x_1| \le M|t_2 - t_1|$ must hold for every pair of points $t_1, t_2 \in I$ with images $x_1 = \varphi(t_1)$, $x_2 = \varphi(t_2)$.

Now let $\{I_i\}$ be a covering of E_t by intervals such that $\sum_i |I_i| < \varepsilon$. Without loss of generality we may assume that $I_i = I_i \cap I \subset I$.

The collection $\{\varphi(I_i)\}$ of sets $\varphi(I_i)$ obviously forms a covering of $E_x = \varphi(E_t)$. If t_i is the center of the interval I_i, then by the estimate just given for the possible change in distances under the mapping φ, the entire set $\varphi(I_i)$ can be covered by the interval $\widetilde{I}_i$ with center $x_i = -\varphi(t_i)$ whose linear dimensions are M times those of the interval I_i. Since $|\widetilde{I}_i| = M^n |I_i|$, and $\varphi(E_t) \subset \bigcup_i \widetilde{I}_i$, we have obtained a covering of $\varphi(E_t) = E_x$ by intervals whose total volume is less than $M^n \varepsilon$. Assertion a) is now established.

Assertion b) follows from a) if we take into account the fact that $\overline{E}_t$ (and hence by what has been proved, $\overline{E}_x = \varphi(\overline{E}_t)$ also) is a set of Lebesgue measure zero and that $\overline{E}_t$ (and hence also $\overline{E}_x$) is a compact set. Indeed, by Lemma 3 of Sect. 11.1 every compact set that is of Lebesgue measure zero also has Jordan measure zero.

Finally, assertion c) is an immediate consequence of b), if we recall the definition of a measurable set and the fact that interior points of E_t map to interior points of its image $E_x = \varphi(E_t)$ under a diffeomorphism, so that $\partial E_x = \varphi(\partial E_t)$. $\square$

Corollary *Under the hypotheses of the theorem the integral on the right-hand side of formula* (11.10) *exists.*

Proof Since $|\det \varphi'(t)| \ne 0$ in D_t, it follows that $\operatorname{supp}(f \circ \varphi \cdot |\det \varphi'|) = \operatorname{supp}(f \circ \varphi) = \varphi^{-1}(\operatorname{supp} f)$ is a compact subset in D_t. Hence the points at which the function

$f \circ \varphi \cdot |\det \varphi'| \chi_{D_t}$ in $\mathbb{R}^n$ is discontinuous have nothing to do with the function χ_{D_t}, but are the pre-images of points of discontinuity of f in D_x. But $f \in \mathcal{R}(D_x)$, and therefore the set E_x of points of discontinuity of f in D_x is a set of Lebesgue measure zero. But then by assertion a) of the lemma the set $E_t = \varphi^{-1}(E_x)$ has measure zero. By Lebesgue's criterion, we can now conclude that $f \circ \varphi \cdot |\det \varphi'| \chi_{D_t}$ is integrable on any interval $I_t \supset D_t$.   □

11.5.3 The One-Dimensional Case

Lemma 2 a) *If $\varphi : I_t \to I_x$ is a diffeomorphism of a closed interval $I_t \subset \mathbb{R}^1$ onto a closed interval $I_x \subset \mathbb{R}^1$ and $f \in \mathcal{R}(I_x)$, then $f \circ \varphi \cdot |\varphi'| \in \mathcal{R}(I_t)$ and*

$$\int_{I_x} f(x)\,dx = \int_{I_t} \left(f \circ \varphi \cdot |\varphi'|\right)(t)\,dt. \tag{11.11}$$

b) *Formula (11.10) holds in $\mathbb{R}^1$.*

Proof Although we essentially already know assertion a) of this lemma, we shall use the Lebesgue criterion for the existence of an integral, which is now at our disposal, to give a short proof here that is independent of the proof given in Part 1.

Since $f \in \mathcal{R}(I_x)$ and $\varphi : I_t \to I_x$ is a diffeomorphism, the function $f \circ \varphi |\varphi'|$ is bounded on I_t. Only the pre-images of points of discontinuity of f on I_x can be discontinuities of the function $f \circ \varphi |\varphi'|$. By Lebesgue's criterion, the latter form a set of measure zero. The image of this set under the diffeomorphism $\varphi^{-1} : I_x \to I_t$, as we saw in the proof of Lemma 1, has measure zero. Therefore $f \circ \varphi |\varphi'| \in \mathcal{R}(I_t)$.

Now let P_x be a partition of the closed interval I_x. Through the mapping φ^{-1} it induces a partition P_t of the closed interval I_t, and it follows from the uniform continuity of the mappings φ and φ^{-1} that $\lambda(P_x) \to 0 \Leftrightarrow \lambda(P_t) \to 0$. We now write the Riemann sums for the partitions P_x and P_t with distinguished points $\xi_i = \varphi(\tau_i)$:

$$\sum_i f(\xi_i)|x_i - x_{i-1}| = \sum_i f \circ \varphi(\tau_i)|\varphi(t_i) - \varphi(t_{i-1})| =$$

$$= \sum_i f \circ \varphi(\tau_i)|\varphi'(\tau_i)||t_i - t_{i-1}|,$$

and the points ξ_i can be assumed chosen just so that $\xi_i = \varphi(\tau_i)$, where τ_i is the point obtained by applying the mean-value theorem to the difference $\varphi(t_i) - \varphi(t_{i-1})$.

Since both integrals in (11.11) exist, the choice of the distinguished points in the Riemann sums can be made to suit our convenience without affecting the limit. Hence from the equalities just written for the Riemann sums, we find (11.11) for the integrals in the limit as $\lambda(P_x) \to 0 (\lambda(P_t) \to 0)$.

Assertion b) of Lemma 2 follows from Eq. (11.11). We first note that in the one-dimensional case $|\det\varphi'| = |\varphi'|$. Next, the compact set supp f can easily be covered by a finite system of closed intervals contained in D_x, no two of which have common interior points. The integral of f over D_x then reduces to the sum of the integrals of f over the intervals of this system, and the integral of $f \circ \varphi|\varphi'|$ over D_t reduces to the sum of the integrals over the intervals that are the pre-images of the intervals in this system. Applying Eq. (11.11) to each pair of intervals that correspond under the mapping φ and then adding, we obtain (11.10).   □

Remark 1 The formula for change of variable that we proved previously had the form

$$\int_{\varphi(\alpha)}^{\varphi(\beta)} f(x)\,dx = \int_{\alpha}^{\beta} \big((f \circ \varphi) \cdot \varphi'\big)(t)\,dt, \tag{11.12}$$

where φ was any smooth mapping of the closed interval $[\alpha, \beta]$ onto the interval with endpoints $\varphi(\alpha)$ and $\varphi(\beta)$. Formula (11.12) contains the derivative φ' itself rather than its absolute value $|\varphi'|$. The reason is that on the left-hand side it is possible that $\varphi(\beta) < \varphi(\alpha)$.

However, if we observe that the relations

$$\int_I f(x)\,dx = \begin{cases} \int_a^b f(x)\,dx, & \text{if } a \le b, \\ -\int_a^b f(x)\,dx, & \text{if } a > b, \end{cases}$$

hold, it becomes clear that when φ is a diffeomorphism formulas (11.11) and (11.12) differ only in appearance; in essence they are the same.

Remark 2 It is interesting to note (and we shall certainly make use of this observation) that if $\varphi : I_t \to I_x$ is a diffeomorphism of closed intervals, then the formulas

$$\overline{\int}_{I_x} f(x)\,dx = \overline{\int}_{I_t} \big(f \circ \varphi|\varphi'|\big)(t)\,dt,$$

$$\underline{\int}_{I_x} f(x)\,dx = \underline{\int}_{I_t} \big(f \circ \varphi|\varphi'|\big)(t)\,dt,$$

for the upper and lower integrals of real-valued functions are always valid.

Given that fact, we may take as established that in the one-dimensional case formula (11.10) remains valid for any bounded function f if the integrals in it are understood as upper or lower Darboux integrals.

Proof We shall assume temporarily that f is a nonnegative function bounded by a constant M.

Again, as in the proof of assertion a) of Lemma 2, one may take partitions P_x and P_t of the intervals I_x and I_t respectively that correspond to each other under the mapping φ and write the following estimates, in which ε is the maximum oscillation of φ on intervals of the partition P_t:

$$\sum_i \sup_{x \in \Delta x_i} f(x)|x_i - x_{i-1}| \le$$

$$\le \sum_i \sup_{t \in \Delta t_i} f(\varphi(t)) \sup_{t \in \Delta t_i} |\varphi'(t)||t_i - t_{i-1}| \le$$

$$\le \sum_i \sup_{t \in \Delta t_i} \left(f(\varphi(t)) \cdot \sup_{t \in \Delta t_i} |\varphi'(t)| \right) |\Delta t_i| \le$$

$$\le \sum_i \sup_{t \in \Delta t_i} \left(f(\varphi(t)) \right) \left(|\varphi'(t)| + \varepsilon \right) |\Delta t_i| \le$$

$$\le \sum_i \sup_{t \in \Delta t_i} \left(f(\varphi(t))|\varphi'(t)| \right) |\Delta t_i| + \varepsilon \sum_i \sup_{t \in \Delta t_i} f(\varphi(t))|\Delta t_i| \le$$

$$\le \sum_i \sup_{t \in \Delta t_i} \left(f(\varphi(t))|\varphi'(t)| \right) |\Delta t_i| + \varepsilon M |I_t|.$$

Taking account of the uniform continuity of φ we obtain from this the relation

$$\overline{\int}_{I_x} f(x)\,\mathrm{d}x \le \overline{\int}_{I_t} \left(f \circ \varphi |\varphi'| \right)(t)\,\mathrm{d}t$$

as $\lambda(P_t) \to 0$. Applying what has just been proved to the mapping φ^{-1} and the function $f \circ \varphi |\varphi'|$, we obtain the opposite inequality, and thereby establish the first equality in Remark 2 for a nonnegative function. But since any function can be written as $f = \max\{f, 0\} - \max\{-f, 0\}$ (a difference of two nonnegative functions) the equality can be considered to be established in general. The second equality is verified similarly. $\qquad\Box$

From the equalities just proved one can of course obtain once again assertion a) of Lemma 2 for real-valued functions f.

11.5.4 The Case of an Elementary Diffeomorphism in $\mathbb{R}^n$

Let $\varphi : D_t \to D_x$ be a diffeomorphism of a domain $D_t \subset \mathbb{R}^n_t$ onto a domain $D_x \subset \mathbb{R}^n_x$ with $(t^1, \ldots, t^n)$ and $(x^1, \ldots, x^n)$ the coordinates of points $t \in \mathbb{R}^n_t$ and $x \in \mathbb{R}^n_x$ respectively. We recall the following definition.

Definition 2 The diffeomorphism $\varphi : D_t \to D_x$ is *elementary* if its coordinate representation has the form

$$x^1 = \varphi^1(t^1, \ldots, t^n) = t^1,$$

$$\vdots$$

$$x^{k-1} = \varphi^{k-1}(t^1, \ldots, t^n) = t^{k-1},$$

$$x^k = \varphi^k(t^1, \ldots, t^n) = \varphi^k(t^1, \ldots, t^k, \ldots, t^n),$$

$$x^{k+1} = \varphi^k(t^1, \ldots, t^n) = t^{k+1},$$

$$\vdots$$

$$x^n = \varphi^n(t^1, \ldots, t^n) = t^n.$$

Thus only one coordinate is changed under an elementary diffeomorphism (the kth coordinate in this case).

Lemma 3 *Formula* (11.10) *holds for an elementary diffeomorphism.*

Proof Up to a relabeling of coordinates we may assume that we are considering a diffeomorphism φ that changes only the nth coordinate. For convenience we introduce the following notation:

$$(x^1, \ldots, x^{n-1}, x^n) =: (\tilde{x}, x^n); \quad (t^1, \ldots, t^{n-1}, t^n) =: (\tilde{t}, t^n);$$

$$D_{x^n}(\tilde{x}_0) := \{(\tilde{x}, x^n) \in D_x \mid \tilde{x} = \tilde{x}_0\};$$

$$D_{t_n}(\tilde{t}_0) := \{(\tilde{t}, t^n) \in D_t \mid \tilde{t} = \tilde{t}_0\}.$$

Thus $D_{x^n}(\tilde{x})$ and $D_{t^n}(\tilde{t})$ are simply the one-dimensional sections of the sets D_x and D_t respectively by lines parallel to the nth coordinate axis. Let I_x be an interval in $\mathbb{R}^n_x$ containing D_x. We represent I_x as the direct product $I_x = I_{\tilde{x}} \times I_{x^n}$ of an $(n-1)$-dimensional interval $I_{\tilde{x}}$ and a closed interval I_{x^n} of the nth coordinate axis. We give a similar representation $I_t = I_{\tilde{t}} \times I_{t^n}$ for a fixed interval I_t in $\mathbb{R}^n_t$ containing D_t.

Using the definition of the integral over a set, Fubini's theorem, and Remark 2, we can write

$$\int_{D_x} f(x)\,dx = \int_{I_x} f \cdot \chi_{D_x}(x)\,dx = \int_{I_{\tilde{x}}} d\tilde{x} \int_{I_{x^n}} f \cdot \chi_{D_x}(\tilde{x}, x^n)\,dx^n =$$

$$= \int_{I_{\tilde{x}}} d\tilde{x} \int_{D_{x^n}(\tilde{x})} f(\tilde{x}, x^n)\,dx^n =$$

$$= \int_{I_{\tilde{t}}} d\tilde{t} \int_{D_{t^n}(\tilde{t})} f(\tilde{t}, \varphi^n(\tilde{t}, t^n)) \left| \frac{\partial \varphi^n}{\partial t^n} \right| (\tilde{t}, t^n)\,dt^n =$$

$$= \int_{I_{\tilde{t}}} d\tilde{t} \int_{I_{t^n}} (f \circ \varphi |\det \varphi'| \chi_{D_t})(\tilde{t}, t^n)\,dt^n =$$

$$= \int_{I_t} (f \circ \varphi |\det \varphi'| \chi_{D_t})(t)\,dt = \int_{D_t} (f \circ \varphi |\det \varphi'|)(t)\,dt.$$

In this computation we have used the fact that $\det \varphi' = \frac{\partial \varphi^n}{\partial t^n}$ for the diffeomorphism under consideration. □

11.5.5 Composite Mappings and the Formula for Change of Variable

Lemma 4 *If* $D_\tau \overset{\psi}{\to} D_t \overset{\varphi}{\to} D_x$ *are two diffeomorphisms for each of which formula* (11.10) *for change of variable in the integral holds, then it holds also for the composition* $\varphi \circ \psi : D_\tau \to D_x$ *of these mappings.*

Proof It suffices to recall that $(\varphi \circ \psi)' = \varphi' \circ \psi'$ and that $\det(\varphi \circ \psi)'(\tau) = \det \varphi'(t) \det \psi'(\tau)$, where $t = \varphi(\tau)$. We then have

$$\int_{D_x} f(x)\,\mathrm{d}x = \int_{D_t} \left(f \circ \varphi \big| \det \varphi' \big|\right) \mathrm{d}t =$$

$$= \int_{D_\tau} \left((f \circ \varphi \circ \psi) \big| \det \varphi' \circ \psi \big| \big| \det \psi' \big|\right)(\tau)\,\mathrm{d}\tau =$$

$$= \int_{D_\tau} \left(f \circ (\varphi \circ \psi) \big| \det(\varphi \circ \psi)' \big|\right)(\tau)\,\mathrm{d}\tau. \qquad \square$$

11.5.6 Additivity of the Integral and Completion of the Proof of the Formula for Change of Variable in an Integral

Lemmas 3 and 4 suggest that we might use the local decomposition of any diffeomorphism as a composition of elementary diffeomorphisms (see Proposition 2 from Sect. 8.6.4 of Part 1) and thereby obtain the formula (11.10) in the general case.

There are various ways of reducing the integral over a set to integrals over small neighborhoods of its points. For example, one may use the additivity of the integral. That is the procedure we shall use. On the basis of Lemmas 1, 3, and 4 we now carry out the proof of Theorem 1 on change of variable in a multiple integral.

Proof For each point t of the compact set $K_t = \mathrm{supp}((f \circ \varphi) | \det \varphi'|) \subset D_t$ we construct a $\delta(t)$-neighborhood $U(t)$ of it in which the diffeomorphism φ decomposes into a composition of elementary diffeomorphisms. From the $\frac{\delta(t)}{2}$ neighborhoods $\widetilde{U}(t) \subset U(t)$ of the points $t \in K_t$ we choose a finite covering $\widetilde{U}(t_1), \ldots, \widetilde{U}(t_k)$ of the compact set K_t. Let $\delta = \frac{1}{2}\min\{\delta(t_1), \ldots, \delta(t_k)\}$. Then the closure of any set whose diameter is smaller than δ and which intersects K_t must be contained in at least one of the neighborhoods $\widetilde{U}(t_1), \ldots, \widetilde{U}(t_k)$.

Now let I be an interval containing the set D_t and P a partition of the interval I such that $\lambda(P) < \min\{\delta, d\}$, where δ was found above and d is the distance from

K_t to the boundary of D_t. Let $\mathcal{I} := \{I_i\}$ be the intervals of the partition P that have a nonempty intersection with K_t. It is clear that if $I_i \in \mathcal{I}$, then $I_i \subset D_t$ and

$$\int_{D_t} (f \circ \varphi |\det \varphi'|)(t)\,dt = \int_I ((f \circ \varphi |\det \varphi'|)\chi_{D_t})(t)\,dt =$$

$$= \sum_i \int_{I_i} (f \circ \varphi |\det \varphi'|)(t)\,dt. \qquad (11.13)$$

By Lemma 1 the image $E_i = \varphi(I_i)$ of the intervals I_i is a measurable set. Then the set $E = \bigcup_i E_i$ is also measurable and $\operatorname{supp} f \subset E = \overline{E} \subset D_x$. Using the additivity of the integral, we deduce from this that

$$\int_{D_x} f(x)\,dx = \int_{I_x \supset D_x} f\chi_{D_x}(x)\,dx = \int_{I_x \setminus E} f\chi_{D_x}(x)\,dx + \int_E f\chi_{D_x}(x)\,dx =$$

$$= \int_E f\chi_{D_x}(x)\,dx = \int_E f(x)\,dx = \sum_i \int_{E_i} f(x)\,dx. \qquad (11.14)$$

By construction every interval $I_i \in \mathcal{I}$ is contained in some neighborhood $U(x_j)$ inside which the diffeomorphism φ decomposes into a composition of elementary diffeomorphisms. Hence on the basis of Lemmas 3 and 4 we can write

$$\int_{E_i} f(x)\,dx = \int_{I_i} (f \circ \varphi |\det \varphi'|)(t)\,dt. \qquad (11.15)$$

Comparing relations (11.13), (11.14), and (11.15), we obtain formula (11.10). $\square$

11.5.7 Corollaries and Generalizations of the Formula for Change of Variable in a Multiple Integral

a. Change of Variable Under Mappings of Measurable Sets

Proposition 1 *Let $\varphi : D_t \to D_x$ be a diffeomorphism of a bounded open set $D_t \subset \mathbb{R}^n$ onto a set $D_x \subset \mathbb{R}^n$ of the same type; let E_t and E_x be subsets of D_t and D_x respectively and such that $\overline{E}_t \subset D_t, \overline{E}_x \subset D_x$, and $E_x = \varphi(E_t)$. If $f \in \mathcal{R}(E_x)$, then $f \circ \varphi |\det \varphi'| \in \mathcal{R}(E_t)$, and the following equality holds:*

$$\int_{E_x} f(x)\,dx = \int_{E_t} (f \circ \varphi |\det \varphi'|)(t)\,dt. \qquad (11.16)$$

Proof Indeed,

$$
\int_{E_x} f(x)\,dx = \int_{D_x} (f\chi_{E_x})(x)\,dx = \int_{D_t} \big(((f\chi_{E_x})\circ\varphi)\big|\det\varphi'\big|\big)(t)\,dt =
$$
$$
= \int_{D_t} \big((f\circ\varphi)\big|\det\varphi'\big|\chi_{E_t}\big)(t)\,dt = \int_{E_t} \big((f\circ\varphi)\big|\det\varphi'\big|\big)(t)\,dt.
$$

In this computation we have used the definition of the integral over a set, formula (11.10), and the fact that $\chi_{E_t} = \chi_{E_x}\circ\varphi$. □

b. Invariance of the Integral

We recall that the integral of a function $f : E \to \mathbb{R}$ over a set E reduces to computing the integral of the function $f\chi_E$ over an interval $I \supset E$. But the interval I itself was by definition connected with a Cartesian coordinate system in $\mathbb{R}^n$. We can now prove that all Cartesian systems lead to the same integral.

Proposition 2 *The value of the integral of a function f over a set $E \subset \mathbb{R}^n$ is independent of the choice of Cartesian coordinate system in $\mathbb{R}^n$.*

Proof In fact the transition from one Cartesian coordinate system in $\mathbb{R}^n$ to another Cartesian system has a Jacobian constantly equal to 1 in absolute value. By Proposition 1 this implies the equality

$$
\int_{E_x} f(x)\,dx = \int_{E_t} (f\circ\varphi)(t)\,dt.
$$

But this means that the integral is invariantly defined: if p is a point of E having coordinates $x = (x^1,\dots,x^n)$ in the first system and $t = (t^1,\dots,t^n)$ in the second, and $x = \varphi(t)$ is the transition function from one system to the other, then

$$
f(p) = f_x(x^1,\dots,x^n) = f_t(t^1,\dots,t^n),
$$

where $f_t = f_x \circ \varphi$. Hence we have shown that

$$
\int_{E_x} f_x(x)\,dx = \int_{E_t} f_t(t)\,dt,
$$

where E_x and E_t denote the set E in the x and t coordinates respectively. □

We can conclude from Proposition 2 and Definition 3 of Sect. 11.2 for the (Jordan) measure of a set $E \subset \mathbb{R}^n$ that this measure is independent of the Cartesian coordinate system in $\mathbb{R}^n$, or, what is the same, that Jordan measure is invariant under the group of rigid Euclidean motions in $\mathbb{R}^n$.

c. Negligible Sets

The changes of variable or formulas for transforming coordinates used in practice
sometimes have various singularities (for example, one-to-oneness may fail in some
places, or the Jacobian may vanish, or differentiability may fail). As a rule, these
singularities occur on a set of measure zero and so, to meet the demands of practice,
the following theorem is very useful.

Theorem 2 *Let $\varphi : D_t \to D_x$ be a mapping of a (Jordan) measurable set $D_t \subset \mathbb{R}^n_t$
onto a set $D_x \subset \mathbb{R}^n_x$ of the same type. Suppose that there are subsets S_t and S_x
of D_t and D_x respectively having (Lebesgue) measure zero and such that $D_t \backslash S_t$
and $D_x \backslash S_x$ are open sets and φ maps the former diffeomorphically onto the latter
and with a bounded Jacobian. Then for any function $f \in \mathcal{R}(D_x)$ the function $(f \circ
\varphi)|\det \varphi'|$ also belongs to $\mathcal{R}(D_t \backslash S_t)$ and*

$$\int_{D_x} f(x)\, dx = \int_{D_t \backslash S_t} \big((f \circ \varphi)|\det \varphi'|\big)(t)\, dt. \tag{11.17}$$

If, in addition, the quantity $|\det \varphi'|$ is defined and bounded in D_t, then

$$\int_{D_x} f(x)\, dx = \int_{D_t} \big((f \circ \varphi)|\det \varphi'|\big)(t)\, dt. \tag{11.18}$$

Proof By Lebesgue's criterion the function f can have discontinuities in D_x and
hence also in $D_x \backslash S_x$ only on a set of measure zero. By Lemma 1, the image of this
set of discontinuities under the mapping $\varphi^{-1} : D_x \backslash S_x \to D_t \backslash S_t$ is a set of measure
zero in $D_t \backslash S_t$. Thus the relation $(f \circ \varphi)|\det \varphi'| \in \mathcal{R}(D_t \backslash S_t)$ will follow immedi-
ately from Lebesgue's criterion for integrability if we establish that the set $D_t \backslash S_t$ is
measurable. The fact that this is indeed a Jordan measurable set will be a by-product
of the reasoning below.

By hypothesis $D_x \backslash S_x$ is an open set, so that $(D_x \backslash S_x) \cap \partial S_x = \varnothing$. Hence $\partial S_x \subset
\partial D_x \cup S_x$ and consequently $\partial D_x \cup S_x = \partial D_x \cup \overline{S}_x$, where $\overline{S}_x = S_x \cup \partial S_x$ is the
closure of S_x in $\mathbb{R}^n_x$. As a result, $\partial D_x \cup S_x$ is a closed bounded set, that is, it is com-
pact in $\mathbb{R}^n$, and, being the union of two sets of measure zero, is itself of Lebesgue
measure zero. From Lemma 3 of Sect. 11.1 we know that then the set $\partial D_x \cup S_x$
(and along with it, S_x) has measure zero, that is, for every $\varepsilon > 0$ there exists a finite
covering $I_1, \ldots, I_k$ of this set by intervals such that $\sum_{i=1}^{k} |I_i| < \varepsilon$. Hence it follows,
in particular, that the set $D_x \backslash S_x$ (and similarly the set $D_t \backslash S_t$) is Jordan measurable:
indeed, $\partial (D_x \backslash S_x) \subset \partial D_x \cup \partial S_x \subset \partial D_x \cup S_x$.

The covering $I_1, \ldots, I_k$ can obviously also be chosen so that every point $x \in
\partial D_x \backslash S_x$ is an interior point of at least one of the intervals of the covering. Let
$U_x = \bigcup_{i=1}^{k} I_i$. The set U_x is measurable, as is $V_x = D_x \backslash U_x$. By construction the set
V_x is such that $\overline{V}_x \subset D_x \backslash S_x$ and for every measurable set $E_x \subset D_x$ containing the

compact set $\overline{V}_x$ we have the estimate

$$\left| \int_{D_x} f(x)\,\mathrm{d}x - \int_{E_x} f(x)\,\mathrm{d}x \right| = \left| \int_{D_x \setminus E_x} f(x)\,\mathrm{d}x \right| \le$$

$$\le M\mu(D_x \setminus E_x) < M \cdot \varepsilon, \qquad (11.19)$$

where $M = \sup_{x \in D_x} f(x)$.

The pre-image $\overline{V}_t = \varphi^{-1}(\overline{V}_x)$ of the compact set $\overline{V}_x$ is a compact subset of $D_t \setminus S_t$. Reasoning as above, we can construct a measurable compact set W_t subject to the conditions $\overline{V}_t \subset W_t \subset D_t \setminus S_t$ and having the property that the estimate

$$\left| \int_{D_t \setminus S_t} \big((f \circ \varphi) \big| \det \varphi' \big| \big)(t)\,\mathrm{d}t - \int_{E_t} \big((f \circ \varphi) \big| \det \varphi' \big| \big)(t)\,\mathrm{d}t \right| < \varepsilon \qquad (11.20)$$

holds for every measurable set E_t such that $W_t \subset E_t \subset D_t \setminus S_t$.

Now let $E_x = \varphi(E_t)$. Formula (11.16) holds for the sets $E_x \subset D_x \setminus S_x$ and $E_t \subset D_t \setminus S_t$ by Lemma 1. Comparing relations (11.16), (11.19), and (11.20) and taking account of the arbitrariness of the quantity $\varepsilon > 0$, we obtain (11.17).

We now prove the last assertion of Theorem 2. If the function $(f \circ \varphi)|\det \varphi'|$ is defined on the entire set D_t, then, since $D_t \setminus S_t$ is open in $\mathbb{R}_t^n$, the entire set of discontinuities of this function in D_t consists of the set A of points of discontinuity of $(f \circ \varphi)|\det \varphi'||_{D_t \setminus S_t}$ (the restriction of the original function to $D_t \setminus S_t$) and perhaps a subset B of $S_t \cup \partial D_t$.

As we have seen, the set A is a set of Lebesgue measure zero (since the integral on the right-hand side of (11.17) exists), and since $S_t \cup \partial D_t$ has measure zero, the same can be said of B. Hence it suffices to know that the function $(f \circ \varphi)|\det \varphi'|$ is bounded on D_t; it will then follow from the Lebesgue criterion that it is integrable over D_t. But $|f \circ \varphi|(t) \le M$ on D_t, so that the function $(f \circ \varphi)|\det \varphi'|$ is bounded on S_t, given that the function $|\det \varphi'|$ is bounded on S_t by hypothesis. As for the set $D_t \setminus S_t$, the function $(f \circ \varphi)|\det \varphi'|$ is integrable over it and hence bounded. Thus, the function $(f \circ \varphi)|\det \varphi'|$ is integrable over D_t. But the sets D_t and $D_t \setminus S_t$ differ only by the measurable set S_t, whose measure, as has been shown, is zero. Therefore, by the additivity of the integral and the fact that the integral over S_t is zero, we can conclude that the right-hand sides of (11.17) and (11.18) are indeed equal in this case. $\square$

Example The mapping of the rectangle $I = \{(r, \varphi) \in \mathbb{R}^2 \mid 0 \le r \le R \wedge 0 \le \varphi \le 2\pi\}$ onto the disk $K = \{(x, y) \in \mathbb{R}^2 \mid x^2 + y^2 \le R^2\}$ given by the formulas

$$x = r \cos \varphi, \qquad y = r \sin \varphi, \qquad (11.21)$$

is not a diffeomorphism: the entire side of the rectangle I on which $r = 0$ maps to the single point $(0, 0)$ under this mapping; the images of the points $(r, 0)$ and $(r, 2\pi)$ are the same. However, if we consider, for example, the sets $I \setminus \partial I$ and $K \setminus E$, where E is the union of the boundary ∂K of the disk K and the radius ending at $(0, R)$, then the restriction of the mapping (11.21) to the domain $I \setminus \partial I$ turns out to be a diffeomorphism of it onto the domain $K \setminus E$. Hence by Theorem 2, for any function $f \in \mathcal{R}(K)$ we can write

$$\iint_K f(x, y) \, dx \, dy = \iint_I f(r \cos \varphi, r \sin \varphi) r \, dr \, d\varphi$$

and, applying Fubini's theorem

$$\iint_K f(x, y) \, dx \, dy = \int_0^{2\pi} d\varphi \int_0^R f(r \cos \varphi, r \sin \varphi) r \, dr.$$

Relations (11.21) are the well-known formulas for transition from polar coordinates to Cartesian coordinates in the plane.

What has been said can naturally be developed and extended to the polar (spherical) coordinates in $\mathbb{R}^n$ that we studied in Part 1, where we also exhibited the Jacobian of the transition from polar coordinates to Cartesian coordinates in a space $\mathbb{R}^n$ of any dimension.

11.5.8 Problems and Exercises

1. a) Show that Lemma 1 is valid for any smooth mapping $\varphi : D_t \to D_x$ (also see Problem 8 below in this connection).

b) Prove that if D is an open set in $\mathbb{R}^m$ and $\varphi \in C^{(1)}(D, \mathbb{R}^n)$, then $\varphi(D)$ is a set of measure zero in $\mathbb{R}^n$ when $m < n$.

2. a) Verify that the measure of a measurable set E and the measure of its image $\varphi(E)$ under a diffeomorphism φ are connected by the relation $\mu(\varphi(E)) = \theta \mu(E)$, where $\theta \in [\inf_{t \in E} |\det \varphi'(t)|, \sup_{t \in E} |\det \varphi'(t)|]$.

b) In particular, if E is a connected set, there is a point $\tau \in E$ such that $\mu(\varphi(E)) = |\det \varphi'(\tau)| \mu(E)$.

3. a) Show that if formula (11.10) holds for the function $f \equiv 1$, then it holds in general.

b) Carry out the proof of Theorem 1 again, but for the special case $f \equiv 1$, simplifying it for this special situation.

4. Without using Remark 2, carry out the proof of Lemma 3, assuming Lemma 2 is known and that two integrable functions that differ only on a set of measure zero have the same integral.

5. Instead of the additivity of the integral and the accompanying analysis of the measurability of sets, one can use another device for localization when reducing formula (11.10) to its local version (that is to the verification of the formula for a small neighborhood of the points of the domain being mapped). This device is based on the linearity of the integral.

a) If the smooth functions $e_1, \ldots, e_k$ are such that $0 \le e_i \le 1$, $i = 1, \ldots, k$, and $\sum_i^k e_i(x) \equiv 1$ on D_x, then $\int_{D_x} (\sum_{i=1}^k e_i f)(x) \, dx = \int_{D_x} f(x) \, dx$ for every function $f \in \mathcal{R}(D_x)$.

b) If $\operatorname{supp} e_i$ is contained in the set $U \subset D_x$, then $\int_{D_x} (e_i f)(x) \, dx = \int_U (e_i f)(x)(dx)$.

c) Taking account of Lemmas 3 and 4 and the linearity of the integral, one can derive formula (11.10) from a) and b), if for every open covering $\{U_\alpha\}$ of the compact set $K = \operatorname{supp} f \subset D_x$ we construct a set of smooth functions $e_1, \ldots, e_k$ in D_x such that $0 \le e_i \le 1$, $i = 1, \ldots, k$, $\sum_{i=1}^k e_i \equiv 1$ on K, and for every function $e_i \in \{e_i\}$ there is a set $U_{\alpha_i} \in \{U_\alpha\}$ such that $\operatorname{supp} e_i \subset U_{\alpha_i}$.

In that case the set of functions $\{e_i\}$ is said to be a *partition of unity on the compact set K subordinate to the covering $\{U_\alpha\}$.*
6. This problem contains a scheme for constructing the partition of unity discussed in Problem 5.

a) Construct a function $f \in C^{(\infty)}(\mathbb{R}, \mathbb{R})$ such that $f|_{[-1,1]} \equiv 1$ and $\operatorname{supp} f \subset [-1 - \delta, 1 + \delta]$, where $\delta > 0$.

b) Construct a function $f \in C^{(\infty)}(\mathbb{R}^n, \mathbb{R})$ with the properties indicated in a) for the unit cube in $\mathbb{R}^n$ and its δ-dilation.

c) Show that for every open covering of the compact set $K \subset \mathbb{R}^n$ there exists a smooth partition of unity on K subordinate to this covering.

d) Extending c), construct a $C^{(\infty)}$-partition of unity in $\mathbb{R}^n$ subordinate to a locally finite open covering of the entire space. (A covering is *locally finite* if every point of the set that is covered, in this case $\mathbb{R}^n$, has a neighborhood that intersects only a finite number of the sets in the covering. For a partition of unity containing an infinite number of functions $\{e_i\}$ we impose the requirement that every point of the set on which this partition is constructed belongs to the support of at most finitely many of the functions $\{e_i\}$. Under this hypothesis no questions arise as to the meaning of the equality $\sum_i e_i \equiv 1$; more precisely, there are no questions as to the meaning of the sum on the left-hand side.)

7. One can obtain a proof of Theorem 1 that is slightly different from the one given above and relies on the possibility of decomposing only a linear mapping into a composition of elementary mappings. Such a proof is closer to the heuristic considerations in Sect. 11.5.1 and is obtained by proving the following assertions.

a) Verify that under elementary linear mappings $L : \mathbb{R}^n \to \mathbb{R}^n$ of the form

$$\left(x^1, \ldots, x^k, \ldots, x^n\right) \mapsto \left(x^1, \ldots, x^{k-1}, \lambda x^k, x^{k+1}, \ldots, x^n\right),$$

$\lambda \ne 0$, and

$$\left(x^1, \ldots, x^k, \ldots, x^n\right) \mapsto \left(x^1, \ldots, x^{k-1}, x^k + x^j, \ldots, x^n\right)$$

the relation $\mu(L(E)) = |\det L'| \mu(E)$ holds for every measurable set $E \subset \mathbb{R}^n$; then show that this relation holds for every linear transformation $L : \mathbb{R}^n \to \mathbb{R}^n$. (Use Fubini's theorem and the possibility of decomposing a linear mapping into a composition of the elementary mappings just exhibited.)

b) Show that if $\varphi : D_t \to D_x$ is a diffeomorphism, then $\mu(\varphi(K)) \le \int_K |\det \varphi'(t)| \, dt$ for every measurable compact set $K \subset D_t$ and its image $\varphi(K)$. (If $a \in D_t$, then $\exists(\varphi'(a))^{-1}$ and in the representation $\varphi(t) = (\varphi'(a) \circ (\varphi'(a))^{-1} \circ \varphi)(t)$ the mapping $\varphi'(a)$ is linear while the transformation $(\varphi'(a))^{-1} \circ \varphi$ is nearly an isometry on a neighborhood of a.)

c) Show that if the function f in Theorem 1 is nonnegative, then $\int_{D_x} f(x) \, dx \le \int_{D_t} ((f \circ \varphi)|\det \varphi'|)(t) \, dt$.

d) Applying the preceding inequality to the function $(f \circ \varphi)|\det \varphi'|$ and the mapping $\varphi^{-1} : D_x \to D_t$, show that formula (11.10) holds for a nonnegative function.

e) By representing the function f in Theorem 1 as the difference of integrable nonnegative functions, prove that formula (11.10) holds.

8. Sard's lemma. *Let D be an open set in $\mathbb{R}^n$, let $\varphi \in C^{(1)}(D, \mathbb{R}^n)$, and let S be the set of critical points of the mapping φ. Then $\varphi(S)$ is a set of (Lebesgue) measure zero.*

We recall that a *critical point of a smooth mapping* φ of a domain $D \subset \mathbb{R}^m$ into $\mathbb{R}^n$ is a point $x \in D$ at which $\operatorname{rank} \varphi'(x) < \min\{m, n\}$. In the case $m = n$, this is equivalent to the condition $\det \varphi'(x) = 0$.

a) Verify Sard's lemma for a linear transformation.

b) Let I be an interval in the domain D and $\varphi \in C^{(1)}(D, \mathbb{R}^n)$. Show that there exists a function $\alpha(h)$, $\alpha : \mathbb{R}^n \to \mathbb{R}$ such that $\alpha(h) \to 0$ as $h \to 0$ and $|\varphi(x + h) - \varphi(x) - \varphi'(x)h| \le \alpha(h)|h|$ for every $x, x + h \in I$.

c) Using b), estimate the deviation of the image $\varphi(I)$ of the interval I under the mapping φ from the same image under the linear mapping $L(x) = \varphi(a) + \varphi'(a)(x - a)$, where $a \in I$.

d) Based on a), b), and c), show that if S is the set of critical points of the mapping φ in the interval I, then $\varphi(S)$ is a set of measure zero.

e) Now finish the proof of Sard's lemma.

f) Using Sard's lemma, show that in Theorem 1 it suffices to require that the mapping φ be a one-to-one mapping of class $C^{(1)}(D_t, D_x)$.

We remark that the version of Sard's lemma given here is a simple special case of a theorem of Sard and Morse, according to which the assertion of the lemma holds even if $D \subset \mathbb{R}^m$ and $\varphi \in C^{(k)}(D, \mathbb{R}^n)$, where $k = \max\{m - n + 1, 1\}$. The quantity k here, as an example of Whitney shows, cannot be decreased for any pair of numbers m and n.

In geometry Sard's lemma is known as the assertion that if $\varphi : D \to \mathbb{R}^n$ is a smooth mapping of an open set $D \subset \mathbb{R}^m$ into $\mathbb{R}^n$, then for almost all points $x \in \varphi(D)$, the complete pre-image $\varphi^{-1}(x) = M_x$ in D is a surface (manifold) of codimension n in $\mathbb{R}^m$ (that is, $m - \dim M_x = n$ for almost all $x \in D$).

9. Suppose we consider an arbitrary mapping $\varphi \in C^{(1)}(D_t, D_x)$ such that $\det \varphi'(t) \ne 0$ in D_t instead of the diffeomorphism φ of Theorem 1. Let $n(x) = \operatorname{card}\{t \in \operatorname{supp}(f \circ \varphi) \mid \varphi(t) = x\}$, that is, $n(x)$ is the number of points of the support of the function $f \circ \varphi$ that map to the point $x \in D_x$ under $\varphi : D_t \to D_x$. The following formula holds:

$$\int_{D_x} (f \cdot n)(x)\, dx = \int_{D_t} \left((f \circ \varphi) \left| \det \varphi' \right| \right)(t)\, dt.$$

a) What is the geometric meaning of this formula for $f \equiv 1$?

b) Prove this formula for the special mapping of the annulus $D_t = \{t \in \mathbb{R}^2_t \mid 1 < |t| < 2\}$ onto the annulus $D_x = \{x \in \mathbb{R}^2_x \mid 1 < |x| < 2\}$ given in polar coordinates (r, φ) and (ρ, θ) in the planes $\mathbb{R}^2_x$ and $\mathbb{R}^2_t$ respectively by the formulas $r = \rho$, $\varphi = 2\theta$.

c) Now try to prove the formula in general.

11.6 Improper Multiple Integrals

11.6.1 Basic Definitions

Definition 1 An *exhaustion* of a set $E \subset \mathbb{R}^m$ is a sequence of measurable sets $\{E_n\}$ such that $E_n \subset E_{n+1} \subset E$ for any $n \in \mathbb{N}$ and $\bigcup_{n=1}^{\infty} E_n = E$.

Lemma *If $\{E_n\}$ is an exhaustion of a measurable set E, then*:

a) $\lim_{n\to\infty} \mu(E_n) = \mu(E)$;

b) *for every function $f \in \mathcal{R}(E)$ the function $f|_{E_n}$ also belongs to $\mathcal{R}(E_n)$, and*

$$\lim_{n\to\infty} \int_{E_n} f(x)\, dx = \int_E f(x)\, dx.$$

Proof Since $E_n \subset E_{n+1} \subset E$, it follows that $\mu(E_n) \leq \mu(E_{n+1}) \leq \mu(E)$ and $\lim_{n\to\infty} \mu(E_n) \leq \mu(E)$. To prove a) we shall show that the inequality $\lim_{n\to\infty} \mu(E_n) \geq \mu(E)$ also holds.

The boundary ∂E of E has content zero, and hence can be covered by a finite number of open intervals of total content less than any preassigned number $\varepsilon > 0$. Let Δ be the union of all these open intervals. Then the set $E \cup \Delta =: \widetilde{E}$ is open in $\mathbb{R}^m$ and by construction $\widetilde{E}$ contains the closure of E and $\mu(\widetilde{E}) \leq \mu(E) + \mu(\Delta) < \mu(E) + \varepsilon$.

For every set E_n of the exhaustion $\{E_n\}$ the construction just described can be repeated with the value $\varepsilon_n = \varepsilon/2^n$. We then obtain a sequence of open sets $\widetilde{E}_n = E_n \cup \Delta_n$ such that $E_n \subset \widetilde{E}_n$, $\mu(\widetilde{E}_n) \leq \mu(E_n) + \mu(\Delta_n) < \mu(E_n) + \varepsilon_n$, and $\bigcup_{n=1}^{\infty} \widetilde{E}_n \supset \bigcup_{n=1}^{\infty} E_n \supset E$.

The system of open sets $\Delta, \widetilde{E}_1, \widetilde{E}_2, \ldots$, forms an open covering of the compact set $\overline{E}$.

Let $\Delta, \widetilde{E}_1, \widetilde{E}_2, \ldots, \widetilde{E}_k$ be a finite covering of $\overline{E}$ extracted from this covering. Since $E_1 \subset E_2 \subset \cdots \subset E_k$, the sets $\Delta, \Delta_1, \ldots, \Delta_k, E_k$ also form a covering of $\overline{E}$ and hence

$$\mu(E) \leq \mu(\overline{E}) \leq \mu(E_k) + \mu(\Delta) + \mu(\Delta_1) + \cdots + \mu(\Delta_k) < \mu(E_k) + 2\varepsilon.$$

It follows from this that $\mu(E) \le \lim_{n \to \infty} \mu(E_n)$.

b) The relation $f|_E \in \mathcal{R}(E_n)$ is well known to us and follows from Lebesgue's criterion for the existence of the integral over a measurable set. By hypothesis $f \in \mathcal{R}(E)$, and so there exists a constant M such that $|f(x)| \le M$ on E. From the additivity of the integral and the general estimate for the integral we obtain

$$\left| \int_E f(x)\, dx - \int_{E_n} f(x)\, dx \right| = \left| \int_{E \setminus E_n} f(x)\, dx \right| \le M\mu(E \setminus E_n).$$

From this, together with what was proved in a), we conclude that b) does indeed hold. □

Definition 2 Let $\{E_n\}$ be an exhaustion of the set E and suppose the function $f: E \to \mathbb{R}$ is integrable on the sets $E_n \in \{E_n\}$. If the limit

$$\int_E f(x)\, dx := \lim_{n \to \infty} \int_{E_n} f(x)\, dx$$

exists and has a value independent of the choice of the sets in the exhaustion of E, this limit is called the *improper integral of f over E*.

The integral sign on the left in this last equality is usually written for any function defined on E, but we say that the integral *exists* or *converges* if the limit in Definition 2 exists. If there is no common limit for all exhaustions of E, we say that the integral of f over E does not exist, or that the integral *diverges*.

The purpose of Definition 2 is to extend the concept of integral to the case of an unbounded integrand or an unbounded domain of integration.

The symbol introduced to denote an improper integral is the same as the symbol for an ordinary integral, and that fact makes the following remark necessary.

Remark 1 If E is a measurable set and $f \in \mathcal{R}(E)$, then the integral of f over E in the sense of Definition 2 exists and has the same value as the proper integral of f over E.

Proof This is precisely the content of assertion b) in the lemma above. □

The set of all exhaustions of any reasonably rich set is immense, and we do not use all exhaustions. The verification that an improper integral converges is often simplified by the following proposition.

Proposition 1 *If a function $f: E \to \mathbb{R}$ is nonnegative and the limit in Definition 2 exists for even one exhaustion $\{E_n\}$ of the set E, then the improper integral of f over E converges.*

Proof Let $\{E_k'\}$ be a second exhaustion of E into elements on which f is integrable. The sets $E_n^k := E_k' \cap E_n$, $n = 1, 2, \ldots$ form an exhaustion of the set E_k', and so it follows from part b) of the lemma that

$$\int_{E'_k} f(x)\,dx = \lim_{n\to\infty} \int_{E_n^k} f(x)\,dx \le \lim_{n\to\infty} \int_{E_n} f(x)\,dx = A.$$

Since $f \ge 0$ and $E'_k \subset E'_{k+1} \subset E$, it follows that

$$\exists \lim_{k\to\infty} \int_{E'_k} f(x)\,dx = B \le A.$$

But there is symmetry between the exhaustions $\{E_n\}$ and $\{E'_k\}$, so that $A \le B$ also, and hence $A = B$. $\qquad\qquad\square$

Example 1 Let us find the improper integral $\iint_{\mathbb{R}^2} e^{-(x^2+y^2)}\,dx\,dy$.

We shall exhaust the plane $\mathbb{R}^2$ by the sequence of disks $E_n = \{(x, y) \in \mathbb{R}^2 \mid x^2 + y^2 < n^2\}$. After passing to polar coordinates we find easily that

$$\iint_{E_n} e^{-(x^2+y^2)}\,dx\,dy = \int_0^{2\pi} d\varphi \int_0^n e^{-r^2} dr = \pi\left(1 - e^{-n^2}\right) \to \pi$$

as $n \to \infty$.

By Proposition 1 we can now conclude that this integral converges and equals π.

One can derive a useful corollary from this result if we now consider the exhaustion of the plane by the squares $E'_n = \{(x, y) \in \mathbb{R}^2 \mid |x| \le n \wedge |y| \le n\}$. By Fubini's theorem

$$\iint_{E'_n} e^{-(x^2+y^2)}\,dx\,dy = \int_{-n}^n dy \int_{-n}^n e^{-(x^2+y^2)}\,dx = \left(\int_{-n}^n e^{-t^2}\,dt\right)^2.$$

By Proposition 1 this last quantity must tend to π as $n \to \infty$. Thus, following Euler and Poisson, we find that

$$\int_{-\infty}^{+\infty} e^{-x^2}\,dx = \sqrt{\pi}.$$

Some additional properties of Definition 2 of an improper integral, which are not completely obvious at first glance, will be given below in Remark 3.

11.6.2 The Comparison Test for Convergence of an Improper Integral

Proposition 2 *Let f and g be functions defined on the set E and integrable over exactly the same measurable subsets of it, and suppose $|f(x)| \le g(x)$ on E. If the improper integral $\int_E g(x)\,dx$ converges, then the integrals $\int_E |f|(x)\,dx$ and $\int_E f(x)\,dx$ also converge.*

Proof Let $\{E_n\}$ be an exhaustion of E on whose elements both g and f are integrable. It follows from the Lebesgue criterion that the function $|f|$ is integrable on the sets E_n, $n \in \mathbb{N}$, and so we can write

$$\int_{E_{n+k}} |f|(x)\,\mathrm{d}x - \int_{E_n} |f|(x)\,\mathrm{d}x = \int_{E_{n+k}\setminus E_n} |f|(x)\,\mathrm{d}x \le$$

$$\le \int_{E_{n+k}\setminus E_n} g(x)\,\mathrm{d}x = \int_{E_{n+k}} g(x)\,\mathrm{d}x - \int_{E_n} g(x)\,\mathrm{d}x,$$

where k and n are any natural numbers. When we take account of Proposition 1 and the Cauchy criterion for the existence of a limit of a sequence, we conclude that the integral $\int_E |f|(x)\,\mathrm{d}x$ converges.

Now consider the functions $f_+ := \frac{1}{2}(|f| + f)$ and $f_- := \frac{1}{2}(|f| - f)$. Obviously $0 \le f_+ \le |f|$ and $0 \le f_- \le |f|$. By what has just been proved, the improper integrals of f_+ and f_- over E both converge. But $f = f_+ - f_-$, and hence the improper integral of f over the same set converges as well (and is equal to the difference of the integrals of f_+ and f_-). $\qquad\square$

In order to make effective use of Proposition 2 in studying the convergence of improper integrals, it is useful to have a store of standard functions for comparison. In this connection we consider the following example.

Example 2 In the deleted n-dimensional ball of radius 1, $B \subset \mathbb{R}^n$ with its center at 0 removed, consider the function $1/r^\alpha$, where $r = d(0, x)$ is the distance from the point $x \in B\setminus 0$ to the point 0. Let us determine the values of $\alpha \in \mathbb{R}$ for which the integral of $r^{-\alpha}$ over the domain $B\setminus 0$ converges. To do this we construct an exhaustion of the domain by the annular regions $B(\varepsilon) = \{x \in B \mid \varepsilon < d(0, x) < 1\}$.

Passing to polar coordinates with center at 0, by Fubini's theorem, we obtain

$$\int_{B(\varepsilon)} \frac{\mathrm{d}x}{r^\alpha(x)} = \int_S f(\varphi)\,\mathrm{d}\varphi \int_\varepsilon^1 \frac{r^{n-1}\,\mathrm{d}r}{r^\alpha} = c \int_\varepsilon^1 \frac{\mathrm{d}r}{r^{\alpha-n+1}},$$

where $\mathrm{d}\varphi = \mathrm{d}\varphi_1 \ldots \mathrm{d}\varphi_{n-1}$ and $f(\varphi)$ is a certain product of sines of the angles $\varphi_1, \ldots, \varphi_{n-2}$ that appears in the Jacobian of the transition to polar coordinates in $\mathbb{R}^n$, while c is the magnitude of the integral over s, which depends only on n, not on r and ε.

As $\varepsilon \to +0$ the value just obtained for the integral over $B(\varepsilon)$ will have a finite limit if $\alpha < n$. In all other cases this last integral tends to infinity as $\varepsilon \to +0$.

Thus we have shown that the function $\frac{1}{d^\alpha(0,x)}$, where d is the distance to the point 0, can be integrated in a deleted neighborhood of 0 only when $\alpha < n$, where n is the dimension of the space.

Similarly one can show that outside the ball B, that is, in a neighborhood of infinity, this same function is integrable in the improper sense only for $\alpha > n$.

Example 3 Let $I = \{x \in \mathbb{R}^n \mid 0 \le x^i \le 1, i = 1, \ldots, n\}$ be the n-dimensional cube and I_k the k-dimensional face of it defined by the conditions $x^{k+1} = \cdots = x^n = 0$.

On the set $I \setminus I_k$ we consider the function $\frac{1}{d^\alpha(x)}$, where $d(x)$ is the distance from $x \in I \setminus I_k$ to the face I_k. Let us determine the values of $\alpha \in \mathbb{R}$ for which the integral of this function over $I \setminus I_k$ converges.

We remark that if $x = (x^1, \ldots, x^k, x^{k+1}, \ldots, x^n)$ then

$$d(x) = \sqrt{(x^{k+1})^2 + \cdots + (x^n)^2}.$$

Let $I(\varepsilon)$ be the cube I from which the ε-neighborhood of the face I_k has been removed. By Fubini's theorem

$$\int_{I(\varepsilon)} \frac{dx}{d^\alpha(x)} = \int_{I_k} dx^1 \ldots dx^k \int_{I_{n-k}(\varepsilon)} \frac{dx^{k+1} \cdots dx^n}{((x^{k+1})^2 + \cdots + (x^n)^2)^{\alpha/2}} = \int_{I_{n-k}(\varepsilon)} \frac{du}{|u|^\alpha},$$

where $u = (x^{k+1}, \ldots, x^n)$ and $I_{n-k}(\varepsilon)$ is the face $I_{n-k} \subset \mathbb{R}^{n-k}$ from which the ε-neighborhood of 0 has been removed.

But it is clear on the basis of the experience acquired in Example 1 that the last integral converges only for $\alpha < n - k$. Hence the improper integral under consideration converges only for $\alpha < n - k$, where k is the dimension of the face near which the function may increase without bound.

Remark 2 In the proof of Proposition 2 we verified that the convergence of the integral $|f|$ implies the convergence of the integral of f. It turns out that the converse is also true for an improper integral in the sense of Definition 2, which was not the case previously when we studied improper integrals on the line. In the latter case, we distinguished absolute and nonabsolute (conditional) convergence of an improper integral. To understand right away the essence of the new phenomenon that has arisen in connection with Definition 2, consider the following example.

Example 4 Let the function $f : \mathbb{R}_+ \to \mathbb{R}$ be defined on the set $\mathbb{R}_+$ of nonnegative numbers by the following conditions: $f(x) = \frac{(-1)^{n-1}}{n}$, if $n - 1 \leq x < n$, $n \in \mathbb{N}$.

Since the series $\sum_{n=1}^{\infty} \frac{(-1)^{n-1}}{n}$ converges, the integral $\int_0^A f(x) \, dx$ has a limit as $A \to \infty$ equal to the sum of this series.

However, this series does not converge absolutely, and one can make it divergent to $+\infty$, for example, by rearranging its terms. The partial sums of the new series can be interpreted as the integrals of the function f over the union E_n of the closed intervals on the real line corresponding to the terms of the series. The sets E_n, taken all together, however, form an exhaustion of the domain $\mathbb{R}_+$ on which f is defined.

Thus the improper integral $\int_0^\infty f(x) \, dx$ of the function $f : \mathbb{R}_+ \to \mathbb{R}$ exists in its earlier sense, but not in the sense of Definition 2.

We see that the condition in Definition 2 that the limit be independent of the choice of the exhaustion is equivalent to the independence of the sum of a series on the order of summation. The latter, as we know, is exactly equivalent to absolute convergence.

In practice one nearly always has to consider only special exhaustions of the following type. Let a function $f : D \to \mathbb{R}$ defined in the domain D be unbounded in a neighborhood of some set $E \subset \partial D$. We then remove from D the points lying in the ε-neighborhood of E and obtain a domain $D(\varepsilon) \subset D$. As $\varepsilon \to 0$ these domains generate an exhaustion of D. If the domain is unbounded, we can obtain an exhaustion of it by taking the D-complements of neighborhoods of infinity. These are the special exhaustions we mentioned earlier and studied in the one-dimensional case, and it is these special exhaustions that lead directly to the generalization of the notion of Cauchy principal value of an improper integral to the case of a space of any dimension, which we discussed earlier when studying improper integrals on the line.

11.6.3 Change of Variable in an Improper Integral

In conclusion we obtain the formula for change of variable in improper integrals, thereby making a valuable, although very simple, supplement to Theorems 1 and 2 of Sect. 11.5.

Theorem 1 *Let $\varphi : D_t \to D_x$ be a diffeomorphism of the open set $D_t \subset \mathbb{R}^n_t$ onto the set $D_x \subset \mathbb{R}^n_x$ of the same type, and let $f : D_x \to \mathbb{R}$ be integrable on all measurable compact subsets of D_x. If the improper integral $\int_{D_x} f(x)\, dx$ converges, then the integral $\int_{D_t} ((f \circ \varphi)|\det \varphi'|)(t)\, dt$ also converges and has the same value.*

Proof The open set $D_t \subset \mathbb{R}^n_t$ can be exhausted by a sequence of compact sets E^k_t, $k \in \mathbb{N}$, contained in $\mathbb{N}$, each of which is the union of a finite number of intervals in $\mathbb{R}^n_t$ (in this connection, see the beginning of the proof of Lemma 1 in Sect. 11.5). Since $\varphi : D_t \to D_x$ is a diffeomorphism, the exhaustion E^k_x of D_x, where $E^k_x = \varphi(E^k_t)$, corresponds to the exhaustion $\{E^k_t\}$ of D_t. Here the sets $E^k_x = \varphi(E^k_t)$ are measurable compact sets in D_x (measurability follows from Lemma 1 of Sect. 11.5). By Proposition 1 of Sect. 11.5 we can write

$$\int_{E^k_x} f(x)\, dx = \int_{E^k_t} ((f \circ \varphi)|\det \varphi'|)(t)\, dt.$$

The left-hand side of this equality has a limit by hypothesis as $k \to \infty$. Hence the right-hand side also has the same limit. □

Remark 3 By the reasoning just given we have verified that the integral on the right-hand side of the last equality has the same limit for any exhaustion D_t of the given special type. It is this proven part of the theorem that we shall be using. But formally, to complete the proof of the theorem in accordance with Definition 2 it is necessary to verify that this limit exists for every exhaustion of the domain D_t. We leave this (not entirely elementary) proof to the reader as an excellent exercise. We remark only that one can already deduce the convergence of the improper integral of $|f \circ \varphi||\det \varphi'|$ over the set D_t (see Problem 7).

Theorem 2 *Let* $\varphi : D_t \to D_x$ *be a mapping of the open sets* D_t *and* D_x. *Assume that there are subsets* S_t *and* S_x *of measure zero contained in* D_t *and* D_x *respectively such that* $D_t \setminus S_t$ *and* $D_x \setminus S_x$ *are open sets and* φ *is a diffeomorphism of the former onto the latter. Under these hypotheses, if the improper integral* $\int_{D_x} f(x)\,dx$ *converges, then the integral* $\int_{D_t \setminus S_t} ((f \circ \varphi) |\det \varphi'|)(t)\,dt$ *also converges to the same value. If in addition* $|\det \varphi'|$ *is defined and bounded on compact subsets of* D_t, *then* $(f \circ \varphi) |\det \varphi'| xs$ *improperly integrable over the set* D_t, *and the following equality holds:*

$$\int_{D_x} f(x)\,dx = \int_{D_t \setminus S_t} \big((f \circ \varphi) |\det \varphi'| \big)(t)\,dt.$$

Proof The assertion is a direct corollary of Theorem 1 and Theorem 2 of Sect. 11.5, provided we take account of the fact that when finding an improper integral over an open set one may restrict consideration to exhaustions that consist of measurable compact sets (see Remark 3). $\qquad\square$

Example 5 Let us compute the integral $\iint_{x^2+y^2<1} \frac{dx\,dy}{(1-x^2-y^2)^\alpha}$, which is an improper integral when $\alpha > 0$, since the integrand is unbounded in that case in a neighborhood of the disk $x^2 + y^2 = 1$.

Passing to polar coordinates, we obtain from Theorem 2

$$\iint_{x^2+y^2<1} \frac{dx\,dy}{(1-x^2-y^2)^\alpha} = \iint_{\substack{0<\varphi<2\pi \\ 0<r<1}} \frac{r\,dr\,d\varphi}{(1-r^2)^\alpha}.$$

For $\alpha > 0$ this last integral is also improper, but, since the integrand is nonnegative, it can be computed as the limit over the special exhaustion of the rectangle $I = \{(r, \varphi) \in \mathbb{R}^2 \mid 0 < \varphi < 2\pi \wedge 0 < r < 1\}$ by the rectangles $I_n = \{(r, \varphi) \in \mathbb{R}^2 \mid 0 < \varphi < 2\pi \wedge 0 < r < 1 - \frac{1}{n}\}, n \in \mathbb{N}$. Using Fubini's theorem, we find that

$$\iint_{\substack{0<\varphi<2\pi \\ 0<r<1}} \frac{r\,dr\,d\varphi}{(1-r^2)^\alpha} = \lim_{n \to \infty} \int_0^{2\pi} d\varphi \int_0^{1-\frac{1}{n}} \frac{r\,dr}{(1-r^2)^\alpha} = \frac{\pi}{1-\alpha}.$$

By the same considerations, one can deduce that the original integral diverges for $\alpha \geq 1$.

Example 6 Let us show that the integral $\iint_{|x|+|y|\geq 1} \frac{dx\,dy}{|x|^p+|y|^q}$ converges only under the condition $\frac{1}{p} + \frac{1}{q} < 1$.

Proof In view of the obvious symmetry it suffices to consider the integral only over the domain D in which $x \geq 0$, $y \geq 0$ and $x + y \geq 1$.

It is clear that the simultaneous conditions $p > 0$ and $q > 0$ are necessary for the integral to converge. Indeed, if $p \leq 0$ for example, we would obtain the following estimate for the integral over the rectangle $I_A = \{(x, y) \in \mathbb{R}^2 \mid 1 \leq x \leq A \wedge 0 \leq$

$y \leq 1\}$ alone, which is contained in D:

$$\iint_{I_A} \frac{dx\,dy}{|x|^p + |y|^q} = \int_1^A dx \int_0^1 \frac{dy}{|x|^p + |y|^q} \geq \int_1^A dx \int_0^1 \frac{dy}{1 + |y|^q} =$$

$$= (A - 1) \int_0^1 \frac{dy}{1 + |y|^q},$$

which shows that as $A \to \infty$, this integral increases without bound. Thus from now on we may assume that $p > 0$ and $q > 0$.

The integrand has no singularities in the bounded portion of the domain D, so that studying the convergence of this integral is equivalent to studying the convergence of the integral of the same function over, for example, the portion G of the domain D where $x^p + y^q \geq a > 0$. The number a can be assumed sufficiently large that the curve $x^p + y^q = a$ lies in D for $x \geq 0$ and $y \geq 0$.

Passing to generalized curvilinear coordinates φ using the formulas

$$x = \left(r \cos^2 \varphi\right)^{1/p}, \qquad y = \left(r \sin^2 \varphi\right)^{1/q},$$

by Theorem 2 we obtain

$$\iint_G \frac{dx\,dy}{|x|^p + |y|^q} = \frac{2}{p \cdot q} \iint_{\substack{0 < \varphi < \pi/2 \\ a \leq r < \infty}} \left(r^{\frac{1}{p}+\frac{1}{q}-2} \cos^{\frac{2}{p}-1} \varphi \sin^{\frac{2}{q}-1} \varphi\right) dr\,d\varphi.$$

Using the exhaustion of the domain $\{(r, \varphi) \in \mathbb{R}^2 \mid 0 < \varphi < \pi/2 \wedge a \leq r < \infty\}$ by intervals $I_{\varepsilon A} = \{(r, \varphi) \in \mathbb{R}^2 \mid 0 < \varepsilon \leq \varphi \leq \pi/2 - \varepsilon \wedge a \leq r \leq A\}$ and applying Fubini's theorem, we obtain

$$\iint_{\substack{0 < \varphi < \pi/2 \\ a \leq r < \infty}} \left(r^{\frac{1}{p}+\frac{1}{q}-2} \cos^{\frac{2}{p}-1} \varphi \sin^{\frac{2}{q}-1} \varphi\right) dr\,d\varphi =$$

$$= \lim_{\varepsilon \to 0} \int_\varepsilon^{\pi/2-\varepsilon} \cos^{\frac{2}{p}-1} \varphi \sin^{\frac{2}{q}-1} \varphi\,d\varphi \lim_{A \to \infty} \int_a^A r^{\frac{1}{p}+\frac{1}{q}-2}\,dr.$$

Since $p > 0$ and $q > 0$, the first of these limits is necessarily finite and the second is finite only when $\frac{1}{p} + \frac{1}{q} < 1$. $\qquad \square$

11.6.4 Problems and Exercises

1. Give conditions on p and q under which the integral $\iint_{0 < |x|+|y| \leq 1} \frac{dx\,dy}{|x|^p + |y|^q}$ converges.

2. a) Does the limit $\lim_{A \to \infty} \int_0^A \cos x^2\,dx$ exist?

b) Does the integral $\int_{\mathbb{R}^1} \cos x^2\,dx$ converge in the sense of Definition 2?

c) By verifying that

$$\lim_{n\to\infty} \iint_{|x|\le n} \sin\left(x^2 + y^2\right) dx\, dy = \pi$$

and

$$\lim_{n\to\infty} \iint_{x^2+y^2\le 2\pi n} \sin\left(x^2 + y^2\right) dx\, dy = 0$$

verify that the integral of $\sin(x^2 + y^2)$ over the plane $\mathbb{R}^2$ diverges.

3. a) Compute the integral $\int_0^1 \int_0^1 \int_0^1 \frac{dx\, dy\, dz}{x^p\, y^q\, z^r}$.

b) One must be careful when applying Fubini's theorem to improper integrals (but of course one must also be careful when applying it to proper integrals). Show that the integral $\iint_{x\ge 1, y\ge 1} \frac{x^2-y^2}{(x^2+y^2)^2}\, dx\, dy$ diverges, while both of the iterated integrals $\int_1^\infty dx \int_1^\infty \frac{x^2-y^2}{(x^2+y^2)^2}\, dy$ and $\int_1^\infty dy \int_1^\infty \frac{x^2-y^2}{(x^2+y^2)^2}\, dx$ converge.

c) Prove that if $f \in C(\mathbb{R}^2, \mathbb{R})$ and $f \ge 0$ in $\mathbb{R}^2$, then the existence of either of the iterated integrals $\int_{-\infty}^\infty dx \int_{-\infty}^\infty f(x, y)\, dy$ and $\int_{-\infty}^\infty dy \int_{-\infty}^\infty f(x, y)\, dx$ implies that the integral $\iint_{\mathbb{R}^2} f(x, y)\, dx\, dy$ converges to the value of the iterated integral in question.

4. Show that if $f \in C(\mathbb{R}, \mathbb{R})$, then

$$\lim_{h\to 0} \frac{1}{\pi} \int_{-1}^1 \frac{h}{h^2 + x^2} f(x)\, dx = f(0).$$

5. Let D be a bounded domain in $\mathbb{R}^n$ with a smooth boundary and S a smooth k-dimensional surface contained in the boundary of D. Show that if the function $f \in C(D, \mathbb{R})$ admits the estimate $|f| < \frac{1}{d^{n-k-\varepsilon}}$, where $d = d(S, x)$ is the distance from $x \in D$ to S and $\varepsilon > 0$, then the integral of f over D converges.

6. As a supplement to Remark 1 show that it remains valid even if the set E is not assumed to be measurable.

7. Let D be an open set in $\mathbb{R}^n$ and let the function $f : D \to \mathbb{R}$ be integrable over any measurable compact set contained in D.

a) Show that if the improper integral of the function $|f|$ over D diverges, then there exists an exhaustion $\{E_n\}$ of D such that each set E_n is an *elementary compact set*, consisting of a finite number of n-dimensional intervals and $\iint_{E_n} |f|(x)\, dx \to +\infty$ as $n \to \infty$.

b) Verify that if the integral of f over a set converges while the integral of $|f|$ diverges, then the integrals of $f_+ = \frac{1}{2}(|f| + f)$ and $f_- = \frac{1}{2}(|f| - f)$ over the set both diverge.

c) Show that the exhaustion $\{E_n\}$ obtained in a) can be distributed in such a way that $\int_{E_{n+1}\setminus E_n} f_+(x)\, dx > \int_{E_n} |f|(x)\, dx$ for all $n \in \mathbb{N}$.

d) Using lower Darboux sums, show that if $\int_E f_+(x)\, dx > A$, then there exists an elementary compact set $F \subset E$ consisting of a finite number of intervals such that $\int_F f(x)\, dx > A$.

e) Deduce from c) and d) that there exists an elementary compact set $F_n \subset E_{n+1} \backslash E_n$ for which $\int_{F_n} f(x)\,dx > \int_{E_n} |f|(x)\,dx + n$.

f) Show using e) that the sets $G_n = F_n \cap E_n$ are elementary compact sets (that is, they consist of a finite number of intervals) contained in D that, taken together, constitute an exhaustion of D, and for which the relation $\int_{G_n} f(x)\,dx \to +\infty$ as $n \to \infty$ holds.

Thus, if the integral of $|f|$ diverges, then the integral of f (in the sense of Definition 2) also diverges.

8. Carry out the proof of Theorem 2 in detail.

9. We recall that if $x = (x^1, \ldots, x^n)$ and $\xi = (\xi^1, \ldots, \xi^n)$, then $\langle x, \xi \rangle = x^1 \xi^1 + \cdots + x^n \xi^n$ is the standard inner product in $\mathbb{R}^n$. Let $A = (a_{ij})$ be a symmetric $n \times n$ matrix of complex numbers. We denote by $\operatorname{Re} A$ the matrix with elements $\operatorname{Re} a_{ij}$. Writing $\operatorname{Re} A \geq 0$ (resp. $\operatorname{Re} A > 0$) means that $\langle (\operatorname{Re} A)x, x \rangle \geq 0$ (resp. $\langle (\operatorname{Re} A)x, x \rangle > 0$) for every $x \in \mathbb{R}^n$, $x \neq 0$.

a) Show that if $\operatorname{Re} A \geq 0$, then for $\lambda > 0$ and $\xi \in \mathbb{R}^n$ we have

$$\int_{\mathbb{R}^n} \exp\left(-\frac{\lambda}{2}\langle Ax, x \rangle - i\langle x, \xi \rangle\right) dx =$$
$$= \left(\frac{2\pi}{\lambda}\right)^{n/2} (\det A)^{-1/2} \exp\left(-\frac{1}{2\lambda}\langle A^{-1}\xi, \xi \rangle\right).$$

Here the branch of $\sqrt{\det A}$ is chosen as follows:

$$(\det A)^{-1/2} = |\det A|^{-1/2} \exp(-i \operatorname{Ind} A),$$

$$\operatorname{Ind} A = \frac{1}{2} \sum_{j=1}^{n} \arg \mu_j(A), \qquad \left|\arg \mu_j(A)\right| \leq \frac{\pi}{2},$$

where $\mu_j(A)$ are the eigenvalues of A.

b) Let A be a real-valued symmetric nondegenerate $(n \times n)$ matrix. Then for $\xi \in \mathbb{R}^n$ and $\lambda > 0$ we have

$$\int_{\mathbb{R}^n} \exp\left(i\frac{\lambda}{2}\langle Ax, x \rangle - i\langle x, \xi \rangle\right) dx =$$
$$= \left(\frac{2\pi}{\lambda}\right)^{n/2} |\det A|^{-1/2} \exp\left(-\frac{i}{2\lambda}\langle A^{-1}\xi, \xi \rangle\right) \exp\left(\frac{i\pi}{4} \operatorname{sgn} A\right).$$

Here $\operatorname{sgn} A$ is the signature of the matrix, that is,

$$\operatorname{sgn} A = \nu_+(A) - \nu_-(A),$$

where $\nu_+(A)$ is the number of positive eigenvalues of A and $\nu_-(A)$ the number of negative eigenvalues.

Chapter 12
Surfaces and Differential Forms in $\mathbb{R}^n$

In this chapter we discuss the concepts of surface, boundary of a surface, and consistent orientation of a surface and its boundary; we derive a formula for computing the area of a surface lying in $\mathbb{R}^n$; and we give some elementary information on differential forms. Mastery of these concepts is very important in working with line and surface integrals, to which the next chapter is devoted.

12.1 Surfaces in $\mathbb{R}^n$

The standard model for a k-dimensional surface is $\mathbb{R}^k$.

Definition 1 A *surface of dimension k* (or *k-dimensional surface* or *k-dimensional manifold*) in $\mathbb{R}^n$ is a subset $S \subset \mathbb{R}^n$ each point of which has a neighborhood[1] in S homeomorphic[2] to $\mathbb{R}^k$.

Definition 2 The mapping $\varphi : \mathbb{R}^k \to U \subset S$ provided by the homeomorphism referred to in the definition of a surface is called a *chart* or a *local chart* of the surface S, $\mathbb{R}^k$ is called the *parameter domain*, and U is the *range* or *domain of action of the chart on the surface S*.

A local chart introduces curvilinear coordinates in U by assigning to the point $x = \varphi(t) \in U$ the set of numbers $t = (t^1, \ldots, t^k) \in \mathbb{R}^k$. It is clear from the definition that the set of objects S described by the definition does not change if $\mathbb{R}^k$ is replaced

---

[1] As before, a neighborhood of a point $x \in S \subset \mathbb{R}^n$ in S is a set $Us(x) = S \cap U(x)$, where $U(x)$ is a neighborhood of x in $\mathbb{R}^n$. Since we shall be discussing only neighborhoods of a point on a surface in what follows, we shall simplify the notation where no confusion can arise by writing U or $U(x)$ instead of $U_S(x)$.

[2] On $S \subset \mathbb{R}^n$ and hence also on $U \subset S$ there is a unique metric induced from $\mathbb{R}^n$, so that one can speak of a topological mapping of U into $\mathbb{R}^k$.

© Springer-Verlag Berlin Heidelberg 2016
V.A. Zorich, *Mathematical Analysis II*, Universitext,
DOI 10.1007/978-3-662-48993-2_4

in it by any topological space homeomorphic to $\mathbb{R}^k$. Most often the standard parameter region for local charts is assumed to be an open cube I^k or an open ball B^k in $\mathbb{R}^k$. But this makes no substantial difference.

To carry out certain analogies and in order to make a number of the following constructions easier to visualize, we shall as a rule take a cube I^k as the canonical parameter domain for local charts on a surface. Thus a chart

$$\varphi : I^k \to U \subset S \tag{12.1}$$

gives a local parametric equation $x = \varphi(t)$ for the surface $S \subset \mathbb{R}^n$, and the k-dimensional surface itself thus has the local structure of a deformed standard k-dimensional interval $I^k \subset \mathbb{R}^n$.

The parametric definition of a surface is especially important for computational purposes, as will become clear below. Sometimes one can define the entire surface by a single chart. Such a surface is usually called *elementary*. For example, the graph of a continuous function $f : I^k \to \mathbb{R}$ in $\mathbb{R}^{k+1}$ is an elementary surface. However, elementary surfaces are more the exception than the rule. For example, our ordinary two-dimensional terrestrial sphere cannot be defined by only one chart. An atlas of the surface of the Earth must contain at least two charts (see Problem 3 at the end of this section).

In accordance with this analogy we adopt the following definition.

Definition 3 A set $A(S) := \{\varphi_i : I_i^k \to U_i, i \in \mathbb{N}\}$ of local charts of a surface S whose domains of action together cover the entire surface (that is, $S = \bigcup_i U_i$) is called an *atlas of the surface* S.

The union of two atlases of the same surface is obviously also an atlas of the surface.

If no restrictions are imposed on the mappings (12.1), the local parametrizations of the surface, except that they must be homeomorphisms, the surface may be situated very strangely in $\mathbb{R}^n$. For example, it can happen that a surface homeomorphic to a two-dimensional sphere, that is, a topological sphere, is contained in $\mathbb{R}^3$, but the region it bounds is not homeomorphic to a ball (the so-called *Alexander horned sphere*).[3]

To eliminate such complications, which have nothing to do with the questions considered in analysis, we defined a *smooth k-dimensional surface* in $\mathbb{R}^n$ in Sect. 8.7 to be a set $S \subset \mathbb{R}^n$ such that for each $x_0 \in S$ there exists a neighborhood $U(x_0)$ in $\mathbb{R}^n$ and a diffeomorphism $\psi : U(x_0) \to I^n = \{t \in \mathbb{R}^n \mid |t| < 1, i = 1, \ldots, n\}$ under which the set $U_S(x_0) := S \cap U(x_0)$ maps into the cube $I^k = I^n \cap \{t \in \mathbb{R}^n \mid t^{k+1} = \cdots = t^n = 0\}$.

It is clear that a surface that is smooth in this sense is a surface in the sense of Definition 1, since the mappings $x = \psi^{-1}(t^1, \ldots, t^k, 0, \ldots, 0) = \varphi(t^1, \ldots, t^k)$

---

[3] An example of the surface described here was constructed by the American topologist J.W. Alexander (1888–1977).

obviously define a local parametrization of the surface. The converse, as follows from the example of the horned sphere mentioned above, is generally not true, if the mappings φ are merely homeomorphisms. However, if the mappings (12.1) are sufficiently regular, the concept of a surface is actually the same in both the old and new definitions.

In essence this has already been shown by Example 8 in Sect. 8.7, but considering the importance of the question, we give a precise statement of the assertion and recall how the answer is obtained.

Proposition *If the mapping* (12.1) *belongs to class* $C^{(1)}(I^k, \mathbb{R}^n)$ *and has maximal rank at each point of the cube* I^k, *there exists a number* $\varepsilon > 0$ *and a diffeomorphism* $\varphi_\varepsilon : I_\varepsilon^n \to \mathbb{R}^n$ *of the cube* $I_\varepsilon^n := \{t \in \mathbb{R}^n \mid |t^i| \leq \varepsilon_i, i = 1, \ldots, n\}$ *of dimension* n *in* $\mathbb{R}^n$ *such that* $\varphi|_{I^k \cap I_\varepsilon^n} = \varphi_\varepsilon|_{I^k \cap I_\varepsilon^n}$.

In other words, it is asserted that under these hypotheses the mappings (12.1) are locally the restrictions of diffeomorphisms of the full-dimensional cubes I_ε^n to the k-dimensional cubes $I_\varepsilon^k = I^k \cap I_\varepsilon^n$.

Proof Suppose for definiteness that the first k of the n coordinate functions $x^k = \varphi^i(t^1, \ldots, t^k)$, $i = 1, \ldots, n$, of the mapping $x = \varphi(t)$ are such that $\det(\frac{\partial \varphi^i}{\partial t^j})(0) \neq 0$, $i, j = 1, \ldots, k$. Then by the implicit function theorem the relations

$$\begin{cases} x^1 = \varphi^1(t^1, \ldots, t^k), \\ \vdots \\ x^k = \varphi^k(t^1, \ldots, t^k), \\ x^{k+1} = \varphi^{k+1}(t^1, \ldots, t^k), \\ \vdots \\ x^n = \varphi^n(t^1, \ldots, t^k) \end{cases}$$

near the point $(t_0, x_0) = (0, \varphi(0))$ are equivalent to relations

$$\begin{cases} t^1 = f^1(x^1, \ldots, x^k), \\ \vdots \\ t^k = f^k(x^1, \ldots, x^k), \\ x^{k+1} = f^{k+1}(x^1, \ldots, x^k), \\ \vdots \\ x^n = f^n(x^1, \ldots, x^k). \end{cases}$$

In this case the mapping

$$
\begin{cases}
t^1 = f^1(x^1, \ldots, x^k), \\
\;\vdots \\
t^k = f^k(x^1, \ldots, x^k), \\
t^{k+1} = x^{k+1} - f^{k+1}(x^1, \ldots, x^k), \\
\;\vdots \\
t^n = x^n - f^n(x^1, \ldots, x^k)
\end{cases}
$$

is a diffeomorphism of a full-dimensional neighborhood of the point $x_0 \in \mathbb{R}^n$. As φ_ε we can now take the restriction to some cube I_ε^n of the diffeomorphism inverse to it.   □

By a change of scale, of course, one can arrange to have $\varepsilon = 1$ and a unit cube I_ε^n in the last diffeomorphism.

Thus we have shown that for a smooth surface in $\mathbb{R}^n$ one can adopt the following definition, which is equivalent to the previous one.

Definition 4 The k-dimensional surface in $\mathbb{R}^n$ introduced by Definition 1 is *smooth* (*of class* $C^{(m)}$, $m \geq 1$) if it has an atlas whose local charts are smooth mappings (of class $C^{(m)}$, $m \geq 1$) and have rank k at each point of their domains of definition.

We remark that the condition on the rank of the mappings (12.1) is essential. For example, the analytic mapping $\mathbb{R} \ni t \mapsto (x^1, x^2) \in \mathbb{R}^2$ defined by $x^1 = t^2$, $x^2 = t^3$ defines a curve in the plane $\mathbb{R}^2$ having a cusp at $(0, 0)$. It is clear that this curve is not a smooth one-dimensional surface in $\mathbb{R}^2$, since the latter must have a tangent (a one-dimensional tangent plane) at each point.[4]

Thus, in particular one should not conflate the concept of a smooth path of class $C^{(m)}$ and the concept of a smooth curve of class $C^{(m)}$.

In analysis, as a rule, we deal with rather smooth parametrizations (12.1) of rank k. We have verified that in this case Definition 4 adopted here for a smooth surface agrees with the one considered earlier in Sect. 8.7. However, while the previous definition was intuitive and eliminated certain unnecessary complications immediately, the well-known advantage of Definition 4 of a surface, in accordance with Definition 1, is that it can easily be extended to the definition of an abstract manifold, not necessarily embedded in $\mathbb{R}^n$. For the time being, however, we shall be interested only in surfaces in $\mathbb{R}^n$.

Let us consider some examples of such surfaces.

---

[4]For the tangent plane see Sect. 8.7.

Example 1 We recall that if $F^i \in C^{(m)}(\mathbb{R}^n, \mathbb{R})$, $i = 1, \ldots, n - k$, is a set of smooth functions such that the system of equations

$$
\begin{cases}
F^1(x^1, \ldots, x^k, x^{k+1}, \ldots, x^n) = 0, \\
\vdots \\
F^{n-k}(x^1, \ldots, x^k, x^{k+1}, \ldots, x^n) = 0
\end{cases}
\tag{12.2}
$$

has rank $n - k$ at each point in the set S of its solutions, then either this system has no solutions at all or the set of its solutions forms a k-dimensional $C^{(m)}$-smooth surface S in $\mathbb{R}^n$.

Proof We shall verify that if $S \neq \varnothing$, then S does indeed satisfy Definition 4. This follows from the implicit function theorem, which says that in some neighborhood of each point $x_0 \in S$ the system (12.2) is equivalent, up to a relabeling of the variables, to a system

$$
\begin{cases}
x^{k+1} = f^{k+1}(x^1, \ldots, x^k), \\
\vdots \\
x^n = f^n(x^1, \ldots, x^k)
\end{cases}
$$

where $f^{k+1}, \ldots, f^n \in C^{(m)}$. By writing this last system as

$$
\begin{cases}
x^1 = t^1, \\
\vdots \\
x^k = t^k, \\
x^{k+1} = f^{k+1}(t^1, \ldots, t^k), \\
\vdots \\
x^n = f^n(t^1, \ldots, t^k),
\end{cases}
$$

we arrive at a parametric equation for the neighborhood of the point $x_0 \in S$ on S. By an additional transformation one can obviously turn the domain into a canonical domain, for example, into I^k and obtain a standard local chart (12.1). □

Example 2 In particular, the sphere defined in $\mathbb{R}^n$ by the equation

$$
(x^1)^2 + \cdots + (x^n)^2 = r^2 \quad (r > 0)
\tag{12.3}
$$

is an $(n - 1)$-dimensional smooth surface in $\mathbb{R}^n$ since the set S of solutions of Eq. (12.3) is obviously nonempty and the gradient of the left-hand side of (12.3) is nonzero at each point of S.

When $n = 2$, we obtain the circle in $\mathbb{R}^2$ given by

$$\left(x^1\right)^2 + \left(x^2\right)^2 = r^2,$$

which can easily be parametrized locally by the polar angle θ using the polar coordinates

$$\begin{cases} x^1 = r \cos\theta, \\ x^2 = r \sin\theta. \end{cases}$$

For fixed $r > 0$ the mapping $\theta \mapsto (x^1, x^2)(\theta)$ is a diffeomorphism on every interval of the form $\theta_0 < \theta < \theta_0 + 2\pi$, and two charts (for example, those corresponding to values $\theta_0 = 0$ and $\theta_0 = -\pi$) suffice to produce an atlas of the circle. We could not get by with one canonical chart (12.1) here because a circle is compact, in contrast to $\mathbb{R}^1$ or $I^1 = B^1$, and compactness is invariant under topological mappings.

Polar (spherical) coordinates can also be used to parametrize the two-dimensional sphere

$$\left(x^1\right)^2 + \left(x^2\right)^2 + \left(x^3\right)^2 = r^2$$

in $\mathbb{R}^3$. Denoting by ψ the angle between the direction of the vector (x^1, x^2, x^3) and the positive x^3-axis (that is, $0 \le \psi \le \pi$) and by φ the polar angle of the projection of the radius-vector (x^1, x^2, x^3) onto the (x^1, x^2)-plane, we obtain

$$\begin{cases} x^3 = r \cos\psi, \\ x^2 = r \sin\psi \sin\varphi, \\ x^1 = r \sin\psi \cos\varphi. \end{cases}$$

In general polar coordinates $(r, \theta_1, \ldots, \theta_{n-1})$ in $\mathbb{R}^n$ are introduced via the relations

$$\begin{cases} x^1 = r \cos\theta_1, \\ x^2 = r \sin\theta_1 \cos\theta_2, \\ \vdots \\ x^{n-1} = r \sin\theta_1 \sin\theta_2 \cdot \ldots \cdot \sin\theta_{n-2} \cos\theta_{n-1}, \\ x^n = r \sin\theta_1 \sin\theta_2 \cdot \ldots \cdot \sin\theta_{n-1} \sin\theta_{n-1}. \end{cases} \tag{12.4}$$

We recall the Jacobian

$$J = r^{n-1} \sin^{n-2}\theta_1 \sin^{n-3}\theta_2 \cdot \ldots \cdot \sin\theta_{n-2} \tag{12.5}$$

for the transition (12.4) from polar coordinates $(r, \theta_1, \ldots, \theta_{n-1})$ to Cartesian coordinates $(x^1, \ldots, x^n)$ in $\mathbb{R}^n$. It is clear from the expression for the Jacobian that it is nonzero if, for example, $0 < \theta_i < \pi$, $i = 1, \ldots, n-2$, and $r > 0$. Hence, even without invoking the simple geometric meaning of the parameters $\theta_1, \ldots, \theta_{n-1}$, one can

Fig. 12.1

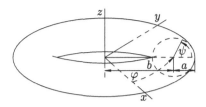

guarantee that for a fixed $r > 0$ the mapping $(\theta_1, \ldots, \theta_{n-1}) \mapsto (x^1, \ldots, x^n)$, being the restriction of a local diffeomorphism $(r, \theta_1, \ldots, \theta_{n-1}) \mapsto (x^1, \ldots, x^n)$ is itself a local diffeomorphism. But the sphere is homogeneous under the group of orthogonal transformations of $\mathbb{R}^n$, so that the possibility of constructing a local chart for a neighborhood of any point of the sphere now follows.

Example 3 The cylinder

$$\left(x^1\right)^2 + \cdots + \left(x^k\right)^2 = r^2 \quad (r > 0),$$

for $k < n$ is an $(n - 1)$-dimensional surface in $\mathbb{R}^n$ that is the direct product of the $(k - 1)$-dimensional sphere in the plane of the variables $(x^1, \ldots, x^k)$ and the $(n - k)$-dimensional plane of the variables $(x^{k+1}, \ldots, x^n)$.

A local parametrization of this surface can obviously be obtained if we take the first $k - 1$ of the $n - 1$ parameters $(t^1, \ldots, t^{n-1})$ to be the polar coordinates $\theta_1, \ldots, \theta_{k-1}$ of a point of the $(k - 1)$-dimensional sphere in $\mathbb{R}^k$ and set $t^k, \ldots, t^{n-1}$ equal to $x^{k+1}, \ldots, x^n$ respectively.

Example 4 If we take a curve (a one-dimensional surface) in the plane $x = 0$ of $\mathbb{R}^3$ endowed with Cartesian coordinates (x, y, z), and the curve does not intersect the z-axis, we can rotate the curve about the z-axis and obtain a 2-dimensional surface. The local coordinates can be taken as the local coordinates of the original curve (the meridian) and, for example, the angle of revolution (a local coordinate on a parallel of latitude).

In particular, if the original curve is a circle of radius a with center at $(b, 0, 0)$, for $a < b$ we obtain the two-dimensional torus (Fig. 12.1). Its parametric equation can be represented in the form

$$\begin{cases} x = (b + a \cos \psi) \cos \varphi, \\ y = (b + a \cos \psi) \sin \varphi, \\ z = a \sin \psi, \end{cases}$$

where ψ is the angular parameter on the original circle – the meridian – and φ is the angle parameter on a parallel of latitude.

It is customary to refer to any surface homeomorphic to the torus of revolution just constructed as a *torus* (more precisely, a *two-dimensional torus*). As one can see, a two-dimensional torus is the direct product of two circles. Since a circle can

Fig. 12.2

be obtained from a closed interval by gluing together (identifying) its endpoints, a torus can be obtained from the direct product of two closed intervals (that is, a rectangle) by gluing the opposite sides together at corresponding points (Fig. 12.2).

In essence, we have already made use of this device earlier when we established that the configuration space of a double pendulum is a two-dimensional torus, and that a path on the torus corresponds to a motion of the pendulum.

Example 5 If a flexible ribbon (rectangle) is glued along the arrows shown in Fig. 12.3a, one can obtain an annulus (Fig. 12.3c) or a cylindrical surface (Fig. 12.3b), which are the same from a topological point of view. (These two surfaces are homeomorphic.) But if the ribbon is glued together along the arrows shown in Fig. 12.4a, we obtain a surface in $\mathbb{R}^3$ (Fig. 12.4b) called *a Möbius band*.[5]

Local coordinates on this surface can be naturally introduced using the coordinates on the plane in which the original rectangle lies.

Example 6 Comparing the results of Examples 4 and 5 in accordance with the natural analogy, one can now prescribe how to glue a rectangle (Fig. 12.5a) that combines elements of the torus and elements of the Möbius band. But, just as it was necessary to go outside $\mathbb{R}^2$ in order to glue the Möbius band without tearing or self-intersections, the gluing prescribed here cannot be carried out in $\mathbb{R}^3$. However, this can be done in $\mathbb{R}^4$, resulting in a surface in $\mathbb{R}^4$ usually called the *Klein bottle*.[6] An attempt to depict this surface has been undertaken in Fig. 12.5b.

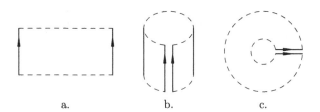

a.   b.   c.

Fig. 12.3

---

[5]A.F. Möbius (1790–1868) – German mathematician and astronomer.

[6]F.Ch. Klein (1849–1925) – outstanding German mathematician, the first to make a rigorous investigation of non-Euclidean geometry. An expert in the history of mathematics and one of the organizers of the "Encyclopädie der mathematischen Wisaenschaftm".

Fig. 12.4

a.   b.

Fig. 12.5

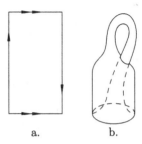

a.   b.

This last example gives some idea of how a surface can be intrinsically described more easily than the same surface lying in a particular space $\mathbb{R}^n$. Moreover, many important surfaces (of different dimensions) originally arise not as subsets of $\mathbb{R}^n$, but, for example, as the phase spaces of mechanical systems or the geometric image of continuous transformation groups of automorphisms, as the quotient spaces with respect to groups of automorphisms of the original space, and so on, and so forth. We confine ourselves for the time being to these introductory remarks, waiting to make them more precise until Chap. 15, where we shall give a general definition of a surface not necessarily lying in $\mathbb{R}^n$. But already at this point, before the definition has even been given, we note that by a well-known theorem of Whitney[7] any k-dimensional surface can be mapped homeomorphically onto a surface lying in $\mathbb{R}^{2k+1}$. Hence in considering surfaces in $\mathbb{R}^n$ we really lose nothing from the point of view of topological variety and classification. These questions, however, are somewhat off the topic of our modest requirements in geometry.

12.1.1 Problems and Exercises

1. For each of the sets E_α given by the conditions

$$E_\alpha = \left\{ (x, y) \in \mathbb{R}^2 \mid x^2 - y^2 = \alpha \right\},$$

$$E_\alpha = \left\{ (x, y, z) \in \mathbb{R}^3 \mid x^2 - y^2 = \alpha \right\},$$

$$E_\alpha = \left\{ (x, y, z) \in \mathbb{R}^3 \mid x^2 + y^2 - z^2 = \alpha \right\},$$

$$E_\alpha = \left\{ z \in \mathbb{C} \mid \left| z^2 - 1 \right| = \alpha \right\},$$

---

[7]H. Whitney (1907–1989) – American topologist, one of the founders of the theory of fiber bundles.

depending on the value of the parameter $\alpha \in \mathbb{R}$, determine

    a) whether E_α is a surface;
    b) if so, what the dimension of E_α is;
    c) whether E_α is connected.

2. Let $f : \mathbb{R}^n \to \mathbb{R}^n$ be a smooth mapping satisfying the condition $f \circ f = f$.

    a) Show that the set $f(\mathbb{R}^n)$ is a smooth surface in $\mathbb{R}^n$.
    b) By what property of the mapping f is the dimension of this surface determined?

3. Let $e_0, e_1, \ldots, e_n$ be an orthonormal basis in the Euclidean space $\mathbb{R}^{n+1}$, let $x = x^0 e_0 + x^1 e_1 + \cdots + x^n e_n$, let $\{x\}$ be the point $(x^0, x^1, \ldots, x^n)$, and let $e_1, \ldots, e_n$ be a basis in $\mathbb{R}^n \subset \mathbb{R}^{n+1}$.
The formulas

$$\psi_1 = \frac{x - x^0 e_0}{1 - x^0} \quad \text{for } x \neq e_0, \qquad \psi_2 = \frac{x - x^0 e_0}{1 + x^0} \quad \text{for } x \neq -e_0$$

define the stereographic projections

$$\psi_1 : S^n \backslash \{e_0\} \to \mathbb{R}^n, \qquad \psi : S^n \backslash \{-e_0\} \to \mathbb{R}^n$$

from the points $\{e_0\}$ and $\{-e_0\}$ respectively.

    a) Determine the geometric meaning of these mappings.
    b) Verify that if $t \in \mathbb{R}^n$ and $t \neq 0$, then $(\psi_2 \circ \psi_1^{-1})(t) = \frac{t}{|t|^2}$, where $\psi_1^{-1} = (\psi_1|_{S_n \backslash \{e_0\}})^{-1}$.
    c) Show that the two charts $\psi_1^{-1} = \varphi_1 : \mathbb{R}^n \to S^n \backslash \{e_0\}$ and $\psi_2^{-1} = \varphi_2 : \mathbb{R}^n \to S^n \backslash \{-e_0\}$ form an atlas of the sphere $S^n \subset \mathbb{R}^{n+1}$.
    d) Prove that every atlas of the sphere must have at least two charts.

12.2 Orientation of a Surface

We recall first of all that the transition from one frame $\mathbf{e}_1, \ldots, \mathbf{e}_n$ in $\mathbb{R}^n$ to a second frame $\tilde{\mathbf{e}}_1, \ldots, \tilde{\mathbf{e}}_n$ is effected by means of the square matrix obtained from the expansions $\tilde{\mathbf{e}}_j = a^i_j \mathbf{e}_i$. The determinant of this matrix is always nonzero, and the set of all frames divides into two equivalence classes, each class containing all possible frames such that for any two of them the determinant of the transition matrix is positive. Such equivalence classes are called *orientation classes of frames* in $\mathbb{R}^n$.

    To define an orientation means to fix one of these orientation classes. Thus, the *oriented space* $\mathbb{R}^n$ is the space $\mathbb{R}^n$ itself together with a fixed orientation class of frames. To specify the orientation class it suffices to exhibit any of the frames in it, so that one can also say that the oriented space $\mathbb{R}^n$ is $\mathbb{R}^n$ together with a fixed frame in it.

Fig. 12.6

A frame in $\mathbb{R}^n$ generates a coordinate system in $\mathbb{R}^n$, and the transition from one such coordinate system to another is effected by the matrix (a_i^j) that is the transpose of the matrix (a_j^i) that connects the two frames. Since the determinants of these two matrices are the same, everything that was said above about orientation can be repeated on the level of *orientation classes of coordinate systems in* $\mathbb{R}^n$, placing in one class all the coordinate systems such that the transition matrix between any two systems in the same class has a positive Jacobian.

Both of these essentially identical approaches to describing the concept of an orientation in $\mathbb{R}^n$ will also manifest themselves in describing the orientation of a surface, to which we now turn.

We recall, however, another connection between coordinates and frames in the case of curvilinear coordinate systems, a connection that will be useful in what is to follow.

Let G and D be diffeomorphic domains lying in two copies of the space $\mathbb{R}^n$ endowed with Cartesian coordinates $(x^1, \ldots, x^n)$ and $(t^1, \ldots, t^n)$ respectively. A diffeomorphism $\varphi : D \to G$ can be regarded as the introduction of curvilinear coordinates $(t^1, \ldots, t^n)$ into the domain G via the rule $x = \varphi(t)$, that is, the point $x \in G$ is endowed with the Cartesian coordinates $(t^1, \ldots, t^n)$ of the point $t = \varphi^{-1}(x) \in D$. If we consider a frame $\mathbf{e}_1, \ldots, \mathbf{e}_n$ of the tangent space $T\mathbb{R}_t^n$ at each point $t \in D$ composed of the unit vectors along the coordinate directions, a field of frames arises in D, which can be regarded as the translations of the orthogonal frame of the original space $\mathbb{R}^n$ containing D, parallel to itself, to the points of D. Since $\varphi : D \to G$ is a diffeomorphism, the mapping $\varphi'(t) : TD_t \to TG_{x=\varphi(t)}$ of tangent spaces effected by the rule $TD_t \ni \mathbf{e} \mapsto \varphi'(t)\mathbf{e} = \boldsymbol{\xi} \in TG_x$, is an isomorphism of the tangent spaces at each point t. Hence from the frame $\mathbf{e}_1, \ldots, \mathbf{e}_n$ in TD_t we obtain a frame $\boldsymbol{\xi}_1 = \varphi'(t)\mathbf{e}_1, \ldots, \boldsymbol{\xi}_n = \varphi'(t)\mathbf{e}_n$ in TG_x, and the field of frames on D transforms into a field of frames on G (see Fig. 12.6). Since $\varphi \in C^{(1)}(D, G)$, the vector field $\boldsymbol{\xi}(x) = \boldsymbol{\xi}(\varphi(t)) = \varphi'(t)\mathbf{e}(t)$ is continuous in G if the vector field $\mathbf{e}(t)$ is continuous in D. Thus every continuous field of frames (consisting of n continuous vector fields) transforms under a diffeomorphism to a continuous field of frames. Now let us consider a pair of diffeomorphisms $\varphi_i : D_i \to G$, $i = 1, 2$, which introduce two systems of curvilinear coordinates $(t_1^1, \ldots, t_1^n)$ and $(t_2^1, \ldots, t_2^n)$ into the same domain G. The mutually inverse diffeomorphisms $\varphi_2^{-1} \circ \varphi_1 : D_1 \to D_2$ and $\varphi_1^{-1} \circ \varphi_2 : D_2 \to D_1$ provide mutual transitions between these coordinate systems. The Jacobians of these mappings at corresponding points of D_1 and D_2 are mutually inverse to each other and consequently have the same sign. If the domain G (and together with it D_1 and D_2) is connected, then by the continuity and nonvanishing of the Jacobians under consideration, they have the same sign at all points of the domains D_1 and D_2 respectively.

Hence the set of all curvilinear coordinate systems introduced in a connected domain G by this method divide into exactly two equivalence classes when each class is assigned systems whose mutual transitions are effected with a positive Jacobian. Such equivalence classes are called the *orientation classes of curvilinear coordinate systems* in G.

To define an orientation in G means by definition to fix an orientation class of its curvilinear coordinate systems.

It is not difficult to verify that curvilinear coordinate systems belonging to the same orientation class generate continuous fields of frames in G (as described above) that are in the same orientation class of the tangent space TG_x at each point $x \in G$. It can be shown in general that, if G is connected, the continuous fields of frames on G divide into exactly two equivalence classes if each class is assigned the fields whose frames belong to the same orientation class of frames of the space TG_x at each point $x \in G$ (in this connection, see Problems 3 and 4 at the end of this section).

Thus the same orientation of a domain G can be defined in two completely equivalent ways: by exhibiting a curvilinear coordinate system in G, or by defining any continuous field of frames in G, all belonging to the same orientation class as the field of frames generated by this coordinate system.

It is now clear that the orientation of a connected domain G is completely determined if a frame that orients TG_x is prescribed at even one point $x \in G$. This circumstance is widely used in practice. If such an *orienting frame* is defined at some point $x_0 \in G$, and a curvilinear coordinate system $\varphi : D \to G$ is taken in G, then after constructing the frame induced by this coordinate system in TG_{x_0}, we compare it with the orienting frame in TG_x. If the two frames both belong to the same orientation class of TG_{x_0}, we regard the curvilinear coordinates as defining the same orientation on G as the orienting frame. Otherwise, we regard them as defining the opposite orientation.

If G is an open set, not necessarily connected, since what has just been said is applicable to any connected component of G, it is necessary to define an orienting frame in each component of G in order to orient G. Hence, if there are m components, the set G admits 2^m different orientations.

What has just been said about the orientation of a domain $G \subset \mathbb{R}^n$ can be repeated verbatim if instead of the domain G we consider a smooth k-dimensional surface S in $\mathbb{R}^n$ defined by a single chart (see Fig. 12.7). In this case the curvilinear coordinate systems on S also divide naturally into two orientation classes in accordance with the sign of the Jacobian of their mutual transition transformations; fields of frames also arise on S; and the orientation can also be defined by an orienting frame in some tangent plane TS_{x_0} to S.

The only new element that arises here and requires verification is the implicitly occurring proposition that follows.

Fig. 12.7

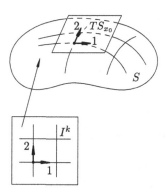

Proposition 1 *The mutual transitions from one curvilinear coordinate system to another on a smooth surface $S \subset \mathbb{R}^n$ are diffeomorphisms of the same degree of smoothness as the charts of the surface.*

Proof In fact, by the proposition in Sect. 12.1, we can regard any chart $I^k \to U \subset S$ locally as the restriction to $I^k \cap O(t)$ of a diffeomorphism $\mathcal{F} : O(t) \to O(x)$ from some n-dimensional neighborhood $O(t)$ of the point $t \in I^k \subset \mathbb{R}^n$ to an n-dimensional neighborhood $O(x)$ of $x \in S \subset \mathbb{R}^n$, $\mathcal{F}$ being of the same degree of smoothness as φ. If now $\varphi_1 : I_1^k \to U_1$ and $\varphi_2 : I_2^k \to U_2$ are two such charts, then the action of the mapping $\varphi_2^{-1} \circ \varphi_1$ (the transition from the first coordinate system to the second) which arises in the common domain of action can be represented locally as $\varphi_2^{-1} \circ \varphi_1(t^1, \dots, t^k) = \mathcal{F}_2^{-1} \circ \mathcal{F}(t^1, \dots, t^k, 0, \dots, 0)$, where $\mathcal{F}_1$ and $\mathcal{F}_2$ are the corresponding diffeomorphisms of the n-dimensional neighborhoods. $\quad\square$

We have studied all the essential components of the concept of an orientation of a surface using the example of an elementary surface defined by a single chart. We now finish up this business with the final definitions relating to the case of an arbitrary smooth surface in $\mathbb{R}^n$.

Let S be a smooth k-dimensional surface in $\mathbb{R}^n$, and let $\varphi_i : I_i^k \to U_i$, $\varphi_j : I_j^k \to U_j$ be two local charts of the surface S whose domains of action intersect, that is, $U_i \cap U_j \neq \varnothing$. Then between the sets $I_{ij}^k = \varphi_i^{-1}(U_j)$ and $I_{ji}^k = \varphi_j^{-1}(U_i)$, as was just proved, there are natural mutually inverse diffeomorphisms $\varphi_{ij} : I_{ij}^k \to I_{ji}^k$ and $\varphi_{ji} : I_{ji}^k \to I_{ij}^k$ that realize the transition from one local curvilinear coordinate system on S to the other.

Definition 1 Two local charts of a surface are *consistent* if their domains of action either do not intersect, or have a nonempty intersection for which the mutual transitions are effected by diffeomorphisms with positive Jacobian in their common domain of action.

Definition 2 An atlas of a surface is an *orienting atlas of the surface* if it consists of pairwise consistent charts.

Definition 3 A surface is *orientable* if it has an orienting atlas. Otherwise it is *nonorientable*.

In contrast to domains of $\mathbb{R}^n$ or elementary surfaces defined by a single chart, an arbitrary surface may turn out to be nonorientable.

Example 1 The Möbius band, as one can verify (see Problems 2 and 3 at the end of this section), is a nonorientable surface.

Example 2 The Klein bottle is also a nonorientable surface, since it contains a Möbius band. This last fact can be seen immediately from the construction of the Klein bottle shown in Fig. 12.5.

Example 3 A circle and in general a k-dimensional sphere are orientable, as can be proved by exhibiting directly an atlas of the sphere consisting of consistent charts (see Example 2 of Sect. 12.1).

Example 4 The two-dimensional torus studied in Example 4 of Sect. 12.1 is also an orientable surface. Indeed, using the parametric equations of the torus exhibited in Example 4 of Sect. 12.1, one can easily exhibit an orienting atlas for it.

We shall not go into detail, since a more visualizable method of controlling the orientability of sufficiently simple surfaces will be exhibited below, making it easy to verify the assertions in Examples 1–4.

The formal description of the concept of orientation of a surface will be finished if we add Definitions 4 and 5 below to Definitions 1, 2, and 3.

Two orienting atlases of a surface are *equivalent* if their union is also an orienting atlas of the surface.

This relation is indeed an equivalence relation between orienting atlases of an orientable surface.

Definition 4 An equivalence class of orienting atlases of a surface under this relation is called an *orientation class of atlases* or simply an *orientation of the surface*.

Definition 5 An *oriented surface* is a surface with a fixed orientation class of atlases (that is, a fixed orientation of the surface).

Thus *orienting a surface* means exhibiting a particular orientation class of orienting atlases of the surface by some means or other.

Some special manifestations of the following proposition are already familiar to us.

Proposition 2 *There exist precisely two orientations on a connected orientable surface.*

Fig. 12.8

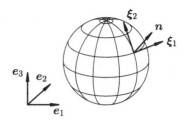

They are usually called *opposite orientations*.

The proof of Proposition 2 will be given in Sect. 15.2.3.

If an orientable surface is connected, an orientation of it can be defined by specifying any local chart of the surface or an orienting frame in any of its tangent planes. This fact is widely used in practice.

When a surface has more than one connected component, such a local chart or frame is naturally to be exhibited in each component.

The following way of defining an orientation of a surface embedded in a space that already carries an orientation is widely used in practice. Let S be an orientable $(n-1)$-dimensional surface embedded in the Euclidean space $\mathbb{R}^n$ with a fixed orienting frame $\mathbf{e}_1, \ldots, \mathbf{e}_n$ in $\mathbb{R}^n$. Let TS_x be the $(n-1)$-dimensional plane tangent to S at $x \in S$, and $\mathbf{n}$ the vector orthogonal to TS_x, that is, the vector normal to the surface S at x. If we agree that for the given vector $\mathbf{n}$ the frame $\boldsymbol{\xi}_1, \ldots, \boldsymbol{\xi}_{n-1}$ is to be chosen in TS_x so that the frames $(\mathbf{e}_1, \ldots, \mathbf{e}_n)$ and $(\mathbf{n}, \boldsymbol{\xi}_1, \ldots, \boldsymbol{\xi}_{n-1}) = (\tilde{\mathbf{e}}_1, \ldots, \tilde{\mathbf{e}}_n)$ belong to the same orientation class on $\mathbb{R}^n$, then, as one can easily see, such frames $(\boldsymbol{\xi}_1, \ldots, \boldsymbol{\xi}_n)$ of the plane TS_x will themselves all turn out to belong to the same orientation class for this plane. Hence in this case defining an orientation class for TS_x and along with it an orientation on a connected orientable surface can be done by defining the normal vector $\mathbf{n}$ (Fig. 12.8).

It is not difficult to verify (see Problem 4) that the orientability of an $(n-1)$-dimensional surface embedded in the Euclidean space $\mathbb{R}^n$ is equivalent to the existence of a continuous field of nonzero normal vectors on the surface.

Hence, in particular, the orientability of the sphere and the torus follow obviously, as does the nonorientability of the Möbius band, as was stated in Examples 1–4.

In geometry the connected $(n-1)$-dimensional surfaces in the Euclidean space $\mathbb{R}^n$ on which there exists a (single-valued) continuous field of unit normal vectors are called *two-sided*.

Thus, for example, a sphere, torus, or plane in $\mathbb{R}^3$ is a two-sided surface, in contrast to the Möbius band, which is a *one-sided surface* in this sense.

To finish our discussion of the concept of orientation of a surface, we make several remarks on the practical use of this concept in analysis.

In the computations that are connected in analysis with oriented surfaces in $\mathbb{R}^n$ one usually finds first some local parametrization of the surface S without bothering about orientation. In some tangent plane TS_x to the surface one then constructs a frame $\boldsymbol{\xi}_1, \ldots, \boldsymbol{\xi}_{n-1}$ consisting of (velocity) vectors tangent to the coordinate lines of a chosen curvilinear coordinate system, that is, the orienting frame induced by this coordinate system.

If the space $\mathbb{R}^n$ has been oriented and an orientation of S has been defined by a field of normal vectors, one chooses the vector **n** of the given field at the point x and compares the frame $\mathbf{n}, \boldsymbol{\xi}_1, \ldots, \boldsymbol{\xi}_{n-1}$ with the frame $\mathbf{e}_1, \ldots, \mathbf{e}_n$ that orients the space. If these are in the same orientation class, the local chart defines the required orientation of the surface in accordance with our convention. If these two frames are inconsistent, the chosen chart defines an orientation of the surface opposite to the one prescribed by the normal **n**.

It is clear that when there is a local chart of an $(n-1)$-dimensional surface, one can obtain a local chart of the required orientation (the one prescribed by the fixed normal vector **n** to the two-sided hypersurface embedded in the oriented space $\mathbb{R}^n$) by a simple change in the order of the coordinates.

In the one-dimensional case, in which a surface is a curve, the orientation is more often defined by the tangent vector to the curve at some point; in that case we often say *the direction of motion along the curve* rather than "the orientation of the curve".

If an orienting frame has been chosen in $\mathbb{R}^2$ and a closed curve is given, the *positive direction of circuit* around the domain D bounded by the curve is taken to be the direction such that the frame **n**, **v**, where **n** is the exterior normal to the curve with respect to D and **v** is the velocity of the motion, is consistent with the orienting frame in $\mathbb{R}^2$.

This means, for example, that for the traditional frame drawn in the plane a positive circuit is "counterclockwise", in which the domain is always "on the left".

In this connection the orientation of the plane itself or of a portion of the plane is often defined by giving the positive direction along some closed curve, usually a circle, rather than a frame in $\mathbb{R}^2$.

Defining such a direction amounts to exhibiting the direction of shortest rotation from the first vector in the frame until it coincides with the second, which is equivalent to defining an orientation class of frames on the plane.

12.2.1 Problems and Exercises

1. Is the atlas of the sphere exhibited in Problem 3c) of Sect. 12.1 an orienting atlas of the sphere?

2. a) Using Example 4 of Sect. 12.1, exhibit an orienting atlas of the two-dimensional torus.

b) Prove that there does not exist an orienting atlas for the Möbius band.

c) Show that under a diffeomorphism $f : D \to \tilde{D}$ an orientable surface $S \subset D$ maps to an orientable surface $\tilde{S} \subset \tilde{D}$.

3. a) Verify that the curvilinear coordinate systems on a domain $G \subset \mathbb{R}^n$ belonging to the same orientation class generate continuous fields of frames in G that determine frames of the same orientation class on the space TG_x at each point $x \in G$.

b) Show that in a connected domain $G \subset \mathbb{R}^n$ the continuous fields of frames divide into exactly two orientation classes.

c) Use the example of the sphere to show that a smooth surface $S \subset \mathbb{R}^n$ may be orientable even those there is no continuous field of frames in the tangent spaces to S.

d) Prove that on a connected orientable surface one can define exactly two different orientations.

4. a) A subspace $\mathbb{R}^{n-1}$ has been fixed, a vector $\mathbf{v} \in \mathbb{R}^n \backslash \mathbb{R}^{n-1}$ has been chosen, along with two frames $(\boldsymbol{\xi}_1, \ldots, \boldsymbol{\xi}_{n-1})$ and $(\widetilde{\boldsymbol{\xi}}_1, \ldots, \widetilde{\boldsymbol{\xi}}_{n-1})$ of the subspace $\mathbb{R}^{n-1}$. Verify that these frames belong to the same orientation class of frames of $\mathbb{R}^{n-1}$ if and only if the frames $(\mathbf{v}, \boldsymbol{\xi}_1, \ldots, \boldsymbol{\xi}_{n-1})$ and $(\mathbf{v}, \widetilde{\boldsymbol{\xi}}_1, \ldots, \widetilde{\boldsymbol{\xi}}_{n-1})$ define the same orientation on $\mathbb{R}^n$.

b) Show that a smooth hypersurface $S \subset \mathbb{R}^n$ is orientable if and only if there exists a continuous field of unit normal vectors to S. Hence, in particular, it follows that a two-sided surface is orientable.

c) Show that if $\operatorname{grad} F \neq 0$, then the surface defined by $F(x^1, \ldots, x^m) = 0$ is orientable (assuming that the equation has solutions).

d) Generalize the preceding result to the case of a surface defined by a system of equations.

e) Explain why not every smooth two-dimensional surface in $\mathbb{R}^3$ can be defined by an equation $F(x, y, z) = 0$, where F is a smooth surface having no critical points (a surface for which $\operatorname{grad} F \neq 0$ at all points).

12.3 The Boundary of a Surface and Its Orientation

12.3.1 Surfaces with Boundary

Let $\mathbb{R}^k$ be a Euclidean space of dimension k endowed with Cartesian coordinates $t^1, \ldots, t^k$. Consider the half-space $H^k := \{t \in \mathbb{R}^k \mid t^1 \leq 0\}$ of the space $\mathbb{R}^k$. The hyperplane $\partial H^k := \{t \in \mathbb{R}^k \mid t^1 = 0\}$ will be called the *boundary* of the half-space H^k.

We remark that the set $\mathring{H}^k := H^k \backslash \partial H^k$, that is, the open part of H^k, is an elementary k-dimensional surface. The half-space H^k itself does not formally satisfy the definition of a surface because of the presence of the boundary points from ∂H^k. The set H^k is the standard model for surfaces with boundary, which we shall now describe.

Definition 1 A set $S \subset \mathbb{R}^n$ is a (k-dimensional) *surface with boundary* if every point $x \in S$ has a neighborhood U in S homeomorphic either to $\mathbb{R}^k$ or to H^k.

Definition 2 If a point $x \in U$ corresponds to a point of the boundary ∂H^k under the homeomorphism of Definition 1, then x is called a *boundary point* of the surface (with boundary) S and of its neighborhood U. The set of all such boundary points is called the *boundary of the surface S*.

As a rule, the boundary of a surface S will be denoted ∂S. We note that for $k = 1$ the space ∂H^k consists of a single point. Hence, preserving the relation $\partial H^k = \mathbb{R}^{k-1}$, we shall from now on take $\mathbb{R}^0$ to consist of a single point and regard $\partial \mathbb{R}^0$ as the empty set.

We recall that under a homeomorphism $\varphi_{ij} : G_i \to G_j$ of the domain $G_i \subset \mathbb{R}^k$ onto the domain $G_j \subset \mathbb{R}^k$ the interior points of G_i map to interior points of the image $\varphi_{ij}(G_i)$ (this is a theorem of Brouwer). Consequently, the concept of a boundary point of the surface is independent of the choice of the local chart, that is, the concept is well defined.

Formally Definition 1 includes the case of the surface described in Definition 1 of Sect. 12.1. Comparing these definitions, we see that if S has no boundary points, we return to our previous definition of a surface, which can now be regarded as the definition of a surface without boundary. In this connection we note that the term "surface with boundary" is normally used when the set of boundary points is nonempty.

The concept of a smooth surface S (of class $C^{(m)}$) with boundary can be introduced, as for surfaces without boundary, by requiring that S have an atlas of charts of the given smoothness class. When doing this we assume that for charts of the form $\varphi : H^k \to U$ the partial derivatives of φ are computed at points of the boundary ∂H^k only over the domain H^k of definition of the mapping φ, that is, these derivatives are sometimes one-sided, and that the Jacobian of the mapping φ is nonzero throughout H^k.

Since $\mathbb{R}^k$ can be mapped to the cube $I^k = \{t \in \mathbb{R}^k \mid |t^i| < 1, i = 1, \ldots, k\}$ by a diffeomorphism of class $C^{(\infty)}$ and in such a way that H^k maps to the portion I_H^k of the cube I^k defined by the additional condition $t^1 \le 0$, it is clear that in the definition of a surface with boundary (even a smooth one) we could have replaced $\mathbb{R}^k$ by I^k and H^k by I_H^k or by the cube $\widetilde{I}^k$ with one of its faces attached: $\widetilde{I}^{k-1} := \{t \in \mathbb{R}^k \mid t^1 = 1, |t^i| < 1, i = 2, \ldots, k\}$, which is obviously a cube of dimension one less.

Taking account of this always-available freedom in the choice of canonical local charts of a surface, comparing Definitions 1 and 2 and Definition 1 of Sect. 12.1, we see that the following proposition holds.

Proposition 1 *The boundary of a k-dimensional surface of class $C^{(m)}$ is itself a surface of the same smoothness class, and is a surface without boundary having dimension one less than the dimension of the original surface with boundary.*

Proof Indeed, if $A(S) = \{(H^k, \varphi_i, U_i)\} \cup \{(\mathbb{R}^k, \varphi_j, U_j)\}$ is an atlas for the surface S with boundary, then $A(\partial S) = \{(\mathbb{R}^{k-1}, \varphi_i|_{\partial H^k = \mathbb{R}^{k-1}}, \partial U_i)\}$ is obviously an atlas of the same smoothness class for ∂S.   □

We now give some simple examples of surfaces with boundary.

Example 1 A closed n-dimensional ball $\overline{B}^n$ in $\mathbb{R}^n$ is an n-dimensional surface with boundary. Its boundary $\partial \overline{B}^n$ is the $(n-1)$-dimensional sphere (see Figs. 12.8 and 12.9a). The ball $\overline{B}^n$, which is often called in analogy with the two-dimensional case an *n-dimensional disk*, can be homeomorphically mapped to half of an

Fig. 12.9

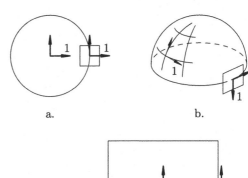

a.   b.

Fig. 12.10

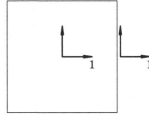

Fig. 12.11

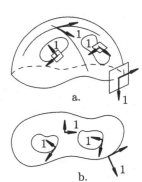

a.

b.

n-dimensional sphere whose boundary is the equatorial $(n-1)$-dimensional sphere (Fig. 12.9b).

Example 2 The closed cube $\overline{I}^n$ in $\mathbb{R}^n$ can be homeomorphically mapped to the closed ball $\partial \overline{B}^n$ along rays emanating from its center. Consequently $\overline{I}^n$, like $\overline{B}^n$ is an n-dimensional surface with boundary, which in this case is formed by the faces of the cube (Fig. 12.10). We note that on the edges, which are the intersections of the faces, it is obvious that no mapping of the cube onto the ball can be regular (that is, smooth and of rank n).

Example 3 If the Möbius band is obtained by gluing together two opposite sides of a closed rectangle, as described in Example 5 of Sect. 12.1, the result is obviously a surface with boundary in $\mathbb{R}^3$, and the boundary is homeomorphic to a circle (to be sure, the circle is not knotted in $\mathbb{R}^3$).

Under the other possible gluing of these sides the result is a cylindrical surface whose boundary consists of two circles. This surface is homeomorphic to the usual planar annulus (see Fig. 12.3 and Example 5 of Sect. 12.1). Figures 12.11a, 12.11b,

Fig. 12.12

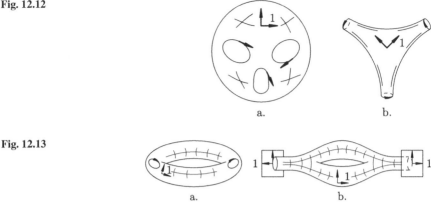

a.   b.

Fig. 12.13

a.   b.

12.12a, 12.12b, 12.13a, and 12.13b, which we will use below, show pairwise home-omorphic surfaces with boundary embedded in $\mathbb{R}^2$ and $\mathbb{R}^3$. As one can see, the boundary of a surface may be disconnected, even when the surface itself is connected.

12.3.2 Making the Orientations of a Surface and Its Boundary Consistent

If an orienting orthoframe $\mathbf{e}_1, \ldots, \mathbf{e}_k$ that induces Cartesian coordinates $x^1, \ldots, x^k$ is fixed in $\mathbb{R}^k$, the vectors $\mathbf{e}_2, \ldots, \mathbf{e}_k$ define an orientation on the boundary $\partial H^k = \mathbb{R}^{k-1}$ of $\partial H^k = \{x \in \mathbb{R}^k \mid x^1 \leq 0\}$ which is regarded as the orientation of the half-space H^k consistent with the orientation of the half-space H^k given by the frame $\mathbf{e}_1, \ldots, \mathbf{e}_k$.

In the case $k = 1$ where $\partial H^k = \mathbb{R}^{k-1} = \mathbb{R}^0$ is a point, a special convention needs to be made as to how to orient the point. By definition, the point is oriented by assigning a sign $+$ or $-$ to it. In the case $\partial H^1 = \mathbb{R}^0$, we take $(\mathbb{R}^0, +)$, or more briefly $+\mathbb{R}^0$.

We now wish to determine what is meant in general by consistency of the orientation of a surface and its boundary. This is very important in carrying out computations connected with surface integrals, which will be discussed below.

We begin by verifying the following general proposition.

Proposition 2 *The boundary ∂S of a smooth orientable surface S is itself a smooth orientable surface (although possibly not connected).*

Proof After we take account of Proposition 1 all that remains is to verify that ∂S is orientable. We shall show that if $A(S) = \{(H^k, \varphi_i, U_i)\} \cup \{(\mathbb{R}^k, \varphi_j, U_j)\}$ is an orienting atlas for a surface S with boundary, then the atlas $A(\partial S) = \{(\mathbb{R}^{k-1}, \varphi_i|_{\partial H^k = \mathbb{R}^{k-1}}, \partial U_i)\}$ of the boundary also consists of pairwise consistent charts. To do this it obviously suffices to verify that if $\tilde{t} = \psi(t)$ is a diffeomorphism with positive Jacobian from an H^k-neighborhood $U_{H^k}(t_0)$ of the point t_0 in ∂H^k onto an H^k-neighborhood $\tilde{U}_{H^k}(\tilde{t}_0)$ of the point $\tilde{t}_0 \in \partial H^k$, then the mapping $\psi|_{\partial U_{H^k}(t_0)}$ from the H^k-neighborhood $U_{\partial H^k}(t_0) = \partial U_{H^k}(t_0)$ of $t_0 \in \partial H^k$ onto the H^k-neighborhood $\tilde{U}_{\partial H^k}(\tilde{t}_0) = \partial \tilde{U}_{H^k}(\tilde{t}_0)$ of $\tilde{t}_0 = \psi(t_0) \in \partial H^k$ also has a positive Jacobian.

We remark that at each point $t_0 = (0, t_0^2, \ldots, t_0^k) \in \partial H^k$ the Jacobian J of the mapping ψ has the form

$$
J(t_0) = \begin{vmatrix} \frac{\partial \psi^1}{\partial t^1} & 0 & \cdots & 0 \\ \frac{\partial \psi^2}{\partial t^1} & \frac{\partial \psi^2}{\partial t^2} & \cdots & \frac{\partial \psi^2}{\partial t^k} \\ \vdots & \vdots & \ddots & \vdots \\ \frac{\partial \psi^k}{\partial t^1} & \frac{\partial \psi^k}{\partial t^2} & \cdots & \frac{\partial \psi^k}{\partial t^k} \end{vmatrix} = \frac{\partial \psi^1}{\partial t^1} \begin{vmatrix} \frac{\partial \psi^2}{\partial t^2} & \cdots & \frac{\partial \psi^2}{\partial t^k} \\ \vdots & \ddots & \vdots \\ \frac{\partial \psi^k}{\partial t^2} & \cdots & \frac{\partial \psi^k}{\partial t^k} \end{vmatrix},
$$

since for $t^1 = 0$ we must also have $\tilde{t}^1 = \psi^1(0, t^2, \ldots, t^k) \equiv 0$ (boundary points map to boundary points under a diffeomorphism). It now remains only to remark that when $t^1 < 0$ we must also have $\tilde{t} = \psi^1(t^1, t^2, \ldots, t^k) < 0$ (since $\tilde{t} = \psi(t) \in H^k$), so that the value of $\frac{\partial \psi^1}{\partial t^1}(0, t^2, \ldots, t^k)$ cannot be negative. By hypothesis $J(t_0) > 0$, and since $\frac{\partial \psi^1}{\partial t^1}(0, t^2, \ldots, t^k) > 0$ it follows from the equality given above connecting the determinants that the Jacobian of the mapping $\psi|_{\partial U_{H^k}} = \psi(0, t^2, \ldots, t^k)$ is positive. $\square$

We note that the case of a one-dimensional surface ($k = 1$) in Proposition 2 and Definition 3 below must be handled by a special convention in accordance with the convention adopted at the beginning of this subsection.

Definition 3 If $A(S) = \{(H^k, \varphi_i, U_i)\} \cap \{(\mathbb{R}^k, \varphi_j, U_j)\}$ is an orienting atlas of standard local charts of the surface S with boundary ∂S, then $A(\partial S) = \{(\mathbb{R}^{k-1}, \varphi_i|_{\partial H^k = \mathbb{R}^{k-1}}, \partial U_i)\}$ is an orienting atlas for the boundary. The orientation of ∂S that it defines is said to be the orientation *consistent with the orientation of the surface*.

To finish our discussion of orientation of the boundary of an orientable surface, we make two useful remarks.

Remark 1 In practice, as already noted above, an orientation of a surface embedded in $\mathbb{R}^n$ is often defined by a frame of tangent vectors to the surface. For that reason, the verification of the consistency of the orientation of the surface and its boundary in this case can be carried out as follows. Take a k-dimensional plane TS_{x_0} tangent to the smooth surface S at the point $x_0 \in \partial S$. Since the local structure of S near x_0 is the same as the structure of the half-space H^k near $0 \in \partial H^k$, directing the first vector of the orthoframe $\boldsymbol{\xi}_1, \boldsymbol{\xi}_2, \ldots, \boldsymbol{\xi}_k \in TS_{x_0}$ along the normal to ∂S and in the direction exterior to the local projection of S on TS_{x_0}, we obtain a frame $\boldsymbol{\xi}_2, \ldots, \boldsymbol{\xi}_k$ in the

$(k-1)$-dimensional plane $T\partial S_{x_0}$ tangent to ∂S at x_0, which defines an orientation of $T\partial S_{x_0}$, and hence also of ∂S, consistent with orientation of the surface S defined by the given frame $\boldsymbol{\xi}_1, \boldsymbol{\xi}_2, \ldots, \boldsymbol{\xi}_k$.

Figures 12.9–12.12 show the process and the result of making the orientations of a surface and its boundary consistent using a simple example.

We note that this scheme presumes that it is possible to translate a frame that defines the orientation of S to different points of the surface and its boundary, which, as examples show, may be disconnected.

Remark 2 In the oriented space $\mathbb{R}^k$ we consider the half-space $H_-^k = H^k = \{x \in \mathbb{R}^k \mid x^1 \le 0\}$ and $H_+^k = \{x \in \mathbb{R}^k \mid x^1 \ge 0\}$ with the orientation induced from $\mathbb{R}^k$. The hyperplane $\Gamma = \{x \in \mathbb{R}^k \mid x^1 = 0\}$ is the common boundary of H_-^k and H_+^k. It is easy to see that the orientations of the hyperplane Γ consistent with the orientations of H_-^k and H_+^k are opposite to each other. This also applies to the case $k = 1$, by convention.

Similarly, if an oriented k-dimensional surface is cut by some $(k-1)$-dimensional surface (for example, a sphere intersected by its equator), two opposite orientations arise on the intersection, induced by the parts of the original surface adjacent to it.

This observation is often used in the theory of surface integrals.

In addition, it can be used to determine the orientability of a piecewise-smooth surface.

We begin by giving the definition of such a surface.

Definition 4 (Inductive definition of a piecewise-smooth surface) We agree to call a point a *zero-dimensional* surface of any smoothness class.

A *piecewise smooth one-dimensional surface* (piecewise smooth curve) is a curve in $\mathbb{R}^n$ which breaks into smooth one-dimensional surfaces (curves) when a finite or countable number of zero-dimensional surfaces are removed from it.

A surface $S \subset \mathbb{R}^n$ of dimension k is *piecewise smooth* if a finite or countable number of piecewise smooth surfaces of dimension at most $k-1$ can be removed from it in such a way that the remainder decomposes into smooth k-dimensional surfaces S_i (with boundary or without).

Example 4 The boundary of a plane angle and the boundary of a square are piecewise-smooth curves.

The boundary of a cube or the boundary of a right circular cone in $\mathbb{R}^3$ are two-dimensional piecewise-smooth surfaces.

Let us now return to the orientation of a piecewise-smooth surface.

A point (zero-dimensional surface), as already pointed out, is by convention oriented by ascribing the sign $+$ or $-$ to it. In particular, the boundary of a closed interval $[a, b] \subset \mathbb{R}$, which consists of the two points a and b is by convention consistent with the orientation of the closed interval from a to b if the orientation is $(a, -)$, $(b, +)$, or, in another notation, $-a, +b$.

Now let us consider a k-dimensional piecewise smooth surface $S \subset \mathbb{R}^n$ $(k > 0)$.

We assume that the two smooth surfaces S_{i_1} and S_{i_2} in Definition 4 are oriented and abut each other along a smooth portion Γ of a $(k-1)$-dimensional surface (edge). Orientations then arise on Γ, which is a boundary, consistent with the orientations of S_{i_1} and S_{i_2}. If these two orientations are opposite on every edge $\Gamma \subset \overline{S}_{i_1} \cap \overline{S}_{i_2}$, the original orientations of S_{i_1} and S_{i_2} are considered *consistent*. If $\overline{S}_{i_1} \cap \overline{S}_{i_2}$ is empty or has dimension less than $(k-1)$, all orientations of S_{i_1} and S_{i_2} are consistent.

Definition 5 A piecewise-smooth k-dimensional surface $(k > 0)$ will be considered *orientable* if up to a finite or countable number of piecewise-smooth surfaces of dimension at most $(k-1)$ it is the union of smooth orientable surfaces S_i any two of which have a mutually consistent orientation.

Example 5 The surface of a three-dimensional cube, as one can easily verify, is an orientable piecewise-smooth surface. In general, all the piecewise-smooth surfaces exhibited in Example 4 are orientable.

Example 6 The Möbius band can easily be represented as the union of two orientable smooth surfaces that abut along a piece of the boundary. But these surfaces cannot be oriented consistently. One can verify that the Möbius band is not an orientable surface, even from the point of view of Definition 5.

12.3.3 Problems and Exercises

1. a) Is it true that the boundary of a surface $S \subset \mathbb{R}^n$ is the set $\overline{S} \backslash S$, where $\overline{S}$ is the closure of S in $\mathbb{R}^n$?

b) Do the surfaces $S_1 = \{(x, y) \in \mathbb{R}^2 \mid 1 < x^2 + y^2 < 2\}$ and $S_2 = \{(x, y) \mid 0 < x^2 + y^2\}$ have a boundary?

c) Give the boundary of the surfaces $S_1 = \{(x, y) \in \mathbb{R}^2 \mid 1 \le x^2 + y^2 < 2\}$ and $S_2 = \{(x, y) \in \mathbb{R}^2 \mid 1 \le x^2 + y^2\}$.

2. Give an example of a nonorientable surface with an orientable boundary.

3. a) Each face $I^k = \{x \in \mathbb{R}^k \mid |x^i| < 1, i = 1, \ldots, k\}$ is parallel to the corresponding $(k-1)$-dimensional coordinate hyperplane in $\mathbb{R}^k$, so that one may consider the same frame and the same coordinate system in the face as in the hyperplane. On which faces is the resulting orientation consistent with the orientation of the cube I^k induced by the orientation of $\mathbb{R}^k$, and on which is it not consistent? Consider successively the cases $k = 2$, $k = 3$, and $k = n$.

b) The local chart $(t^1, t^2) \mapsto (\sin t^2 \cos t^2, \sin t^2 \sin t^2, \cos t^1)$ acts in a certain domain of the hemisphere $S = \{(x, y, z) \in \mathbb{R}^3 \mid x^2 + y^2 + z^2 = 1 \wedge z > 0\}$, and the local chart $t \mapsto (\cos t, \sin t, 0)$ acts in a certain domain of the boundary ∂S of this hemisphere. Determine whether these charts give a consistent orientation of the surface S and its boundary ∂S.

c) Construct the field of frames on the hemisphere S and its boundary ∂S induced by the local charts shown in b).

d) On the boundary ∂S of the hemisphere S exhibit a frame that defines the orientation of the boundary consistent with the orientation of the hemisphere given in c).

e) Define the orientation of the hemisphere S obtained in c) using a normal vector to $S \subset \mathbb{R}^3$.

4. a) Verify that the Möbius band is not an orientable surface even from the point of view of Definition 5.

b) Show that if S is a smooth surface in $\mathbb{R}^n$, determining its orientability as a smooth surface and as a piecewise-smooth surface are equivalent processes.

5. a) We shall say that a set $S \subset \mathbb{R}^n$ is a k-dimensional surface with boundary if for each point $x \in S$ there exists a neighborhood $U(x) \in \mathbb{R}^n$ and a diffeomorphism $\psi : U(x) \to I^n$ of this neighborhood onto the standard cube $I^n \subset \mathbb{R}^n$ under which $\psi(S \cap U(x))$ coincides either with the cube $I^k = \{t \in I^n \mid t^{k+1} = \cdots = t^n = 0\}$ or with a portion of it $I^k \cap \{t \in \mathbb{R}^n \mid t^k \le 0\}$ that is a k-dimensional open interval with one of its faces attached.

Based on what was said in Sect. 12.1 in the discussion of the concept of a surface, show that this definition of a surface with boundary is equivalent to Definition 1.

b) Is it true that if $f \in C^{(l)}(H^k, \mathbb{R})$, where $H^k = \{x \in \mathbb{R}^k \mid x^1 \le 0\}$, then for every point $x \in \partial H^k$ one can find a neighborhood of it $U(x)$ in $\mathbb{R}^k$ and a function $\mathcal{F} \in C^{(l)}(U(x), \mathbb{R})$ such that $\mathcal{F}|_{H^k \cap U(x)} = f|_{H^k \cap U(x)}$?

c) If the definition given in part a) is used to describe a smooth surface with boundary, that is, we regard ψ as a smooth mapping of maximal rank, will this definition of a smooth surface with boundary be the same as the one adopted in Sect. 12.3?

12.4 The Area of a Surface in Euclidean Space

We now turn to the problem of defining the area of a k-dimensional piecewise-smooth surface embedded in the Euclidean space $\mathbb{R}^n$, $n \ge k$.

We begin by recalling that if $\boldsymbol{\xi}_1, \dots, \boldsymbol{\xi}_k$ are k vectors in Euclidean space $\mathbb{R}^k$, then the volume $V(\boldsymbol{\xi}_1, \dots, \boldsymbol{\xi}_k)$ of the parallelepiped spanned by these vectors as edges can be computed as the determinant

$$V(\boldsymbol{\xi}_1, \dots, \boldsymbol{\xi}_k) = \det\left(\xi_i^j\right) \tag{12.6}$$

of the matrix $J = (\xi_i^j)$ whose rows are formed by the coordinates of these vectors in some orthonormal basis $\mathbf{e}_1, \dots, \mathbf{e}_k$ of $\mathbb{R}^k$. We note, however, that in actual fact formula (12.6) gives the so-called *oriented volume of the parallelepiped* rather than simply the volume. If $V \ne 0$, the value of V given by (12.6) is positive or negative according as the frames $\mathbf{e}_1, \dots, \mathbf{e}_k$ and $\boldsymbol{\xi}_1, \dots, \boldsymbol{\xi}_k$ belong to the same or opposite orientation classes of $\mathbb{R}^k$.

Fig. 12.14

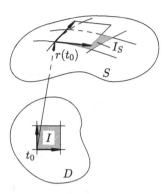

We now remark that the product JJ^* of the matrix J and its transpose J^* has elements that are none other than the matrix $G = (g_{ij})$ of pairwise inner products $g_{ij} = \langle \boldsymbol{\xi}_i, \boldsymbol{\xi}_j \rangle$ of these vectors, that is, the *Gram matrix*[8] of the system of vectors $\boldsymbol{\xi}_1, \ldots, \boldsymbol{\xi}_k$. Thus

$$\det G = \det(JJ^*) = \det J \det J^* = (\det J)^2, \tag{12.7}$$

and hence the nonnegative value of the volume $V(\boldsymbol{\xi}_1, \ldots, \boldsymbol{\xi}_k)$ can be obtained as

$$V(\boldsymbol{\xi}_1, \ldots, \boldsymbol{\xi}_k) = \sqrt{\det(\langle \boldsymbol{\xi}_i, \boldsymbol{\xi}_j \rangle)}. \tag{12.8}$$

This last formula is convenient in that it is essentially coordinate-free, containing only a set of geometric quantities that characterize the parallelepiped under consideration. In particular, if these same vectors $\boldsymbol{\xi}_1, \ldots, \boldsymbol{\xi}_k$ are regarded as embedded in n-dimensional Euclidean space $\mathbb{R}^n$ ($n \geq k$), formula (12.8) for the k-dimensional volume (or k-dimensional surface area) of the parallelepiped they span remains unchanged.

Now let $\mathbf{r} : D \to S \subset \mathbb{R}^n$ be a k-dimensional smooth surface S in the Euclidean space $\mathbb{R}^n$ defined in parametric form $\mathbf{r} = \mathbf{r}(t^1, \ldots, t^k)$, that is, as a smooth vector-valued function $\mathbf{r}(t) = (x^1, \ldots, x^n)(t)$ defined in the domain $D \subset \mathbb{R}^k$. Let $\mathbf{e}_1, \ldots, \mathbf{e}_k$ be the orthonormal basis in $\mathbb{R}^k$ that generates the coordinate system $(t^1, \ldots, t^k)$. After fixing a point $t_0 = (t_0^1, \ldots, t_0^k) \in D$, we take the positive numbers $h^1, \ldots, h^k$ to be so small that the parallelepiped I spanned by the vectors $h^i \mathbf{e}_i \in TD_{t_0}$, $i = 1, \ldots, k$, attached at the point t_0 is contained in D.

Under the mapping $D \to S$ a figure I_S on the surface S, which we may provisionally call a curvilinear parallelepiped, corresponds to the parallelepiped I (see Fig. 12.14, which corresponds to the case $k = 2$, $n = 3$). Since

$$\mathbf{r}(t_0^1, \ldots, t_0^{i-1}, t_0^i + h^i, t_0^{i+1}, \ldots, t_0^k) - \mathbf{r}(t_0^1, \ldots, t_0^{i-1}, t_0^i, t_0^{i+1}, \ldots, t_0^k)$$

$$= \frac{\partial \mathbf{r}}{\partial t^i}(t_0)h^i + o(h^i),$$

---

[8]See the footnote on p. 497.

a displacement in $\mathbb{R}^n$ from $\mathbf{r}(t_0)$ that can be replaced, up to $o(h^i)$, by the partial differential $\frac{\mathbf{r}}{\partial t^i}(t_0)h^i =: \dot{\mathbf{r}}_i h^i$ as $h^i \to 0$ corresponds to displacement from t_0 by $h^i \mathbf{e}_i$. Thus, for small values of h^i, $i = 1, \ldots, k$, the curvilinear parallelepiped I_S differs only slightly from the parallelepiped spanned by the vectors $h^1 \dot{\mathbf{r}}_1, \ldots, h^i \dot{\mathbf{r}}_k$ tangent to the surface S at $\mathbf{r}(t_0)$. Assuming on that basis that the volume ΔV of the curvilinear parallelepiped I_S must also be close to the volume of the standard parallelepiped just exhibited, we find the approximate formula

$$\Delta V \approx \sqrt{\det(g_{ij})(t_0)} \Delta t^1 \cdot \ldots \cdot \Delta t^k, \tag{12.9}$$

where we have set $g_{ij}(t_0) = \langle \dot{\mathbf{r}}_i, \dot{\mathbf{r}}_j \rangle (t_0)$ and $\Delta t^i = h^i$, $i, j = 1, \ldots, k$.

If we now tile the entire space $\mathbb{R}^k$ containing the parameter domain D with k-dimensional parallelepipeds of small diameter d, take the ones that are contained in D, compute an approximate value of the k-dimensional volume of their images using formula (12.9), and then sum the resulting values, we arrive at the quantity

$$\sum_\alpha \sqrt{\det(g_{ij})(t_\alpha)} \Delta t^1 \cdot \ldots \cdot \Delta t^k,$$

which can be regarded as an approximation to the k-dimensional volume or area of the surface S under consideration, and this approximate value should become more precise as $d \to 0$. Thus we adopt the following definition.

Definition 1 The *area* (or *k-dimensional volume*) of a smooth k-dimensional surface S given parametrically by $D \ni t \to \mathbf{r}(t) \in S$ and embedded in the Euclidean space $\mathbb{R}^n$ is the quantity

$$V_k(S) := \int_D \sqrt{\det(\langle \dot{\mathbf{r}}_i, \dot{\mathbf{r}}_j \rangle)} \, dt^1 \cdots dt^k. \tag{12.10}$$

Let us see how formula (12.10) looks in the cases that we already know about.

For $k = 1$ the domain $D \subset \mathbb{R}^1$ is an interval with certain endpoints a and b $(a < b)$ on the line $\mathbb{R}^1$, and S is a curve in $\mathbb{R}^n$ in this case. Thus for $k = 1$ formula (12.10) becomes the formula

$$V_1(S) = \int_a^b |\dot{\mathbf{r}}(t)| \, dt = \int_a^b \sqrt{(\dot{x}^1)^2 + \cdots + (\dot{x}^n)^2}(t) \, dt$$

for computing the length of a curve.

If $k = n$, then S is an n-dimensional domain in $\mathbb{R}^n$ diffeomorphic to D. In this case the Jacobian matrix $J = x'(t)$ of the mapping $D \ni (t^1, \ldots, t^n) = t \mapsto \mathbf{r}(t) = (x^1, \ldots, x^n)(t) \in S$ is a square matrix. Now using relation (12.7) and the formula for change of variable in a multiple integral, one can write

$$V_n(S) = \int_D \sqrt{\det G(t)} \, dt = \int_D |\det x'(t)| \, dt = \int_S dx = V(S).$$

That is, as one should have expected, we have arrived at the volume of the domain S in $\mathbb{R}^n$.

We note that for $k = 2$, $n = 3$, that is, when S is a two-dimensional surface in $\mathbb{R}^3$, one often replaces the standard notation $g_{ij} = \langle \dot{\mathbf{r}}_i, \dot{\mathbf{r}}_j \rangle$ by the following: $\sigma := V_2(S)$, $E := g_{11} = \langle \dot{\mathbf{r}}_1, \dot{\mathbf{r}}_1 \rangle$, $F := g_{12} = g_{21} = \langle \dot{\mathbf{r}}_1, \dot{\mathbf{r}}_2 \rangle$, $G := g_{22} = \langle \dot{\mathbf{r}}_2, \dot{\mathbf{r}}_2 \rangle$; and one writes u, v respectively instead of t^1, t^2. In this notation formula (12.10) assumes the form

$$\sigma = \iint_D \sqrt{EG - F^2}\, du\, dv.$$

In particular, if $u = x$, $v = y$, and the surface S is the graph of a smooth real-valued function $z = f(x, y)$ defined in a domain $D \subset \mathbb{R}^2$, then, as one can easily compute,

$$\sigma = \iint_D \sqrt{1 + \left(f_x'\right)^2 + \left(f_y'\right)^2}\, dx\, dy.$$

We now return once again to Definition 1 and make a number of remarks that will be useful later.

Remark 1 Definition 1 makes sense only when the integral on the right-hand side of (12.10) exists. It demonstrably exists, for example, if D is a Jordan-measurable domain and $\mathbf{r} \in C^{(1)}(\overline{D}, \mathbb{R}^n)$.

Remark 2 If the surface S in Definition 1 is partitioned into a finite number of surfaces $S_1, \ldots, S_m$ with piecewise smooth boundaries, the same kind of partition of the domain D into domains $D_1, \ldots, D_m$ corresponding to these surfaces will correspond to this partition. If the surface S had area in the sense of Eq. (12.10), then the quantities

$$V_k(S_\alpha) = \int_{D_\alpha} \sqrt{\det\langle \dot{\mathbf{r}}_i, \dot{\mathbf{r}}_j \rangle(t)}\, dt$$

are defined for each value of $\alpha = 1, \ldots, m$.

By the additivity of the integral, it follows that

$$V_k(S) = \sum_\alpha V_k(S_\alpha).$$

We have thus established that the area of a k-dimensional surface is additive in the same sense as the ordinary multiple integral.

Remark 3 This last remark allows us to exhaust the domain D when necessary, and thereby to extend the meaning of the formula (12.10), in which the integral may now be interpreted as an improper integral.

Remark 4 More importantly, the additivity of area can be used to define the area of an arbitrary smooth or even piecewise smooth surface (not necessarily given by a single chart).

Definition 2 Let S be an arbitrary piecewise smooth k-dimensional surface in $\mathbb{R}^n$. If, after a finite or countable number of piecewise smooth surfaces of dimension at most $k - 1$ are removed, it breaks up into a finite or countable number of smooth parametrized surfaces $S_1, \ldots, S_m, \ldots$, we set

$$V_k(S) := \sum_\alpha V_k(S_\alpha).$$

The additivity of the multiple integral makes it possible to verify that the quantity $V_k(S)$ so defined is independent of the way in which the surface S is partitioned into smooth pieces $S_1, \ldots, S_m, \ldots$, each of which is contained in the range of some local chart of the surface S.

We further remark that it follows easily from the definitions of smooth and piecewise smooth surfaces that the partition of S into parametrized pieces, as described in Definition 2, is always possible, and can even be done while observing the natural additional requirement that the partition be *locally finite*. The latter means that any compact set $K \subset S$ can intersect only a finite number of the surfaces $S_1, \ldots, S_m, \ldots$. This can be expressed more vividly in another way: every point of S must have a neighborhood that intersects at most a finite number of the sets $S_1, \ldots, S_m, \ldots$.

Remark 5 The basic formula (12.10) contains a system of curvilinear coordinates $t^1, \ldots, t^k$. For that reason, it is natural to verify that the quantity $V_k(S)$ defined by (12.10) (and thereby also the quantity $V_k(S)$ from Definition 2) is invariant under a diffeomorphic transition $\widetilde{D} \ni (\widetilde{t}^1, \ldots, \widetilde{t}^k) \mapsto t = (t^1, \ldots, t^k) \in D$ to new curvilinear coordinates $\widetilde{t}^1, \ldots, \widetilde{t}^k$ varying in the domain $\widetilde{D} \subset \mathbb{R}^k$.

Proof For the verification it suffices to remark that the matrices

$$G = (g_{ij}) = \left(\left\langle \frac{\partial \mathbf{r}}{\partial t^i}, \frac{\partial \mathbf{r}}{\partial t^j} \right\rangle \right) \quad \text{and} \quad \widetilde{G} = (\widetilde{g}_{ij}) = \left(\left\langle \frac{\partial \mathbf{r}}{\partial \widetilde{t}^i}, \frac{\partial \mathbf{r}}{\partial \widetilde{t}^j} \right\rangle \right)$$

at corresponding points of the domains D and $\widetilde{D}$ are connected by the relation $\widetilde{G} = J^* G J$, where $J = (\frac{\partial t^j}{\partial \widetilde{t}^i})$ is the Jacobian matrix of the mapping $\widetilde{D} \ni \widetilde{t} \mapsto t \in D$ and J^* is the transpose of the matrix J. Thus, $\det \widetilde{G}(\widetilde{t}) = \det G(t)(\det J)^2(\widetilde{t})$, from which it follows that

$$\int_D \sqrt{\det G(t)}\, dt = \int_{\widetilde{D}} \sqrt{\det G(t(\widetilde{t}))}\, |J(\widetilde{t})|\, d\widetilde{t} = \int_{\widetilde{D}} \sqrt{\det \widetilde{G}(\widetilde{t})}\, d\widetilde{t}. \qquad \square$$

Thus, we have given a definition of the k-dimensional volume or area of a k-dimensional piecewise-smooth surface that is independent of the choice of coordinate system.

Remark 6 We precede the remark with a definition.

Definition 3 A set E embedded in a k-dimensional piecewise-smooth surface S is *a set of k-dimensional measure zero* or *has area zero* in the Lebesgue sense if for every $\varepsilon > 0$ it can be covered by a finite or countable system $S_1, \ldots, S_m, \ldots$ of (possibly intersecting) surfaces $S_\alpha \subset S$ such that $\sum_\alpha V_k(S_\alpha) < \varepsilon$.

As one can see, this is a verbatim repetition of the definition of a set of measure zero in $\mathbb{R}^k$.

It is easy to see that in the parameter domain D of any local chart $\varphi : D \to S$ of a piecewise-smooth surface S the set $\varphi^{-1}(E) \subset D \subset \mathbb{R}^k$ of k-dimensional measure zero corresponds to such a set E. One can even verify that this is the characteristic property of sets $E \subset S$ of measure zero.

In the practical computation of areas and the surface integrals introduced below, it is useful to keep in mind that if a piecewise-smooth surface $\widetilde{S}$ has been obtained from a piecewise-smooth surface S by removing a set E of measure zero from S, then the areas of $\widetilde{S}$ and S are the same.

The usefulness of this remark lies in the fact that it is often easy to remove such a set of measure zero from a piecewise-smooth surface in such a way that the result is a smooth surface $\widetilde{S}$ defined by a single chart. But then the area of $\widetilde{S}$ and hence the area of S also can be computed directly by formula (12.10).

Let us consider some examples.

Example 1 The mapping $]0, 2\pi[\ni t \mapsto (R\cos t, R\sin t) \in \mathbb{R}^2$ is a chart for the arc $\widetilde{S}$ of the circle $x^2 + y^2 = R^2$ obtained by removing the single point $E = (R, 0)$ from that circle. Since E is a set of measure zero on S, we can write

$$V_1(S) = V_1(\widetilde{S}) = \int_0^{2\pi} \sqrt{R^2 \sin^2 t + R^2 \cos^2 t}\, dt = 2\pi R.$$

Example 2 In Example 4 of Sect. 12.1 we exhibited the following parametric representation of the two-dimensional torus S in $\mathbb{R}^3$:

$$\mathbf{r}(\varphi, \psi) = \big((b + a\cos\psi)\cos\varphi, (b + a\cos\psi)\sin\varphi, a\sin\psi\big).$$

In the domain $D = \{(\varphi, \psi) \mid 0 < \varphi < 2\pi, 0 < \psi < 2\pi\}$ the mapping $(\varphi, \psi) \mapsto \mathbf{r}(\varphi, \psi)$ is a diffeomorphism. The image $\widetilde{S}$ of the domain D under this diffeomorphism differs from the torus by the set E consisting of the coordinate line $\varphi = 2\pi$ and the line $\psi = 2\pi$. The set E thus consists of one parallel of latitude and one meridian of longitude of the torus, and, as one can easily see, has measure zero. Hence the area of the torus can be found by formula (12.10) starting from this parametric representation, considered within the domain D.

Let us carry out the necessary computations:

$$\dot{\mathbf{r}}_\varphi = \big(-(b + a\cos\psi)\sin\varphi, (b + a\cos\psi)\cos\varphi, 0\big),$$

$$\dot{\mathbf{r}}_\psi = (-a\sin\psi)\cos\varphi, -a\sin\psi\sin\varphi, a\cos\psi),$$

$$g_{11} = \langle \dot{\mathbf{r}}_\varphi, \dot{\mathbf{r}}_\varphi \rangle = (b + a\cos\psi)^2,$$

$$g_{12} = g_{21} = \langle \dot{\mathbf{r}}_\varphi, \dot{\mathbf{r}}_\psi \rangle = 0,$$

$$g_{22} = \langle \dot{\mathbf{r}}_\psi, \dot{\mathbf{r}}_\psi \rangle = a^2,$$

$$\det G = \begin{vmatrix} g_{11} & g_{12} \\ g_{21} & g_{22} \end{vmatrix} = a^2 (b + a\cos\psi)^2.$$

Consequently,

$$V_2(S) = V_2(\widetilde{S}) = \int_0^{2\pi} d\varphi \int_0^{2\pi} a(b + a\cos\psi)\, d\psi = 4\pi^2 ab.$$

In conclusion we note that the method indicated in Definition 2 can now be used to compute the areas of piecewise-smooth curves and surfaces.

12.4.1 Problems and Exercises

1. a) Let P and $\widetilde{P}$ be two hyperplanes in the Euclidean space $\mathbb{R}^n$, D a subdomain of P, and $\widetilde{D}$ the orthogonal projection of D on the hyperplane $\widetilde{P}$. Show that the $(n-1)$-dimensional areas of D and $\widetilde{D}$ are connected by the relation $\sigma(\widetilde{D}) = \sigma(D)\cos\alpha$, where α is the angle between the hyperplanes P and $\widetilde{P}$.

b) Taking account of the result of a), give the geometric meaning of the formula $d\sigma = \sqrt{1 + (f_x')^2 + (f_y')^2}\, dx\, dy$ for the element of area of the graph of a smooth function $z = f(x, y)$ in three-dimensional Euclidean space.

c) Show that if the surface S in Euclidean space $\mathbb{R}^3$ is defined as a smooth vector-valued function $\mathbf{r} = \mathbf{r}(u, v)$ defined in a domain $D \subset \mathbb{R}^2$, then the area of the surface S can be found by the formula

$$\sigma(S) = \iint_D \left| [\mathbf{r}_u', \mathbf{r}_v'] \right| du\, dv,$$

where $[\mathbf{r}_u', \mathbf{r}_v']$ is the vector product of $\frac{\partial \mathbf{r}}{\partial u}$ and $\frac{\partial \mathbf{r}}{\partial v}$.

d) Verify that if the surface $S \subset \mathbb{R}^3$ is defined by the equation $F(x, y, z) = 0$ and the domain U of the surface S projects orthogonally in a one-to-one manner onto the domain D of the xy-plane, we have the formula

$$\sigma(U) = \iint_D \frac{\operatorname{grad} F}{|F_z'|}\, dx\, dy.$$

2. Find the area of the spherical rectangle formed by two parallels of latitude and two meridians of longitude of the sphere $S \subset \mathbb{R}^3$.

3. a) Let (r, φ, h) be cylindrical coordinates in $\mathbb{R}^3$. A smooth curve lying in the plane $\varphi = \varphi_0$ and defined there by the equation $\mathbf{r} = \mathbf{r}(s)$, where s is the arc length parameter, is revolved about the h-axis. Show that the area of the surface obtained by revolving the piece of this curve corresponding to the closed interval $[s_1, s_2]$ of variation of the parameter s can be found by the formula

$$\sigma = 2\pi \int_{s_1}^{s_2} r(s)\,ds.$$

b) The graph of a smooth nonnegative function $y = f(x)$ defined on a closed interval $[a, b] \subset \mathbb{R}_+$ is revolved about the x-axis, then about the y-axis. In each of these cases, write the formula for the area of the corresponding surface of revolution as an integral over the closed interval $[a, b]$.

4. a) The center of a ball of radius 1 slides along a smooth closed plane curve of length L. Show that the area of the surface of the tubular body thereby formed is $2\pi \cdot 1 \cdot L$.

b) Based on the result of part a), find the area of the two-dimensional torus obtained by revolving a circle of radius a about an axis lying in the plane of the circle and lying at distance $b > a$ from its center.

5. Describe the helical surface defined in Cartesian coordinates (x, y, z) in $\mathbb{R}^3$ by the equation

$$y = x \tan \frac{z}{h} = 0, \quad |z| \le \frac{\pi}{2}h,$$

and find the area of the portion of it for which $r^2 < x^2 + y^2 \le R^2$.

6. a) Show that the area Ω_{n-1} of the unit sphere in $\mathbb{R}^n$ is $\frac{2(\sqrt{\pi})^n}{\Gamma(\frac{n}{2})}$, where $\Gamma(a) = \int_0^\infty e^{-x} x^{a-1}\,dx$. (In particular, if n is even, then $\Gamma(\frac{n}{2}) = (\frac{n-2}{2})!$, while if n is odd, $\Gamma(\frac{n}{2}) = \frac{(n-2)!!}{2^{\frac{n-1}{2}}} \sqrt{\pi}$.)

b) By verifying that the volume $V_n(r)$ of the ball of radius r in $\mathbb{R}^n$ is $\frac{(\sqrt{\pi})^n}{\Gamma(\frac{n+2}{2})} r^n$, show that $\frac{dV_n}{dr}\big|_{r=1} = \Omega_{n-1}$.

c) Find the limit as $n \to \infty$ of the ratio of the area of the hemisphere $\{x \in \mathbb{R}^n \mid |x| = 1 \wedge x^n > 0\}$ to the area of its orthogonal projection on the plane $x^n = 0$.

d) Show that as $n \to \infty$, the majority of the volume of the n-dimensional ball is concentrated in an arbitrarily small neighborhood of the boundary sphere, and the majority of the area of the sphere is concentrated in an arbitrarily small neighborhood of its equator.

e) Show that the following beautiful corollary on *concentration phenomena* follows from the observation made in d).

A regular function that is continuous on a sphere of large dimension is nearly constant on it (recall *pressure* in thermodynamics).

Specifically, let us consider, for example, functions satisfying a Lipschitz condition with a fixed constant. Then for any $\varepsilon > 0$ and $\delta > 0$ there exists N such that for $n > N$ and any function $f : S^n \to \mathbb{R}$ there exists a value c with the following properties: the area of the set on which the value of f differs from c by more than ε is at most δ times the area of the whole sphere.

7. a) Let $x_1, \ldots, x_k$ be a system of vectors in Euclidean space $\mathbb{R}^n$, $n \geq k$. Show that the Gram determinant of this system can be represented as

$$\det(\langle x_i, x_j \rangle) = \sum_{1 \leq i_1 < \cdots < i_k \leq n} P^2_{i_1 \cdots i_k},$$

where

$$P_{i_1 \cdots i_k} = \det \begin{pmatrix} x_1^{i_1} & \cdots & x_1^{i_k} \\ \vdots & \ddots & \vdots \\ x_k^{i_1} & \cdots & x_k^{i_k} \end{pmatrix}.$$

b) Explain the geometric meaning of the quantities $P_{i_1 \cdots i_k}$ from a) and state the result of a) as *the Pythagorean theorem for measures of arbitrary dimension k, $1 \leq k \leq n$.*

c) Now explain the formula

$$\sigma = \int_D \sqrt{\sum_{1 \leq i_1 < \cdots < i_k \leq n} \det^2 \begin{pmatrix} \frac{\partial x^{i_1}}{\partial t^1} & \cdots & \frac{\partial x^{i_1}}{\partial t^k} \\ \vdots & \ddots & \vdots \\ \frac{\partial x^{i_k}}{\partial t^1} & \cdots & \frac{\partial x^{i_k}}{\partial t^k} \end{pmatrix}} \, dt^1 \cdots dt^k$$

for the area of a k-dimensional surface given in the parametric form $x = x(t^1, \ldots, t^k)$, $t \in D \subset \mathbb{R}^k$.

8. a) Verify that the quantity $V_k(S)$ in Definition 2 really is independent of the method of partitioning the surface S into smooth pieces $S_1, \ldots, S_m, \ldots$.

b) Show that a piecewise-smooth surface S admits the locally finite partition into pieces $S_1, \ldots, S_m, \ldots$ described in Definition 2.

c) Show that a set of measure 0 can always be removed from a piecewise-smooth surface S so as to leave a smooth surface $\widetilde{S} = S \setminus E$ that can be described by a single standard local chart $\varphi : I \to S$.

9. The length of a curve, like the high-school definition of the circumference of a circle, is often defined as the limit of the lengths of suitably inscribed broken lines. The limit is taken as the length of the links in the inscribed broken lines tend to zero. The following simple example, due to H. Schwarz, shows that the analogous procedure in an attempt to define the area of even a very simple smooth surface in terms of the areas of polyhedral surfaces "inscribed" in it, may lead to an absurdity.

In a cylinder of radius R and height H we inscribe a polyhedron as follows. Cut the cylinder into m equal cylinders each of height H/m by means of horizontal planes. Break each of the $m+1$ circles of intersection (including the upper and lower bases of the original cylinder) into n equal parts so that the points of division on each circle lie beneath the midpoints of the points of division of the circle immediately above. We now take a pair of division points of each circle and the point lying directly above or below the midpoint of the arc whose endpoints they are.

These three points form a triangle, and the set of all such triangles forms a poly-hedral surface inscribed in the original cylindrical surface (the lateral surface of a right circular cylinder). In shape this polyhedron resembles the calf of a boot that has been crumpled like an accordion. For that reason it is often called the *Schwarz boot*.

a) Show that if m and n are made to tend to infinity in such a way that the ratio n^2/m tends to zero, then the area of the polyhedral surface just constructed will increase without bound, even though the dimensions of each of its faces (each triangle) tend to zero.

b) If n and m tend to infinity in such a way that the ratio m/n^2 tends to some finite limit p, the area of the polyhedral surfaces will tend to a finite limit, which may be larger than, smaller than, or (when $p = 0$) equal to the area of the original cylindrical surface.

c) Compare the method of introducing the area of a smooth surface described here with what was just done above, and explain why the results are the same in the one-dimensional case, but in general not in the two-dimensional case. What are the conditions on the sequence of inscribed polyhedral surfaces that guarantee that the two results will be the same?

10. *The isoperimetric inequality.* Let $V(E)$ denote the volume of a set $E \subset \mathbb{R}^n$, and $A + B$ the (vector) sum of the sets $A, B \subset \mathbb{R}^n$. (The sum in the sense of Minkowski is meant. See Problem 4 in Sect. 11.2.)

Let B be a ball of radius h. Then $A + B =: A_h$ is the h-neighborhood of the set A. The quantity

$$\lim_{h \to 0} \frac{V(A_h) - V(A)}{h} =: \mu_+(\partial A)$$

is called the *Minkowski outer area of the boundary ∂A of A.*

a) Show that if ∂A is a smooth or sufficiently regular surface, then $\mu_+(\partial A)$ equals the usual area of the surface ∂A.

b) Using the Brunn–Minkowski inequality (Problem 4 of Sect. 11.2), obtain now the classical *isoperimetric inequality in $\mathbb{R}^n$*:

$$\mu_+(\partial A) \geq n v^{\frac{1}{n}} V^{\frac{n-1}{n}}(A) =: \mu(S_A);$$

here V is the volume of the unit ball in $\mathbb{R}^n$, and $\mu(S_A)$ the area of the $((n-1)$-dimensional) surface of the ball having the same volume as A.

The isoperimetric inequality means that a solid $A \subset \mathbb{R}^n$ has boundary area $\mu_+(\partial A)$ not less than that of a ball of the same volume.

12.5 Elementary Facts About Differential Forms

We now give an elementary description of the convenient mathematical machinery known as differential forms, paying particular attention here to its algorithmic potential rather than the details of the theoretical constructions, which will be discussed in Chap. 15.

12.5.1 Differential Forms: Definition and Examples

Having studied algebra, the reader is well acquainted with the concept of a linear form, and we have already made extensive use of that concept in constructing the differential calculus. In that process we encountered mostly symmetric forms. In the present subsection we will be discussing skew-symmetric (anti-symmetric) forms.

We recall that a form $L : X^k \to Y$ of degree or order k defined on ordered sets $\boldsymbol{\xi}_1, \ldots, \boldsymbol{\xi}_k$ of vectors of a vector space X and assuming values in a vector space Y is *skew-symmetric* or *anti-symmetric* if the value of the form changes sign when any pair of its arguments are interchanged, that is,

$$L(\boldsymbol{\xi}_1, \ldots, \boldsymbol{\xi}_i, \ldots, \boldsymbol{\xi}_j, \ldots, \boldsymbol{\xi}_k) = -L(\boldsymbol{\xi}_1, \ldots, \boldsymbol{\xi}_j, \ldots, \boldsymbol{\xi}_i, \ldots, \boldsymbol{\xi}_k).$$

In particular, if $\boldsymbol{\xi}_i = \boldsymbol{\xi}_j$ then the value of the form will be zero, regardless of the other vectors.

Example 1 The vector (cross) product $[\boldsymbol{\xi}_1, \boldsymbol{\xi}_2]$ of two vectors in $\mathbb{R}^3$ is a skew-symmetric bilinear form with values in $\mathbb{R}^3$.

Example 2 The oriented volume $V(\boldsymbol{\xi}_1, \ldots, \boldsymbol{\xi}_k)$ of the parallelepiped spanned by the vectors $\boldsymbol{\xi}_1, \ldots, \boldsymbol{\xi}_k$ of $\mathbb{R}^k$, defined by Eq. (12.6) of Sect. 12.4, is a skew-symmetric real-valued k-form on $\mathbb{R}^k$.

For the time being we shall be interested only in real-valued k-forms (the case $Y = \mathbb{R}$), even though everything that will be discussed below is applicable to the more general situation, for example, when Y is the field $\mathbb{C}$ of complex numbers.

A linear combination of skew-symmetric forms of the same degree is in turn a skew-symmetric form, that is, the skew-symmetric forms of a given degree constitute a vector space.

In addition, in algebra one introduces the *exterior product* $\wedge$ of skew-symmetric forms, which assigns to an ordered pair A^p, B^q of such forms (of degrees p and q respectively) a skew-symmetric form $A^p \wedge B^q$ of degree $p + q$. This operation is

associative: $(A^p \wedge B^q) \wedge C^r = A^p \wedge (B^q \wedge C^r)$,
distributive: $(A^p + B^p) \wedge C^q = A^p \wedge C^q + B^p \wedge C^q$,
skew-commutative: $A^p \wedge B^q = (-1)^{pq} B^q \wedge A^p$.

In particular, in the case of 1-forms A and B, we have anticommutativity $A \wedge B = -B \wedge A$, for the operations, like the anticommutativity of the vector product shown in Example 1. The exterior product of forms is in fact a generalization of the vector product.

Without going into the details of the definition of the exterior product, we take as known for the time being the properties of this operation just listed and observe that in the case of the exterior product of 1-forms $L_1, \ldots, L_k \in \mathcal{L}(\mathbb{R}^n, \mathbb{R})$ the result $L_1 \wedge \cdots \wedge L_k$ is a k-form that assumes the value

$$
L_1 \wedge \cdots \wedge L_k(\boldsymbol{\xi}_1, \ldots, \boldsymbol{\xi}_k) =
\begin{vmatrix}
L_1(\boldsymbol{\xi}_1) & \cdots & L_k(\boldsymbol{\xi}_1) \\
\vdots & \ddots & \vdots \\
L_1(\boldsymbol{\xi}_k) & \cdots & L_k(\boldsymbol{\xi}_k)
\end{vmatrix}
= \det\bigl(L_j(\boldsymbol{\xi}_i)\bigr) \qquad (12.11)
$$

on the set of vectors $\boldsymbol{\xi}_1, \ldots, \boldsymbol{\xi}_k$.

If relation (12.11) is taken as the definition of the left-hand side, it follows from properties of determinants that in the case of linear forms A, B, and C, we do indeed have $A \wedge B = -B \wedge A$ and $(A + B) \wedge C = A \wedge C + B \wedge C$.

Let us now consider some examples that will be useful below.

Example 3 Let $\pi^i \in \mathcal{L}(\mathbb{R}^n, \mathbb{R})$, $i = 1, \ldots, n$, be the projections. More precisely, the linear function $\pi^i : \mathbb{R}^n \to \mathbb{R}$ is such that on each vector $\boldsymbol{\xi} = (\xi^1, \ldots, \xi^n) \in \mathbb{R}^n$ it assumes the value $\pi^i(\boldsymbol{\xi}) = \xi^i$ of the projection of that vector on the corresponding coordinate axis. Then, in accordance with formula (12.11) we obtain

$$
\pi^{i_1} \wedge \cdots \wedge \pi^{i_k}(\boldsymbol{\xi}_1, \ldots, \boldsymbol{\xi}_k) =
\begin{vmatrix}
\xi_1^{i_1} & \cdots & \xi_1^{i_k} \\
\vdots & \ddots & \vdots \\
\xi_k^{i_1} & \cdots & \xi_k^{i_k}
\end{vmatrix} . \qquad (12.12)
$$

Example 4 The Cartesian coordinates of the vector product $[\boldsymbol{\xi}_1, \boldsymbol{\xi}_2]$ of the vectors $\boldsymbol{\xi}_1 = (\xi_1^1, \xi_1^2, \xi_1^3)$ and $\boldsymbol{\xi}_2 = (\xi_2^1, \xi_2^2, \xi_2^3)$ in the Euclidean space $\mathbb{R}^3$, as is known, are defined by the equality

$$
[\boldsymbol{\xi}_1, \boldsymbol{\xi}_2] = \left(
\begin{vmatrix} \xi_1^2 & \xi_1^3 \\ \xi_2^2 & \xi_2^3 \end{vmatrix},
\begin{vmatrix} \xi_1^3 & \xi_1^1 \\ \xi_2^3 & \xi_2^1 \end{vmatrix},
\begin{vmatrix} \xi_1^1 & \xi_1^2 \\ \xi_2^1 & \xi_2^2 \end{vmatrix}
\right).
$$

Thus, in accordance with the result of Example 3 we can write

$$
\pi^1\bigl([\boldsymbol{\xi}_1, \boldsymbol{\xi}_2]\bigr) = \pi^2 \wedge \pi^3(\boldsymbol{\xi}_1, \boldsymbol{\xi}_2),
$$

$$
\pi^2\bigl([\boldsymbol{\xi}_1, \boldsymbol{\xi}_2]\bigr) = \pi^3 \wedge \pi^1(\boldsymbol{\xi}_1, \boldsymbol{\xi}_2),
$$

$$
\pi^3\bigl([\boldsymbol{\xi}_1, \boldsymbol{\xi}_2]\bigr) = \pi^1 \wedge \pi^2(\boldsymbol{\xi}_1, \boldsymbol{\xi}_2).
$$

Example 5 Let $f : D \to \mathbb{R}$ be a function that is defined in a domain $D \subset \mathbb{R}^n$ and differentiable at $x_0 \in D$. As is known, the differential $\mathrm{d}f(x_0)$ of the function at a point is a linear function defined on displacement vectors $\boldsymbol{\xi}$ from that point. More precisely, on vectors of the tangent space TD_{x_0} to D (or $\mathbb{R}^n$) at the point under consideration. We recall that if $x^1, \ldots, x^n$ are the coordinates in $\mathbb{R}^n$ and $\boldsymbol{\xi} = (\xi^1, \ldots, \xi^n)$, then

$$\mathrm{d}f(x_0)(\boldsymbol{\xi}) = \frac{\partial f}{\partial x^1}(x_0)\xi^1 + \cdots + \frac{\partial f}{\partial x^n}(x_0)\xi^n = D_{\boldsymbol{\xi}}f(x_0).$$

In particular $\mathrm{d}x^i(\boldsymbol{\xi}) = \xi^i$, or, more formally, $\mathrm{d}x^i(x_0)(\boldsymbol{\xi}) = \xi^i$. If $f_1, \ldots, f_k$ are real-valued functions defined in G and differentiable at the point $x_0 \in G$, then in accordance with (12.11) we obtain

$$\mathrm{d}f_1 \wedge \cdots \wedge \mathrm{d}f_k(\boldsymbol{\xi}_1, \ldots, \boldsymbol{\xi}_k) = \begin{vmatrix} \mathrm{d}f_1(\boldsymbol{\xi}_1) & \cdots & \mathrm{d}f_k(\boldsymbol{\xi}_1) \\ \vdots & \ddots & \vdots \\ \mathrm{d}f_1(\boldsymbol{\xi}_k) & \cdots & \mathrm{d}f_k(\boldsymbol{\xi}_k) \end{vmatrix} \tag{12.13}$$

at the point x_0 for the set $\boldsymbol{\xi}_1, \ldots, \boldsymbol{\xi}_k$ of vectors in the space TG_{x_0}; and, in particular,

$$\mathrm{d}x^{i_1} \wedge \cdots \wedge \mathrm{d}x^{i_k}(\boldsymbol{\xi}_1, \ldots, \boldsymbol{\xi}_k) = \begin{vmatrix} \xi_1^{i_1} & \cdots & \xi_1^{i_k} \\ \vdots & \ddots & \vdots \\ \xi_k^{i_1} & \cdots & \xi_k^{i_k} \end{vmatrix}. \tag{12.14}$$

In this way skew-symmetric forms of degree k defined on the space $TD_{x_0} \approx T\mathbb{R}^n_{x_0} \approx \mathbb{R}^n$ have been obtained from the linear forms $\mathrm{d}f_1, \ldots, \mathrm{d}f_k$ defined on this space.

Example 6 If $f \in C^{(1)}(D, \mathbb{R})$, where D is a domain in $\mathbb{R}^n$, then the differential $\mathrm{d}f(x)$ of the functions f is defined at any point $x \in D$, and this differential, as has been stated, is a linear function $\mathrm{d}f(x) : TD_x \to T\mathbb{R}_{f(x)} \approx \mathbb{R}$ on the tangent space TD_x to D at x. In general the form $\mathrm{d}f(x) = f'(x)$ varies in passage from one point to another in D. Thus a smooth scalar-valued function $f : D \to \mathbb{R}$ generates a linear form $\mathrm{d}f(x)$ at each point, or, as we say, generates a *field* of linear *forms* in D, defined on the corresponding tangent spaces TD_x.

Definition 1 We shall say that a real-valued *differential p-form* ω is defined in the domain $D \subset \mathbb{R}^n$ if a skew-symmetric form $\omega(x) : (TD_x)^p \to \mathbb{R}$ is defined at each point $x \in D$.

The number p is usually called the *degree* or *order* of ω. In this connection the p-form ω is often denoted ω^p.

Thus, the field of the differential $\mathrm{d}f$ of a smooth function $f : D \to \mathbb{R}$ considered in Example 6 is a differential 1-form in D, and $\omega = \mathrm{d}x^{i_1} \wedge \cdots \wedge \mathrm{d}x^{i_p}$ is the simplest example of a differential form of degree p.

Example 7 Suppose a vector field $D \subset \mathbb{R}^n$ is defined, that is, a vector $\mathbf{F}(x)$ is attached to each point $x \in D$. When there is a Euclidean structure in $\mathbb{R}^n$ this vector field generates the following differential 1-form $\omega_{\mathbf{F}}^1$ in D.

If $\boldsymbol{\xi}$ is a vector attached to $x \in D$, that is, $\boldsymbol{\xi} \in TD_x$, we set

$$\omega_{\mathbf{F}}^1(x)(\boldsymbol{\xi}) = \langle \mathbf{F}(x), \boldsymbol{\xi} \rangle.$$

It follows from properties of the inner product that $\omega_{\mathbf{F}}^1(x) = \langle \mathbf{F}(x), \cdot \rangle$ is indeed a linear form at each point $x \in D$.

Such differential forms arise very frequently. For example, if $\mathbf{F}$ is a continuous force field in D and $\boldsymbol{\xi}$ an infinitesimal displacement vector from the point $x \in D$, the element of work corresponding to this displacement, as is known from physics, is defined precisely by the quantity $\langle \mathbf{F}(x), \boldsymbol{\xi} \rangle$.

Thus a force field $\mathbf{F}$ in a domain D of the Euclidean space $\mathbb{R}^n$ naturally generates a differential 1-form $\omega_{\mathbf{F}}^1$ in D, which it is natural to call the *work form of the field* $\mathbf{F}$ in this case.

We remark that in Euclidean space the differential df of a smooth function $f : D \to \mathbb{R}$ in the domain $D \subset \mathbb{R}^n$ can also be regarded as the 1-form generated by a vector field, in this case the field $\mathbf{F} = \operatorname{grad} f$. In fact, by definition $\operatorname{grad} f$ is such that $df(x)(\boldsymbol{\xi}) = \langle \operatorname{grad} f(x), \boldsymbol{\xi} \rangle$ for every vector $\boldsymbol{\xi} \in TD_x$.

Example 8 A vector field $\mathbf{V}$ defined in a domain D of the Euclidean space $\mathbb{R}^n$ can also be regarded as a differential form $\omega_{\mathbf{V}}^{n-1}$ of degree $n - 1$. If at a point $x \in D$ we take the vector field $\mathbf{V}(x)$ and $n - 1$ additional vectors $\boldsymbol{\xi}_1, \ldots, \boldsymbol{\xi}_n \in TD_x$ attached to the point x, then the oriented volume of the parallelepiped spanned by the vectors $\mathbf{V}(x), \boldsymbol{\xi}_1, \ldots, \boldsymbol{\xi}_{n-1}$, which is the determinant of the matrix whose rows are the co-ordinates of these vectors, will obviously be a skew-symmetric $(n - 1)$-form with respect to the variables $\boldsymbol{\xi}_1, \ldots, \boldsymbol{\xi}_{n-1}$.

For $n = 3$ the form $\omega_{\mathbf{V}}^2$ is the usual scalar triple product $(\mathbf{V}(x), \boldsymbol{\xi}_1, \boldsymbol{\xi}_2)$ of vectors, one of which $\mathbf{V}(x)$ is given, resulting in a skew-symmetric 2-form $\omega_{\mathbf{V}}^2 = (\mathbf{V}, \cdot, \cdot)$.

For example, if a steady flow of a fluid is taking place in the domain D and $\mathbf{V}(x)$ is the velocity vector at the point $x \in D$, the quantity $(\mathbf{V}(x), \boldsymbol{\xi}_1, \boldsymbol{\xi}_2)$ is the element of volume of the fluid passing through the (parallelogram) area spanned by the small vectors $\boldsymbol{\xi}_1 \in TD_x$ and $\boldsymbol{\xi}_2 \in TD_x$ in unit time. By choosing different vectors $\boldsymbol{\xi}_1$ and $\boldsymbol{\xi}_2$, we shall obtain areas (parallelograms) of different configuration, differently situated in space, all having one vertex at x. For each such area there will be, in general, a different value $(\mathbf{V}(x), \boldsymbol{\xi}_1, \boldsymbol{\xi}_2)$ of the form $\omega_{\mathbf{V}}^2(x)$. As has been stated, this value shows how much fluid has flowed through the surface in unit time, that is, it characterizes the flux across the chosen element of area. For that reason we often call the form $\omega_{\mathbf{V}}^2$ (and indeed its multidimensional analogue $\omega_{\mathbf{V}}^{n-1}$) the *flux form of the vector field* $\mathbf{V}$ in D.

12.5.2 Coordinate Expression of a Differential Form

Let us now investigate the coordinate expression of skew-symmetric algebraic and differential forms and show, in particular, that every differential k-form is in a certain sense a linear combination of standard differential forms of the form (12.14).

To abbreviate the notation, we shall assume summation over the range of allowable values for indices that occur as both superscripts and subscripts (as we did earlier in similar situations).

Let L be a k-linear form in $\mathbb{R}^n$. If a basis $\mathbf{e}_1, \ldots, \mathbf{e}_n$ is fixed in $\mathbb{R}^n$, then each vector $\boldsymbol{\xi} \in \mathbb{R}^n$ gets a coordinate representation $\boldsymbol{\xi} = \xi^i \mathbf{e}_i$ in that basis, and the form L acquires the coordinate expression

$$L(\boldsymbol{\xi}_1, \ldots, \boldsymbol{\xi}_k) = L\big(\xi_1^{i_1} \mathbf{e}_{i_1}, \ldots, \xi_k^{i_k} \mathbf{e}_{i_k}\big) = L(\mathbf{e}_{i_1}, \ldots, \mathbf{e}_{i_k}) \xi_1^{i_1} \cdots \xi_k^{i_k}. \qquad (12.15)$$

The numbers $a_{i_1, \ldots, i_k} = L(\mathbf{e}_{i_1}, \ldots, \mathbf{e}_{i_k})$ characterize the form L completely if the basis in which they have been obtained is known. These numbers are obviously symmetric or skew-symmetric with respect to their indices if and only if the form L possesses the corresponding type of symmetry.

In the case of a skew-symmetric form L the coordinate representation can be transformed slightly. To make the direction of that transformation clear and natural, let us consider the special case of (12.15) that occurs when L is a skew-symmetric 2-form in $\mathbb{R}^3$. Then for the vectors $\boldsymbol{\xi}_1 = \xi_1^{i_1} \mathbf{e}_{i_1}$ and $\boldsymbol{\xi}_2 = \xi_2^{i_2} \mathbf{e}_{i_2}$, where $i_1, i_2 = 1, 2, 3$, we obtain

$$
\begin{aligned}
L(\boldsymbol{\xi}_1, \boldsymbol{\xi}_2) &= L\big(\xi_1^{i_1} \mathbf{e}_{i_1}, \xi_2^{i_2} \mathbf{e}_{i_2}\big) = L(\mathbf{e}_{i_1}, \mathbf{e}_{i_2}) \xi_1^{i_1} \xi_2^{i_2} = \\
&= L(\mathbf{e}_1, \mathbf{e}_1)\xi_1^1 \xi_2^1 + L(\mathbf{e}_1, \mathbf{e}_2)\xi_1^1 \xi_2^2 + L(\mathbf{e}_1, \mathbf{e}_3)\xi_1^1 \xi_2^3 + \\
&\quad + L(\mathbf{e}_2, \mathbf{e}_1)\xi_1^2 \xi_2^1 + L(\mathbf{e}_2, \mathbf{e}_2)\xi_1^2 \xi_2^2 + L(\mathbf{e}_2, \mathbf{e}_3)\xi_1^2 \xi_2^3 + \\
&\quad + L(\mathbf{e}_3, \mathbf{e}_1)\xi_1^3 \xi_2^1 + L(\mathbf{e}_3, \mathbf{e}_2)\xi_1^3 \xi_2^2 + L(\mathbf{e}_3, \mathbf{e}_3)\xi_1^3 \xi_2^3 = \\
&= L(\mathbf{e}_1, \mathbf{e}_2)\big(\xi_1^1 \xi_2^2 - \xi_1^2 \xi_2^1\big) + L(\mathbf{e}_1, \mathbf{e}_3)\big(\xi_1^1 \xi_2^3 - \xi_1^3 \xi_2^1\big) + \\
&\quad + L(\mathbf{e}_2, \mathbf{e}_3)\big(\xi_1^2 \xi_2^3 - \xi_1^3 \xi_2^2\big) = \sum_{1 \le i_1 < i_2 \le 3} L(\mathbf{e}_{i_1}, \mathbf{e}_{i_2}) \begin{vmatrix} \xi_1^{i_1} & \xi_1^{i_2} \\ \xi_2^{i_1} & \xi_2^{i_2} \end{vmatrix},
\end{aligned}
$$

where the summation extends over all combinations of indices i_1 and i_2 that satisfy the inequalities written under the summation sign.

Similarly in the general case we can also obtain the following representation for a skew-symmetric form L:

$$L(\boldsymbol{\xi}_1, \ldots, \boldsymbol{\xi}_k) = \sum_{1 \le i_1 < \cdots < i_k \le n} L(\mathbf{e}_{i_1}, \ldots, \mathbf{e}_{i_k}) \begin{vmatrix} \xi_1^{i_1} & \cdots & \xi_1^{i_k} \\ \vdots & \ddots & \vdots \\ \xi_k^{i_1} & \cdots & \xi_k^{i_k} \end{vmatrix}. \qquad (12.16)$$

Then, in accordance with formula (12.12) this last equality can be rewritten as

$$L(\boldsymbol{\xi}_1, \ldots, \boldsymbol{\xi}_k) = \sum_{1 \le i_1 < \cdots < i_k \le i_n} L(\mathbf{e}_{i_1}, \ldots, \mathbf{e}_{i_k}) \pi^{i_1} \wedge \cdots \wedge \pi^{i_k}(\boldsymbol{\xi}_1, \ldots, \boldsymbol{\xi}_k).$$

Thus, any skew-symmetric form L can be represented as a linear combination

$$L = \sum_{1 \le i_1 < \cdots < i_k \le i_n} a_{i_1 \cdots i_k} \pi^{i_1} \wedge \cdots \wedge \pi^{i_k} \tag{12.17}$$

of the k-forms $\pi^{i_1} \wedge \cdots \wedge \pi^{i_k}$, which are the exterior product formed from the elementary 1-forms $\pi^1, \ldots, \pi^n$ in $\mathbb{R}^n$.

Now suppose that a differential k-form ω is defined in some domain $D \subset \mathbb{R}^n$ along with a curvilinear coordinate system $x^1, \ldots, x^n$. At each point $x \in D$ we fix the basis $\mathbf{e}_1(x), \ldots, \mathbf{e}_n(x)$ of the space TD_x, formed from the unit vectors along the coordinate axes. (For example, if $x^1, \ldots, x^n$ are Cartesian coordinates in $\mathbb{R}^n$, then $\mathbf{e}_1(x), \ldots, \mathbf{e}_n(x)$ is simply the frame $\mathbf{e}_1, \ldots, \mathbf{e}_n$ in $\mathbb{R}^n$ translated parallel to itself from the origin to x.) Then at each point $x \in D$ we find by formulas (12.14) and (12.16) that

$$\omega(x)(\boldsymbol{\xi}_1, \ldots, \boldsymbol{\xi}_k) =$$
$$= \sum_{1 \le i_1 < \cdots < i_k \le n} \omega\big(\mathbf{e}_{i_1}(x), \ldots, \mathbf{e}_{i_k}(x)\big) \, dx^{i_1} \wedge \cdots \wedge dx^{i_k}(\boldsymbol{\xi}_1, \ldots, \boldsymbol{\xi}_k)$$

or

$$\omega(x) = \sum_{1 \le i_1 < \cdots < i_k \le n} a_{i_1 \cdots i_k}(x) \, dx^{i_1} \wedge \cdots \wedge dx^{i_k}. \tag{12.18}$$

Thus, every differential k-form is a combination of the elementary k-forms $dx^{i_1} \wedge \cdots \wedge dx^{i_k}$ formed from the differentials of the coordinates. As a matter of fact, that is the reason for the term "differential form".

The coefficients $a_{i_1 \cdots i_k}(x)$ of the linear combination (12.18) generally depend on the point x, that is, they are functions defined in the domain in which the form ω^k is given.

In particular, we have long known the expansion of the differential

$$df(x) = \frac{\partial f}{\partial x^1}(x) \, dx^1 + \cdots + \frac{\partial f}{\partial x^n}(x) \, dx^n, \tag{12.19}$$

and, as can be seen from the equalities

$$\langle \mathbf{F}, \boldsymbol{\xi} \rangle = \big\langle F^{i_1} \mathbf{e}_{i_1}(x), \xi^{i_2} \mathbf{e}_{i_2}(x) \big\rangle =$$
$$= \big\langle \mathbf{e}_{i_1}(x), \mathbf{e}_{i_2}(x) \big\rangle F^{i_1}(x) \xi^{i_2} = g_{i_1 i_2}(x) F^{i_1}(x) \xi^{i_2} =$$
$$= g_{i_1 i_2}(x) F^{i_1}(x) \, dx^{i_2}(\boldsymbol{\xi}),$$

the expansion

$$\omega_{\mathbf{F}}^1(x) = \langle \mathbf{F}(x), \cdot \rangle = \big(g_{i_1 i}(x) F^{i_1}(x)\big) \, dx^i = a_i(x) \, dx^i \tag{12.20}$$

also holds. In Cartesian coordinates this expansion looks especially simple:

$$\omega_{\mathbf{F}}^1(x) = \langle \mathbf{F}(x), \cdot \rangle = \sum_{i=1}^{n} F^i(x)\,dx^i. \tag{12.21}$$

Next, the following equality holds in $\mathbb{R}^3$:

$$\omega_{\mathbf{V}}^2(x)(\boldsymbol{\xi}_1, \boldsymbol{\xi}_2) = \begin{vmatrix} V^1(x) & V^2(x) & V^3(x) \\ \xi_1^1 & \xi_1^2 & \xi_1^3 \\ \xi_2^1 & \xi_2^2 & \xi_2^3 \end{vmatrix} =$$

$$= V^1(x)\begin{vmatrix} \xi_1^2 & \xi_1^3 \\ \xi_2^2 & \xi_2^3 \end{vmatrix} + V^2(x)\begin{vmatrix} \xi_1^3 & \xi_1^1 \\ \xi_2^3 & \xi_2^1 \end{vmatrix} + V^3(x)\begin{vmatrix} \xi_1^1 & \xi_1^2 \\ \xi_2^1 & \xi_2^2 \end{vmatrix},$$

from which it follows that

$$\omega_{\mathbf{V}}^2(x) = V^1(x)\,dx^2 \wedge dx^3 + V^2(x)\,dx^3 \wedge dx^1 + V^3(x)\,dx^1 \wedge dx^2. \tag{12.22}$$

Similarly, expanding the determinant of order n for the form $\omega_{\mathbf{V}}^{n-1}$ by minors along the first row, we obtain the expansion

$$\omega_{\mathbf{V}}^{n-1} = \sum_{i=1}^{n-1}(-1)^{i+1}V^i(x)\,dx^1 \wedge \cdots \wedge \widehat{dx^i} \wedge \cdots \wedge dx^n, \tag{12.23}$$

where the sign $\frown$ stands over the differential that is to be omitted in the indicated term.

12.5.3 The Exterior Differential of a Form

All that has been said up to now about differential forms essentially involved each individual point x of the domain of definition of the form and had a purely algebraic character. The operation of (exterior) differentiation of such forms is specific to analysis.

Let us agree from now on to define the 0-*forms* in a domain to be functions $f : D \to \mathbb{R}$ defined in that domain.

Definition 2 The (*exterior*) *differential* of a 0-form f, when f is a differentiable function, is the usual differential df of that function.

If a differential p-form ($p \geq 1$) defined in a domain $D \subset \mathbb{R}^n$

$$\omega(x) = a_{i_1 \cdots i_p}(x)\,dx^{i_1} \wedge \cdots \wedge dx^{i_p}$$

has differentiable coefficients $a_{i_1 \cdots i_p}(x)$, then its *(exterior) differential* is the form

$$d\omega(x) = da_{i_1 \cdots i_p}(x) \wedge dx^{i_1} \wedge \cdots \wedge dx^{i_p}.$$

Using the expansion (12.19) for the differential of a function, and relying on the distributivity of the exterior product of 1-forms, which follows from relation (12.11), we conclude that

$$d\omega(x) = \frac{\partial a_{i_1 \cdots i_p}}{\partial x^i}(x)\, dx^i \wedge dx^{i_1} \wedge \cdots \wedge dx^{i_p} =$$

$$= \alpha_{i i_1 \cdots i_p}(x)\, dx^i \wedge dx^{i_1} \wedge \cdots \wedge dx^{i_p},$$

that is, the (exterior) differential of a p-form ($p \geq 0$) is always a form of degree $p + 1$.

We note that Definition 1 given above for a differential p-form in a domain $D \subset \mathbb{R}^n$, as one can now understand, is too general, since it does not in any way connect the forms $\omega(x)$ corresponding to different points of the domain D. In actuality, the only forms used in analysis are those whose coordinates $a_{i_1 \cdots i_p}(x)$ in a coordinate representation are sufficiently regular (most often infinitely differentiable) functions in the domain D. The *order of smoothness of the form* ω in the domain $D \subset \mathbb{R}^n$ is customarily characterized by the smallest order of smoothness of its coefficients. The totality of all forms of degree $p \geq 0$ with coefficients of class $C^{(\infty)}(D, \mathbb{R})$ is most often denoted $\Omega^p(D, \mathbb{R})$ or Ω^p.

Thus the operation of differentiation of forms that we have defined effects a mapping $d : \Omega^p \to \Omega^{p+1}$.

Let us consider several useful specific examples.

Example 9 For a 0-form $\omega = f(x, y, z)$ – a differentiable function – defined in a domain $D \subset \mathbb{R}^3$, we obtain

$$d\omega = \frac{\partial f}{\partial x}\, dx + \frac{\partial f}{\partial y}\, dy + \frac{\partial f}{\partial z}\, dz.$$

Example 10 Let

$$\omega(x, y) = P(x, y)\, dx + Q(x, y)\, dy$$

be a differential 1-form in a domain D of $\mathbb{R}^2$ endowed with coordinates (x, y). Assuming that P and Q are differentiable in D, by Definition 2 we obtain

$$d\omega(x, y) = dP \wedge dx + dQ \wedge dy =$$

$$= \left(\frac{\partial P}{\partial x}\, dx + \frac{\partial P}{\partial y}\, dy \right) \wedge dx + \left(\frac{\partial Q}{\partial x}\, dx + \frac{\partial Q}{\partial y}\, dy \right) \wedge dy =$$

$$= \frac{\partial P}{\partial y}\, dy \wedge dx + \frac{\partial Q}{\partial x}\, dx \wedge dy = \left(\frac{\partial Q}{\partial x} - \frac{\partial P}{\partial y} \right)(x, y)\, dx \wedge dy.$$

Example 11 For a 1-form

$$\omega = P\,dx + Q\,dy + R\,dz$$

defined in a domain D in $\mathbb{R}^3$ we obtain

$$d\omega = \left(\frac{\partial R}{\partial y} - \frac{\partial Q}{\partial z}\right)dy \wedge dz + \left(\frac{\partial P}{\partial z} - \frac{\partial R}{\partial x}\right)dz \wedge dx + \left(\frac{\partial Q}{\partial x} - \frac{\partial P}{\partial y}\right)dx \wedge dy.$$

Example 12 Computing the differential of the 2-form

$$\omega = P\,dy \wedge dz + Q\,dz \wedge dx + R\,dx \wedge dy,$$

where P, Q, and R are differentiable in the domain $D \subset \mathbb{R}^3$, leads to the relation

$$d\omega = \left(\frac{\partial P}{\partial x} + \frac{\partial Q}{\partial y} + \frac{\partial R}{\partial z}\right)dx \wedge dy \wedge dz.$$

If (x^1, x^2, x^3) are Cartesian coordinates in the Euclidean space $\mathbb{R}^3$ and $x \mapsto f(x), x \mapsto \mathbf{F}(x) = (F^1, F^2, F^3)(x)$, and $x \mapsto \mathbf{V} = (V^1, V^2, V^3)(x)$ are smooth scalar and vector fields in the domain $D \subset \mathbb{R}^3$, then along with these fields, we often consider the respective vector fields

$$\operatorname{grad} f = \left(\frac{\partial f}{\partial x^1}, \frac{\partial f}{\partial x^2}, \frac{\partial f}{\partial x^3}\right) - \text{the } \textit{gradient} \text{ of } f, \tag{12.24}$$

$$\operatorname{curl} \mathbf{F} = \left(\frac{\partial F^3}{\partial x^2} - \frac{\partial F^2}{\partial x^3}, \frac{\partial F^1}{\partial x^3} - \frac{\partial F^3}{\partial x^1}, \frac{\partial F^2}{\partial x^1} - \frac{\partial F^1}{\partial x^2}\right) - \text{the } \textit{curl} \text{ of } \mathbf{F}, \tag{12.25}$$

and the scalar field

$$\operatorname{div} \mathbf{V} = \frac{\partial V^1}{\partial x^1} + \frac{\partial V^2}{\partial x^2} = \frac{\partial V^3}{\partial x^3} - \text{the } \textit{divergence} \text{ of } \mathbf{V}. \tag{12.26}$$

We have already mentioned the gradient of a scalar field earlier. Without dwelling on the physical content of the curl and divergence of a vector field at the moment, we note only the connections that these classical operators have with the operation of differentiating forms.

In the oriented Euclidean space $\mathbb{R}^3$ there is a one-to-one correspondence between vector fields and 1- and 2-forms:

$$\mathbf{F} \leftrightarrow \omega_{\mathbf{F}}^1 = \langle \mathbf{F}, \cdot \rangle, \qquad \mathbf{V} \leftrightarrow \omega_{\mathbf{V}}^2(\mathbf{V}, \cdot, \cdot).$$

We remark also that every 3-form in the domain $D \subset \mathbb{R}^3$ has the form $\rho(x^1, x^2, x^3)\,dx^1 \wedge dx^2 \wedge dx^3$. Taking this circumstance into account, one can introduce the following definitions for $\operatorname{grad} f$, $\operatorname{curl} \mathbf{F}$, and $\operatorname{div} \mathbf{V}$:

$$f \mapsto \omega^0(= f) \mapsto d\omega^0(= df) = \omega_{\mathbf{g}}^1 \mapsto \mathbf{g} := \operatorname{grad} f, \tag{12.24'}$$

$$\mathbf{F} \mapsto \omega_{\mathbf{F}}^1 \mapsto d\omega_{\mathbf{F}}^1 = \omega_{\mathbf{r}}^2 \mapsto \mathbf{r} := \operatorname{curl} \mathbf{F}, \qquad (12.25')$$

$$\mathbf{V} \mapsto \omega_{\mathbf{V}}^2 \mapsto d\omega_{\mathbf{V}}^2 = \omega_\rho^3 \mapsto \rho := \operatorname{div} \mathbf{V}. \qquad (12.26')$$

Examples 9, 11, and 12 show that when we do this in Cartesian coordinates, we arrive at the expressions (12.24), (12.25), and (12.26) above for grad f, curl $\mathbf{F}$, and div $\mathbf{V}$. Thus these operators in field theory can be regarded as concrete manifestations of the operation of differentiation of exterior forms, which is carried out in a single manner on forms of any degree. More details on the gradient, curl, and divergence will be given in Chap. 14.

12.5.4 Transformation of Vectors and Forms Under Mappings

Let us consider in more detail what happens with functions (0-forms) under a mapping of their domains.

Let $\varphi : U \to V$ be a mapping of the domain $U \subset \mathbb{R}^m$ into the domain $V \subset \mathbb{R}^n$. Under the mapping φ each point $t \in U$ maps to a definite point $x = \varphi(t)$ of the domain V.

If a function f is defined on V, then, because of the mapping $\varphi : U \to V$ a function $\varphi^* f$ naturally arises on the domain U, defined by the relation

$$(\varphi^* f)(t) := f(\varphi(t)),$$

that is, to find the value of $\varphi^* f$ at a point $t \in U$ one must send t to the point $x = \varphi(t) \in V$ and compute the value of f at that point.

Thus, if the domain U maps to the domain V under the mapping $\varphi : U \to V$, then the set of functions defined on V maps (in the opposite direction) to the set of functions defined on U under the correspondence $f \mapsto \varphi^* f$ just defined.

In other words, we have shown that a mapping $\varphi^* : \Omega^0(V) \to \Omega^0(U)$ transforming 0-forms defined on V into 0-forms defined on U naturally arises from a mapping $\varphi : U \to V$.

Now let us consider the general case of transformation of forms of any degree.

Let $\varphi : U \to V$ be a smooth mapping of a domain $U \subset \mathbb{R}_t^m$ into a domain $V \subset \mathbb{R}_x^n$, and $\varphi'(t) : TU_t \to TV_{x=\varphi(t)}$ the mapping of tangent spaces corresponding to φ, and let ω be a p-form in the domain V. Then one can assign to ω the p-form $\varphi^* \omega$ in the domain U defined at $t \in U$ on the set of vectors $\boldsymbol{\tau}_1, \ldots, \boldsymbol{\tau}_p \in TU_t$ by the equality

$$\varphi^* \omega(t)(\boldsymbol{\tau}_1, \ldots, \boldsymbol{\tau}_p) := \omega(\varphi(t))(\varphi_1' \boldsymbol{\tau}_1, \ldots, \varphi_p' \boldsymbol{\tau}_p). \qquad (12.27)$$

Thus to each smooth mapping $\varphi : U \to V$ there corresponds a mapping $\omega^* : \Omega^p(V) \to \Omega^p(U)$ that transforms forms defined on V into forms defined on U. It obviously follows from (12.27) that

$$\varphi^* (\omega' + \omega'') = \varphi^* (\omega') + \varphi^* (\omega''), \qquad (12.28)$$

$$\varphi^*(\lambda\omega) = \lambda\varphi^*\omega, \quad \text{if } \lambda \in \mathbb{R}. \tag{12.29}$$

Recalling the rule $(\psi \circ \varphi)' = \psi' \circ \varphi'$ for differentiating the composition of the mappings $\varphi : U \to V$, $\psi : V \to W$, we conclude in addition from (12.27) that

$$(\psi \circ \varphi)^* = \varphi^* \circ \psi^* \tag{12.30}$$

(the natural reverse path: the composition of the mappings)

$$\psi^* : \Omega^p(W) \to \Omega^p(V), \qquad \varphi^* : \Omega^p(V) \to \Omega^p(U).$$

Now let us consider how to carry out the transformation of forms in practice.

Example 13 In the domain $V \subset \mathbb{R}_x^n$ let us take the 2-form $\omega = dx^{i_1} \wedge dx^{i_2}$. Let $x^i = x^i(t^1, \ldots, t^m)$, $i = 1, \ldots, n$, be the coordinate expression for the mapping $\varphi : U \to V$ of a domain $U \subset \mathbb{R}_t^m$ into V.

We wish to find the coordinate representation of the form $\varphi^*\omega$ in U. We take a point $t \in U$ and vectors $\boldsymbol{\tau}_1, \boldsymbol{\tau}_2 \in TU_t$. The vectors $\boldsymbol{\xi}_1 = \varphi'(t)\boldsymbol{\tau}_1$ and $\boldsymbol{\xi}_2 = \varphi'(t)\boldsymbol{\tau}_2$ correspond to them in the space $TV_{x=\varphi(t)}$. The coordinates $(\xi_1^1, \ldots, \xi_1^n)$ and $(\xi_2^1, \ldots, \xi_2^n)$ of these vectors can be expressed in terms of the coordinates $(\tau_1^1, \ldots, \tau_1^m)$ and $(\tau_2^1, \ldots, \tau_2^m)$ of $\boldsymbol{\tau}_1$ and $\boldsymbol{\tau}_2$ using the Jacobian matrix via the formulas

$$\xi_1^i = \frac{\partial x^i}{\partial t^j}(t)\tau_1^j, \quad \xi_2^i = \frac{\partial x^i}{\partial t^j}(t)\tau_2^j, \quad i = 1, \ldots, n.$$

(The summation on j runs from 1 to m.)

Thus,

$$\varphi^*\omega(t)(\boldsymbol{\tau}_1, \boldsymbol{\tau}_2) := \omega\big(\varphi(t)\big)(\boldsymbol{\xi}_1, \boldsymbol{\xi}_2) = dx^{i_1} \wedge dx^{i_2}(\boldsymbol{\xi}_1, \boldsymbol{\xi}_2) =$$

$$= \begin{vmatrix} \xi_1^{i_1} & \xi_1^{i_2} \\ \xi_2^{i_1} & \xi_2^{i_2} \end{vmatrix} = \begin{vmatrix} \frac{\partial x^{i_1}}{\partial t^{j_1}}\tau_1^{j_1} & \frac{\partial x^{i_2}}{\partial t^{j_2}}\tau_1^{j_2} \\ \frac{\partial x^{i_1}}{\partial t^{j_1}}\tau_2^{j_1} & \frac{\partial x^{i_2}}{\partial t^{j_2}}\tau_2^{j_2} \end{vmatrix} =$$

$$= \sum_{j_1,j_2=1}^m \frac{\partial x^{i_1}}{\partial t^{j_1}} \frac{\partial x^{i_2}}{\partial t^{j_2}} \begin{vmatrix} \tau_1^{j_1} & \tau_1^{j_2} \\ \tau_2^{j_1} & \tau_2^{j_2} \end{vmatrix} =$$

$$= \sum_{j_1,j_2=1}^m \frac{\partial x^{i_1}}{\partial t^{j_1}} \frac{\partial x^{i_2}}{\partial t^{j_2}} dt^{j_1} \wedge dt^{j_2}(\boldsymbol{\tau}_1, \boldsymbol{\tau}_2) =$$

$$= \sum_{1 \le j_1 < j_2 \le m} \left(\frac{\partial x^{i_1}}{\partial t^{j_1}} \frac{\partial x^{i_2}}{\partial t^{j_2}} - \frac{\partial x^{i_1}}{\partial t^{j_2}} \frac{\partial x^{i_2}}{\partial t^{j_1}} \right) dt^{j_1} \wedge dt^{j_2}(\boldsymbol{\tau}_1, \boldsymbol{\tau}_2) =$$

$$= \sum_{1 \le j_1 < j_2 \le m} \begin{vmatrix} \frac{\partial x^{i_1}}{\partial t^{j_1}} & \frac{\partial x^{i_2}}{\partial t^{j_1}} \\ \frac{\partial x^{i_1}}{\partial t^{j_2}} & \frac{\partial x^{i_2}}{\partial t^{j_2}} \end{vmatrix} (t)\, dt^{j_1} \wedge dt^{j_2}(\boldsymbol{\tau}_1, \boldsymbol{\tau}_2).$$

Consequently, we have shown that

$$\varphi^*\left(dx^{i_1} \wedge dx^{i_2}\right) = \sum_{1 \le i_1 < i_2 \le m} \frac{\partial(x^{i_1}, x^{i_2})}{\partial(t^{j_1}, t^{j_2})}(t) \, dt^{j_1} \wedge dt^{j_2}.$$

If we use properties (12.28) and (12.29) for the operation of transformation of forms[9] and repeat the reasoning of the last example, we obtain the following equality:

$$\varphi^*\left(\sum_{1 \le i_1 < \cdots < i_p \le n} a_{i_1,\ldots,i_p}(x) \, dx^{i_1} \wedge \cdots \wedge dx^{i_p} \right) =$$

$$= \sum_{\substack{1 \le i_1 < \cdots < i_p \le n \\ 1 \le j_1 < \cdots < j_p \le m}} a_{i_1,\ldots,i_p}\left(x(t)\right) \frac{\partial(x^{i_1}, \ldots, x^{i_p})}{\partial(t^{j_1}, \ldots, t^{j_p})} \, dt^{j_1} \wedge \cdots \wedge dt^{j_p}. \quad (12.31)$$

We remark that if we make the formal change of variable $x = x(t)$ in the form that is the argument of φ^* on the left, express the differentials $dx^1, \ldots, dx^n$ in terms of the differentials $dt^1, \ldots, dt^m$, and gather like terms in the resulting expression, using the properties of the exterior product, we obtain precisely the right-hand side of Eq. (12.31).

Indeed, for each fixed choice of indices $i_1, \ldots, i_p$ we have

$$a_{i_1,\ldots,i_p}(x) \, dx^{i_1} \wedge \cdots \wedge dx^{i_p} =$$

$$= a_{i_1,\ldots,i_p}\left(x(t)\right) \left(\frac{\partial x^{i_1}}{\partial t^{j_1}} dt^{j_1} \right) \wedge \cdots \wedge \left(\frac{\partial x^{i_p}}{\partial t^{j_p}} dt^{j_p} \right) =$$

$$= a_{i_1,\ldots,i_p}\left(x(t)\right) \frac{\partial x^{i_1}}{\partial t^{j_1}} \cdot \ldots \cdot \frac{\partial x^{i_p}}{\partial t^{j_p}} dt^{j_1} \wedge \cdots \wedge dt^{j_p} =$$

$$= \sum_{1 < j_1 < \cdots j_p \le m} a_{i_1,\ldots,i_p}\left(x(t)\right) \frac{\partial(x^{i_1}, \ldots, x^{i_p})}{\partial(t^{j_1}, \ldots, t^{j_p})} dt^{j_1} \wedge \cdots \wedge dt^{j_p}.$$

Summing such equalities over all ordered sets $1 \le i_1 < \cdots < i_p \le n$, we obtain the right-hand side of (12.31).

Thus we have proved the following proposition, of great technical importance.

Proposition *If a differential form ω is defined in a domain $V \subset \mathbb{R}^n$ and $\varphi : U \to V$ is a smooth mapping of a domain $U \subset \mathbb{R}^m$ into V, then the coordinate expression*

---

[9]If (12.29) is used pointwise, one can see that

$$\varphi^*\left(a(x)\omega\right) = a\left(\varphi(t)\right)\varphi^*\omega.$$

of the form $\varphi^*\omega$ can be obtained from the coordinate expression

$$\sum_{1\leq i_1<\cdots<i_p\leq n} a_{i_1,\ldots,i_p}(x)\,\mathrm{d}x^{i_1}\wedge\cdots\wedge\mathrm{d}x^{i_p}$$

of the form ω by the direct change of variable $x = \varphi(t)$ (with subsequent transformations in accordance with the properties of the exterior product).

Example 14 In particular, if $m = n = p$, relation (12.31) reduces to the equality

$$\varphi^*\big(\mathrm{d}x^1\wedge\cdots\wedge\mathrm{d}x^n\big) = \det\varphi'(t)\,\mathrm{d}t^1\wedge\cdots\wedge\mathrm{d}t^n. \tag{12.32}$$

Hence, if we write $f(x)\,\mathrm{d}x^1\wedge\cdots\wedge\mathrm{d}x^n$ in a multiple integral instead of $f(x)\,\mathrm{d}x^1\cdots\mathrm{d}x^n$, the formula

$$\int_{V=\varphi(U)} f(x)\,\mathrm{d}x = \int_U f\big(\varphi(t)\big)\det\varphi'(t)\,\mathrm{d}t$$

for change of variable in a multiple integral via an orientation-preserving diffeomorphism (that is, when $\det\varphi'(t) > 0$) could be obtained automatically by the formal substitution $x = \varphi(t)$, just as happened in the one-dimensional case, and it could be given the following form:

$$\int_{\varphi(U)}\omega = \int_U \varphi^*\omega. \tag{12.33}$$

We remark in conclusion that if the degree p of the form ω in the domain $V \subset \mathbb{R}_x^n$ is larger than the dimension m of the domain $U \subset \mathbb{R}^m$ that is mapped into V via $\varphi : U \to V$, then the form $\varphi^*\omega$ on U corresponding to ω is obviously zero. Thus the mapping $\varphi^* : \Omega^p(V) \to \Omega^p(U)$ is not necessarily injective in general.

On the other hand, if $\varphi : U \to V$ has a smooth inverse $\varphi^{-1} : V \to U$, then by (12.30) and the equalities $\varphi^{-1}\circ\varphi = e_U$, $\varphi\circ\varphi^{-1} = e_V$, we find that $(\varphi)^*\circ(\varphi^{-1})^* = e_U^*$ and $(\varphi^{-1})^*\circ\varphi^* = e_V^*$. And, since e_U^* and e_V^* are the identity mappings on $\Omega^p(U)$ and $\Omega^p(V)$ respectively, the mappings $\varphi^* : \Omega^p(V) \to \Omega^p(U)$ and $(\varphi^{-1})^* : \Omega^p(U) \to \Omega^p(V)$, as one would expect, turn out to be inverses of each other. That is, in this case, the mapping $\varphi^* : \Omega^p(V) \to \Omega^p(U)$ is bijective.

We note finally that along with the properties (12.28)–(12.30) the mapping φ^* that transfers forms, as one can verify, also satisfies the relation

$$\varphi^*(\mathrm{d}\omega) = \mathrm{d}\big(\varphi^*\omega\big). \tag{12.34}$$

This theoretically important equality shows in particular that the operation of differentiation of forms, which we defined in coordinate notation, is actually independent of the coordinate system in which the differentiable form ω is written. This will be discussed in more detail in Chap. 15.

12.5.5 Forms on Surfaces

Definition 3 We say that a *differential p-form* ω is defined on a smooth surface $S \subset \mathbb{R}^n$ if a p-form $\omega(x)$ is defined on the vectors of the tangent plane TS_x to S at each point $x \in S$.

Example 15 If the smooth surface S is contained in the domain $D \subset \mathbb{R}^n$ in which a form ω is defined, then, since the inclusion $TS_x \subset TD_x$ holds at each point $x \in S$, one can consider the restriction of $\omega(x)$ to TS_x. In this way a form $\omega|_S$ arises, which it is natural to call the *restriction of ω to S*.

As we know, a surface can be defined parametrically, either locally or globally. Let $\varphi : U \to S = \varphi(U) \subset D$ be a parametrized smooth surface in the domain D and ω a form on D. Then we can transfer the form ω to the domain U of parameters and write $\varphi^*\omega$ in coordinate form in accordance with the algorithm given above. It is clear that the form $\varphi^*\omega$ in U obtained in this way coincides with the form $\varphi^*(\omega|_S)$.

We remark that, since $\varphi'(t) : TU_t \to TS_x$ is an isomorphism between TU_t and TS_x at every point $t \in U$, we can transfer forms both from S to U and from U to S, and so just as the smooth surfaces themselves are usually defined locally or globally by parameters, the forms on them, in the final analysis, are usually defined in the parameter domains of local charts.

Example 16 Let $\omega_{\mathbf{V}}^2$ be the flux form considered in Example 8, generated by the velocity field $\mathbf{V}$ of a flow in the domain D of the oriented Euclidean space $\mathbb{R}^3$. If S is a smooth oriented surface in D, one may consider the restriction $\omega_{\mathbf{V}}^2|_S$ of the form $\omega_{\mathbf{V}}^2$ to S. The form $\omega_{\mathbf{V}}^2|_S$ so obtained characterizes the flux across each element of the surface S.

If $\varphi : I \to S$ is a local chart of the surface S, then, making the change of variable $x = \varphi(t)$ in the coordinate expression (12.22) for the form $\omega_{\mathbf{V}}^2$, we obtain the coordinate expression for the form $\varphi^*\omega_{\mathbf{V}}^2 = \varphi^*(\omega_{\mathbf{V}}^2|_S)$, which is defined on the square I, in these local coordinates of the surface.

Example 17 Let $\omega_{\mathbf{F}}^1$ be the work form considered in Example 7, generated by the force field $\mathbf{F}$ acting in a domain D of Euclidean space. Let $\varphi : I \to \varphi(I) \subset D$ be a smooth path (φ is not necessarily a homeomorphism). Then, in accordance with the general principle of restriction and transfer of forms, a form $\varphi^*\omega_{\mathbf{F}}^1$ arises on the closed interval I, whose coordinate representation $a(t)\,dt$ can be obtained by the change of variable $x = \varphi(t)$ in the coordinate expression (12.21) for the form $\omega_{\mathbf{F}}^1$.

12.5.6 Problems and Exercises

1. Compute the values of the differential forms ω in $\mathbb{R}^n$ given below on the indicated sets of vectors:

a) $\omega = x^2\,\mathrm{d}x^1$ on the vector $\boldsymbol{\xi} = (1, 2, 3) \in T\mathbb{R}_{(3,2,1)}$.

b) $\omega = \mathrm{d}x^1 \wedge \mathrm{d}x^3 + x^1\,\mathrm{d}x^2 \wedge \mathrm{d}x^4$ on the ordered pair of vectors $\boldsymbol{\xi}_1, \boldsymbol{\xi}_2 \in T\mathbb{R}^4_{(1,0,0,0)}$.

c) $\omega = \mathrm{d}f$, where $f = x^1 + 2x^2 + \cdots + nx^n$, and $\boldsymbol{\xi} = (1, -, 1, \ldots, (-1)^{n-1}) \in T\mathbb{R}^n_{(1,1,\ldots,1)}$.

2. a) Verify that the form $\mathrm{d}x^{i_1} \wedge \cdots \wedge \mathrm{d}x^{i_k}$ is identically zero if the indices $i_1, \ldots, i_k$ are not all distinct.

b) Explain why there are no nonzero skew-symmetric forms of degree $p > n$ on an n-dimensional vector space.

c) Simplify the expression for the form

$$2\,\mathrm{d}x^1 \wedge \mathrm{d}x^3 \wedge \mathrm{d}x^2 + 3\,\mathrm{d}x^3 \wedge \mathrm{d}x^1 \wedge \mathrm{d}x^2 - \mathrm{d}x^2 \wedge \mathrm{d}x^3 \wedge \mathrm{d}x^1.$$

d) Remove the parentheses and gather like terms:

$$\left(x^1\,\mathrm{d}x^2 + x^2\,\mathrm{d}x^1\right) \wedge \left(x^3\,\mathrm{d}x^1 \wedge \mathrm{d}x^2 + x^2\,\mathrm{d}x^1 \wedge \mathrm{d}x^3 + x^1\,\mathrm{d}x^2 \wedge \mathrm{d}x^3\right).$$

e) Write the form $\mathrm{d}f \wedge \mathrm{d}g$, where $f = \ln(1 + |x|^2)$, $g = \sin|x|$, and $x = (x^1, x^2, x^3)$ as a linear combination of the forms $\mathrm{d}x^{i_1} \wedge \mathrm{d}x^{i_2}$, $1 \leq i_1 < i_2 \leq 3$.

f) Verify that in $\mathbb{R}^n$

$$\mathrm{d}f^1 \wedge \cdots \wedge \mathrm{d}f^n(x) = \det\left(\frac{\partial f^i}{\partial x^j}\right)(x)\,\mathrm{d}x^1 \wedge \cdots \wedge \mathrm{d}x^n.$$

g) Carry out all the computations and show that for $1 \leq k \leq n$

$$\mathrm{d}f^1 \wedge \cdots \wedge \mathrm{d}f^k = \sum_{1 \leq i_1 < i_2 < \cdots < i_k \leq n} \det \begin{vmatrix} \dfrac{\partial f^1}{\partial x^{i_1}} & \cdots & \dfrac{\partial f^1}{\partial x^{i_k}} \\ \dfrac{\partial f^k}{\partial x^{i_1}} & \cdots & \dfrac{\partial f^k}{\partial x^{i_k}} \end{vmatrix} \mathrm{d}x^{i_1} \wedge \cdots \wedge \mathrm{d}x^{i_k}.$$

3. a) Show that a form α of even degree commutes with any form β, that is, $\alpha \wedge \beta = \beta \wedge \alpha$.

b) Let $\omega = \sum_{i=1}^n \mathrm{d}p_i \wedge \mathrm{d}q^i$ and $\omega^n = \omega \wedge \cdots \wedge \omega$ (n factors). Verify that $\omega^n = n!\,\mathrm{d}p_1 \wedge \mathrm{d}q^1 \wedge \cdots \wedge \mathrm{d}p_n \wedge \mathrm{d}q^n = (-1)^{\frac{n(n-1)}{2}}\,\mathrm{d}p_1 \wedge \cdots \wedge \mathrm{d}p_n \wedge \mathrm{d}q^1 \wedge \cdots \wedge \mathrm{d}q^n$.

4. a) Write the form $\omega = \mathrm{d}f$, where $f(x) = (x^1)^2 + (x^2)^2 + \cdots + (x^n)^2$, as a combination of the forms $\mathrm{d}x^1, \ldots, \mathrm{d}x^n$ and find the differential $\mathrm{d}\omega$ of ω.

b) Verify that $\mathrm{d}^2 f \equiv 0$ for any function $f \in C^{(2)}(D, \mathbb{R})$, where $\mathrm{d}^2 = \mathrm{d} \circ \mathrm{d}$, and d is exterior differentiation.

c) Show that if the coefficients $a_{i_1,\ldots,i_k}$ of the form $\omega = a_{i_1,\ldots,i_k}(x)\,\mathrm{d}x^{i_1} \wedge \cdots \wedge \mathrm{d}x^{i_k}$ belongs to the class $C^{(2)}(D, \mathbb{R})$, then $\mathrm{d}^2\omega \equiv 0$ in the domain D.

d) Find the exterior differential of the form $\frac{y\,\mathrm{d}x - x\,\mathrm{d}y}{x^2 + y^2}$ in its domain of definition.

5. If the product $\mathrm{d}x^1 \cdots \mathrm{d}x^n$ in the multiple integral $\int_D f(x)\,\mathrm{d}x^1 \cdots \mathrm{d}x^n$ is interpreted as the form $\mathrm{d}x^1 \wedge \cdots \wedge \mathrm{d}x^n$, then, by the result of Example 14, we have the possibility of formally obtaining the integrand in the formula for change of variable in a multiple integral. Using this recommendation, carry out the following changes of variable from Cartesian coordinates:

    a) to polar coordinates in $\mathbb{R}^2$,
    b) to cylindrical coordinates in $\mathbb{R}^3$,
    c) to spherical coordinates in $\mathbb{R}^3$.

6. Find the restriction of the following forms:

    a) dx^i to the hyperplane $x^i = 1$.
    b) $dx \wedge dy$ to the curve $x = x(t)$, $y = y(t)$, $a < t < b$.
    c) $dx \wedge dy$ to the plane in $\mathbb{R}^3$ defined by the equation $x = c$.
    d) $dy \wedge dz + dz \wedge dx + dx \wedge dy$ to the faces of the standard unit cube in $\mathbb{R}^3$.

    e) $\omega_i = dx^1 \wedge \cdots \wedge dx^{i-1} \wedge \widehat{dx^i} \wedge dx^{i+1} \wedge \cdots \wedge dx^n$ to the faces of the standard unit cube in $\mathbb{R}^n$. The symbol $\frown$ stands over the differential dx^i that is to be omitted in the product.

7. Express the restriction of the following forms to the sphere of radius R with center at the origin in spherical coordinates on $\mathbb{R}^3$:

    a) dx,
    b) dy,
    c) $dy \wedge dz$.

8. The mapping $\varphi : \mathbb{R}^2 \to \mathbb{R}^2$ is given in the form $(u, v) \mapsto (u \cdot v, 1) = (x, y)$. Find:

    a) $\varphi^*(dx)$,
    b) $\varphi^*(dy)$,
    c) $\varphi^*(y\,dx)$.

9. Verify that the exterior differential $d : \Omega^p(D) \to \Omega^{p+1}(D)$ has the following properties:

    a) $d(\omega_1 + \omega_2) = d\omega_1 + d\omega_2$,
    b) $d(\omega_1 \wedge \omega_2) = d\omega_1 \wedge \omega_2 + (-1)^{\deg \omega_1} \omega_1 \wedge d\omega_2$, where $\deg \omega_1$ is the degree of the form ω_1.
    c) $\forall \omega \in \Omega^p \ d(d\omega) = 0$.
    d) $\forall f \in \Omega^0 \ df = \sum_{i=1}^n \frac{\partial f}{\partial x^i}\, dx^i$.
    Show that there is only one mapping $d : \Omega^p(D) \to \Omega^{p+1}(D)$ having properties a), b), c), and d).

10. Verify that the mapping $\varphi^* : \Omega^p(V) \to \Omega^p(U)$ corresponding to a mapping $\varphi : U \to V$ has the following properties:

    a) $\varphi^*(\omega_1 + \omega_2) = \varphi^*\omega_1 + \varphi^*\omega_2$.
    b) $\varphi^*(\omega_1 \wedge \omega_2) = \varphi^*\omega_1 \wedge \varphi^*\omega_2$.
    c) $d\varphi^*\omega = \varphi^* d\omega$.
    d) If there is a mapping $\psi : V \to W$, then $(\psi \circ \varphi)^* = \varphi^* \circ \psi^*$.

11. Show that a smooth k-dimensional surface is orientable if and only if there exists a k-form on it that is not degenerate at any point.

Chapter 13
Line and Surface Integrals

13.1 The Integral of a Differential Form

13.1.1 The Original Problems, Suggestive Considerations, Examples

a. The Work of a Field

Let $\mathbf{F}(x)$ be a continuous force field acting in a domain G of the Euclidean space $\mathbb{R}^n$. The displacement of a test particle in the field is accompanied by work. We ask how we can compute the work done by the field in moving a unit test particle along a given trajectory, more precisely, a smooth path $\gamma : I \to \gamma(I) \subset G$.

We have already touched on this problem when we studied the applications of the definite integral. For that reason we can merely recall the solution of the problem at this point, noting certain elements of the construction that will be useful in what follows.

It is known that in a constant field $\mathbf{F}$ the displacement by a vector $\boldsymbol{\xi}$ is associated with an amount of work $\langle \mathbf{F}, \boldsymbol{\xi} \rangle$.

Let $t \mapsto \mathbf{x}(t)$ be a smooth mapping $\gamma : I \to G$ defined on the closed interval $I = \{t \in \mathbb{R} \mid a \le t \le b\}$.

We take a sufficiently fine partition of the closed interval $[a, b]$. Then on each interval $I_i = \{t \in I \mid t_{i-1} \le t \le t_i\}$ of the partition we have the equality $\mathbf{x}(t) - \mathbf{x}(t_i) \approx \mathbf{x}'(t)(t_i - t_{i-1})$ up to infinitesimals of higher order. To the displacement vector $\tau_i = t_{i+1} - t_i$ from the point t_i (Fig. 13.1) there corresponds a displacement of $\mathbf{x}(t_i)$ in $\mathbb{R}^n$ by the vector $\Delta \mathbf{x}_i = \mathbf{x}_{i+1} - \mathbf{x}_i$, which can be regarded as equal to the tangent vector $\boldsymbol{\xi}_i = \dot{\mathbf{x}}(t_i)\tau_i$ to the trajectory at $\mathbf{x}(t_i)$ with the same precision. Since the field $\mathbf{F}(\mathbf{x})$ is continuous, it can be regarded a locally constant, and for that reason we can compute the work ΔA_i corresponding to the (time) interval I_i with small relative error as

$$\Delta A_i \approx \langle \mathbf{F}(x_i),\ \boldsymbol{\xi}_i \rangle$$

© Springer-Verlag Berlin Heidelberg 2016
V.A. Zorich, *Mathematical Analysis II*, Universitext,
DOI 10.1007/978-3-662-48993-2_5

Fig. 13.1

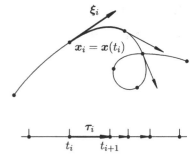

or

$$\Delta A_i \approx \big\langle \mathbf{F}\big(\mathbf{x}(t_i)\big), \dot{\mathbf{x}}(t_i)\tau_i\big\rangle.$$

Hence,

$$A = \sum_i \Delta A_i \approx \sum_i \big\langle \mathbf{F}\big(\mathbf{x}(t_i)\big), \dot{\mathbf{x}}(t_i)\big\rangle \Delta t_i$$

and so, passing to the limit as the partition of the closed interval I is refined, we find that

$$A = \int_a^b \big\langle \mathbf{F}\big(\mathbf{x}(t)\big), \dot{\mathbf{x}}(t)\big\rangle \mathrm{d}t. \tag{13.1}$$

If the expression $\langle \mathbf{F}(\mathbf{x}(t)), \dot{\mathbf{x}}(t)\rangle \,\mathrm{d}t$ is rewritten as $\langle \mathbf{F}(\mathbf{x}), \mathrm{d}\mathbf{x}\rangle$, then, assuming the coordinates in $\mathbb{R}^n$ are Cartesian coordinates, we can give this expression the form $F^1 \,\mathrm{d}x^1 + \cdots + F^n \,\mathrm{d}x^n$, after which we can write (13.1) as

$$A = \int_\gamma F^1 \,\mathrm{d}x^1 + \cdots + F^n \,\mathrm{d}x^n \tag{13.2}$$

or as

$$A = \int_\gamma \omega_{\mathbf{F}}^1. \tag{13.2'}$$

Formula (13.1) provides the precise meaning of the integrals of the work 1-form along the path γ written in formulas (13.2) and (13.2′).

Example 1 Consider the force field $\mathbf{F} = (-\frac{y}{x^2+y^2}, \frac{x}{x^2+y^2})$ defined at all points of the plane $\mathbb{R}^2$ except the origin. Let us compute the work of this field along the curve γ_1 defined as $x = \cos t$, $y = \sin t$, $0 \le t \le 2\pi$, and along the curve defined by $x = 2 + \cos t$, $y = \sin t$, $0 \le t \le 2\pi$. According to formulas (13.1), (13.2), and (13.2′), we find

$$\int_{\gamma_1} \omega_{\mathbf{F}}^1 = \int_{\gamma_1} -\frac{y}{x^2+y^2}\,\mathrm{d}x + \frac{x}{x^2+y^2}\,\mathrm{d}y =$$

$$= \int_0^{2\pi} \left(-\frac{\sin t \cdot (-\sin t)}{\cos^2 t + \sin^2 t} + \frac{\cos t \cdot \cos t}{\cos^2 t + \sin^2 t} \right) dt = 2\pi$$

and

$$\int_{\gamma_2} \omega_F^1 = \int_{\gamma_2} \frac{-y\,dx + x\,dy}{x^2 + y^2} = \int_0^{2\pi} \frac{-\sin t(-\sin t) + (2 + \cos t)(\cos t)}{(2 + \cos t)^2 + \sin^2 t}\,dt =$$

$$= \int_0^{2\pi} \frac{1 + 2\cos t}{5 + 4\cos t}\,dt = \int_0^{\pi} \frac{1 + 2\cos t}{5 + 4\cos t}\,dt + \int_{\pi}^0 \frac{1 + 2\cos(2\pi - u)}{5 + 4\cos(2\pi - u)}\,du =$$

$$= \int_0^{\pi} \frac{1 + 2\cos t}{5 + 4\cos t}\,dt - \int_0^{\pi} \frac{1 + 2\cos u}{5 + 4\cos u}\,du = 0.$$

Example 2 Let $\mathbf{r}$ be the radius vector of a point $(x, y, z) \in \mathbb{R}^3$ and $r = |\mathbf{r}|$. Suppose a force field $\mathbf{F} = f(r)\mathbf{r}$ is defined everywhere in $\mathbb{R}^3$ except at the origin. This is a so-called *central force field*. Let us find the work of $\mathbf{F}$ on a path $\gamma : [0, 1] \to \mathbb{R}^3 \backslash 0$. Using (13.2), we find

$$\int_{\gamma} f(r)(x\,dx + y\,dy + z\,dz) = \frac{1}{2} \int_{\gamma} f(r)\,d(x^2 + y^2 + z^2) =$$

$$= \frac{1}{2} \int_0^1 f(r(t))\,dr^2(t) = \frac{1}{2} \int_0^1 f(\sqrt{u(t)})\,du(t) =$$

$$= \frac{1}{2} \int_{r_0^2}^{r_1^2} f(\sqrt{u})\,du = \Phi(r_0, r_1).$$

Here, as one can see, we have set $x^2(t) + y^2(t) + z^2(t) = r^2(t)$, $r^2(t) = u(t)$, $r_0 = r(0)$, and $r_1 = r(1)$.

Thus in any central field the work on a path γ has turned out to depend only on the distances r_0 and r_1 of the beginning and end of the path from the center 0 of the field.

In particular, for the gravitational field $\frac{1}{r^3}\mathbf{r}$ of a unit point mass located at the origin, we obtain

$$\Phi(r_0, r_1) = \frac{1}{2} \int_{r_0^2}^{r_1^2} \frac{1}{u^{3/2}}\,du = \frac{1}{r_0} - \frac{1}{r_1}.$$

b. The Flux Across a Surface

Suppose there is a steady flow of liquid (or gas) in a domain G of the oriented Euclidean space $\mathbb{R}^3$ and that $x \mapsto \mathbf{V}(x)$ is the velocity field of this flow. In addition, suppose that a smooth oriented surface S has been chosen in G. For definiteness we shall suppose that the orientation of S is given by a field of normal vectors. We ask how to determine the (volumetric) outflow or flux of fluid across the surface S. More

Fig. 13.2

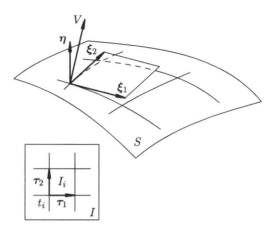

precisely, we ask how to find the volume of fluid that flows across the surface S per unit time in the direction indicated by the orienting field of normals to the surface.

To solve the problem, we remark that if the velocity field of the flow is constant and equal to $\mathbf{V}$, then the flow per unit time across a parallelogram Π spanned by vectors $\boldsymbol{\xi}_1$ and $\boldsymbol{\xi}_2$ equals the volume of the parallelepiped constructed on the vectors $\mathbf{V}, \boldsymbol{\xi}_1, \boldsymbol{\xi}_2$. If $\boldsymbol{\eta}$ is normal to Π and we seek the flux across Π in the direction of $\boldsymbol{\eta}$, it equals the scalar triple product $(\mathbf{V}, \boldsymbol{\xi}_1, \boldsymbol{\xi}_2)$, provided $\boldsymbol{\eta}$ and the frame $\boldsymbol{\xi}_1, \boldsymbol{\xi}_2$ give Π the same orientation (that is, if $\boldsymbol{\eta}, \boldsymbol{\xi}_1, \boldsymbol{\xi}_2$ is a frame having the given orientation in $\mathbb{R}^3$). If the frame $\boldsymbol{\xi}_1, \boldsymbol{\xi}_2$ gives the orientation opposite to the one given by $\boldsymbol{\eta}$ in Π, then the flow in the direction of $\boldsymbol{\eta}$ is $-(\mathbf{V}, \boldsymbol{\xi}_1, \boldsymbol{\xi}_2)$.

We now return to the original statement of the problem. For simplicity let us assume that the entire surface S admits a smooth parametrization $\varphi : I \to S \subset G$, where I is a two-dimensional interval in the plane $\mathbb{R}^2$. We partition I into small intervals I_i (Fig. 13.2). We approximate the image $\varphi(I_i)$ of each such interval by the parallelogram spanned by the images $\boldsymbol{\xi}_1 = \varphi'(t_i)\boldsymbol{\tau}_1$ and $\boldsymbol{\xi}_2 = \varphi'(t_i)\boldsymbol{\tau}_2$ of the displacement vectors $\boldsymbol{\tau}_1, \boldsymbol{\tau}_2$ along the coordinate directions. Assuming that $\mathbf{V}(x)$ varies by only a small amount inside the piece of surface $\varphi(I_i)$ and replacing $\varphi(I_i)$ by this parallelogram, we may assume that the flux $\Delta\mathcal{F}_i$ across the piece $\varphi(I_i)$ of the surface is equal, with small relative error, to the flux of a constant velocity field $\mathbf{V}(x_i) = \mathbf{V}(\varphi(t_i))$ across the parallelogram spanned by the vectors $\boldsymbol{\xi}_1, \boldsymbol{\xi}_2$.

Assuming that the frame $\boldsymbol{\xi}_1, \boldsymbol{\xi}_2$ gives the same orientation on S as $\boldsymbol{\eta}$, we find

$$\Delta\mathcal{F}_i \approx \big(\mathbf{V}(x_i), \boldsymbol{\xi}_1, \boldsymbol{\xi}_2\big).$$

Summing the elementary fluxes, we obtain

$$\mathcal{F} = \sum_i \Delta\mathcal{F}_i \approx \sum_i \omega_{\mathbf{V}}^2(x_i)(\boldsymbol{\xi}_1, \boldsymbol{\xi}_2),$$

where $\omega_{\mathbf{V}}^2(x) = (\mathbf{V}(x), \cdot, \cdot)$ is the flux 2-form (studied in Example 8 of Sect. 12.5). If we pass to the limit, taking ever finer partitions P of the interval I, it is natural to assume that

$$\mathcal{F} := \lim_{\lambda(P) \to 0} \sum \omega_{\mathbf{V}}^2(x_i)(\boldsymbol{\xi}_1, \boldsymbol{\xi}_2) =: \int_S \omega_{\mathbf{V}}^2. \tag{13.3}$$

This last symbol is the integral of the 2-form $\omega_{\mathbf{V}}^2$ over the oriented surface S.

Recalling (formula (12.22) of Sect. 12.5) the coordinate expression for the flux form $\omega_{\mathbf{V}}^2$ in Cartesian coordinates, we may now also write

$$\mathcal{F} = \int_S V^1 \, dx^2 \wedge dx^3 + V^2 \, dx^3 \wedge dx^1 + V^3 \, dx^1 \wedge dx^2. \tag{13.4}$$

We have discussed here only the general principle for solving this problem. In essence all we have done is to give the precise definition (13.3) of the flux $\mathcal{F}$ and introduced certain notation (13.3) and (13.4); we have still not obtained any effective computational formula similar to formula (13.1) for the work.

We remark that formula (13.1) can be obtained from (13.2) by replacing $x^1, \ldots, x^n$ with the functions $(x^1, \ldots, x^n)(t) = x(t)$ that define the path γ. We recall (Sect. 12.5) that such a substitution can be interpreted as the transfer of the form ω defined in G to the closed interval $I = [a, b]$.

In a completely analogous way, a computational formula for the flux can be obtained by direct substitution of the parametric equations of the surface into (13.4).

In fact,

$$\omega_{\mathbf{V}}^2(x_i)(\boldsymbol{\xi}_1, \boldsymbol{\xi}_2) = \omega_{\mathbf{V}}\big(\varphi(t_i)\big)\big(\varphi'(t_i)\boldsymbol{\tau}_1, \varphi'(t_i)\boldsymbol{\tau}_2\big) = \big(\varphi^* \omega_{\mathbf{V}}^2\big)(t_i)(\boldsymbol{\tau}_1, \boldsymbol{\tau}_2)$$

and

$$\sum_i \omega_{\mathbf{V}}^2(x_i)(\boldsymbol{\xi}_1, \boldsymbol{\xi}_2) = \sum_i \big(\varphi^* \omega_{\mathbf{V}}^2\big)(t_i)(\boldsymbol{\tau}_1, \boldsymbol{\tau}_2).$$

The form $\varphi^* \omega_{\mathbf{V}}^2$ is defined on a two-dimensional interval $I \subset \mathbb{R}^2$. Any 2-form in I has the form $f(t) \, dt^1 \wedge dt^2$, where f is a function on I depending on the form. Therefore

$$\varphi^* \omega_{\mathbf{V}}^2(t_i)(\boldsymbol{\tau}_1, \boldsymbol{\tau}_2) = f(t_i) \, dt^1 \wedge dt^2(\boldsymbol{\tau}_1, \boldsymbol{\tau}_2).$$

But $dt^1 \wedge dt^2(\boldsymbol{\tau}_1, \boldsymbol{\tau}_2) = \tau_1^1 \cdot \tau_2^2$ is the area of the rectangle I_i spanned by the orthogonal vectors $\boldsymbol{\tau}_1, \boldsymbol{\tau}_2$.

Thus,

$$\sum_i f(t_i) \, dt^1 \wedge dt^2(\boldsymbol{\tau}_1, \boldsymbol{\tau}_2) = \sum_i f(t_i)|I_i|.$$

As the partition is refined we obtain in the limit

$$\int_I f(t) \, dt^1 \wedge dt^2 = \int_I f(t) \, dt^1 \, dt^2, \tag{13.5}$$

where, according to (13.3), the left-hand side contains the integral of the 2-form $\omega^2 = f(t) \, dt^1 \wedge dt^2$ over the elementary oriented surface I, and the right-hand side the integral of the function f over the rectangle I.

It remains only to recall that the coordinate representation $f(t)\,dt^1 \wedge dt^2$ of the form $\varphi^*\omega_{\mathbf{V}}^2$ is obtained from the coordinate expression for the form $\omega_{\mathbf{V}}^2$ by the direct substitution $x = \varphi(t)$, where $\varphi : I \to G$ is a chart of the surface S.

Carrying out this change of variable, we obtain from (13.4)

$$\mathcal{F} = \int_{S=\varphi(I)} \omega_{\mathbf{V}}^2 = \int_{I} \varphi^*\omega_{\mathbf{V}}^2 =$$

$$= \int_{I} \left(V^1\big(\varphi(t)\big) \begin{vmatrix} \frac{\partial x^2}{\partial t^1} & \frac{\partial x^3}{\partial t^1} \\ \frac{\partial x^2}{\partial t^2} & \frac{\partial x^3}{\partial t^2} \end{vmatrix} + V^2\big(\varphi(t)\big) \begin{vmatrix} \frac{\partial x^3}{\partial t^1} & \frac{\partial x^1}{\partial t^1} \\ \frac{\partial x^3}{\partial t^2} & \frac{\partial x^1}{\partial t^2} \end{vmatrix} + \right.$$

$$\left. + V^3\big(\varphi(t)\big) \begin{vmatrix} \frac{\partial x^1}{\partial t^1} & \frac{\partial x^2}{\partial t^1} \\ \frac{\partial x^1}{\partial t^2} & \frac{\partial x^2}{\partial t^2} \end{vmatrix} \right) dt^1 \wedge dt^2.$$

This last integral, as Eq. (13.5) shows, is the ordinary Riemann integral over the rectangle I.

Thus we have found that

$$\mathcal{F} = \int_{I} \begin{vmatrix} V^1(\varphi(t)) & V^2(\varphi(t)) & V^3(\varphi(t)) \\ \frac{\partial\varphi^1}{\partial t^1}(t) & \frac{\partial\varphi^2}{\partial t^1}(t) & \frac{\partial\varphi^3}{\partial t^1}(t) \\ \frac{\partial\varphi^1}{\partial t^2}(t) & \frac{\partial\varphi^2}{\partial t^2}(t) & \frac{\partial\varphi^3}{\partial t^2}(t) \end{vmatrix} dt^1\,dt^2, \tag{13.6}$$

where $x = \varphi(t) = (\varphi^1, \varphi^2, \varphi^3)(t^1, t^2)$ is a chart of the surface S defining the same orientation as the field of normals we have given. If the chart $\varphi : I \to S$ gives S the opposite orientation, Eq. (13.6) does not generally hold. But, as follows from the considerations at the beginning of this subsection, the left- and right-hand sides will differ only in sign in that case.

The final formula (13.6) is obviously merely the limit of the sums of the elementary fluxes $\Delta\mathcal{F}_i \approx (\mathbf{V}(x_i), \boldsymbol{\xi}_1, \boldsymbol{\xi}_2)$ familiar to us, written accurately in the coordinates t^1 and t^2.

We have considered the case of a surface defined by a single chart. In general a smooth surface can be decomposed into smooth pieces S_i having essentially no intersections with one another, and then we can find the flux through S as the sum of the fluxes though the pieces S_i.

Example 3 Suppose a medium is advancing with constant velocity $\mathbf{V} = (1, 0, 0)$. If we take any closed surface in the domain of the flow, then, since the density of the medium does not change, the amount of matter in the volume bounded by this surface must remain constant. Hence the total flux of the medium through such a surface must be zero.

In this case, let us check formula (13.6) by taking S to be the sphere $x^2 + y^2 + z^2 = R^2$.

Up to a set of area zero, which is therefore negligible, this sphere can be defined parametrically

$$x = R \cos \psi \cos \varphi,$$

$$y = R \cos \psi \sin \varphi,$$

$$z = R \sin \psi,$$

where $0 < \varphi < 2\pi$ and $-\pi/2 < \psi < \pi/2$.

After these relations and the relation $\mathbf{V} = (1, 0, 0)$ are substituted in (13.6), we obtain

$$\mathcal{F} = \int_I \begin{vmatrix} \frac{\partial x^1}{\partial t^1} & \frac{\partial x^2}{\partial t^1} \\ \frac{\partial x^1}{\partial t^2} & \frac{\partial x^2}{\partial t^2} \end{vmatrix} d\varphi \, d\psi = R^2 \int_{-\pi/2}^{\pi/2} \cos^2 \psi \, d\psi \int_0^{2\pi} \cos \varphi \, d\varphi = 0.$$

Since the integral equals zero, we have not even bothered to consider whether it was the inward or outward flow we were computing.

Example 4 Suppose the velocity field of a medium moving in $\mathbb{R}^3$ is defined in Cartesian coordinates x, y, z by the equality $\mathbf{V}(x, y, z) = (V^1, V^2, V^3)(x, y, z) = (x, y, z)$. Let us find the flux through the sphere $x^2 + y^2 + z^2 = R^2$ into the ball that it bounds (that is, in the direction of the inward normal) in this case.

Taking the parametrization of the sphere given in the last example, and carrying out the substitution in the right-hand side of (13.6), we find that

$$\int_0^{2\pi} d\varphi \int_{-\pi/2}^{\pi/2} \begin{vmatrix} R \cos \psi \cos \varphi & R \cos \psi \sin \varphi & R \sin \psi \\ -R \cos \psi \sin \varphi & R \cos \psi \cos \varphi & 0 \\ R \sin \psi \cos \varphi & -R \sin \psi \sin \varphi & R \cos \psi \end{vmatrix} d\varphi =$$

$$= \int_0^{2\pi} d\varphi \int_{-\pi/2}^{\pi/2} R^3 \cos \psi \, d\psi = 4\pi R^3.$$

We now check to see whether the orientation of the sphere given by the curvilinear coordinates (φ, ψ) agrees with that given by the inward normal. It is easy to verify that they do not agree. Hence the required flux is given by $\mathcal{F} = -4\pi R^3$.

In this case the result is easy to verify: the velocity vector $\mathbf{V}$ of the flow has magnitude equal to R at each point of the sphere, is orthogonal to the sphere, and points outward. Therefore the outward flux from the inside equals the area of the sphere $4\pi R^2$ multiplied by R. The flux in the opposite direction is then $-4\pi R^3$.

13.1.2 Definition of the Integral of a Form over an Oriented Surface

The solution of the problems considered in Sect. 13.1.1 leads to the definition of the integral of a k-form over a k-dimensional surface.

First let S be a smooth k-dimensional surface in $\mathbb{R}^n$, defined by one standard chart $\varphi : I \to S$. Suppose a k-form ω is defined on S. The integral of the form ω over the parametrized surface $\varphi : I \to S$ is then constructed as follows.

Take a partition P of the k-dimensional standard interval $I \subset \mathbb{R}^n$ induced by partitions of its projections on the coordinate axes (closed intervals). In each interval I_i of the partition P take the vertex t_i having minimal coordinate values and attach to it the k vectors $\boldsymbol{\tau}_1, \ldots, \boldsymbol{\tau}_k$ that go along the direction of the coordinate axes to the k vertices of I_i adjacent to t_i (Fig. 13.2). Find the vectors $\boldsymbol{\xi}_1 = \varphi'(t_i)\boldsymbol{\tau}_1, \ldots, \boldsymbol{\xi}_k = \varphi'(t_i)\boldsymbol{\tau}_k$ of the tangent space $TS_{x_i = \varphi(t_i)}$, then compute $\omega(x_i)(\boldsymbol{\xi}_1, \ldots, \boldsymbol{\xi}_k) =: (\varphi^*\omega)(t_i)(\boldsymbol{\tau}_1, \ldots, \boldsymbol{\tau}_k)$, and form the Riemann sum $\sum_i \omega(x_i)(\boldsymbol{\xi}_1, \ldots, \boldsymbol{\xi}_k)$. Then pass to the limit as the mesh $\lambda(P)$ of the partition tends to zero.

Thus we adopt the following definition:

Definition 1 (Integral of a k-form ω over a given chart $\varphi : I \to S$ of a smooth k-dimensional surface.)

$$\int_S \omega := \lim_{\lambda(P) \to 0} \sum_i \omega(x_i)(\boldsymbol{\xi}_1, \ldots, \boldsymbol{\xi}_k) = \lim_{\lambda(P) \to 0} \sum_i (\varphi^*\omega)(t_i)(\boldsymbol{\tau}_1, \ldots, \boldsymbol{\tau}_k). \tag{13.7}$$

If we apply this definition to the k-form $f(t)\,dt^1 \wedge \cdots \wedge dt^k$ on I (when φ is the identity mapping), we obviously find that

$$\int_I f(t)\,dt^1 \wedge \cdots \wedge dt^k = \int_I f(t)\,dt^1 \cdots dt^k. \tag{13.8}$$

It thus follows from (13.7) that

$$\int_{S=\varphi(I)} \omega = \int_I \varphi^*\omega, \tag{13.9}$$

and the last integral, as Eq. (13.8) shows, reduces to the ordinary multiple integral over the interval I of the function f corresponding to the form $\varphi^*\omega$.

We have derived the important relations (13.8) and (13.9) from Definition 1, but they themselves could have been adopted as the original definitions. In particular, if D is an arbitrary domain in $\mathbb{R}^n$ (not necessarily an interval), then, so as not to repeat the summation procedure, we set

$$\int_D f(t)\,dt^1 \wedge \cdots \wedge dt^k := \int_D f(t)\,dt^1 \cdots dt^k, \tag{13.8'}$$

and for a smooth surface given in the form $\varphi : D \to S$ and a k-form ω on it we set

$$\int_{S=\varphi(D)} \omega := \int_D \varphi^*\omega. \tag{13.9'}$$

If S is an arbitrary piecewise-smooth k-dimensional surface and ω is a k-form defined on the smooth pieces of S, then, representing S as the union $\bigcup_i S_i$ of smooth parametrized surfaces that intersect only in sets of lower dimension, we set

$$\int_S \omega := \sum_i \int_{S_i} \omega. \tag{13.10}$$

In the absence of substantive physical or other problems that can be solved using (13.10), such a definition raises the question whether the magnitude of the integral of the partition $\bigcup_i S_i$ is independent of the choice of the parametrization of its pieces.

Let us verify that this definition is unambiguous.

Proof We begin by considering the simplest case in which S is a domain D_x in $\mathbb{R}^k$ and $\varphi : D_t \to D_x$ is a diffeomorphism of a domain $D_t \subset \mathbb{R}^k$ onto D_x. In $D_x = S$ the k-form ω has the form $f(x)\,dx^1 \wedge \cdots \wedge dx^k$. Then, on the one hand (13.8) implies

$$\int_{D_x} f(x)\,dx^1 \wedge \cdots \wedge dx^k = \int_{D_x} f(x)\,dx^1 \cdots dx^k.$$

On the other hand, by (13.9') and (13.8'),

$$\int_{D_x} \omega := \int_{D_t} \varphi^* \omega = \int_{D_t} f\big(\varphi(t)\big) \det \varphi'(t)\,dt^1 \cdots dt^k.$$

But if $\det \varphi'(t) > 0$ in D_t, then by the theorem on change of variable in a multiple integral we have

$$\int_{D_x = \varphi(D_t)} f(x)\,dx^1 \cdots dx^k = \int_{D_t} f\big(\varphi(t)\big) \det \varphi'(t)\,dt^1 \cdots dt^k.$$

Hence, assuming that there were coordinates $x^1, \ldots, x^k$ in $S = D_x$ and curvilinear coordinates $t^1, \ldots, t^k$ of the same orientation class, we have shown that the value of the integral $\int_S \omega$ is the same, no matter which of these two coordinate systems is used to compute it.

We note that if the curvilinear coordinates $t^1, \ldots, t^k$ had defined the opposite orientation on S, that is, $\det \varphi'(t) < 0$, the right- and left-hand sides of the last equality would have had opposite signs. Thus, one can say that the integral is well-defined only in the case of an oriented surface of integration.

Now let $\varphi_x : D_x \to S$ and $\varphi_t : D_t \to S$ be two parametrizations of the same smooth k-dimensional surface S and ω a k-form on S. Let us compare the integrals

$$\int_{D_x} \varphi_x^* \omega \quad \text{and} \quad \int_{D_t} \varphi_t^* \omega. \tag{13.11}$$

Since $\varphi_t = \varphi_x \circ (\varphi_x^{-1} \circ \varphi_t) = \varphi_x \circ \varphi$, where $\varphi = \varphi_x^{-1} \circ \varphi_t : D_t \to D_x$ is a diffeomorphism of D_t onto D_x, it follows that $\varphi_t^* \omega = \varphi^* (\varphi_x^* \omega)$ (see Eq. (12.30) of Sect. 12.5).

Hence one can obtain the form $\varphi_t^* \omega$ in D_t by the change of variable $x = \varphi(t)$ in the form $\varphi_x^* \omega$. But, as we have just verified, in this case the integrals (13.11) are equal if $\det \varphi'(t) > 0$ and differ in sign if $\det \varphi'(t) < 0$.

Thus it has been shown that if $\varphi_t : D_t \to S$ and $\varphi_x : D_x \to S$ are parametrizations of the surface S belonging to the same orientation class, the integrals (13.11) are equal. The fact that the integral is independent of the choice of curvilinear coordinates on the surface S has now been verified.

The fact that the integral (13.10) over an oriented piecewise-smooth surface S is independent of the method of partitioning $\bigcup_i S_i$ into smooth pieces follows from the additivity of the ordinary multiple integral (it suffices to consider a finer partition obtained by superimposing two partitions and verify that the value of the integral over the finer partition equals the value over each of the two original partitions). $\square$

On the basis of these considerations, it now makes sense to adopt the following chain of formal definitions corresponding to the construction of the integral of a form explained in Definition 1.

Definition 1' (Integral of a form over an oriented surface $S \subset \mathbb{R}^n$.)

a) If the form $f(x) \, dx^1 \wedge \cdots \wedge dx^k$ is defined in a domain $D \subset \mathbb{R}^k$, then

$$\int_D f(x) \, dx^1 \wedge \cdots \wedge dx^k := \int_D f(x) \, dx^1 \cdots dx^k.$$

b) If $S \subset \mathbb{R}^n$ is a smooth k-dimensional oriented surface, $\varphi : D \to S$ is a parametrization of it, and ω is a k-form on S, then

$$\int_S \omega := \pm \int_D \varphi^* \omega,$$

where the $+$ sign is taken if the parametrization φ agrees with the given orientation of S and the $-$ sign in the opposite case.

c) If S is a piecewise-smooth k-dimensional oriented surface in $\mathbb{R}^n$ and ω is a k-form on S (defined where S has a tangent plane), then

$$\int_S \omega := \sum_i \int_{S_i} \omega,$$

where $S_1, \ldots, S_m, \ldots$ is a decomposition of S into smooth parametrizable k-dimensional pieces intersecting at most in piecewise-smooth surfaces of smaller dimension.

We see in particular that changing the orientation of a surface leads to a change in the sign of the integral.

13.1.3 Problems and Exercises

1. a) Let x, y be Cartesian coordinates on the plane $\mathbb{R}^2$. Exhibit the vector field whose work form is $\omega = -\frac{y}{x^2+y^2}\,dx + \frac{x}{x^2+y^2}\,dy$.

b) Find the integral of the form ω in a) along the following paths γ_i:

$$[0, \pi] \ni t \xrightarrow{\gamma_1} (\cos t, \sin t) \in \mathbb{R}^2; \qquad [0, \pi] \ni t \xrightarrow{\gamma_2} (\cos t, -\sin t) \in \mathbb{R}^2;$$

γ_3 consists of a motion along the closed intervals joining the points $(1, 0)$, $(1, 1)$, $(-1, 1)$, $(-1, 0)$ in that order; γ_4 consists of a motion along the closed intervals joining $(1, 0)$, $(1, -1)$, $(-1, -1)$, $(-1, 0)$ in that order.

2. Let f be a smooth function in the domain $D \subset \mathbb{R}^n$ and γ a smooth path in D with initial point $p_0 \in D$ and terminal point $p_1 \in D$. Find the integral of the form $\omega = df$ over γ.

3. a) Find the integral of the form $\omega = dy \wedge dz + dz \wedge dx$ over the boundary of the standard unit cube in $\mathbb{R}^3$ oriented by an outward-pointing normal.

b) Exhibit a velocity field for which the form ω in a) is the flux form.

4. a) Let x, y, z be Cartesian coordinates in $\mathbb{R}^n$. Exhibit a velocity field for which the flux form is

$$\omega = \frac{x\,dy \wedge dz + y\,dz \wedge dx + z\,dx \wedge dy}{(x^2 + y^2 + z^2)^{3/2}}.$$

b) Find the integral of the form ω in a) over the sphere $x^2 + y^2 + z^2 = R^2$ oriented by the outward normal.

c) Show that the flux of the field $\frac{(x,y,z)}{(x^2+y^2+z^2)^{3/2}}$ across the sphere $(x-2)^2 + y^2 + z^2 = 1$ is zero.

d) Verify that the flux of the field in c) across the torus whose parametric equations are given in Example 4 of Sect. 12.1 is also zero.

5. It is known that the pressure P, volume V, and temperature T of a given quantity of a substance are connected by an equation $f(P, V, T) = 0$, called the *equation of state* in thermodynamics. For example, for one mole of an ideal gas the equation of state is given by Clapeyron's formula $\frac{PV}{T} - R = 0$, where R is the universal gas constant.

Since P, V, T are connected by the equation of state, knowing any pair of them, one can theoretically determine the remaining one. Hence the state of any system can be characterized, for example, by points (V, P) of the plane $\mathbb{R}^2$ with coordinates V, P. Then the evolution of the state of the system as a function of time will correspond to some path γ in this plane.

Suppose the gas is located in a cylinder in which a frictionless piston can move. By changing the position of the piston, we can change the state of the gas enclosed by the piston and the cylinder walls at the cost of doing mechanical work. Conversely, by changing the state of the gas (heating it, for example) we can force the gas to do mechanical work (lifting a weight by expanding, for example). In this

Fig. 13.3

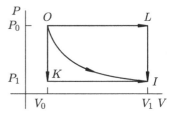

problem and in Problems 6, 7, and 8 below, all processes are assumed to take place so slowly that the temperature and pressure are able to average out at each particular instant of time; thus at each instant of time the system satisfies the equation of state. These are the so-called *quasi-static processes*.

a) Let γ be a path in the *VP*-plane corresponding to a quasi-static transition of the gas enclosed by the piston and the cylinder walls from state V_0, P_0 to V_1, P_1. Show that the quantity A of mechanical work performed on this path is defined by the line integral $A = \int_\gamma P \, dV$.

b) Find the mechanical work performed by one mole of an ideal gas in passing from the state V_0, P_0 to state V_1, P_1 along each of the following paths (Fig. 13.3): γ_{OLI}, consisting of the isobar OL ($P = P_0$) followed by the isochore LI ($V = V_1$); γ_{OKI}, consisting of the isochore OK ($V = V_0$) followed by the isobar KI ($P = P_1$); γ_{OI}, consisting of the isotherm $T = $ const (assuming that $P_0 V_0 = P_1 V_1$).

c) Show that the formula obtained in a) for the mechanical work performed by the gas enclosed by the piston and the cylinder walls is actually general, that is, it remains valid for the work of a gas enclosed in any deformable container.

6. The quantity of heat acquired by a system in some process of varying its states, like the mechanical work performed by the system (see Problem 5), depends not only on the initial and final states of the system, but also on the transition path. An important characteristic of a substance and the thermodynamic process performed by (or on) it is its *heat capacity*, the ratio of the heat acquired by the substance to the change in its temperature. A precise definition of heat capacity can be given as follows. Let $\mathbf{x}$ be a point in the plane of states F (with coordinates V, P or V, T or P, T) and $\mathbf{e} \in TF_x$ a vector indicating the direction of displacement from the point $\mathbf{x}$. Let t be a small parameter. Let us consider the displacement from the state $\mathbf{x}$ to the state $\mathbf{x} + t\mathbf{e}$ along the closed interval in the plane F whose endpoints are these states. Let $\Delta Q(\mathbf{x}, t\mathbf{e})$ be the heat acquired by the substance in this process and $\Delta T(\mathbf{x}, t\mathbf{e})$ the change in the temperature of the substance.

The *heat capacity* $C = C(\mathbf{x}, \mathbf{e})$ of the substance (or system) corresponding to the state $\mathbf{x}$ and the direction $\mathbf{e}$ of displacement from that state is

$$C(\mathbf{x}, t\mathbf{e}) = \lim_{t \to 0} \frac{\Delta Q(\mathbf{x}, t\mathbf{e})}{\Delta T(\mathbf{x}, t\mathbf{e})}.$$

In particular, if the system is thermally insulated, its evolution takes place without any exchange of heat with the surrounding medium. This is a so-called *adiabatic process*. The curve in the plane of states F corresponding to such a process is called an *adiabatic*. Hence, zero heat capacity of the system corresponds to displacement from a given state $\mathbf{x}$ along an adiabatic.

Infinite heat capacity corresponds to displacement along an isotherm $T = \text{const}$.

The heat capacities at constant volume $C_V = C(\mathbf{x}, \mathbf{e}_V)$ and at constant pressure $C_P = C(\mathbf{x}, \mathbf{e}_P)$, which correspond respectively to displacement along an isochore $V = \text{const}$ and an isobar $P = \text{const}$, are used particularly often. Experiment shows that in a rather wide range of states of a given mass of substance, each of the quantities C_V and C_P can be considered practically constant. The heat capacity corresponding to one mole of a given substance is customarily called the *molecular* heat capacity and is denoted (in contrast to the others) by upper case letters rather than lower case. We shall assume that we are dealing with one mole of a substance.

Between the quantity ΔQ of heat acquired by the substance in the process, the change ΔU in its internal energy, and the mechanical work ΔA it performs, the law of conservation of energy provides the connection $\Delta Q = \Delta U + \Delta A$. Thus, under a small displacement $t\mathbf{e}$ from state $\mathbf{x} \in F$ the heat acquired can be found as the value of the form $\delta Q := \mathrm{d}U + P\,\mathrm{d}V$ at the point $\mathbf{x}$ on the vector $t\mathbf{e} \in TF_x$ (for the formula $P\,\mathrm{d}V$ for the work see Problem 5c)). Hence if T and V are regarded as the coordinates of the state and the displacement parameter (in a nonisothermal direction) is taken as T, then we can write

$$C = \lim_{t \to 0} \frac{\Delta Q}{\Delta T} = \frac{\partial U}{\partial T} + \frac{\partial U}{\partial V} \cdot \frac{\mathrm{d}V}{\mathrm{d}T} + P\frac{\mathrm{d}V}{\mathrm{d}T}.$$

The derivative $\frac{\mathrm{d}V}{\mathrm{d}T}$ determines the direction of displacement from the state $\mathbf{x} \in F$ in the plane of states with coordinates T and V. In particular, if $\frac{\mathrm{d}V}{\mathrm{d}T} = 0$ then the displacement is in the direction of the isochore $V = \text{const}$, and we find that $C_V = \frac{\partial U}{\partial T}$. If $P = \text{const}$, then $\frac{\mathrm{d}V}{\mathrm{d}T} = (\frac{\partial V}{\partial T})_{P=\text{const}}$. (In the general case $V = V(P, T)$ is the equation of state $f(P, V, T) = 0$ solved for V.) Hence

$$C_P = \left(\frac{\partial U}{\partial T}\right)_V + \left(\left(\frac{\partial U}{\partial V}\right)_T + P\right)\left(\frac{\partial V}{\partial T}\right)_P,$$

where the subscripts P, V, and T on the right-hand side indicate the parameter of state that is fixed when the partial derivative is taken. Comparing the resulting expressions for C_V and C_P, we see that

$$C_P - C_V = \left(\left(\frac{\partial U}{\partial V}\right)_T + P\right)\left(\frac{\partial V}{\partial T}\right)_P.$$

By experiments on gases (the Joule[1]–Thomson experiments) it was established and then postulated in the model of an ideal gas that its internal energy depends only on the temperature, that is, $(\frac{\partial U}{\partial V})_T = 0$. Thus for an ideal gas $C_P - C_V = P(\frac{\partial V}{\partial T})_P$.

---

[1] G.P. Joule (1818–1889) – British physicist who discovered the law of thermal action of a current and also determined, independently of Mayer, the mechanical equivalent of heat.

Taking account of the equation $PV = RT$ for one mole of an ideal gas, we obtain the relation $C_P - C_V = R$ from this, known as *Mayer's equation*[2] in thermodynamics.

The fact that the internal energy of a mole of gas depends only on temperature makes it possible to write the form δQ as

$$\delta Q = \frac{\partial U}{\partial T}\,dT + P\,dV = C_V\,dT + P\,dV.$$

To compute the quantity of heat acquired by a mole of gas when its state varies over the path γ one must consequently find the integral of the form $C_V\,dT + P\,dV$ over γ. It is sometimes convenient to have this form in terms of the variables V and P. If we use the equation of state $PV = RT$ and the relation $C_P - C_V = R$, we obtain

$$\delta Q = C_P\frac{P}{R}\,dV + C_V\frac{V}{R}\,dP.$$

a) Write the formula for the quantity Q of heat acquired by a mole of gas as its state varies along the path γ in the plane of states F.

b) Assuming the quantities C_P and C_V are constant, find the quantity Q corresponding to the paths γ_{OL1}, γ_{OK1}, and γ_{OI} in Problem 5b).

c) Find (following Poisson) the *equation of the adiabatic* passing through the point (V_0, P_0) in the plane of states F with coordinates V and P. (Poisson found that $PV^{C_P/C_V} = $ const on an adiabatic. The quantity C_P/C_V is the *adiabatic constant* of the gas. For air $C_P/C_V \approx 1.4$.) Now compute the work one must do in order to confine a thermally isolated mole of air in the state (V_0, P_0) to the volume $V_1 = \frac{1}{2}V_0$.

7. We recall that a *Carnot cycle*[3] of variation in the state of the working body of a heat engine (for example, the gas under the piston in a cylinder) consists of the following (Fig. 13.4). There are two energy-storing bodies, a heater and a cooler (for example, a steam boiler and the atmosphere) maintained at constant temperatures T_1 and T_2 respectively ($T_1 > T_2$). The working body (gas) of this heat engine, having temperature T_1 in State 1, is brought into contact with the heater, and by decreasing the external pressure along an isotherm, expands quasi-statically and moves to State 2. In the process the engine borrows a quantity of heat Q_1 from the heater and performs mechanical work A_{12} against the external pressure. In State 2 the gas is thermally insulated and forced to expand quasi-statically to State 3, until its temperature reaches T_2, the temperature of the cooler. In this process the engine also performs a certain quantity of work A_{23} against the external pressure. In State 3 the gas is brought into contact with the cooler and compressed isothermically to State 4 by increasing the pressure. In this process work is done on the gas (the gas itself performs negative work A_{34}), and the gas gives up a certain quantity of heat Q_2 to the

---

[2]J.P. Mayer (1814–1878) – German scholar, a physician by training; he stated the law of conservation and transformation of energy and found the mechanical equivalent of heat.

[3]N.L.S. Carnot (1796–1832) – French engineer, one of the founders of thermodynamics.

Fig. 13.4

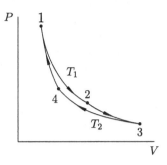

cooler. State 4 is chosen so that it is possible to return from it to State 1 by a quasi-static compression along an adiabatic. Thus the gas is returned to State 1. In the process it is necessary to perform some work on the gas (and the gas itself performs negative work A_{41}). As a result of this cyclic process (a Carnot cycle) the internal energy of the gas (the working body of the engine) obviously does not change (after all, we have returned to the initial state). Therefore the work performed by the engine is $A = A_{12} + A_{23} + A_{34} + A_{41} = Q_1 - Q_2$.

The heat Q_1 acquired from the heater went only partly to perform the work A. It is natural to call the quantity $\eta = \frac{A}{Q_1} = \frac{Q_1 - Q_2}{Q_1}$ the *efficiency* of the heat engine under consideration.

a) Using the results obtained in a) and c) of Problem 6, show that the equality $\frac{Q_1}{T_1} = \frac{Q_2}{T_2}$ holds for a Carnot cycle.

b) Now prove the following theorem, the first of Carnot's two famous theorems. *The efficiency of a heat engine working along a Carnot cycle depends only on the temperatures T_1 and T_2 of the heater and cooler.* (It is independent of the structure of the engine or the form of its working body.)

8. Let γ be a closed path in the plane of states F of the working body of an arbitrary heat engine (see Problem 7) corresponding to a closed cycle of work performed by it. The quantity of heat that the working body (a gas, for example) exchanges with the surrounding medium and the temperature at which the heat exchange takes place are connected by the *Clausius inequality* $\int_\gamma \frac{\delta Q}{T} \leq 0$. Here δQ is the heat exchange form mentioned in Problem 6.

a) Show that for a Carnot cycle (see Problem 7), the Clausius inequality becomes equality.

b) Show that if the work cycle γ can be run in reverse, then the Clausius inequality becomes equality.

c) Let γ_1 and γ_2 be the parts of the path γ on which the working body of a heat engine acquires heat from without and imparts it to the surrounding medium respectively. Let T_1 be the maximal temperature of the working body on γ_1 and T_2 its minimal temperature on γ_2. Finally, let Q_1 be the heat acquired on γ_1 and Q_2 the heat given up on γ_2. Based on Clausius' inequality, show that $\frac{Q_2}{Q_1} \leq \frac{T_2}{T_1}$.

d) Obtain the estimate $\eta \leq \frac{T_1 - T_2}{T_1}$ for the efficiency of any heat engine (see Problem 7). This is *Carnot's second theorem*. (Estimate separately the efficiency of a steam engine in which the maximal temperature of the steam is at most 150 °C, that is, $T_1 = 423$ K, and the temperature of the cooler – the surrounding medium – is of the order 20 °C, that is $T_2 = 291$ K.)

e) Compare the results of Problems 7b) and 8d) and verify that a heat engine working in a Carnot cycle has the maximum possible efficiency for given values of T_1 and T_2.

9. The differential equation $\frac{dy}{dx} = \frac{f(x)}{g(y)}$ is said to have variables separable. It is usually rewritten in the form $g(y)\,dy = f(x)\,dx$, in which "the variables are separated," then "solved" by equating the primitives $\int g(y)\,dy = \int f(x)\,dx$. Using the language of differential forms, now give a detailed mathematical explanation for this algorithm.

13.2 The Volume Element. Integrals of First and Second Kind

13.2.1 The Mass of a Lamina

Let S be a lamina in Euclidean space $\mathbb{R}^n$. Assume that we know the density $\rho(x)$ (per unit area) of the mass distribution on S. We ask how one can determine the total mass of S.

In order to solve this problem it is necessary first of all to take account of the fact that the surface density $\rho(x)$ is the limit of the ratio Δm of the quantity of mass on a portion of the surface in a neighborhood of x to the area $\Delta\sigma$ of that same portion of the surface, as the neighborhood is contracted to x.

By breaking S into small pieces S_i and assuming that ρ is continuous on S, we can find the mass of S_i, neglecting the variation of ρ within each small piece, from the relation

$$\Delta m_i \approx \rho(x_i)\Delta\sigma_i,$$

in which $\Delta\sigma_i$ is the area of the surface S_i and $x_i \in S_i$.

Summing these approximate equalities and passing to the limit as the partition is refined, we find that

$$m = \int_S \rho\,d\sigma. \tag{13.12}$$

The symbol for integration over the surface S here obviously requires some clarification so that computational formulas can be derived from it.

We note that the statement of the problem itself shows that the left-hand side of Eq. (13.12) is independent of the orientation of S, so that the integral on the right-hand side must have the same property. At first glance this appears to contrast with the concept of an integral over a surface, which was discussed in detail in Sect. 13.1. The answer to the question that thus arises is concealed in the definition of the surface element $d\sigma$, to whose analysis we now turn.

13.2.2 The Area of a Surface as the Integral of a Form

Comparing Definition 1 of Sect. 13.1 for the integral of a form with the construction that led us to the definition of the area of a surface (Sect. 12.4), we see that the area of a smooth k-dimensional surface S embedded in the Euclidean space $\mathbb{R}^n$ and given parametrically by $\varphi : D \to S$, is the integral of a form Ω, which we shall provisionally call the volume element on the surface S. It follows from relation (12.10) of Sect. 12.4 that Ω (more precisely $\varphi^* \Omega$) has the form

$$\omega = \sqrt{\det(g_{ij})}(t)\,dt^1 \wedge \cdots \wedge dt^k, \tag{13.13}$$

in the curvilinear coordinates $\varphi : D \to S$ (that is, when transferred to the domain D). Here $g_{ij}(t) = \langle \frac{\partial \varphi}{\partial t^i}, \frac{\partial \varphi}{\partial t^j} \rangle$, $i, j = 1, \ldots, k$.

To compute the area of S over a domain $\tilde{D}$ in a second parametrization $\tilde{\varphi} : \tilde{D} \to S$, one must correspondingly integrate the form

$$\tilde{\omega} = \sqrt{\det(\tilde{g}_{ij})}(\tilde{t})\,d\tilde{t}^1 \wedge \cdots \wedge d\tilde{t}^k, \tag{13.14}$$

where $\tilde{g}_{ij}(\tilde{t}) = \langle \frac{\partial \varphi}{\partial \tilde{t}^i}, \frac{\partial \varphi}{\partial \tilde{t}^j} \rangle$, $i, j = 1, \ldots, k$.

We denote by ψ the diffeomorphism $\varphi^{-1} \circ \tilde{\varphi} : \tilde{D} \to D$ that provides the change from $\tilde{t}$ coordinates to t coordinates on S. Earlier we have computed (see Remark 5 of Sect. 12.4) that

$$\sqrt{\det(\tilde{g}_{ij})}(\tilde{t}) = \sqrt{\det(g_{ij})}(t) \cdot \left| \det \psi'(t) \right|. \tag{13.15}$$

At the same time, it is obvious that

$$\psi^* \omega = \sqrt{\det(g_{ij})}\big(\psi(\tilde{t})\big), \det \psi'(\tilde{t})\,d\tilde{t}^1 \wedge \cdots \wedge d\tilde{t}^k. \tag{13.16}$$

Comparing the equalities (13.13)–(13.16), we see that $\psi^* \omega = \tilde{\omega}$ if $\det \psi'(\tilde{t}) > 0$ and $\psi^* \omega = -\tilde{\omega}$ if $\det \psi'(\tilde{t}) < 0$. If the forms ω and $\tilde{\omega}$ were obtained from the same form Ω on S through the transfers φ^* and $\tilde{\varphi}^*$, then we must always have the equality $\psi^*(\varphi^* \Omega) = \tilde{\varphi}^* \Omega$ or, what is the same, $\psi^* \omega = \tilde{\omega}$.

We thus conclude that the forms on the parametrized surface S that one must integrate in order to obtain the areas of the surface are different: they differ in sign if the parametrizations define different orientations on S; these forms are equal for parametrizations that belong to the same orientation class for the surface S.

Thus the volume element Ω on S must be determined not only by the surface S embedded in $\mathbb{R}^n$, but also by the orientation of S.

This might appear paradoxical: in our intuition, the area of a surface should not depend on its orientation!

But after all, we arrived at the definition of the area of a parametrized surface via an integral, the integral of a certain form. Hence, if the result of our computations is to be independent of the orientation of the surface, it follows that we must integrate different forms when the orientation is different.

Let us now turn these considerations into precise definitions.

13.2.3 The Volume Element

Definition 1 If $\mathbb{R}^k$ is an oriented Euclidean space with inner product $\langle , \rangle$, the *volume element* on $\mathbb{R}^k$ corresponding to a particular orientation and the inner product $\langle , \rangle$ is the skew-symmetric k-form that assumes the value 1 on an orthonormal frame of some orientation class.

The value of the k-form on the frame $\mathbf{e}_1, \ldots, \mathbf{e}_k$ obviously determines this form.

We remark also that the form Ω is determined not by an individual orthonormal frame, but only by its orientation class.

Proof In fact, if $\mathbf{e}_1, \ldots, \mathbf{e}_k$ and $\tilde{\mathbf{e}}_1, \ldots, \tilde{\mathbf{e}}_k$ are two such frames in the same orientation class, then the transition matrix O from the second basis to the first is an orthogonal matrix with $\det O = 1$. Hence

$$\Omega(\mathbf{e}_1, \ldots, \mathbf{e}_k) = \det O \cdot \Omega(\tilde{\mathbf{e}}_1, \ldots, \tilde{\mathbf{e}}_k) = \Omega(\tilde{\mathbf{e}}_1, \ldots, \tilde{\mathbf{e}}_k). \qquad \square$$

If an orthonormal basis $\mathbf{e}_1, \ldots, \mathbf{e}_k$ is fixed in $\mathbb{R}^k$ and $\pi^1, \ldots, \pi^k$ are the projections of $\mathbb{R}^k$ on the corresponding coordinate axes, obviously $\pi^1 \wedge \cdots \wedge \pi^k(\mathbf{e}_1, \ldots, \mathbf{e}_k) = 1$ and

$$\Omega = \pi^1 \wedge \cdots \wedge \pi^k.$$

Thus,

$$\Omega(\boldsymbol{\xi}_1, \ldots, \boldsymbol{\xi}_k) = \begin{vmatrix} \xi_1^1 & \cdots & \xi_1^k \\ \vdots & \ddots & \vdots \\ \xi_k^1 & \cdots & \xi_k^k \end{vmatrix}.$$

This is the oriented volume of the parallelepiped spanned by the ordered set of vectors $\boldsymbol{\xi}_1, \ldots, \boldsymbol{\xi}_k$.

Definition 2 If the smooth k-dimensional oriented surface S is embedded in a Euclidean space $\mathbb{R}^n$, then each tangent plane TS_x to S has an orientation consistent with the orientation of S and an inner product induced by the inner product in $\mathbb{R}^n$; hence there is a volume element $\Omega(x)$. The k-form Ω that arises on S in this way is the *volume element on S* induced by the embedding of S in $\mathbb{R}^n$.

Definition 3 The *area of an oriented smooth surface* is the integral over the surface of the volume element corresponding to the orientation chosen for the surface.

This definition of area, stated in the language of forms and made precise, is of course in agreement with Definition 1 of Sect. 12.4, which we arrived at by consideration of a smooth k-dimensional surface $S \subset \mathbb{R}^n$ defined in parametric form.

Proof Indeed, the parametrization orients the surface and all its tangent planes TS_x. If $\boldsymbol{\xi}_1, \ldots, \boldsymbol{\xi}_k$ is a frame of a fixed orientation class in TS_x, it follows from Definitions 2 and 3 for the volume element Ω that $\Omega(x)(\boldsymbol{\xi}_1, \ldots, \boldsymbol{\xi}_k) > 0$. But then (see Eq. (12.7) of Sect. 12.4)

$$\Omega(x)(\boldsymbol{\xi}_1, \ldots, \boldsymbol{\xi}_k) = \sqrt{\det(\langle \boldsymbol{\xi}_i, \boldsymbol{\xi}_j \rangle)}. \tag{13.17}$$

$\square$

We note that the form $\Omega(x)$ itself is defined on any set $\boldsymbol{\xi}_1, \ldots, \boldsymbol{\xi}_k$ of vectors in TS_x, but Eq. (13.17) holds only on frames of a given orientation class in TS_x.

We further note that the volume element is defined only on an oriented surface, so that it makes no sense, for example, to talk about the volume element on a Möbius band in $\mathbb{R}^3$, although it does make sense to talk about the volume element of each orientable piece of this surface.

Definition 4 Let S be a k-dimensional piecewise-smooth surface (orientable or not) in $\mathbb{R}^n$, and $S_1, \ldots, S_m, \ldots$ a finite or countable number of smooth parametrized pieces of it intersecting at most in surfaces of dimension not larger than $k - 1$ and such that $S = \bigcup_i S_i$.

The *area* (or *k-dimensional volume*) of S is the sum of the areas of the surfaces S_i.

In this sense we can speak of the area of a Möbius band in $\mathbb{R}^3$ or, what is the same, try to find its mass if it is a material surface with matter having unit density.

The fact that Definition 4 is unambiguous (that the area obtained is independent of the partition $S_1, \ldots, S_m, \ldots$ of the surface) can be verified by traditional reasoning.

13.2.4 Expression of the Volume Element in Cartesian Coordinates

Let S be a smooth hypersurface (of dimension $n - 1$) in an oriented Euclidean space $\mathbb{R}^n$ endowed with a continuous field of unit normal vectors $\boldsymbol{\eta}(x)$, $x \in S$, which orients it. Let V be the n-dimensional volume in $\mathbb{R}^n$ and Ω the $(n-1)$-dimensional volume element on S.

If we take a frame $\boldsymbol{\xi}_1, \ldots, \boldsymbol{\xi}_{n-1}$ in the tangent space TS_x from the orientation class determined by the unit normal $\mathbf{n}(x)$ to TS_x, we can obviously write the following equality:

$$V(x)(\boldsymbol{\eta}, \boldsymbol{\xi}_1, \ldots, \boldsymbol{\xi}_{n-1}) = \Omega(x)(\boldsymbol{\xi}_1, \ldots, \boldsymbol{\xi}_{n-1}). \tag{13.18}$$

Proof This fact follows from the fact that under the given hypotheses both sides are nonnegative and equal in magnitude because the volume of the parallelepiped spanned by $\boldsymbol{\eta}, \boldsymbol{\xi}_1, \ldots, \boldsymbol{\xi}_{n-1}$ is the area of the base $\Omega(x)(\boldsymbol{\xi}_1, \ldots, \boldsymbol{\xi}_{n-1})$ multiplied by the height $|\boldsymbol{\eta}| = 1$. $\square$

But,

$$V(x)(\boldsymbol{\eta}, \boldsymbol{\xi}_1, \ldots, \boldsymbol{\xi}_{n-1}) =$$

$$= \begin{vmatrix} \eta^1 & \cdots & \eta^n \\ \xi_1^1 & \cdots & \xi_1^n \\ \vdots & \ddots & \vdots \\ \xi_{n-1}^1 & \cdots & \xi_{n-1}^n \end{vmatrix} =$$

$$= \sum_{i=1}^n (-1)^{i-1} \eta^i(x) \, dx^1 \wedge \cdots \wedge \widehat{dx^i} \wedge \cdots \wedge dx^n (\boldsymbol{\xi}_1, \ldots, \boldsymbol{\xi}_{n-1}).$$

Here the variables $x^1, \ldots, x^n$ are Cartesian coordinates in the orthonormal basis $\mathbf{e}_1, \ldots, \mathbf{e}_n$ that defines the orientation, and the frown over the differential dx^i indicates that it is to be omitted.

Thus we obtain the following coordinate expression for the volume element on the oriented hypersurface $S \subset \mathbb{R}^n$:

$$\Omega = \sum_{i=1}^n (-1)^{i-1} \eta^i(x) \, dx^1 \wedge \cdots \wedge \widehat{dx^i} \wedge \cdots \wedge dx^n (\boldsymbol{\xi}_1, \ldots, \boldsymbol{\xi}_{n-1}). \qquad (13.19)$$

At this point it is worthwhile to remark that when the orientation of the surface is reversed, the direction of the normal $\boldsymbol{\eta}(x)$ reverses, that is, the form Ω is replaced by the new form $-\Omega$.

It follows from the same geometric considerations that for a fixed value of $i \in \{1, \ldots, n\}$

$$\langle \boldsymbol{\eta}(x), \mathbf{e}_i \rangle \Omega(\boldsymbol{\xi}_1, \ldots, \boldsymbol{\xi}_{n-1}) = V(x)(\mathbf{e}_i, \boldsymbol{\xi}_1, \ldots, \boldsymbol{\xi}_{n-1}). \qquad (13.20)$$

This last equality means that

$$\eta^i(x)\Omega(x) = (-1)^{i-1} dx^1 \wedge \cdots \wedge \widehat{dx^i} \wedge \cdots \wedge dx^n (\boldsymbol{\xi}_1, \ldots, \boldsymbol{\xi}_{n-1}). \qquad (13.21)$$

For a two-dimensional surface S in $\mathbb{R}^n$ the volume element is most often denoted $d\sigma$ or dS. These symbols should not be interpreted as the differentials of some forms σ and S; they are only symbols. If x, y, z are Cartesian coordinates on $\mathbb{R}^3$, then in this notation relations (13.19) and (13.21) can be written as follows:

$$d\sigma = \cos\alpha_1 \, dy \wedge dz + \cos\alpha_2 \, dz \wedge dx + \cos\alpha_3 \, dx \wedge dy,$$

$$\cos\alpha_1 \, d\sigma = dy \wedge dz, \quad \text{(oriented areas of the projections}$$

$$\cos\alpha_2 \, d\sigma = dz \wedge dx, \quad \text{on the coordinate planes).}$$

$$\cos\alpha_3 \, d\sigma = dx \wedge dy,$$

Here $(\cos\alpha_1, \cos\alpha_2, \cos\alpha_3)(x)$ are the direction cosines (coordinates) of the unit normal vector $\eta(x)$ to S at the point $x \in S$. In these equalities (as also in (13.19) and (13.21)) it would of course have been more correct to place the restriction sign $|_S$ on the right-hand side so as to avoid misunderstanding. But, in order not to make the formulas cumbersome, we confine ourselves to this remark.

13.2.5 Integrals of First and Second Kind

Integrals of type (13.12) arise in a number of problems, a typical representative of which is the problem considered above of determining the mass of a surface whose density is known. These integrals are often called integrals over a surface or integrals of first kind.

Definition 5 The *integral of a function ρ over an oriented surface S* is the integral

$$\int_S \rho\Omega \tag{13.22}$$

of the differential form $\rho\Omega$, where Ω is the volume element on S (corresponding to the orientation of S chosen in the computation of the integral).

It is clear that the integral (13.22) so defined is independent of the orientation of S, since a reversal of the orientation is accompanied by a corresponding replacement of the volume element.

We emphasize that it is not really a matter of integrating a function, but rather integrating a form $\rho\Omega$ of special type over the surface S with the volume element defined on it.

Definition 6 If S is a piecewise-smooth (orientable or non-orientable) surface and ρ is a function on S, then the *integral* (13.22) *of ρ over the surface S* is the sum $\sum_i \int_{S_i} \rho\Omega$ of the integrals of ρ over the parametrized pieces $S_1, \ldots, S_m, \ldots$ of the partition of S described in Definition 4.

The integral (13.22) is usually called a *surface integral of first kind*.

For example, the integral (13.12), which expresses the mass of the surface S in terms of the density ρ of the mass distribution over the surface, is such an integral.

To distinguish integrals of first kind, which are independent of the orientation of the surface, we often refer to integrals of forms over an oriented surface as *surface integrals of second kind*.

We remark that, since all skew-symmetric forms on a vector space whose degrees are equal to the dimension of the space are multiples of one another, there is a connection $\omega = \rho \Omega$ between any k-form ω defined on a k-dimensional orientable surface S and the volume element Ω on S. Here ρ is some function on S depending on ω. Hence

$$\int_S \omega = \int_S \rho \Omega.$$

That is, every integral of second kind can be written as a suitable integral of first kind.

Example 1 The integral (13.2′) of Sect. 13.1, which expresses the work on the path $\gamma : [a, b] \to \mathbb{R}^n$, can be written as the integral of first kind

$$\int_\gamma \langle \mathbf{F}, \mathbf{e} \rangle \, ds, \qquad\qquad (13.23)$$

where s is arc length on γ, ds is the element of length (a 1-form), and $\mathbf{e}$ is a unit velocity vector containing all the information about the orientation of γ. From the point of view of the physical meaning of the problem solved by the integral (13.23), it is just as informative as the integral (13.1) of Sect. 13.1.

Example 2 The flux (13.3) of Sect. 13.1 of the velocity field $\mathbf{V}$ across a surface $S \subset \mathbb{R}^n$ oriented by unit normals $\mathbf{n}(x)$ can be written as the surface integral of first kind

$$\int_S \langle \mathbf{V}, \mathbf{n} \rangle \, d\sigma. \qquad\qquad (13.24)$$

The information about the orientation of S here is contained in the direction of the field of normals $\mathbf{n}$.

The geometric and physical content of the integrand in (13.24) is just as transparent as the corresponding meaning of the integrand in the final computational formula (13.6) of Sect. 13.1.

For the reader's information we note that quite frequently one encounters the notation $d\mathbf{s} := \mathbf{e} \, ds$ and $d\boldsymbol{\sigma} := \mathbf{n} \, d\sigma$, which introduce a vector element of length and a vector element of area. In this notation the integrals (13.23) and (13.24) have the form

$$\int_\gamma \langle \mathbf{F}, d\mathbf{s} \rangle \quad \text{and} \quad \int_S \langle \mathbf{V}, d\boldsymbol{\sigma} \rangle,$$

which are very convenient from the point of view of physical interpretation. For brevity the inner product $\langle \mathbf{A}, \mathbf{B} \rangle$ of the vectors $\mathbf{A}$ and $\mathbf{B}$ is often written $\mathbf{A} \cdot \mathbf{B}$.

Example 3 Faraday's law[4] asserts that the electromotive force arising in a closed conductor Γ in a variable magnetic field **B** is proportional to the rate of variation of the flux of the magnetic field across a surface S bounded by Γ. Let **E** be the electric field intensity. A precise statement of Faraday's law can be given as the equality

$$\oint_\Gamma \mathbf{E}\cdot d\mathbf{s} = -\frac{\partial}{\partial t}\int_S \mathbf{B}\cdot d\boldsymbol{\sigma}.$$

The circle in the integration sign over Γ is an additional reminder that the integral is being taken over a closed curve. The work of the field over a closed curve is often called the *circulation of the field* along this curve. Thus by Faraday's law the circulation of the electric field intensity generated in a closed conductor by a variable magnetic field equals the rate of variation of the flux of the magnetic field across a surface S bounded by Γ, taken with a suitable sign.

Example 4 Ampère's law[5]

$$\oint_\Gamma \mathbf{B}\cdot d\mathbf{s} = \frac{1}{\varepsilon_0 c^2}\int_S \mathbf{j}\cdot d\boldsymbol{\sigma}$$

(where **B** is the magnetic field intensity, **j** is the current density vector, and ε_0 and c are dimensioning constants) asserts that the circulation of the intensity of a magnetic field generated by an electric current along a contour Γ is proportional to the strength of the current flowing across the surface S bounded by the contour.

We have studied integrals of first and second kind. The reader might have noticed that this terminological distinction is very artificial. In reality we know how to integrate, and we do integrate, only differential forms. No integral is ever taken of anything else (if the integral is to claim independence of the choice of the coordinate system used to compute it).

13.2.6 Problems and Exercises

1. Give a formal proof of Eqs. (13.18) and (13.20).
2. Let γ be a smooth curve and ds the element of arc length on γ.

a) Show that

$$\left|\int_\gamma f(s)\,ds\right| \le \int_\gamma |f(s)|\,ds$$

for any function f on γ for which both integrals are defined.

---

[4]M. Faraday (1791–1867) – outstanding British physicist, creator of the concept of an electromagnetic field.

[5]A.M. Ampère (1775–1836) – French physicist and mathematician, one of the founders of modern electrodynamics.

b) Verify that if $|f(s)| \leq M$ on γ and l is the length of γ, then

$$\left| \int_\gamma f(x)\,\mathrm{d}s \right| \leq Ml.$$

c) State and prove assertions analogous to a) and b) in the general case for an integral of first kind taken over a k-dimensional smooth surface.

3. a) Show that the coordinates (x_0^1, x_0^2, x_0^3) of the center of masses distributed with linear density $\rho(x)$ along the curve γ should be given by the relations

$$x_0^i \int_\gamma \rho(x)\,\mathrm{d}s = \int_\gamma x^i \rho(x)\,\mathrm{d}s, \quad i = 1, 2, 3.$$

b) Write the equation of a helix in $\mathbb{R}^3$ and find the coordinates of the center of mass of a piece of this curve, assuming that the mass is distributed along the curve with constant density equal to 1.

c) Exhibit formulas for the center of masses distributed over a surface S with surface density ρ and find the center of masses that are uniformly distributed over the surface of a hemisphere.

d) Exhibit the formulas for the moment of inertia of a mass distributed with density ρ over the surface S.

e) The tire on a wheel has mass 30 kg and the shape of a torus of outer diameter 1 m and inner diameter 0.5 m. When the wheel is being balanced, it is placed on a balancing lathe and rotated to a velocity corresponding to a speed of the order of 100 km/hr, then stopped by brake pads rubbing against a steel disk of diameter 40 cm and width 2 cm. Estimate the temperature to which the disk would be heated if all the kinetic energy of the spinning tire went into heating the disk when the wheel was stopped. Assume that the heat capacity of steel is $c = 420$ J/(kg-K).

4. a) Show that the gravitational force acting on a point mass m_0 located at (x_0, y_0, z_0) due to a material curve γ having linear density ρ is given by the formula

$$F = Gm_0 \int_\gamma \frac{\rho}{|\mathbf{r}|^3} \mathbf{r}\,\mathrm{d}s,$$

where G is the gravitational constant and $\mathbf{r}$ is the vector with coordinates $(x - x_0, y - y_0, z - z_0)$.

b) Write the corresponding formula in the case when the mass is distributed over a surface S.

c) Find the gravitational field of a homogeneous material line.

d) Find the gravitational field of a homogeneous material sphere. (Exhibit the field both outside the ball bounded by the sphere and inside the ball.)

e) Find the gravitational field created in space by a homogeneous material ball (consider both exterior and interior points of the ball).

f) Regarding the Earth as a liquid ball, find the pressure in it as a function of the distance from the center. (The radius of the Earth is 6400 km, and its average density is 6 g/cm^3.)

5. Let γ_1 and γ_2 be two closed conductors along which currents J_1 and J_2 respectively are flowing. Let ds_1 and ds_2 be the vector elements of these conductors corresponding to the directions of current in them. Let the vector $\mathbf{R}_{12}$ be directed from ds_1 to ds_2, and $\mathbf{R}_{21} = -\mathbf{R}_{12}$.

According to the *Biot–Savart law*[6] the force $d\mathbf{F}_{12}$ with which the first element acts on the second is

$$d\mathbf{F}_{12} = \frac{J_1 J_2}{c_0^2 |\mathbf{R}_{12}|^2}[ds_2, [ds_1, \mathbf{R}_{12}]],$$

where the brackets denote the vector product of the vectors and c_0 is a dimensioning constant.

a) Show that, on the level of an abstract differential form, it could happen that $d\mathbf{F}_{12} \neq -d\mathbf{F}_{21}$ in the differential Biot–Savart formula, that is, "the reaction is not equal and opposite to the action."

b) Write the (integral) formulas for the total forces $\mathbf{F}_{12}$ and $\mathbf{F}_{21}$ for the interaction of the conductors γ_1 and γ_2 and show that $\mathbf{F}_{12} = -\mathbf{F}_{21}$.

6. *The co-area formula (the Kronrod–Federer formula).*

Let M^m and N^n be smooth surfaces of dimensions m and n respectively, embedded in a Euclidean space of high dimension (M^m and N^n may also be abstract Riemannian manifolds, but that is not important at the moment). Suppose that $m \geq n$.

Let $f : M^m \to N^n$ be a smooth mapping. When $m > n$, the mapping $df(x) : T_x M^m \to T_{f(x)} N^n$ has a nonempty kernel $\ker df(x)$. Let us denote by $T_x^\perp M^m$ the orthogonal complement of $\ker df(x)$, and by $J(f, x)$ the Jacobian of the mapping $df(x)|_{T_x^\perp M^m} : T_x^\perp M^m \to T_{f(x)} N^n$. If $m = n$, then $J(f, x)$ is the usual Jacobian.

Let $dv_k(p)$ denote the volume element on a k-dimensional surface at the point p. We shall assume that $v_0(E) = \text{card } E$, where $v_k(E)$ is the k-volume of E.

a) Using Fubini's theorem and the rank theorem (on the local canonical form of a smooth mapping) if necessary, prove the following formula of Kronrod and Federer: $\int_{M^m} J(f, x)\,dv_m(x) = \int_{N^n} v_{m-n}(f^{-1}(y))\,dv_n(y)$.

b) Show that if A is a measurable subset of M^m, then

$$\int_A J(f, x)\,dv_m(x) = \int_{N^n} v_{m-n}\big(A \cap f^{-1}(y)\big)\,dv_n(y).$$

This is the general Kronrod–Federer formula.

c) Prove the following strengthening of Sard's theorem (which in its simplest version asserts that the image of the set of critical points of a smooth mapping has measure zero). (See Problem 8 of Sect. 11.5.)

Suppose as before that $f : M^m \to N^n$ is a smooth mapping and K is a compact set in M^m on which rank $df(x) < n$ for all $x \in K$.

Then $\int_{N^n} v_{m-n}(K \cap f^{-1}(y))\,dv_n(y) = 0$. Use this result to obtain in addition the simplest version of Sard's theorem stated above.

---

[6] Biot (1774–1862), Savart (1791–1841) – French physicists.

d) Verify that if $f : D \to \mathbb{R}$ and $u : D \to \mathbb{R}$ are smooth functions in a regular domain $D \subset \mathbb{R}^n$ and u has no critical points in D, then

$$\int_D f \, dv = \int_{\mathbb{R}} dt \int_{u^{-1}(t)} f \frac{d\sigma}{|\nabla u|}.$$

e) Let $V_f(t)$ be the measure (volume) of the set $\{x \in D \mid f(x) > t\}$, and let the function f be nonnegative and bounded in the domain D.

Show that $\int_D f \, dv = -\int_{\mathbb{R}} t \, dV_f(t) = \int_0^\infty V_f(t) \, dt$.

f) Let $\varphi \in C^{(1)}(\mathbb{R}, \mathbb{R}_+)$ and $\varphi(0) = 0$, while $f \in C^{(1)}(D, \mathbb{R})$ and $V_{|f|}(t)$ is the measure of the set $\{x \in D \mid |f(x)| > t\}$. Verify that $\int_D \varphi \circ f \, dv = \int_0^\infty \varphi'(t) V_{|f|}(t) \, dt$.

13.3 The Fundamental Integral Formulas of Analysis

The most important formula of analysis is the Newton–Leibniz formula (fundamental theorem of calculus). In the present section we shall obtain the formulas of Green, Gauss–Ostrogradskii, and Stokes, which on the one hand are an extension of the Newton–Leibniz formula, and on the other hand, taken together, constitute the most-used part of the machinery of integral calculus.

In the first three subsections of this section, without striving for generality in our statements, we shall obtain the three classical integral formulas of analysis using visualizable material. They will be reduced to one general Stokes formula in the fourth subsection, which can be read formally independently of the others.

13.3.1 Green's Theorem

Green's[7] theorem is the following.

Proposition 1 *Let $\mathbb{R}^2$ be the plane with a fixed coordinate grid x, y, and let $\overline{D}$ be a compact domain in this plane bounded by piecewise-smooth curves. Let P and Q be smooth functions in the closed domain $\overline{D}$. Then the following relation holds:*

$$\iint_{\overline{D}} \left(\frac{\partial Q}{\partial x} - \frac{\partial P}{\partial y} \right) dx \, dy = \int_{\partial \overline{D}} P \, dx + Q \, dy, \qquad (13.25)$$

in which the right-hand side contains the integral over the boundary $\partial \overline{D}$ of the domain $\overline{D}$ oriented consistently with the orientation of the domain $\overline{D}$ itself.

---

[7]G. Green (1793–1841) – British mathematician and mathematical physicist. Newton's grave in Westminster Abbey is framed by five smaller gravestones with brilliant names: Faraday, Thomson (Lord Kelvin), Green, Maxwell, and Dirac.

Fig. 13.5

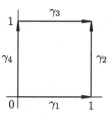

We shall first consider the simplest version of (13.25) in which $\overline{D}$ is the square $I = \{(x, y) \in \mathbb{R}^2 \mid 0 \le x \le 1, 0 \le y \le 1\}$ and $Q \equiv 0$ in I. Then Green's theorem reduces to the equality

$$\iint_I \frac{\partial P}{\partial y}\, dx\, dy = -\int_{\partial I} P\, dx, \tag{13.26}$$

which we shall prove.

Proof Reducing the double integral to an iterated integral and applying the fundamental theorem of calculus, we obtain

$$\iint_{\overline{D}} \frac{\partial P}{\partial y}\, dx\, dy = \int_0^1 dx \int_0^1 \frac{\partial P}{\partial y}\, dy =$$

$$= \int_0^1 \big(P(x, 1) - P(x, 2)\big)\, dx = -\int_0^1 P(x, 0)\, dx + \int_0^1 P(x, 1)\, dx.$$

The proof is now finished. What remains is a matter of definitions and interpretation of the relation just obtained. The point is that the difference of the last two integrals is precisely the right-hand side of relation (13.26).

Indeed, the piecewise-smooth curve ∂I breaks into four pieces (Fig. 13.5), which can be regarded as parametrized curves

$$\gamma_1 : [0, 1] \to \mathbb{R}^2, \quad \text{where } x \xmapsto{\gamma_1} (x, 0),$$

$$\gamma_2 : [0, 1] \to \mathbb{R}^2, \quad \text{where } y \xmapsto{\gamma_2} (1, y),$$

$$\gamma_3 : [0, 1] \to \mathbb{R}^2, \quad \text{where } x \xmapsto{\gamma_3} (x, 1),$$

$$\gamma_4 : [0, 1] \to \mathbb{R}^2, \quad \text{where } y \xmapsto{\gamma_4} (0, y).$$

By definition of the integral of the 1-form $\omega = P\, dx$ over a curve

$$\int_{\gamma_1} P(x, y)\, dx := \int_{[0,1]} \gamma_1^* \big(P(x, y)\, dx\big) := \int_0^1 P(x, 0)\, dx,$$

$$\int_{\gamma_2} P(x, y)\, dx := \int_{[0,1]} \gamma_2^* \big(P(x, y)\, dx\big) := \int_0^1 0\, dy = 0,$$

$$\int_{\gamma_3} P(x, y) \, dx := \int_{[0,1]} \gamma_3^* \big(P(x, y) \, dx \big) := \int_0^1 P(x, 1) \, dx,$$

$$\int_{\gamma_4} P(x, y) \, dx := \int_{[0,1]} \gamma_4^* \big(P(x, y) \, dx \big) := \int_0^1 0 \, dy = 0,$$

and, in addition, by the choice of the orientation of the boundary of the domain, taking account of the orientations of $\gamma_1, \gamma_2, \gamma_3, \gamma_4$, it is obvious that

$$\int_{\partial I} \omega = \int_{\gamma_1} \omega + \int_{\gamma_2} \omega + \int_{-\gamma_3} \omega + \int_{-\gamma_4} \omega = \int_{\gamma_1} \omega + \int_{\gamma_2} \omega - \int_{\gamma_3} \omega - \int_{\gamma_4} \omega,$$

where $-\gamma_i$ is the curve γ_i taken with the orientation opposite to the one defined by γ_i.

Thus Eq. (13.26) is now verified.   □

It can be verified similarly that

$$\iint_I \frac{\partial Q}{\partial x} \, dx \, dy = \int_{\partial I} Q \, dy. \tag{13.27}$$

Adding (13.26) and (13.27), we obtain Green's formula

$$\iint_I \left(\frac{\partial Q}{\partial x} - \frac{\partial P}{\partial y} \right) dx \, dy = \int_{\partial I} P \, dx + Q \, dy \tag{13.25'}$$

for the square I.

We remark that the asymmetry of P and Q in Green's formula (13.25) and in Eqs. (13.26) and (13.27) comes from the asymmetry of x and y: after all, x and y are ordered, and it is that ordering that gives the orientation in $\mathbb{R}^2$ and in I.

In the language of forms, the relation (13.25') just proved can be rewritten as

$$\int_I d\omega = \int_{\partial I} \omega, \tag{13.25''}$$

where ω is an arbitrary smooth form on I. The integrand on the right-hand side here is the restriction of the form ω to the boundary ∂I of the square I.

The proof of relation (13.26) just given admits an obvious generalization: If D_y is not a square, but a "curvilinear quadrilateral" whose lateral sides are vertical closed intervals (possibly degenerating to a point) and whose other two sides are the graphs of piecewise-smooth functions $\varphi_1(x) \le \varphi_2(x)$ over the closed interval $[a, b]$ of the x-axis, then

$$\iint_{D_y} \frac{\partial P}{\partial y} \, dx \, dy = - \int_{\partial D_y} P \, dx. \tag{13.26'}$$

Fig. 13.6

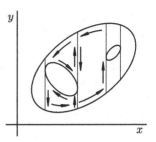

Similarly, if there is such a "quadrilateral" D_x with respect to the y-axis, that is, having two horizontal sides, then for it we have the equality

$$\iint_{D_x} \frac{\partial Q}{\partial x} \, dx \, dy = \int_{\partial D_x} Q \, dy. \qquad (13.27')$$

Now let us assume that the domain $\overline{D}$ can be cut into a finite number of domains of type D_y (Fig. 13.6). Then a formula of the form (13.26′) also holds for that region $\overline{D}$.

Proof In fact, by additivity, the double integral over the domain $\overline{D}$ is the sum of the integrals over the pieces of type D_y into which $\overline{D}$ is divided. Formula (13.26′) holds for each such piece, that is, the double integral over that piece equals the integral of $P \, dx$ over the oriented boundary of the piece. But adjacent pieces induce opposite orientations on their common boundary, so that when the integrals over the boundaries are added, all that remains after cancellation is the integral over the boundary $\partial \overline{D}$ of the domain $\overline{D}$ itself.   □

Similarly, if $\overline{D}$ admits a partition into domains of type D_x, an equality of type (13.27′) holds for it.

We agree to call domains that can be cut both into pieces of type D_x and into pieces of type D_y *elementary domains*. In fact, this class is sufficiently rich for all practical applications.

By writing both relations (13.26′) and (13.27′) for a simple domain, we obtain (13.25) by adding them.

Thus, Green's theorem is proved for simple domains.

We shall not undertake any further sharpenings of Green's formula at this point (on this account see Problem 2 below), but rather demonstrate a second, very fruitful line of reasoning that one may pursue after establishing Eqs. (13.25′) and (13.25″).

Suppose the domain C has been obtained by a smooth mapping $\varphi : I \to C$ of the square I. If ω is a smooth 1-form on C, then

$$\int_C d\omega := \int_I \varphi^* \, d\omega = \int_I d\varphi^* \, \omega \overset{!}{=} \int_{\partial I} \varphi^* \omega =: \int_{\partial C} \omega. \qquad (13.28)$$

The exclamation point here distinguishes the equality we have already proved (see (13.25″)); the extreme terms in these equalities are definitions or direct consequences of them; the remaining equality, the second from the left, results from the fact that exterior differentiation is independent of the coordinate system.

Hence Green's formula also holds for the domain C.

Finally, if it is possible to cut any oriented domain $\overline{D}$ into a finite number of domains of the same type as C, the considerations already described involving the mutual cancellation of the integrals over the portions of the boundaries of the C_i inside $\overline{D}$ imply that

$$\int_{\overline{D}} d\omega = \sum_i \int_{C_i} d\omega = \sum_i \int_{\partial C_i} \omega = \int_{\partial \overline{D}} \omega, \qquad (13.29)$$

that is, Green's formula also holds for $\overline{D}$.

It can be shown that every domain with a piecewise-smooth boundary belongs to this last class of domains, but we shall not do so, since we shall describe below (Chap. 15) a useful technical device that makes it possible to avoid such geometric complications, replacing them by an analytic problem that is comparatively easy to solve.

Let us consider some examples of the use of Green's formula.

Example 1 Let us set $P = -y$, $Q = x$ in (13.25). We then obtain

$$\int_{\partial D} -y \, dx + x \, dy = \int_D 2 \, dx \, dy = 2\sigma(D),$$

where $\sigma(D)$ is the area of D. Using Green's formula one can thus obtain the following expression for the area of a domain on the plane in terms of line integrals over the oriented boundary of the domain:

$$\sigma(D) = \frac{1}{2} \int_{\partial D} -y \, dx + x \, dy = - \int_{\partial D} y \, dx = \int_{\partial D} x \, dy.$$

It follows in particular from this that the work $A = \int_\gamma P \, dV$ performed by a heat engine in changing the state of its working substance over a closed cycle γ equals the area of the domain bounded by the curve γ in the PV-plane of states (see Problem 5 of Sect. 13.1).

Example 2 Let $\overline{B} = \{(x, y) \in \mathbb{R}^2 \mid x^2 + y^2 \le 1\}$ be the closed disk in the plane. We shall show that any smooth mapping $f : \overline{B} \to \overline{B}$ of the closed disk into itself has at least one fixed point (that is, a point $p \in \overline{B}$ such that $f(p) = p$).

Proof Assume that the mapping f has no fixed points. Then for every point $p \in \overline{B}$ the ray with initial point $f(p)$ passing through the point p and the point $\varphi(p) \in \partial B$ where this ray intersects the circle bounding $\overline{B}$ are uniquely determined. Thus a mapping $\varphi : \overline{B} \to \partial \overline{B}$ would arise, and it is obvious that the restriction of this mapping to the boundary would be the identity mapping. Moreover, it would have the same smoothness as the mapping f itself. We shall show that no such mapping φ can exist.

In the domain $\mathbb{R}^2 \backslash 0$ (the plane with the origin omitted) let us consider the form $\omega = \frac{-y\,dx + x\,dy}{x^2 + y^2}$ that we encountered in Sect. 13.1. It can be verified immediately that $d\omega = 0$. Since $\partial\overline{B} \subset \mathbb{R}^2 \backslash 0$, given the mapping $\varphi : \overline{B} \to \partial\overline{B}$, one could obtain a form $\varphi^*\omega$ on $\overline{B}$, and $d\varphi^*\omega = \varphi^*(d\omega) = \varphi^*0 = 0$. Hence by Green's formula

$$\int_{\partial\overline{B}} \varphi^*\omega = \int_{\overline{B}} d\varphi^*\omega = 0.$$

But the restriction of φ to $\partial\overline{B}$ is the identity mapping, and so

$$\int_{\partial\overline{B}} \varphi^*\omega = \int_{\partial\overline{B}} \omega.$$

This last integral, as was verified in Example 1 of Sect. 13.1, is nonzero. This contradiction completes the proof of the assertion. $\square$

This assertion is of course valid for a ball of any dimension (see Example 5 below). It also holds not only for smooth mappings, but for all continuous mappings $f : \mathbb{B} \to \mathbb{B}$. In this general form it is called the *Brouwer fixed-point theorem*.[8]

13.3.2 The Gauss–Ostrogradskii Formula

Just as Green's formula connects the integral over the boundary of a plane domain with a corresponding integral over the domain itself, the Gauss–Ostrogradskii formula given below connects the integral over the boundary of a three-dimensional domain with an integral over the domain itself.

Proposition 2 *Let $\mathbb{R}^3$ be three-dimensional space with a fixed coordinate system x, y, z and $\overline{D}$ a compact domain in $\mathbb{R}^3$ bounded by piecewise-smooth surfaces. Let $P, Q,$ and R be smooth functions in the closed domain $\overline{D}$.*
    Then the following relation holds:

$$\boxed{\begin{aligned} \iiint_{\overline{D}} \left(\frac{\partial P}{\partial x} + \frac{\partial Q}{\partial y} + \frac{\partial R}{\partial z} \right) dx\,dy\,dz = \\ = \iint_{\partial\overline{D}} P\,dy \wedge dz + Q\,dz \wedge dx + R\,dx \wedge dy. \end{aligned}} \tag{13.30}$$

---

[8]L.E.J. Brouwer (1881–1966) – well-known Dutch mathematician. A number of fundamental theorems of topology are associated with his name, as well as an analysis of the foundations of mathematics that leads to the philosophico-mathematical concepts called intuitionism.

Fig. 13.7

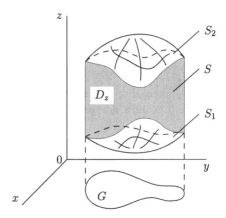

The Gauss–Ostrogradskii formula (13.30) can be derived by repeating the derivation of Green's formula step by step with obvious modifications. So as not to do a verbatim repetition, let us begin by considering not a cube in $\mathbb{R}^3$, but the domain D_x shown in Fig. 13.7, which is bounded by a lateral cylindrical surface S with generator parallel to the z-axis and two caps S_1 and S_2 which are the graphs of piecewise-smooth functions φ_1 and φ_2 defined in the same domain $G \subset \mathbb{R}^2_{xy}$. We shall verify that the relation

$$\iiint_{D_z} \frac{\partial R}{\partial z}\, dx\, dy\, dz = \iint_{\partial D_z} R\, dx \wedge dy \qquad (13.31)$$

holds for D_z.

Proof

$$\iiint_{D_z} \frac{\partial R}{\partial z}\, dx\, dy\, dz =$$

$$= \iint_G dx\, dy \int_{\varphi_1(x,y)}^{\varphi_2(x,y)} \frac{\partial R}{\partial z}\, dz =$$

$$= \iint_G \big(R(x, y, \varphi_2(x, y)) - R(x, y, \varphi_1(x, y))\big)\, dx\, dy =$$

$$= -\iint_G \big(R(x, y, \varphi_1(x, y))\big)\, dx\, dy + \iint_G \big(R(x, y, \varphi_2(x, y))\big)\, dx\, dy.$$

The surfaces S_1 and S_2 have the following parametrizations:

$$S_1 : (x, y) \longmapsto (x, y, \varphi_1(x, y)),$$
$$S_2 : (x, y) \longmapsto (x, y, \varphi_2(x, y)).$$

The curvilinear coordinates (x, y) define the same orientation on S_2 that is induced by the orientation of the domain D_z, and the opposite orientation on S_1.

Hence if S_1 and S_2 are regarded as pieces of the boundary of D_z oriented as indicated in Proposition 2, these last two integrals can be interpreted as integrals of the form $R \, dx \wedge dy$ over S_1 and S_2.

The cylindrical surface S has a parametric representation $(t, z) \mapsto (x(t), y(t), z)$, so that the restriction of the form $R \, dx \wedge dy$ to S equals zero, and so consequently, its integral over S is also zero.

Thus relation (13.31) does indeed hold for the domain D_z.   □

If the oriented domain $\overline{D}$ can be cut into a finite number of domains of the type D_z, then, since adjacent pieces induce opposite orientations on their common boundary, the integrals over these pieces will cancel out, leaving only the integral over the boundary $\partial \overline{D}$.

Consequently, formula (13.31) also holds for domains that admit this kind of partition into domains of type D_z.

Similarly, one can introduce domains D_y and D_x whose cylindrical surfaces have generators parallel to the y-axis or x-axis respectively and show that if a domain $\overline{D}$ can be divided into domains of type D_y or D_x, then the relations

$$\iiint_{\overline{D}} \frac{\partial Q}{\partial y} \, dx \, dy \, dz = \iint_{\partial \overline{D}} Q \, dz \wedge dx, \tag{13.32}$$

$$\iiint_{\overline{D}} \frac{\partial P}{\partial x} \, dx \, dy \, dz = \iint_{\partial \overline{D}} P \, dy \wedge dz. \tag{13.33}$$

Thus, if $\overline{D}$ is a *simple domain*, that is, a domain that admits each of the three types of partitions just described into domains of types D_x, D_y, and D_z, then, by adding (13.31), (13.32), and (13.33), we obtain (13.30) for $\overline{D}$.

For the reasons given in the derivation of Green's theorem, we shall not undertake the description of the conditions for a domain to be simple or any further sharpening of what has been proved (in this connection see Problem 8 below or Example 12 in Sect. 17.5).

We note, however, that in the language of forms, the Gauss–Ostrogradskii formula can be written in coordinate-free form as follows:

$$\int_{\overline{D}} d\omega = \int_{\partial \overline{D}} \omega, \tag{13.30'}$$

where ω is a smooth 2-form in $\overline{D}$.

Since formula (13.30') holds for the cube $I = I^3 = \{(x, y, z) \in \mathbb{R}^3 \mid 0 \leq 1 \leq 1, 0 \leq y \leq 1, 0 \leq z \leq 1\}$, as we have shown, its extension to more general classes of domains can of course be carried out using the standard computations (13.28) and (13.29).

Example 3 (The law of Archimedes) Let us compute the buoyant force of a homogeneous liquid on a body D immersed in it. We choose the Cartesian coordinates x, y, z in $\mathbb{R}^3$ so that the xy-plane is the surface of the liquid and the z-axis is directed out of the liquid. A force $\rho g z \mathbf{n} \, d\sigma$ is acting on an element $d\sigma$ of the surface S

of D located at depth z, where ρ is the density of the liquid, g is the acceleration of gravity, and $\mathbf{n}$ is a unit outward normal to the surface at the point of the surface corresponding to $d\sigma$. Hence the resultant force can be expressed by the integral

$$\mathbf{F} = \iint_S \rho g z \mathbf{n} \, d\sigma.$$

If $\mathbf{n} = \mathbf{e}_x \cos \alpha_x + \mathbf{e}_y \cos \alpha_y + \mathbf{e}_z \cos \alpha_z$, then $\mathbf{n} \, d\sigma = \mathbf{e}_x \, dy \wedge dz + \mathbf{e}_y \, dz \wedge dx + \mathbf{e}_z \, dx \wedge dy$ (see Sect. 13.2.4). Using the Gauss–Ostrogradskii formula (13.30), we thus find that

$$\mathbf{F} = \mathbf{e}_x \rho g \iint_S z \, dy \wedge dz + \mathbf{e}_y \rho g \iint_S z \, dz \wedge dx + \mathbf{e}_z \rho g \iint_S z \, dx \wedge dy =$$

$$= \mathbf{e}_x \rho g \iiint_D 0 \, dx \, dy \, dz + \mathbf{e}_y \rho g \iiint_D 0 \, dx \, dy \, dz +$$

$$+ \mathbf{e}_z \rho g \iiint_D dx \, dy \, dz = \rho g V \mathbf{e}_z,$$

where V is the volume of the body D. Hence $P = \rho g V$ is the weight of a volume of the liquid equal to the volume occupied by the body. We have arrived at Archimedes' law: $\mathbf{F} = P \mathbf{e}_z$.

Example 4 Using the Gauss–Ostrogradskii formula (13.30), one can give the following formulas for the volume $V(D)$ of a body D bounded by a surface ∂D.

$$V(D) = \frac{1}{3} \iint_{\partial D} x \, dy \wedge dz + y \, dz \wedge dx + z \, dx \wedge dy =$$

$$= \iint_{\partial D} x \, dy \wedge dz = \iint_{\partial D} y \, dz \wedge dx = \iint_{\partial D} z \, dx \wedge dy.$$

13.3.3 Stokes' Formula in $\mathbb{R}^3$

Proposition 3 *Let S be an oriented piecewise-smooth compact two-dimensional surface with boundary ∂S embedded in a domain $G \subset \mathbb{R}^3$, in which a smooth 1-form $\omega = P \, dx + Q \, dy + R \, dz$ is defined. Then the following relation holds:*

$$\int_{\partial S} P \, dx + Q \, dy + R \, dz = \iint_S \left(\frac{\partial R}{\partial y} - \frac{\partial Q}{\partial z} \right) dy \wedge dz +$$

$$+ \left(\frac{\partial P}{\partial z} - \frac{\partial R}{\partial x} \right) dz \wedge dx + \left(\frac{\partial Q}{\partial x} - \frac{\partial P}{\partial y} \right) dx \wedge dy,$$

$$(13.34)$$

Fig. 13.8

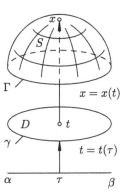

where the orientation of the boundary ∂S is chosen consistently with the orientation of the surface S.

In other notation, this means that

$$\int_S d\omega = \int_{\partial\omega} \omega. \tag{13.34'}$$

Proof If C is a standard parametrized surface $\varphi : I \to C$ in $\mathbb{R}^3$, where I is a square in $\mathbb{R}^2$, relation (13.34) follows from Eqs. (13.28) taking account of what has been proved for the square and Green's formula.

If the orientable surface S can be cut into elementary surfaces of this type, then relation (13.34) is also valid for it, as follows from Eqs. (13.29) with $\overline{D}$ replaced by S.   □

As in the preceding cases, we shall not prove at this point that, for example, a piecewise-smooth surface admits such a partition.

Let us show what this proof of formula (13.34) would look like in coordinate notation. To avoid expressions that are really too cumbersome, we shall write out only the first, main part of its two expressions, and with some simplifications even in that. To be specific, let us introduce the notation x^1, x^2, x^3 for the coordinates of a point $x \in \mathbb{R}^3$ and verify only that

$$\int_{\partial S} P(x)\,dx^1 = \iint_S \frac{\partial P}{\partial x^2}\,dx^2 \wedge dx^1 + \frac{\partial P}{\partial x^3}\,dx^3 \wedge dx^1,$$

since the other two terms on the left-hand side of (13.34) can be studied similarly. For simplicity we shall assume that S can be obtained by a smooth mapping $x = x(t)$ of a domain D in the plane $\mathbb{R}^2$ of the variables t^1, t^2 and bounded by a smooth curve $\gamma = \partial D$ parametrized via a mapping $t = t(\tau)$ by the points of the closed interval $\alpha \le \tau \le \beta$ (Fig. 13.8). Then the boundary $\Gamma = \partial S$ of the surface S can be written as $x = x(t(\tau))$, where τ ranges over the closed interval $[\alpha, \beta]$. Using the definition of the integral over a curve, Green's formula for a plane domain D, and the definition of the integral over a parametrized surface, we find successively

$$\int_\Gamma P(x)\,\mathrm{d}x^1 := \int_\alpha^\beta P\big(x(t(\tau))\big)\left(\frac{\partial x^1}{\partial t^1}\frac{\mathrm{d}t^1}{\mathrm{d}\tau} + \frac{\partial x^1}{\partial t^2}\frac{\mathrm{d}t^2}{\mathrm{d}\tau}\right)\mathrm{d}\tau =$$

$$= \int_\gamma \left(P\big(x(t)\big)\frac{\partial x^1}{\partial t^1}\right)\mathrm{d}t^1 + \left(P\big(x(t)\big)\frac{\partial x^1}{\partial t^2}\right)\mathrm{d}t^2 \overset{!}{=}$$

$$\overset{!}{=} \iint_D \left[\frac{\partial}{\partial t^1}\left(P\frac{\partial x^1}{\partial t^2}\right) - \frac{\partial}{\partial t^2}\left(P\frac{\partial x^1}{\partial t^1}\right)\right]\mathrm{d}t^1 \wedge \mathrm{d}t^2 =$$

$$= \iint_D \left(\frac{\partial P}{\partial t^1}\frac{\partial x^1}{\partial t^2} - \frac{\partial P}{\partial t^2}\frac{\partial x^1}{\partial t^1}\right)\mathrm{d}t^1 \wedge \mathrm{d}t^2 =$$

$$= \iint_D \sum_{i=1}^3 \left(\frac{\partial P}{\partial x^i}\frac{\partial x^i}{\partial t^1}\frac{\partial x^1}{\partial t^2} - \frac{\partial P}{\partial x^i}\frac{\partial x^i}{\partial t^2}\frac{\partial x^1}{\partial t^1}\right)\mathrm{d}t^1 \wedge \mathrm{d}t^2 =$$

$$= \iint_D \left[\left(\frac{\partial P}{\partial x^2}\frac{\partial x^2}{\partial t^1} + \frac{\partial P}{\partial x^3}\frac{\partial x^3}{\partial t^1}\right)\frac{\partial x^1}{\partial t^2} -\right.$$

$$\left. - \left(\frac{\partial P}{\partial x^2}\frac{\partial x^2}{\partial t^2} + \frac{\partial P}{\partial x^3}\frac{\partial x^3}{\partial t^2}\right)\frac{\partial x^1}{\partial t^1}\right]\mathrm{d}t^1 \wedge \mathrm{d}t^2 =$$

$$= \iint_D \left(\frac{\partial P}{\partial x^2}\begin{vmatrix}\frac{\partial x^2}{\partial t^1} & \frac{\partial x^2}{\partial t^2}\\ \frac{\partial x^1}{\partial t^1} & \frac{\partial x^1}{\partial t^2}\end{vmatrix} + \frac{\partial P}{\partial x^3}\begin{vmatrix}\frac{\partial x^3}{\partial t^1} & \frac{\partial x^3}{\partial t^2}\\ \frac{\partial x^1}{\partial t^1} & \frac{\partial x^1}{\partial t^2}\end{vmatrix}\right)\mathrm{d}t^1 \wedge \mathrm{d}t^2 =$$

$$=: \iint_S \left(\frac{\partial P}{\partial x^2}\,\mathrm{d}x^2 \wedge \mathrm{d}x^1 + \frac{\partial P}{\partial x^3}\,\mathrm{d}x^3 \wedge \mathrm{d}x^1\right).$$

The colon here denotes equality by definition, and the exclamation point denotes a transition that uses the Green's formula already proved. The rest consists of identities.

Using the basic idea of the proof of formula (13.34′), we have thus verified directly (without invoking the relation $\varphi^*\mathrm{d} = \mathrm{d}\varphi^*$, but essentially proving it for the case under consideration) that formula (13.34) does indeed hold for a simple parametrized surface. We have carried out the reasoning formally only for the term $P\,\mathrm{d}x$, but it is clear that the same thing could also be done for the other two terms in the 1-form in the integrand on the left-hand side of (13.34).

13.3.4 The General Stokes Formula

Despite the differences in the external appearance of formulas (13.25), (13.30), and (13.34), their coordinate-free expressions (13.25″), (13.29), (13.30′), and (13.34′) turn out to be identical. This gives grounds for supposing that we have been dealing with particular manifestations of a general rule, which one can now easily guess.

Proposition 4 *Let S be an oriented piecewise smooth k-dimensional compact surface with boundary ∂S in the domain $G \subset \mathbb{R}^n$, in which a smooth $(k-1)$-form ω is defined.*

Then the following relation holds:

$$\boxed{\int_S d\omega = \int_{\partial S} \omega,} \tag{13.35}$$

in which the orientation of the boundary ∂S is that induced by the orientation of S.

Proof Formula (13.35) can obviously be proved by the same general computations (13.28) and (13.29) as Stokes' formula (13.34′) provided it holds for a standard k-dimensional interval $I^k = \{x = (x^1, \dots, x^k) \in \mathbb{R}^k \mid 0 \le x^i \le 1, i = 1, \dots, k\}$. Let us verify that (13.35) does indeed hold for I^k.

Since a $(k-1)$-form on I^k has the form $\omega = \sum_i a_i(x)\,dx^1 \wedge \cdots \wedge \widehat{dx^i} \wedge \cdots \wedge dx^k$ (summation over $i = 1, \dots, k$, with the differential dx^i omitted), it suffices to prove (13.35) for each individual term. Let $\omega = a(x)\,dx^1 \wedge \cdots \wedge \widehat{dx^i} \wedge \cdots \wedge dx^k$. Then $d\omega = (-1)^{i-1} \frac{\partial a}{\partial x^i}(x)\,dx^1 \wedge \cdots \wedge dx^i \wedge \cdots \wedge dx^k$. We now carry out the computation:

$$\int_{I^k} d\omega = \int_{I^k} (-1)^{i-1} \frac{\partial a}{\partial x^i}(x)\,dx^1 \wedge \cdots \wedge dx^k =$$

$$= (-1)^{i-1} \int_{I^{k-1}} dx^1 \cdots \widehat{dx^i} \cdots dx^k \int_0^1 \frac{\partial a}{\partial x^i}(x)\,dx^i =$$

$$= (-1)^{i-1} \int_{I^{k-1}} \big(a\big(x^1, \dots, x^{i-1}, 1, x^{i+1}, \dots, x^k\big) -$$

$$- a\big(x^1, \dots, x^{i-1}, 0, x^{i+1}, \dots x^k\big)\big)\,dx^1 \cdots \widehat{dx^i} \cdots dx^k =$$

$$= (-1)^{i-1} \int_{I^{k-1}} a\big(t^1, \dots, t^{i-1}, 1, t^i, \dots, t^{k-1}\big)\,dt^1 \cdots dt^{k-1} +$$

$$+ (-1)^i \int_{I^{k-1}} a\big(t^1, \dots, t^{i-1}, 0, t^i, \dots, t^{k-1}\big)\,dt^1 \cdots dt^{k-1}.$$

Here I^{k-1} is the same as I^k in $\mathbb{R}^k$, only it is a $(k-1)$-dimensional interval in $\mathbb{R}^{k-1}$. In addition, we have relabeled the variables $x^1 = t^1, \dots, x^{i-1} = t^{i-1}, x^{i+1} = t^i, \dots, x^k = t^{k-1}$.

The mappings

$$I^{k-1} \ni t = \big(t^1, \dots, t^{k-1}\big) \longmapsto \big(t^1, \dots, t^{i-1}, 1, t^i, \dots, t^{k-1}\big) \in I^k,$$

$$I^{k-1} \ni t = \big(t^1, \dots, t^{k-1}\big) \longmapsto \big(t^1, \dots, t^{i-1}, 0, t^i, \dots, t^{k-1}\big) \in I^k$$

are parametrizations of the upper and lower faces Γ_{i1} and Γ_{i0} of the interval I^k respectively orthogonal to the x^i axis. These coordinates define the same frame $\mathbf{e}_1, \dots, \mathbf{e}_{i-1}, \mathbf{e}_{i+1}, \dots, \mathbf{e}_k$ orienting the faces and differing from the frame $\mathbf{e}_1, \dots, \mathbf{e}_k$

of $\mathbb{R}^k$ in the absence of $\mathbf{e}_i$. On Γ_{i1} the vector $\mathbf{e}_i$ is the exterior normal to I^k, as the vector $-\mathbf{e}_i$ is for the face Γ_{i0}. The frame $\mathbf{e}_i, \mathbf{e}_1, \ldots, \mathbf{e}_{i-1}, \mathbf{e}_{i+1}, \ldots, \mathbf{e}_k$ becomes the frame $\mathbf{e}_1, \ldots, \mathbf{e}_k$ after $i - 1$ inter-changes of adjacent vectors, that is, the agreement or disagreement of the orientations of these frames is determined by the sign of $(-1)^{i-1}$. Thus, this parametrization defines an orientation on Γ_{i1} consistent with the orientation of I^k if taken with the corrective coefficient $(-1)^{i-1}$ (that is, not changing the orientation when i is odd, but changing it when i is even).

Analogous reasoning shows that for the face Γ_{i0} it is necessary to take a corrective coefficient $(-1)^i$ to the orientation defined by this parametrization of the face Γ_{i0}.

Thus, the last two integrals (together with the coefficients in front of them) can be interpreted respectively as the integrals of the form ω over the faces Γ_{i1} and Γ_{i0} of I^k with the orientation induced by the orientation of I^k.

We now remark that on each of the remaining faces of I^k one of the coordinates $x^1, \ldots, x^{i-1}, x^{i+1}, \ldots, x^k$ is constant. Hence the differential corresponding to it is equal to zero on such a face. Thus, the form $d\omega$ is identically equal to zero and its integral equals zero over all faces except Γ_{i0} and Γ_{i1}.

Hence we can interpret the sum of the integrals over these two faces as the integral of the form ω over the entire boundary ∂I^k of the interval I^k oriented in consistency with the orientation of the interval I^k itself.

The formula

$$\int_{I_k} d\omega = \int_{\partial I^k} \omega,$$

and along with it formula (13.35), is now proved.   □

As one can see, formula (13.35) is a corollary of the Newton–Leibniz formula (fundamental theorem of calculus), Fubini's theorem, and a series of definitions of such concepts as surface, boundary of a surface, orientation, differential form, differentiation of a differential form, and transference of forms.

Formulas (13.25), (13.30), and (13.34), the formulas of Green, Gauss–Ostrogradskii, and Stokes respectively, are special cases of the general formula (13.35). Moreover, if we interpret a function f defined on a closed interval $[a, b] \subset \mathbb{R}$ as a 0-form ω, and the integral of a 0-form over an oriented point as the value of the function at that point taken with the sign of the orientation of the point, then the Newton–Leibniz formula itself can be regarded as an elementary (but independent) version of (13.35). Consequently, the fundamental relation (13.35) holds in all dimensions $k \geq 1$.

Formula (13.35) is often called the *general Stokes formula*. As historical information, we quote here some lines from the preface of M. Spivak to his book (cited in the bibliography below):

The first statement of the Theorem[9] appears as a postscript to a letter, dated July 2, 1850, from Sir William Thomson (Lord Kelvin) to Stokes. It appeared publicly as question 8 on

---

[9] The classical Stokes theorem (13.34) is meant.

the Smith's Prize Examination for 1854. This competitive examination, which was taken annually by the best mathematics students at Cambridge University, was set from 1849 to 1882 by Professor Stokes; by the time of his death the result was known universally as Stokes' theorem. At least three proofs were given by his contemporaries: Thomson published one, another appeared in Thomson and Tait's *Treatise on Natural Philosophy*, and Maxwell provided another in *Electricity and Magnetism*. Since this time the name of Stokes has been applied to much more general results, which have figured so prominently in the development of certain parts of mathematics that Stokes' theorem may be considered a case study in the value of generalization.

We note that the modern language of differential forms originates with Élie Cartan,[10] but the form (13.35) for the general Stokes' formula for surfaces in $\mathbb{R}^n$ seems to have been first proposed by Poincaré. For domains in n-dimensional space $\mathbb{R}^n$ Ostrogradskii already knew the formula, and Leibniz wrote down the first differential forms.

Thus it is not an accident that the general Stokes formula (13.35) is sometimes called the Newton–Leibniz–Green–Gauss–Ostrogradskii–Stokes Poincaré formula. One can conclude from what has been said that this is by no means its full name.

Let us use this formula to generalize the result of Example 2.

Example 5 Let us show that every smooth mapping $f : \overline{B} \to \overline{B}$ of a closed ball $\overline{B} \subset \mathbb{R}^m$ into itself has at least one fixed point.

Proof If the mapping f had no fixed points, then, as in Example 2, one could construct a smooth mapping $\varphi : \overline{B} \to \partial\overline{B}$ that is the identity on the sphere $\partial\overline{B}$. In the domain $\mathbb{R}^m \backslash 0$, we consider the vector field $\frac{\mathbf{r}}{|\mathbf{r}|^m}$, where $\mathbf{r}$ is the radius-vector of the point $x = (x^1, \dots, x^m) \in \mathbb{R}^m \backslash 0$, and the flux form

$$\omega = \left\langle \frac{\mathbf{r}}{|\mathbf{r}|^m}, \mathbf{n} \right\rangle \Omega = \sum_{i=1}^m \frac{(-1)^{i-1} x^i \, dx^1 \wedge \cdots \wedge \widehat{dx^i} \wedge \cdots \wedge dx^m}{((x^1)^2 + \cdots + (x^m)^2)^{m/2}}$$

corresponding to this field (see formula (13.19) of Sect. 13.2). The flux of such a field across the boundary of the ball $\overline{B} = \{x \in \mathbb{R} \mid |x| = 1\}$ in the direction of the outward normal to the sphere $\partial\overline{B}$ is obviously equal to the area of the sphere $\partial\overline{B}$, that is, $\int_{\partial\overline{B}} \omega \neq 0$. But, as one can easily verify by direct computation, $d\omega = 0$ in $\mathbb{R}^m \backslash 0$, from which, by using the general Stokes formula, as in Example 2, we find that

$$\int_{\partial\overline{B}} \omega = \int_{\partial\overline{B}} \varphi^* \omega = \int_{\overline{B}} d\varphi^* \, \omega = \int_{\overline{B}} \varphi^* \, d\omega = \int_{\overline{B}} \varphi^* 0 = 0.$$

This contradiction finishes the proof.   □

---

[10] Élie Cartan (1869–1951) – outstanding French geometer.

13.3.5 Problems and Exercises

1. a) Does Green's formula (13.25) change if we pass from the coordinate system x, y to the system y, x?

   b) Does formula (13.25″) change in this case?

2. a) Prove that formula (13.25) remains valid if the functions P and Q are continuous in a closed square I, their partial derivatives $\frac{\partial P}{\partial y}$ and $\frac{\partial Q}{\partial x}$ are continuous at interior points of I, and the double integrals exist, even if as improper integrals (13.25′).

   b) Verify that if the boundary of a compact domain D consists of piecewise-smooth curves, then under assumptions analogous to those in a), formula (13.25) remains valid.

3. a) Verify the proof of (13.26′) in detail.

   b) Show that if the boundary of a compact domain $D \subset \mathbb{R}^2$ consists of a finite number of smooth curves having only a finite number of points of inflection, then D is a simple domain with respect to any pair of coordinate axes.

   c) Is it true that if the boundary of a plane domain consists of smooth curves, then one can choose the coordinate axes in $\mathbb{R}^2$ such that it is a simple domain relative to them?

4. a) Show that if the functions P and Q in Green's formula are such that $\frac{\partial Q}{\partial x} - \frac{\partial P}{\partial y} = 1$, then the area $\sigma(D)$ of the domain D can be found using the formula $\sigma(D) = \int_{\partial D} P \, dx + Q \, dy$.

   b) Explain the geometric meaning of the integral $\int_\gamma y \, dx$ over some (possibly nonclosed) curve in the plane with Cartesian coordinates x, y. Starting from this, give a new interpretation of the formula $\sigma(D) = - \int_{\partial D} y \, dx$.

   c) As a check on the preceding formula, use it to find the area of the domain

$$D = \left\{ (x, y) \in \mathbb{R}^2 \,\middle|\, \frac{x^2}{a^2} + \frac{y^2}{b^2} \leq 1 \right\}.$$

5. a) Let $x = x(t)$ be a diffeomorphism of the domain $D_t \subset \mathbb{R}_t^2$ onto the domain $D_x \subset \mathbb{R}_x^2$. Using the results of Problem 4 and the fact that a line integral is independent of the admissible change in the parametrization of the path, prove that

$$\int_{D_x} dx = \int_{D_t} |x'(t)| \, dt,$$

where $dx = dx^1 \, dx^2$, $dt = dt^1 \, dt^2$, $|x'(t)| = \det x'(t)$.

   b) From a) derive the formula

$$\int_{D_x} f(x) \, dx = \int_{D_t} f(x(t)) \left| \det x'(t) \right| dt$$

for change of variable in a double integral.

6. Let $f(x, y, t)$ be a smooth function satisfying the condition $(\frac{\partial f}{\partial x})^2 + (\frac{\partial f}{\partial y})^2 \neq 0$ in its domain of definition. Then for each fixed value of the parameter t the equation $f(x, y, t) = 0$ defines a curve γ_t in the plane $\mathbb{R}^2$. Then a family of curves $\{\gamma_t\}$ depending on the parameter t arises in the plane. A smooth curve $\Gamma \subset \mathbb{R}^2$ defined by parametric equations $x = x(t)$, $y = y(t)$, is the *envelope of the family of curves* $\{\gamma_t\}$ if the point $x(t_0)$, $y(t_0)$ lies on the corresponding curve γ_{t_0} and the curves Γ and γ_{t_0} are tangent at that point, for every value of t_0 in the common domain of definition of $\{\gamma_t\}$ and the functions $x(t)$, $y(t)$.

a) Assuming that x, y are Cartesian coordinates in the plane, show that the functions $x(t), y(t)$ that define the envelope must satisfy the system of equations

$$\begin{cases} f(x, y, t) = 0, \\ \dfrac{\partial f}{\partial t}(x, y, t) = 0, \end{cases}$$

and from the geometric point of view the envelope itself is the boundary of the projection (shadow) of the surface $f(x, y, t) = 0$ of $\mathbb{R}^3_{(x,y,t)}$ on the plane $\mathbb{R}^2_{(x,y)}$.

b) A family of lines $x \cos\alpha + y \sin\alpha - p(\alpha) = 0$ is given in the plane with Cartesian coordinates x and y. The role of the parameter is played here by the polar angle α. Give the geometric meaning of the quantity $p(\alpha)$, and find the envelope of this family if $p(\alpha) = c + a\cos\alpha + b\sin\alpha$, where a, b, and c are constants.

c) Describe the accessible zone of a shell that can be fired from an adjustable cannon making any angle $\alpha \in [0, \pi/2]$ to the horizon.

d) Show that if the function $p(\alpha)$ of b) is 2π-periodic, then the corresponding envelope Γ is a closed curve.

e) Using Problem 4, show that the length L of the closed curve Γ obtained in d) can be found by the formula

$$L = \int_0^{2\pi} p(\alpha) \, d\alpha.$$

(Assume that $p \in C^{(2)}$.)

f) Show also that the area σ of the region bounded by the closed curve Γ obtained in d) can be computed as

$$\sigma = \frac{1}{2} \int_0^{2\pi} \left(p^2 - \dot{p}^2 \right)(\alpha) \, d\alpha, \qquad \dot{p}(\alpha) = \frac{dp}{d\alpha}(\alpha).$$

7. Consider the integral $\int_\gamma \frac{\cos(\mathbf{r}, \mathbf{n})}{r} \, ds$, in which γ is a smooth curve in $\mathbb{R}^2$, $\mathbf{r}$ is the radius-vector of the point $(x, y) \in \gamma$, $r = |\mathbf{r}| = \sqrt{x^2 + y^2}$, $\mathbf{n}$ is the unit normal vector to γ at (x, y) varying continuously along γ, and ds is arc length on the curve. This integral is called *Gauss' integral*.

a) Write Gauss' integral in the form of a flux $\int_\gamma \langle \mathbf{V}, \mathbf{n} \rangle \, ds$ of the plane vector field $\mathbf{V}$ across the curve γ.

b) Show that in Cartesian coordinates x and y Gauss' integral has the form $\pm \int_\gamma \frac{-y\,dx+x\,dy}{x^2+y^2}$ familiar to us from Example 1 of Sect. 13.1, where the choice of sign is determined by the choice of the field of normals $\mathbf{n}$.

c) Compute Gauss' integral for a closed curve γ that encircles the origin once and for a curve γ bounding a domain that does not contain the origin.

d) Show that $\frac{\cos(\mathbf{r},\mathbf{n})}{r}\,ds = d\varphi$, where φ is the polar angle of the radius-vector $\mathbf{r}$, and give the geometric meaning of the value of Gauss' integral for a closed curve and for an arbitrary curve $\gamma \subset \mathbb{R}^2$.

8. In deriving the Gauss–Ostrogradskii formula we assumed that D is a simple domain and the functions P, Q, R belong to $C^{(1)}(\overline{D}, \mathbb{R})$. Show by improving the reasoning that formula (13.30) holds if D is a compact domain with piecewise smooth boundary, $P, Q, R \in C(\overline{D}, \mathbb{R})$, $\frac{\partial P}{\partial x}, \frac{\partial Q}{\partial y}, \frac{\partial R}{\partial z} \in C(D, \mathbb{R})$, and the triple integral converges, even if it is an improper integral.

9. a) If the functions P, Q, and R in formula (13.30) are such that $\frac{\partial P}{\partial x} + \frac{\partial Q}{\partial y} + \frac{\partial R}{\partial z} = 1$, then the volume $V(D)$ of the domain D can be found by the formula

$$V(D) = \iint_{\partial D} P\,dy \wedge dz + Q\,dz \wedge dx + R\,dx \wedge dy.$$

b) Let $f(x,t)$ be a smooth function of the variables $x \in D_X \subset \mathbb{R}^n_x$, $t \in D_t \subset \mathbb{R}^n_t$ and $\frac{\partial f}{\partial x} = (\frac{\partial f}{\partial x^1}, \dots, \frac{\partial f}{\partial x^n}) \neq 0$. Write the system of equations that must be satisfied by the $(n-1)$-dimensional surface in $\mathbb{R}^n_x$ that is the envelope of the family of surfaces $\{S_t\}$ defined by the conditions $f(x,t) = 0$, $t \in D_t$ (see Problem 6).

c) Choosing a point on the unit sphere as the parameter t, exhibit a family of planes in $\mathbb{R}^3$ depending on the parameter t whose envelope is the ellipsoid $\frac{x^2}{a^2} + \frac{y^2}{b^2} + \frac{z^2}{c^2} = 1$.

d) Show that if a closed surface S is the envelope of a family of planes

$$\cos \alpha_1(t)x + \cos \alpha(t)y + \cos \alpha_3(t)z - p(t) = 0,$$

where $\alpha_1, \alpha_2, \alpha_3$ are the angles formed by the normal to the plane and the coordinate axes and the parameter t is a variable point of the unit sphere $S^2 \subset \mathbb{R}^3$, then the area σ of the surface S can be found by the formula $\sigma = \int_{S^2} p(t)\,d\sigma$.

e) Show that the volume of the body bounded by the surface S considered in d) can be found by the formula $V = \frac{1}{3}\int_S p(t)\,d\sigma$.

f) Test the formula given in e) by finding the volume of the ellipsoid $\frac{x^2}{a^2} + \frac{y^2}{b^2} + \frac{z^2}{c^2} \leq 1$.

g) What does the n-dimensional analogue of the formulas in d) and e) look like?

10. a) Using the Gauss–Ostrogradskii formula, verify that the flux of the field $\mathbf{r}/r^3$ (where $\mathbf{r}$ is the radius-vector of the point $x \in \mathbb{R}^3$ and $r = |\mathbf{r}|$) across a smooth surface S enclosing the origin and homeomorphic to a sphere equals the flux of the same field across an arbitrarily small sphere $|x| = \varepsilon$.

b) Show that the flux in a) is 4π.

c) Interpret Gauss' integral $\int_S \frac{\cos(\mathbf{r},\mathbf{n})}{r} \, ds$ in $\mathbb{R}^3$ as the flux of the field $\mathbf{r}/r^3$ across the surface S.

d) Compute Gauss' integral over the boundary of a compact domain $D \subset \mathbb{R}^3$, considering both the case when D contains the origin in its interior and the case when the origin lies outside D.

e) Comparing Problems 7 and 10a)–d), give an n-dimensional version of Gauss' integral and the corresponding vector field. Give an n-dimensional statement of problems a)–d) and verify it.

11. a) Show that a closed rigid surface $S \subset \mathbb{R}^3$ remains in equilibrium under the action of a uniformly distributed pressure on it. (By the principles of statics the problem reduces to verifying the equalities $\iint_S \mathbf{n} \, d\sigma = 0$, $\iint_S [\mathbf{r}, \mathbf{n}] \, d\sigma = 0$, where $\mathbf{n}$ is a unit normal vector, $\mathbf{r}$ is the radius-vector, and $[\mathbf{r}, \mathbf{n}]$ is the vector product of $\mathbf{r}$ and $\mathbf{n}$.)

b) A solid body of volume V is completely immersed in a liquid having specific gravity 1. Show that the complete static effect of the pressure of the liquid on the body reduces to a single force $\mathbf{F}$ of magnitude V directed vertically upward and attached to the center of mass C of the solid domain occupied by the body.

12. Let $\Gamma : I^k \to D$ be a smooth (not necessarily homeomorphic) mapping of an interval $I^k \subset \mathbb{R}^k$ into a domain D of $\mathbb{R}^n$, in which a k-form ω is defined. By analogy with the one-dimensional case, we shall call a mapping Γ a k-cell or k-path and by definition set $\int_\Gamma \omega = \int_{I^k} \Gamma^* \omega$. Study the proof of the general Stokes formula and verify that it holds not only for k-dimensional surfaces but also for k-cells.

13. Using the generalized Stokes formula, prove by induction the formula for change of variable in a multiple integral (the idea of the proof is shown in Problem 5a)).

14. *Integration by parts in a multiple integral.*

Let D be a bounded domain in $\mathbb{R}^m$ with a regular (smooth or piecewise smooth) boundary ∂D oriented by the outward unit normal $\mathbf{n} = (n^1, \ldots, n^m)$.

Let f, g be smooth functions in $\overline{D}$.

a) Show that

$$\int_D \partial_i f \, dv = \int_{\partial D} f n^i \, d\sigma.$$

b) Prove the following formula for integration by parts:

$$\int_D (\partial_i f) g \, dv = \int_{\partial D} f g n^i \, d\sigma - \int_D f (\partial_i g) \, dv.$$

Chapter 14
Elements of Vector Analysis and Field Theory

14.1 The Differential Operations of Vector Analysis

14.1.1 Scalar and Vector Fields

In field theory we consider functions $x \mapsto T(x)$ that assign to each point x of a given domain D a special object $T(x)$ called a *tensor*. If such a function is defined in a domain D, we say that a *tensor field* is defined in D. We do not intend to give the definition of a tensor at this point: that concept will be studied in algebra and differential geometry. We shall say only that numerical functions $D \ni x \mapsto f(x) \in \mathbb{R}$ and vector-valued functions $\mathbb{R}^n \supset D \ni x \mapsto V(x) \in T\mathbb{R}^n_x \approx \mathbb{R}^n$ are special cases of tensor fields and are called *scalar fields* and *vector fields* respectively in D (we have used this terminology earlier).

A differential p-form ω in D is a function $\mathbb{R}^n \supset D \ni x \mapsto \omega(x) \in \mathcal{L}((\mathbb{R}^n)^p, \mathbb{R})$ which can be called a *field of forms* of degree p in D. This also is a special case of a tensor field.

At present we are primarily interested in scalar and vector fields in domains of the oriented Euclidean space $\mathbb{R}^n$. These fields play a major role in many applications of analysis in natural science.

14.1.2 Vector Fields and Forms in $\mathbb{R}^3$

We recall that in the Euclidean vector space $\mathbb{R}^3$ with inner product $\langle \, , \, \rangle$ there is a correspondence between linear functionals $A : \mathbb{R}^3 \to \mathbb{R}$ and vectors $\mathbf{A} \in \mathbb{R}^3$ consisting of the following: Each such functional has the form $A(\xi) = \langle \mathbf{A}, \xi \rangle$, where $\mathbf{A}$ is a completely definite vector in $\mathbb{R}^3$.

If the space is also oriented, each skew-symmetric bilinear functional $B: \mathbb{R}^3 \times \mathbb{R}^3 \to \mathbb{R}$ can be uniquely written in the form $B(\xi_1, \xi_2) = (\mathbf{B}, \xi_1, \xi_2)$, where $\mathbf{B}$ is a completely definite vector in $\mathbb{R}^3$ and $(\mathbf{B}, \xi_1, \xi_2)$, as always, is the scalar triple

© Springer-Verlag Berlin Heidelberg 2016
V.A. Zorich, *Mathematical Analysis II*, Universitext,
DOI 10.1007/978-3-662-48993-2_6

product of the vectors $\mathbf{B}$, $\boldsymbol{\xi}_1$, and $\boldsymbol{\xi}_2$, or what is the same, the value of the volume element on these vectors. Thus, in the oriented Euclidean vector space $\mathbb{R}^3$ one can associate with each vector a linear or bilinear form, and defining the linear or bilinear form is equivalent to defining the corresponding vector in $\mathbb{R}^3$.

If there is an inner product in $\mathbb{R}^3$, it also arises naturally in each tangent space $T\mathbb{R}_x^3$ consisting of the vectors attached to the point $\mathbb{R}^3$, and the orientation of $\mathbb{R}^3$ orients each space $T\mathbb{R}_x^3$.

Hence defining a 1-form $\omega^1(x)$ or a 2-form $\omega^2(x)$ in $T\mathbb{R}_x^3$ under the conditions just listed is equivalent to defining some vector $\mathbf{A}(x) \in T\mathbb{R}_x^3$ corresponding to the form $\omega^1(x)$ or a vector $\mathbf{B}(x) \in T\mathbb{R}_x^3$ corresponding to the form $\omega^2(x)$.

Consequently, defining a 1-form ω^1 or a 2-form ω^2 in a domain D of the oriented Euclidean space $\mathbb{R}^3$ is equivalent to defining the vector field $\mathbf{A}$ or $\mathbf{B}$ in D corresponding to the form.

In explicit form, this correspondence amounts to the following:

$$\omega_{\mathbf{A}}^1(x)(\boldsymbol{\xi}) = \langle \mathbf{A}(x), \boldsymbol{\xi} \rangle, \tag{14.1}$$

$$\omega_{\mathbf{B}}^2(x)(\boldsymbol{\xi}_1, \boldsymbol{\xi}_2) = (\mathbf{B}(x), \boldsymbol{\xi}_1, \boldsymbol{\xi}_2), \tag{14.2}$$

where $\mathbf{A}(x)$, $\mathbf{B}(x)$, $\boldsymbol{\xi}$, $\boldsymbol{\xi}_1$, and $\boldsymbol{\xi}_2$ belong to TD_x.

Here we see the work form $\omega^1 = \omega_{\mathbf{A}}^1$ of the vector field $\mathbf{A}$ and the flux form $\omega^2 = \omega_{\mathbf{B}}^2$ of the vector field $\mathbf{B}$, which are already familiar to us.

To a scalar field $f : D \to \mathbb{R}$, we can assign a 0-form and a 3-form in D as follows:

$$\omega_f^0 = f, \tag{14.3}$$

$$\omega_f^3 = f \, dV, \tag{14.4}$$

where dV is the volume element in the oriented Euclidean space $\mathbb{R}^3$.

In view of the correspondences (14.1)–(14.4), definite operations on vector and scalar fields correspond to operations on forms. This observation, as we shall soon verify, is very useful technically.

Proposition 1 *To a linear combination of forms of the same degree there corresponds a linear combination of the vector and scalar fields corresponding to them.*

Proof Proposition 1 is of course obvious. However, let us write out the full proof, as an example, for 1-forms:

$$\alpha_1 \omega_{\mathbf{A}_1}^1 + \alpha_2 \omega_{\mathbf{A}_2}^1 = \alpha_1 \langle \mathbf{A}_1, \cdot \rangle + \alpha \langle \mathbf{A}_2, \cdot \rangle =$$

$$= \langle \alpha_1 \mathbf{A}_1 + \alpha_2 \mathbf{A}_2, \cdot \rangle = \omega_{\alpha_1 \mathbf{A}_1 + \alpha_2 \mathbf{A}_2}^2. \qquad \square$$

It is clear from the proof that α_1 and α_2 can be regarded as functions (not necessarily constant) in the domain D in which the forms and fields are defined.

As an abbreviation, let us agree to use, along with the symbols $\langle , \rangle$ and $[,]$ for the inner product and the vector product of vectors $\mathbf{A}$ and $\mathbf{B}$ in $\mathbb{R}^3$, the alternative notation $\mathbf{A} \cdot \mathbf{B}$ and $\mathbf{A} \times \mathbf{B}$ wherever convenient.

Proposition 2 *If* **A**, **B**, $\mathbf{A}_1$, *and* $\mathbf{B}_1$ *are vector fields in the oriented Euclidean space* $\mathbb{R}^3$, *then*

$$\omega_{\mathbf{A}_1}^1 \wedge \omega_{\mathbf{A}_2}^1 = \omega_{\mathbf{A}_1 \times \mathbf{A}_2}^2, \tag{14.5}$$

$$\omega_{\mathbf{A}}^1 \wedge \omega_{\mathbf{B}}^2 = \omega_{\mathbf{A} \cdot \mathbf{B}}^3. \tag{14.6}$$

In other words, the vector product $\mathbf{A}_1 \times \mathbf{A}_2$ of fields $\mathbf{A}_1$ and $\mathbf{A}_2$ that generate 1-forms corresponds to the exterior product of the 1-forms they generate, since it generates the 2-form that results from the product.

In the same sense the inner product of the vector fields **A** and **B** that generate a 1-form $\omega_{\mathbf{A}}^1$ and a 2-form $\omega_{\mathbf{B}}^2$ corresponds to the exterior product of these forms.

Proof To prove these assertions, fix an orthonormal basis in $\mathbb{R}^3$ and the Cartesian coordinates x^1, x^2, x^3 corresponding to it.

In Cartesian coordinates

$$\omega_{\mathbf{A}}^1(x)(\boldsymbol{\xi}) = \mathbf{A}(x) \cdot \boldsymbol{\xi} = \sum_{i=1}^{3} A^i(x)\xi^i = \sum_{i=1}^{3} A^i(x)\,dx^i(\boldsymbol{\xi}),$$

that is,

$$\omega_{\mathbf{A}}^1 = A^1\,dx^1 + A^2\,dx^2 + A^3\,dx^3, \tag{14.7}$$

and

$$\omega_{\mathbf{B}}^2(x)(\boldsymbol{\xi}_1, \boldsymbol{\xi}_2) = \begin{vmatrix} B^1(x) & B^2(x) & B^3(x) \\ \xi_1^1 & \xi_1^2 & \xi_1^3 \\ \xi_2^1 & \xi_2^2 & \xi_2^3 \end{vmatrix} =$$

$$= \left(B^1(x)\,dx^2 \wedge dx^3 + B^2\,dx^3 \wedge dx^1 + B^3(x)\,dx^1 \wedge dx^2\right)(\boldsymbol{\xi}_1, \boldsymbol{\xi}_2),$$

that is,

$$\omega_{\mathbf{B}}^2 = B^1\,dx^2 \wedge dx^3 + B^2\,dx^3 \wedge dx^1 + B^3\,dx^1 \wedge dx^2. \tag{14.8}$$

Therefore in Cartesian coordinates, taking account of expressions (14.7) and (14.8), we obtain

$$\omega_{\mathbf{A}_1}^1 \wedge \omega_{\mathbf{A}_2}^1 = \left(A_1^1\,dx^1 + A_1^2\,dx^2 + A_1^3\,dx^3\right) \wedge \left(A_2^1\,dx^1 + A_2^2\,dx^2 + A_2^3\,dx^3\right) =$$

$$= \left(A_1^2 A_2^3 - A_1^3 A_2^2\right)dx^2 \wedge dx^3 + \left(A_1^3 A_2^1 - A_1^1 A_2^3\right)dx^3 \wedge dx^1 +$$

$$+ \left(A_1^1 A_2^2 - A_1^2 A_2^1\right)dx^1 \wedge dx^2 =$$

$$= \omega_{\mathbf{B}}^2,$$

where $\mathbf{B} = \mathbf{A}_1 \times \mathbf{A}_2$.

Coordinates were used in this proof only to make it easier to find the vector $\mathbf{B}$ of the corresponding 2-form. The equality (14.5) itself, of course, is independent of the coordinate system.

Similarly, multiplying Eqs. (14.7) and (14.8), we obtain

$$\omega_{\mathbf{A}}^1 \wedge \omega_{\mathbf{B}}^2 = \left(A^1 B^1 + A^2 B^2 + A^3 B^3\right) dx^1 \wedge dx^2 \wedge dx^3 = \omega_\rho^3.$$

In Cartesian coordinates $dx^1 \wedge dx^2 \wedge dx^3$ is the volume element in $\mathbb{R}^3$, and the sum of the pairwise products of the coordinates of the vectors $\mathbf{A}$ and $\mathbf{B}$, which appears in parentheses just before the 3-form, is the inner product of these vectors at the corresponding points of the domain, from which it follows that $\rho(x) = \mathbf{A}(x) \cdot \mathbf{B}(x)$. □

14.1.3 The Differential Operators grad, curl, div, and ∇

Definition 1 To the exterior differentiation of 0-forms (functions), 1-forms, and 2-forms in oriented Euclidean space $\mathbb{R}^3$ there correspond respectively the operations of finding the *gradient* (grad) of a scalar field and the *curl* and *divergence* (div) of a vector field. These operations are defined by the relations

$$d\omega_f^0 =: \omega_{\text{grad} f}^1, \tag{14.9}$$

$$d\omega_{\mathbf{A}}^1 =: \omega_{\text{curl} \mathbf{A}}^2, \tag{14.10}$$

$$d\omega_{\mathbf{B}}^1 =: \omega_{\text{div} \mathbf{B}}^3. \tag{14.11}$$

By virtue of the correspondence between forms and scalar and vector fields in $\mathbb{R}^3$ established by Eqs. (14.1)–(14.4), relations (14.9)–(14.11) are unambiguous definitions of the operations grad, curl, and div, performed on scalar and vector fields respectively. These operations, the *operators of field theory* as they are called, correspond to the single operation of exterior differentiation of forms, but applied to forms of different degree.

Let us give right away the explicit form of these operators in Cartesian coordinates x^1, x^2, x^3 in $\mathbb{R}^3$.

As we have explained, in this case

$$\omega_f^0 = f, \tag{14.3'}$$

$$\omega_{\mathbf{A}}^1 = A^1 dx^1 + A^2 dx^2 + A^3 dx^3, \tag{14.7'}$$

$$\omega_{\mathbf{B}}^2 = B^1 dx^2 \wedge dx^3 + B^2 dx^3 \wedge dx^1 + B^3 dx^1 \wedge dx^2, \tag{14.8'}$$

$$\omega_\rho^3 = \rho \, dx^1 \wedge dx^2 \wedge dx^3. \tag{14.4'}$$

Since

$$\omega_{\text{grad} f}^1 := d\omega_f^0 = df = \frac{\partial f}{\partial x^1} dx^1 + \frac{\partial f}{\partial x^2} dx^2 + \frac{\partial f}{\partial x^3} dx^3,$$

it follows from (14.7′) that in these coordinates

$$\operatorname{grad} f = \mathbf{e}_1 \frac{\partial f}{\partial x^1} + \mathbf{e}_2 \frac{\partial f}{\partial x^2} + \mathbf{e}_3 \frac{\partial f}{\partial x^3}, \tag{14.9′}$$

where $\mathbf{e}_1, \mathbf{e}_2, \mathbf{e}_3$ is a fixed orthonormal basis of $\mathbb{R}^3$.

Since

$$\omega^2_{\operatorname{curl}\mathbf{A}} := \mathrm{d}\omega^1_{\mathbf{A}} = \mathrm{d}\big(A^1\,\mathrm{d}x^1 + A^2\,\mathrm{d}x^2 + A^3\,\mathrm{d}x^3\big) =$$

$$= \left(\frac{\partial A^3}{\partial x^2} - \frac{\partial A^2}{\partial x^3}\right)\mathrm{d}x^2 \wedge \mathrm{d}x^3 + \left(\frac{\partial A^1}{\partial x^3} - \frac{\partial A^3}{\partial x^1}\right)\mathrm{d}x^3 \wedge \mathrm{d}x^1 +$$

$$+ \left(\frac{\partial A^2}{\partial x^1} - \frac{\partial A^1}{\partial x^2}\right)\mathrm{d}x^1 \wedge \mathrm{d}x^2,$$

it follows from (14.8′) that in Cartesian coordinates

$$\operatorname{curl}\mathbf{A} = \mathbf{e}_1\left(\frac{\partial A^3}{\partial x^2} - \frac{\partial A^2}{\partial x^3}\right) + \mathbf{e}_2\left(\frac{\partial A^1}{\partial x^3} - \frac{\partial A^3}{\partial x^1}\right) + \mathbf{e}_3\left(\frac{\partial A^2}{\partial x^1} - \frac{\partial A^1}{\partial x^2}\right). \tag{14.10′}$$

As an aid to memory this last relation is often written in symbolic form as

$$\operatorname{curl}\mathbf{A} = \begin{vmatrix} \mathbf{e}_1 & \mathbf{e}_2 & \mathbf{e}_3 \\ \frac{\partial}{\partial x^1} & \frac{\partial}{\partial x^2} & \frac{\partial}{\partial x^3} \\ A^1 & A^2 & A^3 \end{vmatrix}. \tag{14.10″}$$

Next, since

$$\omega^3_{\operatorname{div}\mathbf{B}} := \mathrm{d}\omega^2_{\mathbf{B}} = \mathrm{d}\big(B^1\,\mathrm{d}x^2 \wedge \mathrm{d}x^3 + B^2\,\mathrm{d}x^3 \wedge \mathrm{d}x^1 + B^3\,\mathrm{d}x^1 \wedge \mathrm{d}x^2\big) =$$

$$= \left(\frac{\partial B^1}{\partial x^1} + \frac{\partial B^2}{\partial x^2} + \frac{\partial B^3}{\partial x^3}\right)\mathrm{d}x^1 \wedge \mathrm{d}x^2 \wedge \mathrm{d}x^3,$$

it follows from (14.4′) that in Cartesian coordinates

$$\operatorname{div}\mathbf{B} = \frac{\partial B^1}{\partial x^1} + \frac{\partial B^2}{\partial x^2} + \frac{\partial B^3}{\partial x^3}. \tag{14.11′}$$

One can see from the formulas (14.9′), (14.10′), and (14.11′) just obtained that grad, curl, and div are linear differential operations (operators). The grad operator is defined on differentiable scalar fields and assigns vector fields to the scalar fields. The curl operator is also vector-valued, but is defined on differentiable vector fields. The div operator is defined on differentiable vector fields and assigns scalar fields to them.

We note that in other coordinates these operators will have expressions that are in general different from those obtained above in Cartesian coordinates. We shall discuss this point in Sect. 14.1.5 below.

We remark also that the vector field curl $\mathbf{A}$ is sometimes called the *rotation* of $\mathbf{A}$ and written rot $\mathbf{A}$.

As an example of the use of these operators we write out the famous[1] system of equations of Maxwell,[2] which describe the state of the components of an electromagnetic field as functions of a point $x = (x^1, x^2, x^3)$ in space and time t.

Example 1 (The Maxwell equations for an electromagnetic field in a vacuum)

$$1.\ \operatorname{div} \mathbf{E} = \frac{\rho}{\varepsilon_0}. \qquad 2.\ \operatorname{div} \mathbf{B} = 0.$$

$$3.\ \operatorname{curl} \mathbf{E} = -\frac{\partial \mathbf{B}}{\partial t}. \qquad 4.\ \operatorname{curl} \mathbf{B} = \frac{\mathbf{j}}{\varepsilon_0 c^2} + \frac{1}{c^2}\frac{\partial \mathbf{E}}{\partial t}. \qquad (14.12)$$

Here $\rho(x, t)$ is the electric charge density (the quantity of charge per unit volume), $\mathbf{j}(x, t)$ is the electrical current density vector (the rate at which charge is flowing across a unit area), $\mathbf{E}(x, t)$ and $\mathbf{B}(x, t)$ are the electric and magnetic field intensities respectively, and ε_0 and c are dimensioning constants (and in fact c is the speed of light in a vacuum).

In mathematical and especially in physical literature, along with the operators grad, curl, and div, wide use is made of the symbolic differential operator nabla proposed by Hamilton (the *Hamilton operator*)[3]

$$\nabla = \mathbf{e}_1 \frac{\partial}{\partial x^1} + \mathbf{e}_2 \frac{\partial}{\partial x^2} + \mathbf{e}_3 \frac{\partial}{\partial x^3}, \qquad (14.13)$$

where $\{\mathbf{e}_1, \mathbf{e}_2, \mathbf{e}_3\}$ is an orthonormal basis of $\mathbb{R}^3$ and x^1, x^2, x^3 are the corresponding Cartesian coordinates.

By definition, applying the operator ∇ to a scalar field f (that is, to a function), gives the vector field

$$\nabla f = \mathbf{e}_1 \frac{\partial f}{\partial x^1} + \mathbf{e}_2 \frac{\partial f}{\partial x^2} + \mathbf{e}_3 \frac{\partial f}{\partial x^3},$$

which coincides with the field (14.9'), that is, the nabla operator is simply the grad operator written in a different notation.

---

[1] On this subject the famous American physicist and mathematician R. Feynman (1918–1988) writes, with his characteristic acerbity, "From a long view of the history of mankind – seen from, say, ten thousand years from now – there can be little doubt that the most significant event of the 19th century will be judged as Maxwell's discovery of the laws of electrodynamics. The American Civil War will pale into provincial insignificance in comparison with this important scientific event of the same decade." Richard R. Feynman, Robert B. Leighton, and Matthew Sands, *The Feynman Lectures on Physics: Mainly Electromagnetism and Matter*, Addison-Wesley, Reading, MA, 1964.

[2] J.C. Maxwell (1831–1879) – outstanding Scottish physicist; he created the mathematical theory of the electromagnetic field and is also famous for his research in the kinetic theory of gases, optics and mechanics.

[3] W.R. Hamilton (1805–1865) – famous Irish mathematician and specialist in mechanics; he stated the variational principle (Hamilton's principle) and constructed a phenomenological theory of optic phenomena; he was the creator of the theory of quaternions and the founder of vector analysis (in fact, the term "vector" is due to him).

Using, however, the vector form in which ∇ is written, Hamilton proposed a system of formal operations with it that imitates the corresponding algebraic operations with vectors.

Before we illustrate these operations, we note that in dealing with ∇ one must adhere to the same principles and cautionary rules as in dealing with the usual differentiation operator $D = \frac{d}{dx}$. For example, $\varphi D f$ equals $\varphi \frac{df}{dx}$ and not $\frac{d}{dx}(\varphi f)$ or $f \frac{d\varphi}{dx}$. Thus, the operator operates on whatever is placed to the right of it; left multiplication in this case plays the role of a coefficient, that is, φD is the new differential operator $\varphi \frac{d}{dx}$, not the function $\frac{d\varphi}{dx}$. Moreover, $D^2 = D \cdot D$, that is, $D^2 f = D(Df) = \frac{d}{dx}(\frac{d}{dx} f) = \frac{d^2}{dx^2} f$.

If we now, following Hamilton, deal with ∇ as if it were a vector field defined in Cartesian coordinates, then, comparing relations (14.13), (14.9′), (14.10″), and (14.11′), we obtain

$$\operatorname{grad} f = \nabla f, \tag{14.14}$$

$$\operatorname{curl} \mathbf{A} = \nabla \times \mathbf{A}, \tag{14.15}$$

$$\operatorname{div} \mathbf{B} = \nabla \cdot \mathbf{B}. \tag{14.16}$$

In this way the operators grad, curl, and div, can be written in terms of the Hamilton operator and the vector operations in $\mathbb{R}^3$.

Example 2 Only the curl and div operators occurred in writing out the Maxwell equations (14.12). Using the principles for dealing with $\nabla = \operatorname{grad}$, we rewrite the Maxwell equations as follows, to compensate for the absence of grad in them:

$$
\begin{array}{ll}
1.\ \nabla \cdot \mathbf{E} = \dfrac{\rho}{\varepsilon_0}. & 2.\ \nabla \cdot \mathbf{B} = 0. \\[2mm]
3.\ \nabla \times \mathbf{E} = -\dfrac{\partial \mathbf{B}}{\partial t}. & 4.\ \nabla \times \mathbf{B} = \dfrac{\mathbf{j}}{\varepsilon_0 c^2} + \dfrac{1}{c^2} \dfrac{\partial \mathbf{E}}{\partial t}.
\end{array}
\tag{14.12'}
$$

14.1.4 Some Differential Formulas of Vector Analysis

In the oriented Euclidean space $\mathbb{R}^3$ we have established the connection (14.1)–(14.4) between forms on the one hand and vector and scalar fields on the other. This connection enabled us to associate corresponding operators on fields with exterior differentiation (see formulas (14.5), (14.6), and (14.9)–(14.11)).

This correspondence can be used to obtain a number of basic differential formulas of vector analysis.

For example, the following relations hold:

$$\operatorname{curl}(f\mathbf{A}) = f \operatorname{curl} \mathbf{A} - \mathbf{A} \times \operatorname{grad} f, \tag{14.17}$$

$$\operatorname{div}(f\mathbf{A}) = \mathbf{A} \cdot \operatorname{grad} f + f \operatorname{div} \mathbf{A}, \tag{14.18}$$

$$\operatorname{div}(\mathbf{A} \times \mathbf{B}) = \mathbf{B} \cdot \operatorname{curl} \mathbf{A} - \mathbf{A} \cdot \operatorname{curl} \mathbf{B}. \tag{14.19}$$

Proof We shall verify this last equality:

$$\omega^3_{\operatorname{div}\mathbf{A}\times\mathbf{B}} = d\omega^2_{\mathbf{A}\times\mathbf{B}} = d\left(\omega^1_\mathbf{A}\wedge\omega^1_\mathbf{B}\right) = d\omega^1_\mathbf{A}\wedge\omega^1_\mathbf{B} - \omega^1_\mathbf{A}\wedge d\omega^1_\mathbf{B} =$$

$$= \omega^2_{\operatorname{curl}\mathbf{A}}\wedge\omega^1_\mathbf{B} - \omega^1_\mathbf{A}\wedge\omega^2_{\operatorname{curl}\mathbf{B}} = \omega^3_{\mathbf{B}\cdot\operatorname{curl}\mathbf{A}} - \omega^3_{\mathbf{A}\cdot\operatorname{curl}\mathbf{B}} = \omega^3_{\mathbf{B}\cdot\operatorname{curl}\mathbf{A}-\mathbf{A}\cdot\operatorname{curl}\mathbf{B}}.$$

The first two relations are verified similarly. Of course, the verification of all these equalities can also be carried out by direct differentiation in coordinates. □

If we take account of the relation $d^2\omega = 0$ for any form ω, we can also assert that the following equalities hold:

$$\operatorname{curl}\operatorname{grad} f = \mathbf{0}, \qquad (14.20)$$

$$\operatorname{div}\operatorname{curl}\mathbf{A} = 0. \qquad (14.21)$$

Proof Indeed:

$$\omega^2_{\operatorname{curl}\operatorname{grad} f} = d\omega^1_{\operatorname{grad} f} = d(d\omega^0_f) = d^2\omega^0_f = 0,$$
$$\omega^3_{\operatorname{div}\operatorname{curl}\mathbf{A}} = d\omega^2_{\operatorname{curl}\mathbf{A}} = d(d\omega^1_\mathbf{A}) = d^2\omega^1_\mathbf{A} = 0.$$
□

In formulas (14.17)–(14.19) the operators grad, curl, and div are applied once, while (14.20) and (14.21) involve the second-order operators obtained by successive execution of two of the three original operations. Besides the rules given in (14.20) and (14.21), one can also consider other combinations of these operators:

$$\operatorname{grad}\operatorname{div}\mathbf{A}, \quad \operatorname{curl}\operatorname{curl}\mathbf{A}, \quad \operatorname{div}\operatorname{grad} f. \qquad (14.22)$$

The operator div grad is applied, as one can see, to a scalar field. This operator is denoted Δ (Delta) and is called the *Laplace operator*[4] or *Laplacian*.

It follows from (14.9′) and (14.11′) that in Cartesian coordinates

$$\Delta f = \frac{\partial^2 f}{\partial(x^1)^2} + \frac{\partial^2 f}{\partial(x^2)^2} + \frac{\partial^2 f}{\partial(x^3)^2}. \qquad (14.23)$$

Since the operator Δ acts on numerical functions, it can be applied component-wise to the coordinates of vector fields $\mathbf{A} = \mathbf{e}_1 A^1 + \mathbf{e}_2 A^2 + \mathbf{e}_3 A^3$, where $\mathbf{e}_1$, $\mathbf{e}_2$, and $\mathbf{e}_3$ are an orthonormal basis in $\mathbb{R}^3$. In that case

$$\Delta\mathbf{A} = \mathbf{e}_1\Delta A^1 + \mathbf{e}_2\Delta A^2 + \mathbf{e}_3\Delta A^3.$$

Taking account of this last equality, we can write the following relation for the triple of second-order operators (14.22):

$$\operatorname{curl}\operatorname{curl}\mathbf{A} = \operatorname{grad}\operatorname{div}\mathbf{A} - \Delta\mathbf{A}, \qquad (14.24)$$

---

[4]P.S. Laplace (1749–1827) – famous French astronomer, mathematician, and physicist; he made fundamental contributions to the development of celestial mechanics, the mathematical theory of probability, and experimental and mathematical physics.

whose proof we shall not take the time to present (see Problem 2 below). The equality (14.24) can serve as the definition of $\Delta \mathbf{A}$ in any coordinate system, not necessarily orthogonal.

Using the language of vector algebra and formulas (14.14)–(14.16), we can write all the second-order operators (14.20)–(14.22) in terms of the Hamilton operator ∇:

$$\operatorname{curl} \operatorname{grad} f = \nabla \times \nabla f = 0,$$

$$\operatorname{div} \operatorname{curl} \mathbf{A} = \nabla \cdot (\nabla \times \mathbf{A}) = 0,$$

$$\operatorname{grad} \operatorname{div} \mathbf{A} = \nabla (\nabla \cdot \mathbf{A}),$$

$$\operatorname{curl} \operatorname{curl} \mathbf{A} = \nabla \times (\nabla \times \mathbf{A}),$$

$$\operatorname{div} \operatorname{grad} f = \nabla \cdot \nabla f.$$

From the point of view of vector algebra the vanishing of the first two of these operators seems completely natural.

The last equality means that the following relation holds between the Hamilton operator ∇ and the Laplacian Δ:

$$\Delta = \nabla^2.$$

14.1.5 *Vector Operations in Curvilinear Coordinates

a.

Just as, for example, the sphere $x^2 + y^2 + z^2 = a^2$ has a particularly simple equation $R = a$ in spherical coordinates, vector fields $x \mapsto \mathbf{A}(x)$ in $\mathbb{R}^3$ (or $\mathbb{R}^n$) often assume a simpler expression in a coordinate system that is not Cartesian. For that reason we now wish to find explicit formulas from which one can find grad, curl, and div in a rather extensive class of curvilinear coordinates.

But first it is necessary to be precise as to what is meant by the coordinate expression for a field $\mathbf{A}$ in a curvilinear coordinate system.

We begin with two introductory examples of a descriptive character.

Example 3 Suppose we have a fixed Cartesian coordinate system x^1, x^2 in the Euclidean plane $\mathbb{R}^2$. When we say that a vector field $(A^1, A^2)(x)$ is defined in $\mathbb{R}^2$, we mean that some vector $\mathbf{A}(x) \in T\mathbb{R}^2_x$ is connected with each point $x = (x^1, x^2) \in \mathbb{R}^2$, and in the basis of $T\mathbb{R}^2_x$ consisting of the unit vectors $\mathbf{e}_1(x)$, $\mathbf{e}_2(x)$ in the coordinate directions we have the expansion $\mathbf{A}(x) = A^1(x)\mathbf{e}_1(x) + A^2(x)\mathbf{e}_2(x)$ (see Fig. 14.1). In this case the basis $\{\mathbf{e}_1(x), \mathbf{e}_2(x)\}$ of $T\mathbb{R}^2_x$ is essentially independent of x.

Example 4 In the case when polar coordinates (r, φ) are defined in the same plane $\mathbb{R}^2$, at each point $x \in \mathbb{R}^2 \setminus 0$ one can also attach unit vectors $\mathbf{e}_1(x) = \mathbf{e}_r(x)$, $\mathbf{e}_2 = \mathbf{e}_\varphi(x)$ (Fig. 14.2) in the coordinate directions. They also form a basis in $T\mathbb{R}^2_x$

Fig. 14.1

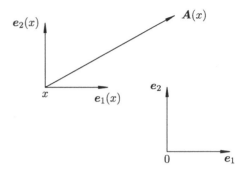

Fig. 14.2

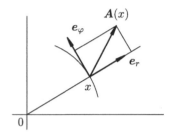

with respect to which one can expand the vector $\mathbf{A}(x)$ of the field $\mathbf{A}$ attached to x : $\mathbf{A}(x) = A^1(x)\mathbf{e}_1(x) + A^2(x)\mathbf{e}_2(x)$. It is then natural to regard the ordered pair of functions $(A^1, A^2)(x)$ as the expression for the field $\mathbf{A}$ in polar coordinates.

Thus, if $(A^1, A^2)(x) \equiv (1, 0)$, this is a field of unit vectors in $\mathbb{R}^2$ pointing radially away from the center 0.

The field $(A^1, A^2)(x) \equiv (0, 1)$ can be obtained from the preceding field by rotating each vector in it counterclockwise by the angle $\pi/2$.

These are not constant fields in $\mathbb{R}^2$, although the components of their coordinate representation are constant. The point is that the basis in which the expansion is taken varies synchronously with the vector of the field in a transition from one point to another.

It is clear that the components of the coordinate representation of these fields in Cartesian coordinates would not be constant at all. On the other hand, a truly constant field (consisting of a vector translated parallel to itself to all points of the plane) which does have constant components in a Cartesian coordinate system, would have variable components in polar coordinates.

b.

After these introductory considerations, let us consider more formally the problem of defining vector fields in curvilinear coordinate systems.

We recall first of all that a curvilinear coordinate system t^1, t^2, t^3 in a domain $D \subset \mathbb{R}^3$ is a diffeomorphism $\varphi : D_t \to D$ of a domain D_t in the Euclidean parameter space $\mathbb{R}_t^3$ onto the domain D, as a result of which each point $x = \varphi(t) \in D$ acquires the Cartesian coordinates t^1, t^2, t^3 of the corresponding point $t \in D_t$.

Since φ is a diffeomorphism, the tangent mapping $\varphi'(t) : T\mathbb{R}_t^3 \to T\mathbb{T}_{x=\varphi(t)}^3$ is a vector-space isomorphism. To the canonical basis $\boldsymbol{\xi}_1(t) = (1, 0, 0)$, $\boldsymbol{\xi}_2(t) = (0, 1, 0)$, $\boldsymbol{\xi}_3(t) = (0, 0, 1)$ of $T\mathbb{R}_t^3$ corresponds the basis of $T\mathbb{R}_{x=\varphi(t)}^3$ consisting of the vectors $\boldsymbol{\xi}_i(x) = \varphi'(t)\boldsymbol{\xi}_i(t) = \frac{\partial \varphi(t)}{\partial t^i}$, $i = 1, 2, 3$, giving the coordinate directions. To the expansion $\mathbf{A}(x) = \alpha_1 \boldsymbol{\xi}_1(x) + \alpha_2 \boldsymbol{\xi}_2(x) + \alpha_3 \boldsymbol{\xi}_3(x)$ of any vector $\mathbf{A}(x) \in T\mathbb{R}_x^3$ in this basis there corresponds the same expansion $\mathbf{A}(t) = \alpha_1 \boldsymbol{\xi}_1(t) + \alpha_2 \boldsymbol{\xi}_2(t) + \alpha_3 \boldsymbol{\xi}_3(t)$ (with the same components $\alpha_1, \alpha_2, \alpha_3$!) of the vector $\mathbf{A}(t) = (\varphi')^{-1}\mathbf{A}(x)$ in the canonical basis $\boldsymbol{\xi}_1(t), \boldsymbol{\xi}_2(t), \boldsymbol{\xi}_3(t)$ in $T\mathbb{R}_t^3$. In the absence of a Euclidean structure in $\mathbb{R}^3$, the numbers $\alpha_1, \alpha_2, \alpha_3$ would be the most natural coordinate expression for the vector $\mathbf{A}(x)$ connected with this curvilinear coordinate system.

c.

However, adopting such a coordinate representation would not be quite consistent with what we agreed to in Example 4. The point is that the basis $\boldsymbol{\xi}_1(x), \boldsymbol{\xi}_2(x), \boldsymbol{\xi}_3(x)$ in $T\mathbb{R}_x^3$ corresponding to the canonical basis $\boldsymbol{\xi}_1(t), \boldsymbol{\xi}_2(t), \boldsymbol{\xi}_3(t)$ in $T\mathbb{R}_t^3$, although it consists of vectors in the coordinate directions, is not at all required to consist of *unit vectors* in those directions, that is, in general $\langle \boldsymbol{\xi}_i, \boldsymbol{\xi}_i \rangle(x) \neq 1$.

We shall now take account of this circumstance which results from the presence of a Euclidean structure in $\mathbb{R}^3$ and consequently in each vector space $T\mathbb{R}_x^3$ also.

Because of the isomorphism $\varphi'(t) : T\mathbb{R}_t^3 \to T\mathbb{R}_{x=\varphi(t)}^3$ we can transfer the Euclidean structure of $T\mathbb{R}_x^3$ into $T\mathbb{R}_t^3$ by setting $\langle \tau_1, \tau_2 \rangle := \langle \varphi'\tau_1, \varphi'\tau_2 \rangle$ for every pair of vectors $\tau_1, \tau_2 \in T\mathbb{R}_t^3$. In particular, we obtain from this the following expression for the square of the length of a vector:

$$\langle \tau, \tau \rangle = \langle \varphi'(t)\tau, \varphi'(t)\tau \rangle = \left\langle \frac{\partial \varphi(t)}{\partial t^i} \tau^i, \frac{\partial \varphi(t)}{\partial t^j} \tau^j \right\rangle =$$

$$= \left\langle \frac{\partial \varphi}{\partial t^i}, \frac{\partial \varphi}{\partial t^j} \right\rangle (t)\tau^i \tau^j = \langle \boldsymbol{\xi}_i, \boldsymbol{\xi}_j \rangle(t)\tau^i \tau^i = g_{ij}(t)\,\mathrm{d}t^i(\tau)\,\mathrm{d}t^j(\tau).$$

The quadratic form

$$\mathrm{d}s^2 = g_{ij}(t)\,\mathrm{d}t^i\,\mathrm{d}t^j \tag{14.25}$$

whose coefficients are the pairwise inner products of the vectors in the canonical basis determines the inner product on $T\mathbb{R}_t^3$ completely. If such a form is defined at each point of a domain $D_t \subset \mathbb{R}_t^3$, then, as is known from geometry, one says that a *Riemannian metric* is defined in this domain. A Riemannian metric makes it possible to introduce a Euclidean structure in each tangent space $T\mathbb{R}_t^3(t \in D_t)$ within the context of rectilinear coordinates t^1, t^2, t^3 in $\mathbb{R}_t^3$, corresponding to the "curved" embedding $\varphi : D_t \to D$ of the domain D_t in the Euclidean space $\mathbb{R}^3$.

If the vectors $\boldsymbol{\xi}_i(x) = \varphi'(t)\boldsymbol{\xi}_i(t) = \frac{\partial\varphi}{\partial t^i}(t)$, $i = 1, 2, 3$, are orthogonal in $T\mathbb{R}_x^3$, then $g_{ij}(t) = 0$ for $i \neq j$. This means that we are dealing with a *triorthogonal coordinate grid*. In terms of the space $T\mathbb{R}_t^3$ it means that the vectors $\boldsymbol{\xi}_i(t)$, $i = 1, 2, 3$, in the canonical basis are mutually orthogonal in the sense of the inner product in $T\mathbb{R}_t^3$ defined by the quadratic form (14.25). In what follows, for the sake of simplicity, we shall consider only triorthogonal curvilinear coordinate systems. For them, as has been noted, the quadratic form (14.25) has the following special form:

$$ds^2 = E_1(t)(dt^1)^2 + E_2(t)(dt^2)^2 + E_3(t)(dt^3)^2, \tag{14.26}$$

where $E_i(t) = g_{ii}(t)$, $i = 1, 2, 3$.

Example 5 In Cartesian coordinates (x, y, z), cylindrical coordinates (r, φ, z), and spherical coordinates (R, φ, θ) on Euclidean space $\mathbb{R}^3$ the quadratic form (14.25) has the respective forms

$$ds^2 = dx^2 + dy^2 + dz^2 = \tag{14.26$'$}$$

$$= dr^2 + r^2\,d\varphi^2 + dz^2 = \tag{14.26$''$}$$

$$= dR^2 + R^2\cos^2\theta\,d\varphi^2 + R^2\,d\theta^2. \tag{14.26$'''$}$$

Thus, each of these coordinate systems is a triorthogonal system in its domain of definition.

The vectors $\boldsymbol{\xi}_1(t)$, $\boldsymbol{\xi}_2(t)$, $\boldsymbol{\xi}_3(t)$ of the canonical basis $(1, 0, 0)$, $(0, 1, 0)$, $(0, 0, 1)$ in $T\mathbb{R}_t^3$, like the vectors $\boldsymbol{\xi}_i(x) \in T\mathbb{R}_x^3$ corresponding to them, have the following norm:[5] $|\boldsymbol{\xi}_i| = \sqrt{g_{ii}}$. Hence the unit vectors (in the sense of the square-norm of a vector) in the coordinate directions have the following coordinate representation for the triorthogonal system (14.26):

$$\mathbf{e}_1(t) = \left(\frac{1}{\sqrt{E_1}}, 0, 0\right), \qquad \mathbf{e}_2(t) = \left(0, \frac{1}{\sqrt{E_2}}, 0\right), \qquad \mathbf{e}_3(t) = \left(0, 0, \frac{1}{\sqrt{E_3}}\right). \tag{14.27}$$

Example 6 It follows from formulas (14.27) and the results of Example 5 that for Cartesian, cylindrical, and spherical coordinates, the triples of unit vectors along the coordinate directions have respectively the following forms:

$$\mathbf{e}_x = (1, 0, 0), \qquad \mathbf{e}_y = (0, 1, 0), \qquad\qquad \mathbf{e}_z = (0, 0, 1); \tag{14.27$'$}$$

$$\mathbf{e}_r = (1, 0, 0), \qquad \mathbf{e}_\varphi = \left(0, \frac{1}{r}, 0\right), \qquad\quad \mathbf{e}_z = (0, 0, 1); \tag{14.27$''$}$$

---

[5]In the triorthogonal system (14.26) we have $|\boldsymbol{\xi}_i| = \sqrt{E_i} = H_i$, $i = 1, 2, 3$. The quantities H_1, H_2, H_3 are usually called the *Lamé' coefficient* or *Lamé' parameters*. G. Lamé (1795–1870) French engineer, mathematician, and physicist.

$$\mathbf{e}_R = (1, 0, 0), \qquad \mathbf{e}_\varphi = \left(0, \frac{1}{R\cos\theta}, 0\right), \qquad \mathbf{e}_\theta = \left(0, 0, \frac{1}{R}\right). \qquad (14.27''')$$

Examples 3 and 4 considered above assumed that the vector of the field was expanded in a basis consisting of *unit vectors* along the coordinate directions. Hence the vector $\mathbf{A}(t) \in T\mathbb{R}_t^3$ corresponding to the vector $\mathbf{A}(x) \in T\mathbb{R}_x^3$ of the field should be expanded in the basis $\mathbf{e}_1(t)$, $\mathbf{e}_2(t)$, $\mathbf{e}_3(t)$ consisting of unit vectors in the coordinate directions, rather than in the canonical basis $\boldsymbol{\xi}_1(t)$, $\boldsymbol{\xi}_2(t)$, $\boldsymbol{\xi}_3(t)$.

Thus, abstracting from the original space $\mathbb{R}^3$, one can assume that a Riemannian metric (14.25) or (14.26) and a vector field $t \mapsto \mathbf{A}(t)$ are defined in the domain $D_t \subset \mathbb{R}_t^3$ and that the coordinate representation $(A^1, A^2, A^3)(t)$ of $\mathbf{A}(t)$ at each point $t \in D_t$ is obtained from the expansion of the vector $\mathbf{A}(t) = A^i(t)\mathbf{e}_i(t)$ of the field corresponding to this point with respect to unit vectors along the coordinate axes.

d.

Let us now investigate forms. Under the diffeomorphism $\varphi : D_t \to D$ every form in D automatically transfers to the domain D_t. This transfer, as we know, occurs at each point $x \in D$ from the space $T\mathbb{R}_x^3$ into the corresponding space $T\mathbb{R}_t^3$. Since we have transferred the Euclidean structure into $T\mathbb{R}_t^3$ from $T\mathbb{R}_x^3$, it follows from the definition of the transfer of vectors and forms that, for example, to a given form $\omega_\mathbf{A}^1(x) = \langle \mathbf{A}(x), \cdot \rangle$ defined in $T\mathbb{R}_x^3$ there corresponds exactly the same kind of form $\omega_\mathbf{A}^1(t) = \langle \mathbf{A}(t), \cdot \rangle$ in $T\mathbb{R}_t^3$, where $\mathbf{A}(x) = \varphi'(t)\mathbf{A}(t)$. The same can be said of forms of the type $\omega_\mathbf{B}^2$ and ω_ρ^3, to say nothing of forms ω_f^0 – that is, functions.

After these clarifications, the rest of our study can be confined to the domain $D_t \subset \mathbb{R}^3$, abstracting from the original space $\mathbb{R}^3$ and assuming that a Riemannian metric (14.25) is defined in D_t and that scalar fields f, ρ and vector fields $\mathbf{A}$, $\mathbf{B}$ are defined in D_t along with the forms ω_f^0, $\omega_\mathbf{A}^1$, $\omega_\mathbf{B}^2$, ω_ρ^3, which are defined at each point $t \in D_t$ in accordance with the Euclidean structure on $T\mathbb{R}_t^3$ defined by the Riemannian metric.

Example 7 The volume element dV in curvilinear coordinates t^1, t^2, t^3, as we know, has the form

$$dV = \sqrt{\det g_{ij}}(t) \, dt^1 \wedge dt^2 \wedge dt^3.$$

For a triorthogonal system

$$dV = \sqrt{E_1 E_2 E_3}(t) \, dt^1 \wedge dt^2 \wedge dt^3. \qquad (14.28)$$

In particular, in Cartesian, cylindrical, and spherical coordinates, respectively, we obtain

$$dV = dx \wedge dy \wedge dz = \qquad\qquad (14.28')$$

$$= r \, dr \wedge d\varphi \wedge dz = \qquad\qquad (14.28'')$$

$$= R^2 \cos\theta \, dR \wedge d\varphi \wedge d\theta. \tag{14.28'''}$$

What has just been said enables us to write the form $\omega_\rho^3 = \rho \, dV$ in different curvilinear coordinate systems.

e.

Our main problem (now easily solvable) is, knowing the expansion $\mathbf{A}(t) = A^i(t)\mathbf{e}_i(t)$ for a vector $\mathbf{A}(t) \in T\mathbb{R}_t^3$ with respect to the unit vectors $\mathbf{e}_i(t) \in T\mathbb{R}_t^3$, $i = 1, 2, 3$, of the triorthogonal coordinate system determined by the Riemannian metric (14.26), to find the expansion of the forms $\omega_\mathbf{A}^1(t)$ and $\omega_\mathbf{B}^2(t)$ in terms of the canonical 1-forms dt^i and the canonical 2-forms $dt^i \wedge dt^j$ respectively.

Since all the reasoning applies at every given point t, we shall abbreviate the notation by suppressing the letter t that shows that the vectors and forms are attached to the tangent space at t.

Thus, $\mathbf{e}_1$, $\mathbf{e}_2$, $\mathbf{e}_3$ is a basis in $T\mathbb{R}_t^3$ consisting of the unit vectors (14.27) along the coordinate directions, and $\mathbf{A} = A^1\mathbf{e}_1 + A^2\mathbf{e}_2 + A^3\mathbf{e}_3$ is the expansion of $\mathbf{A} \in T\mathbb{R}_t^3$ in that basis.

We remark first of all that formula (14.27) implies that

$$dt^j(\mathbf{e}_i) = \frac{1}{\sqrt{E_i}}\delta_j^i, \quad \text{where } \delta_j^i = \begin{cases} 0, & \text{if } i \neq j, \\ 1, & \text{if } i = j, \end{cases} \tag{14.29}$$

$$dt^i \wedge dt^j(\mathbf{e}_k, \mathbf{e}_l) = \frac{1}{\sqrt{E_i E_j}}\delta_{kl}^{ij}, \quad \text{where } \delta_{kl}^{ij} = \begin{cases} 0, & \text{if } (i, j) \neq (k, l), \\ 1, & \text{if } (i, j) = (k, l). \end{cases} \tag{14.30}$$

f.

Thus, if $\omega_\mathbf{A}^1 := \langle \mathbf{A}, \cdot \rangle = a_1 \, dt^1 + a_2 \, dt^2 + a_3 \, dt^3$, then on the one hand

$$\omega_\mathbf{A}^1(\mathbf{e}_i) = \langle \mathbf{A}, \mathbf{e}_i \rangle = A^i,$$

and on the other hand, as one can see from (14.29),

$$\omega_\mathbf{A}^1(\mathbf{e}_i) = \left(a_1 \, dt^1 + a_2 \, dt^2 + a_3 \, dt^3\right)(\mathbf{e}_i) = a_i \cdot \frac{1}{\sqrt{E_i}}.$$

Consequently, $a_i = A^i \sqrt{E_i}$, and we have found the expansion

$$\omega_\mathbf{A}^1 = A^1\sqrt{E_1} \, dt^1 + A^2\sqrt{E_2} \, dt^2 + A^3\sqrt{E_3} \, dt^3 \tag{14.31}$$

for the form $\omega_\mathbf{A}^1$ corresponding to the expansion $\mathbf{A} = A^1\mathbf{e}_1 + A^2\mathbf{e}_2 + A^3\mathbf{e}_3$ of the vector $\mathbf{A}$.

Example 8 Since in Cartesian, spherical, and cylindrical coordinates we have respectively

$$\mathbf{A} = A_x \mathbf{e}_x + A_y \mathbf{e}_y + A_z \mathbf{e}_z =$$
$$= A_r \mathbf{e}_r + A_\varphi \mathbf{e}_\varphi + A_z \mathbf{e}_z =$$
$$= A_R \mathbf{e}_R + A_\varphi \mathbf{e}_\varphi + A_\theta \mathbf{e}_\theta,$$

as follows from the results of Example 6,

$$\omega_{\mathbf{A}}^1 = A_x \, \mathrm{d}x + A_y \, \mathrm{d}y + A_z \, \mathrm{d}z = \tag{14.31'}$$
$$= A_r \, \mathrm{d}r + A_\varphi r \, \mathrm{d}\varphi + A_z \, \mathrm{d}z = \tag{14.31''}$$
$$= A_R \, \mathrm{d}R + A_\varphi R \cos\varphi \, \mathrm{d}\varphi + A_\theta R \, \mathrm{d}\theta. \tag{14.31'''}$$

g.

Now let $\mathbf{B} = B^1 \mathbf{e}_1 + B^2 \mathbf{e}_2 + B^3 \mathbf{e}_3$ and $\omega_{\mathbf{B}}^2 = b_1 \, \mathrm{d}t^2 \wedge \mathrm{d}t^3 + b_2 \, \mathrm{d}t^3 \wedge \mathrm{d}t^1 + b_3 \, \mathrm{d}t^1 \wedge \mathrm{d}t^2$. Then, on the one hand,

$$\omega_{\mathbf{B}}^2(\mathbf{e}_2, \mathbf{e}_3) := \mathrm{d}V(\mathbf{B}, \mathbf{e}_2, \mathbf{e}_3) =$$

$$= \sum_{i-1}^{3} B^i \, \mathrm{d}V(\mathbf{e}_i, \mathbf{e}_2, \mathbf{e}_3) = B^1 \cdot (\mathbf{e}_1, \mathbf{e}_2, \mathbf{e}_3) = B^1,$$

where $\mathrm{d}V$ is the volume element in $T\mathbb{R}_t^3$ see (14.28) and (14.27)).

On the other hand, by (14.30) we obtain

$$\omega_{\mathbf{B}}^2(\mathbf{e}_2, \mathbf{e}_3) = \left(b_1 \, \mathrm{d}t^2 \wedge \mathrm{d}t^3 + b_2 \, \mathrm{d}t^3 \wedge \mathrm{d}t^1 + b_3 \, \mathrm{d}t^1 \wedge \mathrm{d}t^2\right)(\mathbf{e}_2, \mathbf{e}_3) =$$

$$= b_1 \, \mathrm{d}t^2 \wedge \mathrm{d}t^3(\mathbf{e}_2, \mathbf{e}_3) = \frac{b_1}{\sqrt{E_2 E_3}}.$$

Comparing these results, we conclude that $b_1 = B^1 \sqrt{E_2 E_3}$. Similarly, we verify that $b_2 = B^2 \sqrt{E_1 E_3}$ and $b_3 = B^3 \sqrt{E_1 E_2}$.

Thus we have found the representation

$$\omega_{\mathbf{B}}^2 = B^1 \sqrt{E_2 E_3} \, \mathrm{d}t^2 \wedge \mathrm{d}t^3 + B^2 \sqrt{E_3 E_1} \, \mathrm{d}t^3 \wedge \mathrm{d}t^1 + B^3 \sqrt{E_1 E_2} \, \mathrm{d}t^1 \wedge \mathrm{d}t^2 =$$

$$= \sqrt{E_1 E_2 E_3} \left(\frac{B^1}{\sqrt{E_1}} \, \mathrm{d}t^2 \wedge \mathrm{d}t^3 + \frac{B^2}{\sqrt{E_2}} \, \mathrm{d}t^3 \wedge \mathrm{d}t^1 + \frac{B^3}{\sqrt{E_3}} \, \mathrm{d}t^1 \wedge \mathrm{d}t^2 \right) \tag{14.32}$$

of the form $\omega_{\mathbf{B}}^2$ corresponding to the vector $\mathbf{B} = B^1 \mathbf{e}_1 + B^2 \mathbf{e}_2 + B^3 \mathbf{e}_3$.

Example 9 Using the notation introduced in Example 8 and formulas (14.26′), (14.26″) and (14.26‴), we obtain in Cartesian, cylindrical, and spherical coordinates respectively

$$\omega_{\mathbf{B}}^2 = B_x \, dy \wedge dz + B_y \, dz \wedge dx + B_z \, dx \wedge dy = \tag{14.32′}$$

$$= B_r r \, d\varphi \wedge dz + B_\varphi \, dz \wedge dr + B_z r \, dr \wedge d\varphi = \tag{14.32″}$$

$$= B_R R^2 \cos\theta \, d\varphi \wedge d\theta + B_\varphi R \, d\theta \wedge dR + B_\theta R \cos\theta \, dR \wedge d\varphi. \tag{14.32‴}$$

h.

We add further that on the basis of (14.28) we can write

$$\omega_\rho^3 = \rho\sqrt{E_1 E_2 E_3} \, dt^1 \wedge dt^2 \wedge dt^3. \tag{14.33}$$

Example 10 In particular, for Cartesian, cylindrical, and spherical coordinates respectively, formula (14.33) has the following forms:

$$\omega_\rho^3 = \rho \, dx \wedge dy \wedge dz = \tag{14.33′}$$

$$= \rho r \, dr \wedge d\varphi \wedge dz = \tag{14.33″}$$

$$= \rho R^2 \cos\theta \, dR \wedge d\varphi \wedge d\theta. \tag{14.33‴}$$

Now that we have obtained formulas (14.31)–(14.33), it is easy to find the coordinate representation of the operators grad, curl, and div in a triorthogonal curvilinear coordinate system using Definitions (14.9)–(14.11).

Let grad $f = A^1 \mathbf{e}_1 + A^2 \mathbf{e}_2 + A^3 \mathbf{e}_3$. Using the definitions, we write

$$\omega_{\text{grad}\,f}^1 := d\omega_f^0 := df := \frac{\partial f}{\partial t^1} \, dt^1 + \frac{\partial f}{\partial t^2} \, dt^2 + \frac{\partial f}{\partial t^3} \, dt^3.$$

From this, using formula (14.31), we conclude that

$$\text{grad}\, f = \frac{1}{\sqrt{E_1}} \frac{\partial f}{\partial t^1} \mathbf{e}_1 + \frac{1}{\sqrt{E_2}} \frac{\partial f}{\partial t^2} \mathbf{e}_2 + \frac{1}{\sqrt{E_3}} \frac{\partial f}{\partial t^3} \mathbf{e}_3. \tag{14.34}$$

Example 11 In Cartesian, cylindrical, and spherical coordinates respectively,

$$\text{grad}\, f = \frac{\partial f}{\partial x} \mathbf{e}_x + \frac{\partial f}{\partial y} \mathbf{e}_y + \frac{\partial f}{\partial z} \mathbf{e}_z = \tag{14.34′}$$

$$= \frac{\partial f}{\partial r} \mathbf{e}_r + \frac{1}{r} \frac{\partial f}{\partial \varphi} \mathbf{e}_\varphi + \frac{\partial f}{\partial z} \mathbf{e}_z = \tag{14.34″}$$

$$= \frac{\partial f}{\partial R} \mathbf{e}_R + \frac{1}{R \cos\theta} \frac{\partial f}{\partial \varphi} \mathbf{e}_\varphi + \frac{1}{R^2} \frac{\partial f}{\partial \theta} \mathbf{e}_\theta. \tag{14.34‴}$$

Suppose given a field $\mathbf{A}(t) = (A^1\mathbf{e}_1 + A^2\mathbf{e}_2 + A^3\mathbf{e}_3)(t)$. Let us find the coordinates B^1, B^2, B^3 of the field $\operatorname{curl}\mathbf{A}(t) = \mathbf{B}(t) = (B^1\mathbf{e}_1 + B^2\mathbf{e}_2 + B^3\mathbf{e}_3)(t)$.

Based on the definition (14.10) and formula (14.31), we obtain

$$\omega^2_{\operatorname{curl}\mathbf{A}} := d\omega^1_{\mathbf{A}} = d\left(A^1\sqrt{E_1}\,dt^1 + A^2\sqrt{E_2}\,dt^2 + A^3\sqrt{E_3}\,dt^3\right) =$$

$$= \left(\frac{\partial A^3\sqrt{E_3}}{\partial t^2} - \frac{\partial A^2\sqrt{E_2}}{\partial t^3}\right)dt^2 \wedge dt^3 +$$

$$+ \left(\frac{\partial A^1\sqrt{E_1}}{\partial t^3} - \frac{\partial A^3\sqrt{E_3}}{\partial t^1}\right)dt^3 \wedge dt^1 +$$

$$+ \left(\frac{\partial A^2\sqrt{E_2}}{\partial t^1} - \frac{\partial A^1\sqrt{E_1}}{\partial t^2}\right)dt^1 \wedge dt^2.$$

On the basis of (14.32) we now conclude that

$$B^1 = \frac{1}{\sqrt{E_2 E_3}}\left(\frac{\partial A^3\sqrt{E_3}}{\partial t^2} - \frac{\partial A^2\sqrt{E_2}}{\partial t^3}\right),$$

$$B^2 = \frac{1}{\sqrt{E_3 E_1}}\left(\frac{\partial A^1\sqrt{E_1}}{\partial t^3} - \frac{\partial A^3\sqrt{E_3}}{\partial t^1}\right),$$

$$B^3 = \frac{1}{\sqrt{E_1 E_2}}\left(\frac{\partial A^2\sqrt{E_2}}{\partial t^1} - \frac{\partial A^1\sqrt{E_1}}{\partial t^2}\right),$$

that is,

$$\operatorname{curl}\mathbf{A} = \frac{1}{\sqrt{E_1 E_2 E_3}}\begin{vmatrix} \sqrt{E_1}\mathbf{e}_1 & \sqrt{E_2}\mathbf{e}_2 & \sqrt{E_3}\mathbf{e}_3 \\ \frac{\partial}{\partial t^1} & \frac{\partial}{\partial t^2} & \frac{\partial}{\partial t^3} \\ \sqrt{E_1}A^1 & \sqrt{E_2}A^2 & \sqrt{E_3}A^3 \end{vmatrix}. \tag{14.35}$$

Example 12 In Cartesian, cylindrical, and spherical coordinates respectively

$$\operatorname{curl}\mathbf{A} = \left(\frac{\partial A_z}{\partial y} - \frac{\partial A_y}{\partial z}\right)\mathbf{e}_x + \left(\frac{\partial A_x}{\partial z} - \frac{\partial A_z}{\partial x}\right)\mathbf{e}_y + \left(\frac{\partial A_y}{\partial x} - \frac{\partial A_x}{\partial y}\right)\mathbf{e}_z = \tag{14.35'}$$

$$= \frac{1}{r}\left(\frac{\partial A_z}{\partial \varphi} - \frac{\partial r A_\varphi}{\partial z}\right)\mathbf{e}_r + \left(\frac{\partial A_r}{\partial z} - \frac{\partial A_z}{\partial r}\right)\mathbf{e}_\varphi + \frac{1}{r}\left(\frac{\partial r A_\varphi}{\partial r} - \frac{\partial A_r}{\partial \varphi}\right)\mathbf{e}_z =$$

$$\tag{14.35''}$$

$$= \frac{1}{R\cos\theta}\left(\frac{\partial A_\theta}{\partial \varphi} - \frac{\partial A_\varphi\cos\theta}{\partial \theta}\right)\mathbf{e}_R + \frac{1}{R}\left(\frac{\partial A_R}{\partial \theta} - \frac{\partial R A_\theta}{\partial R}\right)\mathbf{e}_\varphi +$$

$$+ \frac{1}{R}\left(\frac{\partial R A_\varphi}{\partial R} - \frac{1}{\cos\theta}\frac{\partial A_R}{\partial \varphi}\right)\mathbf{e}_\theta. \tag{14.35'''}$$

i.

Now suppose given a field $\mathbf{B}(t) = (B^1\mathbf{e}_1 + B^2\mathbf{e}_2 + B^3\mathbf{e}_3)(t)$. Let us find an expression for div $\mathbf{B}$.

Starting from the definition (14.11) and formula (14.32), we obtain

$$\omega_{\mathrm{div}\mathbf{B}} := \mathrm{d}\omega_{\mathbf{B}}^2 = \mathrm{d}(B^1\sqrt{E_2E_3}\,\mathrm{d}t^2 \wedge \mathrm{d}t^3 +$$
$$+ B^2\sqrt{E_3E_1}\,\mathrm{d}t^3 \wedge \mathrm{d}t^1 + B^3\sqrt{E_1E_2}\,\mathrm{d}t^1 \wedge \mathrm{d}t^2 =$$
$$= \left(\frac{\partial\sqrt{E_2E_3}B^1}{\partial t^1} + \frac{\partial\sqrt{E_3E_1}B^2}{\partial t^2} + \frac{\partial\sqrt{E_1E_2}B^3}{\partial t^3}\right)\mathrm{d}t^1 \wedge \mathrm{d}t^2 \wedge \mathrm{d}t^3.$$

On the basis of formula (14.33) we now conclude that

$$\mathrm{div}\,\mathbf{B} = \frac{1}{\sqrt{E_1E_2E_3}}\left(\frac{\partial\sqrt{E_2E_3}B^1}{\partial t^1} + \frac{\partial\sqrt{E_3E_1}B^2}{\partial t^2} + \frac{\partial\sqrt{E_1E_2}B^3}{\partial t^3}\right). \qquad (14.36)$$

In Cartesian, cylindrical, and spherical coordinates respectively, we obtain

$$\mathrm{div}\,\mathbf{B} = \frac{\partial B_x}{\partial x} + \frac{\partial B_y}{\partial y} + \frac{\partial B_z}{\partial z} = \qquad\qquad (14.36')$$

$$= \frac{1}{r}\left(\frac{\partial r B_r}{\partial r} + \frac{\partial B_\varphi}{\partial \varphi}\right) + \frac{\partial B_z}{\partial z} = \qquad\qquad (14.36'')$$

$$= \frac{1}{R^2\cos\theta}\left(\frac{\partial R^2\cos\theta\,B_R}{\partial R} + \frac{\partial R B_\varphi}{\partial \varphi} + \frac{\partial R\cos\theta\,B_\theta}{\partial \theta}\right). \qquad (14.36''')$$

j.

Relations (14.34) and (14.36) can be used to obtain an expression for the Laplacian $\Delta = \mathrm{div}\,\mathrm{grad}$ in an arbitrary triorthogonal coordinate system:

$$\Delta f = \mathrm{div}\,\mathrm{grad}\,f =$$
$$= \mathrm{div}\left(\frac{1}{\sqrt{E_1}}\frac{\partial f}{\partial t^1}\mathbf{e}_1 + \frac{1}{\sqrt{E_2}}\frac{\partial f}{\partial t^2}\mathbf{e}_2 + \frac{1}{\sqrt{E_3}}\frac{\partial f}{\partial t^3}\mathbf{e}_3\right) =$$
$$= \frac{1}{\sqrt{E_1E_2E_3}}\left(\frac{\partial}{\partial t^1}\left(\sqrt{\frac{E_2E_3}{E_1}}\frac{\partial f}{\partial t^1}\right) + \right.$$
$$\left. + \frac{\partial}{\partial t^2}\left(\sqrt{\frac{E_3E_1}{E_2}}\frac{\partial f}{\partial t^2}\right) + \frac{\partial}{\partial t^3}\left(\sqrt{\frac{E_1E_2}{E_3}}\frac{\partial f}{\partial t^3}\right)\right). \qquad (14.37)$$

Example 13 In particular, for Cartesian, cylindrical, and spherical coordinates, we obtain respectively

$$\Delta f = \frac{\partial^2 f}{\partial x^2} + \frac{\partial^2 f}{\partial y^2} + \frac{\partial^2 f}{\partial z^2} = \tag{14.37$'$}$$

$$= \frac{1}{r} \frac{\partial}{\partial r}\left(r \frac{\partial f}{\partial r}\right) + \frac{1}{r^2} \frac{\partial^2 f}{\partial \varphi^2} + \frac{\partial^2 f}{\partial z^2} = \tag{14.37$''$}$$

$$= \frac{1}{R^2} \frac{\partial}{\partial R}\left(R^2 \frac{\partial f}{\partial R}\right) + \frac{1}{R^2 \cos^2 \theta} \frac{\partial^2 f}{\partial \varphi^2} + \frac{1}{R^2 \cos \theta} \frac{\partial}{\partial \theta}\left(\cos \theta \frac{\partial f}{\partial \theta}\right). \tag{14.37$'''$}$$

14.1.6 Problems and Exercises

1. The operators grad, curl, and div and the algebraic operations.
Verify the following relations:
for grad:

a) $\nabla(f + g) = \nabla f + \nabla g$,
b) $\nabla(f \cdot g) = f \nabla g + g \nabla f$,
c) $\nabla(\mathbf{A} \cdot \mathbf{B}) = (\mathbf{B} \cdot \nabla)\mathbf{A} + (\mathbf{A} \cdot \nabla)\mathbf{B} + \mathbf{B} \times (\nabla \times \mathbf{A}) + \mathbf{A} \times (\nabla \times \mathbf{B})$,
d) $\nabla(\frac{1}{2}\mathbf{A}^2) = (\mathbf{A} \cdot \nabla)\mathbf{A} + \mathbf{A} \times (\nabla \times \mathbf{A})$;

for curl:

e) $\nabla \times (f\mathbf{A}) = f\nabla \times \mathbf{A} + \nabla f \times \mathbf{A}$,
f) $\nabla \times (\mathbf{A} \times \mathbf{B}) = (\mathbf{B} \cdot \nabla)\mathbf{A} - (\mathbf{A} \cdot \nabla)\mathbf{B} + (\nabla \cdot \mathbf{B})\mathbf{A} - (\nabla \cdot \mathbf{A})\mathbf{B}$;

for div:

g) $\nabla \cdot (f\mathbf{A}) = \nabla f \cdot \mathbf{A} + f\nabla \cdot \mathbf{A}$,
h) $\nabla \cdot (\mathbf{A} \times \mathbf{B}) = \mathbf{B} \cdot (\nabla \times \mathbf{A}) - \mathbf{A} \cdot (\nabla \times \mathbf{B})$

and rewrite them in the symbols grad, curl, and div.
(Hints. $\mathbf{A} \cdot \nabla = A^1 \frac{\partial}{\partial x^1} + A^2 \frac{\partial}{\partial x^2} + A^3 \frac{\partial}{\partial x^3}$; $\mathbf{B} \cdot \nabla \neq \nabla \cdot \mathbf{B}$; $\mathbf{A} \times (\mathbf{B} \times \mathbf{C}) = \mathbf{B}(\mathbf{A} \cdot \mathbf{C}) - \mathbf{C}(\mathbf{A} \cdot \mathbf{B})$.)

2. a) Write the operators (14.20)–(14.22) in Cartesian coordinates.
b) Verify relations (14.20) and (14.21) by direct computation.
c) Verify formula (14.24) in Cartesian coordinates.
d) Write formula (14.24) in terms of ∇ and prove it, using the formulas of vector algebra.

3. From the system of Maxwell equations in Example 2 deduce that $\nabla \cdot \mathbf{j} = -\frac{\partial \rho}{\partial t}$.

4. a) Exhibit the Lamé parameters H_1, H_2, H_3 of Cartesian, cylindrical, and spherical coordinates in $\mathbb{R}^3$.
b) Rewrite formulas (14.28), (14.34)–(14.37), using the Lamé parameters.

5. Write the field $\mathbf{A} = \operatorname{grad} \frac{1}{r}$, where $r = \sqrt{x^2 + y^2 + z^2}$ in

a) Cartesian coordinates x, y, z;

b) cylindrical coordinates;

c) spherical coordinates.

d) Find curl $\mathbf{A}$ and div $\mathbf{A}$.

6. In cylindrical coordinates (r, φ, z) the function f has the form $\ln \frac{1}{r}$. Write the field $\mathbf{A} = \operatorname{grad} f$ in

a) Cartesian coordinates;

b) cylindrical coordinates;

c) spherical coordinates.

d) Find curl $\mathbf{A}$ and div $\mathbf{A}$.

7. Write the formula for transformation of coordinates in a fixed tangent space $T\mathbb{R}^3_p$, $p \in \mathbb{R}^3$, when passing from Cartesian coordinates in $\mathbb{R}^3$ to

a) cylindrical coordinates;

b) spherical coordinates;

c) an arbitrary triorthogonal curvilinear coordinate system.

d) Applying the formulas obtained in c) and formulas (14.34)–(14.37), verify directly that the vector fields grad f, curl $\mathbf{A}$, and the quantities div $\mathbf{A}$ and Δf are invariant relative to the choice of the coordinate system in which they are computed.

8. The space $\mathbb{R}^3$, being a rigid body, revolves about a certain axis with constant angular velocity ω. Let $\mathbf{v}$ be the field of linear velocities of the points at a fixed instant of time.

a) Write the field $\mathbf{v}$ in the corresponding cylindrical coordinates.

b) Find curl $\mathbf{v}$.

c) Indicate how the field curl $\mathbf{v}$ is directed relative to the axis of rotation.

d) Verify that $|\operatorname{curl} \mathbf{v}| = 2\omega$ at each point of space.

e) Interpret the geometric meaning of curl $\mathbf{v}$ and the geometric meaning of the constancy of this vector at all points of space for the situation in d).

14.2 The Integral Formulas of Field Theory

14.2.1 The Classical Integral Formulas in Vector Notation

a. Vector Notation for the Forms ω_A^1 and ω_B^2

In the preceding chapter we noted (see Sect. 13.2, formulas (13.23) and (13.24)) that the restriction of the work form $\omega_{\mathbf{F}}^1$ of a field $\mathbf{F}$ to an oriented smooth curve (path) γ or the restriction of the flux form $\omega_{\mathbf{V}}^2$ of a field $\mathbf{V}$ to an oriented surface S can be written respectively in the following forms:

$$\omega_{\mathbf{F}}^1|_\gamma = \langle \mathbf{F}, \mathbf{e} \rangle \, ds, \qquad \omega_{\mathbf{V}}^2|_S = \langle \mathbf{V}, \mathbf{n} \rangle \, d\sigma,$$

where **e** is the unit vector that orients γ, codirectional with the velocity vector of the motion along γ, ds is the element (form) of arc length on γ, **n** is the unit normal vector to S that orients the surface, and $d\sigma$ is the element (form) of area on S.

In vector analysis we often use the vector element of length of a curve $\mathbf{ds} := \mathbf{e}\, ds$ and the vector element of area on a surface $\boldsymbol{d\sigma} := \mathbf{n}\, d\sigma$. Using this notation, we can now write:

$$\omega_{\mathbf{A}}^1|_\gamma = \langle \mathbf{A}, \mathbf{e}\rangle\, ds = \langle \mathbf{A}, \mathbf{ds}\rangle = \mathbf{A}\cdot\mathbf{ds}, \tag{14.38}$$

$$\omega_{\mathbf{B}}^2|_S = \langle \mathbf{B}, \mathbf{n}\rangle\, d\sigma = \langle \mathbf{B}, \boldsymbol{d\sigma}\rangle = \mathbf{B}\cdot\boldsymbol{d\sigma}. \tag{14.39}$$

b. The Newton–Leibniz Formula

Let $f \in C^{(1)}(D, \mathbb{R})$, and let $\gamma : [a, b] \to D$ be a path in the domain D.

Applied to the 0-form ω_f^0, Stokes' formula

$$\int_{\partial\gamma} \omega_f^0 = \int_\gamma d\omega_f^0$$

means, on the one hand, the equality

$$\int_{\partial\gamma} f = \int_\gamma df,$$

which agrees with the classical formula

$$f\big(\gamma(b)\big) - f\big(\gamma(a)\big) = \int_a^b df\big(\gamma(t)\big)$$

of Newton–Leibniz (the fundamental theorem of calculus). On the other hand, by definition of the gradient, it means that

$$\int_{\partial\gamma} \omega_f^0 = \int_\gamma \omega_{\text{grad } f}^1. \tag{14.40}$$

Thus, using relation (14.38), we can rewrite the Newton–Leibniz formula as

$$\boxed{f\big(\gamma(b)\big) - f\big(\gamma(a)\big) = \int_\gamma (\text{grad } f)\cdot\mathbf{ds}.} \tag{14.40'}$$

In this form it means that

the increment of a function on a path equals the work done by the gradient of the function on the path.

This is a very convenient and informative notation. In addition to the obvious deduction that the work of the field grad f along a path γ depends only on the endpoints of the path, the formula enables us to make a somewhat more subtle observation. To be specific, motion over a level surface $f = c$ of f takes place without any work being done by the field grad f since in this case grad $f \cdot d\boldsymbol{\sigma} = 0$. Then, as the left-hand side of the formula shows, the work of the field grad f depends not even on the initial and final points of the path but only on the level surfaces of f to which they belong.

c. Stokes' Formula

We recall that the work of a field on a closed path is called the *circulation of the field on that path*. To indicate that the integral is taken over a closed path, we often write $\oint_\gamma \mathbf{F} \cdot d\mathbf{s}$ rather than the traditional notation $\int_\gamma \mathbf{F} \cdot d\mathbf{s}$. If γ is a curve in the plane, we often use the symbols $\oint_\gamma$ and $\oint_\gamma$, in which the direction of traversal of the curve γ is indicated.

The term *circulation* is also used when speaking of the integral over some finite set of closed curves. For example, it might be the integral over the boundary of a compact surface with boundary.

Let $\mathbf{A}$ be a smooth vector field in a domain D of the oriented Euclidean space $\mathbb{R}^3$ and S a (piecewise) smooth oriented compact surface with boundary in D. Applied to the 1-form $\omega_{\mathbf{A}}^1$, taking account of the definition of the curl of a vector field, Stokes' formula means the equality

$$\int_{\partial S} \omega_{\mathbf{A}}^1 = \int_S \omega_{\mathrm{curl}\,\mathbf{A}}^2 . \tag{14.41}$$

Using relation (14.39), we can rewrite (14.41) as the classical Stokes formula

$$\boxed{\oint_{\partial S} \mathbf{A} \cdot d\mathbf{s} = \iint_S (\mathrm{curl}\,\mathbf{A}) \cdot d\boldsymbol{\sigma}.} \tag{14.41'}$$

In this notation it means that

> the circulation of a vector field on the boundary of a surface equals the flux of the curl of the field across the surface.

As always, the orientation chosen on ∂S is the one induced by the orientation of S.

d. The Gauss–Ostrogradskii Formula

Let V be a compact domain of the oriented Euclidean space $\mathbb{R}^3$ bounded by a (piecewise-) smooth surface ∂V, the boundary of V. If $\mathbf{B}$ is a smooth field in V, then in accordance with the definition of the divergence of a field, Stokes' formula yields the equality

$$\int_{\partial V} \omega_{\mathbf{B}}^2 = \int_V \omega_{\mathrm{div}\,\mathbf{B}}^3 . \tag{14.42}$$

Using relation (14.39) and the notation $\rho \, \mathrm{d}V$ for the form ω_ρ^3 in terms of the volume element $\mathrm{d}V$ in $\mathbb{R}^3$, we can rewrite Eq. (14.42) as the classical Gauss–Ostrogradskii formula

$$\boxed{\iint_{\partial V} \mathbf{B} \cdot \mathrm{d}\boldsymbol{\sigma} = \iiint_V \operatorname{div} \mathbf{B} \, \mathrm{d}V.} \qquad (14.42')$$

In this form it means that

the flux of a vector field across the boundary of a domain equals the integral of the divergence of the field over the domain itself.

e. Summary of the Classical Integral Formulas

In sum, we have arrived at the following vector notation for the three classical integral formulas of analysis:

$$\int_{\partial \gamma} f = \int_\gamma (\nabla f) \cdot \mathrm{d}\mathbf{s} \quad \text{(the Newton–Leibniz formula)}, \qquad (14.40'')$$

$$\int_{\partial S} \mathbf{A} \cdot \mathrm{d}\mathbf{s} = \int_S (\nabla \times \mathbf{A}) \cdot \mathrm{d}\boldsymbol{\sigma} \quad \text{(Stokes' formula)}, \qquad (14.41'')$$

$$\int_{\partial V} \mathbf{B} \cdot \mathrm{d}\boldsymbol{\sigma} = \int_V (\nabla \cdot \mathbf{B}) \, \mathrm{d}V \quad \text{(the Gauss–Ostrogradskii formula)}. \qquad (14.42'')$$

14.2.2 The Physical Interpretation of div, curl, and grad

a. The Divergence

Formula (14.42') can be used to explain the physical meaning of $\operatorname{div} \mathbf{B}(x)$ – the divergence of the vector field $\mathbf{B}$ at a point x in the domain V in which the field is defined. Let $V(x)$ be a neighborhood of x (for example, a ball) contained in V. We permit ourselves to denote the volume of this neighborhood by the same symbol $V(x)$ and its diameter by the letter d.

By the mean-value theorem and the formula (14.42') we obtain the following relation for the triple integral

$$\iint_{\partial V(x)} \mathbf{B} \cdot \mathrm{d}\boldsymbol{\sigma} = \operatorname{div} \mathbf{B}(x') V(x),$$

where x' is a point in the neighborhood $V(x)$. If $d \to 0$, then $x' \to x$, and since $\mathbf{B}$ is a smooth field, we also have $\operatorname{div} \mathbf{B}(x') \to \operatorname{div} \mathbf{B}(x)$. Hence

$$\operatorname{div} \mathbf{B}(x) = \lim_{d \to 0} \frac{\iint_{\partial V(x)} \mathbf{B} \cdot \mathrm{d}\boldsymbol{\sigma}}{V(x)}. \qquad (14.43)$$

Let us regard $\mathbf{B}$ as the velocity field for a flow (of liquid or gas). Then, by the law of conservation of mass, a flux of this field across the boundary of the domain V or, what is the same, a volume of the medium diverging across the boundary of the domain, can arise only when there are sinks or sources (including those associated with a change in the density of the medium). The flux is equal to the total power of all these factors, which we shall collectively call "sources", in the domain $V(x)$. Hence the fraction on the right-hand side of (14.43) is the mean intensity (per unit volume) of sources in the domain $V(x)$, and the limit of that quantity, that is, $\operatorname{div}\mathbf{B}(x)$, is the specific intensity (per unit volume) of the source at the point x. But the limit of the ratio of the total amount of some quantity in the domain $V(x)$ to the volume of that domain as $d \to 0$ is customarily called the *density* of that quantity at x, and the density as a function of a point is usually called the *density of the distribution* of the given quantity in a portion of space.

Thus, we can interpret the divergence $\operatorname{div}\mathbf{B}$ of a vector field B as the density of the distribution of sources in the domain of the flow, that is, in the domain of definition of the field $\mathbf{B}$.

Example 1 If, in particular, $\operatorname{div}\mathbf{B} \equiv 0$, that is, there are no sources, then the flux across the boundary of the region must be zero: the amount flowing in equals the amount flowing out. And, as formula (14.42$'$) shows, this is indeed the case.

Example 2 A point electric charge of magnitude q creates an electric field in space. Suppose the charge is located at the origin. By *Coulomb's law*[6] the intensity $\mathbf{E} = E(x)$ of the field at the point $x \in \mathbb{R}^3$ (that is, the force acting on a unit test charge at the point x) can be written as

$$\mathbf{E} = \frac{q}{4\pi\varepsilon_0}\frac{\mathbf{r}}{|\mathbf{r}|^3},$$

where ε_0 is a dimensioning constant and $\mathbf{r}$ is the radius-vector of the point x.

The field $\mathbf{E}$ is defined at all points different from the origin. In spherical coordinates $\mathbf{E} = \frac{q}{4\pi\varepsilon_0}\frac{1}{R^2}\mathbf{e}_R$, so that by formula (14.36$'''$) of the preceding section, one can see immediately that $\operatorname{div}\mathbf{E} = 0$ everywhere in the domain of definition of the field $\mathbf{E}$.

Hence, if we take any domain V not containing the origin, then by formula (14.42$'$) the flux of $\mathbf{E}$ across the boundary ∂V of V is zero.

Let us now take the sphere $S_R = \{x \in \mathbb{R}^3 \mid |x| = R\}$ of radius R with center at the origin and find the outward flux (relative to the ball bounded by the sphere) of $\mathbf{E}$ across this surface. Since the vector $\mathbf{e}_R$ is itself the unit outward normal to the sphere, we find

$$\int_{S_R}\mathbf{E}\cdot\mathrm{d}\boldsymbol{\sigma} = \int_{S_R}\frac{q}{4\pi\varepsilon_0}\frac{1}{R^2}\,\mathrm{d}\sigma = \frac{q}{4\pi\varepsilon_0 R^2}\cdot 4\pi R^2 = \frac{q}{\varepsilon_0}.$$

---

[6]Ch.O. Coulomb (1736–1806) – French physicist. He discovered experimentally the law (Coulomb's law) of interaction of charges and magnetic fields using a torsion balance that he invented himself.

Thus, up to the dimensioning constant ε_0, which depends on the choice of the system of physical units, we have found the amount of charge in the volume bounded by the sphere.

We remark that under the hypotheses of Example 2 just studied the left-hand side of formula (14.42') is well-defined on the sphere $\partial V = S_R$, but the integrand on the right-hand side is defined and equal to zero everywhere in the ball V except at one point – the origin. Nevertheless, the computations show that the integral on the right-hand side of (14.42') cannot be interpreted as the integral of a function that is identically zero.

From the formal point of view one could dismiss the need to study this situation by saying that the field $\mathbf{E}$ is not defined at the point $0 \in V$, and hence we do not have the right to speak about the equality (14.42'), which was proved for smooth fields defined in the entire domain V of integration. However, the physical interpretation of (14.42') as the law of conservation of mass shows that, when suitably interpreted, it ought to be valid always.

Let us study the indeterminacy of $\operatorname{div} \mathbf{E}$ at the origin in Example 2 more attentively to see what is causing it. Formally the original field $\mathbf{E}$ is not defined at the origin, but, if we seek $\operatorname{div} \mathbf{E}$ from formula (14.43), then, as Example 2 shows, we would have to assume that $\operatorname{div} \mathbf{E}(0) = +\infty$. Hence the integrand on the right-hand side of (14.42) would be a "function" equal to zero everywhere except at one point, where it is equal to infinity. This corresponds to the fact that there are no charges at all outside the origin, and we somehow managed to put the entire charge q into a space of volume zero – into the single point 0, at which the charge density naturally became infinite. Here we are encountering the so-called Dirac[7] δ-function (delta-function).

The densities of physical quantities are needed ultimately so that one can find the values of the quantities themselves by integrating the density. For that reason there is no need to define the δ-function at each individual point; it is more important to define its integral. If we assume that physically the "function" $\delta_{x_0}(x) = \delta(x_0; x)$ must correspond to the density of a distribution, for example the distribution of mass in space, for which the entire mass, equal to 1 in magnitude, is concentrated at the single point x_0, it is natural to set

$$\int_V \delta(x_0, x)\, dV = \begin{cases} 1, & \text{when } x_0 \in V, \\ 0, & \text{when } x_0 \notin V. \end{cases}$$

Thus, from the point of view of a mathematical idealization of our ideas of the possible distribution of a physical quantity (mass, charge, and the like) in space, we must assume that its distribution density is the sum of an ordinary finite function corresponding to a continuous distribution of the quantity in space and a certain set

---

[7]P.A.M. Dirac (1902–1984) – British theoretical physicist, one of the founders of quantum mechanics. More details on the Dirac δ-function will be given in Sects. 17.4.4 and 17.5.4.

of singular "functions" (of the same type as the Dirac δ-function) corresponding to a concentration of the quantity at individual points of space.

Hence, starting from these positions, the results of the computations in Example 2 can be expressed as the single equality $\operatorname{div} \mathbf{E} = \frac{q}{\varepsilon_0} \delta(0; x)$. Then, as applied to the field $\mathbf{E}$, the integral on the right-hand side of (14.42′) is indeed equal either to q/ε_0 or to 0, according as the domain V contains the origin (and the point charge concentrated there) or not.

In this sense one can assert (following Gauss) that the flux of electric field intensity across the surface of a body equals (up to a factor depending on the units chosen) the sum of the electric charges contained in the body. In this same sense one must interpret the electric charge density ρ in the Maxwell equations considered in Sect. 14.1 (formula (14.12)).

b. The Curl

We begin our study of the physical meaning of the curl with an example.

Example 3 Suppose the entire space, regarded as a rigid body, is rotating with constant angular speed ω about a fixed axis (let it be the x-axis). Let us find the curl of the field $\mathbf{v}$ of linear velocities of the points of space. (The field is being studied at any fixed instant of time.)

In cylindrical coordinates (r, φ, z) we have the simple expression $\mathbf{v}(r, \varphi, z) = \omega r \mathbf{e}_\varphi$. Then by formula (14.35″) of Sect. 14.1, we find immediately that $\operatorname{curl} \mathbf{v} = 2\omega \mathbf{e}_z$. That is, $\operatorname{curl} \mathbf{v}$ is a vector directed along the axis of rotation. Its magnitude 2ω equals the angular velocity of the rotation, up to the coefficient 2, and the direction of the vector, taking account of the orientation of the whole space $\mathbb{R}^3$, completely determines the direction of rotation.

The field described in Example 3 in the small resembles the velocity field of a funnel (sink) or the field of the vorticial motion of air in the neighborhood of a tornado (also a sink, but one that drains upward). Thus, the curl of a vector field at a point characterizes the degree of vorticity of the field in a neighborhood of the point.

We remark that the circulation of a field over a closed contour varies in direct proportion to the magnitude of the vectors in the field, and, as one can verify using the same Example 3, it can also be used to characterize the vorticity of the field. Only now, to describe completely the vorticity of the field in a neighborhood of a point, it is necessary to compute the circulation over contours lying in three different planes. Let us now carry out this program.

We take a disk $S_i(x)$ with center at the point x and lying in a plane perpendicular to the ith coordinate axis, $i = 1, 2, 3$. We orient $S_i(x)$ using a normal, which we take to be the unit vector $\mathbf{e}_i$ along this coordinate axis. Let d be the diameter of $S_i(x)$.

From formula (14.41′) for a smooth field **A** we find that

$$(\operatorname{curl} \mathbf{A}) \cdot \mathbf{e}_i = \lim_{d \to 0} \frac{\oint_{\partial S_i(x)} \mathbf{A} \cdot d\mathbf{s}}{S_i(x)}, \qquad (14.44)$$

where $S_i(x)$ denotes the area of the disk under discussion. Thus the circulation of the field **A** over the boundary ∂S_i per unit area in the plane orthogonal to the ith coordinate axis characterizes the ith component of curl **A**.

To clarify still further the meaning of the curl of a vector field, we recall that every linear transformation of space is a composition of dilations in three mutually perpendicular directions, translation of the space as a rigid body, and rotation as a rigid body. Moreover, every rotation can be realized as a rotation about some axis. Every smooth deformation of the medium (flow of a liquid or gas, sliding of the ground, bending of a steel rod) is locally linear. Taking account of what has just been said and Example 3, we can conclude that if there is a vector field that describes the motion of a medium (the velocity field of the points in the medium), then the curl of that field at each point gives the instantaneous axis of rotation of a neighborhood of the point, the magnitude of the instantaneous angular velocity, and the direction of rotation about the instantaneous axis. That is, the curl characterizes completely the rotational part of the motion of the medium. This will be made slightly more precise below, where it will be shown that the curl should be regarded as a sort of density for the distribution of local rotations of the medium.

c. The Gradient

We have already said quite a bit about the gradient of a scalar field, that is, about the gradient of a function. Hence at this point we shall merely recall the main things.

Since $\omega^1_{\operatorname{grad} f}(\boldsymbol{\xi}) = \langle \operatorname{grad} f, \boldsymbol{\xi} \rangle = d f(\boldsymbol{\xi}) = D_{\boldsymbol{\xi}} f$, where $D_{\boldsymbol{\xi}} f$ is the derivative of the function f with respect to the vector $\boldsymbol{\xi}$, it follows that $\operatorname{grad} f$ is orthogonal to the level surfaces of f, and at each point it points in the direction of most rapid increase in the values of the function. Its magnitude $|\operatorname{grad} f|$ gives the rate of that growth (per unit of length in the space in which the argument varies).

The significance of the gradient as a density will be discussed below.

14.2.3 Other Integral Formulas

a. Vector Versions of the Gauss–Ostrogradskii Formula

The interpretation of the curl and gradient as vector densities, analogous to the interpretation (14.43) of the divergence as a density, can be obtained from the following classical formulas of vector analysis, connected with the Gauss–Ostrogradskii formula.

$$\int_V \nabla \cdot \mathbf{B}\, dV = \int_{\partial V} d\boldsymbol{\sigma} \cdot B \quad \text{(the divergence theorem)}, \qquad (14.45)$$

$$\int_V \nabla \times \mathbf{A}\, dV = \int_{\partial V} d\boldsymbol{\sigma} \times \mathbf{A} \quad \text{(the curl theorem)}, \qquad (14.46)$$

$$\int_V \nabla f\, dV = \int_{\partial V} d\boldsymbol{\sigma}\, f \quad \text{(the gradient theorem)}. \qquad (14.47)$$

The first of these three relations coincides with (14.42′) up to notation and is the Gauss–Ostrogradskii formula. The vector equalities (14.46) and (14.47) follow from (14.45) if we apply that formula to each component of the corresponding vector field.

Retaining the notation $V(x)$ and d used in Eq. (14.43), we obtain from formulas (14.45)–(14.47) in a unified manner,

$$\nabla \cdot \mathbf{B}(x) = \lim_{d \to 0} \frac{\int_{\partial V(x)} d\boldsymbol{\sigma} \cdot \mathbf{B}}{V(x)}, \qquad (14.43′)$$

$$\nabla \times \mathbf{A}(x) = \lim_{d \to 0} \frac{\int_{\partial V(x)} d\boldsymbol{\sigma} \times \mathbf{A}}{V(x)}, \qquad (14.48)$$

$$\nabla f(x) = \lim_{d \to 0} \frac{\int_{\partial V(x)} d\boldsymbol{\sigma}\, f}{V(x)}. \qquad (14.49)$$

The right-hand sides of (14.45)–(14.47) can be interpreted respectively as the scalar flux of the vector field $\mathbf{B}$, the vector flux of the vector field $\mathbf{A}$, and the vector flux of the scalar field f across the surface ∂V bounding the domain V. Then the quantities div $\mathbf{B}$, curl $\mathbf{A}$, and grad f on the left-hand sides of Eqs. (14.43′), (14.48), and (14.49) can be interpreted as the corresponding source densities of these fields.

We remark that the right-hand sides of Eqs. (14.43′), (14.48), and (14.49) are independent of the coordinate system. From these we can once again derive the invariance of the gradient, curl, and divergence.

b. Vector Versions of Stokes' Formula

Just as formulas (14.45)–(14.47) were the result of combining the Gauss–Ostrogradskii formula with the algebraic operations on vector and scalar fields, the following triple of formulas can be obtained by combining these same operations with the classical Stokes formula (which appears as the first of the three relations).

Let S be a (piecewise-) smooth compact oriented surface with a consistently oriented boundary ∂S, let $d\sigma$ be the vector element of area on S, and ds the vector element of length on ∂S. Then for smooth fields $\mathbf{A}$, $\mathbf{B}$, and f, the following relations hold:

$$\int_S d\sigma \cdot (\nabla \times \mathbf{A}) = \int_{\partial S} ds \cdot \mathbf{A}, \tag{14.50}$$

$$\int_S (d\sigma \times \nabla) \times \mathbf{B} = \int_{\partial S} ds \times \mathbf{B}, \tag{14.51}$$

$$\int_S d\sigma \times \nabla f = \int_{\partial S} ds\, f. \tag{14.52}$$

Formulas (14.51) and (14.52) follow from Stokes' formula (14.50). We shall not take time to give the proofs.

c. Green's Formulas

If S is a surface and $\mathbf{n}$ a unit normal vector to S, then the derivative $D_{\mathbf{n}} f$ of the function f with respect to $\mathbf{n}$ is usually denoted $\frac{\partial f}{\partial n}$ in field theory. For example, $\langle \nabla f, d\sigma \rangle = \langle \nabla f, \mathbf{n} \rangle\, d\sigma = \langle \operatorname{grad} f, \mathbf{n} \rangle\, d\sigma = D_{\mathbf{n}} f\, d\sigma = \frac{\partial f}{\partial n}\, d\sigma$. Thus, $\frac{\partial f}{\partial n}\, d\sigma$ is the flux of $\operatorname{grad} f$ across the element of surface $d\sigma$.

In this notation we can write the following formulas of Green, which are very widely used in analysis:

$$\int_V \nabla f \cdot \nabla g\, dV + \int_V g\nabla^2 f\, dV = \int_{\partial V} (g\nabla f) \cdot d\sigma \left(= \int_{\partial V} g\frac{\partial f}{\partial n}\, d\sigma \right), \tag{14.53}$$

$$\int_V \left(g\nabla^2 f - f\nabla^2 g \right) dV =$$

$$= \int_{\partial V} (g\nabla f - f\nabla g) \cdot d\sigma \left(= \int_{\partial V} \left(g\frac{\partial f}{\partial n} - f\frac{\partial g}{\partial n} \right) d\sigma \right). \tag{14.54}$$

In particular, if we set $f = g$ in (14.53) and $g \equiv 1$ in (14.54), we find respectively,

$$\int_V |\nabla f|^2\, dV + \int_V f\Delta f\, dV = \int_{\partial V} f\Delta f \cdot d\sigma \left(= \int_{\partial V} f\frac{\partial f}{\partial n}\, d\sigma \right), \tag{14.53'}$$

$$\int_V \Delta f\, dV = \int_{\partial V} \nabla f \cdot d\sigma \left(= \int_{\partial V} \frac{\partial f}{\partial n}\, d\sigma \right). \tag{14.54'}$$

This last equality is often called *Gauss' theorem*. Let us prove, for example, the second of Eqs. (14.53) and (14.54):

Proof

$$\int_{\partial V} (g \nabla f - f \nabla g) \cdot d\boldsymbol{\sigma} = \int_V \nabla \cdot (g \nabla f - f \nabla g) \, dV =$$

$$= \int_V \left(\nabla g \cdot \nabla f + g \nabla^2 f - \nabla f \cdot \nabla g - f \nabla^2 g \right) dV =$$

$$= \int_V \left(g \nabla^2 f - f \nabla^2 g \right) dV = \int_V (g \Delta f - f \Delta g) \, dV.$$

In this formula we have used the Gauss–Ostrogradskii formula and the relation $\nabla \cdot (\varphi \mathbf{A}) = \nabla \varphi \cdot \mathbf{A} + \varphi \nabla \cdot \mathbf{A}$.   □

14.2.4 Problems and Exercises

1. Using the Gauss–Ostrogradskii formula (14.45), prove relations (14.46) and (14.47).

2. Using Stokes' formula (14.50), prove relations (14.51) and (14.52).

3. a) Verify that formulas (14.45), (14.46), and (14.47) remain valid for an un-bounded domain V if the integrands in the surface integrals are of order $O(\frac{1}{r^3})$ as $r \to \infty$. (Here $r = |\mathbf{r}|$, and $\mathbf{r}$ is the radius-vector in $\mathbb{R}^3$.)

b) Determine whether formulas (14.50), (14.51), and (14.52) remain valid for a noncompact surface $S \subset \mathbb{R}^3$ if the integrands in the line integrals are of order $O(\frac{1}{r^2})$ as $r \to \infty$.

c) Give examples showing that for unbounded surfaces and domains Stokes' formula (14.41′) and the Gauss–Ostrogradskii formula (14.42′) are in general not true.

4. a) Starting from the interpretation of the divergence as a source density, explain why the second of the Maxwell equations (formula (14.12) of Sect. 14.1) implies that there are no point sources in the magnetic field (that is, there are no magnetic charges).

b) Using the Gauss–Ostrogradskii formula and the Maxwell equations (formula (14.12) of Sect. 14.1), show that no rigid configuration of test charges (for example a single charge) can be in a stable equilibrium state in the domain of an electrostatic field that is free of the (other) charges that create the field. (It is assumed that no forces except those exerted by the field act on the system.) This fact is known as *Earnshaw's theorem*.

5. If an electromagnetic field is steady, that is, independent of time, then the system of Maxwell equations (formula (14.12) of Sect. 14.1) decomposes into two indepen-dent parts – the *electrostatic equations* $\nabla \cdot \mathbf{E} = \frac{\rho}{\varepsilon_0}$, $\nabla \times \mathbf{E} = 0$, and the *magnetostatic equations* $\nabla \times \mathbf{B} = \frac{\mathbf{j}}{\varepsilon_0 c^2}$, $\nabla \cdot \mathbf{B} = 0$.

The equation $\nabla \cdot \mathbf{E} = \rho/\varepsilon_0$, where ρ is the charge density, transforms via the Gauss–Ostrogradskii formula into $\int_S \mathbf{E} \cdot d\boldsymbol{\sigma} = Q/\varepsilon_0$, where the left-hand side is the flux of the electric field intensity across the closed surface S and the right-hand side is the sum Q of the charges in the domain bounded by S, divided by the dimensioning constant ε_0. In electrostatics this relation is usually called *Gauss' law*. Using Gauss' law, find the electric field $\mathbf{E}$

a) created by a uniformly charged sphere, and verify that outside the sphere it is the same as the field of a point charge of the same magnitude located at the center of the sphere;

b) of a uniformly charged line;

c) of a uniformly charged plane;

d) of a pair of parallel planes uniformly charged with charges of opposite sign;

e) of a uniformly charged ball.

6. a) Prove Green's formula (14.53).

b) Let f be a *harmonic function* in the bounded domain V (that is, f satisfies Laplace's equation $\Delta f = 0$ in V). Show, starting from $(14.54')$ that the flux of the gradient of this function across the boundary of the domain V is zero.

c) Verify that a harmonic function in a bounded connected domain is determined up to an additive constant by the values of its normal derivative on the boundary of the domain.

d) Starting from $(14.53')$, prove that if a harmonic function in a bounded domain vanishes on the boundary, it is identically zero throughout the domain.

e) Show that if the values of two harmonic functions are the same on the boundary of a bounded domain, then the functions are equal in the domain.

f) Starting from (14.53), verify the following *principle of Dirichlet. Among all continuous differentiable functions in a domain assuming prescribed values on the boundary, a harmonic function in the region is the only one that minimizes the Dirichlet integral (that is, the integral of the squared-modulus of the gradient over the domain).*

7. a) Let $r(p,q) = |p - q|$ be the distance between the points p and q in the Euclidean space $\mathbb{R}^3$. By fixing p, we obtain a function $r_p(q)$ of $q \in \mathbb{R}^3$. Show that $\Delta r_p^{-1}(q) = 4\pi\delta(p; q)$, where δ is the δ-function.

b) Let g be harmonic in the domain V. Setting $f = 1/r_p$ in (14.54) and taking account of the preceding result, we obtain

$$4\pi g(p) = \int_S \left(g\nabla\frac{1}{r_p} - \frac{1}{r_p}\nabla g \right) \cdot d\boldsymbol{\sigma}.$$

Prove this equality precisely.

c) Deduce from the preceding equality that if S is a sphere of radius R with center at p, then

$$g(p) = \frac{1}{4\pi R^2} \int_S g\, d\sigma.$$

This is the so-called *mean-value theorem* for harmonic functions.

d) Starting from the preceding result, show that if B is the ball bounded by the sphere S considered in part c) and $V(B)$ is its volume, then

$$g(p) = \frac{1}{V(B)} \int_B g \, dV.$$

e) If p and q are points of the Euclidean plane $\mathbb{R}^2$, then along with the function $\frac{1}{r_p}$ considered in a) above (corresponding to the potential of a charge located at p), we now take the function $\ln \frac{1}{r_p}$ (corresponding to the potential of a uniformly charged line in space). Show that $\Delta \ln \frac{1}{r_p} = 2\pi \delta(p; q)$, where $\delta(p; q)$ is now the δ-function in $\mathbb{R}^2$.

f) By repeating the reasoning in a), b), c), and d), obtain the mean-value theorem for functions that are harmonic in plane regions.

8. *Cauchy's multi-dimensional mean-value theorem.*

The classical mean-value theorem for the integral ("Lagrange's theorem") asserts that if the function $f : D \to \mathbb{R}$ is continuous on a compact, measurable, and connected set $D \subset \mathbb{R}^n$ (for example, in a domain), then there exists a point $\xi \in D$ such that

$$\int_D f(x) \, dx = f(\xi) \cdot |D|,$$

where $|D|$ is the measure (volume) of D.

a) Now let $f, g \in C(D, \mathbb{R})$, that is, f and g are continuous real-valued functions in D. Show that the following theorem ("Cauchy's theorem") holds: *There exists $\xi \in D$ such that*

$$g(\xi) \int_D f(x) \, dx = f(\xi) \int_D g(x) \, dx.$$

b) Let D be a compact domain with smooth boundary ∂D and $\mathbf{f}$ and $\mathbf{g}$ two smooth vector fields in D. Show that there exists a point $\xi \in D$ such that

$$\operatorname{div} \mathbf{g}(\xi) \cdot \underset{\partial D}{\operatorname{Flux}} \mathbf{f} = \operatorname{div} \mathbf{f}(\xi) \cdot \underset{\partial D}{\operatorname{Flux}} \mathbf{g},$$

where $\underset{\partial D}{\operatorname{Flux}}$ is the flux of a vector field across the surface ∂D.

14.3 Potential Fields

14.3.1 The Potential of a Vector Field

Definition 1 Let $\mathbf{A}$ be a vector field in the domain $D \subset \mathbb{R}^n$. A function $U : D \to \mathbb{R}$ is called a *potential of the field* $\mathbf{A}$ if $\mathbf{A} = \operatorname{grad} U$ in D.

Definition 2 A field that has a potential is called a *potential field*.

Since the partial derivatives of a function determine the function up to an additive constant in a connected domain, the potential is unique in such a domain up to an additive constant.

We briefly mentioned potentials in the first part of this course. Now we shall discuss this important concept in somewhat more detail. In connection with these definitions we note that when different force fields are studied in physics, the potential of a field $\mathbf{F}$ is usually defined as a function U such that $\mathbf{F} = -\operatorname{grad} U$. This potential differs from the one given in Definition 1 only in sign.

Example 1 At a point of space having radius-vector $\mathbf{r}$ the intensity $\mathbf{F}$ of the gravitational field due to a point mass M located at the origin can be computed from Newton's law as

$$\mathbf{F} = -GM\frac{\mathbf{r}}{r^3},\tag{14.55}$$

where $r = |\mathbf{r}|$.

This is the force with which the field acts on a unit mass at this point of space. The gravitational field (14.55) is a potential field. Its potential in the sense of Definition 1 is the function

$$U = GM\frac{1}{r}.\tag{14.56}$$

Example 2 At a point of space having radius-vector $\mathbf{r}$ the intensity $\mathbf{E}$ of the electric field due to a point charge q located at the origin can be computed from Coulomb's law

$$\mathbf{E} = \frac{q}{4\pi\varepsilon_0}\frac{\mathbf{r}}{r^3}$$

Thus such an electrostatic field, like the gravitational field, is a potential field. Its potential φ in the sense of physical terminology is defined by the relation

$$\varphi = \frac{q}{4\pi\varepsilon_0}\frac{1}{r}.$$

14.3.2 Necessary Condition for Existence of a Potential

In the language of differential forms the equality $\mathbf{A} = \operatorname{grad} U$ means that $\omega_{\mathbf{A}}^1 = d\omega_U^0 = dU$, from which it follows that

$$d\omega_{\mathbf{A}}^1 = 0,\tag{14.57}$$

since $d^2\omega_U^0 = 0$. This is a necessary condition for the field $\mathbf{A}$ to be a potential field.

In Cartesian coordinates this condition can be expressed very simply. If $\mathbf{A} = (A^1, \ldots, A^n)$ and $\mathbf{A} = \operatorname{grad} U$, then $A^i = \frac{\partial U}{\partial x^i}$, $i = 1, \ldots, n$, and if the potential U is sufficiently smooth (for example, if its second-order partial derivatives are continuous), we must have

$$\frac{\partial A^i}{\partial x^j} = \frac{\partial A^j}{\partial x^i}, \quad i, j = 1, \ldots, n, \tag{14.57$'$}$$

which simply means that the mixed partial derivatives are equal in both orders:

$$\frac{\partial^2 U}{\partial x^i \partial x^j} = \frac{\partial^2 U}{\partial x^j \partial x^i}.$$

In Cartesian coordinates $\omega_{\mathbf{A}}^1 = \sum_{i=1}^n A^i \, dx^i$, and therefore the equalities (14.57) and (14.57$'$) are indeed equivalent in this case.

In the case of $\mathbb{R}^3$ we have $d\omega_{\mathbf{A}}^1 = \omega_{\operatorname{curl} \mathbf{A}}^2$, so that the necessary condition (14.57) can be rewritten as

$$\operatorname{curl} \mathbf{A} = \mathbf{0},$$

which corresponds to the relation $\operatorname{curl} \operatorname{grad} U = \mathbf{0}$, which we already know.

Example 3 The field $\mathbf{A} = (x, xy, xyz)$ in Cartesian coordinates in $\mathbb{R}^3$ cannot be a potential field, since, for example, $\frac{\partial xy}{\partial x} \neq \frac{\partial x}{\partial y}$.

Example 4 Consider the field $\mathbf{A} = (A_x, A_y)$ given by

$$\mathbf{A} = \left(-\frac{y}{x^2 + y^2}, \frac{x}{x^2 + y^2} \right), \tag{14.58}$$

defined in Cartesian coordinates at all points of the plane except the origin. The necessary condition for a field to be a potential field $\frac{\partial A_x}{\partial y} = \frac{\partial A_y}{\partial x}$ is fulfilled in this case. However, as we shall soon verify, this field is not a potential field in its domain of definition.

Thus the necessary condition (14.57), or, in Cartesian coordinates (14.57$'$), is in general not sufficient for a field to be a potential field.

14.3.3 Criterion for a Field to be Potential

Proposition 1 *A continuous vector field* $\mathbf{A}$ *in a domain* $D \subset \mathbb{R}^n$ *is a potential field in* D *if and only if its circulation (work) around every closed curve* γ *contained in* D *is zero*:

$$\oint_\gamma \mathbf{A} \cdot d\mathbf{s} = 0. \tag{14.59}$$

Proof Necessity. Suppose $\mathbf{A} = \operatorname{grad} U$. Then by the Newton–Leibniz formula (Formula (14.40$'$) of Sect. 14.2),

$$\oint_\gamma \mathbf{A} \cdot d\mathbf{s} = U\big(\gamma(b)\big) - U\big(\gamma(a)\big),$$

where $\gamma : [a, b] \to D$. If $\gamma(a) = \gamma(b)$, that is, when the path γ is closed, it is obvious that the right-hand side of this last equality vanishes, and hence the left-hand side does also.

Sufficiency. Suppose condition (5) holds. Then the integral over any (not necessarily closed) path in D depends only on its initial and terminal points, not on the path joining them. Indeed, if γ_1 and γ_2 are two paths having the same initial and terminal points, then, traversing first γ_1, then $-\gamma_2$ (that is, traversing γ_2 in the opposite direction), we obtain a closed path γ whose integral, by (14.59), equals zero, but is also the difference of the integrals over γ_1 and γ_2. Hence these last two integrals really are equal.

We now fix some point $x_0 \in D$ and set

$$U(x) = \int_{x_0}^x \mathbf{A} \cdot d\mathbf{s}, \tag{14.60}$$

where the integral on the right is the integral over any path in D from x_0 to x. We shall verify that the function U so defined is the required potential for the field $\mathbf{A}$. For convenience, we shall assume that a Cartesian coordinate system $(x^1, \ldots, x^n)$ has been chosen in $\mathbb{R}^n$. Then $\mathbf{A} \cdot d\mathbf{s} = A^1 dx^1 + \cdots + A^n dx^n$. If we move away from x along a straight line in the direction $h\mathbf{e}_i$, where $\mathbf{e}_i$ is the unit vector along the x^i-axis, the function U receives an increment equal to

$$U(x + h\mathbf{e}_i) - U(x) = \int_{x^i}^{x^i + h^i} A^i\big(x^1, \ldots, x^{i-1}, t, x^{i+1}, \ldots, x^n\big) dt,$$

equal to the integral of the form $\mathbf{A} \cdot d\mathbf{s}$ over this path from x to $x + h\mathbf{e}_i$. By the continuity of A and the mean-value theorem, this last equality can be written as

$$U(x + h\mathbf{e}_i) - U(x) = A^i\big(x^1, \ldots, x^{i-1}, x^i + \theta h, x^{i+1}, \ldots, x^n\big)h,$$

where $0 \leq \theta \leq 1$. Dividing this last equality by h and letting h tend to zero, we find

$$\frac{\partial U}{\partial x^i}(x) = A^i(x),$$

that is, $\mathbf{A} = \operatorname{grad} U$. $\qquad\square$

Remark 1 As can be seen from the proof, a sufficient condition for a field to be a potential field is that (14.59) hold for smooth paths or, for example, for broken lines whose links are parallel to the coordinate axes.

We now return to Example 4. Earlier (Example 1 of Sect. 8.1) we computed that the circulation of the field (14.58) over the circle $x^2 + y^2 = 1$ traversed once in the counterclockwise direction was $2\pi (\neq 0)$.

Thus, by Proposition 1 we can conclude that the field (14.58) is not a potential field in the domain $\mathbb{R}^2 \backslash 0$.

But surely, for example,

$$\operatorname{grad} \arctan \frac{y}{x} = \left(-\frac{y}{x^2 + y^2}, \frac{x}{x^2 + y^2} \right),$$

and it would seem that the function $\arctan \frac{y}{x}$ is a potential for (14.58). What is this, a contradiction?! There is no contradiction as yet, since the only correct conclusion that one can make in this situation is that the function $\arctan \frac{y}{x}$ is not defined in the entire domain $\mathbb{R}^2 \backslash 0$. And that is indeed the case: Take for example, the points on the y-axis. But then, you may say, we could consider the function $\varphi(x, y)$, the polar angular coordinate of the point (x, y). That is practically the same thing as $\arctan \frac{y}{x}$, but $\varphi(x, y)$ is also defined for $x = 0$, provided the point (x, y) is not at the origin. Throughout the domain $\mathbb{R}^2 \backslash 0$ we have

$$d\varphi = -\frac{y}{x^2 + y^2} \, dx + \frac{x}{x^2 + y^2} \, dy.$$

However, there is still no contradiction, although the situation is now more delicate. Please note that in fact φ is not a continuous single-valued function of a point in the domain $\mathbb{R}^2 \backslash 0$. As a point encircles the origin counterclockwise, its polar angle, varying continuously, will have increased by 2π when the point returns to its starting position. That is, we arrive at the original point with a new value of the function, different from the one we began with. Consequently, we must give up either the continuity or the single-valuedness of the function φ in the domain $\mathbb{R}^2 \backslash 0$.

In a small neighborhood (not containing the origin) of each point of the domain $\mathbb{R}^2 \backslash 0$ one can distinguish a continuous single-valued branch of the function φ. All such branches differ from one another by an additive constant, a multiple of 2π. That is why they all have the same differential and can all serve locally as potentials of the field (14.58). Nevertheless, the field (14.58) has no potential in the entire domain $\mathbb{R}^2 \backslash 0$.

The situation studied in Example 4 turns out to be typical in the sense that the necessary condition (14.57) or (14.57$'$) for the field **A** to be a potential field is locally also sufficient. The following proposition holds.

Proposition 2 *If the necessary condition for a field to be a potential field holds in a ball, then the field has a potential in that ball.*

Proof For the sake of intuitiveness we first carry out the proof in the case of a disk $D = \{(x, y) \in \mathbb{R}^2 \mid x^2 + y^2 < r\}$ in the plane $\mathbb{R}^2$. One can arrive at the point (x, y) of the disk from the origin along two different two-link broken lines γ_1 and γ_2 with

Fig. 14.3

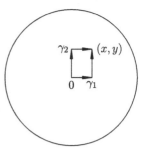

links parallel to the coordinate axes (see Fig. 14.3). Since D is a convex domain, the entire rectangle I bounded by these lines is contained in D.

By Stokes' formula, taking account of condition (14.57), we obtain

$$\int_{\partial I} \omega_{\mathbf{A}}^1 = \int_I d\omega_{\mathbf{A}}^1 = 0.$$

By the remark to Proposition 1 we can conclude from this that the field $\mathbf{A}$ is a potential field in D. Moreover, by the proof of sufficiency in Proposition 1, the function (14.60) can again be taken as the potential, the integral being interpreted as the integral over a broken line from the center to the point in question with links parallel to the axes. In this case the independence of the choice of path γ_1, γ_2 for such an integral followed immediately from Stokes' formula for a rectangle.

In higher dimensions it follows from Stokes' formula for a two-dimensional rectangle that replacing two adjacent links of the broken line by two links forming the sides of a rectangle parallel to the original does not change the value of the integral over the path. Since one can pass from one broken-line path to any other broken-line path leading to the same point by a sequence of such reconstructions, the potential is unambiguously defined in the general case.   □

14.3.4 Topological Structure of a Domain and Potentials

Comparing Example 4 and Proposition 2, one can conclude that when the necessary condition (14.57) for a field to be a potential field holds, the question whether it is always a potential field depends on the (topological) structure of the domain in which the field is defined. The following considerations (here and in Sect. 14.3.5 below) give an elementary idea as to exactly how the characteristics of the domain bring this about.

It turns out that if the domain D is such that every closed path in D can be contracted to a point of the domain without going outside the domain, then the necessary condition (14.57) for a field to be a potential field in D is also sufficient. We shall call such domains *simply connected* below. A ball is a simply connected domain (and that is why Proposition 2 holds). But the punctured plane $\mathbb{R}^2 \backslash 0$ is not simply connected, since a path that encircles the origin cannot be contracted

Fig. 14.4

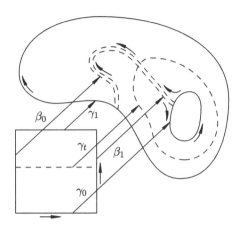

to a point without going outside the region. This is why not every field in $\mathbb{R}^2\backslash 0$ satisfying (14.57′), as we saw in Example 4, is necessarily a potential field in $\mathbb{R}^2\backslash 0$.

We now turn from the general description to precise formulations. We begin by stating clearly what we mean we speak of deforming or contracting a path.

Definition 3 A *homotopy* (or *deformation*) in D from a closed path $\gamma_0 : [0, 1] \to D$ to a closed path $\gamma_1 : [0, 1] \to D$ is a continuous mapping $\Gamma : I^2 \to D$ of the square $I^2 = \{(t^1, t^2) \in \mathbb{R}^2 \mid 0 \le t^i \le 1, i = 1, 2\}$ into D such that $\Gamma(t^1, 0) = \gamma_0(t^1)$, $\Gamma(t^1, 1) = \gamma_1(t^1)$, and $\Gamma(0, t^2) = \Gamma(1, t^2)$ for all $t^1, t^2 \in [0, 1]$.

Thus a homotopy is a mapping $\Gamma : I^2 \to D$ (Fig. 14.4). If the variable t^2 is regarded as time, according to Definition 3 at each instant of time $t = t^2$ we have a closed path $\Gamma(t^1, t) = \gamma_t$ (Fig. 14.4).[8] The change in this path with time is such that at the initial instant $t = t^2 = 0$ it coincides with γ_0 and at time $t = t^2 = 1$ it becomes γ_1.

Since the condition $\gamma_t(0) = \Gamma(0, t) = \Gamma(1, t) = \gamma_t(1)$, which means that the path γ_t is closed, holds at all times $t \in [0, 1]$, the mapping $\Gamma : I^2 \to D$ induces the same mappings $\beta_0(t^1) := \Gamma(t^1, 0) = \Gamma(t^1, 1) =: \beta_1(t^1)$ on the vertical sides of the square I^2.

The mapping Γ is a formalization of our intuitive picture of gradually deforming γ_0 to γ_1.

It is clear that time can be allowed to run backwards, and then we obtain the path γ_0 from γ_1.

Definition 4 Two closed paths are *homotopic* in a domain if they can be obtained from each other by a homotopy in that domain, that is a homotopy can be constructed in that domain from one to the other.

---

[8]Orienting arrows are shown along certain curves in Fig. 14.4. These arrows will be used a little later; for the time being the reader should not pay any attention to them.

Remark 2 Since the paths we have to deal with in analysis are as a rule paths of integration, we shall consider only smooth or piecewise-smooth paths and smooth or piecewise-smooth homotopies among them, without noting this explicitly.

For domains in $\mathbb{R}^n$ one can verify that the presence of a continuous homotopy of (piecewise-) smooth paths guarantees the existence of (piecewise-) smooth homotopies of these paths.

Proposition 3 *If the* 1*-form* $\omega_{\mathbf{A}}^1$ *in the domain* D *is such that* $d\omega_{\mathbf{A}}^1 = 0$, *and the closed paths* γ_0 *and* γ_1 *are homotopic in* D, *then*

$$\int_{\gamma_0} \omega_{\mathbf{A}}^1 = \int_{\gamma_1} \omega_{\mathbf{A}}^1.$$

Proof Let $\Gamma : I^2 \to D$ be a homotopy from γ_0 to γ_1 (see Fig. 14.4). If I_0 and I_1 are the bases of the square I^2 and J_0 and J_1 its vertical sides, then by definition of a homotopy of closed paths, the restrictions of Γ to I_0 and I_1 coincide with γ_0 and γ_1 respectively, and the restrictions of Γ to J_0 and J_1 give some paths β_0 and β_1 in D. Since $\Gamma(0, t^2) = \Gamma(1, t^2)$, the paths β_0 and β_1 are the same. As a result of the change of variables $x = \Gamma(t)$, the form $\omega_{\mathbf{A}}^1$ transfers to the square I^2 as some 1-form $\omega = \Gamma^* \omega_{\mathbf{A}}^1$. In the process $d\omega = d\Gamma^* \omega_{\mathbf{A}}^1 = \Gamma^* d\omega_{\mathbf{A}}^1 = 0$, since $d\omega_{\mathbf{A}}^1 = 0$. Hence, by Stokes' formula

$$\int_{\partial I^2} \omega = \int_{I^2} d\omega = 0.$$

But

$$\int_{\partial I^2} \omega = \int_{I_0} \omega + \int_{J_1} \omega - \int_{I_1} \omega - \int_{J_0} \omega =$$

$$= \int_{\gamma_0} \omega_{\mathbf{A}}^1 + \int_{\beta_1} \omega_{\mathbf{A}}^1 - \int_{\gamma_2} \omega_{\mathbf{A}}^1 - \int_{\beta_0} \omega_{\mathbf{A}}^1 = \int_{\gamma_0} \omega_{\mathbf{A}}^1 - \int_{\gamma_1} \omega_{\mathbf{A}}^1. \qquad \square$$

Definition 5 A domain is *simply connected* if every closed path in it is homotopic to a point (that is, a constant path).

Thus simply connected domains are those in which every closed path can be contracted to a point.

Proposition 4 *If a field* **A** *defined in a simply connected domain* D *satisfies the necessary condition* (14.57) *or* (14.57′) *to be a potential field, then it is a potential field in* D.

Proof By Proposition 1 and Remark 1 it suffices to verify that Eq. (14.59) holds for every smooth path γ in D. The path γ is by hypothesis homotopic to a constant path whose support consists of a single point. The integral over such a one-point path is obviously zero. But by Proposition 3 the integral does not change under a homotopy, and so Eq. (14.59) must hold for γ. □

Remark 3 Proposition 4 subsumes Proposition 2. However, since we had certain applications in mind, we considered it useful to give an independent constructive proof of Proposition 2.

Remark 4 Proposition 2 was proved without invoking the possibility of a smooth homotopy of smooth paths.

14.3.5 Vector Potential. Exact and Closed Forms

Definition 6 A field $\mathbf{A}$ is a *vector potential* for a field $\mathbf{B}$ in a domain $D \subset \mathbb{R}^3$ if the relation $\mathbf{B} = \operatorname{curl} \mathbf{A}$ holds in the domain.

If we recall the connection between vector fields and forms in the oriented Euclidean space $\mathbb{R}^3$ and also the definition of the curl of a vector field, the relation $\mathbf{B} = \operatorname{curl} \mathbf{A}$ can be rewritten as $\omega_{\mathbf{B}}^2 = d\omega_{\mathbf{A}}^1$. It follows from this relation that $\omega_{\operatorname{div} \mathbf{B}}^3 = d\omega_{\mathbf{B}}^2 = d^2\omega_{\mathbf{A}}^1 = 0$. Thus we obtain the necessary condition

$$\operatorname{div} \mathbf{B} = 0, \tag{14.61}$$

which the field $\mathbf{B}$ must satisfy in D in order to have a vector potential, that is, in order to be the curl of a vector field $\mathbf{A}$ in that domain.

A field satisfying condition (14.61) is often, especially in physics, called a *solenoidal field*.

Example 5 In Sect. 14.1 we wrote out the system of Maxwell equations. The second equation of this system is exactly Eq. (14.61). Thus, the desire naturally arises to regard a magnetic field $\mathbf{B}$ as the curl of some vector field $\mathbf{A}$ – the vector potential of $\mathbf{B}$. When solving the Maxwell equations, one passes to exactly such a vector potential.

As can be seen from Definitions 1 and 6, the questions of the scalar and vector potential of vector fields (the latter question being posed only in $\mathbb{R}^3$) are special cases of the general question as to when a differential p-form ω^p is the differential $d\omega^{p-1}$ of some form ω^{p-1}.

Definition 7 A differential form ω^p is *exact* in a domain D if there exists a form ω^{p-1} in D such that $\omega^p = d\omega^{p-1}$.

If the form ω^p is exact in D, then $d\omega^p = d^2\omega^{p-1} = 0$. Thus the condition

$$d\omega = 0 \tag{14.62}$$

is a necessary condition for the form ω to be exact.

Fig. 14.5

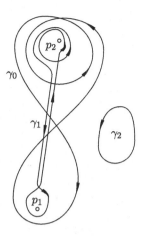

As we have already seen (Example 4), not every form satisfying this condition is exact. For that reason we make the following definition.

Definition 8 The differential form ω is *closed* in a domain D if it satisfies condition (14.62) there.

The following theorem holds.

Theorem (Poincaré's lemma) *If a form is closed in a ball, then it is exact there.*

Here we are talking about a ball in $\mathbb{R}^n$ and a form of any order, so that Proposition 2 is an elementary special case of this theorem.

The Poincaré lemma can also be interpreted as follows: The necessary condition (14.62) for a form to be exact is also locally sufficient, that is, every point of a domain in which (14.62) holds has a neighborhood in which ω is exact.

In particular, if a vector field **B** satisfies condition (14.61), it follows from the Poincaré lemma that at least locally it is the curl of some vector field **A**.

We shall not take the time at this point to prove this important theorem (those who wish to do so can read it in Chap. 15). We prefer to conclude by explaining in general outline the connection between the problem of the exactness of closed forms and the topology of their domains of definition (based on information about 1-forms).

Example 6 Consider the plane $\mathbb{R}^2$ with two points p_1 and p_2 removed (Fig. 14.5), and the paths γ_0, γ_1, and γ_2 whose supports are shown in the figure. The path γ_2 can be contracted to a point inside D, and therefore if a closed form ω is given in D, its integral over γ_2 is zero. The path γ_0 cannot be contracted to a point, but without changing the value of the integral of the form, it can be homotopically converted into the path γ_1.

The integral over γ_1 obviously reduces to the integral over one cycle enclosing the point p_1 clockwise and the double of the integral over a cycle enclosing p_2 counterclockwise. If T_1 and T_2 are the integrals of the form ω over small circles enclosing the points p_1 and p_2 and traversed, say, counterclockwise, one can see that the integral of the form ω over any closed path in D will be equal to $n_1 T_1 + n_2 T_2$, where n_1 and n_2 are certain integers indicating how many times we have encircled each of the holes p_1 and p_2 in the plane $\mathbb{R}^2$ and in which direction.

Circles c_1 and c_2 enclosing p_1 and p_2 serve as a sort of basis in which every closed path $\gamma \subset D$ has the form $\gamma = n_1 c_1 + n_2 c_2$, up to a homotopy, which has no effect on the integral. The quantities $\int_{c_i} \omega = T_i$ are called the *cyclic constants* or the *periods* of the integral. If the domain is more complicated and there are k independent elementary cycles, then in agreement with the expansion $\gamma = n_1 c_1 + \cdots + n_k c_k$, it results that $\int_{\gamma} \omega = n_1 T_1 + \cdots + n_k T_k$. It turns out that for any set $T_1, \ldots, T_k$ of numbers in such a domain one can construct a closed 1-form that will have exactly that set of periods. (This is a special case of de Rham's theorem – see Chap. 15.)

For the sake of visualization, we have resorted here to considering a plane domain, but everything that has been said can be repeated for any domain $D \subset \mathbb{R}^n$.

Example 7 In an anchor ring (the solid domain in $\mathbb{R}^3$ enclosed by a torus) all closed paths are obviously homotopic to a circle that encircles the hole a certain number of times. This circle serves as the unique non-constant basic cycle c.

Moreover, everything that has just been said can be repeated for paths of higher dimension. If instead of one-dimensional closed paths – mappings of a circle or, what is the same, mappings of the one-dimensional sphere – we take mappings of a k-dimensional sphere, introduce the concept of homotopy for them, and examine how many mutually nonhomotopic mappings of the k-dimensional sphere into a given domain $D \subset \mathbb{R}^n$ exist, the result is a certain characteristic of the domain D which is formalized in topology as the so-called kth homotopy group of D and denoted $\pi_k(D)$. If all the mappings of the k-dimensional sphere into D are homotopic to a constant mapping, the group $\pi_k(D)$ is considered trivial. (It consists of the identity element alone.) It can happen that $\pi_1(D)$ is trivial and $\pi_2(D)$ is not.

Example 8 If D is taken to be the space $\mathbb{R}^3$ with the point 0 removed, obviously every closed path in D can be contracted to a point, but a sphere enclosing the point 0 cannot be homotopically converted to a point.

It turns out that the homotopy group $\pi_k(D)$ has less to do with the periods of a closed k-form than the so-called *homology group* $H_k(D)$. (See Chap. 15.)

Example 9 From what has been said we can conclude that, for example, in the domain $\mathbb{R}^3 \backslash 0$ every closed 1-form is exact ($\mathbb{R}^3 \backslash 0$ is a simply connected domain), but not very closed 2-form is exact. In the language of vector fields, this means that every irrotational field $\mathbf{A}$ in $\mathbb{R}^3 \setminus 0$ is the gradient of a function, but not every source-free field $\mathbf{B}(\operatorname{div} \mathbf{B} = 0)$ is the curl of some field in this domain.

Example 10 To balance Example 9 we take the anchor ring. For the anchor ring the group $\pi_1(D)$ is not trivial (see Example 7), but $\pi_2(D)$ is trivial, since every mapping $f : S^2 \to D$ of the two-sphere into D can be contracted to a constant mapping (any image of a sphere can be contracted to a point). In this domain not every irrotational field is a potential field, but every source-free field is the curl of some field.

14.3.6 Problems and Exercises

1. Show that every central field $\mathbf{A} = f(r)\mathbf{r}$ is a potential field.
2. Let $\mathbf{F} = -\operatorname{grad} U$ be a potential force field. Show that the stable equilibrium positions of a particle in such a field are the minima of the potential U of that field.
3. For an electrostatic field $\mathbf{E}$ the Maxwell equations (formula (14.12) of Sect. 14.1), as already noted, reduce to the pair of equations $\nabla \cdot \mathbf{E} = \frac{\rho}{\varepsilon_0}$ and $\nabla \times \mathbf{E} = 0$.

The condition $\nabla \times \mathbf{E} = 0$ means, at least locally, that $\mathbf{E} = -\operatorname{grad} \varphi$. The field of a point charge is a potential field, and since every electric field is the sum (or integral) of such fields, it is always a potential field. Substituting $\mathbf{E} = -\nabla \varphi$ in the first equation of the electrostatic field, we find that its potential satisfies *Poisson's equation*[9] $\Delta \varphi = \frac{\rho}{\varepsilon_0}$. The potential φ determines the field completely, so that describing $\mathbf{E}$ reduces to finding the function φ, the solution of the Poisson equation.

Knowing the potential of a point charge (Example 2), solve the following problem.

a) Two charges $+q$ and $-q$ are located at the points $(0, 0, -d/2)$ and $(0, 0, d/2)$ in $\mathbb{R}^3$ with Cartesian coordinates (x, y, z). Show that at distances that are large relative to d the potential of the electrostatic field has the form

$$\varphi = \frac{1}{4\pi\varepsilon_0} \frac{z}{r^3} qd + o\!\left(\frac{1}{r^3}\right),$$

where r is the absolute value of the radius-vector $\mathbf{r}$ of the point (x, y, z).

b) Moving very far away from the charges is equivalent to moving the charges together, that is, decreasing the distance d. If we now fix the quantity $qd =: p$ and decrease d, then in the limit we obtain the function $\varphi = \frac{1}{4\pi\varepsilon_0} \frac{z}{r^3} p$ in the domain $\mathbb{R}^3 \setminus 0$. It is convenient to introduce the vector $\mathbf{p}$ equal to p in absolute value and directed from $-q$ to $+q$. We call the pair of charges $-q$ and $+q$ and the construction obtained by the limiting procedure just described a *dipole*, and the vector $\mathbf{p}$ the

---

[9]S.D. Poisson (1781–1849) – French scientist, specializing in mechanics and physics; his main work was on theoretical and celestial mechanics, mathematical physics, and probability theory. The Poisson equation arose in his research into gravitational potential and attraction by spheroids.

dipole moment. The function φ obtained in the limit is called the *dipole potential.*
Find the asymptotics of the dipole potential as one moves away from the dipole
along a ray forming angle θ with the direction of the dipole moment.

c) Let φ_0 be the potential of a unit point charge and φ_1 the dipole potential
having dipole moment $\mathbf{p}_1$. Show that $\varphi_1 = -(\mathbf{p}_1 \cdot \nabla)\varphi_0$.

d) We can repeat the construction with the limiting passage that we carried out
for a pair of charges in obtaining the dipole for the case of four charges (more precisely, for two dipoles with moments $\mathbf{p}_1$ and $\mathbf{p}_2$) and obtain a *quadrupole* and a
corresponding potential. In general we can obtain a *multipole of order j* with poten-

tial $\varphi_j = (-1)^j (\mathbf{p}_j \cdot \nabla)(\mathbf{P}_{j-1} \cdot \nabla) \cdots (\mathbf{p}_1 \cdot \nabla)\varphi_0 = \sum_{i+k+l=j} Q^j_{ikl} \frac{\partial^j \varphi_0}{\partial x^i \partial_{yk} \partial_{zl}}$, where

Q^j_{ikl} are the so-called *components of the multipole moment.* Carry out the computations and verify the formula for the potential of a multipole in the case of a
quadrupole.

e) Show that the main term in the asymptotics of the potential of a cluster of
charges with increasing distance from the cluster is $\frac{1}{4\pi\varepsilon_0}\frac{Q}{r}$, where Q is the total
charge of the cluster.

f) Show that the main term of the asymptotics of the potential of an electrically
neutral body consisting of charges of opposite signs (for example, a molecule) at a
distance that is large compared to the dimensions of the body is $\frac{1}{4\pi\varepsilon_0}\frac{\mathbf{p}\cdot\mathbf{e}_r}{r^2}$. Here $\mathbf{e}_r$
is a unit vector directed from the body to the observer; $\mathbf{p} = \sum q_i \mathbf{d}_i$, where q_i is the
magnitude of the ith charge and $\mathbf{d}_i$ is its radius-vector. The origin is chosen at some
point of the body.

g) The potential of any cluster of charges at a great distance from the cluster
can be expanded (asymptotically) in functions of multipole potential type. Show
this using the example of the first two terms of such a potential (see d), e), and
f)).

4. Determine whether the following domains are simply connected.

a) the disk $\{(x, y) \in \mathbb{R}^2 \mid x^2 + y^2 < 1\}$;
b) the disk with its center removed $\{(x, y) \in \mathbb{R}^2 \mid 0 < x^2 + y^2 < 1\}$;
c) a ball with its center removed $\{(x, y, z) \in \mathbb{R}^3 \mid 0 < x^2 + y^2 + z^2 < 1\}$;
d) an annulus $\{(x, y) \in \mathbb{R}^2 \mid \frac{1}{2} < x^2 + y^2 < 1\}$;
e) a spherical annulus $\{(x, y, z) \in \mathbb{R}^3 \mid \frac{1}{2} < x^2 + y^2 + z^2 < 1\}$;
f) an anchor ring in $\mathbb{R}^3$.

5. a) Give the definition of homotopy of paths with endpoints fixed.

b) Prove that a domain is simply connected if and only if every two paths in
it having common initial and terminal points are homotopic in the sense of the
definition given in part a).

6. Show that

a) every continuous mapping $f : S^1 \to S^2$ of a circle S^1 (a one-dimensional
sphere) into a two-dimensional sphere S^2 can be contracted in S^2 to a point (a
constant mapping);

b) every continuous mapping $f : S^2 \to S^1$ is also homotopic to a single point;

c) every mapping $f : S^1 \to S^1$ is homotopic to a mapping $\varphi \mapsto n\varphi$ for some $n \in \mathbb{Z}$, where φ is the polar angle;

d) every mapping of the sphere S^2 into an anchor ring is homotopic to a mapping to a single point;

e) every mapping of a circle S^1 into an anchor ring is homotopic to a closed path encircling the hole in the anchor ring n times, for some $n \in \mathbb{Z}$.

7. In the domain $\mathbb{R}^3 \backslash 0$ (three-dimensional space with the point 0 removed) construct:

a) a closed but not exact 2-form;

b) a source-free vector field that is not the curl of any vector field in that domain.

8. a) Can there be closed, but not exact forms of degree $p < n - 1$ in the domain $D = \mathbb{R}^n \backslash 0$ (the space $\mathbb{R}^n$ with the point 0 removed)?

b) Construct a closed but not exact form of degree $p = n - 1$ in $D = \mathbb{R}^n \backslash 0$.

9. If a 1-form ω is closed in a domain $D \subset \mathbb{R}^n$, then by Proposition 2 every point $x \in D$ has a neighborhood $U(x)$ inside which ω is exact. From now on ω is assumed to be a closed form.

a) Show that if two paths $\gamma_i : [0, 1] \to D$, $i = 1, 2$, have the same initial and terminal points and differ only on an interval $[\alpha, \beta] \subset [0, 1]$ whose image under either of the mappings γ_i is contained inside the same neighborhood $U(x)$, then $\int_{\gamma_1} \omega = \int_{\gamma_1} \omega$.

b) Show that for every path $[0, 1] \ni t \mapsto \gamma(t) \in D$ one can find a number $\delta > 0$ such that if the path $\tilde{\gamma}$ has the same initial and terminal point as γ and differs from γ at most by δ, that is $\max_{0 \le t \le 1} |\overline{\gamma}(t) - \gamma(t)| \le \delta$, then $\int_{\tilde{\gamma}} \omega = \int_\gamma \omega$.

c) Show that if two paths γ_1 and γ_2 with the same initial and terminal points are homotopic in D as paths with fixed endpoints, then $\int_{\gamma_1} \omega = \int_{\gamma_2} \omega$ for any closed form ω in D.

10. a) It will be proved below that every continuous mapping $\Gamma : I^2 \to D$ of the square I^2 can be uniformly approximated with arbitrary accuracy by a smooth mapping (in fact by a mapping with polynomial components). Deduce from this that if the paths γ_1 and γ_2 in the domain D are homotopic, then for every $\varepsilon > 0$ there exist smooth mutually homotopic paths $\tilde{\gamma}_1$ and $\tilde{\gamma}_2$ such that $\max_{0 \le t \le 1} |\tilde{\gamma}_i(t) - \gamma_i(t)| \le \varepsilon$, $i = 1, 2$.

b) Using the results of Example 9, show now that if the integrals of a closed form in D over smooth homotopic paths are equal, then they are equal for any paths that are homotopic in this domain (regardless of the smoothness of the homotopy). The paths themselves, of course, are assumed to be as regular as they need to be for integration over them.

11. a) Show that if the forms ω^p, ω^{p-1}, and $\tilde{\omega}^{p-1}$ are such that $\omega^p = d\omega^{p-1} = d\tilde{\omega}^{p-1}$, then (at least locally) one can find a form ω^{p-2} such that $\tilde{\omega}^{p-1} = \omega^{p-1} + d\omega^{p-2}$. (The fact that any two forms that differ by the differential of a form have the same differential obviously follows from the relation $d^2\omega = 0$.)

b) Show that the potential φ of an electrostatic field (Problem 3) is determined up to an additive constant, which is fixed if we require that the potential tend to zero at infinity.

12. The Maxwell equations (formula (14.12) of Sect. 14.1) yield the following pair of magnetostatic equations: $\nabla \cdot \mathbf{B} = 0$, $\nabla \times \mathbf{B} = -\frac{\mathbf{j}}{\varepsilon_0 c^2}$. The first of these shows that at least locally, $\mathbf{B}$ has a vector potential $\mathbf{A}$, that is, $\mathbf{B} = \nabla \times \mathbf{A}$.

a) Describe the amount of arbitrariness in the choice of the potential $\mathbf{A}$ of the magnetic field $\mathbf{B}$ (see Problem 11a)).

b) Let x, y, z be Cartesian coordinates in $\mathbb{R}^3$. Find potentials $\mathbf{A}$ for a uniform magnetic field $\mathbf{B}$ directed along the z-axis, each satisfying one of the following additional requirements: the field $\mathbf{A}$ must have the form $(0, A_y, 0)$; the field $\mathbf{A}$ must have the form $(A_x, 0, 0)$; the field A must have the form $(A_x, A_y, 0)$; the field $\mathbf{A}$ must be invariant under rotations about the z-axis.

c) Show that the choice of the potential $\mathbf{A}$ satisfying the additional requirement $\nabla \cdot \mathbf{A} = 0$ reduces to solving Poisson's equation; more precisely, to finding a scalar-valued function ψ satisfying the equation $\Delta \psi = f$ for a given scalar-valued function f.

d) Show that if the potential $\mathbf{A}$ of a static magnetic field $\mathbf{B}$ is chosen so that $\nabla \cdot \mathbf{A} = 0$, it will satisfy the vector Poisson equation $\Delta \mathbf{A} = -\frac{\mathbf{j}}{\varepsilon_0 c^2}$. Thus, invoking the potential makes it possible to reduce the problem of finding electrostatic and magnetostatic fields to solving Poisson's equation.

13. The following *theorem of Helmholtz*[10] is well known: *Every smooth field* $\mathbf{F}$ *in a domain D of oriented Euclidean space* $\mathbb{R}^3$ *can be decomposed into a sum* $\mathbf{F} = \mathbf{F}_1 + \mathbf{F}_2$ *of an irrotational field* $\mathbf{F}_1$ *and a solenoidal field* $\mathbf{F}_2$. Show that the construction of such a decomposition can be reduced to solving a certain Poisson equation.

14. Suppose a given mass of a certain substance passes from a state characterized thermodynamically by the parameters V_0, $P_0(T_0)$ into the state V, P, (T). Assume that the process takes place slowly (quasi-statically) and over a path γ in the plane of states (with coordinates V, P). It can be proved in thermodynamics that the quantity $S = \int_\gamma \frac{\delta Q}{T}$, where δQ is the heat exchange form, depends only on the initial point (V_0, P_0) and the terminal point (V, P) of the path, that is, after one of these points is fixed, for example (V_0, P_0), S becomes a function of the state (V, P) of the system. This function is called the *entropy* of the system.

a) Deduce from this that the form $\omega = \frac{\delta Q}{T}$ is exact, and that $\omega = dS$.

b) Using the form of δQ given in Problem 6 of Sect. 13.1 for an ideal gas, find the entropy of an ideal gas.

---

[10]H.L.F. Helmholtz (1821–1894) – German physicist and mathematician; one of the first to discover the general law of conservation of energy. Actually, he was the first to make a clear distinction between the concepts of force and energy.

14.4 Examples of Applications

To show the concepts we have introduced in action, and also to explain the physical meaning of the Gauss–Ostrogradskii–Stokes formula as a conservation law, we shall examine here some illustrative and important equations of mathematical physics.

14.4.1 The Heat Equation

We are studying the scalar field $T = T(x, y, z, t)$ of the temperature of a body being observed as a function of the point (x, y, z) of the body and the time t. As a result of heat transfer between various parts of the body the field T may vary. However, this variation is not arbitrary; it is subject to a particular law which we now wish to write out explicitly.

Let D be a certain three-dimensional part of the observed body bounded by a surface S. If there are no heat sources inside S, a change in the internal energy of the substance in D can occur only as the result of heat transfer, that is, in this case by the transfer of energy across the boundary S of D.

By computing separately the variation in internal energy in the volume D and the flux of energy across the surface S, we can use the law of conservation of energy to equate these two quantities and obtain the needed relation.

It is known that an increase in the temperature of a homogeneous mass m by ΔT requires energy $cm\Delta T$, where c is the specific heat capacity of the substance under consideration. Hence if our field T changes by $\Delta T = T(x, y, z, t + \Delta t) - T(x, y, z, t)$ over the time interval Δt, the internal energy in D will have changed by an amount

$$\iiint_D c\rho\,\Delta T\,\mathrm{d}V, \tag{14.63}$$

where $\rho = \rho(x, y, z)$ is the density of the substance.

It is known from experiments that over a wide range of temperatures the quantity of heat flowing across a distinguished area $\mathrm{d}\boldsymbol{\sigma} = \mathbf{n}\,\mathrm{d}\sigma$ per unit time as the result of heat transfer is proportional to the flux $-\operatorname{grad} T \cdot \mathrm{d}\boldsymbol{\sigma}$ of the field $-\operatorname{grad} T$ across that area (the gradient is taken with respect to the spatial variables x, y, z). The coefficient of proportionality k depends on the substance and is called its *coefficient of thermal conductivity*. The negative sign in front of $\operatorname{grad} T$ corresponds to the fact that the energy flows from hotter parts of the body to cooler parts. Thus, the energy flux (up to terms of order $o(\Delta t)$)

$$\Delta t \iint_S -k \operatorname{grad} T \cdot \mathrm{d}\boldsymbol{\sigma} \tag{14.64}$$

takes place across the boundary S of D in the direction of the external normal over the time interval Δt.

Equating the quantity (14.63) to the negative of the quantity (14.64), dividing by Δt, and passing to the limit as $\Delta t \to 0$, we obtain

$$\iiint_D c\rho \frac{\partial T}{\partial t}\, dV = \iint_S k\, \text{grad}\, T \cdot d\boldsymbol{\sigma}. \qquad (14.65)$$

This equality is the equation for the function T. Assuming T is sufficiently smooth, we transform (14.65) using the Gauss–Ostrogradskii formula:

$$\iiint_D c\rho \frac{\partial T}{\partial t}\, dV = \iiint_D \text{div}(k\, \text{grad}\, T)\, dV.$$

Hence, since D is arbitrary, it follows obviously that

$$c\rho \frac{\partial T}{\partial t} = \text{div}(k\, \text{grad}\, T). \qquad (14.66)$$

We have now obtained the differential version of the integral equation (14.65).

If there were heat sources (or sinks) in D whose intensities have density $F(x, y, z, t)$, instead of (14.65) we would write the equality

$$\iiint_D c\rho \frac{\partial T}{\partial t}\, dV = \iint_S k\, \text{grad}\, T \cdot d\boldsymbol{\sigma} + \iiint_D F\, dV, \qquad (14.65')$$

and then instead of (14.66) we would have the equation

$$c\rho \frac{\partial T}{\partial t} = \text{div}(k\, \text{grad}\, T) + F. \qquad (14.66')$$

If the body is assumed isotropic and homogeneous with respect to its heat conductivity, the coefficient k in (14.66) will be constant, and the equation will transform to the canonical form

$$\frac{\partial T}{\partial t} = a^2 \Delta T + f, \qquad (14.67)$$

where $f = \frac{F}{c\rho}$ and $a^2 = \frac{k}{c\rho}$ is the *coefficient of thermal diffusivity*. The equation (14.67) is usually called the *heat equation*.

In the case of steady-state heat transfer, in which the field T is independent of time, this equation becomes *Poisson's equation*

$$\Delta T = \varphi, \qquad (14.68)$$

where $\varphi = -\frac{1}{a^2} f$; and if in addition there are no heat sources in the body, the result is *Laplace's equation*

$$\Delta T = 0. \qquad (14.69)$$

The solutions of Laplace's equation, as already noted, are called *harmonic functions*. In the thermophysical interpretation, harmonic functions correspond to steady-state temperature fields in a body in which the heat flows occur without any sinks or sources inside the body itself, that is, all sources are located outside the body. For example, if we maintain a steady temperature distribution $T|_{\partial V} = \tau$ over the boundary ∂V of a body, then the temperature field in the body V will eventually stabilize in the form of a harmonic function T. Such an interpretation of the solutions of the Laplace equation (14.69) enables us to predict a number of properties of harmonic functions. For example, one must presume that a harmonic function in V cannot have local maxima inside the body; otherwise heat would only flow away from these hotter portions of the body, and they would cool off, contrary to the assumption that the field is stationary.

14.4.2 The Equation of Continuity

Let $\rho = \rho(x, y, z, t)$ be the density of a material medium that fills a space being observed and $\mathbf{v} = \mathbf{v}(x, y, z, t)$ the velocity field of motion of the medium as function of the point of space (x, y, z) and the time t.

From the law of conservation of mass, using the Gauss–Ostrogradskii formula, we can find an interconnection between these quantities.

Let D be a domain in the space being observed bounded by a surface S. Over the time interval Δt the quantity of matter in D varies by an amount

$$\iiint_D \big(\rho(x, y, z, t + \Delta t) - \rho(x, y, z, t) \big) \, dV.$$

Over this small time interval Δt, the flow of matter across the surface S in the direction of the outward normal to S is (up to $o(\Delta t)$)

$$\Delta t \cdot \iint_S \rho \mathbf{v} \cdot d\boldsymbol{\sigma}.$$

If there were no sources or sinks in D, then by the law of conservation of matter, we would have

$$\iiint_D \Delta \rho \, dV = -\Delta t \iint_S \rho \mathbf{v} \cdot d\boldsymbol{\sigma}$$

or, in the limit as $\Delta t \to 0$

$$\iiint_D \frac{\partial \rho}{\partial t} \, dV = - \iint_S \rho \mathbf{v} \cdot d\boldsymbol{\sigma}.$$

Applying the Gauss–Ostrogradskii formula to the right-hand side of this equality and taking account of the fact that D is an arbitrary domain, we conclude that the following relation must hold for sufficiently smooth functions ρ and $\mathbf{v}$:

$$\frac{\partial \rho}{\partial t} = -\operatorname{div}(\rho \mathbf{v}), \tag{14.70}$$

called the *equation of continuity* of a continuous medium.

In vector notation the equation of continuity can be written as

$$\frac{\partial \rho}{\partial t} + \nabla \cdot (\rho v) = 0, \tag{14.70'}$$

or, in more expanded form,

$$\frac{\partial \rho}{\partial t} + v \cdot \nabla \rho + \rho \nabla \cdot v = 0. \tag{14.70''}$$

If the medium is incompressible (a liquid), the volumetric outflow of the medium across a closed surface S must be zero:

$$\iint_S v \cdot d\sigma = 0,$$

from which (again on the basis of the Gauss–Ostrogradskii formula) it follows that for an incompressible medium

$$\text{div}\, v = 0. \tag{14.71}$$

Hence, for an incompressible medium of variable density (a mixture of water and oil) Eq. (14.70'') becomes

$$\frac{\partial \rho}{\partial t} + v \cdot \nabla \rho = 0. \tag{14.72}$$

If the medium is also homogeneous, then $\nabla \rho = 0$ and therefore $\frac{\partial \rho}{\partial t} = 0$.

14.4.3 The Basic Equations of the Dynamics of Continuous Media

We shall now derive the equations of the dynamics of a continuous medium moving in space. Together with the functions ρ and v already considered, which will again denote the density and the velocity of the medium at a given point (x, y, z) of space and at a given instant t of time, we consider the pressure $p = p(x, y, z, t)$ as a function of a point of space and time.

In the space occupied by the medium we distinguish a domain D bounded by a surface S and consider the forces acting on the distinguished volume of the medium at a fixed instant of time.

Certain force fields (for example, gravitation) are acting on each element $\rho\, dV$ of mass of the medium. These fields create the so-called *mass forces*. Let $\mathbf{F} = \mathbf{F}(x, y, z, t)$ be the density of the external fields of mass force. Then a force $\mathbf{F}\rho\, dV$ acts on the element from the direction of these fields. If this element has an acceleration $\mathbf{a}$ at a given instant of time, then by Newton's second law, this is equivalent to the presence of another mass force called inertia, equal to $-\mathbf{a}\rho\, dV$.

Finally, on each element $d\boldsymbol{\sigma} = \mathbf{n}\, d\sigma$ of the surface S there is a surface tension due to the pressure of the particles of the medium near those in D, and this surface force equals $-p\, d\boldsymbol{\sigma}$ (where $\mathbf{n}$ is the outward normal to S).

By d'Alembert's principle, at each instant during the motion of any material system, all the forces applied to it, including inertia, are in mutual equilibrium, that is, the force required to balance them is zero. In our case, this means that

$$\iiint_D (\mathbf{F} - \mathbf{a})\rho \, dV - \iint_S p \, d\boldsymbol{\sigma} = 0. \tag{14.73}$$

The first term in this sum is the equilibrant of the mass and inertial forces, and the second is the equilibrant of the pressure on the surface S bounding the volume. For simplicity we shall assume that we are dealing with an ideal (nonviscous) fluid or gas, in which the pressure on the surface $d\boldsymbol{\sigma}$ has the form $p \, d\boldsymbol{\sigma}$, where the number p is independent of the orientation of the area in the space.

Applying formula (14.47) from Sect. 14.2, we find by (14.73) that

$$\iiint_D (\mathbf{F} - \mathbf{a})\rho \, dV - \iiint_D \text{grad } p \, dv = \mathbf{0},$$

from which, since the domain D is arbitrary, it follows that

$$\rho\mathbf{a} = \rho\mathbf{F} - \text{grad } p. \tag{14.74}$$

In this local form the equation of motion of the medium corresponds perfectly to Newton's law of motion for a material particle.

The acceleration $\mathbf{a}$ of a particle of the medium is the derivative $\frac{d\mathbf{v}}{dt}$ of the velocity $\mathbf{v}$ of the particle. If $x = x(t)$, $y = y(t)$, $z = z(t)$ is the law of motion of a particle in space and $\mathbf{v} = \mathbf{v}(x, y, z, t)$ is the velocity field of the medium, then for each individual particle we obtain

$$\mathbf{a} = \frac{d\mathbf{v}}{dt} = \frac{\partial\mathbf{v}}{\partial t} + \frac{\partial\mathbf{v}}{\partial x}\frac{dx}{dt} + \frac{\partial\mathbf{v}}{\partial y}\frac{dy}{dt} + \frac{\partial\mathbf{v}}{\partial z}\frac{dz}{dt}$$

or

$$\mathbf{a} = \frac{\partial\mathbf{v}}{\partial t} + (\mathbf{v} \cdot \nabla)\mathbf{v}.$$

Thus the equation of motion (14.74) assumes the following form

$$\frac{d\mathbf{v}}{dt} = \mathbf{F} - \frac{1}{\rho}\text{grad } p \tag{14.75}$$

or

$$\frac{\partial\mathbf{v}}{\partial t} + (\mathbf{v} \cdot \nabla)\mathbf{v} = \mathbf{F} - \frac{1}{\rho}\nabla p. \tag{14.76}$$

Equation (14.76) is usually called *Euler's hydrodynamic equation*.

The vector equation (14.76) is equivalent to a system of three scalar equations for the three components of the vector $\mathbf{v}$ and the pair of functions ρ and p.

Thus, Euler's equation does not completely determine the motion of an ideal continuous medium. To be sure, it is natural to adjoin to it the equation of continuity (14.70), but even then the system is underdetermined.

To make the motion of the medium determinate one must also add to Eqs. (14.70) and (14.76) some information on the thermodynamic state of the medium (for example, the equation of state $f(p, \rho, T) = 0$ and the equation for heat transfer). The reader may obtain some idea of what these relations can yield in the final subsection of this section.

14.4.4 The Wave Equation

We now consider the motion of a medium corresponding to the propagation of an acoustic wave. It is clear that such a motion is also subject to Eq. (14.76); this equation can be simplified due to the specifics of the phenomenon.

Sound is an alternating state of rarefaction and compression of a medium, the deviation of the pressure from its mean value in a sound wave being very small – of the order of 1 %. Therefore acoustic motion consists of small deviations of the elements of volume of the medium from the equilibrium position at small velocities. However, the rate of propagation of the disturbance (wave) through the medium is comparable with the mean velocity of motion of the molecules of the medium and usually exceeds the rate of heat transfer between the different parts of the medium under consideration. Thus, an acoustic motion of a volume of gas can be regarded as small oscillations about the equilibrium position occurring without heat transfer (an adiabatic process).

Neglecting the term $(\mathbf{v} \cdot \nabla)\mathbf{v}$ in the equation of motion (14.76) in view of the small size of the macroscopic velocities $\mathbf{v}$, we obtain the equality

$$\rho \frac{\partial \mathbf{v}}{\partial t} = \rho \mathbf{F} - \nabla p.$$

If we neglect the term of the form $\frac{\partial \rho}{\partial t}\mathbf{v}$ for the same reason, the last equality reduces to the equation

$$\frac{\partial}{\partial t}(\rho \mathbf{v}) = \rho \mathbf{F} - \nabla p.$$

Applying the operator ∇ (on x, y, z coordinates) to it, we obtain

$$\frac{\partial}{\partial t}(\nabla \cdot \rho \mathbf{v}) = \nabla \cdot \rho \mathbf{F} - \Delta p.$$

Using the equation of continuity (14.70′) and introducing the notation $\nabla \cdot \rho \mathbf{F} = -\Phi$, we arrive at the equation

$$\frac{\partial^2 \rho}{\partial t^2} = \Phi + \Delta p. \tag{14.77}$$

If we can neglect the influence of the exterior fields, Eq. (14.77) reduces to the relation

$$\frac{\partial^2 \rho}{\partial t^2} = \Delta p \qquad (14.78)$$

between the density and pressure in the acoustic medium. Since the process is adiabatic, the equation of state $f(p, \rho, T) = 0$ reduces to a relation $\rho = \psi(p)$, from which it follows that $\frac{\partial^2 \rho}{\partial t^2} = \psi'(p)\frac{\partial^2 p}{\partial t^2} + \psi''(p)(\frac{\partial p}{\partial t})^2$ Since the pressure oscillations are small in an acoustic wave, one may assume that $\psi'(p) \equiv \psi'(p_0)$, where p_0 is the equilibrium pressure. Then $\psi'' = 0$ and $\frac{\partial^2 \rho}{\partial t^2} \approx \psi'(p)\frac{\partial^2 p}{\partial t^2}$. Taking this into account, from (14.78) we finally obtain

$$\frac{\partial^2 p}{\partial t^2} = a^2 \Delta p, \qquad (14.79)$$

where $a = (\psi'(p_0))^{-1/2}$. This equation describes the variation in pressure in a medium in a state of acoustic motion. Equation (14.79) describes the simplest wave process in a continuous medium. It is called the *homogeneous wave equation*. The quantity a has a simple physical meaning: it is the speed of propagation of an acoustic disturbance in the medium, that is, the speed of sound in it (see Problem 4).

In the case of forced oscillations, when certain forces are acting on each element of volume of the medium, the three-dimensional density of whose distribution is given, Eq. (14.79) is replaced by the relation

$$\frac{\partial^2 p}{\partial t^2} = a^2 \Delta p + f \qquad (14.80)$$

corresponding to Eq. (14.77), which for $f \not\equiv 0$ is called the *inhomogeneous wave equation*.

14.4.5 Problems and Exercises

1. Suppose the velocity field $\mathbf{v}$ of a moving continuous medium is a potential field. Show that if the medium is incompressible, the potential φ of the field $\mathbf{v}$ is a harmonic function, that is, $\Delta\varphi = 0$ (see (14.71)).

2. a) Show that Euler's equation (14.76) can be rewritten as

$$\frac{\partial \mathbf{v}}{\partial t} + \mathrm{grad}\left(\frac{1}{2}\mathbf{v}^2\right) - \mathbf{v} \times \mathrm{curl}\,\mathbf{v} = \mathbf{F} - \frac{1}{\rho}\,\mathrm{grad}\,p$$

(see Problem 1 of Sect. 14.1).

b) Verify on the basis of the equation of a) that an irrotational flow ($\mathrm{curl}\,\mathbf{v} = \mathbf{0}$) of a homogeneous incompressible liquid can occur only in a potential field $\mathbf{F}$.

c) It turns out (Lagrange's theorem) that if at some instant the flow in a potential field $\mathbf{F} = \mathrm{grad}\,U$ is irrotational, then it always has been and always will be irrotational. Such a flow consequently is at least locally a potential flow, that is, $\mathbf{v} = \operatorname{grad} \varphi$. Verify that for a potential flow of a homogeneous incompressible liquid taking place in a potential field $\mathbf{F}$, the following relation holds at each instant of time:

$$\operatorname{grad}\left(\frac{\partial \varphi}{\partial t} + \frac{v^2}{2} + \frac{p}{\rho} - U\right) = 0.$$

d) Derive the so-called *Cauchy integral* from the equality just obtained:

$$\frac{\partial \varphi}{\partial t} + \frac{v^2}{2} + \frac{p}{\rho} - U = \Phi(t),$$

a relation that asserts that the left-hand side is independent of the spatial coordinates.

e) Show that if the flow is also steady-state, that is, the field $\mathbf{v}$ is independent of time, the following relation holds

$$\frac{v^2}{2} + \frac{p}{\rho} - U = \text{const},$$

called the *Bernoulli integral*.

3. A flow whose velocity field has the form $\mathbf{v} = (v_x, v_y, 0)$ is naturally called *plane-parallel* or simply a *planar flow*.

a) Show that the conditions $\operatorname{div} \mathbf{v} = 0$, $\operatorname{curl} \mathbf{v} = 0$ for a flow to be incompressible and irrotational have the following forms:

$$\frac{\partial v_x}{\partial x} + \frac{\partial v_y}{\partial y} = 0, \qquad \frac{\partial v_x}{\partial y} - \frac{\partial v_y}{\partial x} = 0.$$

b) Show that these equations at least locally guarantee the existence of functions $\psi(x, y)$ and $\varphi(x, y)$ such that $(-v_y, v_x) = \operatorname{grad} \psi$ and $(v_x, v_y) = \operatorname{grad} \varphi$.

c) Verify that the level curves $\varphi = c_1$ and $\psi = c_2$ of these functions are orthogonal and show that in the steady-state flow the curves $\psi = c$ coincide with the trajectories of the moving particles of the medium. It is for that reason that the function ψ is called the *current function*, in contrast to the function φ, which is the *velocity potential*.

d) Show, assuming that the functions φ and ψ are sufficiently smooth, that they are both harmonic functions and satisfy the *Cauchy–Riemann equations*:

$$\frac{\partial \varphi}{\partial x} = \frac{\partial \psi}{\partial y}, \qquad \frac{\partial \varphi}{\partial y} = -\frac{\partial \psi}{\partial x}.$$

Harmonic functions satisfying the Cauchy–Riemann equations are called *conjugate harmonic functions*.

e) Verify that the function $f(z) = (\varphi + i\psi)(x, y)$, where $z = x + iy$, is a differentiable function of the complex variable z. This determines the connection of the planar problems of hydrodynamics with the theory of functions of a complex variable.

4. Consider the elementary version $\frac{\partial^2 p}{\partial t^2} = a^2 \frac{\partial^2 p}{\partial x^2}$ of the wave equation (14.79). This is the case of a plane wave in which the pressure depends only on the x-coordinate of the point (x, y, z) of space.

a) By making the change of variable $u = x - at$, $v = x + at$, reduce this equation to the form $\frac{\partial^2 p}{\partial u \partial v} = 0$ and show that the general form of the solution of the original equation is $p = f(x + at) + g(x - at)$, where f and g are arbitrary functions of class $C^{(2)}$.

b) Interpret the solution just obtained as two waves $f(x)$ and $g(x)$ propagating left and right along the x-axis with velocity a.

c) Assuming that the quantity a is the velocity of propagation of a disturbance even in the general case (14.79), and taking account of the relation $a = (\psi'(p_0))^{-1/2}$, find, following Newton, the velocity c_N of sound in air, assuming that the temperature in an acoustic wave is constant, that is, assuming that the process of acoustic oscillation is isothermic. (The equation of state is $\rho = \frac{\mu p}{RT}$, $R = 8.31 \frac{J}{\text{deg·mole}}$ is the universal gas constant, and $\mu = 28.8 \frac{g}{\text{mole}}$ is the molecular weight of air. Carry out the computation for air at a temperature of 0 °C, that is, $T = 273$ K. Newton found that $c_N = 280$ m/s.)

d) Assuming that the process of acoustic vibrations is adiabatic, find, following Laplace, the velocity c_L of sound in air, and thereby sharpen Newton's result c_N. (In an adiabatic process $p = c\rho^{\gamma}$. This is Poisson's formula from Problem 6 of Sect. 13.1. Show that if $c_N = \sqrt{\frac{p}{\rho}}$, then $c_L = \sqrt{\gamma \frac{p}{\varrho}}$. For air $\gamma \approx 1.4$. Laplace found $c_L = 330$ m/s, which is in excellent agreement with experiment.)

5. Using the scalar and vector potentials one can reduce the Maxwell equations ((14.12) of Sect. 14.1) to the wave equation (more precisely, to several wave equations of the same type). By solving this problem, you will verify this statement.

a) It follows from the equation $\nabla \cdot \mathbf{B} = 0$ that at least locally $\mathbf{B} = \nabla \times \mathbf{A}$, where $\mathbf{A}$ is the vector potential of the field $\mathbf{B}$.

b) Knowing that $\mathbf{B} = \nabla \times \mathbf{A}$, show that the equation $\nabla \times \mathbf{E} = -\frac{\partial \mathbf{B}}{\partial t}$ implies that at least locally there exists a scalar function φ such that $\mathbf{E} = -\nabla \varphi - \frac{\partial \mathbf{A}}{\partial t}$.

c) Verify that the fields $\mathbf{E} = -\nabla \varphi - \frac{\partial \mathbf{A}}{\partial t}$ and $\mathbf{B} = \nabla \times \mathbf{A}$ do not change if instead of φ and $\mathbf{A}$ we take another pair of potentials $\tilde{\varphi}$ and $\tilde{\mathbf{A}}$ such that $\tilde{\varphi} = \varphi - \frac{\partial \psi}{\partial t}$ and $\tilde{\mathbf{A}} = \mathbf{A} + \nabla \psi$, where ψ is an arbitrary function of class $C^{(2)}$.

d) The equation $\nabla \cdot \mathbf{E} = \frac{\rho}{\varepsilon_0}$ implies the first relation $-\nabla^2 \varphi - \frac{\partial}{\partial t} \nabla \cdot \mathbf{A} = \frac{\rho}{\partial \varepsilon_0}$ between the potentials φ and $\mathbf{A}$.

e) The equation $c^2 \nabla \times \mathbf{B} - \frac{\mathbf{E}}{\partial t} = \frac{\mathbf{j}}{\partial \varepsilon_0}$ implies the second relation

$$-c^2 \nabla^2 \mathbf{A} + c^2 \nabla(\nabla \cdot \mathbf{A}) + \frac{\partial}{\partial t} \nabla \varphi + \frac{\partial^2 \mathbf{A}}{\partial t^2} = \frac{\mathbf{j}}{\varepsilon_0}$$

between the potentials φ and $\mathbf{A}$.

f) Using c), show that by solving the auxiliary wave equation $\Delta \psi + f = \frac{1}{c^2} \frac{\partial^2 \psi}{\partial t^2}$, without changing the fields $\mathbf{E}$ and $\mathbf{B}$ one can choose the potentials φ and $\mathbf{A}$ so that they satisfy the additional (so-called *gauge*) condition $\nabla \cdot \mathbf{A} = -\frac{1}{c^2} \frac{\partial \varphi}{\partial t}$.

g) Show that if the potentials φ and $\mathbf{A}$ are chosen as stated in f), then the required inhomogeneous wave equations

$$\frac{\partial^2 \varphi}{\partial t^2} = c^2 \Delta \varphi + \frac{\rho c^2}{\varepsilon_0}, \qquad \frac{\partial^2 \mathbf{A}}{\partial t^2} = c^2 \Delta \mathbf{A} + \frac{\mathbf{j}}{\varepsilon_0}$$

for the potentials φ and $\mathbf{A}$ follow from d) and e). By finding φ and $\mathbf{A}$, we also find the fields $\mathbf{E} = \nabla \varphi$, $\mathbf{B} = \nabla \times \mathbf{A}$.

Chapter 15
*Integration of Differential Forms on Manifolds

15.1 A Brief Review of Linear Algebra

15.1.1 The Algebra of Forms

Let X be a vector space and $F^k : X^k \to \mathbb{R}$ a real-valued k-form on X. If $e_1, \ldots, e_n$ is a basis in X and $x_1 = x^{i_1} e_{i_1}, \ldots, x_k = x^{i_k} e_{i_k}$ is the expansion of the vectors $x_1, \ldots, x_k \in X$ with respect to this basis, then by the linearity of F^k, with respect to each argument

$$F^k(x_1, \ldots, x_k) = F^k\left(x^{i_1} e_{i_1}, \ldots, x^{i_k} e_{i_k}\right) =$$
$$= F^k(e_{i_1}, \ldots, e_{i_k}) x^{i_1} \cdot \ldots \cdot x^{i_k} = a_{i_1 \ldots i_k} x^{i_1} \cdot \ldots \cdot x^{i_k}. \quad (15.1)$$

Thus, after a basis is given in X, one can identify the k-form $F^k : X^k \to \mathbb{R}$ with the set of numbers $a_{i_1 \ldots i_k} = F^k(e_{i_1}, \ldots, e_{i_k})$.

If $\tilde{e}_1, \ldots, \tilde{e}_n$ is another basis in X and $\tilde{a}_{j_1 \ldots j_k} = F^k(\tilde{e}_{j_1}, \ldots, \tilde{e}_{j_k})$, then, setting $\tilde{e}_j = c^i_j e_i$, $j = 1, \ldots, n$, we find the (tensor) law

$$\tilde{a}_{j_1 \ldots j_k} = F^k\left(c^{i_1}_{j_1} e_{i_1}, \ldots, c^{i_k}_{j_k} e_{i_k}\right) = a_{i_1 \ldots i_k} c^{i_1}_{j_1} \cdot \ldots \cdot c^{i_k}_{j_k} \quad (15.2)$$

for transformation of the number sets $a_{i_1 \ldots i_k}, \tilde{a}_{j_1 \ldots j_k}$ corresponding to the same form F^k.

The set $\mathcal{F}^k := \{F^k : X^k \to \mathbb{R}\}$ of k-forms on a vector space X is itself a vector space relative to the standard operations

$$\left(F^k_1 + F^k_2\right)(x) := F^k_1(x) + F^k_2(x), \quad (15.3)$$

$$\left(\lambda F^k\right)(x) := \lambda F^k(x) \quad (15.4)$$

of addition of k-forms and multiplication of a k-form by a scalar.

© Springer-Verlag Berlin Heidelberg 2016
V.A. Zorich, *Mathematical Analysis II*, Universitext,
DOI 10.1007/978-3-662-48993-2_7

For forms F^k and F^l of arbitrary degrees k and l the following *tensor product* operation $\otimes$ is defined:

$$\left(F^k \otimes F^l\right)(x_1, \ldots, x_k, x_{k+1}, \ldots, x_{k+l}) :=$$
$$= F^k(x_1, \ldots, x_k) F^l(x_{k+1}, \ldots, x_{k+l}). \tag{15.5}$$

Thus $F^k \otimes F^l$ is a form F^{k+l} of degree $k+l$. The following relations are obvious:

$$\left(\lambda F^k\right) \otimes F^l = \lambda\left(F^k \otimes F^l\right), \tag{15.6}$$

$$\left(F_1^k + F_2^k\right) \otimes F^l = F_1^k \otimes F^l + F_2^k \otimes F^l, \tag{15.7}$$

$$F^k \otimes \left(F_1^l + F_2^l\right) = F^k \otimes F_1^l + F^k \otimes F_2^l, \tag{15.8}$$

$$\left(F^k \otimes F^l\right) \otimes F^m = F^k \otimes \left(F^l \otimes F^m\right). \tag{15.9}$$

Thus the set $\mathcal{F} = \{\mathcal{F}^k\}$ of forms on the vector space X is a graded algebra $\mathcal{F} = \bigoplus_k \mathcal{F}^k$ with respect to these operations, in which the vector-space operations are carried out inside each space $\mathcal{F}^k$ occurring in the direct sum, and if $F^k \in \mathcal{F}^k$, $F^l \in \mathcal{F}^l$, then $F^k \otimes F^l \in \mathcal{F}^{k+l}$.

Example 1 Let X^* be the dual space to X (consisting of the linear functionals on X) and $e^1, \ldots, e^n$ the basis of X^* dual to the basis $e_1, \ldots, e_n$ in X, that is, $e^i(e_j) = \delta_j^i$.

Since $e^i(x) = e^i(x^j e_j) = x^j e^i(e_j) = x^j \delta_j^i = x^i$, taking account of (15.1) and (15.9), we can write any k-form $F^k : X^k \to \mathbb{R}$ as

$$F^k = a_{i_1 \ldots i_k} e^{i_1} \otimes \cdots \otimes e^{i_k}. \tag{15.10}$$

15.1.2 The Algebra of Skew-Symmetric Forms

Let us now consider the space Ω^k of skew-symmetric forms in $\mathcal{F}^k$, that is, $\omega \in \Omega^k$ if the equality

$$\omega(x_1, \ldots, x_i, \ldots, x_j, \ldots, x_k) = -\omega(x_1, \ldots, x_j, \ldots, x_i, \ldots, x_k)$$

holds for any distinct indices $i, j \in \{1, \ldots, n\}$.

From any form $F^k \in \mathcal{F}^k$ one can obtain a skew-symmetric form using the operation $A : \mathcal{F} \to \Omega^k$ of *alternation*, defined by the relation

$$A F^k(x_1, \ldots, x_k) := \frac{1}{k!} F^k(x_{i_1}, \ldots, x_{i_k}) \delta_{1 \ldots k}^{i_1 \ldots i_k}, \tag{15.11}$$

where

$$
\delta_{1\ldots k}^{i_1\ldots i_k} =
\begin{cases}
1, & \text{if the permutation } \begin{pmatrix} i_1 & \cdots & i_k \\ 1 & \cdots & k \end{pmatrix} \text{ is even,} \\[2mm]
-1, & \text{if the permutation } \begin{pmatrix} i_1 & \cdots & i_k \\ 1 & \cdots & k \end{pmatrix} \text{ is odd,} \\[2mm]
0, & \text{if } \begin{pmatrix} i_1 & \cdots & i_k \\ 1 & \cdots & k \end{pmatrix} \text{ is not a permutation.}
\end{cases}
$$

If F^k is a skew-symmetric form, then, as one can see from (15.11), $AF^k = F^k$. Thus $A(AF^k) = AF^k$ and $A\omega = \omega$ if $\omega \in \Omega^k$. Hence $A : \mathcal{F}^k \to \Omega^k$ is a mapping of $\mathcal{F}^k$ onto Ω^k.

Comparing Definitions (15.3), (15.4), and (15.11), we obtain

$$A\bigl(F_1^k + F_2^k\bigr) = AF_1^k + AF_2^k, \tag{15.12}$$

$$A\bigl(\lambda F^k\bigr) = \lambda AF^k. \tag{15.13}$$

Example 2 Taking account of relations (15.12) and (15.13), we find by (15.10) that

$$AF^k = a_{i_1\ldots i_k} A\bigl(e^{i_1} \otimes \cdots \otimes e^{i_k}\bigr),$$

so that it is of interest to find $A(e^{i_1} \otimes \cdots \otimes e^{i_k})$.

From Definition (15.11), taking account of the relation $e^i(x) = x^i$, we find

$$
A\bigl(e^{j_1} \otimes \cdots \otimes e^{j_k}\bigr)(x_1,\ldots,x_k) =
$$

$$
= \frac{1}{k!} e^{j_1}(x_{i_1}) \cdot \ldots \cdot e^{j_k}(x_{i_k}) \delta_{1\ldots k}^{i_1\ldots i_k} =
$$

$$
= \frac{1}{k!} x_{i_1}^{j_1} \cdot \ldots \cdot x_{i_k}^{j_k} \delta_{1\ldots k}^{i_1\ldots i_k} = \frac{1}{k!}
\begin{vmatrix}
x_1^{j_1} & \cdots & x_1^{j_k} \\
\vdots & \ddots & \vdots \\
x_k^{j_1} & \cdots & x_k^{j_k}
\end{vmatrix}. \tag{15.14}
$$

The tensor product of skew-symmetric forms is in general not skew-symmetric, so that we introduce the following *exterior product* in the class of skew-symmetric forms:

$$\omega^k \wedge \omega^l := \frac{(k+l)!}{k!\,l!} A\bigl(\omega^k \otimes \omega^l\bigr). \tag{15.15}$$

Thus $\omega^k \wedge \omega^l$ is a skew-symmetric form ω^{k+l} of degree $k + l$.

Example 3 Based on the result (15.14) of Example 2, we find by Definition (15.15) that

$$e^{i_1} \wedge e^{i_2}(x_1, x_2) = \frac{2!}{1!1!} A\left(e^{i_1} \otimes e^{i_2}\right)(x_1, x_2) =$$

$$= \begin{vmatrix} e^{i_1}(x_1) & e^{i_2}(x_1) \\ e^{i_1}(x_2) & e^{i_2}(x_2) \end{vmatrix} = \begin{vmatrix} x_1^{i_1} & x_1^{i_2} \\ x_2^{i_1} & x_2^{i_2} \end{vmatrix}. \tag{15.16}$$

Example 4 Using the equality obtained in Example 3, relation (15.14), and the definitions (15.11) and (15.15), we can write

$$e^{i_1} \wedge \left(e^{i_2} \wedge e^{i_3}\right)(x_1, x_2, x_3) =$$

$$= \frac{(1+2)!}{1!2!} A\left(e^{i_1} \otimes \left(e^{i_2} \otimes e^{i_3}\right)\right)(x_1, x_2, x_3) =$$

$$= \frac{3!}{1!2!} e^{i_1}(x_{j_1})\left(e^{i_2} \wedge e^{i_3}\right)(x_{j_2}, x_{j_3}) \delta^{j_1 j_2 j_3}_{1\,2\,3} = \frac{1}{2!} x_{j_1}^{i_1} \begin{vmatrix} x_{j_2}^{i_2} & x_{j_2}^{i_3} \\ x_{j_3}^{i_2} & x_{j_3}^{i_3} \end{vmatrix} \delta^{j_1 j_2 j_3}_{1\,2\,3} =$$

$$= x_1^{i_1} \begin{vmatrix} x_2^{i_2} & x_2^{i_3} \\ x_3^{i_2} & x_3^{i_3} \end{vmatrix} - x_2^{i_1} \begin{vmatrix} x_1^{i_2} & x_1^{i_3} \\ x_3^{i_2} & x_3^{i_3} \end{vmatrix} + x_3^{i_1} \begin{vmatrix} x_1^{i_2} & x_1^{i_3} \\ x_2^{i_2} & x_2^{i_3} \end{vmatrix} =$$

$$= \begin{vmatrix} x_1^{i_1} & x_1^{i_2} & x_1^{i_3} \\ x_2^{i_1} & x_2^{i_2} & x_2^{i_3} \\ x_3^{i_1} & x_3^{i_2} & x_3^{i_3} \end{vmatrix}.$$

A similar computation shows that

$$e^{i_1} \wedge \left(e^{i_2} \wedge e^{i_3}\right) = \left(e^{i_1} \wedge e^{i_2}\right) \wedge e^{i_3}. \tag{15.17}$$

Using the expansion of the determinant along a column, we conclude by induction that

$$e^{i_1} \wedge \cdots \wedge e^{i_k}(x_1, \ldots, x_k) = \begin{vmatrix} e^{i_1}(x_1) & \cdots & e^{i_k}(x_1) \\ \vdots & \ddots & \vdots \\ e^{i_1}(x_k) & \cdots & e^{i_k}(x_k) \end{vmatrix}, \tag{15.18}$$

and, as one can see from the computations just carried out, formula (15.18) holds for any 1-forms $e^{i_1}, \ldots, e^{i_k}$ (not just the basis forms of the space X^*).

Taking the properties of the tensor product and the alternation operation listed above into account, we obtain the following properties of the exterior product of skew-symmetric forms:

$$\left(\omega_1^k + \omega_2^k\right) \wedge \omega^l = \omega_1^k \wedge \omega^l + \omega_2^k \wedge \omega^l, \tag{15.19}$$

$$\left(\lambda \omega^k\right) \wedge \omega^l = \lambda\left(\omega^k \wedge \omega^l\right), \tag{15.20}$$

$$\omega^k \wedge \omega^l = (-1)^{kl} \omega^l \wedge \omega^k, \tag{15.21}$$

$$\left(\omega^k \wedge \omega^l\right) \wedge \omega^m = \omega^k \wedge \left(\omega^l \wedge \omega^m\right). \tag{15.22}$$

Proof Equalities (15.19) and (15.20) follow obviously from relations (15.6)–(15.8) and (15.12) and (15.13).

From relations (15.10)–(15.14) and (15.17), for every skew-symmetric form $\omega = a_{i_1 \ldots i_k} e^{i_1} \otimes \cdots \otimes e^{i_k}$ we obtain

$$\omega = A\omega = a_{i_1 \ldots i_k} A\left(e^{i_1} \otimes \cdots \otimes e^{i_k}\right) = \frac{1}{k!} a_{i_1 \ldots i_k} e^{i_1} \wedge \cdots \wedge e^{i_k}.$$

Using the equalities (15.19) and (15.20) we see that it now suffices to verify (15.21) and (15.22) for the forms $e^{i_1} \wedge \cdots \wedge e^{i_k}$.

Associativity (15.22) for such forms was already established by (15.17).

We now obtain (15.21) immediately from (15.18) and the properties of determinants for these particular forms.   □

Along the way we have shown that every form $\omega \in \Omega^k$ can be represented as

$$\omega = \sum_{1 \le i_1 < i_2 < \cdots < i_k \le n} a_{i_1 \ldots i_k} e^{i_1} \wedge \cdots \wedge e^{i_k}. \tag{15.23}$$

Thus, the set $\Omega = \{\Omega^k\}$ of skew-symmetric forms on the vector space X relative to the linear vector-space operations (15.3) and (15.4) and the exterior multiplication (15.15) is a graded algebra $\Omega = \bigoplus_{k=0}^{\dim X} \Omega^k$. The vector-space operations on Ω are carried out inside each vector space Ω^k, and if $\omega^k \in \Omega^k$, $\omega^l \in \Omega^l$, then $\omega^k \wedge \omega^l \in \Omega^{k+1}$.

In the direct sum $\oplus \Omega^k$ the summation runs from zero the dimension of the space X, since the skew-symmetric forms $\omega^k : X^k \to \mathbb{R}$ of degree larger than the dimension of X are necessarily identically zero, as one can see by (15.21) (or from relations (15.23) and (15.8)).

15.1.3 Linear Mappings of Vector Spaces and the Adjoint Mappings of the Conjugate Spaces

Let X and Y be vector spaces over the field $\mathbb{R}$ of real numbers (or any other field, so long as it is the same field for both X and Y), and let $l : X \to Y$ be a linear mapping of X into Y, that is, for every $x, x_1, x_2 \in X$ and every $\lambda \in \mathbb{R}$,

$$l(x_1 + x_2) = l(x_1) + l(x_2) \quad \text{and} \quad l(\lambda x) = \lambda l(x). \tag{15.24}$$

A linear mapping $l : X \to Y$ naturally generates its adjoint mapping $l^* : \mathcal{F}_Y \to \mathcal{F}_X$ from the set of linear functionals on $Y(\mathcal{F}_Y)$ into the analogous set $\mathcal{F}_X$. If F_Y^k is a k-form on Y, then by definition

$$(l^* F_Y^k)(x_1, \ldots, x_k) := F_Y^k(lx_1, \ldots, lx_k). \tag{15.25}$$

It can be seen by (15.24) and (15.25) that $l^* F_Y^k$ is a k-form F_X^k on X, that is, $l^*(\mathcal{F}_Y^k) \subset \mathcal{F}_X^k$. Moreover, if the form F_Y^k was skew-symmetric, then $(l^* F_Y^k) = F_X^k$ is also skew-symmetric, that is, $l^*(\Omega_Y^k) \subset \Omega_X^k$. Inside each vector space $\mathcal{F}_Y^k$ and Ω_Y^k the mapping l^* is obviously linear, that is,

$$l^*(F_1^k + F_2^k) = l^* F_1^k + l^* F_2^k \quad \text{and} \quad l^*(\lambda F^k) = \lambda l^* F^k. \tag{15.26}$$

Now comparing definition (15.25) with the definitions (15.5), (15.11), and (15.15) of the tensor product, alternation, and exterior product of forms, we conclude that

$$l^*(F^p \otimes F^q) = (l^* F^p) \otimes (l^* F^q), \tag{15.27}$$

$$l^*(AF^p) = A(l^* F^p), \tag{15.28}$$

$$l^*(\omega^p \wedge \omega^q) = (l^* \omega^p) \wedge (l^* \omega^q). \tag{15.29}$$

Example 5 Let $e_1, \ldots, e_m$ be a basis in X, $\tilde{e}_1, \ldots, \tilde{e}_n$ a basis in Y, and $l(e_i) = c_i^j \tilde{e}_j$, $i \in \{1, \ldots, m\}$, $j \in \{1, \ldots, n\}$. If the k-form F_Y^k has the coordinate representation

$$F_Y^k(y_1, \ldots, y_k) = b_{j_1 \ldots j_k} y_1^{j_1} \cdot \ldots \cdot y_k^{j_k}$$

in the basis $\tilde{e}_1, \ldots, \tilde{e}_n$, where $b_{j_1 \ldots j_k} = F_Y^k(\tilde{e}_{j_1}, \ldots, \tilde{e}_{j_k})$, then

$$(l^* F_Y^k)(x_1, \ldots, x_k) = a_{i_1 \ldots i_k} x_1^{i_1} \cdot \ldots \cdot x_k^{i_k},$$

where $a_{i_1 \ldots i_k} = b_{j_1 \ldots j_k} C_{i_1}^{j_1} \cdot \ldots \cdot C_{i_k}^{j_k}$, since

$$a_{i_1 \ldots i_k} =: (l^* F_Y^k)(e_{i_1}, \ldots, e_{i_k}) := F_Y^k(le_{i_1}, \ldots, le_{i_k}) =$$

$$= F_Y^k(C_{i_1}^{j_1} \tilde{e}_{j_1}, \ldots, C_{i_k}^{j_k} \tilde{e}_{j_k}) = F_Y^k(\tilde{e}_{j_1}, \ldots, \tilde{e}_{j_k}) C_{i_1}^{j_1} \cdot \ldots \cdot C_{i_k}^{j_k}.$$

Example 6 Let $e^1, \ldots, e^m$ and $\tilde{e}^1, \ldots, \tilde{e}^n$ be the bases of the conjugate spaces X^* and Y^* dual to the bases in Example 5. Under the hypotheses of Example 5 we obtain

$$(l^* \tilde{e}^j)(x) = (l^* \tilde{e}^j)(x^i e_i) = \tilde{e}^j(x^i le_i) = x^i \tilde{c}^j(c_i^k \tilde{e}_k) =$$

$$= x^i c_i^k \tilde{e}^j(\tilde{e}_k) = x^i c_i^k \delta_k^j = c_i^j x^i = c_i^j e^i(x).$$

Example 7 Retaining the notation of Example 6 and taking account of relations (15.22) and (15.29), we now obtain

$$l^* \left(\tilde{e}^{j_1} \wedge \cdots \wedge \tilde{e}^{j_k} \right) = l^* \tilde{e}^{j_1} \wedge \cdots \wedge \tilde{l}^* e^{j_k} =$$

$$= \left(c_{i_1}^{j_1} e^{i_1} \right) \wedge \cdots \wedge \left(c_{i_k}^{j_k} e^{i_k} \right) = c_{i_1}^{j_1} \cdot \ldots \cdot c_{i_k}^{j_k} e^{i_1} \wedge \cdots \wedge e^{i_k} =$$

$$= \sum_{1 \le i_1 < \cdots < i_k \le m} \begin{vmatrix} c_{i_1}^{j_1} & \cdots & c_{i_1}^{j_k} \\ \vdots & \ddots & \vdots \\ c_{i_k}^{j_1} & \cdots & c_{i_k}^{j_k} \end{vmatrix} e^{i_1} \wedge \cdots \wedge e^{i_k}.$$

Keeping Eq. (15.26) in mind, we can conclude from this that

$$l^* \left(\sum_{1 \le j_1 < \cdots < j_k \le n} b_{j_1 \ldots j_k} \tilde{e}^{j_1} \wedge \cdots \wedge \tilde{e}^{j_k} \right) =$$

$$= \sum_{\substack{1 \le i_1 < \cdots < i_k \le m \\ 1 \le j_1 < \cdots < j_k \le m}} b_{j_1 \ldots j_k} \begin{vmatrix} c_{i_1}^{j_1} & \cdots & c_{i_1}^{j_k} \\ \vdots & \ddots & \vdots \\ c_{i_k}^{j_1} & \cdots & c_{i_k}^{j_k} \end{vmatrix} e^{i_1} \wedge \cdots \wedge e^{i_k} =$$

$$= \sum_{1 \le i_1 < \cdots < i_k \le m} a_{i_1 \ldots i_k} e^{i_1} \wedge \cdots \wedge e^{i_k}.$$

15.1.4 Problems and Exercises

1. Show by examples that in general

a) $F^k \otimes F^l \ne F^l \otimes F^k$;

b) $A(F^k \otimes F^l) \ne A F^k \otimes A F^l$;

c) if $F^k, F^l \in \Omega$, then it is not always true that $F^k \otimes F^l \in \Omega$.

2. a) Show that if $e_1, \ldots, e_n$ is a basis of the vector space X and the linear functionals $e^1, \ldots, e^n$ on X (that is elements of the conjugate space X^*) are such that $e^j(e_i) = \delta_i^j$, then $e^1, \ldots, e^n$ is a basis in X^*.

b) Verify that one can always form a basis of the space $\mathcal{F}^k = \mathcal{F}^k(X)$ from k-forms of the form $e^{i_1} \otimes \cdots \otimes e^{i_k}$, and find the dimension $(\dim \mathcal{F}^k)$ of this space, knowing that $\dim X = n$.

c) Verify that one can always form a basis of the space Ω^k from forms of the form $e^{i_1} \wedge \cdots \wedge e^{i_k}$, and find $\dim \Omega^k$ knowing that $\dim X = n$.

d) Show that if $\Omega = \bigoplus_{k=0}^{k=n} \Omega^k$, then $\dim \Omega = 2^n$.

3. The *exterior (Grassmann)*[1] *algebra* G over a vector space X and a field P (usually denoted $\bigwedge(X)$ in agreement with the symbol $\wedge$ for the multiplication operation

---

[1] H. Grassmann (1809–1877) – German mathematician, physicist and philologist; in particular, he created the first systematic theory of multidimensional and Euclidean vector spaces and gave the definition of the inner product of vectors.

in G) is defined as the associative algebra with identity 1 having the following properties:

1^0 G is generated by the identity and X, that is, any subalgebra of G containing 1 and X is equal to G;

2^0 $x \wedge x = 0$ for every vector $x \in X$;

3^0 $\dim G = 2^{\dim X}$.

a) Show that if $e_1, \ldots, e_n$ is a basis in X, then the set 1, $e_1, \ldots, e_n, e_1 \wedge e_2, \ldots, e_{n-1} \wedge e_n, \ldots, e_1 \wedge \cdots \wedge e_n$ of elements of G of the form $e_{i_1} \wedge \cdots \wedge e_{i_k} = e_I$, where $I = \{i_1 < \cdots < i_k\} \subset \{1, 2, \ldots, n\}$, forms a basis in G.

b) Starting from the result in a) one can carry out the following formal construction of the algebra $G = \bigwedge(X)$.

For the subsets $I = \{i_1, \ldots, i_k\}$ of $\{1, 2, \ldots, n\}$ shown in a) we form the formal elements e_I, (by identifying $e_{\{i\}}$ with e_i, and $e_\varnothing$ with 1), which we take as a basis of the vector space G over the field P. We define multiplication in G by the formula

$$\left(\sum_I a_I e_I\right)\left(\sum_J b_J e_J\right) = \sum_{I,J} a_I b_J \varepsilon(I, J) e_{I \cup J},$$

where $\varepsilon(I, J) = \operatorname{sgn} \prod_{i \in I, j \in J}(j - i)$. Verify that the Grassmann algebra $\bigwedge(X)$ is obtained in this way.

c) Prove the uniqueness (up to isomorphism) of the algebra $\bigwedge(X)$.

d) Show that the algebra $\bigwedge(X)$ is graded: $\bigwedge(X) = \bigoplus_{k=0}^{k=n} \bigwedge^k(X)$, where $\bigwedge^k(X)$ is the linear span of the elements of the form $e_{i_1} \wedge \cdots \wedge e_{i_k}$; here if $a \in \bigwedge^p(X)$ and $b \in \bigwedge^q(X)$, then $a \wedge b \in \bigwedge^{p+q}(X)$. Verify that $a \wedge b = (-1)^{pq} b \wedge a$.

4. a) Let $A : X \to Y$ be a linear mapping of X into Y. Show that there exists a unique homomorphism $\bigwedge(A) : \bigwedge(X) \to \bigwedge(Y)$ from $\bigwedge(X)$ into $\bigwedge(Y)$ that agrees with A on the subspace $\bigwedge'(X) \subset \bigwedge(X)$ identified with X.

b) Show that the homomorphism $\bigwedge(A)$ maps $\bigwedge^k(X)$ into $\bigwedge^k(Y)$. The restriction of $\bigwedge(A)$ to $\bigwedge^k(X)$ is denoted by $\bigwedge^k(A)$.

c) Let $\{e_i : i = 1, \ldots, m\}$ be a basis in X and $\{e_j : j = 1, \ldots, n\}$ a basis in Y, and let the matrix (a_j^i) correspond to the operator A in these bases. Show that if $\{e_I : I \subset \{1, \ldots, m\}\}, \{e_J : J \subset \{1, \ldots, n\}\}$ are the corresponding bases of the spaces $\bigwedge(X)$ and $\bigwedge(Y)$, then the matrix of the operator $\bigwedge^k(A)$ has the form $a_J^I = \det(a_j^i)$, $i \in I$, $j \in J$, where $\operatorname{card} I = \operatorname{card} J = k$.

d) Verify that if $A : X \to Y$, $B : Y \to Z$ are linear operators, then the equality $\bigwedge(B \circ A) = \bigwedge(B) \circ \bigwedge(A)$ holds.

15.2 Manifolds

15.2.1 Definition of a Manifold

Definition 1 A Hausdorff topological space whose topology has a countable base[2] is called an *n-dimensional manifold* if each of its points has a neighborhood U homeomorphic either to all of $\mathbb{R}^n$ or to the half-space $H^n = \{x \in \mathbb{R}^n \mid x^1 \leq 0\}$.

Definition 2 A mapping $\varphi : \mathbb{R}^n \to U \subset M$ (or $\varphi : H^n \to U \subset M$) that realizes the homeomorphism of Definition 1 is a *local chart of the manifold* M, $\mathbb{R}^n$ (or H^n) is called the *parameter domain*, and U the *range* of the chart on the manifold M.

A local chart endows each point $x \in U$ with the coordinates of the point $t = \varphi^{-1}(x) \in \mathbb{R}^n$ corresponding to it. Thus, a local coordinate system is introduced in the region U; for that reason the mapping φ, or, in more expanded notation, the pair (U, φ) is a map of the region U in the ordinary meaning of the term.

Definition 3 A set of charts whose ranges taken together cover the entire manifold is called an *atlas* of the manifold.

Example 1 The sphere $S^2 = \{x \in \mathbb{R}^3 \mid |x| = 1\}$ is a two-dimensional manifold. If we interpret S^2 as the surface of the Earth, then an atlas of geographical maps will be an atlas of the manifold S^2.

The one-dimensional sphere $S^1 = \{x \in \mathbb{R}^2 \mid |x| = 1\}$ – a circle in $\mathbb{R}^2$ – is obviously a one-dimensional manifold. In general, the sphere $S^n = \{x \in \mathbb{R}^{n+1} \mid |x| = 1\}$ is an n-dimensional manifold. (See Sect. 12.1.)

Remark 1 The object (the manifold M) introduced by Definition 1 obviously does not change if we replace $\mathbb{R}^n$ and H^n by any parameter domains in $\mathbb{R}^n$ homeomorphic to them. For example, such a domain might be the open cube $I^n = \{x \in \mathbb{R}^n \mid 0 < x^i < 1, i = 1, \ldots, n\}$ and the cube with a face attached $\widetilde{I}^n = \{x \in \mathbb{R}^n \mid 0 < x^1 \leq 1, 0 < x^i < 1, i = 2, \ldots, n\}$. Such standard parameter domains are used quite often.

It is also not difficult to verify that the object introduced by Definition 1 does not change if we require only that each point $x \in M$ have a neighborhood U in M homeomorphic to some open subset of the half-space H^n.

Example 2 If X is an m-dimensional manifold with an atlas of charts $\{(U_\alpha, \varphi_\alpha)\}$ and Y is an n-dimensional manifold with atlas $\{(V_\beta, \psi_\beta)\}$, then $X \times Y$ can be regarded as an $(m + n)$-dimensional manifold with the atlas $\{(W_{\alpha\beta}, \chi_{\alpha\beta})\}$, where $W_{\alpha\beta} = U_\alpha \times V_\beta$ and the mapping $\chi_{\alpha\beta} = (\varphi_\alpha, \psi_\beta)$ maps the direct product of the domains of definition of φ_α and ψ_β into $W_{\alpha\beta}$.

---

[2]See Sect. 9.2 and also Remarks 2 and 3 in the present section.

Fig. 15.1

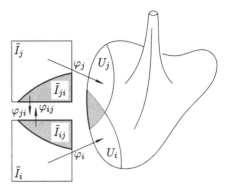

In particular, the two-dimensional torus $T^2 = S^1 \times S^1$ (Fig. 12.1) or the n-dimensional torus $T^n = \underbrace{S^1 \times \cdots \times S^1}_{n \text{ factors}}$ is a manifold of the corresponding dimension.

If the ranges U_i and U_j of two charts (U_i, φ_i) and (U_j, φ_j) of a manifold M intersect, that is, $U_i \cap U_j \neq \varnothing$, mutually inverse homeomorphisms $\varphi_{ij} : I_{ij} \to I_{ji}$ and $\varphi_{ji} : I_{ji} \to I_{ij}$ naturally arise between the sets $I_{ij} = \varphi_i^{-1}(U_j)$ and $I_{ji} = \varphi_j^{-1}(U_i)$. These homeomorphisms are given by $\varphi_{ij} = \varphi_j^{-1} \circ \varphi_i|_{I_{ij}}$ and $\varphi_{ji} = \varphi_i^{-1} \circ \varphi_j|_{I_{ji}}$. These homeomorphisms are often called *changes of coordinates*, since they effect a transition from one local coordinate system to another system of the same kind in their common range $U_i \cap U_j$ (Fig. 15.1).

Definition 4 The number n in Definition 1 is the *dimension of the manifold M* and is usually denoted dim M.

Definition 5 If a point $\varphi^{-1}(x)$ on the boundary ∂H^n of the half-space H^n corresponds to a point $x \in U$ under the homeomorphism $\varphi : H^n \to U$, then x is called a *boundary point of the manifold M* (and of the neighborhood U). The set of all boundary points of a manifold M is called the *boundary* of this manifold and is usually denoted ∂M.

By the topological invariance of interior points (Brouwer's theorem[3]) the concepts of dimension and boundary point of a manifold are unambiguously defined, that is, independent of the particular local charts used in Definitions 4 and 5. We have not proved Brouwer's theorem, but the invariance of interior points under diffeomorphisms is well-known to us (a consequence of the inverse function theorem). Since it is diffeomorphisms that we shall be dealing with, we shall not digress here to discuss Brouwer's theorem.

---

[3]This theorem asserts that under a homeomorphism $\varphi : E \to \varphi(E)$ of a set $E \subset \mathbb{R}^n$ onto a set $\varphi(E) \subset \mathbb{R}^n$ the interior points of E map to interior points of $\varphi(E)$.

Fig. 15.2

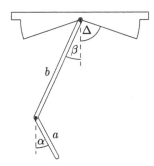

Example 3 The closed ball $\overline{B}^n = \{x \in \mathbb{R}^n \mid |x| \leq 1\}$ or, as we say, the *n-dimensional disk*, is an n-dimensional manifold whose boundary is the $(n-1)$-dimensional sphere $S^{n-1} = \{x \in \mathbb{R}^n \mid |x| = 1\}$.

Remark 2 A manifold M having a nonempty set of boundary points is usually called a *manifold with boundary*, the term *manifold* (in the proper sense of the term) being reserved for manifolds without boundary. In Definition 1 these cases are not distinguished.

Proposition 1 *The boundary ∂M of an n-dimensional manifold with boundary M is an $(n-1)$-dimensional manifold without boundary.*

Proof Indeed, $\partial H^n = \mathbb{R}^{n-1}$, and the restriction to ∂H^n of a chart of the form $\varphi_i : H^n \to U_i$ belonging to an atlas of M generates an atlas of ∂M. □

Example 4 Consider the planar double pendulum (Fig. 15.2) with arm a shorter than arm b, both being free to oscillate, except that the oscillations of b are limited in range by barriers. The configuration of such a system is characterized at each instant of time by the two angles α and β. If there were no constraints, the configuration space of the double pendulum could be identified with the two-dimensional torus $T^2 = S_\alpha^1 \times S_\beta^1$.

Under these constraints, the configuration space of the double pendulum is parametrized by the points of the cylinder $S_\alpha^1 \times I_\beta^1$, where S_α^1 is the circle, corresponding to all possible positions of the arm a, and $I_\beta^1 = \{\beta \in \mathbb{R} \mid |\beta| \leq \Delta\}$ is the interval within which the angle β may vary, characterizing the position of the arm b.

In this case we obtain a manifold with boundary. The boundary of this manifold consists of the two circles $S_\alpha^1 \times \{-\Delta\}$ and $S_\alpha^1 \times \{\Delta\}$, which are the products of the circle S_α^1 and the endpoints $\{-\Delta\}$ and $\{\Delta\}$ of the interval I_β^1.

Remark 3 It can be seen from Example 4 just considered that coordinates sometimes arise naturally on M (α and β in this example), and they themselves induce a topology on M. Hence, in Definition 1 of a manifold, it is not always necessary to require in advance that M have a topology. The essence of the concept of a manifold is that the points of some set M can be parametrized by the points of a set of sub-domains of $\mathbb{R}^n$. A natural connection then arises between the coordinate systems that thereby arise on parts of M, expressed in the mappings of the corresponding domains of $\mathbb{R}^n$. Hence we can assume that M is obtained from a collection of do-mains of $\mathbb{R}^n$ by exhibiting some rule for identifying their points or, figuratively speaking, exhibiting a rule for gluing them together. Thus defining a manifold es-sentially means giving a set of subdomains of $\mathbb{R}^n$ and a rule of correspondence for the points of these subdomains. We shall not take the time to make this any more precise by formalizing the concept of gluing or identifying points, introducing a topology on M, and the like.

Definition 6 A manifold is *compact* (resp. *connected*) if it is compact (resp. con-nected) as a topological space.

The manifolds considered in Examples 1–4 are compact and connected. The boundary of the cylinder $S_\alpha^1 \times I_\beta^1$ in Example 4 consists of two independent cir-cles and is a one-dimensional compact, but not connected, manifold. The boundary $S^{n-1} = \partial \overline{B}^n$ of the n-dimensional disk of Example 3 is a compact manifold, which is connected for $n > 1$ and disconnected (it consists of two points) if $n = 1$.

Example 5 The space $\mathbb{R}^n$ itself is obviously a connected noncompact manifold without boundary, and the half-space H^n provides the simplest example of a con-nected noncompact manifold with boundary. (In both cases the atlas can be taken to consist of the single chart corresponding to the identity mapping.)

Proposition 2 *If a manifold M is connected, it is path connected.*

Proof After fixing a point $x_0 \in M$, consider the set E_{x_0} of points of M that can be joined to x_0 by a path in M. The set E_{x_0}, as one can easily verify from the definition of a manifold, is both open and closed in M. But that means that $E_{x_0} = M$. □

Example 6 If to each real $n \times n$ matrix we assign the point of $\mathbb{R}^{n^2}$ whose coordi-nates are obtained by writing out the elements of the matrix in some fixed order, then the group $GL(n, \mathbb{R})$ of nonsingular $n \times n$ matrices becomes a manifold of di-mension n^2. This manifold is noncompact (the elements of the matrices are not bounded) and nonconnected. This last fact follows from the fact that $GL(n, \mathbb{R})$ con-tains matrices with both positive and negative determinants. The points of $GL(n, \mathbb{R})$ corresponding to two such matrices cannot be joined by a path. (On such a path there would have to be a point corresponding to a matrix whose determinant is zero.)

Example 7 The group $SO(2, \mathbb{R})$ of orthogonal mappings of the plane $\mathbb{R}^2$ having de-terminant equal to 1 consists of matrices of the form $\left(\begin{smallmatrix} \cos\alpha & \sin\alpha \\ -\sin\alpha & \cos\alpha \end{smallmatrix} \right)$ and hence can be regarded as a manifold that is identified with the circle – the domain of variation of the angular parameter α. Thus $SO(2, \mathbb{R})$ is a one-dimensional compact connected manifold. If we also allow reflections about lines in the plane $\mathbb{R}^2$, we obtain the group $O(2, \mathbb{R})$ of all real orthogonal 2×2 matrices. It can be naturally identified with two different circles, corresponding to matrices with determinants $+1$ and -1 respectively. That is, $O(2, \mathbb{R})$ is a one-dimensional compact, but not connected manifold.

Example 8 Let $\mathbf{a}$ be a vector in $\mathbb{R}^2$ and $T_\mathbf{a}$ the group of rigid motions of the plane generated by $\mathbf{a}$. The elements of $T_\mathbf{a}$ are translations by vectors of the form $n\mathbf{a}$, where $n \in \mathbb{Z}$. Under the action of the elements g of the group $T_\mathbf{a}$ each point x of the plane is displaced to a point $g(x)$ of the form $x + n\mathbf{a}$. The set of all points to which a given point $x \in \mathbb{R}^2$ passes under the action of the elements of this group of transformations is called its *orbit*. The property of points of $\mathbb{R}^2$ of belonging to the same orbit is obviously an equivalence relation on $\mathbb{R}^2$, and the orbits are the equivalence classes of this relation. A domain in $\mathbb{R}^2$ containing one point from each equivalence class is called a *fundamental domain* of this group of automorphisms (for a more precise statement see Problem 5d)).

In the present case we can take as a fundamental domain a strip of width $|\mathbf{a}|$ bounded by two parallel lines orthogonal to $\mathbf{a}$. We need only take into account that these lines themselves are obtained from each other through translations by $\mathbf{a}$ and $-\mathbf{a}$ respectively. Inside a strip of width less than $|\mathbf{a}|$ and orthogonal to a there are no equivalent points, so that all orbits having representatives in that strip are endowed uniquely with the coordinates of their representatives. Thus the quotient set $\mathbb{R}^2/T_\mathbf{a}$ consisting of orbits of the group $T_\mathbf{a}$ becomes a manifold. From what was said above about a fundamental domain, one can easily see that this manifold is homeomorphic to the cylinder obtained by gluing the boundary lines of a strip of width $|\mathbf{a}|$ together at equivalent points.

Example 9 Now let $\mathbf{a}$ and $\mathbf{b}$ be a pair of orthogonal vectors of the plane $\mathbb{R}^2$ and $T_{\mathbf{a},\mathbf{b}}$ the group of translations generated by these vectors. In this case a fundamental domain is the rectangle with sides $\mathbf{a}$ and $\mathbf{b}$. Inside this rectangle the only equivalent points are those that lie on opposite sides. After gluing the sides of this fundamental rectangle together, we verify that the resulting manifold $\mathbb{R}^2/T_{\mathbf{a},\mathbf{b}}$ is homeomorphic to the two-dimensional torus.

Example 10 Now consider the group $G_{a,b}$ of rigid motions of the plane $\mathbb{R}^2$ generated by the transformations $a(x, y) = (x + 1, 1 - y)$ and $b(x, y) = (x, y + 1)$.

A fundamental domain for the group $G_{a,b}$ is the unit square whose horizontal sides are identified at points lying on the same vertical line, but whose vertical sides are identified at points symmetric about the center. Thus the resulting manifold $\mathbb{R}^2/G_{a,b}$ turns out to be homeomorphic to the Klein bottle (see Sect. 12.1).

We shall not take time to discuss here the useful and important examples studied in Sect. 12.1.

15.2.2 Smooth Manifolds and Smooth Mappings

Definition 7 An atlas of a manifold is *smooth (of class* $C^{(k)}$ *or analytic)* if all the coordinate-changing functions for the atlas are smooth mappings (diffeomorphisms) of the corresponding smoothness class.

Two atlases of a given smoothness (the same smoothness for both) are *equivalent* if their union is an atlas of this smoothness.

Example 11 An atlas consisting of a single chart can be regarded as having any desired smoothness. Consider in this connection the atlas on the line $\mathbb{R}^1$ generated by the identity mapping $\mathbb{R}^1 \ni x \mapsto \varphi(x) = x \in \mathbb{R}^1$, and a second atlas – generated by any strictly monotonic function $\mathbb{R}^1 \ni x \mapsto \widetilde{\varphi}(x) \in \mathbb{R}^1$, mapping $\mathbb{R}^1$ onto $\mathbb{R}^1$. The union of these atlases is an atlas having smoothness equal to the smaller of the smoothnesses of $\widetilde{\varphi}$ and $\widetilde{\varphi}^{-1}$.

In particular, if $\widetilde{\varphi}(x) = x^3$, then the atlas consisting of the two charts $\{x, x^3\}$ is not smooth, since $\widetilde{\varphi}^{-1}(x) = x^{1/3}$. Using what has just been said, we can construct infinitely smooth atlases in $\mathbb{R}^1$ whose union is an atlas of a preassigned smoothness class $C^{(k)}$.

Definition 8 A *smooth manifold (of class* $C^{(k)}$ *or analytic)* is a manifold M with an equivalence class of atlases of the given smoothness.

After this definition the following terminology is comprehensible: *topological manifold* (of class $C^{(0)}$), $C^{(k)}$*-manifold, analytic manifold.*

To give the entire equivalence class of atlases of a given smoothness on a manifold M it suffices to give any atlas A of this equivalence class. Thus we can assume that a smooth manifold is a pair (M, A), where M is a manifold and A an atlas of the given smoothness on M.

The set of equivalent atlases of a given smoothness on a manifold is often called a *structure of this smoothness on the manifold.* There may be different smooth structures of even the same smoothness on a given topological manifold (see Example 11 and Problem 3).

Let us consider some more examples in which our main attention is directed to the smoothness of the coordinate changes.

Example 12 The one-dimensional manifold $\mathbb{RP}^1$ called the *real projective line,* is the pencil of lines in $\mathbb{R}^2$ passing through the origin, with the natural notion of distance between two lines (measured, for example, by the magnitude of the smaller angle between them). Each line of the pencil is uniquely determined by a nonzero direction vector (x^1, x^2), and two such vectors give the same line if and only if they are collinear. Hence $\mathbb{RP}^1$ can be regarded as a set of equivalence classes of ordered pairs (x^1, x^2) of real numbers. Here at least one of the numbers in the pair must be nonzero, and two pairs are considered equivalent (identified) if they are proportional. The pairs (x^1, x^2) are usually called *homogeneous coordinates* on $\mathbb{RP}^1$. Using the interpretation of $\mathbb{RP}^1$ in homogeneous coordinates, it is easy to construct an atlas of two charts on $\mathbb{RP}^1$. Let U_i, $i = 1, 2$, be the lines (classes of pairs (x^1, x^2)) in $\mathbb{RP}^1$ for which $x^i \neq 0$. To each point (line) $p \in U_1$ there corresponds a unique pair $(1, \frac{x^2}{x^1})$ determined by the number $t_1^2 = \frac{x^2}{x^1}$. Similarly the points of the region U_2 are in one-to-one correspondence with pairs of the form $(\frac{x^1}{x^2}, 1)$ and are determined by the number $t_2^1 = \frac{x^1}{x^2}$. Thus local coordinates arise in U_1 and U_2, which obviously correspond to the topology introduced above on $\mathbb{RP}^1$. In the common range $U_1 \cap U_2$ of these local charts the coordinates they introduce are connected by the relations $t_2^1 = (t_1^2)^{-1}$ and $t_1^2 = (t_2^1)^{-1}$, which shows that the atlas is not only $C^{(\infty)}$ but even analytic.

It is useful to keep in mind the following interpretation of the manifold $\mathbb{RP}^1$. Each line of the original pencil of lines is completely determined by its intersection with the unit circle. But there are exactly two such points, diametrically opposite to each other. Lines are near if and only if the corresponding points of the circle are near. Hence $\mathbb{RP}^1$ can be interpreted as a circle with diametrically opposite points identified (glued together). If we take only a semicircle, there is only one pair of identified points on it, the end-points. Gluing them together, we again obtain a topological circle. Thus $\mathbb{RP}^1$ is homeomorphic to the circle as a topological space.

Example 13 If we now consider the pencil of lines passing through the origin in $\mathbb{R}^3$, or, what is the same, the set of equivalence classes of ordered triples of points (x^1, x^2, x^3) of real numbers that are not all three zero, we obtain the *real projective plane* $\mathbb{RP}^2$. In the regions U_1, U_2, and U_3 where $x^1 \neq 0$, $x^2 \neq 0$, $x^3 \neq 0$ respectively, we introduce local coordinate systems $(1, \frac{x^2}{x^1}, \frac{x^3}{x^1}) = (1, t_1^2, t_1^3) \sim (t_1^2, t_1^3), (\frac{x^1}{x^2}, 1, \frac{x^3}{x^2}) = (t_2^1, 1, t_2^3) \sim (t_2^1, t_2^3)$, and $(\frac{x^1}{x^3}, \frac{x^2}{x^3}, 1) = (t_3^1, t_3^2, 1) \sim (t_3^1, t_3^2)$, which are obviously connected by the relations $t_i^j = (t_j^i)^{-1}$, $t_i^j = t_k^j (t_i^k)^{-1}$, which apply in the common portions of the ranges of these charts.

For example, the transition from (t_1^2, t_1^3) to (t_2^1, t_2^3) in the domain $U_1 \cap U_2$ is given by the formulas

$$t_2^1 = (t_1^2)^{-1}, \qquad t_2^3 = t_1^3 \cdot (t_1^2)^{-1}.$$

The Jacobian of this transformation is $-(t_1^2)^{-3}$, and since $t_1^2 = \frac{x^2}{x^1}$, it is defined and nonzero at points of the set $U_1 \cap U_2$ under consideration.

Thus $\mathbb{RP}^2$ is a two-dimensional manifold having an analytic atlas consisting of three charts.

By the same considerations as in Example 12, where we studied the projective line $\mathbb{RP}^1$, we can interpret the projective plane $\mathbb{RP}^2$ as the two-dimensional sphere $S^2 \subset \mathbb{R}^2$ with antipodal points identified, or as a hemisphere, with diametrically opposite points of its boundary circle identified. Projecting the hemisphere into the plane, we obtain the possibility of interpreting $\mathbb{RP}^2$ as a (two-dimensional) disk with diametrically opposite points of its boundary circle identified.

Example 14 The set of lines in the plane $\mathbb{R}^2$ can be partitioned into two sets: U, the nonvertical lines, and V, the nonhorizontal lines. Each line in U has an equation of the form $y = u_1 x + u_2$, and hence is characterized by the coordinates (u_1, u_2), while each line in V has an equation $x = v_1 y + v_2$ and is determined by coordinates (v_1, v_2). For lines in the intersection $U \cap V$ have the coordinate transformation $v_1 = u_1^{-1}$, $v_2 = -u_2 u_1^{-1}$ and $u_1 = v_1^{-1}$, $u_2 = -v_2 v_1^{-1}$. Thus this set is endowed with an analytic atlas consisting of two charts.

Every line in the plane has an equation $ax + by + c = 0$ and is characterized by a triple of numbers (a, b, c), proportional triples defining the same line. For that reason, it might appear that we are again dealing with the projective plane $\mathbb{RP}^2$ considered in Example 13. However, whereas in $\mathbb{RP}^2$ we admitted any triples of numbers not all zero, now we do not admit triples of the form $(0, 0, c)$ where $c \neq 0$. A single point in $\mathbb{RP}^2$ corresponds to the set of all such triples. Hence the manifold obtained in our present example is homeomorphic to the one obtained from $\mathbb{RP}^2$ by removing one point. If we interpret $\mathbb{RP}^2$ as a disk with diametrically opposite points of the boundary circle identified, then, deleting the center of the circle, we obtain, up to homeomorphism, an annulus whose outer circle is glued together at diametrically opposite points. By a simple incision one can easily show that the result is none other than the familiar Möbius band.

Definition 9 Let M and N be $C^{(k)}$-manifolds. A mapping $f : M \to N$ is *l-smooth* (a $C^{(l)}$-mapping) if the local coordinates of the point $f(x) \in N$ are $C^{(l)}$-functions of the local coordinates of $x \in M$.

This definition has an unambiguous meaning (one that is independent of the choice of local coordinates) if $l \leq k$.

In particular, the smooth mappings of M into $\mathbb{R}^1$ are smooth functions on M, and the smooth mappings of $\mathbb{R}^1$ (or an interval of $\mathbb{R}^1$) into M are smooth paths on M.

Thus the degree of smoothness of a function $f : M \to N$ on a manifold M cannot exceed the degree of smoothness of the manifold itself.

15.2.3 Orientation of a Manifold and Its Boundary

Definition 10 Two charts of a smooth manifold are *consistent* if the transition from the local coordinates in one to the other in their common range is a diffeomorphism whose Jacobian is everywhere positive.

In particular, if the ranges of two local charts have empty intersection, they are considered consistent.

Definition 11 An atlas A of a smooth manifold (M, A) is an *orienting atlas of M* if it consists of pairwise consistent charts.

Definition 12 A manifold is *orientable* if it has an orienting atlas. Otherwise it is *nonorientable*.

Two orienting atlases of a manifold will be regarded as *equivalent* (in the sense of the question of orientation of the manifold considered just now) if their union is also an orienting atlas of the manifold. It is easy to see that this relation really is an equivalence relation.

Definition 13 An equivalence class of orienting atlases of a manifold in the relation just defined is called an *orientation class of atlases of the manifold* or an *orientation of the manifold*.

Definition 14 An *oriented manifold* is a manifold with this class of orientations of its atlases, that is, with a fixed orientation on the manifold.

Thus orienting the manifold means exhibiting (by some means or other) a certain orientation class of atlases on it. To do this, for example, it suffices to exhibit any specific orienting atlas from the orientation class.

Various methods used in practice to define an orientation of manifolds embedded in $\mathbb{R}^n$ are described in Sects. 12.2 and 12.3.

Proposition 3 *A connected manifold is either nonorientable or admits exactly two orientations.*

Proof Let A and $\widetilde{A}$ be two orienting atlases of the manifold M with diffeomorphic transitions from the local coordinates of charts of one to charts of the other. Assume that there is a point $p_0 \in M$ and two charts of these atlases whose ranges U_{i_0} and $\widetilde{U}_{i_0}$ contain p_0; and suppose the Jacobian of the change of coordinates of the charts at points of the parameter space corresponding to the point p_0 is positive. We shall show that then for every point $p \in M$ and any charts of the atlases A and $\widetilde{A}$ whose ranges contain p the Jacobian of the coordinate transformation at corresponding coordinate points is also positive.

We begin by making the obvious observation that if the Jacobian of the transformation is positive (resp. negative) at the point p for any pair of charts containing p in the atlases A and $\widetilde{A}$, then it is positive (resp. negative) at p for any such pair of charts, since inside each given atlas the coordinate transformations occur with positive Jacobian, and the Jacobian of a composition of two mappings is the product of the Jacobians of the individual mappings.

Now let E be the subset of M consisting of the points $p \in M$ at which the coordinate transformations from the charts of one atlas to those of the other have positive Jacobian.

The set E is nonempty, since $p_0 \in E$. The set E is open in M. Indeed, for every point $p \in E$ there exist ranges U_i and $\widetilde{U}_j$ of certain charts of the atlases A and $\widetilde{A}$ containing p. The sets U_i and $\widetilde{U}_j$ are open in M, so that the set $U_i \cap \widetilde{U}_j$ is open in M. On the connected component of the set $U_i \cap \widetilde{U}_j$ containing p, which is open in $U_i \cap \tilde{U}_j$ and in M, the Jacobian of the transformation cannot change sign without vanishing at some point. That is, in some neighborhood of p the Jacobian remains positive, which proves that E is open. But E is also closed in M. This follows from the continuity of the Jacobian of a diffeomorphism and the fact that the Jacobian of a diffeomorphism never vanishes.

Thus E is a nonempty open-closed subset of the connected set M. Hence $E = M$, and the atlases A and $\tilde{A}$ define the same orientation on M.

Replacing one coordinate, say t^1 by $-t^1$ in every chart of the atlas A, we obtain the orienting atlas $-A$ belonging to a different orientation class. Since the Jacobians of the coordinate transformations from an arbitrary chart to the charts of A and $-A$ have opposite signs, every atlas that orients M is equivalent either to A or to $-A$. $\square$

Definition 15 A finite sequence of charts of a given atlas will be called a *chain of charts* if the ranges of any pair of charts having adjacent indices have a nonempty intersection ($U_i \cap U_{i+1} \neq \varnothing$).

Definition 16 A chain of charts is *contradictory* or *disorienting* if the Jacobian of the coordinate transformation from each chart in the chain to the next is positive and the ranges of the first and last charts of the chain intersect, but the coordinate transformation from the last to the first has negative Jacobian.

Proposition 4 *A manifold is orientable if and only if there does not exist a contradictory chain of charts on it.*

Proof Since every manifold decomposes into connected components whose orientations can be defined independently, it suffices to prove Proposition 4 for a connected manifold M.

Necessity. Suppose the connected manifold M is orientable and A is an atlas defining an orientation. From what has been said and Proposition 3, every smooth local chart of the manifold M connected with the charts of the atlas A is either consistent with all the charts of A or consistent with all the charts of $-A$. This can easily be seen from Proposition 3 itself, if we restrict charts of A to the range of the chart we have taken, which can be regarded as a connected manifold oriented by one chart. It follows from this that there is no contradictory chain of charts on M.

Sufficiency. It follows from Definition 1 that there exists an atlas on the manifold consisting of a finite or countable number of charts. We take such an atlas A and number its charts. Consider the chart (U_1, φ_1) and any chart (U_i, φ_i) such that $U_1 \cap U_i \neq \varnothing$. Then the Jacobians of the coordinate transformations φ_{1i} and φ_{i1} are either everywhere negative or everywhere positive in their domains of definition. The Jacobians cannot have values of different signs, since otherwise one could exhibit connected subsets U_- and U_+ in $U_1 \cup U_i$ where the Jacobian is negative and positive respectively, and the chain of charts (U_1, φ_1), (U_+, φ_1), (U_i, φ_i), (U_-, φ_i) would be contradictory.

Thus, changing the sign of one coordinate if necessary in the chart (U_i, φ_i), we could obtain a chart with the same range U_i and consistent with (U_1, φ_1). After that procedure, two charts (U_i, φ_i) and (U_j, φ_j) such that $U_1 \cap U_i \neq \varnothing$, $U_1 \cap U_j \neq \varnothing$, $U_i \cap U_j \neq \varnothing$ are themselves consistent: otherwise we would have constructed a contradictory chain of three charts.

Thus, all the charts of an atlas whose ranges intersect U_1 can now be considered consistent with one another. Taking each of those charts now as the standard, one can adjust the charts of the atlas not covered in the first stage so that they are consistent. No contradictions arise when we do this, since by hypothesis, there are no contradictory chains on the manifold. Continuing this process and taking account of the connectedness of the manifold, we construct on it an atlas consisting of pairwise consistent charts, which proves the orientability of the manifold. $\qquad \square$

This criterion for orientability of the manifold, like the considerations used in its proof, can be applied to the study of specific manifolds. Thus, the manifold $\mathbb{RP}^1$ studied in Example 12 is orientable. From the atlas shown there it is easy to obtain an orienting atlas of $\mathbb{RP}^1$. To do this, it suffices to reverse the sign of the local coordinates of one of the two charts constructed there. However, the orientability of the projective line $\mathbb{RP}^1$ obviously also follows from the fact that the manifold $\mathbb{RP}^1$ is homeomorphic to a circle.

The projective plane $\mathbb{RP}^2$ is nonorientable: every pair of charts in the atlas constructed in Example 13 is such that the coordinate transformations have domains of positivity and domains of negativity of the Jacobian. As we saw in the proof of Proposition 4, it follows from this that a contradictory chain of charts on $\mathbb{RP}^2$ exists.

For the same reason the manifold considered in Example 14 is nonorientable, which, as was noted, is homeomorphic to a Möbius band.

Proposition 5 *The boundary of an orientable smooth n-dimensional manifold is an orientable $(n-1)$-dimensional manifold admitting a structure of the same smoothness as the original manifold.*

Proof The proof of Proposition 5 is a verbatim repetition of the proof of the analogous Proposition 2 of Sect. 12.3.2 for surfaces embedded in $\mathbb{R}^n$. $\qquad \square$

Definition 17 If $A(M) = \{(H^n, \varphi_i, U_i)\} \cup \{(\mathbb{R}^n, \varphi_j, U_j)\}$ is an atlas that orients the manifold M, then the charts $A(\partial M) = \{(\mathbb{R}^{n-1}, \varphi_i|_{\partial H^n = \mathbb{R}^{n-1}}, \partial U_i)\}$ provide an orienting atlas for the boundary ∂M of M. The orientation of the boundary defined by this atlas is called the *orientation of the boundary induced by the orientation of the manifold.*

Important techniques for defining the orientation of a surface embedded in $\mathbb{R}^n$ and the induced orientation of its boundary, which are frequently used in practice, were described in detail in Sects. 12.2 and 12.3.

15.2.4 Partitions of Unity and the Realization of Manifolds as Surfaces in $\mathbb{R}^n$

In this subsection we shall describe a special construction called a *partition of unity*. This construction is often the basic device for reducing global problems to local ones. Later on we shall demonstrate it in deriving Stokes' formula on a manifold, but here we shall use the partition of unity to clarify the possibility of realizing any manifold as a surface in $\mathbb{R}^n$ of sufficiently high dimension.

Lemma *One can construct a function* $f \in C^{(\infty)}(\mathbb{R}, \mathbb{R})$ *on* $\mathbb{R}$ *such that* $f(x) \equiv 0$ *for* $|x| \geq 3$, $f(x) \equiv 1$ *for* $|x| \leq 1$, *and* $0 < f(x) < 1$ *for* $1 < |x| < 3$.

Proof We shall construct one such function using the familiar function $g(x) = \begin{cases} e^{(-1/x^2)} & \text{for } x \neq 0, \\ 0 & \text{for } x=0. \end{cases}$ Previously (see Exercise 2 of Sect. 5.2) we verified that $g \in C^{(\infty)}(\mathbb{R}, \mathbb{R})$ by showing that $g^{(n)}(0) = 0$ for every value $n \in \mathbb{N}$.

In such a case the nonnegative function

$$G(x) = \begin{cases} e^{-(x-1)^{-2}} \cdot e^{-(x+1)^{-2}} & \text{for } |x| < 1, \\ 0 & \text{for } |x| \geq 1 \end{cases}$$

also belongs to $C^{(\infty)}(\mathbb{R}, \mathbb{R})$, and along with it the function

$$F(x) = \int_{-\infty}^{x} G(t)\,dt \Big/ \int_{-\infty}^{+\infty} G(t)\,dt+$$

belongs to this class, since $F'(x) = G(x)/\int_{-\infty}^{\infty} G(t)\,dt$.

The function F is strictly increasing on $[-1, 1]$, $F(x) \equiv 0$ for $x \leq -1$, and $F(x) \equiv 1$ for $x \geq 1$.

We can now take the required function to be

$$f(x) = F(x+2) + F(-x-2) - 1. \qquad \square$$

Remark If $f : \mathbb{R} \to \mathbb{R}$ is the function constructed in the proof of the lemma, then the function

$$\theta(x^1, \ldots, x^n) = f(x^1 - a^1) \cdot \ldots \cdot f(x^n - a^n)$$

defined in $\mathbb{R}^n$ is such that $\theta \in C^{(\infty)}(\mathbb{R}^n, \mathbb{R})$, $0 \leq \theta \leq 1$, at every point $x \in \mathbb{R}^n$, $\theta(x) \equiv 1$ on the interval $I(a) = \{x \in \mathbb{R}^n \mid |x^i - a^i| \leq 1, i = 1, \ldots, n\}$, and the support $\mathrm{supp}\,\theta$ of the function θ is contained in the interval $\tilde{I}(a) = \{x \in \mathbb{R}^n \mid |x^i - a^i| \leq 3, i = 1, \ldots, n\}$.

Definition 18 Let M be a $C^{(k)}$-manifold and X a subset of M. The system $E = \{e_\alpha, \alpha \in A\}$ of functions $e_\alpha \in C^{(k)}(M, \mathbb{R})$ is a $C^{(k)}$ *partition of unity* on X if

1^0 $0 \leq e_\alpha(x) \leq 1$ for every function $e_\alpha \in E$ and every $x \in M$;

2^0 each point $x \in X$ has a neighborhood $U(x)$ in M such that all but a finite number of functions of E are identically zero on $U(x)$;

3^0 $\sum_{e_\alpha \in E} e_\alpha(x) \equiv 1$ on X.

We remark that by condition 2^0 only a finite number of terms in this last sum are nonzero at each point $x \in X$.

Definition 19 Let $\mathcal{O} = \{o_\beta, \beta \in B\}$ be an open covering of $X \subset M$. We say that *the partition of unity* $E = \{e_\alpha, \alpha \in A\}$ on X is *subordinate to the covering* $\mathcal{O}$ if the support of each function in the system E is contained in at least one of the sets of the system $\mathcal{O}$.

Proposition 6 Let $\{(U_i, \varphi_i), i = 1, \ldots, m\}$ be a finite set of charts of some $C^{(k)}$ atlas of the manifold M, whose ranges U_i, $i = 1, \ldots, m$, form a covering of a compact set $K \subset M$. Then there exists a $C^{(k)}$ partition of unity on K subordinate to the covering $\{U_i, i = 1, \ldots, m\}$.

Proof For any point $x_0 \in K$ we first carry out the following construction. We choose successively a domain U_i containing x_0 corresponding to a chart $\varphi_i : \mathbb{R}^n \to U_i$ (or $\varphi_i : H^n \to U_i$), the point $t_0 = \varphi_i^{-1}(x_0) \in \mathbb{R}^n$ (or H^n), the function $\theta(t - t_0)$ (where $\theta(t)$ is the function shown in the remark to the lemma), and the restriction θ_{t_0} of $\theta(t - t_0)$ to the parameter domain of φ_i.

Let I_{t_0} be the intersection of the unit cube centered at $t_0 \in \mathbb{R}^n$ with the parameter domain of φ_i. Actually θ_{t_0} differs from $\theta(t - t_0)$ and I_{t_0} differs from the corresponding unit cube only when the parameter domain of the chart φ_i is the half-space H^n. The open sets $\varphi_i(I_t)$ constructed at each point $x \in K$ and the point $t = \varphi_i^{-1}(x)$, taken for all admissible values of $i = 1, 2, \ldots, m$, form an open covering of the compact set K. Let $\{\varphi_{i_j}(I_{t_j}), j = 1, 2, \ldots, l\}$ be a finite covering of K extracted from it. It is obvious that $\varphi_{i_j}(I_{t_j}) \subset U_{i_j}$. We define on U_{i_j} the function $\widetilde{\theta}_i(x) = \theta_{t_j} \circ \varphi_{i_j}^{-1}(x)$. We then extend $\widetilde{\theta}_j(x)$ to the entire manifold M by setting the function equal to zero outside U_{i_j}. We retain the previous notation $\widetilde{\theta}_j$ for this function extended to M. By construction $\widetilde{\theta}_j \in C^{(k)}(M, \mathbb{R})$, $\operatorname{supp} \widetilde{\theta}_j \subset U_{i_j}, 0 \leq \widetilde{\theta}_j(x) \leq 1$ on M, and $\widetilde{\theta}_j(x) \equiv 1$ on $\varphi_{i_j}(I_{t_j}) \subset U_{i_j}$. Then the functions $e_1(x) = \widetilde{\theta}_1(x)$, $e_2(x) = \widetilde{\theta}_2(x)(1 - \widetilde{\theta}_1(x)), \ldots, e_l(x) = \widetilde{\theta}_l(x) \cdot (1 - \widetilde{\theta}_{l-1}(x)) \cdot \ldots \cdot (1 - \widetilde{\theta}_1(x))$ form the required partition of unity. We shall verify only that $\sum_{j=1}^l e_j(x) \equiv 1$ on K, since the system of functions $\{e_1, \ldots, e_l\}$ obviously satisfies the other conditions required of a partition of unity on K subordinate to the covering $\{U_{i_1}, \ldots, U_{i_l}\} \subset \{U_i, i = 1, \ldots, m\}$. But

$$1 - \sum_{j=1}^l e_j(x) = \left(1 - \widetilde{\theta}_1(x)\right) \cdot \ldots \cdot \left(1 - \widetilde{\theta}_1(x)\right) \equiv 0 \quad \text{on } K,$$

since each point $x \in K$ is covered by some set $\varphi_{i_j}(I_{t_j})$ on which the corresponding function $\widetilde{\theta}_j$ is identically equal to 1. $\qquad \square$

Corollary 1 *If M is a compact manifold and A a $C^{(k)}$ atlas on M, then there exists a finite partition of unity $\{e_1, \ldots, e_l\}$ on M subordinate to a covering of the manifold by the ranges of the charts of A.*

Proof Since M is compact, the atlas A can be regarded as finite. We now have the hypotheses of Proposition 6, if we set $K = M$ in it.   □

Corollary 2 *For every compact set K contained in a manifold M and every open set $G \subset M$ containing K, there exists a function $f : M \to \mathbb{R}$ with smoothness equal to that of the manifold and such that $f(x) \equiv 1$ on K and $\operatorname{supp} f \subset G$.*

Proof Cover each point $x \in K$ by a neighborhood $U(x)$ contained in G and inside the range of some chart of the manifold M. From the open covering $\{U(x), x \in K\}$ of the compact set K extract a finite covering, and construct a partition of unity $\{e_1, \ldots, e_l\}$ on K subordinate to it. The function $f = \sum_{i=1}^{l} e_i$ is the one required. □

Corollary 3 *Every (abstractly defined) compact smooth n-dimensional manifold M is diffeomorphic to some compact smooth surface contained in $\mathbb{R}^N$ of sufficiently large dimension N.*

Proof So as not to complicate the idea of the proof with inessential details, we carry it out for the case of a compact manifold M without boundary. In that case there is a finite smooth atlas $A = \{\varphi_i : I \to U_i, i = 1, \ldots, m\}$ on M, where I is an open n-dimensional cube in $\mathbb{R}^n$. We take a slightly smaller cube I' such that $I' \subset I$ and the set $\{U_i' = \varphi_i(I'), i = 1, \ldots, m\}$ still forms a covering of M. Setting $K = I'$, $G = I$, and $M = \mathbb{R}^n$ in Corollary 2, we construct a function $f \in C^{(\infty)}(\mathbb{R}^n, \mathbb{R})$ such that $f(t) \equiv 1$ for $t \in I'$ and $\operatorname{supp} f \subset I$.

We now consider the coordinate functions $t_i^1(x), \ldots, t_i^n(x)$ of the mappings $\varphi_i^{-1} : U_I \to I$, $i = 1, \ldots, m$, and use them to introduce the following function on M:

$$y_i^k(x) = \begin{cases} (f \circ \varphi_i^{-1})(x) \cdot t_i^k(x) & \text{for } x \in U_i, \\ 0 & \text{for } x \notin U_i, \end{cases}$$

$$i = 1, \ldots, m; \ k = 1, \ldots, n.$$

At every point $x \in M$ the rank of the mapping $M \ni x \mapsto y(x) = (y_1^1, \ldots, y_1^n, \ldots, y_m^1, \ldots, y_m^n)(x) \in \mathbb{R}^{m \cdot n}$ is maximal and equal to n. Indeed, if $x \in U_i'$, then $\varphi_i^{-1}(x) = t \in I'$, $f \circ \varphi_i^{-1}(x) = 1$, and $y_i^k(\varphi_i(t)) = t_i^k$, $k = 1, \ldots, n$.

If finally, we consider the mapping $M \ni x \mapsto Y(x) = (y(x), f \circ \varphi_1^{-1}(x), \ldots, f \circ \varphi_m^{-1}(x)) \in \mathbb{R}^{m \cdot n + m}$, setting $f \circ \varphi_i^{-1}(x) \equiv 0$ outside U_i, $i = 1, \ldots, m$, then this mapping, on the one hand will obviously have the same rank n as the mapping $x \mapsto y(x)$; on the other hand it will be demonstrably a one-to-one mapping of M onto the image of M in $\mathbb{R}^{m \cdot n + m}$. Let us verify this last assertion. Let p, q be different points of M. We find a domain U_i' from the system $\{U_i', i = 1, \ldots, m\}$ covering M that contains the point p. Then $f \circ \varphi_i^{-1}(p) = 1$. If $f \circ \varphi_i^{-1}(q) < 1$, then $Y(p) \neq Y(q)$. If $f \circ \varphi_i^{-1}(q) = 1$, then $p, q \in U_i$, $y_i^k(p) = t^k(p)$, $y_i^k(q) = t^k(q)$, and $t_i^k(p) \neq t_i^k(q)$ for at least one value of $k \in \{1, \ldots, n\}$. That is, $Y(p) \neq Y(q)$ in this case.   □

For information on the general Whitney embedding theorem for an arbitrary manifold as a surface in $\mathbb{R}^n$ the reader may consult the specialized geometric literature.

15.2.5 Problems and Exercises

1. Verify that the object (a *manifold*) introduced by Definition 1 does not change if we require only that each point $x \in M$ have a neighborhood $U(x) \subset M$ homeomorphic to an open subset of the half-space H^n.

2. Show that

a) the manifold $GL(n, \mathbb{R})$ of Example 6 is noncompact and has exactly two connected components;

b) the manifold $SO(n, \mathbb{R})$ (see Example 7) is connected;

c) the manifold $O(n, \mathbb{R})$ is compact and has exactly two connected components.

3. Let (M, A) and $(\widetilde{M}, \widetilde{A})$ be manifolds with smooth structures of the same degree of smoothness $C^{(k)}$ on them. The smooth manifolds (M, A) and $(\widetilde{M}, \widetilde{A})$ (*smooth structures*) are considered *isomorphic* if there exists a $C^{(k)}$ mapping $f : M \to \widetilde{M}$ having a $C^{(k)}$ inverse $f^{-1} : \widetilde{M} \to M$ in the atlases $A, \widetilde{A}$.

a) Show that all structures of the same smoothness on $\mathbb{R}^1$ are isomorphic.

b) Verify the assertions made in Example 11, and determine whether they contradict a).

c) Show that on the circle S^1 (the one-dimensional sphere) any two $C^{(\infty)}$ structures are isomorphic. We note that this assertion remains valid for spheres of dimension not larger than 6, but on S^7, as Milnor[4] has shown, there exist nonisomorphic $C^{(\infty)}$ structures.

4. Let S be a subset of an n-dimensional manifold M such that for every point $x_0 \in S$ there exists a chart $x = \varphi(t)$ of the manifold M whose range U contains x_0, and the k-dimensional surface defined by the relations $t^{k+1} = 0, \ldots, t^n = 0$ corresponds to the set $S \cap U$ in the parameter domain $t = (t^1, \ldots, t^n)$ of φ. In this case S is called a *k-dimensional submanifold of M*.

a) Show that a k-dimensional manifold structure naturally arises on S, induced by the structure of M and having the same smoothness as the manifold M.

b) Verify that the k-dimensional surfaces S in $\mathbb{R}^n$ are precisely the k-dimensional submanifolds of $\mathbb{R}^n$.

---

[4] J. Milnor (b. 1931) – one of the most outstanding modern American mathematicians; his main works are in algebraic topology and the topology of manifolds.

c) Show that under a smooth homeomorphic mapping $f : \mathbb{R}^1 \to T^2$ of the line $\mathbb{R}^1$ into the torus T^2 the image $f(\mathbb{R}^1)$ may be an everywhere dense subset of T^2 and in that case will not be a one-dimensional submanifold of the torus, although it will be an abstract one-dimensional manifold.

d) Verify that the extent of the concept "submanifold" does not change if we consider $S \subset M$ a k-dimensional submanifold of the n-dimensional manifold M when there exists a local chart of the manifold M whose range contains x_0 for every point $x_0 \in S$ and some k-dimensional surface of the space $\mathbb{R}^n$ corresponds to the set $S \cap U$ in the parameter domain of the chart.

5. Let X be a Hausdorff topological space (manifold) and G the group of homeomorphic transformations of X. The group G is a *discrete group of transformations of X* if for every two (possibly equal) points $x_1, x_2 \in X$ there exist neighborhoods U_1 and U_2 of them respectively, such that the set $\{g \in G \mid g(U_1) \cap U_2 \neq \varnothing\}$ is finite.

a) It follows from this that the *orbit* $\{g(x) \in X \mid g \in G\}$ of every point $x \in X$ is discrete, and the *stabilizer* $G_x = \{g \in G \mid g(x) = x\}$ of every point $x \in X$ is finite.

b) Verify that if G is a group of isometries of a metric space, having the two properties in a), then G is a discrete group of transformations of X.

c) Introduce the natural topological space (manifold) structure on the set X/G of orbits of the discrete group G.

d) A closed subset F of the topological space (manifold) X with a discrete group G of transformations is a *fundamental domain of the group G* if it is the closure of an open subset of X and the sets $g(F)$, where $g \in G$, have no interior points in common and form a locally finite covering of X. Show using Examples 8–10 how the quotient space X/G (of orbits) of the group G can be obtained from F by "gluing" certain boundary points.

6. a) Using the construction of Examples 12 and 13, construct n-dimensional projective space $\mathbb{RP}^n$.

b) Show that $\mathbb{RP}^n$ is orientable if n is odd and nonorientable if n is even.

c) Verify that the manifolds $SO(3, \mathbb{R})$ and $\mathbb{RP}^3$ are homeomorphic.

7. Verify that the manifold constructed in Example 14 is indeed homeomorphic to the Möbius band.

8. a) A *Lie group*[5] is a group G endowed with the structure of an analytic manifold such that the mappings $(g_1, g_2) \mapsto g_1 \cdot g_2$ and $g \mapsto g^{-1}$ are analytic mappings of $G \times G$ and G into G. Show that the manifolds in Examples 6 and 7 are Lie groups.

b) A *topological group* (or *continuous group*) is a group G endowed with the structure of a topological space such that the group operations of multiplication and inversion are continuous as mappings $G \times G \to G$, and $G \to G$ in the topology of G. Using the example of the group $\mathbb{Q}$ of rational numbers show that not every topological group is a Lie group.

---

[5]S. Lie (1842–1899) – outstanding Norwegian mathematician, creator of the theory of continuous groups (Lie groups), which is now of fundamental importance in geometry, topology, and the mathematical methods of physics; one of the winners of the International Lobachevskii Prize (awarded in 1897 for his work in applying group theory to the foundations of geometry).

c) Show that every Lie group is a topological group in the sense of the definition given in b).

d) It has been proved[6] that every topological group G that is a manifold is a Lie group (that is, as a manifold G admits an analytic structure in which the group becomes a Lie group). Show that every group manifold (that is, every Lie group) is an orientable manifold.

9. A system of subsets of a topological space is *locally finite* if each point of the space has a neighborhood intersecting only a finite number of sets in the system. In particular, one may speak of a locally finite covering of a space.

A system of sets is said to be a *refinement* of a second system if every set of the first system is contained in at least one of the sets of the second system. In particular it makes sense to speak of one covering of a set being a refinement of another.

a) Show that every open covering of $\mathbb{R}^n$ has a locally finite refinement.

b) Solve problem a) with $\mathbb{R}^n$ replaced by an arbitrary manifold M.

c) Show that there exists a partition of unity on $\mathbb{R}^n$ subordinate to any preassigned open covering of $\mathbb{R}^n$.

d) Verify that assertion c) remains valid for an arbitrary manifold.

15.3 Differential Forms and Integration on Manifolds

15.3.1 The Tangent Space to a Manifold at a Point

We recall that to each smooth path $\mathbb{R} \ni t \overset{\gamma}{\longmapsto} x(t) \in \mathbb{R}^n$ (a motion in $\mathbb{R}^n$) passing through the point $x_0 = x(t_0) \in \mathbb{R}^n$ at time t_0 we have assigned the instantaneous velocity vector $\xi = (\xi^1, \dots, \xi^n) : \xi(t) = \dot{x}(t) = (\dot{x}^1, \dots, \dot{x}^n)(t_0)$. The set of all such vectors ξ attached to the point $x_0 \in \mathbb{R}^n$ is naturally identified with the arithmetic space $\mathbb{R}^n$ and is denoted $T\mathbb{R}^n_{x_0}$ (or $T_{x_0}(\mathbb{R}^n)$). In $T\mathbb{R}^n_{x_0}$ one introduces the same vector operations on elements $\xi \in T\mathbb{R}^n_{x_0}$ as on the corresponding elements of the vector space $\mathbb{R}^n$. In this way a vector space $T\mathbb{R}^n_{x_0}$ arises, called the *tangent space to $\mathbb{R}^n$ at the point $x_0 \in \mathbb{R}^n$*.

Forgetting about motivation and introductory considerations, we can now say that formally $T\mathbb{R}^n_{x_0}$ is a pair $(x_0, \mathbb{R}^n)$ consisting of a point $x_0 \in \mathbb{R}^n$ and a copy of the vector space $\mathbb{R}^n$ attached to it.

Now let M be a smooth n-dimensional manifold with an atlas A of at least $C^{(1)}$ smoothness. We wish to define a tangent vector ξ and a tangent space TM_{p_0} to the manifold M at a point $p_0 \in M$.

To do this we use the interpretation of the tangent vector as the instantaneous velocity of a motion. We take a smooth path $\mathbb{R}^n \ni t \overset{\gamma}{\longmapsto} p(t) \in M$ on the manifold M passing through the point $p_0 = p(t_0) \in M$ at time t_0. The parameters of charts (that

---

[6]This is the solution to Hilbert's fifth problem.

is, local coordinates) of the manifold M will be denoted by the letter x here, with the subscript of the corresponding chart and a superscript giving the number of the co-ordinate. Thus, in the parameter domain of each chart (U_i, φ_i) whose range U_i contains p_0, the path $t \xmapsto{\gamma_i} \varphi_i^{-1} \circ p(t) = x_i(t) \in \mathbb{R}^n$ (or H^n) corresponds to the path γ. This path is smooth by definition of the smooth mapping $\mathbb{R} \ni t \xmapsto{\gamma} p(t) \in M$.

Thus, in the parameter domain of the chart (U_i, φ_i), where φ_i is a mapping $p = \varphi_i(x_i)$, there arises a point $x_i(t_0) = \varphi_i^{-1}(p_0)$ and a vector $\xi_i = \dot{x}_i(t_0) \in T\mathbb{R}^n_{x_i(t_0)}$. In another such chart (U_j, φ_j) these objects will be respectively the point $x_j(t_0) = \varphi_j^{-1}(p_0)$ and the vector $\xi_j = \dot{x}_j(t_0) \in T\mathbb{R}^n_{x_j(t_0)}$. It is natural to regard these as the coordinate expressions in different charts of what we would like to call a tangent vector ξ to the manifold M at the point $p_0 \in M$.

Between the coordinates x_i and x_j there are smooth mutually inverse transition mappings

$$x_i = \varphi_{ji}(x_j), \qquad x_j = \varphi_{ij}(x_i), \tag{15.30}$$

as a result of which the pairs $(x_i(t_0), \xi_i)$, $(x_j(t_0), \xi_j)$ turn out to be connected by the relations

$$x_i(t_0) = \varphi_{ji}(x_j(t_0)), \qquad x_j(t_0) = \varphi_{ij}(x_i(t_0)), \tag{15.31}$$

$$\xi_i = \varphi'_{ji}(x_j(t_0))\xi_j, \qquad \xi_j = \varphi'_{ij}(x_i(t_0))\xi_i. \tag{15.32}$$

Equality (15.32) obviously follows from the formulas

$$\dot{x}_i(t) = \varphi'_{ji}(x_j(t))\dot{x}_j(t), \qquad \dot{x}_j(t) = \varphi'_{ij}(x_i(t))\dot{x}_i(t),$$

obtained from (15.30) by differentiation.

Definition 1 We shall say that a *tangent vector* ξ *to the manifold* M *at the point* $p \in M$ is defined if a vector ξ_i is fixed in each space $T\mathbb{R}^n_{x_i}$ tangent to $\mathbb{R}^n$ at the point x_i corresponding to p in the parameter domain of a chart (U_i, φ_i), where $U_i \ni p$, in such a way that (15.32) holds.

If the elements of the Jacobian matrix φ'_{ji} of the mapping φ_{ji} are written out explicitly as $\frac{\partial x_i^k}{\partial x_j^m}$, we find the following explicit form for the connection between the two coordinate representations of a given vector ξ:

$$\xi_i^k = \sum_{m=1}^n \frac{\partial x_i^k}{\partial x_j^m}\xi_j^m, \quad k = 1, 2, \ldots, n, \tag{15.33}$$

where the partial derivatives are computed at the point $x_j = \varphi_j^{-1}(p)$ corresponding to p.

We denote by TM_p the set of all tangent vectors to the manifold M at the point $p \in M$.

Definition 2 If we introduce a vector-space structure on the set TM_p by identifying TM_p with the corresponding space $T\mathbb{R}^n_{x_i}$ (or $TH^n_{x_i}$), that is, the sum of vectors in TM_p is regarded as the vector whose coordinate representation in $T\mathbb{R}^n_{x_i}$ (or $TH^n_{x_i}$) corresponds to the sum of the coordinate representations of the terms, and multiplication of a vector by a scalar is defined analogously, the vector space so obtained is usually denoted either TM_p or T_pM, and is called the *tangent space to the manifold M at the point $p \in M$*.

It can be seen from formulas (15.32) and (15.33) that the vector-space structure introduced in TM_p is independent of the choice of individual chart, that is, Definition 2 is unambiguous in that sense.

Thus we have now defined the tangent space to a manifold. There are various interpretations of a tangent vector and the tangent space (see Problem 1). For example, one such interpretation is to identify a tangent vector with a linear functional. This identification is based on the following observation, which we make in $\mathbb{R}^n$.

Each vector $\xi \in T\mathbb{R}^n_{x_0}$ is the velocity vector corresponding to some smooth path $x = x(t)$, that is, $\xi = \dot{x}(t)|_{t=t_0}$ with $x_0 = x(t_0)$. This makes it possible to define the derivative $D_\xi f(x_0)$ of a smooth function f defined on $\mathbb{R}^n$ (or in a neighborhood of x_0) with respect to the vector $\xi \in T\mathbb{R}^n_{x_0}$. To be specific,

$$D_\xi f(x_0) := \frac{\mathrm{d}}{\mathrm{d}t}(f \circ x)(t)\big|_{t=t_0}, \qquad (15.34)$$

that is,

$$D_\xi f(x_0) = f'(x_0)\xi, \qquad (15.35)$$

where $f'(x_0)$ is the tangent mapping to f (the differential of f) at a point x_0.

The functional $D_\xi : C^{(1)}(\mathbb{R}^n, \mathbb{R}) \to \mathbb{R}$ assigned to the vector $\xi \in T\mathbb{R}^n_{x_0}$ by the formulas (15.34) and (15.35) is obviously linear with respect to f. It is also clear from (15.35) that for a fixed function f the quantity $D_\xi f(x_0)$ is a linear function of ξ, that is, the sum of the corresponding linear functionals corresponds to a sum of vectors, and multiplication of a functional D_ξ by a number corresponds to multiplying the vector ξ by the same number. Thus there is an isomorphism between the vector space $T\mathbb{R}^n_{x_0}$ and the vector space of corresponding linear functionals D_ξ. It remains only to define the linear functional D_ξ by exhibiting a set of characteristic properties of it, in order to obtain a new interpretation of the tangent space $T\mathbb{R}^n_{x_0}$, which is of course isomorphic to the previous one.

We remark that, in addition to the linearity indicated above, the functional D_ξ possesses the following property:

$$D_\xi(f \cdot g)(x_0) = D_\xi f(x_0) \cdot g(x_0) + f(x_0) \cdot D_\xi g(x_0). \qquad (15.36)$$

This is the law for differentiating a product.

In differential algebra an additive mapping $a \mapsto a'$ of a ring A satisfying the relation $(a \cdot b)' = a' \cdot b + a \cdot b'$ is called *derivation* (more precisely *derivation of the ring A*). Thus the functional $D_\xi : C^{(1)}(\mathbb{R}^n, \mathbb{R})$ is a derivation of the ring $C^{(1)}(\mathbb{R}^n, \mathbb{R})$. But D_ξ is also linear relative to the vector-space structure of $C^{(1)}(\mathbb{R}^n, \mathbb{R})$.

One can verify that a linear functional $l : C^{(\infty)}(\mathbb{R}^n, \mathbb{R}) \to \mathbb{R}$ possessing the properties

$$l(\alpha f + \beta g) = \alpha l(f) + \beta l(g), \quad \alpha, \beta \in \mathbb{R}, \tag{15.37}$$

$$l(f \cdot g) = l(f)g(x_0) + f(x_0)l(g), \tag{15.38}$$

has the form D_ξ, where $\xi \in T\mathbb{R}^n_{x_0}$. Thus the tangent space $T\mathbb{R}^n_{x_0}$ to $\mathbb{R}^n$ at x_0 can be interpreted as a vector space of functionals (derivations) on $C^{(\infty)}(\mathbb{R}^n, \mathbb{R})$ satisfying conditions (15.37) and (15.38).

The functions $D_{e_k} f(x_0) = \frac{\partial}{\partial x^k} f(x)|_{x=x_0}$ that compute the corresponding partial derivative of the function f at x_0 correspond to the basis vectors $e_1, \ldots, e_n$ of the space $T\mathbb{R}^n_{x_0}$. Thus, under the functional interpretation of $T\mathbb{R}^n_{x_0}$ one can say that the functionals $\{\frac{\partial}{\partial x^1}, \ldots, \frac{\partial}{\partial x^n}\}|_{x=x_0}$ form a basis of $T\mathbb{R}^n_{x_0}$.

If $\xi = (\xi^1, \ldots, \xi^n) \in T\mathbb{R}^n_{x_0}$, then the operator D_ξ corresponding to the vector ξ has the form $D_\xi = \xi^k \frac{\partial}{\partial x^k}$.

In a completely analogous manner the tangent vector ξ to an n-dimensional $C^{(\infty)}$ manifold M at a point $p_0 \in M$ can be interpreted (or defined) as the element of the space of derivations l on $C^{(\infty)}(M, \mathbb{R})$ having properties (15.37) and (15.38), x_0 of course being replaced by p_0 in relation (15.38), so that the functional l is connected with precisely the point $p_0 \in M$. Such a definition of the tangent vector ξ and the tangent space TM_{p_0} does not formally require the invocation of any local coordinates, and in that sense it is obviously invariant. In coordinates $(x^1, \ldots, x^n)$ of a local chart (U_i, φ_i) the operator l has the form $\xi_i^1 \frac{\partial}{\partial x_i^1} + \cdots + \xi_i^n \frac{\partial}{\partial x_i^n} = D_{\xi_i}$. The numbers $(\xi_i^1, \ldots, \xi_i^n)$ are naturally called the *coordinates of the tangent vector* $l \in TM_{p_0}$ in coordinates of the chart (U_i, φ_i). By the laws of differentiation, the coordinate representations of the same functional $l \in TM_{p_0}$ in the charts (U_i, φ_i), (U_j, φ_j) are connected by the relations

$$\sum_{k=1}^n \xi_i^k \frac{\partial}{\partial x_i^k} = \sum_{m=1}^n \xi_j^m \frac{\partial}{\partial x_j^m} = \sum_{k=1}^n \left(\sum_{m=1}^n \frac{\partial x_i^k}{\partial x_j^m} \xi_j^m \right) \frac{\partial}{\partial x_i^k}, \tag{15.33'}$$

which of course duplicate (15.33).

15.3.2 Differential Forms on a Manifold

Let us now consider the space T^*M_p conjugate to the tangent space TM_p, that is, T^*M_p is the space of real-valued linear functionals on TM_p.

Definition 3 The space T^*M_p conjugate to the tangent space TM_p to the manifold M at the point $p \in M$ is called the *cotangent space to M at p.*

If the manifold M is a $C^{(\infty)}$ manifold, $f \in C^{(\infty)}(M, \mathbb{R})$, and l_ξ is the derivation corresponding to the vector $\xi \in TM_0$, then for a fixed $f \in C^{(\infty)}(M, \mathbb{R})$ the mapping $\xi \mapsto l_\xi f$ will obviously be an element of the space T^*M_p. In the case $M = \mathbb{R}^n$ we obtain $\xi \mapsto D_\xi f(p) = f'(p)\xi$, so that the resulting mapping $\xi \mapsto l_\xi f$ is naturally called the differential of the function f at p, and is denoted by the usual symbol $df(p)$.

If $T\mathbb{R}^n_{\varphi_\alpha^{-1}(p)}$ (or $TH^n_{\varphi_\alpha^{-1}(p)}$ when $p \in \partial M$) is the space corresponding to the tangent space TM_p in the chart $(U_\alpha, \varphi_\alpha)$ on the manifold M, it is natural to regard the space $T^*\mathbb{R}^n_{\varphi_\alpha^{-1}(p)}$ conjugate to $T\mathbb{R}^j_{\varphi_\alpha^{-1}(p)}$ as the representative of the space T^*M_p in this local chart. In coordinates $(x_\alpha^1, \ldots, x_\alpha^n)$ of a local chart $(U_\alpha, \varphi_\alpha)$ the dual basis $\{dx^1, \ldots, dx^n\}$ in the conjugate space corresponds to the basis $\{\frac{\partial}{\partial x_\alpha^1}, \ldots, \frac{\partial}{\partial x_\alpha^n}\}$ of $T\mathbb{R}^n_{\varphi_\alpha^{-1}(p)}$ (or $TH^n_{\varphi_\alpha^{-1}(p)}$ if $p \in \partial M$). We recall that $dx^i(\xi) = \xi^i$, so that $dx^i(\frac{\partial}{\partial x^j}) = \delta_j^i$. The expressions for these dual bases in another chart (U_β, φ_β) may turn out to be not so simple, for $\frac{\partial}{\partial x_\beta^j} = \frac{\partial x_\alpha^i}{\partial x_\beta^j}\frac{\partial}{\partial x_\alpha^i}$, $dx_\alpha^i = \frac{\partial x_\alpha^i}{\partial x_\beta^j} dx_\beta^j$.

Definition 4 We say that a *differential form* ω^m *of degree* m is defined on an n-dimensional manifold M if a skew-symmetric form $\omega^m(p) : (TM_p)^m \to \mathbb{R}$ is defined on each tangent space TM_p to M, $p \in M$.

In practice this means only that a corresponding m-form $\omega_\alpha(x_\alpha)$, where $x_\alpha = \varphi_\alpha^{-1}(p)$, is defined on each space $T\mathbb{R}^n_{\varphi_\alpha^{-1}(p)}$ (or $TH^n_{\varphi_\alpha^{-1}(p)}$) corresponding to TM_0 in the chart $(U_\alpha, \varphi_\alpha)$ of the manifold M. The fact that two such forms $\omega_\alpha(x_\alpha)$ and $\omega_\beta(x_\beta)$ are representatives of the same form $\omega(p)$ can be expressed by the relation

$$\omega_\alpha(x_\alpha)\big((\xi_1)_\alpha, \ldots, (\xi_m)_\alpha\big) = \omega_\beta(x_\beta)\big((\xi_1)_\beta, \ldots, (\xi_m)_\beta\big), \tag{15.39}$$

in which x_α and x_β are the representatives of the point $p \in M$, and $(\xi_1)_\alpha, \ldots, (\xi_m)_\alpha$ and $(\xi_1)_\beta, \ldots, (\xi_m)_\beta$ are the coordinate representations of the vectors $\xi_1, \ldots, \xi_m \in TM_p$ in the charts $(U_\alpha, \varphi_\alpha)$, (U_β, φ_β) respectively.

In more formal notation this means that

$$x_\alpha = \varphi_{\beta\alpha}(x_\beta), \qquad x_\beta = \varphi_{\alpha\beta}(x_\alpha), \tag{15.31'}$$

$$\xi_\alpha = \varphi'_{\beta\alpha}(x_\beta)\xi_\beta, \qquad \xi_\beta = \varphi'_{\alpha\beta}(x_\alpha)\xi_\alpha, \tag{15.32'}$$

where, as usual, $\varphi_{\beta\alpha}$ and $\varphi_{\alpha\beta}$ are respectively the functions $\varphi_\alpha^{-1} \circ \varphi_\beta$ and $\varphi_\beta^{-1} \circ \varphi_\alpha$ for the coordinate transitions, and the tangent mappings to them $\varphi'_{\beta\alpha} =: (\varphi_{\beta\alpha})_*$, $\varphi'_{\alpha\beta} =: (\varphi_{\alpha\beta})_*$ provide an isomorphism of the tangent spaces to $\mathbb{R}^n$ (or H^n) at the corresponding points x_α and x_β. As stated in Sect. 15.1.3, the adjoint mappings $(\varphi'_{\beta\alpha})^* =: \varphi^*_{\beta\alpha}$ and $(\varphi'_{\alpha\beta})^* =: \varphi^*_{\alpha\beta}$ provide the transfer of the forms, and the relation (15.39) means precisely that

$$\omega_\alpha(x_\alpha) = \varphi^*_{\alpha\beta}(x_\alpha)\omega_\beta(x_\beta), \tag{15.39'}$$

where α and β are indices (which can be interchanged).

The matrix (c_i^j) of the mapping $\varphi'_{\alpha\beta}(x_\alpha)$ is known: $(c_i^j) = (\frac{\partial x_\beta^j}{\partial x_\alpha^i})(x_\alpha)$. Thus, if

$$\omega_\alpha(x_\alpha) = \sum_{1 \le i_1 < \cdots < i_m \le n} a_{i_1,\ldots,i_m}\, dx_\alpha^{i_1} \wedge \cdots \wedge dx_\alpha^{i_m} \tag{15.40}$$

and

$$\omega_\beta(x_\beta) = \sum_{1 \le j_1 < \cdots < j_m \le n} b_{j_1,\ldots,j_m}\, dx_\beta^{j_1} \wedge \cdots \wedge dx_\beta^{j_m}, \tag{15.41}$$

then according to Example 7 of Sect. 15.1 we find that

$$\sum_{1 \le i_1 < \cdots < i_m \le n} a_{i_1 \ldots i_m}\, dx_\alpha^{i_1} \wedge \cdots \wedge dx_\alpha^{i_m} =$$

$$= \sum_{\substack{1 \le i_1 < \cdots < i_m \le n \\ 1 \le j_1 < \cdots < j_m \le n}} b_{j_1 \ldots j_m} \frac{\partial(x_\beta^{j_1}, \ldots, x_\beta^{j_m})}{\partial(x_\alpha^{i_1}, \ldots, x_\alpha^{i_m})}(x_\alpha)\, dx_\alpha^{i_1} \wedge \cdots \wedge dx_\alpha^{i_m}, \tag{15.42}$$

where $\frac{\partial()}{\partial()}$, as always, denotes the determinant of the matrix of corresponding partial derivatives.

Thus different coordinate expressions for the same form ω can be obtained from each other by direct substitution of the variables (expanding the corresponding differentials of the coordinates followed by algebraic transformations in accordance with the laws of exterior products).

If we agree to regard the form ω_α as the transfer of a form ω defined on a manifold to the parameter domain of the chart $(U_\alpha, \varphi_\alpha)$, it is natural to write $\omega_\alpha = \varphi_\alpha^* \omega$ and consider that $\omega_\alpha = \varphi_\alpha^* \circ (\varphi_\beta^{-1})^* \omega_\beta = \varphi_{\alpha\beta}^* \omega_\beta$, where the composition $\varphi_\alpha^* \circ (\varphi_\beta^{-1})^*$ in this case plays the role of a formal elaboration of the mapping $\varphi_{\alpha\beta}^* = (\varphi_\beta^{-1} \circ \varphi_\alpha)^*$.

Definition 5 A differential m-form ω on an n-dimensional manifold M is a $C^{(k)}$ *form* if the coefficients $a_{i_1\ldots i_m}(x_\alpha)$ of its coordinate representation

$$\omega_\alpha = \varphi_\alpha^* \omega = \sum_{1 \le i_1 < \cdots < i_m \le n} a_{i_1\ldots i_m}(x_\alpha)\, dx_\alpha^{i_1} \wedge \cdots \wedge dx_\alpha^{i_m}$$

are $C^{(k)}$ functions in every chart $(U_\alpha, \varphi_\alpha)$ of an atlas that defines a smooth structure on M.

It is clear from (15.42) that Definition 5 is unambiguous if the manifold M itself is a $C^{(k+1)}$ manifold, for example if M is a $C^{(\infty)}$ manifold.

For differential forms defined on a manifold the operations of addition, multiplication by a scalar, and exterior multiplication are naturally defined pointwise. (In particular, multiplication by a function $f : M \to \mathbb{R}$, which by definition is regarded as a form of degree zero, is defined.) The first two of these operations turn the set

Ω_k^m of m-forms of class $C^{(k)}$ on M into a vector space. In the case $k = \infty$ this vector space is usually denoted Ω^m. It is clear that exterior multiplication of forms $\omega^{m_1} \in \Omega_k^{m_1}$ and $\omega^{m_2} \in \Omega_k^{m_2}$ yields a form $\omega^{m_1+m_2} = \omega^{m_1} \wedge \omega^{m_2} \in \Omega_k^{m_1+m_2}$.

15.3.3 The Exterior Derivative

Definition 6 The *exterior differential* is the linear operator $\mathrm{d} : \Omega_k^m \to \Omega_{k-1}^{m+1}$ possessing the following properties:

1^0 On every function $f \in \Omega_k^0$ the differential $\mathrm{d} : \Omega_k^0 \to \Omega_{k-1}^1$ equals the usual differential $\mathrm{d}f$ of this function.

2^0 $\mathrm{d} : (\omega^{m_1} \wedge \omega^{m_2}) = \mathrm{d}\omega^{m_1} \wedge \omega^{m_2} + (-1)^{m_1} \omega^{m_1} \wedge \mathrm{d}\omega^{m_2}$, where $\omega^{m_1} \in \Omega_k^{m_1}$ and $\omega^{m_2} \in \Omega_k^{m_2}$.

3^0 $\mathrm{d}^2 := \mathrm{d} \circ \mathrm{d} = 0$.

This last equality means that $\mathrm{d}(\mathrm{d}\omega)$ is zero for every form ω.

Requirement 3^0 thus presumes that we are talking about forms whose smoothness is at least $C^{(2)}$.

In practice this means that we are considering a $C^{(\infty)}$ manifold M and the operator d mapping Ω^m to Ω^{m+1}.

A formula for computing the operator d in local coordinates of a specific chart (and at the same time the uniqueness of the operator d) follows from the relation

$$\mathrm{d}\left(\sum_{1 \le i_1 < \cdots < i_m \le n} c_{i_1 \dots i_m}(x)\, \mathrm{d}x^{i_1} \wedge \cdots \wedge \mathrm{d}x^{i_m} \right) =$$

$$= \sum_{1 \le i_1 < \cdots < i_m \le n} \mathrm{d}c_{i_1 \dots i_m}(x) \wedge \mathrm{d}x^{i_1} \wedge \cdots \wedge \mathrm{d}x^{i_m} +$$

$$+ \left(\sum_{1 \le i_1 < \cdots < i_m \le n} c_{i_1 \dots, i_m}\, \mathrm{d}\left(\mathrm{d}x^{i_1} \wedge \cdots \wedge \mathrm{d}x^{i_m} \right) = 0 \right). \qquad (15.43)$$

The existence of the operator d now follows from the fact that the operator defined by (15.43) in a local coordinate system satisfies conditions 1^0, 2^0, and 3^0 of Definition 6.

It follows in particular from what has been said that if $\omega_\alpha = \varphi_\alpha^* \omega$ and $\omega_\beta = \varphi_\beta^* \omega$ are the coordinate representations of the same form ω, that is, $\omega_\alpha = \varphi_{\alpha\beta}^* \omega_\beta$, then $\mathrm{d}\omega_\alpha$ and $\mathrm{d}\omega_\beta$ will also be the coordinate representations of the same form $(\mathrm{d}\omega)$, that is, $\mathrm{d}\omega_\alpha = \varphi_{\alpha\beta}^*\, \mathrm{d}\omega_\beta$. Thus the relation $\mathrm{d}(\varphi_{\alpha\beta}^* \omega_\beta) = \varphi_{\alpha\beta}^* (\mathrm{d}\omega_\beta)$ holds, which in abstract form asserts the commutativity

$$\mathrm{d}\varphi^* = \varphi^*\mathrm{d} \qquad (15.44)$$

of the operator d and the operation φ^* that transfers forms.

15.3.4 The Integral of a Form over a Manifold

Definition 7 Let M be an n-dimensional smooth oriented manifold on which the coordinates $x^1, \ldots, x^n$ and the orientation are defined by a single chart $\varphi_x : D_x \to M$ with parameter domain $D_x \subset \mathbb{R}^n$. Then

$$\int_M \omega := \int_{D_x} a(x)\,dx^1 \wedge \cdots \wedge dx^n, \qquad (15.45)$$

where the left-hand side is the usual *integral of the form ω over the oriented manifold M* and the right-hand side is the integral of the function $f(x)$ over the domain D_x.

If $\varphi_t : D_t \to M$ is another atlas of M consisting of a single chart defining the same orientation on M as $\varphi_x : D_x \to M$, then the Jacobian $\det \varphi'(t)$ of the function $x = \varphi(t)$ of the coordinate change is everywhere positive in D_t. The form

$$\varphi^*\big(a(x)\,dx^1 \wedge \cdots \wedge dx^n\big) = a\big(x(t)\big) \det \varphi'(t)\,dt^1 \wedge \cdots \wedge dt^n$$

in D_t corresponds to the form ω. By the theorem on change of variables in a multiple integral we have the equality

$$\int_{D_x} a(x)\,dx^1 \cdots dx^n = \int_{D_t} a\big(x(t)\big) \det \varphi'(t)\,dt^1 \cdots dt^n,$$

which shows that the left-hand side of (15.45) is independent of the coordinate system chosen in M.

Thus, Definition 7 is unambiguous.

Definition 8 The *support* of a form ω defined on a manifold M is the closure of the set of points $x \in M$ where $\omega(x) \neq 0$.

The support of a form ω is denoted by $\operatorname{supp}\omega$. In the case of 0-forms, that is, functions, we have already encountered this concept. Outside the support the coordinate representation of the form in any local coordinate system is the zero form of the corresponding degree.

Definition 9 A form ω defined on a manifold M is *of compact support* if $\operatorname{supp}\omega$ is a compact subset of M.

Definition 10 Let ω be a form of degree n and compact support on an n-dimensional smooth manifold M oriented by the atlas A. Let $\varphi_i : D_i \to U_i$, $\{(U_i, \varphi), i = 1, \ldots, m\}$ be a finite set of charts of the atlas A whose ranges $U_1, \ldots, U_m$ cover $\operatorname{supp}\omega$, and let $e_1, \ldots, e_k$ be a partition of unity subordinate to that covering on $\operatorname{supp}\omega$. Repeating some charts several times if necessary, we can assume that $m = k$, and that $\operatorname{supp} e_i \subset U_i$, $i = 1, \ldots, m$.

The *integral of a form ω of compact support over the oriented manifold M* is the quantity

$$\int_M \omega := \sum_{i=1}^{m} \int_{D_i} \varphi_i^*(e_i\omega), \tag{15.46}$$

where $\varphi_i^*(e_i\omega)$ is the coordinate representation of the form $e_i\omega|_{U_i}$ in the domain D_i of variation of the coordinates of the corresponding local chart.

Let us prove that this definition is unambiguous.

Proof Let $\widetilde{A} = \{\widetilde{\varphi}_j : \widetilde{D}_j \to \widetilde{U}_j\}$ be a second atlas defining the same smooth structure and orientation on M as the atlas A, let $\widetilde{U}_1, \ldots, \widetilde{U}_{\widetilde{m}}$ be the corresponding covering of $\operatorname{supp}\omega$, and let $\widetilde{e}_1, \ldots, \widetilde{e}_{\widetilde{m}}$ a partition of unity on $\operatorname{supp}\omega$ subordinate to this covering. We introduce the functions $f_{ij} = e_i\widetilde{e}_j$, $i = 1, \ldots, m$, $j = 1, \ldots, \widetilde{m}$, and we set $\omega_{ij} = f_{ij}\omega$.

We remark that $\operatorname{supp}\omega_{ij} \subset W_{ij} = U_i \cap \widetilde{U}_j$. From this and from the fact that Definition 7 of the integral over an oriented manifold given by a single chart is unambiguous it follows that

$$\int_{D_i} \varphi_i^*(\omega_{ij}) = \int_{\varphi_i^{-1}(W_{ij})} \varphi_i^*(\omega_{ij}) = \int_{\widetilde{\varphi}_j^{-1}(W_{ij})} \widetilde{\varphi}_j^*(\omega_{ij}) = \int_{\widetilde{D}_j} \widetilde{\varphi}_j^*(\omega_{ij}).$$

Summing these equalities on i from 1 to m and on j from 1 to $\widetilde{m}$, taking account of the relation $\sum_{i=1}^{m} f_{ij} = \widetilde{e}_j$, $\sum_{j=1}^{\widetilde{m}} f_{ij} = e_i$, we find the identities we are interested int. $\qquad\square$

15.3.5 Stokes' Formula

Theorem *Let M be an oriented smooth n-dimensional manifold and ω a smooth differential form of degree $n - 1$ and compact support on M. Then*

$$\int_{\partial M} \omega = \int_M d\omega, \tag{15.47}$$

where the orientation of the boundary ∂M of the manifold M is induced by the orientation of the manifold M. If $\partial M = \varnothing$, then $\int_M d\omega = 0$.

Proof Without loss of generality we may assume that the domains of variation of the coordinates (parameters) of all local charts of the manifold M are either the open cube $I = \{x \in \mathbb{R}^n \mid 0 < x^i < 1, i = 1, \ldots, n\}$, or the cube $\widetilde{I} = \{x \in \mathbb{R}^n \mid 0 < x^1 \leq 1 \wedge 0 < x^i < 1, i = 1, \ldots, n\}$ with one (definite!) face adjoined to the cube I.

By the partition of unity the assertion of the theorem reduces to the case when $\operatorname{supp}\omega$ is contained in the range U of a single chart of the form $\varphi : I \to U$ or

$\varphi : \tilde{I} \to U$. In the coordinates of this chart the form ω has the form

$$\omega = \sum_{i=1}^{n} a_i(x)\,dx^1 \wedge \cdots \wedge \widehat{dx^i} \wedge \cdots \wedge dx^n,$$

where the frown $\frown$, as usual, means that the corresponding factor is omitted.

By the linearity of the integral, it suffices to prove the assertion for one term of the sum:

$$\omega_i = a_i(x)\,dx^1 \wedge \cdots \wedge \widehat{dx^i} \wedge \cdots \wedge dx^n. \tag{15.48}$$

The differential of such a form is the n-form

$$d\omega_i = (-1)^{i-1} \frac{\partial a_i}{\partial x^i}(x)\,dx^1 \wedge \cdots \wedge dx^n. \tag{15.49}$$

For a chart of the form $\varphi : I \to U$ both integrals in (15.47) of the corresponding forms (15.48) and (15.49) are zero: the first because $\operatorname{supp} a_i \subset I$ and the second for the same reason, if we take into account Fubini's theorem and the relation $\int_0^1 \frac{\partial a_i}{\partial x^i}\,dx^i = a_i(1) - a_i(0) = 0$. This argument also covers the case when $\partial M = \varnothing$.

Thus it remains to verify (15.47) for a chart $\varphi : \tilde{I} \to U$.

If $i > 1$, both integrals are also zero for such a chart, which follows from the reasoning given above.

And if $i = 1$, then

$$\int_M d\omega_1 = \int_U d\omega_1 = \int_{\tilde{I}} \frac{\partial a_1}{\partial x^1}(x)\,dx^1 \cdots dx^n =$$

$$= \int_0^1 \cdots \int_0^1 \left(\int_0^1 \frac{\partial a_1}{\partial x^1}(x)\,dx^1 \right) dx^2 \cdots dx^n =$$

$$= \int_0^1 \cdots \int_0^1 a_1(1, x^2, \ldots, x^n)\,dx^2 \cdots dx^n = \int_{\partial U} \omega_1 = \int_{\partial M} \omega_1.$$

Thus formula (15.47) is proved for $n > 1$.

The case $n = 1$ is merely the Newton–Leibniz formula (the fundamental theorem of calculus), if we assume that the endpoints α and β of the oriented interval $[\alpha, \beta]$ are denoted α_- and β_+ and the integral of a 0-form $g(x)$ over such an oriented point is equal to $-g(\alpha)$ and $+g(\beta)$ respectively. □

We now make some remarks on this theorem.

Remark 1 Nothing is said in the statement of the theorem about the smoothness of the manifold M and the form ω. In such cases one usually assumes that each of them is $C^{(\infty)}$. It is clear from the proof of the theorem, however, that formula (15.47) is also true for forms of class $C^{(2)}$ on a manifold M admitting a form of this smoothness.

Remark 2 It is also clear from the proof of the theorem, as in fact it was already from the formula (15.47), that if $\operatorname{supp}\omega$ is a compact set contained strictly inside M, that is, $\operatorname{supp}\omega \cap \partial M = \varnothing$, then $\int_M d\omega = 0$.

Remark 3 If M is a compact manifold, then for every form ω on M the support $\operatorname{supp}\omega$, being a closed subset of the compact set M, is compact. Consequently in this case every form ω on M is of compact support and Eq. (15.47) holds. In particular, if M is a compact manifold without boundary, then the equality $\int_M d\omega = 0$ holds for every smooth form on M.

Remark 4 For arbitrary forms ω (not of compact support) on a manifold that is not itself compact, formula (15.47) is in general not true.

Let us consider, for example, the form $\omega = \frac{x\,dy - y\,dx}{x^2 + y^2}$ in a circular annulus $M = \{(x, y) \in \mathbb{R}^2 \mid 1 \leq x^2 + y^2 \leq 2\}$, endowed with standard Cartesian coordinates. In this case M is a compact two-dimensional oriented manifold, whose boundary ∂M consists of the two circles $C_i = \{(x, y) \in \mathbb{R}^2 \mid x^2 + y^2 = i\}$, $i = 1, 2$. Since $d\omega = 0$, we find by formula (15.47) that

$$0 = \int_M d\omega = \int_{C_2} \omega - \int_{C_1} \omega,$$

where both circles C_1 and C_2 are traversed counterclockwise. We know that

$$\int_{C_1} \omega = \int_{C_2} \omega = 2\pi \neq 0.$$

Hence, if we consider the manifold $\widetilde{M} = M \backslash C_1$, then $\partial \widetilde{M} = C_2$ and

$$\int_{\widetilde{M}} d\omega = 0 \neq 2\pi = \int_{\partial \widetilde{M}} \omega.$$

15.3.6 Problems and Exercises

1. a) We call two smooth paths $\gamma_i : \mathbb{R} \to M$, $i = 1, 2$ on a smooth manifold M *tangent* at a point $p \in M$ if $\gamma_1(0) = \gamma_2(0) = p$ and the relation

$$\left| \varphi^{-1} \circ \gamma_1(t) - \varphi^{-1} \circ \gamma_2(t) \right| = o(t) \quad \text{as } t \to 0 \tag{15.50}$$

holds in each local coordinate system $\varphi : \mathbb{R}^n \to U$ (or $\varphi : H^n \to U$) whose range U contains p. Show that if (15.50) holds in one of these coordinate systems, then it holds in any other local coordinate system of the same type on the smooth manifold M.

b) The property of being tangent at a point $p \in M$ is an equivalence relation on the set of smooth paths on M passing through p. We call an equivalence class a *bundle of tangent paths at* $p \in M$. Establish the one-to-one correspondence exhibited in Sect. 15.3.1 between vectors of TM_p and bundles of tangent paths at the point $p \in M$.

c) Show that if the paths γ_1 and γ_2 are tangent at $p \in M$ and $f \in C^{(1)}(M, \mathbb{R})$, then

$$\frac{\mathrm{d}f \circ \gamma_1}{\mathrm{d}t}(0) = \frac{\mathrm{d}f \circ \gamma_2}{\mathrm{d}t}(0).$$

d) Show how to assign a functional $l = l_\xi (= D_\xi) : C^{(\infty)}(M, \mathbb{R}) \to \mathbb{R}$ possessing properties (15.37) and (15.38), where $x_0 = p$, to each vector $\xi \in TM_p$. A functional possessing these properties is called a *derivation at the point* $p \in M$.

Verify that differentiation l at the point p is a local operation, that is, if $f_1, f_2 \in C^{(\infty)}$ and $f_1(x) \equiv f_2(x)$ in some neighborhood of p, then $lf_1 = lf_2$.

e) Show that if $x^1, \ldots, x^n$ are local coordinates in a neighborhood of the point p, then $l = \sum_{i=1}^n (lx^i) \frac{\partial}{\partial x^i}$, where $\frac{\partial}{\partial x^i}$ is the operation of computing the partial derivative with respect to x^i at the point x corresponding to p. (Hint. Write the function $f|_{U(p)} : M \to \mathbb{R}$ in local coordinates; remember that the expansion $f(x) = f(0) + \sum_{i=1}^n x^i g_i(x)$ holds for the function $f \in C^{(\infty)}(\mathbb{R}^n, \mathbb{R})$, where $g_i \in C^{(\infty)}(\mathbb{R}^n, \mathbb{R})$ and $g_i(0) = \frac{\partial f}{\partial x^i}(0)$, $i = 1, \ldots, n$.)

f) Verify that if M is a $C^{(\infty)}$ manifold, then the vector space of derivations at the point $p \in M$ is isomorphic to the space TM_p tangent to M at p constructed in Sect. 15.3.1.

2. a) If we fix a vector $\xi(p) \in TM_p$ at each point $p \in M$ of a smooth manifold M, we say that a *vector field* is defined *on the manifold* M. Let X be a vector field on M. Since by the preceding problem every vector $X(p) = \xi \in TM_p$ can be interpreted as differentiation at the corresponding point p, from any function $f \in C^{(\infty)}(M, \mathbb{R})$ one can construct a function $Xf(p)$ whose value at every point $p \in M$ can be computed by applying $X(p)$ to f, that is, to differentiating f with respect to the vector $X(p)$ in the field X. A field X on M is *smooth (of class $C^{(\infty)}$)* if for every function $f \in C^{(\infty)}(M, \mathbb{R})$ the function Xf also belongs to $C^{(\infty)}(M, \mathbb{R})$.

Give a local coordinate expression for a vector field and the coordinate definition of a smooth ($C^{(\infty)}$) vector field on a smooth manifold equivalent to the one just given.

b) Let X and Y be two smooth vector fields on the manifold M. For functions $f \in C^{(\infty)}(M, \mathbb{R})$ we construct the functional $[X, Y]f = X(Yf) - Y(Xf)$. Verify that $[X, Y]$ is also a smooth vector field on M. It is called the *Poisson bracket of the vector fields* X *and* Y.

c) Give a Lie algebra structure to the smooth vector fields on a manifold.

3. a) Let X and ω be respectively a smooth vector field and a smooth 1-form on a smooth manifold M. Let ωX denote the application of ω to the vector of the field X at corresponding points of M. Show that ωX is a smooth function on M.

b) Taking account of Problem 2, show that the following relation holds:

$$\mathrm{d}\omega^1(X, Y) = X(\omega^1 Y) - Y(\omega^1 X) - \omega^1([X, Y]),$$

where X and Y and smooth vector fields, $d\omega^1$ is the differential of the form ω^1, and $d\omega^1(X, Y)$ is the application of $d\omega^1$ to pairs of vectors of the fields X and Y attached at the same point.

c) Verify that the relation

$$d\omega(X_1, \ldots, X_{m+1}) =$$

$$= \sum_{i=1}^{m+1}(-1)^{i+1}X_i\omega(X_1, \ldots, \widehat{X_i}, \ldots, X_{m+1}) +$$

$$+ \sum_{1\leq i<j\leq m+1}(-1)^{i+j}\omega\big([X_i, X_j], X_1, \ldots, \widehat{X_i}, \ldots, \widehat{X_j}, \ldots, X_{m+1}\big)$$

holds for the general case of a form ω of order m. Here the frown $\frown$ denotes an omitted term, $[X_i, X_j]$ is the Poisson bracket of the fields X_i and X_j, and $X_i\omega$ represents differentiation of the function $\omega(X_1, \ldots, \widehat{X_i}, \ldots, X_{m+1})$ with respect to the vectors of the field X_i. Since the Poisson bracket is invariantly defined, the resulting relation can be thought of as a rather complicated but *invariant definition of the exterior differential operator* $d : \Omega \to \Omega$.

d) Let ω be a smooth m-form on a smooth n-dimensional manifold M. Let $(\xi_1, \ldots, \xi_{m+1})_i$ be vectors in $\mathbb{R}^n$ corresponding to the vectors $\xi_1, \ldots, \xi_{m+1} \in TM_p$ in the chart $\varphi_i : \mathbb{R}^n \to U \subset M$. We denote by Π_i the parallelepiped formed by the vectors $(\xi_1, \ldots, \xi_{m+1})_i$ in $\mathbb{R}^n$, and let $\lambda\Pi_i$ be the parallelepiped spanned by the vectors $(\lambda\xi_1, \ldots, \lambda\xi_{m+1})_i$. We denote the images $\varphi_i(\Pi_i)$ and $\varphi_i(\lambda\Pi_i)$ of these parallelepipeds in M by Π and $\lambda\Pi$ respectively. Show that

$$d\omega(p)(\xi_1, \ldots, \xi_{m+1}) = \lim_{\lambda\to 0}\frac{1}{\lambda^{n+1}}\int_{\partial(\lambda\Pi)}\omega.$$

4. a) Let $f : M \to N$ be a smooth mapping of a smooth m-dimensional manifold M into a smooth n-dimensional manifold N. Using the interpretation of a tangent vector to a manifold as a bundle of tangent paths (see Problem 1), construct the mapping $f_*(p) : TM_p \to TN_{f(p)}$ induced by f.

b) Show that the mapping f_* is linear and write it in corresponding local coordinates on the manifolds M and N. Explain why $f_*(p)$ is called the *differential of* f at p or the mapping *tangent to* f at that point.

Let f be a diffeomorphism. Verify that $f_*[X, Y] = [f_*X, f_*Y]$. Here X and Y are vector fields on M and $[\cdot, \cdot]$ is their Poisson bracket (see Problem 2).

c) As is known from Sect. 15.1, the tangent mapping $f_*(p) : TM_p \to TN_{q=f(p)}$ of tangent spaces generates the adjoint mapping $f^*(p)$ of the conjugate spaces and in general a mapping of k-forms defined on $TN_{f(p)}$ and TM_p.

Let ω be a k-form on N. The k-form $f^*\omega$ on M is defined by the relation

$$\big(f^*\omega\big)(p)(\xi_1, \ldots, \xi_k) := \omega\big(f(p)\big)(f_*\xi_1, \ldots, f_*\xi_k),$$

where $\xi_1, \ldots, \xi_k \in TM_p$. In this way a mapping $f^* : \Omega^k(N) \to \Omega^k(M)$ arises from the space $\Omega^k(N)$ of k-forms defined on N into the space $\Omega^k(M)$ of k-forms defined on M.

Verify the following properties of the mapping f^*, assuming M and N are $C^{(\infty)}$ manifolds:

1^0 f^* is a linear mapping;
2^0 $f^*(\omega_1 \wedge \omega_2) = f^*\omega_1 \wedge f^*\omega_2$;
3^0 $d \circ f^* = f^* \circ d$, that is $d(f^*\omega) = f^*(d\omega)$;
4^0 $(f_2 \circ f_1)^* = f_1^* \circ f_2^*$.

d) Let M and N be smooth n-dimensional oriented manifolds and $\varphi : M \to N$ a diffeomorphism of M onto N. Show that if ω is an n-form on N with compact support, then

$$\int_{\varphi(M)} \omega = \varepsilon \int_M \varphi^*\omega,$$

where $\varepsilon = \begin{cases} 1, & \text{if } \varphi \text{ preserves orientation,} \\ -1, & \text{if } \varphi \text{ reverses orientation.} \end{cases}$

e) Suppose $A \supset B$. The mapping $i : B \to A$ that assigns to each point $x \in B$ that same point as an element of A is called the *canonical embedding of B in A*.

If ω is a form on a manifold M and M' is a submanifold of M, the canonical embedding $i : M' \to M$ generates a form $i^*\omega$ on M' called the *restriction of ω to M'*. Show that the proper expression of Stokes' formula (15.47) should be

$$\int_M d\omega = \int_{\partial M} i^*\omega,$$

where $i : \partial M \to M$ is the canonical embedding of ∂M in M, and the orientation of ∂M is induced from M.

5. a) Let M be a smooth $(C^{(\infty)})$ oriented n-dimensional manifold and $\Omega_c^n(M)$ the space of smooth $(C^{(\infty)})$ n-forms with compact support on M. Show that there exists a unique mapping $\int_M : \Omega_c^n(M) \to \mathbb{R}$ having the following properties:

1^0 the mapping $\int_M$ is linear;
2^0 if $\varphi : I^n \to U \subset M$ (or $\varphi : \tilde{I}^n \to U \subset M$) is a chart of an atlas defining the orientation of M, $\operatorname{supp}\omega \subset U$, and $\omega = a(x)\, dx^1 \wedge \cdots \wedge dx^n$ in the local coordinates $x^1, \ldots, x^n$ of this chart, then

$$\int_M \omega = \int_{I^n} a(x)\, dx^1, \ldots, dx^n \quad \left(\text{or} \int_M \omega = \int_{\tilde{I}^n} a(x)\, dx^1, \ldots, dx^n \right),$$

where the right-hand side contains the Riemann integral of the function a over the corresponding cube I^n (or $\tilde{I}^n$).

b) Can the mapping just exhibited always be extended to a mapping $\int_M : \Omega^n(M) \to \mathbb{R}$ of all smooth n-forms on M, retaining both of these properties?

c) Using the fact that every open covering of the manifold M has an at most countable locally finite refinement and the fact that there exists a partition of unity subordinate to any such covering (see Problem 9), define the integral of an n-form over an oriented smooth n-dimensional (not necessarily compact) manifold so that it has properties 1^0 and 2^0 above when applied to the forms for which the integral is finite. Show that for this integral formula (15.47) does not hold in general, and give conditions on ω that are sufficient for (15.47) in the case when $M = \mathbb{R}^n$ and in the case when $M = H^n$.

6. a) Using the theorem on existence and uniqueness of the solution of the differential equation $\dot{x} = v(x)$ and also the smooth dependence of the solution on the initial data, show that a smooth bounded vector field $v(x) \in \mathbb{R}^n$ can be regarded as the velocity field of a steady-state flow. More precisely, show that there exists a family of diffeomorphisms $\varphi_t : \mathbb{R}^n \to \mathbb{R}^n$ depending smoothly on the parameter t (time) such that $\varphi_t(x)$ is an integral curve of the equation for each fixed value of $x \in \mathbb{R}^n$, that is, $\frac{\partial \varphi_t(x)}{\partial t} = v(\varphi_t(x))$ and $\varphi_0(x) = x$. The mapping $\varphi_t : \mathbb{R}^n \to \mathbb{R}^n$ obviously characterizes the displacement of the particles of the medium at time t. Verify that the family of mappings $\varphi_t : \mathbb{R}^n \to \mathbb{R}^n$ is a *one-parameter group of diffeomorphisms*, that is, $(\varphi_t)^{-1} = \varphi_{-t}$, and $\varphi_{t_2} \circ \varphi_{t_1} = \varphi_{t_1+t_2}$.

b) Let v be a vector field on $\mathbb{R}^n$ and φ_t a one-parameter group of diffeomorphisms of $\mathbb{R}^n$ generated by v. Verify that the relation

$$\lim_{t \to 0} \frac{1}{t} \left(f(\varphi_t(x)) - f(x) \right) = D_{v(x)} f$$

holds for every smooth function $f \in C^{(\infty)}(\mathbb{R}^n, \mathbb{R})$.

If we introduce the notation $v(f) := D_v f$, in consistency with the notation of Problem 2, and recall that $f \circ \varphi_t =: \varphi_t^* f$, we can write

$$\lim_{t \to 0} \frac{1}{t} (\varphi_t^* f - f)(x) = v(f)(x).$$

c) Differentiation of a smooth form ω of any degree defined in $\mathbb{R}^n$ along the field v is now naturally defined. To be specific, we set

$$v(\omega)(x) := \lim_{t \to 0} \frac{1}{t} (\varphi_t^* \omega - \omega)(x).$$

The form $v(\omega)$ is called the *Lie derivative* of the form ω along the field v and usually denoted $L_v \omega$. Define the Lie derivative $L_X \omega$ of a form ω along the field X on an arbitrary smooth manifold M.

d) Show that the Lie derivative on a $C^{(\infty)}$ manifold M has the following properties.

1^0 L_X is a local operation, that is, if the fields X_1 and X_2 and the forms ω_1 and ω_2 are equal in a neighborhood $U \subset M$ of the point x, then $(L_{X_1}\omega_1)(x) = (L_{X_2}\omega_2)(x)$.

2^0 $L_X \Omega^k(M) \subset \Omega^k(M)$.

3^0 $L_X : \Omega^k(M) \to \Omega^k(M)$ is a linear mapping for every $k = 0, 1, 2, \dots$.

4^0 $L_X(\omega_1 \wedge \omega_2) = (L_X \omega_1) \wedge \omega_2 + \omega_1 \wedge L_X \omega_2$.

5^0 If $f \in \Omega^0(M)$, then $L_X f = df(X) =: Xf$.

6^0 If $f \in \Omega^0(M)$, then $L_X df = d(Xf)$.

e) Verify that the properties 1^0–6^0 determine the operation L_X uniquely.

7. Let X be a vector field and ω a form of degree k on the smooth manifold M.

The *inner product* of the field X and the form ω is the $(k-1)$-form denoted by $i_X\omega$ or $X \rfloor \omega$ and defined by the relation $(i_X\omega)(X_1, \ldots, X_{k-1}) :=$ $\omega(X, X_1, \ldots, X_{k-1})$, where $X_1, \ldots, X_{k-1}$ are vector fields on M. For 0-forms, that is, functions on M, we set $X \rfloor f = 0$.

a) Show that if the form ω (more precisely, $\omega|_U$) has the form

$$\sum_{1 \le i_1 < \cdots < i_k \le n} a_{i_1 \cdots i_k}(x)\, dx^{i_1} \wedge \cdots \wedge dx^{i_k} = \frac{1}{k!} a_{i_1 \ldots i_k} dx^{i_1} \wedge \cdots \wedge dx^{i_k}$$

in the local coordinates $x^1, \ldots, x^n$ of the chart $\varphi : \mathbb{R}^n \to U \subset M$, and $X = X^i \frac{\partial}{\partial x^i}$, then

$$i_X\omega = \frac{1}{(k-1)!} X^i a_{i i_2 \ldots i_k}\, dx^{i_2} \wedge \cdots \wedge dx^{i_k}.$$

b) Verify further that if $df = \frac{\partial f}{\partial x^i} dx^i$, then $i_X\, df = X^i \frac{\partial f}{\partial x^i} = X(f) \equiv D_X f$.

c) Let $X(M)$ be the space of vector fields on the manifold M and $\Omega(M)$ the ring of skew-symmetric forms on M. Show that there exists only one mapping $i :$ $X(M) \times \Omega(M) \to \Omega(M)$ having the following properties:

1^0 i is a local operation, that is, if the fields X_1 and X_2 and the forms ω_1 and ω_2 are equal in a neighborhood U of $x \in M$, then $(i_{X_1}\omega_1)(x) = (i_{X_2}\omega_2)(x)$;

2^0 $i_X(\Omega^k(M)) \subset \Omega^{k-1}(M)$;

3^0 $i_X : \Omega^k(M) \to \Omega^{k-1}(M)$ is a linear mapping;

4^0 if $\omega_1 \in \Omega^{k_1}(M)$ and $\omega_2 \in \Omega^{k_2}(M)$, then $i_X(\omega_1 \wedge \omega_2) = i_X\omega_1 \wedge \omega_2 + (-1)^{k_1}\omega_1 \wedge i_X\omega_2$;

5^0 if $\omega \in \Omega^1(M)$, then $i_X\omega = \omega(X)$, and if $f \in \Omega^0(M)$, then $i_X f = 0$.

8. Prove the following assertions.

a) The operators d, i_X, and L_X (see Problems 6 and 7) satisfy the so-called *homotopy identity*

$$L_X = i_X d + d i_X, \qquad (15.51)$$

where X is any smooth vector field on the manifold.

b) The Lie derivative commutes with d and i_X, that is,

$$L_X \circ d = d \circ L_X, \qquad L_X \circ i_X = i_X \circ L_X.$$

c) $[L_X, i_Y] = i_{[X,Y]}$, $[L_X, L_Y] = L_{[X,Y]}$, where, as always, $[A, B] = A \circ B - B \circ A$ for any operators A and B for which the expression $A \circ B - B \circ A$ is defined. In this case, all brackets $[\,,\,]$ are defined.

d) $L_X f\omega = f L_X \omega + df \wedge i_X\omega$, where $f \in \Omega^0(M)$ and $\omega \in \Omega^k(M)$.

(Hint. Part a) is the main part of the problem. It can be verified, for example, by induction on the degree of the form on which the operators act.)

15.4 Closed and Exact Forms on Manifolds

15.4.1 Poincaré's Theorem

In this section we shall supplement what was said about closed and exact differential forms in Sect. 14.3 in connection with the theory of vector fields in $\mathbb{R}^n$. As before, $\Omega^p(M)$ denotes the space of smooth real-valued forms of degree p on the smooth manifold M and $\Omega(M) = \bigcup_p \Omega^p(M)$.

Definition 1 The form $\omega \in \Omega^p(M)$ is *closed* if $d\omega = 0$.

Definition 2 The form $\omega \in \Omega^p(M)$, $p > 0$, is *exact* if there exists a form $\alpha \in \Omega^{p-1}(M)$ such that $\omega = d\alpha$.

The set of closed p-forms on the manifold M will be denoted $Z^p(M)$, and the set of exact p-forms on M will be denoted $B^p(M)$.

The relation[7] $d(d\omega) = 0$ holds for every form $\omega \in \Omega(M)$, which shows that $Z^p(M) \supset B^p(M)$. We already know from Sect. 14.3 that this inclusion is generally strict.

The important question of the solvability (for α) of the equation $d\alpha = \omega$ given the necessary condition $d\omega = 0$ on the form ω turns out to be closely connected with the topological structure of the manifold M. This statement will be deciphered more completely below.

Definition 3 We shall call a manifold M *contractible* (to the point $x_0 \in M$) or *homotopic to a point* if there exists a smooth mapping $h : M \times I \to M$ where $I = \{t \in \mathbb{R} \mid 0 \leq t \leq 1\}$ such that $h(x, 1) = x$ and $h(x, 0) = x_0$.

Example 1 The space $\mathbb{R}^n$ can be contracted to a point by the mapping $h(x, t) = tx$.

Theorem 1 (Poincaré) *Every closed $(p + 1)$-form $(p \geq 0)$ on a manifold that is contractible to a point is exact.*

Proof The nontrivial part of the proof consists of the following "cylindrical" construction, which remains valid for every manifold M.

Consider the "cylinder", $M \times I$, which is the direct product of M and the closed unit interval I, and the two mappings $j_i : M \to M \times I$, where $j_i(x) = (x, i)$, $i = 0, 1$, which identify M with the bases of the cylinder $M \times I$. Then there naturally arise mappings $j_i^* : \Omega^p(M \times I) \to \Omega^p(M)$, reducing to the replacement of the variable t in a form of $\Omega^p(M \times I)$ by the value i ($= 0, 1$), and, of course, $di = 0$.

---

[7]Depending on the way in which the operator d is introduced this property is either proved, in which case it is called the *Poincaré lemma*, or taken as part of the definition of the operator d.

We construct a linear operator $K : \Omega^{p+1}(M \times I) \to \Omega^p(M)$, which we define on monomials as follows:

$$K\left(a(x,t)\,\mathrm{d}x^{i_1} \wedge \cdots \wedge \mathrm{d}x^{i_{p+1}}\right) := 0,$$

$$K\left(a(x,t)\,\mathrm{d}t \wedge \mathrm{d}x^{i_1} \wedge \cdots \wedge \mathrm{d}x^{i_p}\right) := \left(\int_0^1 a(x,t)\,\mathrm{d}t\right)\mathrm{d}x^{i_1} \wedge \cdots \wedge \mathrm{d}x^{i_p}.$$

The main property of the operator K that we need is that the relation

$$K(\mathrm{d}\omega) + \mathrm{d}(K\omega) = j_1^*\omega - j_0^*\omega \tag{15.52}$$

holds for every form $\omega \in \Omega^{p+1}(M \times I)$.

It suffices to verify this relation for monomials, since all the operators K, d, j_1^*, and j_0^* are linear.

If $\omega = a(x,t)\,\mathrm{d}x^{i_1} \wedge \cdots \wedge \mathrm{d}x^{i_{p+1}}$, then $K\omega = 0$, $\mathrm{d}(K\omega) = 0$, and

$$\mathrm{d}\omega = \frac{\partial a}{\partial t}\,\mathrm{d}t \wedge \mathrm{d}x^{i_1} \wedge \cdots \wedge \mathrm{d}x^{i_{p+1}} + [\text{terms not containing } \mathrm{d}t],$$

$$K(\mathrm{d}\omega) = \left(\int_0^1 \frac{\partial a}{\partial t}\,\mathrm{d}t\right)\mathrm{d}x^{i_1} \wedge \cdots \wedge \mathrm{d}x^{i_{p+1}} =$$

$$= \left(a(x,1) - a(x,0)\right)\mathrm{d}x^{i_1} \wedge \cdots \wedge \mathrm{d}x^{i_{p+1}} = j_1^*\omega - j_0^*\omega,$$

and relation (15.52) is valid.

If $\omega = a(x,t)\,\mathrm{d}t \wedge \mathrm{d}x^{i_1} \wedge \cdots \wedge \mathrm{d}x^{i_p}$, then $j_1^*\omega = j_0^*\omega = 0$. Then

$$K(\mathrm{d}\omega) = K\left(-\sum_{i_0} \frac{\partial a}{\partial x^{i_0}}\,\mathrm{d}t \wedge \mathrm{d}x^{i_0} \wedge \mathrm{d}x^{i_1} \wedge \cdots \wedge \mathrm{d}x^{i_p}\right) =$$

$$= -\sum_{i_0}\left(\int_0^1 \frac{\partial a}{\partial x^{i_0}}\,\mathrm{d}t\right)\mathrm{d}x^{i_0} \wedge \cdots \wedge \mathrm{d}x^{i_p},$$

$$\mathrm{d}(K\omega) = \mathrm{d}\left(\left(\int_0^1 a(x,t)\,\mathrm{d}t\right)\mathrm{d}x^{i_1} \wedge \cdots \wedge \mathrm{d}x^{i_p}\right) =$$

$$= \sum_{i_0} \frac{\partial}{\partial x^{i_0}}\left(\int_0^1 a(x,t)\,\mathrm{d}t\right)\mathrm{d}x^{i_0} \wedge \mathrm{d}x^{i_1} \wedge \cdots \wedge \mathrm{d}x^{i_p} =$$

$$= \sum_{i_0}\left(\int_0^1 \frac{\partial a}{\partial x^{i_0}}\,\mathrm{d}t\right)\mathrm{d}x^{i_0} \wedge \mathrm{d}x^{i_1} \wedge \cdots \wedge \mathrm{d}x^{i_p}.$$

Thus relation (15.52) holds in this case also.[8] Now let M be a manifold that is contractible to the point $x_0 \in M$, let $h : M \times I \to M$ be the mapping in Definition 3,

---

[8] For the justification of the differentiation of the integral with respect to x^{i_0} in this last equality, see, for example, Sect. 17.1.

and let ω be a $(p + 1)$-form on M. Then obviously $h \circ j_1 : M \to M$ is the identity mapping and $h \circ j_0 : M \to x_0$ is the mapping of M to the point x_0, so that $(j_1^* \circ h^*)\omega = \omega$ and $(j_0^* \circ h^*)\omega = 0$. Hence it follows from (15.52) that in this case

$$K\big(\mathrm{d}(h^*\omega)\big) + \mathrm{d}\big(K\left(h^*\omega\right)\big) = \omega. \tag{15.53}$$

If in addition ω is a closed form on M, then, since $\mathrm{d}(h^*\omega) = h^*(\mathrm{d}\omega) = 0$, we find by (15.53) that

$$\mathrm{d}\big(K\left(h^*\omega\right)\big) = \omega.$$

Thus the closed form ω is the exterior derivative of the form $\alpha = K(h^*\omega) \in \Omega^p(M)$, that is, ω is an exact form on M. $\qquad\square$

Example 2 Let A, B, and C be smooth real-valued functions of the variables $x, y, z \in \mathbb{R}^3$. We ask how to solve the following system of equations for P, Q, and R:

$$\begin{cases} \dfrac{\partial R}{\partial y} - \dfrac{\partial Q}{\partial z} = A, \\[2mm] \dfrac{\partial P}{\partial z} - \dfrac{\partial R}{\partial x} = B, \\[2mm] \dfrac{\partial Q}{\partial x} - \dfrac{\partial P}{\partial y} = C. \end{cases} \tag{15.54}$$

An obvious necessary condition for the consistency of the system (15.54) is that the functions A, B, and C satisfy the relation

$$\frac{\partial A}{\partial x} + \frac{\partial B}{\partial y} + \frac{\partial C}{\partial z} = 0,$$

which is equivalent to saying that the form

$$\omega = A\,\mathrm{d}y \wedge \mathrm{d}z + B\,\mathrm{d}z \wedge \mathrm{d}x + C\,\mathrm{d}x \wedge \mathrm{d}y$$

is closed in $\mathbb{R}^3$.

The system (15.54) will have been solved if we find a form

$$\alpha = P\,\mathrm{d}x + Q\,\mathrm{d}y + R\,\mathrm{d}z$$

such that $\mathrm{d}\alpha = \omega$.

In accordance with the recipes explained in the proof of Theorem 1, and taking account of the mapping h constructed in Example 1, we find, after simple computations,

$$\alpha = K\left(h^*\omega\right) = \left(\int_0^1 A(tx, ty, tz)t\,dt\right)(y\,dz - z\,dy) +$$

$$+ \left(\int_0^1 B(tx, ty, tz)t\,dt\right)(z\,dx - x\,dz) +$$

$$+ \left(\int_0^1 C(tx, ty, tz)t\,dt\right)(x\,dy - y\,dx).$$

One can also verify directly that $d\alpha = \omega$.

Remark The amount of arbitrariness in the choice of a form α satisfying the condition $d\alpha = \omega$ is usually considerable. Thus, along with α, any form $\alpha + d\eta$ will obviously also satisfy the same equation.

By Theorem 1 any two forms α and β on a contractible manifold M satisfying $d\alpha = d\beta = \omega$ differ by an exact form. Indeed, $d(\alpha - \beta) = 0$, that is, the form $(\alpha - \beta)$ is closed on M and hence exact, by Theorem 1.

15.4.2 Homology and Cohomology

By Poincaré's theorem every closed form on a manifold is locally exact. But it is by no means always possible to glue these local primitives together to obtain a single form. Whether this can be done depends on the topological structure of the manifold. For example, the closed form in the punctured plane $\mathbb{R}^2 \backslash 0$ given by $\omega = \frac{-y\,dx + x\,dy}{x^2 + y^2}$, studied in Sect. 14.3, is locally the differential of a function $\varphi = \varphi(x, y)$ – the polar angle of the point (x, y). However, extending that function to the domain $\mathbb{R}^2 \backslash 0$ leads to multivaluedness if the closed path over which the extension is carried out encloses the hole – the point 0. The situation is approximately the same with forms of other degrees. "Holes" in manifolds may be of different kinds, not only missing points, but also holes such as one finds in a torus or a pretzel. The structure of manifolds of higher dimensions can be rather complicated. The connection between the structure of a manifold as a topological space and the relationship between closed and exact forms on it is described by the so-called (co)homology groups of the manifold.

The closed and exact real-valued forms on a manifold M form the vector spaces $Z^p(M)$ and $B^p(M)$ respectively, and $Z^p(M) \supset B^p(M)$.

Definition 4 The quotient space

$$H^p(M) := Z^p(M)/B^p(M) \tag{15.55}$$

is called the *p-dimensional cohomology group of the manifold M* (with real coefficients).

Thus, two closed forms $\omega_1, \omega_2 \in Z^p(M)$ lie in the same cohomology class, or are cohomologous, if $\omega_1 - \omega_2 \in B^p(M)$, that is, if they differ by an exact form. The cohomology class of the form $\omega \in Z^p(M)$ will be denoted $[\omega]$.

Since $Z^p(M)$ is the kernel of the operator $d^p : \Omega^p(M) \to \Omega^{p+1}(M)$, and $B^p(M)$ is the image of the operator $d^{p-1} : \Omega^{p-1}(M) \to \Omega^p(M)$, we often write

$$H^p(M) = \operatorname{Ker} d^p / \operatorname{Im} d^{p-1}.$$

Computing cohomologies, as a rule, is difficult. However, certain trivial general observations can be made.

It follows from Definition 4 that if $p > \dim M$, then $H^p(M) = 0$.

It follows from Poincaré's theorem that if M is contractible then $H^p(M) = 0$ for $p > 0$.

On any connected manifold M the group $H^0(M)$ is isomorphic to $\mathbb{R}$, since $H^0(M) = Z^0(M)$, and if $df = 0$ holds for the function $f : M \to \mathbb{R}$ on a connected manifold M, then $f = \text{const}$.

Thus, for example, it results for $\mathbb{R}^n$ that $H^p(\mathbb{R}^n) = 0$ for $p > 0$ and $H^0(\mathbb{R}^n) \sim \mathbb{R}$. This assertion (up to the trivial last relation) is equivalent to Theorem 1 with $M = \mathbb{R}^n$ and is also called Poincaré's theorem.

The so-called homology groups have a more visualizable geometrical relation to the manifold M.

Definition 5 A smooth mapping $c : I^p \to M$ of the p-dimensional cube $I \subset \mathbb{R}^p$ into the manifold M is called a *singular p-cube* on M.

This is a direct generalization of the concept of a smooth path to the case of an arbitrary dimension p. In particular, a singular cube may consist of a mapping of the cube I to a single point.

Definition 6 A *p-chain* (of singular cubes) on a manifold M is any finite formal linear combination $\sum_k \alpha_k c_k$ of singular p-cubes on M with real coefficients.

Like paths, singular cubes that can be obtained from each other by a diffeomorphic change of the parametrization with positive Jacobian are regarded as equivalent and are identified. If such a change of parameter has negative Jacobian, then the corresponding oppositely oriented singular cubes c and c_- are regarded as negatives of each other, and we set $c_- = -c$.

The p-chains on M obviously form a vector space with respect to the standard operations of addition and multiplication by a real number. We denote this space by $C_p(M)$.

Definition 7 The *boundary* ∂I of the p-dimensional cube I^p in $\mathbb{R}^p$ is the $(p-1)$-chain

$$\partial I := \sum_{i=0}^{l} \sum_{j=1}^{p} (-1)^{i+j} c_{ij} \tag{15.56}$$

in $\mathbb{R}^p$, where $c_{ij} : I^{p-1} \to \mathbb{R}^p$ is the mapping of the $(p-1)$-dimensional cube into $\mathbb{R}^p$ induced by the canonical embedding of the corresponding face of I^p in $\mathbb{R}^p$. More precisely, if $I^{p-1} = \{\tilde{x} \in \mathbb{R}^{p-1} \mid 0 \le \tilde{x}^m \le 1, m = 1, \ldots, p-1\}$, then $c_{ij}(\tilde{x}) = (\tilde{x}^1, \ldots, \tilde{x}^{j-1}, i, \tilde{x}^{j+1}, \ldots, \tilde{x}^{p-1}) \in \mathbb{R}^p$.

It is easy to verify that this formal definition of the boundary of a cube agrees completely with the operation of taking the boundary of the standard oriented cube I^p (see Sect. 12.3).

Definition 8 The *boundary* ∂c of the singular p-cube c is the $(p-1)$-chain

$$\partial c := \sum_{i=0}^{1} \sum_{j=1}^{p} (-1)^{i+j} c \circ c_{ij}.$$

Definition 9 The boundary of a p-chain $\sum_k \alpha_k c_k$ on the manifold M is the $(p-1)$-chain

$$\partial \left(\sum_k \alpha_k c_k \right) := \sum_k \alpha_k \partial c_k.$$

Thus on any space of chains $C_p(M)$ we have defined a linear operator

$$\partial = \partial_p : C_p(M) \to C_{p-1}(M).$$

Using relation (15.56), one can verify the relation $\partial(\partial I) = 0$ for the cube. Consequently $\partial \circ \partial = \partial^2 = 0$ in general.

Definition 10 A *p-cycle* on a manifold is a p-chain z for which $\partial z = 0$.

Definition 11 A *boundary p-cycle* on a manifold is a p-chain that is the boundary of some $(p+1)$-chain.

Let $Z_p(M)$ and $B_p(M)$ be the sets of p-cycles and boundary p-cycles on the manifold M. It is clear that $Z_p(M)$ and $B_p(M)$ are vector spaces over the field $\mathbb{R}$ and that $Z_p(M) \supset B_p(M)$.

Definition 12 The quotient space

$$H_p(M) := Z_p(M)/B_p(M) \tag{15.57}$$

is the *p-dimensional homology group of the manifold M* (with real coefficients).

Thus, two cycles $z_1, z_2 \in Z_p(M)$ are in the same homology class, or are *homologous*, if $z_1 - z_2 \in B_p(M)$, that is, they differ by the boundary of some chain. We shall denote the homology class of a cycle $z \in Z_p(M)$ by $[z]$.

As in the case of cohomology, relation (15.57) can be rewritten as

$$H_p(M) = \operatorname{Ker} \partial_p / \operatorname{Im} \partial_{p+1}.$$

Definition 13 If $c : I \to M$ is a singular p-cube and ω is a p-form on the manifold M, then the *integral of the form ω over this singular cube* is

$$\int_c \omega := \int_I c^* \omega. \tag{15.58}$$

Definition 14 If $\sum_k \alpha_k c_k$ is a p-chain and ω is a p-form on the manifold M, the integral of the form over such a chain is interpreted as the linear combination $\sum_k \alpha_k \int_{c_k} \omega$ of the integrals over the corresponding singular cubes.

It follows from Definitions 5–8 and 13–14 that Stokes' formula

$$\int_c d\omega = \int_{\partial c} \omega \tag{15.59}$$

holds for the integral over a singular cube, where c and ω have dimension p and degree $p - 1$ respectively. If we take account of Definition 9, we conclude that Stokes' formula (15.59) is valid for integrals over chains.

Theorem 2 a) *The integral of an exact form over a cycle equals zero.*

b) *The integral of a closed form over the boundary of a chain equals zero.*

c) *The integral of a closed form over a cycle depends only on the cohomology class of the form.*

d) *If the closed p-forms ω_1 and ω_2 and the p-cycles z_1 and z_2 are such that $[\omega_1] = [\omega_2]$ and $[z_1] = [z_2]$, then*

$$\int_{z_1} \omega_1 = \int_{z_2} \omega_2.$$

Proof a) By Stokes' formula $\int_z \omega\, dz = \int_{\partial z} \omega = 0$, since $\partial z = 0$.

b) By Stokes' formula $\int_{\partial c} \omega = \int_c d\omega = 0$, since $d\omega = 0$.

c) follows from b).

d) follows from a).

e) follows from c) and d). $\qquad\qquad\qquad\qquad\qquad\qquad\qquad\qquad\qquad\square$

Corollary *The bilinear mapping $\Omega^p(M) \times C_p(M) \to \mathbb{R}$ defined by $(\omega, c) \mapsto \int_c \omega$ induces a bilinear mapping $Z^p(M) \times Z_p(M) \to \mathbb{R}$ and a bilinear mapping $H^p(M) \times H_p(M) \to \mathbb{R}$. The latter is given by the formula*

$$\big([\omega], [z]\big) \mapsto \int_z \omega, \tag{15.60}$$

where $\omega \in Z^p(M)$ and $z \in Z_p(M)$.

Theorem 3 (de Rham[9]) *The bilinear mapping* $H^p(M) \times H_p(M) \to \mathbb{R}$ *given by* (15.60) *is nondegenerate.*[10]

We shall not take the time to prove this theorem here, but we shall find some reformulations of it that will enable us to present in explicit form some corollaries of it that are used in analysis.

We remark first of all that by (15.60) each cohomology class $[\omega] \in H^p(M)$ can be interpreted as a linear function $[\omega]([z]) = \int_z \omega$. Thus a natural mapping $H^p(M) \to H_p^*(M)$ arises, where $H_p^*(M)$ is the vector space conjugate to $H_p(M)$. The theorem of de Rham asserts that this mapping is an isomorphism, and in this sense $H^p(M) = H_p^*(M)$.

Definition 15 If ω is a closed p-form and z is a p-cycle on the manifold M, then the quantity $\mathrm{per}(z) := \int_z \omega$ is called the *period* (or *cyclic constant*) *of the form* ω *over the cycle* z.

In particular, if the cycle z is homologous to zero, then, as follows from assertion b) of Theorem 2, we have $\mathrm{per}(z) = 0$. For that reason the following connection exists between periods:

$$\left[\sum_k \alpha_k z_k\right] = 0 \Longrightarrow \sum_k \alpha_k \, \mathrm{per}(z_k) = 0, \tag{15.61}$$

that is, if a linear combination of cycles is a boundary cycle, or, what is the same, is homologous to zero, then the corresponding linear combination of periods is zero.

The following two theorems of de Rham hold; taken together, they are equivalent to Theorem 3.

Theorem 4 (de Rham's first theorem) *A closed form is exact if and only if all its periods are zero.*

Theorem 5 (de Rham's second theorem) *If a number* $\mathrm{per}(z)$ *is assigned to each p-cycle $z \in Z_p(M)$ on the manifold M in such a way that condition* (15.61) *holds, then there is a closed p-form ω on M such that* $\int_z \omega = \mathrm{per}(z)$ *for every cycle $z \in Z_p(M)$.*

15.4.3 Problems and Exercises

1. Verify by direct computation that the form α obtained in Example 2 does indeed satisfy the equation $d\alpha = \omega$.

---

[9]G. de Rham (1903–1969) – Belgian mathematician who worked mainly in algebraic topology.

[10]We recall that a bilinear form $L(x, y)$ is nondegenerate if for every fixed nonzero value of one of the variables the resulting linear function of the other variable is not identically zero.

2. a) Prove that every simply-connected domain in $\mathbb{R}^2$ is contractible on itself to a point.

b) Show that the preceding assertion is generally not true in $\mathbb{R}^3$.

3. Analyze the proof of Poincaré's theorem and show that if the smooth mapping $h : M \times I \to M$ is regarded as a family of mappings $h_t : M \to M$ depending on the parameter t, then for every closed form ω on M all the forms $h_t^*\omega, t \in I$, will be in the same cohomology class.

4. a) Let $t \mapsto h_t \in C^{(\infty)}(M, N)$ be a family of mappings of the manifold M into the manifold N depending smoothly on the parameter $t \in I \subset \mathbb{R}$. Verify that for every form $\omega \in \Omega(N)$ the following *homotopy formula* holds:

$$\frac{\partial}{\partial t}\left(h_t^*\omega\right)(x) = \mathrm{d}h_t^*(i_X\omega)(x) + h_t^*(i_X \,\mathrm{d}\omega)(x). \qquad (15.62)$$

Here $x \in M$, X is a vector field on N with $X(x,t) \in TN_{h_t(x)}$, $X(x,t)$ is the velocity vector for the path $t' \mapsto h_{t'}(x)$ at $t' = t$, and the operation i_X of taking the inner product of a form and a vector field is defined in Problem 7 of the preceding section.

b) Obtain the assertion of Problem 3 from formula (15.62).

c) Using formula (15.62), prove Poincaré's theorem (Theorem 1) again.

d) Show that if K is a manifold that is contractible to a point, then $H^p(K \times M) = H^p(M)$ for every manifold M and any integer p.

e) Obtain relation (15.51) of Sect. 15.3 from formula (15.62).

5. a) Show, using Theorem 4, and also by direct demonstration, that if a closed 2-form on the sphere S^2 is such that $\int_{S^2} \omega = 0$, then ω is exact.

b) Show that the group $H^2(S^2)$ is isomorphic to $\mathbb{R}$.

c) Show that $H^1(S^2) = 0$.

6. a) Let $\varphi : S^2 \to S^2$ be the mapping that assigns to each point $x \in S^2$ the antipodal point $-x \in S^2$. Show that there is a one-to-one correspondence between forms on the projective plane $\mathbb{RP}^2$ and forms on the sphere S^2 that are invariant under the mapping φ, that is, $\varphi^*\omega = \omega$.

b) Let us represent $\mathbb{RP}^2$ as the quotient space S^2/Γ, where Γ is the group of transformations of the sphere consisting of the identity mapping and the antipodal mapping φ. Let $\pi : S^2 \to \mathbb{RP}^2 = S^2/\Gamma$ be the natural projection, that is $\pi(x) = \{x, -x\}$. Show that $\pi \circ \varphi = \pi$ and verify that

$$\forall \eta \in \Omega^p\left(S^2\right) \left(\varphi^*\eta = \eta\right) \Longleftrightarrow \exists \omega \in \Omega^p\left(\mathbb{RP}^2\right)\left(\pi^*\omega = \eta\right).$$

c) Now show, using the result of Problem 5a), that $H^2(\mathbb{RP}^2) = 0$.

d) Prove that if the function $f \in C(S^2, \mathbb{R})$ is such that $f(x) - f(-x) \equiv \mathrm{const}$, then $f \equiv 0$. Taking account of Problem 5c), deduce from this that $H^1(\mathbb{RP}^2) = 0$.

7. a) Representing $\mathbb{RP}^2$ as a standard rectangle Π with opposite sides identified as shown by the orienting arrows in Fig. 15.3, show that $\partial\Pi = 2c' - 2c$, $\partial c = P - Q$, and $\partial c' = P - Q$.

Fig. 15.3

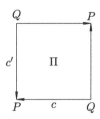

b) Deduce from the observations in the preceding part of the problem that there are no nontrivial 2-cycles on $\mathbb{RP}^2$. Then show by de Rham's theorem that $H^2(\mathbb{RP}^2) = 0$.

c) Show that the only nontrivial 1-cycle on $\mathbb{RP}^2$ (up to a constant factor) is the cycle $c' - c$, and since $c' - c = \frac{1}{2}\partial\Pi$, deduce from de Rham's theorem that $H^1(\mathbb{RP}^2) = 0$.

8. Find the groups $H^0(M)$, $H^1(M)$, and $H^2(M)$ if

a) $M = S^1$ – the circle;
b) $M = T^2$ – the two-dimensional torus;
c) $M = K^2$ – the Klein bottle.

9. a) Prove that diffeomorphic manifolds have isomorphic (co)homology groups of the corresponding dimension.

b) Using the example of $\mathbb{R}^2$ and $\mathbb{RP}^2$, show that the converse is generally not true.

10. Let X and Y be vector spaces over the field $\mathbb{R}$ and $L(x, y)$ a nondegenerate bilinear form $L : X \times Y \to \mathbb{R}$. Consider the mapping $X \to Y^*$ given by the correspondence $X \ni x \mapsto L(x, \cdot) \in Y^*$.

a) Prove that this mapping is injective.

b) Show that for every system $y_1, \ldots, y_k$ of linearly independent vectors in Y there exist vectors $x^1, \ldots, x^k$ such that $x^i(y_j) = L(x^i, y_j) = \delta^i_j$, where $\delta^i_j = 0$ if $i \neq j$ and $\delta^i_j = 1$ if $i = j$.

c) Verify that the mapping $X \to Y^*$ is an isomorphism of the vector spaces X and Y^*.

d) Show that de Rham's first and second theorems together mean that $H^p(M) = H^*_p(M)$ up to isomorphism.

Chapter 16
Uniform Convergence and the Basic Operations of Analysis on Series and Families of Functions

16.1 Pointwise and Uniform Convergence

16.1.1 Pointwise Convergence

Definition 1 We say that *the sequence* $\{f_n; n \in \mathbb{N}\}$ *of functions* $f_n : X \to \mathbb{R}$ *converges at the point* $x \in X$ if the sequence of values at x, $\{f_n(x); n \in \mathbb{N}\}$, converges.

Definition 2 The set of points $E \subset X$ at which the sequence $\{f_n; n \in \mathbb{N}\}$ of functions $f_n : X \to \mathbb{R}$ converges is called the *convergence set of the sequence.*

Definition 3 On the convergence set of the sequence of functions $\{f_n; n \in \mathbb{N}\}$ there naturally arises a function $f : E \to \mathbb{R}$ defined by the relation $f(x) := \lim_{n\to\infty} f_n(x)$. This function is called the *limit function of the sequence* $\{f_n; n \in \mathbb{N}\}$ or the *limit of the sequence* $\{f_n; n \in \mathbb{N}\}$.

Definition 4 If $f : E \to \mathbb{R}$ is the limit of the sequence $\{f_n; n \in \mathbb{N}\}$, we say that *the sequence of functions converges* (or *converges pointwise*) *to* f *on* E.

In this case we write $f(x) = \lim_{n\to\infty} f_n(x)$ on E or $f_n \to f$ on E as $n \to \infty$.

Example 1 Let $X = \{x \in \mathbb{R} \mid x \geq 0\}$ and let the functions $f_n : X \to \mathbb{R}$ be given by the relation $f_n(x) = x^n$, $n \in \mathbb{N}$. The convergence set of this sequence of functions is obviously the closed interval $I = [0, 1]$ and the limit function $f : I \to \mathbb{R}$ is defined by

$$f(x) = \begin{cases} 0, & \text{if } 0 \leq x < 1, \\ 1, & \text{if } x = 1. \end{cases}$$

Example 2 The sequence of functions $f_n(x) = \frac{\sin n^2 x}{n}$ on $\mathbb{R}$ converges on $\mathbb{R}$ to the function $f : \mathbb{R} \to 0$ that is identically 0.

© Springer-Verlag Berlin Heidelberg 2016
V.A. Zorich, *Mathematical Analysis II*, Universitext,
DOI 10.1007/978-3-662-48993-2_8

Example 3 The sequence $f_n(x) = \frac{\sin nx}{n^2}$ also has the identically zero function $f :$ $\mathbb{R} \to 0$ as its limit.

Example 4 Consider the sequence $f_n(x) = 2(n+1)x(1-x^2)^n$ on the closed interval $I = [0, 1]$. Since $nq^n \to 0$ for $|q| < 1$, this sequence tends to zero on the entire closed interval I.

Example 5 Let $m, n \in \mathbb{N}$, and let $f_m(x) := \lim_{n\to\infty}(\cos m!\pi x)^{2n}$. If $m!x$ is an integer, then $f_m(x) = 1$, and if $m!x \notin \mathbb{Z}$, obviously $f_m(x) = 0$.

We shall now consider the sequence $\{f_m; m \in \mathbb{N}\}$ and show that it converges on the entire real line to the Dirichlet function

$$\mathcal{D}(x) = \begin{cases} 0, & \text{if } x \notin \mathbb{Q}, \\ 1, & \text{if } x \in \mathbb{Q}. \end{cases}$$

Indeed, if $x \notin \mathbb{Q}$, then $m!x \notin \mathbb{Z}$, and $f_m(x) = 0$ for every value of $m \in \mathbb{N}$, so that $f(x) = 0$. But if $x = \frac{p}{q}$, where $p \in \mathbb{Z}$ and $q \in \mathbb{N}$, then $m!x \in \mathbb{Z}$ for $m \geq q$, and $f_m(x) = 1$ for all such m, which implies $f(x) = 1$.

Thus $\lim_{m\to\infty} f_m(x) = \mathcal{D}(x)$.

16.1.2   Statement of the Fundamental Problems

Limiting passages are encountered at every step in analysis, and it is often important to know what kind of functional properties the limit function has. The most important properties for analysis are continuity, differentiability, and integrability. Hence it is important to determine whether the limit is a continuous, differentiable, or integrable function if the prelimit functions all have the corresponding property. Here it is especially important to find conditions that are sufficiently convenient in practice and which guarantee that when the functions converge, their derivatives or integrals also converge to the derivative or integral of the limit function.

As the simple examples examined above show, without some additional hypotheses the relation "$f_n \to f$ on $[a, b]$ as $n \to \infty$" does not in general imply either the continuity of the limit function, even when the functions f_n are continuous, or the relation $f_n' \to f'$ or $\int_a^b f_n(x)\,\mathrm{d}x \to \int_a^b f(x)\,\mathrm{d}x$, even when all these derivatives and integrals are defined.

Indeed,

in Example 1 the limit function is discontinuous on $[0, 1]$ although all the prelimit functions are continuous there;

in Example 2 the derivatives $n \cos n^2 x$ of the prelimit functions in general do not converge, and hence cannot converge to the derivative of the limit function, which in this case is identically zero;

in Example 4 we have $\int_0^1 f_n(x)\,\mathrm{d}x = 1$ for every value of $n \in \mathbb{N}$, while $\int_0^1 f(x)\,\mathrm{d}x = 0$;

in Example 5 each of the functions f_m equals zero except at a finite set of points, so that $\int_a^b f_m(x)\,dx = 0$ on every closed interval $[a, b] \subset \mathbb{R}$, while the limit function $\mathcal{D}$ is not integrable on any closed interval of the real line.

At the same time:

in Examples 2, 3, and 4 both the prelimit and the limit functions are continuous;

in Example 3 the limit of the derivatives $\frac{\cos nx}{n}$ of the functions in the sequence $\frac{\sin nx}{n^2}$ does equal the derivative of the limit of that sequence;

in Example 1 we have $\int_0^1 f_n(x)\,dx \to \int_0^1 f(x)\,dx$ as $n \to \infty$.

Our main purpose is to determine the cases in which the limiting passage under the integral or derivative sign is legal.

In this connection, let us consider some more examples.

Example 6 We know that for any $x \in \mathbb{R}$

$$\sin x = x - \frac{1}{3!}x^3 + \frac{1}{5!}x^5 - \cdots + \frac{(-1)^m}{(2m+1)!}x^{2m+1} + \cdots , \qquad (16.1)$$

but after the examples we have just considered, we understand that the relations

$$\sin' x = \sum_{m=0}^{\infty} \left(\frac{(-1)^m}{(2m+1)!}x^{2m+1} \right)', \qquad (16.2)$$

$$\int_a^b \sin x\,dx = \sum_{m=0}^{\infty} \int_a^b \frac{(-1)^m}{(2m+1)!}x^{2m+1}\,dx, \qquad (16.3)$$

require verification in general.

Indeed, if the equality

$$S(x) = a_1(x) + a_2(x) + \cdots + a_m(x) + \cdots$$

is understood in the sense that $S(x) = \lim_{n \to \infty} S_n(x)$, where $S_n(x) = \sum_{m=1}^n a_m(x)$, then by the linearity of differentiation and integration, the relations

$$S'(x) = \sum_{m=1}^{\infty} a_n'(x),$$

$$\int_a^b S(x)\,dx = \sum_{m=1}^{\infty} \int_a^b a_m(x)\,dx$$

are equivalent to

$$S'(x) = \lim_{n \to \infty} S_n'(x),$$

$$\int_a^b S(x)\,dx = \lim_{n \to \infty} \int_a^b S_n(x)\,dx,$$

which we must now look upon with caution.

In this case both relations (16.2) and (16.3) can easily be verified, since it is known that

$$\cos x = 1 - \frac{1}{2!}x^2 + \frac{1}{4!}x^4 - \cdots + \frac{(-1)^m}{(2m)!}x^{2m} + \cdots .$$

However, suppose that Eq. (16.1) is the definition of the function $\sin x$. After all, that was exactly the situation with the definition of the functions $\sin z$, $\cos z$, and e^z for complex values of the argument. At that time we had to get the properties of the new function (its continuity, differentiability, and integrability), as well as the legality of the equalities (16.2) and (16.3) directly from the fact that this function is the limit of the sequence of partial sums of this series.

The main concept by means of which sufficient conditions for the legality of the limiting passages will be derived in Sect. 16.3, is the concept of uniform convergence.

16.1.3 Convergence and Uniform Convergence of a Family of Functions Depending on a Parameter

In our discussion of the statement of the problems above we confined ourselves to consideration of the limit of a sequence of functions. A sequence of functions is the most important special case of a family of functions $f_t(x)$ depending on a parameter t. It arises when $t \in \mathbb{N}$. Sequences of functions thus occupy the same place occupied by the theory of limit of a sequence in the theory of limits of functions. We shall discuss the limit of a sequence of functions and the connected theory of convergence of series of functions in Sect. 16.2. Here we shall discuss only those concepts involving functions depending on a parameter that are basic for everything that follows.

Definition 5 We call a function $(x, t) \mapsto F(x, t)$ of two variables x and t defined on the set $X \times T$ a *family of functions depending on the parameter t* if one of the variables $t \in T$ is distinguished and called the *parameter*.

The set T is called the *parameter set* or *parameter domain*, and the family itself is often written in the form $f_t(x)$ or $\{f_t; t \in T\}$, distinguishing the parameter explicitly.

As a rule, in this book we shall have to consider families of functions for which the parameter domain T is one of the sets $\mathbb{N}$ or $\mathbb{R}$ or $\mathbb{C}$ of natural numbers, real numbers, or complex numbers or subsets of these. In general, however, the set T may be a set of any nature. Thus in Examples 1–5 above we had $T = \mathbb{N}$. In Examples 1–4 we could have assumed without loss of content that the parameter n is any positive number and the limit was taken over the base $n \to \infty$, $n \in \mathbb{R}_+$.

Definition 6 Let $\{f_t : X \to \mathbb{R}; t \in T\}$ be a family of functions depending on a parameter and let $\mathcal{B}$ be a base in the set T of parameter values.

If the limit $\lim_\mathcal{B} f_t(x)$ exists for a fixed value $x \in X$, we say that *the family of functions converges at* x.

The set of points of convergence is called the *convergence set of the family of functions in a given base* $\mathcal{B}$.

Definition 7 We say that *the family of functions converges on the set* $E \subset X$ *over the base* $\mathcal{B}$ if it converges over that base at each point $x \in E$.

The function $f(x) := \lim_\mathcal{B} f_t(x)$ on E is called the *limit function* or the *limit of the family of functions* f_t *on the set* E *over the base* $\mathcal{B}$.

Example 7 Let $f_t(x) = e^{-(x/t)^2}$, $x \in X = \mathbb{R}$, $t \in T = \mathbb{R}\backslash 0$, and let $\mathcal{B}$ be the base $t \to 0$. This family converges on the entire set $\mathbb{R}$, and

$$\lim_{t \to 0} f_t(x) = \begin{cases} 1, & \text{if } x = 0, \\ 0, & \text{if } x \neq 0. \end{cases}$$

We now give two basic definitions.

Definition 8 We say that the family $\{f_t; t \in T\}$ of functions $f_t : X \to \mathbb{R}$ *converges pointwise* (or simply *converges*) *on the set* $E \subset X$ *over the base* $\mathcal{B}$ to the function $f : E \to \mathbb{R}$ if $\lim_\mathcal{B} f_t(x) = f(x)$ at every point $x \in E$.

In this case we shall often write $(f_t \xrightarrow[\mathcal{B}]{} f$ on $E)$.

Definition 9 The family $\{f_t; t \in T\}$ of functions $f_t : X \to \mathbb{R}$ *converges uniformly on the set* $E \subset X$ *over the base* $\mathcal{B}$ to the function $f : E \to \mathbb{R}$ if for every $\varepsilon > 0$ there exists an element B in the base $\mathcal{B}$ such that $|f(x) - f_t(x)| < \varepsilon$ at every value $t \in B$ and at every point $x \in E$.

In this case we shall frequently write $(f_t \underset{\mathcal{B}}{\rightrightarrows} f$ on $E)$.

We give also the formal expression of these important definitions:

$$\left(f_t \xrightarrow[\mathcal{B}]{} f \text{ on } E \right) := \forall \varepsilon > 0 \, \forall x \in E \, \exists B \in \mathcal{B} \, \forall t \in B \, \left(\left| f(x) - f_t(x) \right| < \varepsilon \right),$$

$$\left(f_t \underset{\mathcal{B}}{\rightrightarrows} f \text{ on } E \right) := \forall \varepsilon > 0 \, \exists B \in \mathcal{B} \, \forall x \in E \, \forall t \in B \, \left(\left| f(x) - f_t(x) \right| < \varepsilon \right).$$

The relation between convergence and uniform convergence resembles the relation between continuity and uniform continuity on a set.

To explain better the relationship between convergence and uniform convergence of a family of functions, we introduce the quantity $\Delta_t(x) = |f(x) - f_t(x)|$, which measures the deviation of the value of the function f_t from the value of the function f at the point $x \in E$. Let us consider also the quantity $\Delta_t = \sup_{x \in E} \Delta_t(x)$,

Fig. 16.1

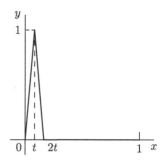

which characterizes, roughly speaking, the maximum deviation (although there may not be a maximum) of the function f_t from the corresponding values of f over all $x \in E$. Thus, at every point $x \in E$ we have $\Delta_t(x) \le \Delta_t$.

In this notation these definitions obviously can be written as follows:

$$\left(f_t \xrightarrow[\mathcal{B}]{} f \text{ on } E \right) := \forall x \in E \quad \left(\Delta_t(x) \to 0 \text{ over } \mathcal{B} \right),$$

$$\left(f_t \underset{\mathcal{B}}{\rightrightarrows} f \text{ on } E \right) := \left(\Delta_t \to 0 \text{ over } B \right).$$

It is now clear that

$$\left(f_t \underset{\mathcal{B}}{\rightrightarrows} f \text{ on } E \right) \Longrightarrow \left(f_t \xrightarrow[\mathcal{B}]{} f \text{ on } E \right),$$

that is, if the family f_t converges uniformly to f on the set E, it converges pointwise to f on that set.

The converse is in general not true.

Example 8 Let us consider the family of functions $f_t : I \to \mathbb{R}$ defined on the closed interval $I = \{x \in \mathbb{R} \mid 0 \le x \le 1\}$ and depending on the parameter $t \in {]0, 1]}$. The graph of the functions $y = f_t(x)$ is shown in Fig. 16.1. It is clear that $\lim_{t \to 0} f_t(x) = 0$ at every point $x \in I$, that is, $f_t \to f \equiv 0$ as $t \to 0$. At the same time $\Delta_t = \sup_{x \in I} |f(x) - f_t(x)| = \sup_{x \in I} |f_t(x)| = 1$, that is, $\Delta_t \not\to 0$ as $t \to 0$, and hence the family converges, but not uniformly.

In such cases we shall say for convenience that the family *converges nonuniformly* to the limit function.

If the parameter t is interpreted as time, then convergence of the family of functions f_t on the set E to the function f means that for any preassigned precision $\varepsilon > 0$ one can exhibit a time t_ε for each point $x \in E$ starting from which (that is, for $t > t_\varepsilon$) the values of all functions f_t at x will differ from $f(x)$ by less than ε.

Uniform convergence means that there is a time t_ε, starting from which (that is, for $t > t_\varepsilon$) the relation $|f(x) - f_t(x)| < \varepsilon$ holds for all $x \in E$.

The figure of a traveling bulge of large deviation depicted in Fig. 16.1 is typical for nonuniform convergence.

Example 9 The sequence of functions $f_n(x) = x^n - x^{2n}$ defined on the closed interval $0 \le x \le 1$, as one can see, converges to zero at each point as $n \to \infty$. To determine whether this convergence is uniform, we find the quantity $\Delta_n = \max_{0 \le x \le 1} |f_n(x)|$. Since $f'_n(x) = nx^{n-1}(1 - 2x^n) = 0$ for $x = 0$ and $x = 2^{-1/n}$, it is clear that $\Delta_n = f_n(2^{-1/n}) = 1/4$. Thus $\Delta_n \not\to 0$ as $n \to \infty$ and our sequence converges to the limit function $f(x) \equiv 0$ nonuniformly.

Example 10 The sequence of functions $f_n = x^n$ on the interval $0 \le x \le 1$ converges to the function

$$f(x) = \begin{cases} 0, & \text{if } 0 \le x < 1, \\ 1, & \text{if } x = 1 \end{cases}$$

nonuniformly, since for each $n \in \mathbb{N}$

$$\Delta_n = \sup_{0 \le x \le 1} |f(x) - f_n(x)| = \sup_{0 \le x < 1} |f(x) - f_n(x)| =$$

$$= \sup_{0 \le x < 1} |f_n(x)| = \sup_{0 \le x < 1} |x^n| = 1.$$

Example 11 The sequence of functions $f_n(x) = \frac{\sin n^2 x}{n}$ studied in Example 2 converges to zero uniformly on the entire set $\mathbb{R}$ as $n \to \infty$, since in this case

$$|f(x) - f_n(x)| = |f_n(x)| = \left| \frac{\sin n^2 x}{n} \right| \le \frac{1}{n},$$

that is, $\Delta_n \le 1/n$, and hence $\Delta_n \to 0$ as $n \to \infty$.

16.1.4 The Cauchy Criterion for Uniform Convergence

In Definition 9 we stated what it means for a family of functions f_t to converge uniformly on a set to a given function on that set. Usually, when the family of functions is defined the limit function is not yet known, so that it makes sense to adopt the following definition.

Definition 10 We shall say that *the family $\{f_t; t \in T\}$ of functions $f_t : X \to \mathbb{R}$ converges on the set $E \subset X$ uniformly over the base $\mathcal{B}$* if it converges on that set and the convergence to the resulting limit function is uniform in the sense of Definition 9.

Theorem (Cauchy criterion for uniform convergence) *Let $\{f_t; t \in T\}$ be a family of functions $f_t : X \to \mathbb{R}$ depending on a parameter $t \in T$, and $\mathcal{B}$ a base in T. A necessary and sufficient condition for the family $\{f_t; t \in T\}$ to converge uniformly on the set $E \subset X$ over the base $\mathcal{B}$ is that for every $\varepsilon > 0$ there exists an element B of the base $\mathcal{B}$ such that $|f_{t_1}(x) - f_{t_2}(x)| < \varepsilon$ for every value of the parameters $t_1, t_2 \in B$ and every point $x \in E$.*

In formal language this means that f_t converges uniformly on E over the base
$\mathcal{B} \Longleftrightarrow \forall \varepsilon > 0 \; \exists B \in \mathcal{B} \; \forall t_1, t_2 \in B \; \forall x \in E \; (|f_{t_1}(x) - f_{t_2}(x)| < \varepsilon).$

Proof Necessity. The necessity of these conditions is obvious, since if $f : E \to \mathbb{R}$
is the limit function and $f_t \rightrightarrows f$ on E over $\mathcal{B}$, there exists an element B in the base
$\mathcal{B}$ such that $|f(x) - f_t(x)| < \varepsilon/2$ for every $t \in B$ and every $x \in E$. Then for every
$t_1, t_2 \in B$ and every $x \in E$ we have

$$\left| f_{t_1}(x) - f_{t_2}(x) \right| \leq \left| f(x) - f_{t_1}(x) \right| + \left| f(x) - f_{t_2}(x) \right| < \varepsilon/2 + \varepsilon/2 = \varepsilon.$$

Sufficiency. For each fixed value of $x \in E$ we can regard $f_t(x)$ as a function of
the variable $t \in T$. If the hypotheses of the theorem hold, then the hypotheses of
the Cauchy convergence criterion for the existence of a limit over the base $\mathcal{B}$ are
fulfilled.

Hence, the family $\{f_t; t \in T\}$ converges at least pointwise to some function $f :$
$E \to \mathbb{R}$ on the set E over the base $\mathcal{B}$.

If we now pass to the limit in the inequality $|f_{t_1}(x) - f_{t_2}(x)| < \varepsilon$, which
is valid for any t_1 and $t_2 \in B$ and every $x \in E$, one can obtain the inequality
$|f(x) - f_{t_2}(x)| \leq \varepsilon$ for every $t_2 \in B$ and every $x \in E$, and this, up to an inessential
relabeling and the change of the strict inequality to the nonstrict, coincides exactly
with the definition of uniform convergence of the family $\{f_t; t \in T\}$ to the function
$f : E \to \mathbb{R}$ on the set E over the base $\mathcal{B}$. $\square$

Remark 1 The definitions of convergence and uniform convergence that we have
given for families of real-valued functions $f_t : X \to \mathbb{R}$ of course remain valid for
families of functions $f_t : X \to Y$ with values in any metric space Y. The natural
modification that one must make in the definitions in this case amounts to replacing
$|f(x) - f_t(x)|$ by $d_Y(f(x), f_t(x))$, where d_Y is the metric in Y.

For normed vector spaces Y, in particular for $Y = \mathbb{C}$ or $Y = \mathbb{R}^m$ or $Y = \mathbb{C}^m$, even
these formal changes are not needed.

Remark 2 The Cauchy criterion of course remains valid for families of functions
$f_t : X \to Y$ with values in a metric space Y provided Y is a complete metric space.
As can be seen from the proofs, the hypothesis that Y be complete is needed only
for the sufficiency part of the criterion.

16.1.5 Problems and Exercises

1. Determine whether the sequences of functions considered in Examples 3–5 converge uniformly.

2. Prove Eqs. (16.2) and (16.3).

3. a) Show that the sequence of functions considered in Example 1 converges uniformly on every closed interval $[0, 1 - \delta] \subset [0, 1]$, but converges nonuniformly on
the interval $[0, 1[$.

b) Show that the same is true for the sequence considered in Example 9.

c) Show that family of functions f_t considered in Example 8 converges uniformly as $t \to 0$ on every closed interval $[\delta, 1] \subset [0, 1]$ but nonuniformly on $[0, 1]$.

d) Investigate the convergence and uniform convergence of the family of functions $f_t(x) = \sin(tx)$ as $t \to 0$ and then as $t \to \infty$.

e) Characterize the convergence of the family of functions $f_t(x) = e^{-tx^2}$ as $t \to +\infty$ on an arbitrary fixed set $E \subset \mathbb{R}$.

4. a) Verify that if a family of functions converges (resp. converges uniformly) on a set, then it also converges (resp. converges uniformly) on any subset of the set.

b) Show that if the family of functions $f_t : X \to \mathbb{R}$ converges (resp. converges uniformly) on a set E over a base $\mathcal{B}$ and $g : X \to \mathbb{R}$ is a bounded function, then the family $g \cdot f_t : X \to \mathbb{R}$ also converges (resp. converges uniformly) on E over the base $\mathcal{B}$.

c) Prove that if the families of functions $f_t : X \to \mathbb{R}$, $g_t : X \to \mathbb{R}$ converge uniformly on $E \subset X$ over the base $\mathcal{B}$, then the family $h_t = \alpha f_t + \beta g_t$, where $\alpha, \beta \in \mathbb{R}$, also converges uniformly on E over $\mathcal{B}$.

5. a) In the proof of the sufficiency of the Cauchy criterion we passed to the limit $\lim_{\mathcal{B}} f_{t_1}(x) = f(x)$ over the base $\mathcal{B}$ in T. But $t_1 \in B$, and $\mathcal{B}$ is a base in T, not in B. Can we pass to this limit in such a way that t_1 remains in B?

b) Explain where the completeness of $\mathbb{R}$ was used in the proof of the Cauchy criterion for uniform convergence of a family of functions $f_t : X \to \mathbb{R}$.

c) Notice that if all the functions of the family $\{f_t : X \to \mathbb{R}; t \in T\}$ are constant, then the theorem proved above is precisely the Cauchy criterion for the existence of the limit of the function $\varphi : T \to \mathbb{R}$ over the base $\mathcal{B}$ in T.

6. Prove that if the family of continuous functions $f_t \in C(I, \mathbb{R})$ on the closed interval $I = \{x \in \mathbb{R} \mid a \le x \le b\}$ converges uniformly on the open interval $]a, b[$, then it converges uniformly on the entire closed interval $[a, b]$.

16.2 Uniform Convergence of Series of Functions

16.2.1 Basic Definitions and a Test for Uniform Convergence of a Series

Definition 1 Let $\{a_n : X \to \mathbb{C}; n \in \mathbb{N}\}$ be a sequence of complex-valued (in particular real-valued) functions. The series $\sum_{n=1}^{\infty} a_n(x)$ *converges* or *converges uniformly on the set* $E \subset X$ if the sequence $\{s_m(x) = \sum_{n=1}^{m} a_n(x); n \in \mathbb{N}\}$ converges or converges uniformly on E.

Definition 2 The function $s_m(x) = \sum_{n=1}^{m} a_n(x)$, as in the case of numerical series, is called the *partial sum* or, more precisely, the *mth partial sum of the series* $\sum_{n=1}^{\infty} a_n(x)$.

Definition 3 The *sum of the series* is the limit of the sequence of its partial sums. Thus, writing

$$s(x) = \sum_{n=1}^{\infty} a_n(x) \quad \text{on } E$$

means that $s_m(x) \to s(x)$ on E as $m \to \infty$, and writing

$$\text{the series } \sum_{n=1}^{\infty} a_n(x) \text{ converges uniformly on } E$$

means that $s_m(x) \rightrightarrows s(x)$ on E as $m \to \infty$.

Investigating the pointwise convergence of a series amounts to investigating the convergence of a numerical series, and we are already familiar with that.

Example 1 Earlier we defined the function $\exp : \mathbb{C} \to \mathbb{C}$ by the relation

$$\exp z := \sum_{n=0}^{\infty} \frac{1}{n!} z^n, \tag{16.4}$$

after first verifying that the series on the right converges for every value $z \in \mathbb{C}$.

In the language of Definitions 1–3 one can now say that the series (16.4) of functions $a_n(z) = \frac{1}{n!} z^n$ converges on the entire complex plane, and the function $\exp z$ is its sum.

By Definitions 1 and 2 just adopted a two-way connection is established between series and their sequences of partial sums: knowing the terms of the series, we obtain the sequence of partial sums, and knowing the sequence of partial sums, we can recover all the terms of the series: the nature of the convergence of the series is identified with the nature of the convergence of its sequence of partial sums.

Example 2 In Example 5 of Sect. 16.1 we constructed a sequence $\{f_m; m \in \mathbb{N}\}$ of functions that converge to the Dirichlet function $\mathcal{D}(x)$ on $\mathbb{R}$. If we set $a_1(x) = f_1(x)$ and $a_n(x) = f_n(x) - f_{n-1}(x)$ for $n > 1$, we obtain a series $\sum_{n=1}^{\infty} a_n(x)$ that will converge on the entire number line, and $\sum_{n=1}^{\infty} a_n(x) = \mathcal{D}(x)$.

Example 3 It was shown in Example 9 of Sect. 16.1 that the sequence of functions $f_n(x) = x^n - x^{2n}$ converges, but nonuniformly, to zero on the closed interval $[0, 1]$. Hence, setting $a_1(x) = f_1(x)$ and $a_n(x) = f_n(x) - f_{n-1}(x)$ for $n > 1$, we obtain a series $\sum_{n=1}^{\infty} a_n(x)$ that converges to zero on the closed interval $[0, 1]$, but converges nonuniformly.

The direct connection between series and sequences of functions makes it possible to restate every proposition about sequences of functions as a corresponding proposition about series of functions.

Thus, in application to the sequence $\{s_n : X \to \mathbb{C}; n \in \mathbb{N}\}$ the Cauchy criterion proved in Sect. 16.1 for uniform convergence of a sequence on a set $E \subset X$ means that

$$\forall \varepsilon > 0 \ \exists N \in \mathbb{N} \ \forall n_1, n_2 > N \ \forall x \in E \ \left(\left| s_{n_1}(x) - s_{n_2}(x) \right| < \varepsilon \right). \tag{16.5}$$

From this, taking account of Definition 1, we obtain the following theorem.

Theorem 1 (Cauchy criterion for uniform convergence of a series) *The series* $\sum_{n=1}^{\infty} a_n(x)$ *converges uniformly on a set* E *if and only if for every* $\varepsilon > 0$ *there exists* $N \in \mathbb{N}$ *such that*

$$\left| a_n(x) + \cdots + a_m(x) \right| < \varepsilon, \tag{16.6}$$

for all natural numbers m, n *satisfying* $m \geq n > N$ *and every point* $x \in E$.

Proof Indeed, setting $n_1 = m$, $n_2 = n - 1$ in (16.5) and assuming that $s_n(x)$ is the partial sum of the series, we obtain inequality (16.6), from which relation (16.5) in turn follows with the same notation and hypotheses of the theorem. □

Remark 1 We did not mention the range of values of the functions $a_n(x)$ in the statement of Theorem 1, taking for granted that it was $\mathbb{R}$ or $\mathbb{C}$. But actually the range of values could obviously be any normed vector space, for example, $\mathbb{R}^n$ or $\mathbb{C}^n$, provided only that the space is complete.

Remark 2 If under the hypotheses of Theorem 1 all the functions $a_n(x)$ are constant, we obtain the familiar Cauchy criterion for convergence of a numerical series $\sum_{n=1}^{\infty} a_n$.

Corollary 1 (Necessary condition for uniform convergence of a series) *A necessary condition for the series* $\sum_{n=1}^{\infty} a_n(x)$ *to converge uniformly on a set* E *is that* $a_n \rightrightarrows 0$ *on* E *as* $n \to \infty$.

Proof This follows from the definition of uniform convergence of a sequence to zero and inequality (16.6) if we set $m = n$ in it. □

Example 4 The series (16.4) converges on the complex plane $\mathbb{C}$ nonuniformly, since $\sup_{z \in \mathbb{C}} \left| \frac{1}{n!} z^n \right| = \infty$ for every $n \in \mathbb{N}$, while by the necessary condition for uniform convergence, the quantity $\sup_{x \in E} |a_n(x)|$ must tend to zero when uniform convergence occurs.

Example 5 The series $\sum_{n=1}^{\infty} \frac{z^n}{n}$, as we know, converges in the unit disk $K = \{z \in \mathbb{C} \mid |z| < 1\}$. Since $\left| \frac{z^n}{n} \right| < \frac{1}{n}$ for $z \in K$, we have $\frac{z^n}{n} \rightrightarrows 0$ on K as $n \to \infty$. The necessary condition for uniform convergence is satisfied; however, this series converges nonuniformly on K. In fact, for any fixed $n \in \mathbb{N}$, by the continuity of the terms of the series, if z is sufficiently close to 1, we can get the inequalities

$$\left| \frac{z^n}{n} + \cdots + \frac{z^{2n}}{2n} \right| > \frac{1}{2} \left| \frac{1}{n} + \cdots + \frac{1}{2n} \right| > \frac{1}{4}.$$

From this we conclude by the Cauchy criterion that the series does not converge uniformly on K.

16.2.2 The Weierstrass M-Test for Uniform Convergence of a Series

Definition 4 The series $\sum_{n=1}^{\infty} a_n(x)$ *converges absolutely* on the set E if the corresponding numerical series converges absolutely at each point $x \in E$.

Proposition 1 *If the series* $\sum_{n=1}^{\infty} a_n(x)$ *and* $\sum_{n=1}^{\infty} b_n(x)$ *are such that* $|a_n(x)| \leq b_n(x)$ *for every* $x \in E$ *and for all sufficiently large indices* $n \in \mathbb{N}$, *then the uniform convergence of the series* $\sum_{n=1}^{\infty} b_n(x)$ *on* E *implies the absolute and uniform convergence of the series* $\sum_{n=1}^{\infty} a_n(x)$ *on the same set* E.

Proof Under these assumptions for all sufficiently large indices n and m (let $n \leq m$) at each point $x \in E$ we have

$$\left|a_n(x) + \cdots + a_m(x)\right| \leq \left|a_n(x)\right| + \cdots + \left|a_m(x)\right| \leq$$
$$\leq b_n(x) + \cdots + b_m(x) = \left|b_n(x) + \cdots + b_m(x)\right|.$$

By the Cauchy criterion and the uniform convergence of the series $\sum_{n=1}^{\infty} b_n(x)$, for each $\varepsilon > 0$ we can exhibit an index $N \in \mathbb{N}$ such that $|b_n(x) + \cdots + b_m(x)| < \varepsilon$ for all $m \geq n > N$ and all $x \in E$. But then it follows from the inequalities just written and the Cauchy criterion that the series $\sum_{n=1}^{\infty} a_n(x)$ and $\sum_{n=1}^{\infty} |a_n(x)|$ both converge uniformly. $\square$

Corollary 2 (Weierstrass' M-test for uniform convergence of a series) *If for the series* $\sum_{n=1}^{\infty} a_n(x)$ *one can exhibit a convergent numerical series* $\sum_{n=1}^{\infty} M_n$ *such that* $\sup_{x \in E} |a_n(x)| \leq M_n$ *for all sufficiently large indices* $n \in \mathbb{N}$, *then the series* $\sum_{n=1}^{\infty} a_n(x)$ *converges absolutely and uniformly on the set* E.

Proof The convergent numerical series can be regarded as a series of constant functions on the set E, which by the Cauchy criterion converges uniformly on E. Hence the Weierstrass test follows from Proposition 1 if we set $b_n(x) = M_n$ in it. $\square$

The Weierstrass M-test is the simplest and at the same time the most frequently used sufficient condition for uniform convergence of a series.

As an example of its application, we prove the following useful fact.

Proposition 2 *If a power series* $\sum_{n=0}^{\infty} c_n (z - z_0)^n$ *converges at a point* $\zeta \neq z_0$, *then it converges absolutely and uniformly in any disk* $K_q = \{z \in \mathbb{C} \mid |z - z_0| < q|\zeta - z_0|\}$, *where* $0 < q < 1$.

Proof By the necessary condition for convergence of a numerical series it follows from the convergence of the series $\sum_{n=0}^{\infty} c_n(\zeta - z_0)^n$ that $c_n(\zeta - z)^n \to 0$ as $n \to \infty$. Hence for all sufficiently large values of $n \in N$ we have the estimates $|c_n(z - z_0)^n| = |c_n(\zeta - z_0)^n| \cdot |\frac{z-z_0}{\zeta-z_0}|^n \leq |c_n(\zeta - z_0)^n| \cdot q^n < q^n$ in the disk K_q. Since the series $\sum_{n=0}^{\infty} q^n$ converges for $|q| < 1$, the estimates $|c_n(z - z_0)^n| < q^n$ and the Weierstrass M-test now imply Proposition 2. $\square$

Comparing this proposition with the Cauchy–Hadamard formula for the radius of convergence of a power series (see Eq. (5.115)), we arrive at the following conclusion.

Theorem 2 (Nature of convergence of a power series) *A power series* $\sum_{n=0}^{\infty} c_n(z - z_0)^n$ *converges in the disk* $K = \{z \in \mathbb{C} \mid |z - z_0| < R\}$ *whose radius of convergence is determined by the Cauchy–Hadamard formula*[1] $R = (\overline{\lim}_{n\to\infty} \sqrt[n]{|c_n|})^{-1}$ *Outside this disk the series diverges. On any closed disk contained in the interior of the disk* K *of convergence of the series, a power series converges absolutely and uniformly.*

Remark 3 As Examples 1 and 5 show, the power series need not converge uniformly on the entire disk K. At the same time, it may happen that the power series does converge uniformly even on the closed disk $\overline{K}$.

Example 6 The radius of convergence of the series $\sum_{n=1}^{\infty} \frac{z^n}{n^2}$ is 1. But if $|z| \leq 1$, then $|\frac{z^n}{n^2}| \leq \frac{1}{n^2}$, and by the Weierstrass M-test this series converges absolutely and uniformly in the closed disk $\overline{K} = \{z \in \mathbb{C} \mid |z| \leq 1\}$.

16.2.3 The Abel–Dirichlet Test

The following pairs of related sufficient conditions for uniform convergence of a series are somewhat more specialized and are essentially connected with the real-valuedness of certain components of the series under consideration. But these conditions are more delicate than the Weierstrass M-test, since they make it possible to investigate series that converge, but nonabsolutely.

Definition 5 The family $\mathcal{F}$ of functions $f : X \to \mathbb{C}$ is *uniformly bounded* on a set $E \subset X$ if there exists a number $M \in \mathbb{R}$ such that $\sup_{x \in E} |f(x)| \leq M$ for every $f \in \mathcal{F}$.

---

[1]In the exceptional case when $\overline{\lim}_{n\to\infty} \sqrt[n]{|c_n|} = \infty$, we take $R = 0$ and the disk K degenerates to the single point z_0.

Definition 6 The sequence of functions $\{b_n : X \to \mathbb{R}; n \in \mathbb{N}\}$ is called *non-decreasing* (resp. *nonincreasing*) *on the set* $E \subset X$ if the numerical sequence $\{b_n(x); n \in \mathbb{N}\}$ is nondecreasing (resp. nonincreasing) for every $x \in E$. Nondecreasing and nonincreasing sequences of functions on set are called *monotonic sequences* on the set.

We recall (if necessary, see Sect. 5.2.3) the following identity, called *Abel's transformation*:

$$\sum_{k=n}^{m} a_k b_k = A_m b_m - A_{n-1} b_n + \sum_{k=n}^{m-1} A_k (b_k - b_{k+1}), \qquad (16.7)$$

where $a_k = A_k - A_{k-1}, k = n, \ldots, m$.

If $b_n, b_{n+1}, \ldots, b_m$ is a monotonic sequence of real numbers, then, even if $a_n, a_{n+1}, \ldots, a_m$ are complex numbers or vectors of a normed space, one can obtain the following estimate, which we need, from the identity (16.7):

$$\left| \sum_{k=n}^{m} a_k b_k \right| \leq 4 \max_{n-1 \leq k \leq m} |A_k| \cdot \max\{|b_n|, |b_m|\}. \qquad (16.8)$$

Proof In fact,

$$|A_m b_m| + |A_{n-1} b_n| + \left| \sum_{k=n}^{m-1} A_k (b_k - b_{k-1}) \right| \leq$$

$$= \max_{n-1 \leq k \leq m} |A_k| \cdot \left(|b_m| + |b_n| + \sum_{k=n}^{m-1} |b_k - b_{k+1}| \right) =$$

$$= \max_{n-1 \leq k \leq m} |A_k| \cdot \left(|b_m| + |b_n| + |b_n - b_m| \right) \leq$$

$$\leq 4 \max_{n-1 \leq k \leq m} |A_k| \cdot \max \left(|b_n|, |b_m| \right).$$

In the equality that occurs in this computation we used the monotonicity of the numerical sequence b_k. $\square$

Proposition 3 (The Abel–Dirichlet test for uniform convergence) *A sufficient condition for uniform convergence on E of a series $\sum_{n=1}^{\infty} a_n(x) b_n(x)$ whose terms are products of complex-valued functions $a_n : X \to \mathbb{C}$ and real-valued functions $b_n : X \to \mathbb{R}$ is that either of the following pairs of hypotheses be satisfied:*

$\alpha_1)$ *the partial sums $s_k(x) = \sum_{n=1}^{k} a_n(x)$ of the series $\sum_{n=1}^{\infty} a_n(x)$ are uniformly bounded on E;*

$\beta_1)$ *the sequence of functions $b_n(x)$ tends monotonically and uniformly to zero on E;*

or

α_2) *the series $\sum_{n=1}^{\infty} a_n(x)$ converges uniformly on E;*
β_2) *the sequence of functions $b_n(x)$ is monotonic and uniformly bounded on E.*

Proof The monotonicity of the sequence $b_n(x)$ allows us to write an estimate analogous to (16.8) for each $x \in E$:

$$\left| \sum_{k=n}^{m} a_k(x) b_k(x) \right| \leq 4 \max_{n-1 \leq k \leq m} |A_k(x)| \cdot \max\{|b_n(x)|, |b_m(x)|\}, \qquad (16.8')$$

where we take $s_k(x) - s_{n-1}(x)$ as $A_k(x)$.

If the hypotheses α_1) and β_1) hold, then, on the one hand, there exists a constant M such that $|A_k(x)| \leq M$ for all $k \in \mathbb{N}$ and all $x \in E$, while on the other hand, for any number $\varepsilon > 0$ we have $\max\{|b_n(x)|, |b_m(x)|\} < \frac{\varepsilon}{4M}$ for all sufficiently large n and m and all $x \in E$. Hence it follows from (16.8) that $|\sum_{k=n}^{m} a_k(x) b_k(x)| < \varepsilon$ for all sufficiently large n and m and all $x \in E$, that is, the Cauchy criterion holds for this series.

In the case of hypotheses α_2) and β_2) the quantity $\max\{|b_n(x)|, |b_m(x)|\}$ is bounded. At the same time, by the uniform convergence of the series $\sum_{n=1}^{\infty} a_n(x)$ and the Cauchy criterion, for every $\varepsilon > 0$ we have $|A_k(x)| = |s_k(x) - s_{n-1}(x)| < \varepsilon$ for all sufficiently large n and $k > n$ and all $x \in E$. Taking this into account, we again conclude from (16.8) that the Cauchy criterion for uniform convergence holds for this series. □

Remark 4 In the case when the functions a_n and b_n are constants Proposition 3 becomes the Abel–Dirichlet criterion for convergence of numerical series.

Example 7 Let us consider the convergence of the series

$$\sum_{n=1}^{\infty} \frac{1}{n^{\alpha}} e^{inx}. \qquad (16.9)$$

Since

$$\left| \frac{1}{n^{\alpha}} e^{inx} \right| = \frac{1}{n^{\alpha}}, \qquad (16.10)$$

the necessary condition for uniform convergence does not hold for the series (16.9) when $\alpha \leq 0$, and it diverges for every $x \in \mathbb{R}$. Thus we shall assume $\alpha > 0$ from now on.

If $\alpha > 1$, we conclude from the Weierstrass M-test and (16.10) that the series (16.9) converges absolutely and uniformly on the entire real line $\mathbb{R}$.

To study the convergence for $0 < \alpha \leq 1$ we use the Abel–Dirichlet test, setting $a_n(x) = e^{inx}$ and $b_n(x) = \frac{1}{n^{\alpha}}$. Since the constant functions $b_n(x)$ are monotonic when $\alpha > 0$ and obviously tend to zero uniformly for $x \in \mathbb{R}$, it remains only to investigate the partial sums of the series $\sum_{n=1}^{\infty} e^{inx}$.

For convenience in citing results below, we shall consider the sums $\sum_{k=0}^{n} e^{ikx}$, which differ from the sums of our series only in the first term, which is 1.

Using the formula for the sum of a finite geometric series and Euler's formula, we obtain successively for $x \neq 2\pi m$, $m \in \mathbb{Z}$,

$$\sum_{k=0}^{n} e^{ikx} = \frac{e^{i(n+1)x} - 1}{e^{ix} - 1} = \frac{\sin \frac{n+1}{2}x}{\sin \frac{x}{2}} \cdot \frac{e^{i\frac{n+1}{2}x}}{e^{i\frac{x}{2}}} =$$

$$= \frac{\sin \frac{n+1}{2}x}{\sin \frac{x}{2}} e^{i\frac{n}{2}x} = \frac{\sin \frac{n+1}{2}x}{\sin \frac{x}{2}} \left(\cos \frac{n}{2}x + i \sin \frac{n}{2}x \right). \tag{16.11}$$

Hence, for every $n \in \mathbb{N}$

$$\left| \sum_{k=0}^{n} e^{ikx} \right| \leq \frac{1}{|\sin \frac{x}{2}|}, \tag{16.12}$$

from which it follows by the Abel–Dirichlet criterion that for $0 < \alpha \leq 1$ the series (16.9) converges uniformly on every set $E \subset \mathbb{R}$ on which $\inf_{x \in E} |\sin \frac{x}{2}| > 0$. In particular the series (16.9) simply converges for every $x \neq 2\pi m$, $m \in \mathbb{Z}$. If $x = 2\pi m$, then $e^{in2\pi m} = 1$, and the series (16.9) becomes the numerical series $\sum_{n=1}^{\infty} \frac{1}{n^{\alpha}}$, which diverges for $0 < \alpha \leq 1$.

We shall show that from what has been said, one can conclude that for $0 < \alpha \leq 1$ the series (16.9) cannot converge uniformly on any set E whose closure contains points of the form $2\pi m$, $m \in \mathbb{Z}$. For definiteness, suppose $0 \in \overline{E}$. The series $\sum_{n=1}^{\infty} \frac{1}{n^{\alpha}}$ diverges for $0 < \alpha \leq 1$. By the Cauchy criterion, there exists $\varepsilon > 0$ such that for every $N \in \mathbb{N}$, no matter how large, one can find numbers $m \geq n > N$ such that $|\frac{1}{n^{\alpha}} + \cdots + \frac{1}{m^{\alpha}}| > \varepsilon_0 > 0$. By the continuity of the functions e^{ikx} on $\mathbb{R}$, it follows that one can choose a point $x \in E$ close enough to 0 so that

$$\left| \frac{e^{inx}}{n^{\alpha}} + \cdots + \frac{e^{imx}}{m^{\alpha}} \right| > \varepsilon_0.$$

But by the Cauchy criterion for uniform convergence this means that the series (16.9) cannot converge uniformly on E.

To supplement what has just been said, we note that, as one can see from (16.10), the series (16.9) converges nonabsolutely for $0 < \alpha \leq 1$.

Remark 5 It is useful for what follows to remark that, separating the real and imaginary parts in (16.11), we obtain the following relations:

$$\sum_{k=0}^{n} \cos kx = \frac{\cos \frac{n}{2}x \cdot \sin \frac{n+1}{2}x}{\sin \frac{x}{2}}, \tag{16.13}$$

$$\sum_{k=0}^{n} \sin kx = \frac{\sin \frac{n}{2}x \cdot \sin \frac{n+1}{2}x}{\sin \frac{x}{2}}, \qquad (16.14)$$

which hold for $x \neq 2\pi m$, $m \in \mathbb{Z}$.

As another example of the use of the Abel–Dirichlet test we prove the following proposition.

Proposition 4 (The so-called second Abel theorem on power series) *If a power series $\sum_{n=0}^{\infty} c_n(z - z_0)^n$ converges at a point $\zeta \in \mathbb{C}$, then it converges uniformly on the closed interval with endpoints z_0 and ζ.*

Proof We represent the points of this interval in the form $z_0 + (\zeta - z_0)t$, where $0 \leq t \leq 1$. Substituting this expression in the power series, we obtain the series $\sum_{n=0}^{\infty} c_n(\zeta - z_0)^n t^n$. By hypothesis, the numerical series $\sum_{n=0}^{\infty} c_n(\zeta - z_0)^n$ converges, and the sequence of functions t^n is monotonic and uniformly bounded on the closed interval $[0, 1]$. Hence conditions $\alpha_2)$ and $\beta_2)$ in the Abel–Dirichlet test are satisfied, and the proposition is proved. □

16.2.4 Problems and Exercises

1. Investigate the nature of the convergence on the sets $E \subset \mathbb{R}$ for different values of the real parameter α in the following series:

a) $\sum_{n=1}^{\infty} \frac{\cos nx}{n^{\alpha}}$.

b) $\sum_{n=1}^{\infty} \frac{\sin nx}{n^{\alpha}}$.

2. Prove that the following series converge uniformly on the indicated sets:

a) $\sum_{n=1}^{\infty} \frac{(-1)^n}{n} x^n$ for $0 \leq x \leq 1$.

b) $\sum_{n=1}^{\infty} \frac{(-1)^n}{n} e^{-nx}$ for $0 \leq x \leq +\infty$.

c) $\sum_{n=1}^{\infty} \frac{(-1)^n}{n+x}$ for $0 \leq x \leq +\infty$.

3. Show that if a *Dirichlet series* $\sum_{n=1}^{\infty} \frac{c_n}{n^x}$ converges at a point $x_0 \in \mathbb{R}$, then it converges uniformly on the set $x \geq x_0$ and absolutely if $x > x_0 + 1$.

4. Verify that the series $\sum_{n=1}^{\infty} \frac{(-1)^{n-1}x^2}{(1+x^2)^n}$ converges uniformly on $\mathbb{R}$, and the series $\sum_{n=1}^{\infty} \frac{x^2}{(1+x^2)^n}$ converges on $\mathbb{R}$, but nonuniformly.

5. a) Using the example of the series from Problem 2, show that the Weierstrass M-test is a sufficient condition but not a necessary one for the uniform convergence of a series.

b) Construct a series $\sum_{n=1}^{\infty} a_n(x)$ with nonnegative terms that are continuous on the closed interval $0 \leq x \leq 1$ and which converges uniformly on that closed interval, while the series $\sum_{n=1}^{\infty} M_n$ formed from the quantities $M = \max_{0 \leq x \leq 1} |a_n(x)|$ diverges.

6. a) State the Abel–Dirichlet test for convergence of a series mentioned in Remark 4.

b) Show that the condition that $\{b_n\}$ be monotonic in the Abel–Dirichlet test can be weakened slightly, requiring only that the sequence $\{b_n\}$ be monotonic up to corrections $\{\beta_n\}$ forming an absolutely convergent series.

7. As a supplement to Proposition 4 shows, following Abel, that if a power series converges at a boundary point of the disk of convergence, its sum has a limit in that disk when the point is approached along any direction not tangential to the boundary circle.

16.3 Functional Properties of a Limit Function

16.3.1 Specifics of the Problem

In this section we shall give answers to the questions posed in Sect. 16.1 as to when the limit of a family of continuous, differentiable, or integrable functions is a function having the same property, and when the limit of the derivatives or integrals of the functions equals the derivative or integral of the limiting function of the family.

To explain the mathematical content of these questions, let us consider, for example, the connection between continuity and passage to the limit.

Let $f_n(x) \to f(x)$ on $\mathbb{R}$ as $n \to \infty$, and suppose that all the functions in the sequence $\{f_n: n \in \mathbb{N}\}$ are continuous at the point $x_0 \in \mathbb{R}$. We are interested in the continuity of the limit function f at the same point x_0. To answer that question, we need to verify the equality $\lim_{x\to x_0} f(x) = f(x_0)$, which in terms of the original sequence can be rewritten as the relation $\lim_{x\to x_0}(\lim_{n\to\infty} f_n(x)) = \lim_{n\to\infty} f_n(x_0)$, or, taking account of the given continuity of f_n at x_0, as the following relation, subject to verification:

$$\lim_{x\to x_0}\left(\lim_{n\to\infty} f_n(x)\right) = \lim_{n\to\infty}\left(\lim_{x\to x_0} f_n(x)\right). \tag{16.15}$$

On the left-hand side here the limit is first taken over the base $n \to \infty$, then over the base $x \to x_0$, while on the right-hand side the limits over the same bases are taken in the opposite order.

When studying functions of several variables we saw that Eq. (16.15) is by no means always true. We also saw this in the examples studied in the two preceding sections, which show that the limit of a sequence of continuous functions is not always continuous.

Differentiation and integration are special operations involving passage to the limit. Hence the question whether we get the same result if we first differentiate (or integrate) the functions of a family, then pass to the limit over the parameter of the family or first find the limit function of the family and then differentiate (or integrate) again reduces to verifying the possibility of changing the order of two limiting passages.

16.3.2 Conditions for Two Limiting Passages to Commute

We shall show that if at least one of two limiting passages is uniform, then the limiting passages commute.

Theorem 1 *Let $\{F_t; t \in T\}$ be a family of functions $F_t : X \to \mathbb{C}$ depending on a parameter t; let $\mathcal{B}_X$ be a base in X and $\mathcal{B}_T$ a base in T. If the family converges uniformly on X over the base $\mathcal{B}_T$ to a function $F : X \to \mathbb{C}$ and the limit $\lim_{\mathcal{B}_X} F_t(x) = A_t$ exists for each $t \in T$, then both repeated limits $\lim_{\mathcal{B}_X}(\lim_{\mathcal{B}_T} F_t(x))$ and $\lim_{\mathcal{B}_T}(\lim_{\mathcal{B}_X} F_t(x))$ exist and the equality*

$$\lim_{\mathcal{B}_X}\left(\lim_{\mathcal{B}_T} F_t(x)\right) = \lim_{\mathcal{B}_T}\left(\lim_{\mathcal{B}_X} F_t(x)\right) \tag{16.16}$$

holds.

This theorem can be conveniently written as the following diagram

$$
\begin{array}{ccc}
F_t(x) & \xRightarrow{\quad} & F(x) \\[2pt]
\mathcal{B}_X \downarrow & \mathcal{B}_T \;\nearrow \atop \nearrow \atop \nearrow \; \exists & \exists \downarrow \mathcal{B}_X \\[2pt]
A_t & \xrightarrow[\mathcal{B}_T]{\quad\quad} & A
\end{array} \tag{16.17}
$$

in which the hypotheses are written above the diagonal and the consequences below it. Equality (16.16) means that this diagram is commutative, that is, the final result A is the same whether the operations corresponding to passage over the upper and right-hand sides are carried out or one first passes down the left-hand side and then to the right over the lower side.

Let us prove this theorem.

Proof Since $F_t \rightrightarrows F$ on X over $\mathcal{B}_T$, by the Cauchy criterion, for every $\varepsilon > 0$ there exists B_T in $\mathcal{B}_T$ such that

$$\left| F_{t_1}(x) - F_{t_2}(x) \right| < \varepsilon \tag{16.18}$$

for every $t_1, t_2 \in B_T$ and every $x \in X$.

Passing to the limit over $\mathcal{B}_X$ in this inequality, we obtain the relation

$$|A_{t_1} - A_{t_2}| \le \varepsilon, \tag{16.19}$$

which holds for every $t_1, t_2 \in B_T$. By the Cauchy criterion for existence of the limit of a function it now follows that A_t has a certain limit A over $\mathcal{B}_T$. We now verify that $A = \lim_{\mathcal{B}_X} F(x)$.

Fixing $t_2 \in B_T$, we find an element B_X in $\mathcal{B}_X$ such that

$$\left| F_{t_2}(x) - A_{t_2} \right| < \varepsilon \tag{16.20}$$

for all $x \in B_X$.

Keeping t_2 fixed, we pass to the limit in (16.18) and (16.19) over $\mathcal{B}_T$ with respect to t_1. We then find

$$\left| F(x) - F_{t_2}(x) \right| \le \varepsilon, \tag{16.21}$$

$$|A - A_{t_2}| \le \varepsilon, \tag{16.22}$$

and (16.22) holds for all $x \in X$.

Comparing (16.20)–(16.22), and using the triangle inequality, we find

$$\left| F(x) - A \right| < 3\varepsilon$$

for every $x \in B_X$. We have thus verified that $A = \lim_{\mathcal{B}_X} F(x)$. $\square$

Remark 1 As the proof shows, Theorem 1 remains valid for functions $F_t : X \to Y$ with values in any complete metric space.

Remark 2 If we add the requirement that the limit $\lim_{\mathcal{B}_T} A_t = A$ exists to the hypotheses of Theorem 1, then, as the proof shows, the equality $\lim_{\mathcal{B}_X} F(x) = A$ can be obtained even without assuming that the space Y of values of the functions $F_t : X \to Y$ is complete.

16.3.3 Continuity and Passage to the Limit

We shall show that if functions that are continuous at a point of a set converge uniformly on that set, then the limit function is also continuous at that point.

Theorem 2 *Let* $\{f_t; t \in T\}$ *be a family of functions* $f_t : X \to \mathbb{C}$ *depending on the parameter* t*; let* $\mathcal{B}$ *be a base in* T*. If* $f_t \rightrightarrows f$ *on* X *over the base* $\mathcal{B}$ *and the functions* f_t *are continuous at* $x_0 \in X$*, then the function* $f : X \to \mathbb{C}$ *is also continuous at that point.*

Proof In this case the diagram (16.17) assumes the following specific form:

$$
\begin{array}{ccc}
f_t(x) & \rightrightarrows^{\mathcal{B}} & f(x) \\
{\scriptstyle x \to x_0}\Big\downarrow & & \Big\downarrow{\scriptstyle x \to x_0} \\
f_t(x_0) & \xrightarrow{\mathcal{B}} & f(x_0)
\end{array}
$$

Here all the limiting passages except the vertical passage on the right are defined by the hypotheses of Theorem 2 itself. The nontrivial conclusion of Theorem 1 that we need is precisely that $\lim_{x \to x_0} f(x) = f(x_0)$. $\square$

Remark 3 We have not said anything specific as to the nature of the set X. In fact it may be any topological space provided the base $x \to x_0$ is defined in it. The values of the functions f_t may lie in any metric space, which, as follows from Remark 2, need not even be complete.

Corollary 1 *If a sequence of functions that are continuous on a set converges uniformly on that set, then the limit function is continuous on the set.*

Corollary 2 *If a series of functions that are continuous on a set converges uniformly on that set, then the sum of the series is also continuous on the set.*

As an illustration of the possible use of these results, consider the following.

Example 1 Abel's method of summing series.
  Comparing Corollary 2 with Abel's second theorem (Proposition 4 of Sect. 16.2), we draw the following conclusion.

Proposition 1 *If a power series $\sum_{n=0}^{\infty} c_n (z - z_0)^n$ converges at a point ζ, it converges uniformly on the closed interval $[z_0, \zeta]$ from z_0 to ζ, and the sum of the series is continuous on that interval.*

In particular, this means that if a numerical series $\sum_{n=0}^{\infty} c_n$ converges, then the power series $\sum_{n=0}^{\infty} c_n x^n$ converges uniformly on the closed interval $0 \le x \le 1$ of the real axis and its sum $s(x) = \sum_{n=0}^{\infty} c_n x^n$ is continuous on that interval. Since $s(1) = \sum_{n=0}^{\infty} c_n$, we can thus assert that if the series $\sum_{n=0}^{\infty} c_n$ converges, then the following equality holds:

$$\sum_{n=0}^{\infty} c_n = \lim_{x \to 1-0} \sum_{n=0}^{\infty} c_n x^n. \tag{16.23}$$

It is interesting that the right-hand side of Eq. (16.23) may have a meaning even when the series on the left diverges in its traditional sense. For example, the series $1 - 1 + 1 - \cdots$ corresponds to the series $x - x^2 + x^3 - \cdots$, which converges to $x/(1+x)$ for $|x| < 1$. As $x \to 1$, this function has the limit $1/2$.
  The method of summing a series known as Abel summation consists of ascribing to the left-hand side of (16.23) the value of the right-hand side if it is defined. We have seen that if the series $\sum_{n=0}^{\infty} c_n$ converges in the traditional sense, then its classical sum will be assigned to it by Abel summation. At the same time, for example, Abel's method assigns to the series $\sum_{n=0}^{\infty} (-1)^n$, which diverges in the traditional sense, the natural average value $1/2$.
  Further questions connected with Example 1 can be found in Problems 5–8 below.

Example 2 Earlier, when discussing Taylor's formula, we showed that the following expansion holds:

$$(1 + x)^\alpha = 1 + \frac{\alpha}{1!}x + \frac{\alpha(\alpha - 1)}{2!}x^2 + \cdots + \frac{\alpha(\alpha - 1)\cdots(\alpha - n + 1)}{n!}x^n + \cdots .$$

$$(16.24)$$

We can verify that for $\alpha > 0$ the numerical series

$$1 + \frac{\alpha}{1!} + \frac{\alpha(\alpha - 1)}{2!} + \cdots + \frac{\alpha(\alpha - 1)\cdots(\alpha - n + 1)}{n!} + \cdots$$

converges. Hence by Abel's theorem, if $\alpha > 0$, the series (16.24) converges uniformly on the closed interval $0 \leq x \leq 1$. But the function $(1 + x)^\alpha$ is continuous at $x = 1$, and so one can assert that if $\alpha > 0$, then Eq. (16.24) holds also for $x = 1$.

In particular, we can assert that for $\alpha > 0$

$$\left(1 - t^2\right)^\alpha = 1 - \frac{\alpha}{1!}t^2 + \frac{\alpha(\alpha - 1)}{2!}t^4 -$$

$$- \cdots + (-1)^n \cdot \frac{\alpha(\alpha - 1)\cdots(\alpha - n + 1)}{n!}t^{2n} + \cdots \quad (16.25)$$

and this series converges to $(1 - t^2)^\alpha$ uniformly on $[-1, 1]$.

Setting $\alpha = \frac{1}{2}$ and $t^2 = 1 - x^2$ in (16.25), for $|x| \leq 1$ we find

$$|x| = 1 - \frac{\frac{1}{2}}{1!}\left(1 - x^2\right) + \frac{\frac{1}{2}\left(\frac{1}{2} - 1\right)}{2!}\left(1 - x^2\right)^2 - \cdots , \quad (16.26)$$

and the series of polynomials on the right-hand side converges to $|x|$ uniformly on the closed interval $[-1, 1]$. Setting $P_n(x) := S_n(x) - S_n(0)$, where $S_n(x)$ is the nth partial sum of the series, we find that for any prescribed tolerance $\varepsilon > 0$ there is a polynomial $P(x)$ such that $P(0) = 0$ and

$$\max_{-1 \leq x \leq 1} \left||x| - P(x)\right| < \varepsilon. \quad (16.27)$$

Let us now return to the general theory.

We have shown that continuity of functions is preserved under uniform passage to the limit. The condition of uniformity in passage to the limit is, however, only a sufficient condition in order that the limit of continuous functions also be a continuous function (see Examples 8 and 9 of Sect. 16.1). At the same time there is a specific situation in which the convergence of continuous functions to a continuous function guarantees that the convergence is uniform.

Proposition 2 (Dini's[2] theorem) *If a sequence of continuous functions on a compact set converges monotonically to a continuous function, then the convergence is uniform.*

---

[2]U. Dini (1845–1918) – Italian mathematician best known for his work in the theory of functions.

Proof For definiteness suppose that f_n is a nondecreasing sequence converging to f. We fix an arbitrary $\varepsilon > 0$, and for every point x of the compact set K we find an index n_x such that $0 \le f(x) - f_{n_x}(x) < \varepsilon$. Since the functions f and f_{n_x} are continuous on K, the inequality $0 \le f(\xi) - f_{n_x}(\xi) < \varepsilon$ holds in some neighborhood $U(x)$ of $x \in K$. From the covering of the compact set K by these neighborhoods one can extract a finite covering $U(x_1), \ldots, U(x_k)$ and then fix the index $n(\varepsilon) = \max\{n_{x_1}, \ldots, n_{x_k}\}$. Then for any $n > n(\varepsilon)$, by the fact that the sequence $\{f_n; n \in \mathbb{N}\}$ is nondecreasing, we have $0 \le f(\xi) - f_n(\xi) < \varepsilon$ at every point $\xi \in K$. $\square$

Corollary 3 *If the terms of the series $\sum_{n=1}^{\infty} a_n(x)$ are nonnegative functions a_n : $K \to \mathbb{R}$ that are continuous on a compact set K and the series converges to a continuous function on K, then it converges uniformly on K.*

Proof The partial sums $s_n(x) = \sum_{k=1}^{n} a_k(x)$ of this series satisfy the hypotheses of Dini's theorem. $\square$

Example 3 We shall show that the sequence of functions $f_n(x) = n(1 - x^{1/n})$ tends to $f(x) = \ln \frac{1}{x}$ as $n \to +\infty$ uniformly on each closed interval $[a, b]$ contained in the interval $0 < x < \infty$.

Proof For fixed $x > 0$ the function $x^t = e^{t \ln x}$ is convex with respect to t, so that the ratio $\frac{x^t - x^0}{t - 0}$ (the slope of the chord) is nonincreasing as $t \to +0$ and tends to $\ln x$.

Hence $f_n(x) \nearrow \ln \frac{1}{x}$ for $x > 0$ as $n \to +\infty$. By Dini's theorem it now follows that the convergence of $f_n(x)$ to $\ln \frac{1}{x}$ is uniform on each closed interval $[a, b] \subset$ $]0, +\infty[$. $\square$

We note that the convergence is obviously not uniform on the interval $0 < x \le 1$, for example, since $\ln \frac{1}{x}$ is unbounded in that interval, while each of the functions $f_n(x)$ is bounded (by a constant depending on n).

16.3.4 Integration and Passage to the Limit

We shall show that if functions that are integrable over a closed interval converge uniformly on that interval, then the limit function is also integrable and its integral over that interval equals the limit of the integrals of the original functions.

Theorem 3 *Let $\{f_t; t \in T\}$ be a family of functions $f_t : [a, b] \to \mathbb{C}$ defined on a closed interval $a \le x \le b$ and depending on the parameter $t \in T$, and let $\mathcal{B}$ be a base in T. If the functions of the family are integrable on $[a, b]$ and $f_t \rightrightarrows f$ on $[a, b]$ over the base $\mathcal{B}$, then the limit function $f : [a, b] \to \mathbb{C}$ is also integrable on $[a, b]$ and*

$$\int_a^b f(x)\,dx = \lim_{\mathcal{B}} \int_a^b f_t(x)\,dx.$$

Proof Let $p = (P, \xi)$ be a partition P of the closed interval $[a, b]$ with distinguished points $\xi = \{\xi_1, \ldots, \xi_n\}$. Consider the Riemann sums $F_t(p) = \sum_{i=1}^{n} f_t(\xi_i) \Delta x_i$, $t \in T$, and $F(p) = \sum_{i=1}^{n} f(\xi_i) \Delta x_i$. Let us estimate the difference $F(p) - F_t(p)$. Since $f_t \rightrightarrows f$ on $[a, b]$ over the base $\mathcal{B}$, for every $\varepsilon > 0$ there exists an element B of $\mathcal{B}$ such that $|f(x) - f_t(x)| < \frac{\varepsilon}{b-a}$ at any $t \in B$ and any point $x \in [a, b]$. Hence for $t \in B$ we have

$$\left| F(p) - F_t(p) \right| = \left| \sum_{i=1}^{n} \left(f(\xi_i) - f_t(\xi_i) \right) \Delta x_i \right| \leq \sum_{i=1}^{n} \left| f(\xi_i) - f_t(\xi_i) \right| \Delta x_i < \varepsilon,$$

and this estimate holds not only for every $t \in B$, but also for every partition p in the set $\mathcal{P} = \{(P, \xi)\}$ of partitions of the closed interval $[a, b]$ with distinguished points. Thus $F_t \rightrightarrows F$ on $\mathcal{P}$ over the base $\mathcal{B}$. Now, taking the traditional base $\lambda(P) \to 0$ in $\mathcal{P}$, we find by Theorem 1 that the following diagram is commutative:

$$
\begin{array}{ccc}
\displaystyle\sum_{i=1}^{n} f_t(\xi_i) \Delta x_i \ =: F_t(p) & \stackrel{\mathcal{B}}{\Longrightarrow} & F(p) := \displaystyle\sum_{i=1}^{n} f(\xi_i) \Delta x_i \\
\lambda(P) \to 0 \ \Big\downarrow & & \exists \ \Big\downarrow \ \lambda(P) \to 0 \\
\displaystyle\int_a^b f_t(x)\, dx =: A_t & \xrightarrow{\ \mathcal{B}\ } & A := \displaystyle\int_a^b f_t(x)\, dx
\end{array}
$$

which proves Theorem 3.   □

Corollary 4 *If the series $\sum_{n=1}^{\infty} f_n(x)$ consisting of integrable functions on a closed interval $[a, b] \subset \mathbb{R}$ converges uniformly on that closed interval, then its sum is also integrable on $[a, b]$ and*

$$\int_a^b \left(\sum_{n=1}^{\infty} f_n(x) \right) dx = \sum_{n=1}^{\infty} \int_a^b f_n(x)\, dx.$$

Example 4 When we write $\frac{\sin x}{x}$ in this example, we shall assume that this ratio equals 1 when $x = 0$.

We have noted earlier that the function $\mathrm{Si}(x) = \int_0^x \frac{\sin t}{t}\, dt$ is not an elementary function. Using the theorem just proved, we can nevertheless obtain a very simple representation of this function as a power series.

To do this, we remark that

$$\frac{\sin t}{t} = \sum_{n=0}^{\infty} \frac{(-1)^n}{(2n+1)!} t^{2n}, \tag{16.28}$$

and the series on the right-hand side converges uniformly on every closed interval $[-a, a] \subset \mathbb{R}$. The uniform convergence of the series follows from the Weierstrass

M-test, since $\frac{|t|^{2n}}{(2n+1)!} \le \frac{a^{2n}}{(2n+1)!}$ for $|t| \le a$, while the numerical series $\sum_{n=0}^{\infty} \frac{a^{2n}}{(2n+1)!}$ converges.

By Corollary 4 we can now write

$$\mathrm{Si}(x) = \int_0^x \left(\sum_{n=0}^{\infty} \frac{(-1)^n}{(2n+1)!} t^{2n} \right) dt =$$

$$= \sum_{n=0}^{\infty} \left(\int_0^x \frac{(-1)^n}{(2n+1)!} t^{2n} \, dt \right) = \sum_{n=0}^{\infty} \frac{(-1)^n x^{2n+1}}{(2n+1)!(2n+1)}.$$

The series just obtained also turns out to converge uniformly on every closed interval of the real line, so that, for any closed interval $[a, b]$ of variation of the argument x and any preassigned absolute error tolerance, one can choose a polynomial – a partial sum of this series – that makes it possible to compute $\mathrm{Si}(x)$ with less than the given error at every point of the closed interval $[a, b]$.

16.3.5 Differentiation and Passage to the Limit

Theorem 4 *Let $\{f_t; t \in T\}$ be a family of functions $f_t : X \to \mathbb{C}$ defined on a convex bounded set X (in $\mathbb{R}$, $\mathbb{C}$, or any other normed space) and depending on the parameter $t \in T$; let B be a base in T. If the functions of the family are differentiable on X, the family of derivatives $\{f_t'; t \in T\}$ converges uniformly on X to a function $\varphi : X \to \mathbb{C}$, and the original family $\{f_t; t \in T\}$ converges at even one point $x_0 \in X$, then it converges uniformly on the entire set X to a differentiable function $f : X \to \mathbb{C}$, and $f' = \varphi$.*

Proof We begin by showing that the family $\{f_t; t \in T\}$ converges uniformly on the set X over the base B. We use the mean-value theorem in the following estimates:

$$\left| f_{t_1}(x) - f_{t_2}(x) \right| \le$$

$$\le \left| \left(f_{t_1}(x) - f_{t_2}(x) \right) - \left(f_{t_1}(x_0) - f_{t_2}(x_0) \right) \right| + \left| f_{t_1}(x_0) - f_{t_2}(x_0) \right| \le$$

$$\le \sup_{\xi \in [x_0, x]} \left| f_{t_1}'(\xi) - f_{t_2}'(\xi) \right| |x - x_0| + \left| f_{t_1}(x_0) - f_{t_2}(x_0) \right| = \Delta(x, t_1, t_2).$$

By hypothesis the family $\{f_t'; t \in T\}$ converges uniformly on X over the base B, and the quantity $f_t(x_0)$ has a limit over the same base as a function of t, while $|x - x_0|$ is bounded for $x \in X$. By the necessity part of the Cauchy criterion for uniform convergence of the family of functions f_t' and the existence of the limit function $f_t(x_0)$, for every $\varepsilon > 0$ there exists B in B such that $\Delta(x, t_1, t_2) < \varepsilon$ for any $t_1, t_2 \in B$ and any $x \in X$. But, by the estimates just written, this means that the family of functions $\{f_t; t \in T\}$ satisfies the hypotheses of the Cauchy criterion and consequently converges on X over the base B to a function $f : X \to \mathbb{C}$.

Again using the mean-value theorem, we now obtain the following estimates:

$$\left|\left(f_{t_1}(x+h) - f_{t_1}(x) - f'_{t_1}(x)h\right) - \left(f_{t_2}(x+h) - f_{t_2}(x) - f'_{t_2}(x)h\right)\right| =$$

$$= \left|(f_{t_1} - f_{t_0})(x+h) - (f_{t_1} - f_{t_2})(x) - (f_{t_1} - f_{t_2})'(x)h\right| \le$$

$$\le \sup_{0<\theta<1} \left|(f_{t_1} - f_{t_2})'(x+\theta h)\right| |h| + \left|(f_{t_1} - f_{t_2})'(x)\right| |h| =$$

$$= \left(\sup_{0<\theta<1} \left|f'_{t_1}(x+\theta h) - f'_{t_2}(x+\theta h)\right| + \left|f'_{t_1}(x) - f'_{t_2}(x)\right| \right) |h|.$$

These estimates, which are valid for $x, x + h \in X$ show, in view of the uniform convergence of the family $\{f'_t; t \in T\}$ on X, that the family $\{F_t; t \in T\}$ of functions

$$F_t(h) = \frac{f_t(x+h) - f_t(x) - f'_t(x)h}{|h|},$$

which we shall consider with a fixed value of $x \in X$, converges over the base $\mathcal{B}$ uniformly with respect to all values of $h \ne 0$ such that $x + h \in X$.

We remark that $F_t(h) \to 0$ as $h \to 0$ since the function f_t is differentiable at the point $x \in X$; and since $f_t \to f$ and $f'_t \to \varphi$ over the base $\mathcal{B}$, we have $F_t(h) \to F(h) = \frac{f(x+h)-f(x)-\varphi(x)h}{|h|}$ over the base $\mathcal{B}$.

Applying Theorem 1, we can now write the commutative diagram

The right-hand limiting passage as $h \to 0$ shows that f is differentiable at $x \in X$ and $f'(x) = \varphi(x)$. □

Corollary 5 *If the series $\sum_{n=1}^{\infty} f_n(x)$ of functions $f_n : X \to \mathbb{C}$ that are differentiable on a bounded convex subset X (contained in $\mathbb{R}$, $\mathbb{C}$, or any other normed vector space) converges at even one point $x \in X$ and the series $\sum_{n=1}^{\infty} f'_n(x)$ converges uniformly on X, then $\sum_{n=1}^{\infty} f_n(x)$ also converges uniformly on X, its sum is differentiable on X, and*

$$\left(\sum_{n=1}^{\infty} f_n(x) \right)'(x) = \sum_{n=1}^{\infty} f'_n(x).$$

This follows from Theorem 4 and the definitions of the sum and uniform convergence of a series, together with the linearity of the operation of differentiation.

Remark 4 The proofs of Theorems 3 and 4, like the theorems themselves and their corollaries, remain valid for functions $f_t : X \to Y$ with values in any complete normed vector space Y. For example, Y may be $\mathbb{R}$, $\mathbb{C}$, $\mathbb{R}^n$, $\mathbb{C}^n$, $C[a, b]$, and so on. The domain of definition X for the functions f_t in Theorem 4 also may be any suitable subset of any normed vector space. In particular, X may be contained in $\mathbb{R}$, $\mathbb{C}$, $\mathbb{R}^n$, or $\mathbb{C}^n$. For real-valued functions of a real argument (under additional convergence requirements) the proofs of these theorems can be made even simpler (see Problem 11).

As an illustration of the use of Theorems 2–4 we shall prove the following proposition, which is widely used in both theory and in specific computations.

Proposition 3 *Let $K \subset \mathbb{C}$ be the convergence disk for a power series $\sum_{n=0}^{\infty} c_n (z - z_0)^n$. If K contains more than just the point z_0, then the sum of the series $f(z)$ is differentiable inside K and*

$$f'(z) = \sum_{n=1}^{\infty} n c_n (z - z_0)^{n-1}. \tag{16.29}$$

Moreover, the function $f(z) : K \to \mathbb{C}$ can be integrated over any path γ : $[0, 1] \to K$, and if $[0, 1] \ni t \overset{\gamma}{\longmapsto} z(t) \in K$, $z(0) = z_0$, and $z(1) = z$, then

$$\int_{\gamma} f(z) \, dz = \sum_{n=0}^{\infty} \frac{c_n}{n+1} (z - z_0)^{n+1}. \tag{16.30}$$

Remark 5 Here $\int_{\gamma} f(z) \, dz := \int_0^1 f(z(t)) z'(t) \, dt$. In particular, if the equality $f(x) = \sum_{n=0}^{\infty} a_n (x - x_0)^n$ holds on an interval $-R < x - x_0 < R$ of the real line $\mathbb{R}$, then

$$\int_{x_0}^{x} f(t) \, dt = \sum_{n=0}^{\infty} \frac{a_n}{n+1} (x - x_0)^{n+1}.$$

Proof Since $\overline{\lim}_{n\to\infty} \sqrt[n-1]{n|c_n|} = \lim_{n\to\infty} \sqrt[n]{|c_n|}$, it follows from the Cauchy–Hadamard formula (Theorem 2 of Sect. 16.2.2 that the power series $\sum_{n=1}^{\infty} n c_n (z - z_0)^{n-1}$ obtained by termwise differentiation of the power series $\sum_{n=0}^{\infty} c_n (z - z_0)^n$, has the same convergence disk K as the original power series. But by Theorem 2 of Sect. 16.2.2 the series $\sum_{n=1}^{\infty} n c_n (z - z_0)^{n-1}$ converges uniformly in any closed disk K_q contained in the interior of K. Since the series $\sum_{n=0}^{\infty} c_n (z - z_0)^n$ obviously converges at $z = z_0$, Corollary 5 is applicable to it, which justifies the equality (16.29). Thus it has now been shown that a power series can be differentiated termwise.

Let us now verify that it can also be integrated termwise.

If $\gamma : [0, 1] \to K$ is a smooth path in K, there exists a closed disk K_q such that $\gamma \subset K_q$ and $K_q \subset K$. On K_q the original power series converges uniformly, so that in the equality

$$f\big(z(t)\big) = \sum_{n=0}^{\infty} c_n \big(z(t) - z_0\big)^n$$

the series of continuous functions on the right-hand side converges uniformly on the closed interval $0 \le t \le 1$ to the continuous function $f(z(t))$. Multiplying this equality by the function $z'(t)$, which is continuous on the closed interval $[0, 1]$, does not violate either the equality itself nor the uniform convergence of the series. Hence by Theorem 3 we obtain

$$\int_0^1 f\big(z(t)\big)z'(t)\,dt = \sum_{n=0}^{\infty} \int_0^1 c_n \big(z(t) - z_0\big)^n z'(t)\,dt.$$

But,

$$\int_0^1 \big(z(t) - z(0)\big)^n z'(t)\,dt = \frac{1}{n+1} \int_0^1 d\big(z(t) - z(0)\big)^{n+1} =$$

$$= \frac{1}{n+1}\big(z(1) - z(0)\big)^{n+1} = \frac{1}{n+1}(z - z_0)^{n+1},$$

and we arrive at Eq. (16.30).   □

Since it is obvious that $c_0 = f(z_0)$ in the expansion $f(z) = \sum_{n=0}^{\infty} c_n(z - z_0)^n$, applying Eq. (16.29) successively, we again obtain the relation $c_n = \frac{f^{(n)}(z_0)}{n!}$, which shows that a power series is uniquely determined by its sum and is the Taylor series of the sum.

Example 5 The *Bessel function* $J_n(x)$, $n \in \mathbb{N}$, is a solution of *Bessel's*[3] *equation*

$$x^2 y'' + xy' + \big(x^2 - n^2\big)y = 0.$$

Let us attempt to solve this equation, for example, for $n = 0$, as a power series $y = \sum_{k=0}^{\infty} c_k x^k$. Applying formula (16.29) successively, after elementary transformations, we arrive at the relation

$$c_1 + \sum_{k=0}^{\infty} \big(k^2 c_k + c_{k-2}\big)x^{k-1} = 0,$$

from which, by the uniqueness of the power series with a given sum, we find

$$c_1 = 0, \qquad k^2 c_k + c_{k-2} = 0, \quad k = 2, 3, \ldots.$$

---

[3]F.W. Bessel (1784–1846) – German astronomer.

From this it is easy to deduce that $c_{2k-1} = 0$, $k \in \mathbb{N}$, and $c_{2k} = (-1)^k \frac{c_0}{(k!)^2 2^{2k}}$. If we assume $J_0(0) = 1$, we arrive at the solution

$$J_0(x) = 1 + \sum_{k=1}^{\infty} (-1)^k \frac{x^{2k}}{(k!)^2 2^{2k}}.$$

This series converges on the entire line $\mathbb{R}$ (and in the entire plane $\mathbb{C}$), so that all the operations carried out above in order to find its specific form are now justified.

Example 6 In Example 5 we sought a solution of an equation as a power series. But if a series is given, using formula (16.29), one can immediately check to see whether it is the solution of a given equation. Thus, by direct computation, one can verify that the function introduced by Gauss

$$F(\alpha, \beta, \gamma, x) = 1 + \sum_{n=1}^{\infty} \frac{\alpha(\alpha+1)\cdots(\alpha+n-1)\beta(\beta+1)\cdots(\beta+n-1)}{\gamma(\gamma+1)\cdots(\gamma+n-1)} x^n$$

(the *hypergeometric series*) is well-defined for $|x| < 1$ and satisfies the so-called *hypergeometric equation*

$$x(x-1)y'' - \left[\gamma - (\alpha+\beta-1)x\right] \cdot y' + \alpha\beta \cdot y = 0.$$

In conclusion we note that, in contrast to Theorems 2 and 3, the hypotheses of Theorem 4 require that the family of derivatives, rather than the original family, converge uniformly. We have already seen (Example 2 of Sect. 16.1) that the sequence of functions $f_n(x) = \frac{1}{n} \sin n^2 x$ converges to the differentiable function $f(x) \equiv 0$ uniformly, while the sequence of derivatives $f_n'(x)$ does not converge to $f'(x)$. The point is that the derivative characterizes the rate of variation of the function, not the size of the values of the function. Even when the function changes by an amount that is small in absolute value, the derivative may formally change very strongly, as happens in the present case of small oscillations with large frequency. This is the circumstance that lies at the basis of Weierstrass' example of a continuous nowhere-differentiable function, which he gave as the series $f(x) = \sum_{n=0}^{\infty} a^n \cos(b^n \pi x)$, which obviously converges uniformly on the entire line $\mathbb{R}$ if $0 < a < 1$. Weierstrass showed that if the parameter b is chosen so as to satisfy the condition $a \cdot b > 1 + \frac{3}{2}\pi$, then on the one hand f will be continuous, being the sum of a uniformly convergent series of continuous functions, while on the other hand, it will not have a derivative at any point $x \in \mathbb{R}$. The rigorous verification of this last assertion is rather taxing, so that those who wish to obtain a simpler example of a continuous function having no derivative may see Problem 5 in Sect. 5.1.

16.3.6 Problems and Exercises

1. Using power series, find a solution of the equation $y''(x) - y(x) = 0$ satisfying the conditions

a) $y(0) = 0$, $y(1) = 1$;

b) $y(0) = 1$, $y(1) = 0$.

2. Find the sum of the series $\sum_{n=1}^{\infty} \frac{x^{n-1}}{n(n+1)}$.

3. a) Verify that the function defined by the series

$$J_n(x) = \sum_{k=0}^{\infty} \frac{(-1)^k}{k!(k+n)!} \left(\frac{x}{2}\right)^{2k+n}$$

is a solution of Bessel's equation of order $n \geq 0$ from Example 5.

b) Verify that the hypergeometric series in Example 6 provides a solution of the hypergeometric equation.

4. Obtain and justify the following expansions, which are suitable for computation, for the complete elliptic integrals of first and second kind with $0 < k < 1$.

$$K(k) = \int_0^{\pi/2} \frac{d\varphi}{\sqrt{1 - k^2 \sin^2 \varphi}} = \frac{\pi}{2} \left(1 + \sum_{n=1}^{\infty} \left(\frac{(2n-1)!!}{(2n)!!}\right)^2 k^{2n}\right);$$

$$E(k) = \int_0^{\pi/2} \sqrt{1 - k^2 \sin^2 \varphi}\, d\varphi = \frac{\pi}{2} \left(1 - \sum_{n=1}^{\infty} \left(\frac{(2n-1)!!}{(2n)!!}\right)^2 \frac{k^{2n}}{2n-1}\right).$$

5. Find

a) $\sum_{k=0}^{n} r^k e^{ik\varphi}$;

b) $\sum_{k=0}^{n} r^k \cos k\varphi$;

c) $\sum_{k=0}^{n} r^k \sin k\varphi$.

Show that the following relations hold for $|r| < 1$:

d) $\sum_{k=0}^{\infty} r^k e^{ik\varphi} = \frac{1}{1 - r\cos\varphi - ir\sin\varphi}$;

e) $\frac{1}{2} + \sum_{k=1}^{\infty} r^k \cos k\varphi = \frac{1}{2} \cdot \frac{1-r^2}{1 - 2r\cos\varphi + r^2}$;

f) $\sum_{k=1}^{\infty} r^k \sin k\varphi = \frac{r\sin\varphi}{1 - 2r\cos\varphi + r^2}$.

Verify that the following equations are true in the sense of Abel summation:

g) $\frac{1}{2} + \sum_{k=1}^{\infty} \cos k\varphi = 0$ if $\varphi \neq 2\pi n$, $n \in \mathbb{Z}$;

h) $\sum_{k=1}^{\infty} \sin k\varphi = \frac{1}{2} \cot \frac{\varphi}{2}$ if $\varphi \neq 2\pi n$, $n \in \mathbb{Z}$.

6. After considering the product of the series

$$(a_0 + a_1 + \cdots)(b_0 + b_1 + \cdots) = (c_0 + c_1 + \cdots),$$

where $c_n = a_0 b_n + a_1 b_{n-1} + \cdots + a_{n-1} b_1 + a_n b_0$, and using Proposition 1, show that if the series $\sum_{n=0}^{\infty} a_n$, $\sum_{n=0}^{\infty} b_n$, and $\sum_{n=0}^{\infty} c_n$ converge respectively to A, B, and C, then $A \cdot B = C$.

7. Let $s_n = \sum_{k=1}^{n} a_k$ and $\sigma_n = \frac{1}{n}\sum_{k=1}^{n} s_k$. The series is *Cesàro*[4] *summable*, more precisely $(c, 1)$-summable to A, if $\lim_{n\to\infty} \sigma_n = A$. In that case we write $\sum_{n=1}^{\infty} a_k = A(c, 1)$.

    a) Verify that $1 - 1 + 1 - 1 + \cdots = \frac{1}{2}(c, 1)$.

    b) Show that $\sigma_n = \sum_{k=1}^{n}(1 - \frac{k-1}{n})a_k$.

    c) Verify that if $\sum_{k=1}^{\infty} a_k = A$ in the usual sense, then $\sum_{k=1}^{\infty} a_k = A(c, 1)$.

    d) The $(c, 2)$-sum of the series $\sum_{k=1}^{\infty} a_k$ is the quantity $\lim_{n\to\infty}\frac{1}{n}(\sigma_1 + \cdots + \sigma_n)$ if this limit exists. In this way one can define the (c, r)-sum of any order r. Show that if $\sum_{k=1}^{\infty} a_k = A(c, r)$, then $\sum_{k=1}^{\infty} a_k = A(c, r + 1)$.

    e) Prove that if $\sum_{k=1}^{\infty} a_k = A(c, 1)$, then the series is also Abel summable to A.

8. a) A *"theorem of Tauberian type"* is the collective description for a class of theorems that make it possible, by introducing various extra hypotheses, to judge the behavior of certain quantities from the behavior of certain of their means. An example of such a theorem involving Cesàro summation of series is the following proposition, which you may attempt to prove following Hardy.[5]

    If $\sum_{n=1}^{\infty} a_n = A(c, 1)$ and $a_n = O(\frac{1}{n})$, then the series $\sum_{n=1}^{\infty} a_n$ converges in the ordinary sense and to the same sum.

    b) Tauber's[6] original theorem relates to Abel summation of series and consists of the following.

    Suppose the series $\sum_{n=1}^{\infty} a_n x^n$ converges for $0 < x < 1$ and $\lim_{x\to 1-0}\sum_{n=1}^{\infty} a_n \times x^n = A$. If $\lim_{n\to\infty}\frac{a_1 + 2a_2 + \cdots + n a_n}{n} = 0$, then the series $\sum_{n=1}^{\infty} \alpha_n$ converges to A in the ordinary sense.

9. It is useful to keep in mind that in relation to the limiting passage under the integral sign there exist theorems that give much freer sufficient conditions for the possibility of such a passage than those made possible by Theorem 3. These theorems constitute one of the major achievements of the so-called Lebesgue integral. In the case when the function is Riemann integrable on a closed interval $[a, b]$, that is, $f \in \mathcal{R}[a, b]$, this function also belongs to the class $\mathcal{L}[a, b]$ of Lebesgue-integrable functions, and the values of the Riemann integral $(R)\int_a^b f(x)\,dx$ of f and the Lebesgue integral $(L)\int_a^b f(x)\,dx$ are the same.

    In general the space $\mathcal{L}[a, b]$ is the completion of $\mathcal{R}[a, b]$ (more precisely, $\widetilde{\mathcal{R}}[a, b]$) with respect to the integral metric), and the integral $(L)\int_a^b$ is the continuation of the linear functional $(R)\int_a^b$ from $\mathcal{R}[a, b]$ to $\mathcal{L}[a, b]$.

    The definitive Lebesgue "dominated convergence" theorem asserts that *if a sequence $\{f_n; n \in \mathbb{N}\}$ of functions $f_n \in \mathcal{L}[a, b]$ is such that there exists a non-*

---

[4]E. Cesàro (1859–1906) – Italian mathematician who studied analysis and geometry.

[5]G.H. Hardy (1877–1947) – British mathematician who worked mainly in number theory and theory of functions.

[6]A. Tauber (b. 1866, year of death unknown) – Austrian mathematician who worked mainly in number theory and theory of functions.

negative function $F \in \mathcal{L}[a, b]$ that majorizes the functions of the sequence, that is, $|f_n(x)| \leq F(x)$ almost everywhere on $[a, b]$, then the convergence $f_n \to f$ at almost all points of the closed interval $[a, b]$ implies that $f \in \mathcal{L}[a, b]$ and $\lim_{n \to \infty} (L) \int_a^b f_n(x) \, dx = (L) \int_a^b f(x)(dx)$.

a) Show by example that even if all the functions of the sequence $\{f_n; n \in \mathbb{N}\}$ are bounded by the same constant M on the interval $[a, b]$, the conditions $f_n \in \mathcal{R}[a, b]$, $n \in \mathbb{N}$ and $f_n \to f$ pointwise on $[a, b]$ still do not imply that $f \in \mathcal{R}[a, b]$. (See Example 5 of Sect. 16.1.)

b) From what has been said about the relation between the integrals $(R) \int_a^b$ and $(L) \int_a^b$ and Lebesgue's theorem show that, under the hypotheses of part a), if it is known that $f \in \mathcal{R}[a, b]$, then $(R) \int_a^b f(x) \, dx = \lim_{n \to \infty} (R) \int_a^b f_n(x) \, dx$. This is a significant strengthening of Theorem 3.

c) In the context of the Riemann integral one can also state the following version of Lebesgue's monotone convergence theorem.

If the sequence $\{f_n; n \in \mathbb{N}\}$ of functions $f_n \in \mathcal{R}[a, b]$ converges to zero monotonically, that is, $0 \leq f_{n+1} \leq f_n$ and $f_n \to 0$ as $n \to \infty$ for every $x \in [a, b]$, then $(R) \int_a^b f_n(x) \, dx \to 0$.

Prove this assertion, using where needed the following useful observation.

d) Let $f \in \mathcal{R}[a, b]$, $|f| \leq M$, and $\int_0^1 f(x) \, dx \geq \alpha > 0$. Then the set $E = \{x \in [0, 1] \mid f(x) \geq \alpha/2\}$ contains a finite number of such intervals the sum of whose lengths (l) is at least $\alpha/(4M)$.

Prove this, using, for example, the intervals of a partition P of the closed interval $[0, 1]$ for which the lower Darboux sum $s(f, P)$ satisfies the relation $0 \leq \int_0^1 f(x) \, dx - s(f, P) < \alpha/4$.

10. a) Show by the examples of Sect. 16.1, that it is not always possible to extract a subsequence that converges uniformly on a closed interval from a sequence of functions that converge pointwise on the interval.

b) It is much more difficult to verify directly that it is impossible to extract a subsequence of the sequence of functions $\{f_n; n \in \mathbb{N}\}$, where $f_n(x) = \sin nx$, that converges at every point of $[0, 2\pi]$. Prove that this is nevertheless the case. (Use the result of Problem 9b) and the circumstance that $\int_0^{2\pi} (\sin n_k x - \sin n_{k+1} x)^2 \, dx = 2\pi \neq 0$ for $n_k < n_{k+1}$.)

c) Let $\{f_n; n \in \mathbb{N}\}$ be a uniformly bounded sequence of functions $f_n \in \mathcal{R}[a, b]$. Let

$$F_n(x) = \int_a^x f_n(t) \, dt \quad (a \leq x \leq b).$$

Show that one can extract a subsequence of the sequence $\{F_n; n \in \mathbb{N}\}$ that converges uniformly on the closed interval $[a, b]$.

11. a) Show that if $f, f_n \in \mathcal{R}([a, b], \mathbb{R})$ and $f_n \rightrightarrows f$ on $[a, b]$ as $n \to \infty$, then for every $\varepsilon > 0$ there exists an integer $N \in \mathbb{N}$ such that

$$\left| \int_a^b (f - f_n)(x)\,\mathrm{d}x \right| < \varepsilon(b - a)$$

for every $n > N$.

b) Let $f_n \in C^{(1)}([a, b], \mathbb{R})$, $n \in \mathbb{N}$. Using the formula $f_n(x) = f_n(x_0) + \int_{x_0}^x f_n'(t)\,\mathrm{d}t$, show that if $f_n' = \varphi$ on $[a, b]$ and there exists a point $x_0 \in [a, b]$ for which the sequence $\{f_n(x_0); n \in \mathbb{N}\}$ converges, then the sequence of functions $\{f_n; n \in \mathbb{N}\}$ converges uniformly on $[a, b]$ to some function $f \in C^{(1)}([a, b], \mathbb{R})$ and $f_n' \rightrightarrows f' = \varphi$.

16.4 *Compact and Dense Subsets of the Space of Continuous Functions

The present section is devoted to more specialized questions, involving the space of continuous functions, which is ubiquitous in analysis. All these questions, like the metric of the space of continuous functions[7] itself, are closely connected with the concept of uniform convergence.

16.4.1 The Arzelà–Ascoli Theorem

Definition 1 A family $\mathcal{F}$ of functions $f : X \to Y$ defined on a set X and assuming values in a metric space Y is *uniformly bounded on X* if the set of values $V = \{y \in Y \mid \exists f \in \mathcal{F} \ \exists x \in X \ (y = f(x))\}$ of the functions in the family is bounded in Y.

For numerical functions or for functions $f : X \to \mathbb{R}^n$, this simply means that there exists a constant $M \in \mathbb{R}$ such that $|f(x)| \leq M$ for all $x \in X$ and all functions $f \in B$.

Definition 1' If the set $V \subset Y$ of values of the functions of the family $\mathcal{F}$ is totally bounded (that is, for every $\varepsilon > 0$ there is a finite ε-grid for V in Y), the family $\mathcal{F}$ is *totally bounded*.

For spaces Y in which the concept of boundedness and total boundedness are the same (for example, for $\mathbb{R}$, $\mathbb{C}$, $\mathbb{R}^n$, and $\mathbb{C}^n$ and in general in the case of a locally compact space Y), the concepts of uniform boundedness and total boundedness are the same.

Definition 2 Let X and Y be metric spaces. A family $\mathcal{F}$ of functions $f : X \to Y$ is *equicontinuous on X* if for every $\varepsilon > 0$ there exists $\delta > 0$ such that

---

[7]If you have not completely mastered the general concepts of Chap. 9, you may assume without any loss of content in the following that the functions discussed always map $\mathbb{R}$ into $\mathbb{R}$ or $\mathbb{C}$ into $\mathbb{C}$, or $\mathbb{R}^m$ into $\mathbb{R}^n$.

$d_Y(f(x_1), f(x_2)) < \varepsilon$ for any function f in the family and any $x_1, x_2 \in X$ such that $d_X(x_1, x_2) < \delta$.

Example 1 The family of functions $\{x^n; n \in \mathbb{N}\}$ is not equicontinuous on $[0, 1]$, but it is equicontinuous on any closed interval of the form $[0, q]$ where $0 < q < 1$.

Example 2 The family of functions $\{\sin nx; n \in \mathbb{N}\}$ is not equicontinuous on any nondegenerate closed interval $[a, b] \subset \mathbb{R}$.

Example 3 If the family $\{f_\alpha : [a, b] \to \mathbb{R}; \alpha \in A\}$ of differentiable functions f_α is such that the family $\{f_\alpha'; \alpha \in A\}$ of their derivatives is uniformly bounded by a constant, then $|f_\alpha(x_2) - f_\alpha(x_1)| \leq M|x_2 - x_1|$, as follows from the mean-value theorem, and hence the original family is equicontinuous on the closed interval $[a, b]$.

The connection of these concepts with uniform convergence of continuous functions is shown by the following lemma.

Lemma 1 *Let K and Y be metric spaces, with K compact. A necessary condition for the sequence $\{f_n; n \in \mathbb{N}\}$ of continuous functions $f_n : K \to Y$ to converge uniformly on K is that the family $\{f_n; n \in \mathbb{N}\}$ be totally bounded and equicontinuous.*

Proof Let $f_n \rightrightarrows f$ on K. By Theorem 2 of Sect. 16.3, we conclude that $f \in C(K, Y)$. It follows from the uniform continuity of f on the compact set K that for every $\varepsilon > 0$ there exists $\delta > 0$ such that $(d_K(x_1, x_2) < \delta \Longrightarrow d_Y(f(x_1), f(x_2)) < \varepsilon)$ for all $x_1, x_2 \in K$. Given the same $\varepsilon > 0$ we can find an index $N \in \mathbb{N}$ such that $d_Y(f(x), f_n(x)) < \varepsilon$ for all $n > N$ and all $x \in X$. Combining these inequalities and using the triangle inequality, we find that $d_K(x_1, x_2) < \delta$ implies $d_Y(f_n(x_1), f_n(x_2)) < 3\varepsilon$ for every $n > N$ and $x_1, x_2 \in K$. Hence the family $\{f_n; n > N\}$ is equicontinuous. Adjoining to this family the equicontinuous family $\{f_1, \ldots, f_N\}$ consisting of a finite number of functions continuous on the compact set K, we obtain an equicontinuous family $\{f_n; n \in \mathbb{N}\}$.

Total boundedness of $\mathcal{F}$ follows from the inequality $d_Y(f(x), f_n(x)) < \varepsilon$, which holds for $x \in K$ and $n > N$, and the fact that $f(K)$ and $\bigcup_{n=1}^{N} f_n(K)$ are compact sets in Y and hence totally bounded in Y. $\qquad\square$

Actually the following general result is true.

Theorem 1 (Arzelà–Ascoli) *Let $\mathcal{F}$ be a family of functions $f : K \to Y$ defined on a compact metric space K with values in a complete metric space Y.*

A necessary and sufficient condition for every sequence $\{f_n \in \mathcal{F}; n \in \mathbb{N}\}$ to contain a uniformly convergent subsequence is that the family $\mathcal{F}$ be totally bounded and equicontinuous.

Proof Necessity. If $\mathcal{F}$ were not a totally bounded family, one could obviously construct a sequence $\{f_n; n \in \mathbb{N}\}$ of functions $f_n \in \mathcal{F}$ that would not be totally bounded and from which (see the lemma) one could not extract a uniformly convergence subsequence.

If $\mathcal{F}$ is not equicontinuous, there exist a number $\varepsilon_0 > 0$, a sequence of functions $\{f_n \in \mathcal{F}; n \in \mathbb{N}\}$, and a sequence $\{(x_n', x_n''); n \in \mathbb{N}\}$ of pairs (x_n', x_n'') of points x_n' and x_n'' that converge to a point $x_0 \in K$ as $n \to \infty$, but $d_Y(f_n(x_n'), f_n(x_n'')) \geq \varepsilon_0 > 0$. Then one could not extract a uniformly convergent subsequence from the sequence $\{f_n; n \in \mathbb{N}\}$: in fact, by Lemma 1, the functions of such a subsequence must form an equicontinuous family.

Sufficiency. We shall assume that the compact set K is infinite, since the assertion is trivial otherwise. We fix a countable dense subset E in K – a sequence $\{x_n \in K; n \in \mathbb{N}\}$. Such a set E is easy to obtain by taking, for example, the union of the points of finite ε-grids in K obtained for $\varepsilon = 1, 1/2, \ldots, 1/n, \ldots$.

Let $\{f_n; n \in \mathbb{N}\}$ be an arbitrary sequence of functions of $\mathcal{F}$.

The sequence $\{f_n(x_1); n \in \mathbb{N}\}$ of values of these functions at the point x_1 is totally bounded in Y by hypothesis. Since Y is a complete space, it is possible to extract from it a convergent subsequence $\{f_{n_k}(x_1); k \in \mathbb{N}\}$. The functions of this sequence, as will be seen, can be conveniently denoted f_n^1, $n \in \mathbb{N}$. The superscript 1 shows that this is the sequence constructed for the point x_1.

From this subsequence we extract a further subsequence $\{f_{n_k}^1; k \in \mathbb{N}\}$ which we denote $\{f_n^2; n \in \mathbb{N}\}$ such that the sequence $\{f_{n_k}^1(x_2); k \in \mathbb{N}\}$ converges.

Continuing this process, we obtain a series $\{f_n^k; n \in \mathbb{N}\}$, $k = 1, 2, \ldots$ of sequences. If we now take the "diagonal" sequence $\{g_n = f_n^n; n \in \mathbb{N}\}$, it will converge at every point of the dense set $E \subset K$, as one can easily see.

We shall show that the sequence $\{g_n; n \in \mathbb{N}\}$ converges at every point of K and that the convergence is uniform on K. To do this, we fix $\varepsilon > 0$ and choose $\delta > 0$ in accordance with Definition 2 of equicontinuity of the family $\mathcal{F}$. Let $E_1 = \{\xi_1, \ldots, \xi_k\}$ be a finite subset of E forming a δ-grid on K. Since the sequences $\{g_n(\xi_i); n \in \mathbb{N}\}$, $i = 1, 2, \ldots, k$, all converge, there exists N such that $d_Y(g_m(\xi_i), g_n(\xi_i)) < \varepsilon$ for $i = 1, 2, \ldots, k$ and all $m, n \geq N$.

For each point $x \in K$ there exists $\xi_j \in E$ such that $d_K(x, \xi_j) < \delta$. By the equicontinuity of the family $\mathcal{F}$, it now follows that $d_Y(g_n(x), g_n(\xi_j)) < \varepsilon$ for every $n \in \mathbb{N}$. Using these inequalities, we now find that

$$d_Y\big(g_m(x), g_n(x)\big) \leq d_Y\big(g_n(x), g_n(\xi_j)\big) + d_Y\big(g_m(\xi_j), g_n(\xi_j)\big) +$$
$$+ d_Y\big(g_m(x), g_m(\xi_j)\big) < \varepsilon + \varepsilon + \varepsilon = 3\varepsilon$$

for all $m, n > N$.

But x was an arbitrary point of the compact set K, so that, by the Cauchy criterion the sequence $\{g_n; n \in \mathbb{N}\}$ indeed converges uniformly on K. $\square$

16.4.2 The Metric Space $C(K, Y)$

One of the most natural metrics on the set $C(K, Y)$ of functions $f : K \to Y$ that are continuous on a compact set K and assume values in a complete metric space Y is the following *metric of uniform convergence*.

$$d(f, g) = \max_{x \in K} d_Y\big(f(x), g(x)\big),$$

where $f, g \in C(K, Y)$, and the maximum exists, since K is compact. The name *metric* comes from the obvious fact that $d(f_n, f) \to 0 \Leftrightarrow f_n \rightrightarrows f$ on K.

Taking account of this last relation, by Theorem 2 of Sect. 16.3 and the Cauchy criterion for uniform convergence we can conclude that the metric space $C(K, Y)$ with the metric of uniform convergence is complete.

We recall that a *precompact subset* of a metric space is a subset such that from every sequence of its points one can extract a Cauchy (fundamental) subsequence. If the original metric space is complete, such a sequence will even be convergent.

The Arzelà–Ascoli theorem gives a description of the precompact subsets of the metric space $C(K, Y)$.

The important theorem we are about to prove gives a description of a large variety of dense subsets of the space $C(K, Y)$. The natural interest of such subsets comes from the fact that one can approximate any continuous function $f : K \to Y$ uniformly with absolute error as small as desired by functions from these subsets.

Example 4 The classical result of Weierstrass, to which we shall often return, and which is generalized by Stone's theorem below, is the following.

Theorem 2 (Weierstrass) *If $f \in C([a, b], \mathbb{C})$, there exists a sequence $\{P_n; n \in \mathbb{N}\}$ of polynomials $P_n : [a, b] \to \mathbb{C}$ such that $P_n \rightrightarrows f$ on $[a, b]$. Here, if $f \in C([a, b], \mathbb{R})$, the polynomials can also be chosen from $C([a, b], \mathbb{R})$.*

In geometric language this means, for example, that the polynomials with real coefficients form an everywhere dense subset of $C([a, b], \mathbb{R})$.

Example 5 Although Theorem 2 still requires a nontrivial proof (given below), one can at least conclude from the uniform continuity of any function $f \in C([a, b], \mathbb{R})$ that the piecewise-linear continuous real-valued functions on the interval $[a, b]$ are a dense subset of $C([a, b], \mathbb{R})$.

Remark 1 We note that if E_1 is everywhere dense in E_2 and E_2 is everywhere dense in E_3, then E_1 is obviously everywhere dense in E_3.

This means, for example, that to prove Theorem 2 it suffices to show that a piecewise linear function can be approximated arbitrarily closely by a polynomial on the given interval.

16.4.3 Stone's Theorem

Before proving the general theorem of Stone, we first give the following proof of Theorem 2 (Weierstrass' theorem) for the case of real-valued functions, which is useful in helping to appreciate what is to follow.

Proof We first remark that if $f, g \in C([a, b], \mathbb{R})$, $\alpha \in \mathbb{R}$, and the functions f and g admit a uniform approximation (with arbitrary accuracy) by polynomials, then the continuous functions $f + g$, $f \cdot g$, and αf also admit such an approximation.

On the closed interval $[-1, 1]$, as was shown in Example 2 of Sect. 16.3, the function $|x|$ admits a uniform approximation by polynomials $P_n(x) = \sum_{k=1}^{n} a_k x^k$. Hence, the corresponding sequence of polynomials $M \cdot P_n(x/M)$ gives a uniform approximation to $|x|$ on the closed interval $|x| \leq M$.

If $f \in C([a, b], \mathbb{R})$ and $M = \max |f(x)|$, it follows from the inequality $||y| - \sum_{k=1}^{n} c_k y^k| < \varepsilon$ for $|y| \leq M$ that $||f(x)| - \sum_{k=1}^{n} c_k f^k(x)| < \varepsilon$ for $a \leq x \leq b$. Hence if f admits a uniform approximation by polynomials on $[a, b]$, then $\sum_{k=1}^{n} c_k f^k$ and $|f|$ also admit such an approximation.

Finally, if f and g admit a uniform approximation by polynomials on the closed interval $[a, b]$, then by what has been said, the functions $\max\{f, g\} = \frac{1}{2}((f + g) + |f - g|)$ and $\min\{f, g\} = \frac{1}{2}((f + g) - |f - g|)$ also admit such an approximation.

Let $a \leq \xi_1 \leq \xi_2 \leq b$, $f(x) \equiv 0$, $g_{\xi_1 \xi_2}(x) = \frac{x - \xi_1}{\xi_2 - \xi_1}$, $h(x) \equiv 1$, $\Phi_{\xi_1 \xi_2} = \max\{f, g_{\xi_1 \xi_2}\}$, and $F_{\xi_1 \xi_2} = \min\{h, \Phi_{\xi_1 \xi_2}\}$. Linear combinations of functions of the form $F_{\xi_1 \xi_2}$ obviously generate the entire set of continuous piecewise-linear functions on the closed interval $[a, b]$, from which, by Example 5, Weierstrass' theorem follows. $\quad\square$

Before stating Stone's theorem, we define some new concepts.

Definition 3 A set A of real- (or complex-)valued functions on a set X is called a *real (or complex) algebra of functions* on X if

$$(f + g) \in A, \quad (f \cdot g) \in A, \quad (\alpha f) \in A$$

when $f, g \in A$ and $\alpha \in \mathbb{R}$ (or $\alpha \in \mathbb{C}$).

Example 6 Let $X \subset \mathbb{C}$. The polynomials $P(z) = c_0 + c_1 z + c_2 z^2 + \cdots + c_n z^n, n \in \mathbb{N}$, obviously form a complex algebra of functions on X.

If we take $X = [a, b] \subset \mathbb{R}$, and take only polynomials with real coefficients, we obtain a real algebra of functions on the closed interval $[a, b]$.

Example 7 The linear combinations of functions e^{nx}, $n = 0, 1, 2, \ldots$ with coefficients in $\mathbb{R}$ or $\mathbb{C}$ also form a (real or complex respectively) algebra on any closed interval $[a, b] \subset \mathbb{R}$.

The same can be said of linear combinations of the functions $\{e^{inx}; n \in \mathbb{Z}\}$.

Definition 4 We shall say that a set S of functions on X *separates points* on X if for every pair of distinct points $x_1, x_2 \in X$ there exists a function $f \in S$ such that $f(x_1) \neq f(x_2)$.

Example 8 The set of functions $\{e^m; n \in \mathbb{N}\}$, and even each individual function in the set, separates points on $\mathbb{R}$.

At the same time, the 2π-periodic functions $\{e^{inx}; n \in \mathbb{Z}\}$ separates points of a closed interval if its length is less than 2π and obviously does not separate the points of an interval of length greater than or equal to 2π.

Example 9 The real polynomials together form a set of functions that separates rates the points of every closed interval $[a, b]$, since the polynomial $P(x) = x$ does that all by itself. What has just been said can be repeated for a set $X \subset \mathbb{C}$ and the set of complex polynomials on X. As a single separating function, one can take $P(z) = z$.

Definition 5 The family $\mathcal{F}$ of functions $f : X \to \mathbb{C}$ *does not vanish on* X (is nondegenerate) if for every point $x_0 \in X$ there is a function $f_0 \in \mathcal{F}$ such that $f_0(x_0) \neq 0$.

Example 10 The family $\mathcal{F} = \{1, x, x^2, \ldots\}$ does not vanish on the closed interval $[0, 1]$, but all the functions of the family $\mathcal{F}_0 = \{x, x^2, \ldots\}$ vanish at $x = 0$.

Lemma 2 *If an algebra A of real (resp. complex) functions on X separates the points of X and does not vanish on X, then for any two distinct points $x_1, x_2 \in X$ and any real (resp. complex) numbers c_1, c_2 there is a function f in A such that $f(x_1) = c_1$ and $f(x_2) = c_2$.*

Proof It obviously suffices to prove the lemma when $c_1 = 0$, $c_2 = 1$ and when $c_1 = 1, c_2 = 0$.

By the symmetry of the hypotheses on x_1 and x_2, we consider only the case $c_1 = 1, c_2 = 0$.

We begin by remarking that A contains a special function s separating the points x_1 and x_2 that, in addition to the condition $s(x_1) \neq s(x_2)$, also satisfies the condition $s(x_1) \neq 0$.

Let $g, h \in A$, $g(x_1) \neq g(x_2)$, $g(x_1) = 0$, and $h(x_1) \neq 0$. There is obviously a number $\lambda \in \mathbb{R} \backslash 0$ such that $\lambda(h(x_1) - h(x_2)) \neq g(x_2)$. The function $s = g + \lambda h$ then has the required properties.

Now, setting $f(x) = \frac{s^2(x) - s(x_2)s(x)}{s^2(x_1) - s(x_1)s(x_2)}$, we obtain a function f in the algebra A satisfying $f(x_1) = 1$ and $f(x_2) = 0$. $\quad\square$

Theorem 3 (Stone[8]) *Let A be an algebra of continuous real-valued functions defined on a compact set K. If A separates the points of K and does not vanish on K, then A is an everywhere-dense subspace of $C(K, \mathbb{R})$.*

---

[8]M.H. Stone (1903–1989) – American mathematician who worked mainly in topology and functional analysis.

Proof Let $\overline{A}$ be the closure of the set $A \subset C(K, \mathbb{R})$ in $C(K, \mathbb{R})$, that is, $\overline{A}$ consists of the continuous functions $f \in C(K, \mathbb{R})$ that can be approximated uniformly with arbitrary precision by functions of A. The theorem asserts that $\overline{A} = C(K, \mathbb{R})$.

Repeating the reasoning in the proof of Weierstrass' theorem, we note that if $f, g \in A$ and $\alpha \in \mathbb{R}$, then the functions $f + g$, $f \cdot g$, αf, $|f|$, $\max\{f, g\}$, $\min\{f, g\}$ also belong to A. By induction we can verify that in general if $f_1, f_2, \ldots, f_n \in A$, then $\max\{f_1, f_2, \ldots, f_n\}$ and $\min\{f_1, f_2, \ldots, f_n\}$ also lie in A.

We now show that for every function $f \in C(K, \mathbb{R})$, every point $x \in K$, and every number $\varepsilon > 0$, there exists a function $g_x \in \overline{A}$ such that $g_x(x) = f(x)$ and $g_x(t) > f(t) - \varepsilon$ for every $t \in K$.

To verify this, for each point $y \in K$ we use Lemma 2 to choose a function $h_y \in A$ such that $h_y(x) = f(x)$ and $h_y(y) = f(y)$. By the continuity of f and h_y on K, there exists an open neighborhood U_y of y such that $h_y(t) > f(t) - \varepsilon$ for every $t \in U_y$. From the covering of the compact set K by the open sets U_y we select a finite covering $\{U_{y_1}, U_{y_2}, \ldots, U_{y_n}\}$. Then the function $g_x = \max\{h_{y_1}, h_{y_2}, \ldots, h_{y_n}\} \in A$ will be the desired function.

Now taking such a function g_x for each point $x \in K$, we remark that by the continuity of g_x and f, there exists an open neighborhood V_x of $x \in K$ such that $g_x(t) < f(t) + \varepsilon$ for every $t \in V_x$. Since K is compact, there exists a finite covering $\{V_{x_1}, V_{x_2}, \ldots, V_{x_m}\}$ by such neighborhoods. The function $g = \min\{g_{x_1}, \ldots, g_{x_m}\}$ belongs to A and by construction, satisfies both inequalities

$$f(t) - \varepsilon < g(t) < f(t) + \varepsilon$$

at every point.

But the number $\varepsilon > 0$ was arbitrary, so that any function $f \in C(K, \mathbb{R})$ can be uniformly approximated on K by functions in A. $\qquad\square$

16.4.4 Problems and Exercises

1. A family $\mathcal{F}$ of functions $f : X \to Y$ defined on the metric space X and assuming values in the metric space Y is *equicontinuous at* $x_0 \in X$ if for every $\varepsilon > 0$ there exists $\delta > 0$ such that $d_X(x, x_0) < \delta$ implies $d_Y(f(x), f(x_0)) < \varepsilon$ for every $f \in \mathcal{F}$.

a) Show that if a family $\mathcal{F}$ of functions $f : X \to Y$ is equicontinuous at $x_0 \in X$, then every function $f \in \mathcal{F}$ is continuous at x_0, although the converse is not true.

b) Prove that if the family $\mathcal{F}$ of functions $f : K \to Y$ is equicontinuous at each point of the compact set K, then it is equicontinuous on K in the sense of Definition 2.

c) Show that if a metric space X is not compact, then equicontinuity of a family $\mathcal{F}$ of functions $f : X \to Y$ at each point $x \in X$ does not imply equicontinuity of $\mathcal{F}$ on X.

For this reason, if the family $\mathcal{F}$ is equicontinuous on a set X in the sense of Definition 2, we often call it *uniformly equicontinuous* on the set. Thus, the relation between equicontinuity at a point and uniform equicontinuity of a family of functions on a set X is the same as that between continuity and uniform continuity of an individual function $f : X \to Y$ on the set X.

d) Let $\omega(f; E)$ be the oscillation of the function $f : X \to Y$ on the set $E \subset X$, and $B(x, \delta)$ the ball of radius δ with center at $x \in X$. What concepts are defined by the following formulas?

$$\forall \varepsilon > 0 \, \exists \delta > 0 \, \forall f \in \mathcal{F} \, \omega\big(f; B(x, \delta)\big) < \varepsilon,$$

$$\forall \varepsilon > 0 \, \exists \delta > 0 \, \forall f \in \mathcal{F} \, \forall x \in X \, \omega\big(f; B(x, \delta)\big) < \varepsilon.$$

e) Show by example that the Arzelà–Ascoli theorem is in general not true if K is not compact: construct a uniformly bounded and equicontinuous sequence $\{f_n; n \in \mathbb{N}\}$ of functions $f_n(x) = \varphi(x + n)$ from which it is not possible to extract a subsequence that converges uniformly on $\mathbb{R}$.

f) Using the Arzelà–Ascoli theorem, solve Problem 10c) from Sect. 16.3.

2. a) Explain in detail why every continuous piecewise-linear function on a closed interval $[a, b]$ can be represented as a linear combination of functions of the form $F_{\xi_1 \xi_2}$ shown in the proof of Weierstrass' theorem.

b) Prove Weierstrass' theorem for complex-valued functions $f : [a, b] \to \mathbb{C}$.

c) The quantity $M_n = \int_a^b f(x) x^n \, dx$ is often called the nth moment of the function $f : [a, b] \to \mathbb{C}$ on the closed interval $[a, b]$. Show that if $f \in C([a, b], \mathbb{C})$ and $M_n = 0$ for all $n \in \mathbb{N}$, then $f(x) \equiv 0$ on $[a, b]$.

3. a) Show that the algebra generated by the pair of functions $\{1, x^2\}$ is dense in the set of all even functions that are continuous on $[-1, 1]$.

b) Solve the preceding problem for the algebra generated by the single function $\{x\}$ and the set of odd functions that are continuous on $[-1, 1]$.

c) Is it possible to approximate every function $f \in C([0, \pi], \mathbb{C})$ uniformly with arbitrary precision by functions in the algebra generated by the pair of functions $\{1, e^{ix}\}$?

d) Answer the preceding question in the case of $f \in C([-\pi, \pi], \mathbb{C})$.

e) Show that the answer to the preceding question is positive if and only if $f(-\pi) = f(\pi)$.

f) Can every function $f \in C([a, b], \mathbb{C})$ be uniformly approximated by linear combinations of the functions $\{1, \cos x, \sin x, \ldots, \cos nx, \sin nx, \ldots\}$ if $[a, b] \subset \,]-\pi, \pi[$?

g) Can any even function $f \in C([-\pi, \pi], \mathbb{C})$ be uniformly approximated by functions of the system $\{1, \cos x, \ldots, \cos nx, \ldots\}$?

h) Let $[a, b]$ be an arbitrary closed interval on the real line $\mathbb{R}$. Show that the algebra generated on $[a, b]$ by any nonvanishing strictly monotonic function $\varphi(x)$ (for example, e^x) is dense in $C([a, b], \mathbb{R})$.

i) For which location of the closed interval $[a, b] \subset \mathbb{R}$ is the algebra generated by $\varphi(x) = x$ dense in $C([a, b], \mathbb{R})$?

4. a) A complex algebra of functions A is *self-adjoint* if it follows from $f \in A$ that $\overline{f} \in A$, where $\overline{f}(x)$ is the value conjugate to $f(x)$. Show that if a complex algebra A is nondegenerate on X and separates the points of X, then, given that A is self-adjoint, one can assert that the subalgebra A_R of real-valued functions in A is also nondegenerate on X and also separates points on X.

b) Prove the following complex version of Stone's theorem.

If a complex algebra A of functions $f : X \to \mathbb{C}$ is nondegenerate on X and separates the points of X, then, given that it is self-adjoint, one can assert that it is dense in $C(X, \mathbb{C})$.

c) Let $X = \{z \in \mathbb{C} \mid |z| = 1\}$ be the unit circle and A the algebra on X generated by the function $e^{i\varphi}$, where φ is the polar angle of the point $z \in \mathbb{C}$. This algebra is nondegenerate on X and separates the points of X, but is not self-adjoint.

Prove that the equalities $\int_0^{2\pi} f(e^{i\varphi})e^{in\varphi}\,d\varphi = 0$, $n \in \mathbb{N}$, must hold for any function $f : X \to \mathbb{C}$ that admits uniform approximation by elements of A. Using this fact, verify that the restriction of the function $f(z) = \bar{z}$ to the circle X is a continuous function on X that does not belong to the closure of the algebra A.

Chapter 17
Integrals Depending on a Parameter

In this chapter the general theorems on families of functions depending on a parameter will be applied to the type of family most frequently encountered in analysis – integrals depending on a parameter.

17.1 Proper Integrals Depending on a Parameter

17.1.1 The Concept of an Integral Depending on a Parameter

An *integral depending on a parameter* is a function of the form

$$F(t) = \int_{E_t} f(x, t)\, dx, \qquad (17.1)$$

where t plays the role of a parameter ranging over a set T, and to each value $t \in T$ there corresponds a set E_t and a function $\varphi_t(x) = f(x, t)$ that is integrable over E_t in the proper or improper sense.

The nature of the set T may be quite varied, but of course the most important cases occur when T is a subset of $\mathbb{R}, \mathbb{C}, \mathbb{R}^n$, or $\mathbb{C}^n$.

If the integral (17.1) is a proper integral for each value of the parameter $t \in T$, we say that the function F in (17.1) is a *proper integral depending on a parameter*.

But if the integral in (17.1) exists only as an improper integral for some or all of the values of $t \in T$, we usually call F an *improper integral depending on a parameter*.

But these are of course merely terminological conventions.

When $x \in \mathbb{R}^m$, $E_t \subset \mathbb{R}^m$, and $m > 1$, we say that we are dealing with a *multiple* (double, triple, and so forth) *integral* (17.1) *depending on a parameter*.

We shall concentrate, however, on the one-dimensional case, which forms the foundation for all generalizations. Moreover, for the sake of simplicity, we shall first take E_t to be intervals of the real line $\mathbb{R}$ independent of the parameter, and we shall assume that the integral (17.1) over these intervals exists as a proper integral.

© Springer-Verlag Berlin Heidelberg 2016
V.A. Zorich, *Mathematical Analysis II*, Universitext,
DOI 10.1007/978-3-662-48993-2_9

17.1.2 Continuity of an Integral Depending on a Parameter

Proposition 1 *Let* $P = \{(x, y) \in \mathbb{R}^2 \mid a \le x \le b \wedge c \le y \le d\}$ *be a rectangle in the plane* $\mathbb{R}^2$. *If the function* $f : P \to \mathbb{R}$ *is continuous, that is, if* $f \in C(P, \mathbb{R})$, *then the function*

$$F(y) = \int_a^b f(x, y) \, dx \qquad (17.2)$$

is continuous at every point $y \in [c, d]$.

Proof It follows from the uniform continuity of the function f on the compact set P that $\varphi_y(x) := f(x, y) \rightrightarrows f(x, y_0) =: \varphi_{y_0}(x)$ on $[a, b]$ as $y \to y_0$, for $y, y_0 \in [c, d]$. For each $y \in [c, d]$ the function $\varphi_y(x) = f(x, y)$ is continuous with respect to x on the closed interval $[a, b]$ and hence integrable over that interval. By the theorem on passage to the limit under an integral sign we can now assert that

$$F(y_0) = \int_a^b f(x, y_0) \, dx = \lim_{y \to y_0} \int_a^b f(x, y) \, dx = \lim_{y \to y_0} F(y). \qquad \square$$

Remark 1 As can be seen from this proof, Proposition 1 on the continuity of the function (17.2) remains valid if we take any compact set K as the set of values of the parameter y, assuming, of course, that $f \in C(I \times K, \mathbb{R})$, where $I = \{x \in \mathbb{R} \mid a \le x \le b\}$.

Hence, in particular, one can conclude that if $f \in C(I \times D, \mathbb{R})$, where D is an open set in $\mathbb{R}^n$, then $F \in C(D, \mathbb{R})$, since every point $y_0 \in D$ has a compact neighborhood $K \subset D$, and the restriction of f to $I \times K$ is a continuous function on the compact set $I \times K$.

We have stated Proposition 1 for real-valued functions, but of course it and its proof remain valid for vector-valued functions, for example, for functions assuming values in $\mathbb{C}$, $\mathbb{R}^m$, or $\mathbb{C}^m$.

Example 1 In the proof of Morse's lemma (see Sect. 8.6, Part 1) we mentioned the following proposition, called Hadamard's lemma.

If a function f belongs to the class $C^{(1)}(U, \mathbb{R})$ *in a neighborhood U of the point* x_0, *then in some neighborhood of* x_0 *it can be represented in the form*

$$f(x) = f(x_0) + \varphi(x)(x - x_0), \qquad (17.3)$$

where φ *is a continuous function and* $\varphi(x_0) = f'(x_0)$.

Equality (17.3) follows easily from the Newton–Leibniz formula

$$f(x_0 + h) - f(x_0) = \int_0^1 f'(x_0 + th) \, dt \cdot h \qquad (17.4)$$

and Proposition 1 applied to the function $F(h) = \int_0^1 f'(x_0 + th) \, dt$. All that remains is to make the substitution $h = x - x_0$ and set $\varphi(x) = F(x - x_0)$.

It is useful to remark that Eq. (17.4) holds for $x_0, h \in \mathbb{R}^n$, where n is not restricted to the value 1. Writing out the symbol f' in more detail, and for simplicity setting $x_0 = 0$, one can write, instead of (17.4)

$$f\left(x^1, \ldots, x^n\right) - f(0, \ldots, 0) = \sum_{i=1}^n \int_0^1 \frac{\partial f}{\partial x^i}\left(tx^1, \ldots, tx^n\right) dt \cdot x^i,$$

and then one should set

$$\varphi(x)x = \sum_{i=1}^n \varphi_i(x)x^i$$

in Eq. (17.3), where $\varphi_i(x) = \int_0^1 \frac{\partial f}{\partial x^i}(tx) \, dt$.

17.1.3 Differentiation of an Integral Depending on a Parameter

Proposition 2 *If the function $f : P \to \mathbb{R}$ is continuous and has a continuous partial derivative with respect to y on the rectangle $P = \{(x, y) \in \mathbb{R}^2 \mid a \leq x \leq b \wedge c \leq y \leq d\}$, then the integral (17.2) belongs to $C^{(1)}([c, d], \mathbb{R})$, and*

$$F'(y) = \int_a^b \frac{\partial f}{\partial y}(x, y) \, dx. \tag{17.5}$$

Formula (17.5) for differentiating the proper integral (17.2) with respect to a parameter is frequently called *Leibniz' formula* or *Leibniz' rule*.

Proof We shall verify directly that if $y_0 \in [c, d]$, then $F'(y_0)$ can be computed by formula (17.5):

$$\left| F(y_0 + h) - F(y_0) - \left(\int_a^b \frac{\partial f}{\partial y}(x, y_0) \, dx \right) h \right| =$$

$$= \left| \int_a^b \left(f(x, y_0 + h) - f(x, y_0) - \frac{\partial f}{\partial y}(x, y_0)h \right) dx \right| \leq$$

$$\leq \int_a^b \left| f(x, y_0 + h) - f(x, y_0) - \frac{\partial f}{\partial y}(x, y_0)h \right| dx \leq$$

$$\leq \int_a^b \sup_{0 < \theta < 1} \left| \frac{\partial f}{\partial y}(x, y_0 + \theta h) - \frac{\partial f}{\partial y}(x, y_0) \right| dx |h| = \varphi(y_0, h) \cdot |h|.$$

By hypothesis $\frac{\partial f}{\partial y} \in C(P, \mathbb{R})$, so that $\frac{\partial f}{\partial y}(x, y) \rightrightarrows \frac{\partial f}{\partial y}(x, y_0)$ on the closed interval $a \leq x \leq b$ as $y \to y_0$, from which it follows that $\varphi(y_0, h) \to 0$ as $h \to 0$. $\qquad \square$

Remark 2 The continuity of the original function f is used in the proof only as a sufficient condition for the existence of all the integrals that appear in the proof.

Remark 3 The proof just given and the form of the mean-value theorem used in it show that Proposition 2 remains valid if the closed interval $[c, d]$ is replaced by any convex compact set in any normed vector space. Here one may obviously assume as well that f takes values in some complete normed vector space.

   In particular – and this is sometimes very useful – formula (17.5) is also applicable to complex-valued functions F of a complex variable and to functions $F(y) = F(y^1, \ldots, y^n)$ of a vector parameter $y = (y^1, \ldots, y^n) \in \mathbb{C}^n$.

   In this case $\frac{\partial f}{\partial y}$ can of course be written coordinatewise as $(\frac{\partial f}{\partial y^1}, \ldots, \frac{\partial f}{\partial y^n})$, and then (17.5) yields the corresponding partial derivatives:

$$\frac{\partial F}{\partial y^i}(y) = \int_a^b \frac{\partial f}{\partial y^i}(x, y^1, \ldots, y^n)\, dx$$

of the function F.

Example 2 Let us verify that the function $u(x) = \int_0^\pi \cos(n\varphi - x \sin\varphi)\, d\varphi$ satisfies Bessel's equation $x^2 u'' + x u' + (x^2 - n^2)u = 0$.

   Indeed, after carrying out the differentiation with formula (17.5) and making simple transformations we find

$$-x^2 \int_0^\pi \sin^2 \varphi \cos(n\varphi - x \sin\varphi)\, d\varphi + x \int_0^\pi \sin\varphi \sin(n\varphi - x \sin\varphi)\, d\varphi +$$

$$+ \left(x^2 - n^2\right) \int_0^\pi \cos(n\varphi - x \sin\varphi)\, d\varphi =$$

$$= - \int_0^\pi \left(\left(x^2 \sin^2 \varphi + n^2 - x^2\right) \cos(n\varphi - x \sin\varphi) - \right.$$

$$\left. - x \sin\varphi \sin(n\varphi - x \sin\varphi)\right) d\varphi =$$

$$= -(n + x \cos\varphi) \sin(n\varphi - x \sin\varphi)\big|_0^\pi = 0.$$

Example 3 The complete elliptic integrals

$$E(k) = \int_0^{\pi/2} \sqrt{1 - k^2 \sin^2 \varphi}\, d\varphi, \qquad K(k) = \int_0^{\pi/2} \frac{d\varphi}{\sqrt{1 - k^2 \sin^2 \varphi}} \qquad (17.6)$$

as functions of the parameter k, $0 < k < 1$, called the *modulus* of the corresponding *elliptic integral*, are connected by the relations

$$\frac{dE}{dk} = \frac{E - K}{k}, \qquad \frac{dK}{dk} = \frac{E}{k(1 - k^2)} - \frac{K}{k}.$$

Let us verify, for example, the first of these. By formula (17.5)

$$\frac{dE}{dk} = -\int_0^{\pi/2} k \sin^2 \varphi \cdot \left(1 - k^2 \sin^2 \varphi\right)^{-1/2} d\varphi =$$

$$= \frac{1}{k} \int_0^{\pi/2} \left(1 - k^2 \sin^2 \varphi\right)^{1/2} d\varphi - \frac{1}{k} \int_0^{\pi/2} \left(1 - k^2 \sin^2 \varphi\right)^{-1/2} d\varphi = \frac{E - K}{k}.$$

Example 4 Formulas (17.5) sometimes make it possible even to compute the integral. Let

$$F(\alpha) = \int_0^{\pi/2} \ln\left(\alpha^2 - \sin^2 \varphi\right) d\varphi \quad (\alpha > 1).$$

According to formula (17.5)

$$F'(\alpha) = \int_0^{\pi/2} \frac{2\alpha \, d\varphi}{\alpha^2 - \sin^2 \varphi} = \frac{\pi}{\sqrt{\alpha^2 - 1}},$$

from which we find $F(\alpha) = \pi \ln(\alpha + \sqrt{\alpha^2 - 1}) + c$.

The constant c is also easy to find, if we note that, on the one hand $F(\alpha) = \pi \ln \alpha + \pi \ln 2 + c + o(1)$ as $\alpha \to +\infty$, and on the other hand, from the definition of $F(\alpha)$, taking account of the equality $\ln(\alpha^2 - \sin^2 \varphi) = 2 \ln \alpha + o(1)$ as $\alpha \to +\infty$, we have $F(\alpha) = \pi \ln \alpha + o(1)$. Hence $\pi \ln 2 + c = 0$ and so $F(\alpha) = \pi \ln \frac{1}{2}(\alpha + \sqrt{\alpha^2 - 1})$.

Proposition 2 can be strengthened slightly.

Proposition 2′ *Suppose the function* $f : P \to \mathbb{R}$ *is continuous and has a continuous partial derivative* $\frac{\partial f}{\partial y}$ *on the rectangle* $P = \{(x, y) \in \mathbb{R}^2 \mid a \leq x \leq b \wedge c \leq y \leq d\}$; *further suppose* $\alpha(y)$ *and* $\beta(y)$ *are continuously differentiable functions on* $[c, d]$ *whose values lie in* $[a, b]$ *for every* $y \in [c, d]$. *Then the integral*

$$F(y) = \int_{\alpha(y)}^{\beta(y)} f(x, y) \, dx \tag{17.7}$$

is defined for every $y \in [c, d]$ *and belongs to* $C^{(1)}([c, d], \mathbb{R})$, *and the following formula holds:*

$$F'(y) = f\left(\beta(y), y\right) \cdot \beta'(y) - f\left(\alpha(y), y\right) \cdot \alpha'(y) + \int_{\alpha(y)}^{\beta(y)} \frac{\partial f}{\partial y}(x, y) \, dx. \tag{17.8}$$

Proof In accordance with the rule for differentiating an integral with respect to the limits of integration, taking account of formula (17.5), we can say that if $\alpha, \beta \in [a, b]$ and $y \in [c, d]$, then the function

$$\Phi(\alpha, \beta, y) = \int_\alpha^\beta f(x, y) \, dx$$

has the following partial derivatives:

$$\frac{\partial \Phi}{\partial \beta} = f(\beta, y), \qquad \frac{\partial \Phi}{\partial \alpha} = -f(\alpha, y), \qquad \frac{\partial \Phi}{\partial y} = \int_\alpha^\beta \frac{\partial f}{\partial y}(x, y) \, dx.$$

Taking account of Proposition 1, we conclude that all the partial derivatives of Φ are continuous in its domain of definition. Hence Φ is continuously differentiable. Formula (17.8) now follows from the chain rule for differentiation of the composite function $F(y) = \Phi(\alpha(y), \beta(y), y)$. $\qquad\qquad\qquad\qquad\qquad\qquad\qquad\square$

Example 5 Let

$$F_n(x) = \frac{1}{(n-1)!} \int_0^x (x-t)^{n-1} f(t) \, dt,$$

where $n \in \mathbb{N}$ and f is a function that is continuous on the interval of integration. Let us verify that $F_n^{(n)}(x) = f(x)$.

For $n = 1$ we have $F_1(x) = \int_0^x f(t) \, dt$ and $F_1'(x) = f(x)$.

By formula (17.8) we find for $n > 1$ that

$$F_n'(x) = \frac{1}{(n-1)!}(x-x)^{n-1} f(x) + \frac{1}{(n-2)!} \int_0^x (x-t)^{n-2} f(t) \, dt = F_{n-1}(x).$$

We now conclude by induction that indeed $F_n^{(n)}(x) = f(x)$ for every $n \in \mathbb{N}$.

17.1.4 Integration of an Integral Depending on a Parameter

Proposition 3 *If the function $f : P \to \mathbb{R}$ is continuous in the rectangle $P = \{(x, y) \in \mathbb{R}^2 \mid a \leq x \leq b \wedge c \leq y \leq d\}$, then the integral (17.2) is integrable over the closed interval $[c, d]$ and the following equality holds*:

$$\int_c^d \left(\int_a^b f(x, y) \, dx \right) dy = \int_a^b \left(\int_c^d f(x, y) \, dy \right) dx. \qquad (17.9)$$

Proof From the point of view of multiple integrals, Eq. (17.9) is an elementary version of Fubini's theorem. However, we shall give a proof of (17.9) that justifies it independently of Fubini's theorem.

Consider the functions

$$\varphi(u) = \int_c^u \left(\int_a^b f(x, y) \, dx \right) dy, \qquad \psi(u) = \int_a^b \left(\int_c^u f(x, y) \, dy \right) dx.$$

Since $f \in C(P, \mathbb{R})$, by Proposition 1 and the continuous dependence of the integral on the upper limit of integration, we conclude that φ and ψ belong to $C([c, d], \mathbb{R})$. Then, by the continuity of the function (17.2), we find that $\varphi'(u) =$

$\int_a^b f(x, u) \, dx$, and finally by formula (17.5) that $\psi'(u) = \int_a^b f(x, u) \, dx$ for $u \in [c, d]$. Thus $\varphi'(u) = \psi'(u)$, and hence $\varphi(u) = \psi(u) + C$ on $[c, d]$. But since $\varphi(c) = \psi(c) = 0$, we have $\varphi(u) = \psi(u)$ on $[c, d]$, from which relation (17.9) follows for $u = d$. $\qquad\square$

17.1.5 Problems and Exercises

1. a) Explain why the function $F(y)$ in (17.2) has the limit $\int_a^b \varphi(x) \, dx$ if the family of functions $\varphi_y(x) = f(x, y)$ depending on the parameter $y \in Y$ and integrable over the closed interval $a \le x \le b$ converges uniformly on that closed interval to a function $\varphi(x)$ over some base $\mathcal{B}$ in Y (for example, the base $y \to y_0$).

b) Prove that if E is a measurable set in $\mathbb{R}^m$ and the function $f : E \times I^n \to \mathbb{R}$ defined on the direct product $E \times I^n = \{(x, t) \in \mathbb{R}^{m+n} \mid x \in E \wedge t \in I^n\}$ of the set E and the n-dimensional interval I^n is continuous, then the function F defined by (17.1) for $E_t = E$ is continuous on I^n.

c) Let $P = \{(x, y) \in \mathbb{R}^2 \mid a \le x \le b \wedge c \le y \le d\}$, and let $f \in C(P, \mathbb{R})$, $\alpha, \beta \in C([c, d], [a, b])$. Prove that in that case the function (17.7) is continuous on the closed interval $[c, d]$.

2. a) Prove that if $f \in C(\mathbb{R}, \mathbb{R})$, then the function $F(x) = \frac{1}{2a} \int_{-a}^a f(x + t) \, dt$ is not only continuous, but also differentiable on $\mathbb{R}$.

b) Find the derivative of this function $F(x)$ and verify that $F \in C^{(1)}(\mathbb{R}, \mathbb{R})$.

3. Using differentiation with respect to the parameter, show that for $|r| < 1$

$$F(r) = \int_0^\pi \ln\left(1 - 2r \cos x + r^2\right) dx = 0.$$

4. Verify that the following functions satisfy Bessel's equation of Example 2:

a) $u = x^n \int_0^\pi \cos(x \cos \varphi) \sin^{2n} \varphi \, d\varphi$;

b) $J_n(x) = \frac{x^n}{(2n-1)!!\pi} \int_{-1}^{+1} (1 - t^2)^{(n-1/2)} \cos xt \, dt$.

c) Show that the functions J_n corresponding to different values of $n \in \mathbb{N}$ are connected by the relation $J_{n+1} = J_{n-1} - 2J_n'$.

5. Developing Example 3 and setting $\tilde{k} := \sqrt{1 - k^2}$, $\tilde{E}(k) := E(\tilde{k})$, $\tilde{K}(k) := K(\tilde{k})$, show, following Legendre, that

a) $\frac{d}{dk}(E\tilde{K} + \tilde{E}K - K\tilde{K}) = 0$.

b) $E\tilde{K} + \tilde{E}K - K\tilde{K} = \pi/2$.

6. Instead of the integral (17.2), consider the integral

$$\mathcal{F}(y) = \int_a^b f(x, y) g(x) \, dx,$$

where g is a function that is integrable over the closed interval $[a, b] (g \in \mathcal{R}[a, b])$.

By repeating the proofs of Propositions 1–3 above verify successively that

a) if the function f satisfies the hypotheses of Proposition 1, then $\mathcal{F}$ is continuous on $[c,d]$ ($\mathcal{F} \in C[c,d]$);

b) if f satisfies the hypotheses of Proposition 2, then $\mathcal{F}$ is continuously differentiable on $[c,d]$ ($\mathcal{F} \in C^{(1)}[c,d]$), and

$$\mathcal{F}'(y) = \int_a^b \frac{\partial y}{\partial y}(x,y)g(x)\,dx;$$

c) if f satisfies the hypotheses of Proposition 3, then $\mathcal{F}$ is integrable over $[c,d]$ ($\mathcal{F} \in \mathcal{R}[c,d]$) and

$$\int_c^d \mathcal{F}(y)\,dy = \int_a^b \left(\int_c^d f(x,y)g(x)\,dy \right) dx.$$

7. *Taylor's formula and Hadamard's lemma.*

a) Show that if f is a smooth function and $f(0) = 0$, then $f(x) = x\varphi(x)$, where φ is a continuous function and $\varphi(0) = f'(0)$.

b) Show that if $f \in C^{(n)}$ and $f^{(k)}(0) = 0$ for $k = 0, 1, \ldots, n-1$, then $f(x) = x^n \varphi(x)$, where φ is a continuous function and $\varphi(0) = \frac{1}{n!} f^{(n)}(0)$.

c) Let f be a $C^{(n)}$ function defined in a neighborhood of 0. Verify that the following version of Taylor's formula with the Hadamard form of the remainder holds:

$$f(x) = f(0) + \frac{1}{1!} f'(0)x + \cdots + \frac{1}{(n-1)!} f^{(n-1)}(0)x^{n-1} + x^n \varphi(x),$$

where φ is a function that is continuous on a neighborhood of zero, and $\varphi(0) = \frac{1}{n!} f^{(n)}(0)$.

d) Generalize the results of a), b), and c) to the case when f is a function of several variables. Write the basic Taylor formula in multi-index notation:

$$f(x) = \sum_{|\alpha|=0}^{n-1} \frac{1}{\alpha!} D^\alpha f(0)x^\alpha + \sum_{|\alpha|=n} x^\alpha \varphi_\alpha(x),$$

and note in addition to what was stated in a), b), and c), that if $f \in C^{(n+p)}$, that $\varphi_\alpha \in C^{(p)}$.

17.2 Improper Integrals Depending on a Parameter

17.2.1 Uniform Convergence of an Improper Integral with Respect to a Parameter

a. Basic Definition and Examples

Suppose that the improper integral

$$F(y) = \int_a^\omega f(x, y) \, dx \qquad (17.10)$$

over the interval $[a, \omega] \subset \mathbb{R}$ converges for each value $y \in Y$. For definiteness we shall assume that the integral (17.10) has only one singularity and that it involves the upper limit of integration (that is, either $\omega = +\infty$ or the function f is unbounded as a function of x in a neighborhood of ω).

Definition We say that *the improper integral* (17.10) depending on the parameter $y \in Y$ *converges uniformly on the set* $E \subset Y$ if for every $\varepsilon > 0$ there exists a neighborhood $U_{[a,\omega[}(\omega)$ of ω in the set $[a, \omega[$ such that the estimate

$$\left| \int_b^\omega f(x, y) \, dx \right| < \varepsilon \qquad (17.11)$$

for the remainder of the integral (17.10) holds for every $b \in U_{[a,\omega[}(\omega)$ and every $y \in E$.

If we introduce the notation

$$F_b(y) := \int_a^b f(x, y) \, dx \qquad (17.12)$$

for a proper integral approximating the improper integral (17.10), the basic definition of this section can be restated (and, as will be seen in what follows, very usefully) in a different form equivalent to the previous one:

uniform convergence of the integral (17.10) *on the set* $E \subset Y$ by definition means that

$$F_b(y) \rightrightarrows F(y) \quad \text{on } E \text{ as } b \to \omega, \ b \in [a, \omega[. \qquad (17.13)$$

Indeed,

$$F(y) = \int_a^\omega f(x, y) \, dx := \lim_{\substack{b \to \omega \\ b \in [a,\omega[}} \int_a^b f(x, y) \, dx = \lim_{\substack{b \to \omega \\ b \in [a,\omega[}} F_b(y),$$

and therefore relation (17.11) can be rewritten as

$$\left| F(y) - F_b(y) \right| < \varepsilon. \qquad (17.14)$$

This last inequality holds for every $b \in U_{[a,b[}(\omega)$ and every $y \in E$, as shown in (17.13).

Thus, relations (17.11), (17.13), and (17.14) mean that if the integral (17.10) converges uniformly on a set E of parameter values, then this improper integral (17.19) can be replaced by a certain proper integral (17.12) depending on the same parameter y with any preassigned precision, simultaneously for all $y \in E$.

Example 1 The integral

$$\int_1^{+\infty} \frac{dx}{x^2 + y^2}$$

converges uniformly on the entire set $\mathbb{R}$ of values of the parameter $y \in \mathbb{R}$, since for every $y \in \mathbb{R}$

$$\int_b^{+\infty} \frac{dx}{x^2 + y^2} \leq \int_b^{+\infty} \frac{dx}{x^2} = \frac{1}{b} < \varepsilon,$$

provided $b > 1/\varepsilon$.

Example 2 The integral

$$\int_0^{+\infty} e^{-xy} \, dx,$$

obviously converges only when $y > 0$. Moreover it converges uniformly on every set $\{y \in \mathbb{R} \mid y \geq y_0 > 0\}$.

Indeed, if $y \geq y_0 > 0$, then

$$0 \leq \int_b^{+\infty} e^{-xy} \, dx = \frac{1}{y} e^{-by} \leq \frac{1}{y_0} e^{-by_0} \to 0 \quad \text{as } b \to +\infty.$$

At the same time, the convergence is not uniform on the entire set $\mathbb{R}_+ = \{y \in \mathbb{R} \mid y > 0\}$. Indeed, negating uniform convergence of the integral (17.10) on a set E means that

$$\exists \varepsilon_0 > 0 \; \forall B \in [a, \omega[\; \exists b \in [B, \omega[\; \exists y \in E \; \left(\left| \int_b^{\omega} f(x, y) \, dx \right| > \varepsilon_0 \right).$$

In the present case ε_0 can be taken as any real number, since

$$\int_b^{+\infty} e^{-xy} \, dx = \frac{1}{y} e^{-by} \to +\infty, \quad \text{as } y \to +0,$$

for every fixed value of $b \in [0, +\infty[$.

Let us consider a less trivial example, which we shall be using below.

Example 3 Let us show that each of the integrals

$$\Phi(x) = \int_0^{+\infty} x^\alpha y^{\alpha+\beta+1} e^{-(1+x)y} \, dy,$$

$$F(y) = \int_0^{+\infty} x^\alpha y^{\alpha+\beta+1} e^{-(1+x)y} \, dx,$$

in which α and β are fixed positive numbers, converges uniformly on the set of nonnegative values of the parameter.

For the remainder of the integral $\Phi(x)$ we find immediately that

$$0 \le \int_b^{+\infty} x^\alpha y^{\alpha+\beta+1} e^{-(1+x)y} \, dy =$$

$$= \int_b^{+\infty} (xy)^\alpha e^{-xy} y^{\beta+1} e^{-y} \, dy < M_\alpha \int_b^{+\infty} y^{\beta+1} e^{-y} \, dy,$$

where $M_\alpha = \max_{0 \le u < +\infty} u^\alpha e^{-u}$. Since this last integral converges, it can be made smaller than any preassigned $\varepsilon > 0$ for sufficiently large values of $b \in \mathbb{R}$. But this means that the integral $\Phi(x)$ converges uniformly.

Let us now consider the remainder of the second integral $F(y)$:

$$0 \le \int_b^{+\infty} x^\alpha y^{\alpha+\beta+1} e^{-(1+x)y} \, dx =$$

$$= y^\beta e^{-y} \int_b^{+\infty} (xy)^\alpha e^{-xy} y \, dx = y^\beta e^{-y} \int_{by}^{+\infty} u^\alpha e^{-u} \, du.$$

Since

$$\int_{by}^{+\infty} u^\alpha e^{-u} \, du \le \int_0^{+\infty} u^\alpha e^{-u} \, du < +\infty,$$

for $y \ge 0$ and $y^\beta e^{-y} \to 0$ as $y \to 0$, for each $\varepsilon > 0$ there obviously exists a number $y_0 > 0$ such that for every $y \in [0, y_0]$ the remainder of the integral will be less than ε even independently of the value of $b \in [0, +\infty[$.

And if $y \ge y_0 > 0$, taking account of the relations $M_\beta = \max_{0 \le y < +\infty} y^\beta e^{-y} < +\infty$ and $0 \le \int_{by}^{+\infty} u^\alpha e^{-u} \, du \le \int_{by_0}^{+\infty} u^\alpha e^{-u} \, du \to 0$ as $b \to +\infty$, we conclude that for all sufficiently large values of $b \in [0, +\infty[$ and simultaneously for all $y \ge y_0 > 0$ the remainder of the integral $F(y)$ can be made less than ε.

Combining the intervals $[0, y_0]$ and $[y_0, +\infty[$, we conclude that indeed for every $\varepsilon > 0$ one can choose a number B such that for every $b > B$ and every $y \ge 0$ the corresponding remainder of the integral $F(y)$ will be less than ε.

b. The Cauchy Criterion for Uniform Convergence of an Integral

Proposition 1 (Cauchy criterion) *A necessary and sufficient condition for the improper integral* (17.10) *depending on the parameter* $y \in Y$ *to converge uniformly on a set* $E \subset Y$ *is that for every* $\varepsilon > 0$ *there exist a neighborhood* $U_{[a,\omega[}(\omega)$ *of the point* *ω such that*

$$\left| \int_{b_1}^{b_2} f(x, y)\, dx \right| < \varepsilon \tag{17.15}$$

for every $b_1, b_2 \in U_{[a,\omega[}(\omega)$ and every $y \in E$.

Proof Inequality (17.15) is equivalent to the relation $|F_{b_2}(y) - F_{b_2}(y)| < \varepsilon$, so that Proposition 1 is an immediate corollary of the form (17.13) for the definition of uniform convergence of the integral (17.10) and the Cauchy criterion for uniform convergence on E of a family of functions $F_b(y)$ depending on the parameter $b \in [a, \omega[$. $\qquad\square$

As an illustration of the use of this Cauchy criterion, we consider the following corollary of it, which is sometimes useful.

Corollary 1 *If the function f in the integral (17.10) is continuous on the set $[a, \omega[\times [c, d]$ and the integral (17.10) converges for every $y \in {]}c, d[$ but diverges for $y = c$ or $y = d$, then it converges nonuniformly on the interval ${]}c, d[$ and also on any set $E \subset {]}c, d[$ whose closure contains the point of divergence.*

Proof If the integral (17.10) diverges at $y = c$, then by the Cauchy criterion for convergence of an improper integral there exists $\varepsilon_0 > 0$ such that in every neighborhood $U_{[a,\omega[}(\omega)$ there exist numbers b_1, b_2 for which

$$\left| \int_{b_1}^{b_2} f(x, c)\, dx \right| > \varepsilon_0. \tag{17.16}$$

The proper integral

$$\int_{b_1}^{b_2} f(x, y)\, dx$$

is in this case a continuous function of the parameter y on the entire closed interval $[c, d]$ (see Proposition 1 of Sect. 17.1), so that for all values of y sufficiently close to c, the inequality

$$\left| \int_{b_1}^{b_2} f(x, y)\, dx \right| > \varepsilon$$

will hold along with the inequality (17.16).

On the basis of the Cauchy criterion for uniform convergence of an improper integral depending on a parameter, we now conclude that this integral cannot converge uniformly on any subset $E \subset {]}c, d[$ whose closure contains the point c.

The case when the integral diverges for $y = d$ is handled similarly. $\qquad\square$

Example 4 The integral

$$\int_0^{+\infty} e^{-tx^2}\,dx$$

converges for $t > 0$ and diverges at $t = 0$, hence it demonstrably converges nonuniformly on every set of positive numbers having 0 as a limit point. In particular, it converges nonuniformly on the whole set $\{t \in \mathbb{R} \mid t > 0\}$ of positive numbers.

In this case, one can easily verify these statements directly:

$$\int_b^{+\infty} e^{-tx^2}\,dx = \frac{1}{\sqrt{t}}\int_{b\sqrt{t}}^{+\infty} e^{-u^2}\,du \to +\infty \quad \text{as } t \to +0.$$

We emphasize that this integral nevertheless converges uniformly on any set $\{t \in \mathbb{R} \mid t \geq t_0 > 0\}$ that is bounded away from 0, since

$$0 < \frac{1}{\sqrt{t}}\int_{b\sqrt{t}}^{+\infty} e^{-u^2}\,du \leq \frac{1}{\sqrt{t_0}}\int_{b\sqrt{t_0}}^{+\infty} e^{-u^2}\,du \to 0 \quad \text{as } b \to +\infty.$$

c. Sufficient Conditions for Uniform Convergence of an Improper Integral Depending on a Parameter

Proposition 2 (The Weierstrass test) *Suppose the functions $f(x, y)$ and $g(x, y)$ are integrable with respect to x on every closed interval $[a, b] \subset [a, \omega[$ for each value of $y \in Y$.*

If the inequality $|f(x, y)| \leq g(x, y)$ holds for each value of $y \in Y$ and every $x \in [a, \omega[$ and the integral

$$\int_a^\omega g(x, y)\,dx$$

converges uniformly on Y, then the integral

$$\int_a^\omega f(x, y)\,dx$$

converges absolutely for each $y \in Y$ and uniformly on Y.

Proof This follows from the estimates

$$\left|\int_{b_1}^{b_2} f(x, y)\,dx\right| \leq \int_{b_1}^{b_2} |f(x, y)|\,dx \leq \int_{b_1}^{b_2} g(x, y)\,dx$$

and Cauchy's criterion for uniform convergence of an integral (Proposition 1). $\quad\square$

The most frequently encountered case of Proposition 2 occurs when the function g is independent of the parameter y. It is this case in which Proposition 2 is usually called the *Weierstrass M-test for uniform convergence of an integral*.

Example 5 The integral

$$\int_0^\infty \frac{\cos \alpha x}{1+x^2}\,dx$$

converges uniformly on the whole set $\mathbb{R}$ of values of the parameter α, since $|\frac{\cos \alpha x}{1+x^2}| \le \frac{1}{1+x^2}$, and the integral $\int_0^\infty \frac{dx}{1+x^2}$ converges.

Example 6 In view of the inequality $|\sin x\, e^{-tx^2}| \le e^{-tx^2}$, the integral

$$\int_0^\infty \sin x\, e^{-tx^2}\,dx,$$

as follows from Proposition 2 and the results of Example 3, converges uniformly on every set of the form $\{t \in \mathbb{R} \mid t \ge t_0 > 0\}$. Since the integral diverges for $t = 0$, on the basis of the Cauchy criterion we conclude that it cannot converge uniformly on any set having zero as a limit point.

Proposition 3 (Abel–Dirichlet test) *Assume that the functions $f(x, y)$ and $g(x, y)$ are integrable with respect to x at each $y \in Y$ on every closed interval $[a, b] \subset [a, \omega[$.*

A sufficient condition for uniform convergence of the integral

$$\int_a^\omega (f \cdot g)(x, y)\,dx$$

on the set Y is that one of the following two pairs of conditions holds:

α_1) *either there exists a constant $M \in \mathbb{R}$ such that*

$$\left| \int_a^b f(x, y)\,dx \right| < M$$

for any $b \in [a, \omega[$ and any $y \in Y$ and
β_1) *for each $y \in Y$ the function $g(x, y)$ is monotonic with respect to x on the interval $[a, \omega[$ and $g(x, y) \rightrightarrows 0$ on Y as $x \to \omega$, $x \in [a, \omega[$,*

or

α_2) *the integral*

$$\int_a^\omega f(x, y)\,dx$$

converges uniformly on the set Y and
β_2) *for each $y \in Y$ the function $g(x, y)$ is monotonic with respect to x on the interval $[a, \omega[$ and there exists a constant $M \in \mathbb{R}$ such that*

$$\left| g(x, y) \right| < M$$

for every $x \in [a, \omega[$ and every $y \in Y$.

Proof Applying the second mean-value theorem for the integral, we write

$$\int_{b_1}^{b_2} (f \cdot g)(x, y)\, dx = g(b_1, y) \int_{b_1}^{\xi} f(x, y)\, dx + g(b_2, y) \int_{\xi}^{b_2} f(x, y)\, dx,$$

where $\xi \in [b_1, b_2]$. If b_1 and b_2 are taken in a sufficiently small neighborhood $U_{[a, \omega[}(\omega)$ of the point ω, then the right-hand side of this equality can be made smaller in absolute value than any prescribed $\varepsilon > 0$, and indeed simultaneously for all values of $y \in Y$. In the case of the first pair of conditions α_1), β_1) this is obvious. In the case of the second pair α_2), β_2), it becomes obvious if we use the Cauchy criterion for uniform convergence of the integral (Proposition 1).

Thus, again invoking the Cauchy criterion, we conclude that the original integral of the product $f \cdot g$ over the interval $[a, \omega[$ does indeed converge uniformly on the set Y of parameter values.   □

Example 7 The integral

$$\int_1^{+\infty} \frac{\sin x}{x^\alpha}\, dx,$$

as follows from the Cauchy criterion and the Abel–Dirichlet test for convergence of improper integrals, converges only for $\alpha > 0$. Setting $f(x, \alpha) = \sin x$, $g(x, \alpha) = x^{-\alpha}$, we see that the pair α_1), β_1) of hypotheses of Proposition 3 holds for $\alpha \geq \alpha_0 > 0$. Consequently, on every set of the form $\{\alpha \in \mathbb{R} \mid \alpha \geq \alpha_0 > 0\}$ this integral converges uniformly. On the set $\{\alpha \in \mathbb{R} \mid \alpha > 0\}$ of positive values of the parameter the integral converges nonuniformly, since it diverges at $\alpha = 0$.

Example 8 The integral

$$\int_0^\infty \frac{\sin x}{x} e^{-xy}\, dx$$

converges uniformly on the set $\{y \in \mathbb{R} \mid y \geq 0\}$.

Proof First of all, on the basis of the Cauchy criterion for convergence of the improper integral one can easily conclude that for $y < 0$ this integral diverges. Now assuming $y \geq 0$ and setting $f(x, y) = \frac{\sin x}{x}$, $g(x, y) = e^{-xy}$, we see that the second pair α_2), β_2) of hypotheses of Proposition 3 holds, from which it follows that this integral converges uniformly on the set $\{y \in \mathbb{R} \mid y \geq 0\}$.   □

Thus we have introduced the concept of uniform convergence of an improper integral depending on a parameter and indicated several of the most important tests for such convergence completely analogous to the corresponding tests for uniform convergence of series of functions. Before passing on, we make two remarks.

Remark 1 So as not to distract the reader's attention from the basic concept of uniform convergence of an integral introduced here, we have assumed throughout that the discussion involves integrating real-valued functions. At the same time, as one can now easily check, these results extend to integrals of vector-valued functions, in particular to integrals of complex-valued functions. Here one need only note that, as always, in the Cauchy criterion one must assume in addition that the corresponding vector space of values of the integrand is complete (this is the case for $\mathbb{R}$, $\mathbb{C}$, $\mathbb{R}^n$, and $\mathbb{C}^n$); and in the Abel–Dirichlet test, as in the corresponding test for uniform convergence of series of functions, the factor in the product $f \cdot g$ that is assumed to be a monotonic function, must of course be real-valued.

Everything that has just been said applies equally to the main results of the following subsections in this section.

Remark 2 We have considered an improper integral (17.10) whose only singularity was at the upper limit of integration. The uniform convergence of an integral whose only singularity is at the lower limit of integration can be defined and studied similarly. If the integral has a singularity at both limits of integration, it can be represented as

$$\int_{\omega_1}^{\omega_2} f(x, y)\, dx = \int_{\omega_1}^{c} f(x, y)\, dx + \int_{c}^{\omega_2} f(x, y)\, dx,$$

where $c \in]\omega_1, \omega_2[$, and regarded as uniformly convergent on a set $E \subset Y$ if both of the integrals on the right-hand side of the equality converge uniformly. It is easy to verify that this definition is unambiguous, that is, independent of the choice of the point $c \in]\omega_1, \omega_2[$.

17.2.2 *Limiting Passage Under the Sign of an Improper Integral and Continuity of an Improper Integral Depending on a Parameter*

Proposition 4 *Let $f(x, y)$ be a family of functions depending on a parameter $y \in Y$ that are integrable, possibly in the improper sense, on the interval $a \leq x < \omega$, and let $\mathcal{B}_Y$ be a base in Y.*

If

a) *for every $b \in [a, \omega[$*

$$f(x, y) \rightrightarrows \varphi(x) \quad \text{on } [a, b] \text{ over the base } \mathcal{B}_Y$$

and

b) *the integral $\int_a^\omega f(x, y)\, dx$ converges uniformly on Y,*

then the limit function φ is improperly integrable on $[a, \omega[$ and the following equality holds:

$$\lim_{\mathcal{B}_Y} \int_a^\omega f(x, y)\, dx = \int_a^\omega \varphi(x)\, dx. \tag{17.17}$$

Proof The proof reduces to checking the following diagram:

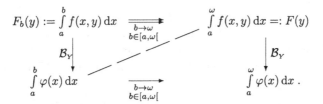

The left vertical limiting passage follows from hypothesis a) and the theorem on passage to the limit under a proper integral sign (see Theorem 3 of Sect. 16.3).

The upper horizontal limiting passage is an expression of hypothesis b).

By the theorem on the commutativity of two limiting passages it follows from this that both limits below the diagonal exist and are equal.

The right-hand vertical limit passage is what stands on the left-hand side of Eq. (17.17), and the lower horizontal limit gives by definition the improper integral on the right-hand side of (17.17). □

The following example shows that condition a) alone is generally insufficient to guarantee Eq. (17.17) in this case.

Example 9 Let $Y = \{y \in \mathbb{R} \mid y > 0\}$ and

$$f(x, y) = \begin{cases} 1/y, & \text{if } 0 \le x \le y, \\ 0, & \text{if } y < x. \end{cases}$$

Obviously, $f(x, y) \rightrightarrows 0$ on the interval $0 \le x < +\infty$ as $y \to +\infty$. At the same time, for every $y \in Y$,

$$\int_0^{+\infty} f(x, y)\, dx = \int_0^y f(x, y)\, dx = \int_0^y \frac{1}{y}\, dx = 1,$$

and therefore Eq. (17.17) does not hold in this case.

Using Dini's theorem (Proposition 2 of Sect. 16.3), we can obtain the following sometimes useful corollary of Proposition 4.

Corollary 2 *Suppose that the real-valued function $f(x, y)$ is nonnegative at each value of the real parameter $y \in Y \subset \mathbb{R}$ and continuous on the interval $a \le x < \omega$.*
If

a) the function $f(x, y)$ is monotonically increasing as y increases and tends to a function $\varphi(x)$ on $[a, \omega[$,
b) $\varphi \in C([a, \omega[, \mathbb{R})$, and
c) the integral $\int_a^\omega \varphi(x)\, dx$ converges,

then Eq. (17.17) holds.

Proof It follows from Dini's theorem that $f(x, y) \rightrightarrows \varphi(x)$ on each closed interval $[a, b] \subset [a, \omega[$.

It follows from the inequalities $0 \le f(x, y) \le \varphi(x)$ and the Weierstrass M-test for uniform convergence that the integral of $f(x, y)$ over the interval $a \le x < \omega$ converges uniformly with respect to the parameter y.

Thus, both hypotheses of Proposition 4 hold, and so Eq. (17.17) holds.   □

Example 10 In Example 3 of Sect. 16.3 we verified that the sequence of functions $f_n(x) = n(1 - x^{1/n})$ is monotonically increasing on the interval $0 < x \le 1$, and $f_n(x) \nearrow \ln \frac{1}{x}$ as $n \to +\infty$.

Hence, by Corollary 2

$$\lim_{n \to \infty} \int_0^1 n\left(1 - x^{1/n}\right) dx = \int_0^1 \ln \frac{1}{x} dx.$$

Proposition 5 *If*

a) *the function $f(x, y)$ is continuous on the set $\{(x, y) \in \mathbb{R}^2 \mid a \le x < \omega \wedge c \le y \le d\}$, and*

b) *the integral $F(y) = \int_a^\omega f(x, y) \, dx$ converges uniformly on $[c, d]$, then the function $F(y)$ is continuous on $[c, d]$.*

Proof It follows from hypothesis a) that for any $b \in [a, \omega[$ the proper integral

$$F_b(y) = \int_a^b f(x, y) \, dx$$

is a continuous function on $[c, d]$ (see Proposition 1 of Sect. 17.1).

By hypothesis b) we have $F_b(y) \rightrightarrows F(y)$ on $[c, d]$ as $b \to \omega$, $b \in [a, \omega[$, from which it now follows that the function $F(y)$ is continuous on $[c, d]$.   □

Example 11 It was shown in Example 8 that the integral

$$F(y) = \int_0^{+\infty} \frac{\sin x}{x} e^{-xy} \, dx \qquad (17.18)$$

converges uniformly on the interval $0 \le y < +\infty$. Hence by Proposition 5 one can conclude that $F(y)$ is continuous on each closed interval $[0, d] \subset [0, +\infty[$, that is, it is continuous on the entire interval $0 \le y < +\infty$. In particular, it follows from this that

$$\lim_{y \to +0} \int_0^{+\infty} \frac{\sin x}{x} e^{-xy} \, dx = \int_0^{+\infty} \frac{\sin x}{x} \, dx. \qquad (17.19)$$

17.2.3 Differentiation of an Improper Integral with Respect to a Parameter

Proposition 6 *If*

a) *the functions* $f(x, y)$ *and* $f'_y(x, y)$ *are continuous on the set* $\{(x, y) \in \mathbb{R}^2 \mid a \leq x < \omega \wedge c \leq y \leq d\}$,

b) *the integral* $\Phi(y) = \int_a^\omega f'_y(x, y) \, dy$ *converges uniformly on the set* $Y = [c, d]$, *and*

c) *the integral* $F(y) = \int_a^\omega f(x, y) \, dx$ *converges for at least one value of* $y_0 \in Y$,

then it converges uniformly on the whole set Y. *Moreover the function* $F(y)$ *is differentiable and the following equality holds*:

$$F'(y) = \int_a^\omega f'_y(x, y) \, dx.$$

Proof By hypothesis a), for every $b \in [a, \omega[$ the function

$$F_b(y) = \int_a^b f(x, y) \, dx$$

is defined and differentiable on the interval $c \leq y \leq d$ and by Leibniz' rule

$$(F_b)'(y) = \int_a^b f'_y(x, y) \, dx.$$

By hypothesis b) the family of functions $(F_b)'(y)$ depending on the parameter $b \in [a, \omega[$ converges uniformly on $[c, d]$ to the function $\Phi(y)$ as $b \to \omega, b \in [a, \omega[$.

By hypothesis c) the quantity $F_b(y_0)$ has a limit as $b \to \omega, b \in [a, \omega[$.

It follows from this (see Theorem 4 of Sect. 16.3) that the family of functions $F_b(y)$ itself converges uniformly on $[c, d]$ to the limiting function $F(y)$ as $b \to \omega$, $b \in [a, \omega[$, the function F is differentiable on the interval $c \leq y \leq d$, and the equality $F'(y) = \Phi(y)$ holds. But this is precisely what was to be proved. $\square$

Example 12 For a fixed value $\alpha > 0$ the integral

$$\int_0^{+\infty} x^\alpha e^{-xy} \, dx$$

converges uniformly with respect to the parameter y on every interval of the form $\{y \in \mathbb{R} \mid y \geq y_0 > 0\}$. This follows from the estimate $0 \leq x^\alpha e^{-xy} < x^\alpha e^{-xy_0} < e^{-x\frac{y_0}{2}}$, which holds for all sufficiently large $x \in \mathbb{R}$.

Hence, by Proposition 6, the function

$$F(y) = \int_0^{+\infty} e^{-xy} \, dx$$

is infinitely differentiable for $y > 0$ and

$$F^{(n)}(y) = (-1)^n \int_0^{+\infty} x^n e^{-xy} \, dx.$$

But $F(y) = \frac{1}{y}$, and therefore $F^{(n)}(y) = (-1)^n \frac{n!}{y^{n+1}}$ and consequently we can conclude that

$$\int_0^{+\infty} x^n e^{-xy} \, dx = \frac{n!}{y^{n+1}}.$$

In particular, for $y = 1$ we obtain

$$\int_0^{+\infty} x^n e^{-x} \, dx = n!.$$

Example 13 Let us compute the *Dirichlet integral*

$$\int_0^{+\infty} \frac{\sin x}{x} \, dx.$$

To do this we return to the integral (17.18), and we remark that for $y > 0$

$$F'(y) = -\int_0^{+\infty} \sin x e^{-xy} \, dx, \tag{17.20}$$

since the integral (17.20) converges uniformly on every set of the form $\{y \in \mathbb{R} \mid y \geq y_0 > 0\}$.

The integral (17.20) is easily computed from the primitive of the integrand, and the result is that

$$F'(y) = -\frac{1}{1 + y^2} \quad \text{for } y > 0,$$

from which it follows that

$$F(y) = -\arctan y + c \quad \text{for } y > 0. \tag{17.21}$$

We have $F(y) \to 0$ as $y \to +\infty$, as can be seen from relation (17.18), so that it follows from (17.21) that $c = \pi/2$. It now results from (17.19) and (17.21) that $F(0) = \pi/2$. Thus,

$$\int_0^{+\infty} \frac{\sin x}{x} \, dx = \frac{\pi}{2}. \tag{17.22}$$

We remark that the relation "$F(y) \to 0$ as $y \to +\infty$" used in deriving (17.22) is not an immediate corollary of Proposition 4, since $\frac{\sin x}{x} e^{-xy} \rightrightarrows 0$ as $y \to +\infty$ only on intervals of the form $\{x \in \mathbb{R} \mid x \geq x_0 > 0\}$, while the convergence is not uniform on intervals of the form $0 < x < x_0$: for $\frac{\sin x}{x} e^{-xy} \to 1$ as $x \to 0$. But for $x_0 > 0$ we have

$$\int_0^\infty \frac{\sin x}{x} e^{-xy} \, dx = \int_0^{x_0} \frac{\sin x}{x} e^{-xy} \, dx + \int_{x_0}^{+\infty} \frac{\sin x}{x} e^{-xy} \, dx$$

and, given $\varepsilon > 0$ we first choose x_0 so close to 0 that $\sin x \geq 0$ for $x \in [0, x_0]$ and

$$0 < \int_0^{x_0} \frac{\sin x}{x} e^{-xy} \, dx < \int_0^{x_0} \frac{\sin x}{x} \, dx < \frac{\varepsilon}{2}$$

for every $y > 0$. Then, after fixing x_0, on the basis of Proposition 4, by letting y tend to $+\infty$, we can make the integral over $[x_0, +\infty[$ also less than $\varepsilon/2$ in absolute value.

17.2.4 Integration of an Improper Integral with Respect to a Parameter

Proposition 7 *If*

a) *the function $f(x, y)$ is continuous on the set $\{(x, y) \in \mathbb{R}^2 \mid a \leq x < \omega \wedge c \leq y \leq d\}$ and*

b) *the integral $F(y) = \int_a^\omega f(x, y) \, dx$ converges uniformly on the closed interval $[c, d]$,*

then the function F is integrable on $[c, d]$ and the following equality holds:

$$\int_c^d dy \int_a^\omega f(x, y) \, dx = \int_a^\omega dx \int_c^d f(x, y) \, dy. \tag{17.23}$$

Proof For $b \in [a, \omega[$, by hypothesis a) and Proposition 3 of Sect. 17.1 for improper integrals one can write

$$\int_c^d dy \int_a^b f(x, y) \, dx = \int_a^b dx \int_c^d f(x, y) \, dy. \tag{17.24}$$

Using hypothesis b) and Theorem 3 of Sect. 16.3 on passage to the limit under an integral sign, we carry out a limiting passage on the left-hand side of (17.24) as $b \to \omega$, $b \in [a, \omega[$ and obtain the left-hand side of (17.23). By the very definition of an improper integral, the right-hand side of (17.23) is the limit of the right-hand side of (17.24) as $b \to \omega$, $b \in [a, \omega[$. Thus, by hypothesis b) we obtain (17.23) from (17.24) as $b \to \omega$, $b \in [a, \omega[$.   □

The following example shows that, in contrast to the reversibility of the order of integration with two proper integrals, condition a) alone is in general not sufficient to guarantee (17.23).

Example 14 Consider the function $f(x, y) = (2 - xy)xy\,e^{-xy}$ on the set $\{(x, y) \in \mathbb{R}^2 \mid 0 \le x < +\infty \land 0 \le y \le 1\}$. Using the primitive $u^2 e^{-u}$ of the function $(2 - u)ue^{-u}$, it is easy to compute directly that

$$0 = \int_0^1 \mathrm{d}y \int_0^{+\infty} (2 - xy)xy\,e^{-xy}\,\mathrm{d}x \ne \int_0^{+\infty} \mathrm{d}x \int_0^1 (2 - xy)xy\,e^{-xy}\,\mathrm{d}y = 1.$$

Corollary 3 *If*

a) *the function $f(x, y)$ is continuous on the set $P = \{(x, y) \in \mathbb{R}^2 \mid a \le x < \omega \land c \le y \le d\}$ and*

b) *nonnegative on P, and*

c) *the integral $F(y) = \int_a^\omega f(x, y)\,\mathrm{d}x$ is continuous on the closed interval $[c, d]$ as a function of y,*

then Eq. (17.23) holds.

Proof It follows from hypothesis a) that for every $b \in [a, \omega[$ the integral

$$F_b(y) = \int_a^b f(x, y)\,\mathrm{d}x$$

is continuous with respect to y on the closed interval $[c, d]$.

It follows from b) that $F_{b_1}(y) \le F_{b_2}(y)$ for $b_1 \le b_2$.

By Dini's theorem and hypothesis c) we now conclude that $F_b \rightrightarrows F$ on $[c, d]$ as $b \to \omega$, $b \in [a, \omega[$.

Thus the hypotheses of Proposition 7 are satisfied and consequently Eq. (17.23) indeed holds in the present case.   □

Corollary 3 shows that Example 14 results from the fact that the function $f(x, y)$ is not of constant sign.

In conclusion we now prove a sufficient condition for two improper integrals to commute.

Proposition 8 *If*

a) *the function $f(x, y)$ is continuous on the set $\{(x, y) \in \mathbb{R}^2 \mid a \le x < \omega \land c \le y < \widetilde{\omega}\}$,*

b) *both integrals*

$$F(y) = \int_a^\omega f(x, y)\,\mathrm{d}x, \qquad \Phi(x) = \int_c^{\widetilde{\omega}} f(x, y)\,\mathrm{d}y$$

converge uniformly, the first with respect to y on any closed interval $[c, d] \subset [c, \widetilde{\omega}]$, the second with respect to x on any closed interval $[a, b] \subset [a, \omega[$, and

c) *at least one of the iterated integrals*

$$\int_c^{\widetilde{\omega}} \mathrm{d}y \int_a^\omega |f|(x, y)\,\mathrm{d}x, \qquad \int_a^\omega \mathrm{d}x \int_c^{\widetilde{\omega}} |f|(x, y)\,\mathrm{d}y$$

converges, then the following equality holds:

$$\int_{c}^{\widetilde{\omega}} dy \int_{a}^{\omega} f(x, y) \, dx = \int_{a}^{\omega} dx \int_{c}^{\widetilde{\omega}} f(x, y) \, dy. \tag{17.25}$$

Proof For definiteness suppose that the second of the two iterated integrals in c) exists.

By condition a) and the first condition in b) one can say by Proposition 7 that Eq. (17.23) holds for the function f for every $d \in [c, \widetilde{\omega}[$.

If we show that the right-hand side of (17.23) tends to the right-hand side of (17.25) as $d \to \widetilde{\omega}$, $d \in [c, \widetilde{\omega}[$, then Eq. (17.25) will have been proved, since the left-hand side will then also exist and be the limit of the left-hand side of Eq. (17.23) by the very definition of an improper integral.

Let us set

$$\Phi_d(x) := \int_{c}^{d} f(x, y) \, dy.$$

The function Φ_d is defined for each fixed $d \in [c, \widetilde{\omega}[$ and, since f is continuous, Φ_d is continuous on the interval $a \leq x < \omega$.

By the second of hypotheses b) we have $\Phi_d(x) \rightrightarrows \Phi(x)$ as $d \to \widetilde{\omega}$, $d \in [c, \widetilde{\omega}[$ on each closed interval $[a, b] \subset [a, \omega[$.

Since $|\Phi_d(x)| \leq \int_{c}^{\widetilde{\omega}} |f|(x, y) \, dy =: G(x)$ and the integral $\int_{a}^{\omega} G(x) \, dx$, which equals the second integral in hypothesis c), converges by hypothesis, we conclude by the Weierstrass M-test for uniform convergence that the integral $\int_{a}^{\omega} \Phi_d(x) \, dx$ converges uniformly with respect to the parameter d.

Thus the hypotheses of Proposition 4 hold, and we can conclude that

$$\lim_{\substack{d \to \widetilde{\omega} \\ d \in [c, \widetilde{\omega}]}} \int \Phi_d(x) \, dx = \int_{a}^{\omega} \Phi(x) \, dx;$$

and that was precisely what remained to be verified. $\qquad\square$

The following example shows that the appearance of the extra hypothesis c) in Proposition 8 in comparison with Proposition 7 is not accidental.

Example 15 Computing the integral

$$\int_{A}^{+\infty} \frac{x^2 - y^2}{(x^2 + y^2)^2} \, dx = -\frac{x}{x^2 + y^2} \Big|_{A}^{+\infty} = \frac{A}{A^2 + y^2} < \frac{1}{A}$$

for $A > 0$ shows at the same time that for every fixed value of $A > 0$ it converges uniformly with respect to the parameter y on the entire set of real numbers $\mathbb{R}$. The same thing could have been said about the integral obtained from this one by replacing dx with dy. The values of these integrals happen to differ only in sign. A direct computation shows that

$$-\frac{\pi}{4} = \int_A^{+\infty} dx \int_A^{+\infty} \frac{x^2 - y^2}{(x^2 + y^2)^2} \, dy \neq \int_A^{+\infty} dy \int_A^{+\infty} \frac{x^2 - y^2}{(x^2 + y^2)^2} \, dx = \frac{\pi}{4}.$$

Example 16 For $\alpha > 0$ and $\beta > 0$ the iterated integral

$$\int_0^{+\infty} dy \int_0^{+\infty} x^\alpha y^{\alpha+\beta-1} e^{-(1+x)y} \, dx = \int_0^{+\infty} y^\beta e^{-y} \, dy \int_0^{+\infty} (xy)^\alpha e^{-(xy)} y \, dx$$

of a nonnegative continuous function exists, as this identity shows: it equals zero for $y = 0$ and $\int_0^{+\infty} y^\beta e^{-y} \, dy \cdot \int_0^{+\infty} u^\alpha e^{-u} \, du$ for $y > 0$. Thus, in this case hypotheses a) and c) of Proposition 8 hold. The fact that both conditions of b) hold for this integral was verified in Example 3. Hence by Proposition 8 we have the equality

$$\int_0^{+\infty} dy \int_0^{+\infty} x^\alpha y^{\alpha+\beta+1} e^{-(1+x)y} \, dx = \int_0^{+\infty} dx \int_0^{+\infty} x^\alpha y^{\alpha+\beta+1} e^{-(1+x)y} \, dy.$$

Just as Corollary 3 followed from Proposition 7, we can deduce the following corollary from Proposition 8.

Corollary 4 *If*

    a) *the function $f(x, y)$ is continuous on the set*

$$P = \{(x, y) \in \mathbb{R}^2 \mid a \leq x < \omega \wedge c \leq y \leq \widetilde{\omega}\}, \quad and$$

    b) *is nonnegative on P, and*
    c) *the two integrals*

$$F(y) = \int_a^\omega f(x, y) \, dx, \qquad \Phi(x) = \int_c^{\widetilde{\omega}} f(x, y) \, dy$$

are continuous functions on $[a, \omega[$ and $[c, \widetilde{\omega}[$ respectively, and
    d) *at least one of the iterated integrals*

$$\int_c^{\widetilde{\omega}} dy \int_a^\omega f(x, y) \, dx, \qquad \int_a^\omega dx \int_a^{\widetilde{\omega}} f(x, y) \, dy,$$

exists, then the other iterated integral also exists and their values are the same.

Proof Reasoning as in the proof of Corollary 3, we conclude from hypotheses a), b), and c) and Dini's theorem that hypothesis b) of Proposition 8 holds in this case. Since $f \geq 0$, hypothesis d) here is the same as hypothesis c) of Proposition 8. Thus all the hypotheses of Proposition 8 are satisfied, and so Eq. (17.24) holds. □

Remark 3 As pointed out in Remark 2, an integral having singularities at both limits of integration reduces to the sum of two integrals, each of which has a singularity at only one limit. This makes it possible to apply the propositions and corollaries just proved to integrals over intervals $]\omega_1, \omega_2[\subset \mathbb{R}$. Here naturally the hypotheses that were satisfied previously on closed intervals $[a, b] \subset [a, \omega[$ must now hold on closed intervals $[a, b] \subset]\omega_1, \omega_2[$.

Example 17 By changing the order of integration in two improper integrals, let us show that

$$\int_0^{+\infty} e^{-x^2}\, dx = \frac{1}{2}\sqrt{\pi}. \tag{17.26}$$

This is the famous *Euler–Poisson integral*.

Proof We first observe that for $y > 0$

$$\mathcal{J} := \int_0^{+\infty} e^{-u^2}\, du = y \int_0^{+\infty} e^{-(xy)^2}\, dx,$$

and that the value of the integral in (17.26) is the same whether it is taken over the half-open interval $[0, +\infty[$ or the open interval $]0, +\infty[$.

Thus,

$$\int_0^{+\infty} y\, e^{-y^2}\, dy \int_0^{+\infty} e^{-(xy)^2}\, dx = \int_0^{+\infty} e^{-y^2}\, dy \int_0^{+\infty} e^{-u^2}\, du = \mathcal{J}^2,$$

and we assume that the integration on y extends over the interval $]0, +\infty[$.

As we shall verify, it is permissible to reverse the order of integration over x and y in this iterated integral, and therefore

$$\mathcal{J}^2 = \int_0^{+\infty} dx \int_0^{+\infty} y\, e^{-(1+x^2)y^2}\, dy = \frac{1}{2} \int_0^{+\infty} \frac{dx}{1+x^2} = \frac{\pi}{4},$$

from which Eq. (17.26) follows.

Let us now justify reversing the order of integration.

The function

$$\int_0^{+\infty} y\, e^{-(1+x^2)y^2}\, dy = \frac{1}{2}\frac{1}{1+x^2}$$

is continuous for $x \geq 0$, and the function

$$\int_0^{+\infty} y\, e^{-(1+x^2)y^2}\, dx = e^{-y^2} \cdot \mathcal{J}$$

is continuous for $y > 0$. Taking account of the general Remark 3, we now conclude from Corollary 4 that this reversal in the order of integration is indeed legal. ☐

17.2.5 Problems and Exercises

1. Let $a = a_0 < a_1 < \cdots < a_n < \cdots < \omega$. We represent the integral (17.10) as the sum of the series $\sum_{n=1}^{\infty} \varphi_n(y)$, where $\varphi_n(y) = \int_{a_{n-1}}^{a_n} f(x, y)\, dx$. Prove that the integral converges uniformly on the set $E \subset Y$ if and only if to each sequence $\{a_n\}$ of this form there corresponds a series $\sum_{n=1}^{\infty} \varphi_n(y)$ that converges uniformly on E.

2. a) In accordance with Remark 1 carry out all the constructions in Sect. 17.2.1 for the case of a complex-valued integrand f.

  b) Verify the assertions in Remark 2.

3. Verify that the function $J_0(x) = \frac{1}{\pi} \int_0^1 \frac{\cos xt}{\sqrt{1-b^2}}\, dt$ satisfies Bessel's equation $y'' + \frac{1}{x} y' + y = 0$.

4. a) Starting from the equality $\int_0^{+\infty} \frac{dy}{x^2+y^2} = \frac{\pi}{2}\frac{1}{x}$, show that $\int_0^{+\infty} \frac{dy}{(x^2+y^2)^2} = \frac{\pi}{2} \cdot \frac{(2n-3)!!}{(2n-2)!!} \cdot \frac{1}{x^{2n-1}}$.

  b) Verify that $\int_0^{+\infty} \frac{dy}{(1+(y^2/n))^n} = \frac{\pi}{2} \frac{(2n-3)!!}{(2n-2)!!} \sqrt{n}$.

  c) Show that $(1 + (y^2/n))^{-n} \searrow e^{-y^2}$ on $\mathbb{R}$ as $n \to +\infty$ and that

$$\lim_{n\to+\infty} \int_0^{+\infty} \frac{dy}{(1+(y^2/n))^n} = \int_0^{+\infty} e^{-y^2}\, dy.$$

  d) Obtain the following formula of Wallis:

$$\lim_{n\to\infty} \frac{(2n-3)!!}{(2n-2)!!} = \frac{1}{\sqrt{\pi}}.$$

5. Taking account of Eq. (17.26), show that

  a) $\int_0^{+\infty} e^{-x^2} \cos 2xy\, dx = \frac{1}{2}\sqrt{\pi} e^{-y^2}$.
  b) $\int_0^{+\infty} e^{-x^2} \sin 2xy\, dx = e^{-y^2} \int_0^y e^{t^2}\, dt$.

6. Assuming $t > 0$, prove the identity

$$\int_0^{+\infty} \frac{e^{-tx}}{1+x^2}\, dx = \int_t^{+\infty} \frac{\sin(x - t)}{x}\, dx,$$

using the fact that both of these integrals, as functions of the parameter t, satisfy the equation $\ddot{y} + y = 1/t$ and tend to zero as $t \to +\infty$.

7. Show that

$$\int_0^1 K(k)\, dk = \int_0^{\pi/2} \frac{\varphi}{\sin \varphi}\, d\varphi \left(= \int_0^1 \frac{\arctan x}{x}\, dx \right),$$

where $K(k) = \int_0^{\pi/2} \frac{d\varphi}{\sqrt{1-k^2 \sin^2 \varphi}}$ is the complete elliptic integral of first kind.

8. a) Assuming that $a > 0$ and $b > 0$ and using the equality

$$\int_0^{+\infty} dx \int_a^b e^{-xy}\, dy = \int_0^{+\infty} \frac{e^{-ax} - e^{-bx}}{x}\, dx,$$

compute this last integral.

  b) For $a > 0$ and $b > 0$ compute the integral

$$\int_0^{+\infty} \frac{e^{-ax} - e^{-bx}}{x} \cos x\, dx.$$

  c) Using the Dirichlet integral (17.22) and the equality

$$\int_0^{+\infty} \frac{dx}{x} \int_a^b \sin xy\, dy = \int_0^{+\infty} \frac{\cos ax - \cos bx}{x^2}\, dx,$$

compute this last integral.

9. a) Prove that for $k > 0$

$$\int_0^{+\infty} e^{-kt} \sin t\, dt \int_0^{+\infty} e^{-tu^2}\, du = \int_0^{+\infty} du \int_0^{+\infty} e^{-(k+u^2)t} \sin t\, dt.$$

  b) Show that the preceding equality remains valid for the value $k = 0$.
  c) Using the Euler–Poisson integral (17.26), verify that

$$\frac{1}{\sqrt{t}} = \frac{2}{\sqrt{\pi}} \int_0^{+\infty} e^{-tu^2}\, du.$$

  d) Using this last equality and the relations

$$\int_0^{+\infty} \sin x^2\, dx = \frac{1}{2} \int_0^{+\infty} \frac{\sin t}{\sqrt{t}}\, dt, \qquad \int_0^{+\infty} \cos x^2\, dx = \frac{1}{2} \int_0^{+\infty} \frac{\cos t}{\sqrt{t}}\, dt,$$

obtain the value $(\frac{1}{2}\sqrt{\frac{\pi}{2}})$ for the Fresnel integrals

$$\int_0^{+\infty} \sin x^2\, dx, \qquad \int_0^{+\infty} \cos x^2\, dx.$$

10. a) Use the equality

$$\int_0^{+\infty} \frac{\sin x}{x}\, dx = \int_0^{+\infty} \sin x\, dx \int_0^{+\infty} e^{-xy}\, dy$$

and, by justifying a reversal in the order of integration in the iterated integral, obtain once again the value of the Dirichlet integral (17.22) found in Example 13.

b) Show that for $\alpha > 0$ and $\beta > 0$

$$\int_0^{+\infty} \frac{\sin \alpha x}{x} \cos \beta x \, dx = \begin{cases} \frac{\pi}{2}, & \text{if } \beta < \alpha, \\ \frac{\pi}{4}, & \text{if } \beta = \alpha, \\ 0, & \text{if } \beta > \alpha. \end{cases}$$

This integral is often called the *Dirichlet discontinuous factor*.

c) Assuming $\alpha > 0$ and $\beta > 0$, verify the equality

$$\int_0^{+\infty} \frac{\sin \alpha x}{x} \frac{\sin \beta x}{x} \, dx = \begin{cases} \frac{\pi}{2}\beta, & \text{if } \beta \le \alpha, \\ \frac{\pi}{2}\alpha, & \text{if } \alpha \le \beta. \end{cases}$$

d) Prove that if the numbers $\alpha, \alpha_1, \ldots, \alpha_n$ are positive and $\alpha > \sum_{i=1}^n \alpha_i$, then

$$\int_0^{+\infty} \frac{\sin \alpha x}{x} \frac{\sin \alpha_1 x}{x} \cdots \frac{\sin \alpha_n x}{x} \, dx = \frac{\pi}{2} \alpha_1 \alpha_2 \cdots \alpha_n.$$

11. Consider the integral

$$\mathcal{F}(y) = \int_a^\omega f(x, y) g(x) \, dx,$$

where g is a locally integrable function $[a, \omega[$ (that is, for each $b \in [a, \omega[\, g|_{[a,b]} \in \mathcal{R}[a, b])$. Let the function f satisfy the various hypotheses a) of Propositions 5–8. If the integrand $f(x, y)$ is replaced by $f(x, y) \cdot g(x)$ in the other hypotheses of these propositions, the results are hypotheses under which, by using Problem 6 of Sect. 17.1 and repeating verbatim the proofs of Propositions 5–8, one can conclude respectively that

a) $\mathcal{F} \in C[c, d]$;

b) $\mathcal{F} \in C^{(1)}[c, d]$, and

$$\mathcal{F}'(y) = \int_a^\omega \frac{\partial f}{\partial y}(x, y) g(x) \, dx;$$

c) $\mathcal{F} \in \mathcal{R}[c, d]$ and

$$\int_c^d \mathcal{F}(y) \, dy = \int_a^\omega \left(\int_c^d f(x, y) g(x) \, dy \right) dx;$$

c) $\mathcal{F}$ is improperly integrable on $[c, \tilde{\omega}[$, and

$$\int_c^{\tilde{\omega}} \mathcal{F}(y) \, dy = \int_a^\omega \left(\int_c^{\tilde{\omega}} f(x, y) g(x) \, dy \right) dx.$$

Verify this.

17.3 The Eulerian Integrals

In this section and the next we shall illustrate the application of the theory developed above to some specific integrals of importance in analysis that depend on a parameter.

Following Legendre, we define the *Eulerian integrals of first and second kinds* respectively as the two special functions that follow:

$$B(\alpha, \beta) := \int_0^1 x^{\alpha-1}(1-x)^{\beta-1}\,dx, \qquad (17.27)$$

$$\Gamma(\alpha) := \int_0^{+\infty} x^{\alpha-1}e^{-x}\,dx. \qquad (17.28)$$

The first of these is called the *beta function*, and the second, which is the most frequently used, is the *gamma function of Euler*.[1]

17.3.1 The Beta Function

a. Domain of Definition

A necessary and sufficient condition for the convergence of the integral (17.27) at the lower limit is that $\alpha > 0$. Similarly, convergence at 1 occurs if and only if $\beta > 0$.

Thus the function $B(\alpha, \beta)$ is defined when both of the following conditions hold simultaneously:

$$\alpha > 0 \quad \text{and} \quad \beta > 0.$$

Remark We are regarding α and β as real numbers here. However, it should be kept in mind that the most complete picture of the properties of the beta and gamma functions and the most profound applications of them involve their extension into the complex parameter domain.

b. Symmetry

Let us verify that

$$B(\alpha, \beta) = B(\beta, \alpha). \qquad (17.29)$$

Proof It suffices to make the change of variable $x = 1 - t$ in the integral (17.27). □

---

[1]L. Euler (1707–1783), a brilliant scientist and above all a mathematician and specialist in mechanics. If one were to select a name, after the names of Newton and Leibniz, for a professional mathematician, that name would likely be pronounced "Euler". Euler's works and ideas still permeate almost all areas of modern mathematics. Swiss by birth, he spent a significant part of his life living and working in Russia, where he was buried.

c. The Reduction Formula

If $\alpha > 1$, the following equality holds:

$$B(\alpha, \beta) = \frac{\alpha - 1}{\alpha + \beta - 1} B(\alpha - 1, \beta). \tag{17.30}$$

Proof Integrating by parts and carrying out some identity transformations for $\alpha > 1$ and $\beta > 0$, we obtain

$$B(\alpha, \beta) = -\frac{1}{\beta} x^{\alpha-1}(1-x)^{\beta}\Big|_0^1 + \frac{\alpha - 1}{\beta} \cdot \int_0^1 x^{\alpha-2}(1-x)^{\beta}\, dx =$$

$$= \frac{\alpha - 1}{\beta} \int_0^1 x^{\alpha-2}\big((1-x)^{\beta-1} - (1-x)^{\beta-1}x\big)\, dx =$$

$$= \frac{\alpha - 1}{\beta} B(\alpha - 1, \beta) - \frac{\alpha - 1}{\beta} B(\alpha, \beta),$$

from which the reduction formula (17.30) follows.   □

Taking account of formula (17.29), we can now write the reduction formula

$$B(\alpha, \beta) = \frac{\beta - 1}{\alpha + \beta - 1} B(\alpha, \beta - 1) \tag{17.30'}$$

on the parameter β, assuming, of course, that $\beta > 1$.

It can be seen immediately from the definition of the beta function that $B(\alpha, 1) = \frac{1}{\alpha}$, and so for $n \in \mathbb{N}$ we obtain

$$B(\alpha, n) = \frac{n - 1}{\alpha + n - 1} \cdot \frac{n - 2}{\alpha + n - 2} \cdot \ldots \cdot \frac{n - (n-1)}{\alpha + n - (n-1)} B(\alpha, 1) =$$

$$= \frac{(n - 1)!}{\alpha(\alpha + 1) \cdot \ldots \cdot (\alpha + n - 1)}. \tag{17.31}$$

In particular, for $m, n \in \mathbb{N}$

$$B(m, n) = \frac{(m - 1)!(n - 1)!}{(m + n - 1)!}. \tag{17.32}$$

d. Another Integral Representation of the Beta Function

The following representation of the beta function is sometimes useful:

$$B(\alpha, \beta) = \int_0^{+\infty} \frac{y^{\alpha-1}}{(1 + y)^{\alpha+\beta}}\, dy. \tag{17.33}$$

Proof This representation can be obtained from (17.27) by the change of variables $x = \frac{y}{1+y}$. □

17.3.2 The Gamma Function

a. Domain of Definition

It can be seen by formula (17.28) that the integral defining the gamma function converges at zero only for $\alpha > 0$, while it converges at infinity for all values of $\alpha \in \mathbb{R}$, due to the presence of the rapidly decreasing factor e^{-x}.

Thus the gamma function is defined for $\alpha > 0$.

b. Smoothness and the Formula for the Derivatives

The gamma function is infinitely differentiable, and

$$\Gamma^{(n)}(\alpha) = \int_0^{+\infty} x^{\alpha-1} \ln^n x \, e^{-x} \, dx. \tag{17.34}$$

Proof We first verify that the integral (17.34) converges uniformly with respect to the parameter α on each closed interval $[a, b] \subset \,]0, +\infty[$ for each fixed value of $n \in \mathbb{N}$.

If $0 < a \leq \alpha$, then (since $x^{\alpha/2} \ln^n x \to 0$ as $x \to +0$) there exists $c_n > 0$ such that

$$\left| x^{\alpha-1} \ln^n x e^{-x} \right| < x^{\frac{a}{2}-1}$$

for $0 < x \leq c_n$. Hence by the Weierstrass M-test for uniform convergence we conclude that the integral

$$\int_0^{c_n} x^{\alpha-1} \ln^n x \, e^{-x} \, dx$$

converges uniformly with respect to α on the interval $[a, +\infty[$.

If $\alpha \leq b < +\infty$, then for $x \geq 1$,

$$\left| x^{\alpha-1} \ln^n x \, e^{-x} \right| \leq x^{b-1} \left| \ln^n x \right| e^{-x},$$

and we conclude similarly that the integral

$$\int_{c_n}^{+\infty} x^{\alpha-1} \ln^n x \, e^{-x} \, dx$$

converges uniformly with respect to α on the interval $]0, b]$.

Combining these conclusions, we find that the integral (17.34) converges uniformly on every closed interval $[a, b] \subset]0, +\infty[$.

But under these conditions differentiation under the integral sign in (17.27) is justified. Hence, on any such closed interval, and hence on the entire open interval $0 < \alpha$, the gamma function is infinitely differentiable and formula (17.34) holds. $\square$

c. The Reduction Formula

The relation

$$\Gamma(\alpha + 1) = \alpha \Gamma(\alpha) \tag{17.35}$$

holds. It is known as the *reduction formula* for the gamma function.

Proof Integrating by parts, we find that for $\alpha > 0$

$$\Gamma(\alpha + 1) := \int_0^{+\infty} x^\alpha e^{-x}\, dx = -x^\alpha e^{-x}\big|_0^{+\infty} + \alpha \int_0^{+\infty} x^{\alpha-1} e^{-x}\, dx =$$

$$= \alpha \int_0^{+\infty} x^{\alpha-1} e^{-x}\, dx = \alpha \Gamma(\alpha). \qquad \square$$

Since $\Gamma(1) = \int_0^{+\infty} e^{-x}\, dx = 1$, we conclude that for $n \in \mathbb{N}$

$$\Gamma(n + 1) = n!. \tag{17.36}$$

Thus the gamma function turns out to be closely connected with the number-theoretic function $n!$.

d. The Euler–Gauss Formula

This is the name usually given to the following equality:

$$\Gamma(\alpha) = \lim_{n \to \infty} n^\alpha \cdot \frac{(n-1)!}{\alpha(\alpha+1) \cdot \ldots \cdot (\alpha+n-1)}. \tag{17.37}$$

Proof To prove this formula, we make the change of variable $x = \ln \frac{1}{u}$ in the integral (17.28), resulting in the following integral representation of the gamma function:

$$\Gamma(\alpha) = \int_0^1 \ln^{\alpha-1}\left(\frac{1}{u}\right) du. \tag{17.38}$$

It was shown in Example 3 of Sect. 16.3 that the sequence of functions $f_n(u) = n(1 - u^{1/n})$ increases monotonically and converges to $\ln(\frac{1}{u})$ on the interval $0 < u < 1$ as $n \to \infty$. Using Corollary 2 of Sect. 17.2 (see also Example 10 of Sect. 17.2), we conclude that for $\alpha \geq 1$

$$\int_0^1 \ln^{\alpha-1}\left(\frac{1}{u}\right) du = \lim_{n\to\infty} n^{\alpha-1} \int_0^1 \left(1 - u^{1/n}\right)^{\alpha-1} du. \qquad (17.39)$$

Making the change of variable $u = v^n$ in the last integral, we find by (17.38), (17.39), (17.27), (17.29), and (17.31) that

$$\Gamma(\alpha) = \lim_{n\to\infty} n^\alpha \int_0^1 v^{n-1}(1-v)^{\alpha-1} dv =$$

$$= \lim_{n\to\infty} n^\alpha B(n, \alpha) = \lim_{n\to\infty} n^\alpha B(\alpha, n) =$$

$$= \lim_{n\to\infty} n^\alpha \cdot \frac{(n-1)!}{\alpha(\alpha+1)\cdot\ldots\cdot(\alpha+n-1)}.$$

Applying the reduction formulas (17.30) and (17.35) to the relation $\Gamma(\alpha) = \lim_{n\to\infty} n^\alpha B(\alpha, n)$ just proved for $\alpha \geq 1$, we verify that formula (17.37) holds for all $\alpha > 0$. $\qquad\qquad\square$

e. The Complement Formula

For $0 < \alpha < 1$ the values α and $1 - \alpha$ of the argument of the gamma function are mutually complementary, so that the equality

$$\Gamma(\alpha) \cdot \Gamma(1 - \alpha) = \frac{\pi}{\sin \pi\alpha} \qquad (0 < \alpha < 1) \qquad (17.40)$$

is called the *complement formula for the gamma function.*

Proof Using the Euler–Gauss formula (17.37) and simple identities, we find that

$$\Gamma(\alpha)\Gamma(1-\alpha) = \lim_{n\to\infty}\left(n^\alpha \frac{(n-1)!}{\alpha(\alpha+1)\cdot\ldots\cdot(\alpha+n-1)}\times\right.$$

$$\left.\times n^{1-\alpha}\frac{(n-1)!}{(1-\alpha)(2-\alpha)\cdot\ldots\cdot(n-\alpha)}\right) =$$

$$= \lim_{n\to\infty}\left(n\frac{1}{\alpha(1+\frac{\alpha}{1})\cdot\ldots\cdot(1+\frac{\alpha}{n-1})}\times\right.$$

$$\left.\times\frac{1}{(1-\frac{\alpha}{1})(1-\frac{\alpha}{2})\cdot\ldots\cdot(1-\frac{\alpha}{n-1})(n-\alpha)}\right) =$$

$$= \frac{1}{\alpha}\lim_{n\to\infty}\frac{1}{(1-\frac{\alpha^2}{1^2})(1-\frac{\alpha^2}{2^2})\cdot\ldots\cdot(1-\frac{\alpha^2}{(n-1)^2})}.$$

Hence for $0 < \alpha < 1$

$$\Gamma(\alpha)\Gamma(1-\alpha) = \frac{1}{\alpha}\prod_{n=1}^{\infty}\frac{1}{1-\frac{\alpha^2}{n^2}}. \tag{17.41}$$

But the following expansion is classical:

$$\sin\pi\alpha = \pi\alpha\prod_{n=1}^{\infty}\left(1-\frac{\alpha^2}{n^2}\right). \tag{17.42}$$

(We shall not take the time to prove this formula just now, since it will be obtained as a simple example of the use of the general theory when we study Fourier series. See Example 6 of Sect. 18.2.)

Comparing relations (17.41) and (17.42), we obtain (17.40).   □

It follows in particular from (17.40) that

$$\Gamma\left(\frac{1}{2}\right) = \sqrt{\pi}. \tag{17.43}$$

We observe that

$$\Gamma\left(\frac{1}{2}\right) = \int_0^{+\infty} x^{-1/2}e^{-x}\,dx = 2\int_0^{+\infty} e^{-u^2}\,du,$$

and thus we again arrive at the Euler–Poisson integral

$$\int_0^{+\infty} e^{-u^2}\,du = \frac{1}{2}\sqrt{\pi}.$$

17.3.3 Connection Between the Beta and Gamma Functions

Comparing formulas (17.32) and (17.36), one may suspect the following connection:

$$B(\alpha,\beta) = \frac{\Gamma(\alpha)\cdot\Gamma(\beta)}{\Gamma(\alpha+\beta)} \tag{17.44}$$

between the beta and gamma functions. Let us prove this formula.

Proof We remark that for $y > 0$

$$\Gamma(\alpha) = y^\alpha \int_0^{+\infty} x^{\alpha-1}e^{-xy}\,dx,$$

and therefore the following equality also holds:

$$\frac{\Gamma(\alpha+\beta)\cdot y^{\alpha-1}}{(1+y)^{\alpha+\beta}} = y^{\alpha-1}\int_0^{+\infty} x^{\alpha+\beta-1}e^{-(1+y)x}\,dx,$$

using which, taking account of (17.33), we obtain

$$\Gamma(\alpha+\beta)\cdot B(\alpha,\beta) = \int_0^{+\infty}\frac{\Gamma(\alpha+\beta)y^{\alpha-1}}{(1+y)^{\alpha+\beta}}\,dy =$$

$$= \int_0^{+\infty}\left(y^{\alpha-1}\int_0^{+\infty} x^{\alpha+\beta-1}e^{-(1+y)x}\,dx\right)dy \overset{!}{=}$$

$$\overset{!}{=} \int_0^{+\infty}\left(\int_0^{+\infty} y^{\alpha-1}x^{\alpha+\beta-1}e^{-(1+y)x}\,dy\right)dx =$$

$$= \int_0^{+\infty}\left(x^{\beta-1}e^{-x}\int_0^{+\infty}(xy)^{\alpha-1}e^{-xy}x\,dy\right)dx =$$

$$= \int_0^{+\infty}\left(x^{\beta-1}e^{-x}\int_0^{+\infty} u^{\alpha-1}e^{-u}\,du\right)dx = \Gamma(\alpha)\cdot\Gamma(\beta).$$

All that remains is to explain the equality distinguished by the exclamation point. But that is exactly what was done in Example 16 of Sect. 17.2. □

17.3.4 Examples

In conclusion let us consider a small group of interconnected examples in which the special functions B and Γ introduced here occur.

Example 1

$$\int_0^{\pi/2}\sin^{\alpha-1}\varphi\cos^{\beta-1}\varphi\,d\varphi = \frac{1}{2}B\left(\frac{\alpha}{2},\frac{\beta}{2}\right). \tag{17.45}$$

Proof To prove this, it suffices to make the change of variable $\sin^2\varphi = x$ in the integral. □

Using formula (17.44), we can express the integral (17.45) in terms of the gamma function. In particular, taking account of (17.43), we obtain

$$\int_0^{\pi/2}\sin^{\alpha-1}\varphi\,d\varphi = \int_0^{\pi/2}\cos^{\alpha-1}\varphi\,d\varphi = \frac{\sqrt{\pi}}{2}\frac{\Gamma(\frac{\alpha}{2})}{\Gamma(\frac{\alpha+1}{2})}. \tag{17.46}$$

Example 2 A one-dimensional ball of radius r is simply an open interval and its (one-dimensional) volume $V_1(r)$ is the length $(2r)$ of that interval. Thus $V_1(r) = 2r$.

If we assume that the $((n-1)$-dimensional) volume of the $(n-1)$-dimensional ball of radius r is expressed by the formula $V_{n-1}(r) = c_{n-1}r^{n-1}$, then, integrating over sections (see Example 3 of Sect. 11.4), we obtain

$$V_n(r) = \int_{-r}^{r} c_{n-1}(r^2 - x^2)^{\frac{n-1}{2}}\,dx = \left(c_{n-1}\int_{-\pi/2}^{\pi/2}\cos^n\varphi\,d\varphi\right)\cdot r^n,$$

that is, $V_n(r) = c_n r^n$, where

$$c_n = 2c_{n-1}\int_0^{\pi/2}\cos^n\varphi\,d\varphi.$$

By relations (17.46) we can rewrite this last equality as

$$c_n = \sqrt{\pi}\,\frac{\Gamma(\frac{n+1}{2})}{\Gamma(\frac{n+2}{2})}c_{n-1},$$

so that

$$c_n = (\sqrt{\pi})^{n-1}\frac{\Gamma(\frac{n+1}{2})}{\Gamma(\frac{n+2}{2})}\cdot\frac{\Gamma(\frac{n}{2})}{\Gamma(\frac{n+1}{2})}\cdot\ldots\cdot\frac{\Gamma(\frac{3}{2})}{\Gamma(\frac{4}{2})}\cdot c_1$$

or, more briefly,

$$c_n = \pi^{\frac{n-1}{2}}\frac{\Gamma(\frac{3}{2})}{\Gamma(\frac{n+2}{2})}c_1.$$

But $c_1 = 2$, and $\Gamma(\frac{3}{2}) = \frac{1}{2}\Gamma(\frac{1}{2}) = \frac{1}{2}\sqrt{\pi}$, so that

$$c_n = \frac{\pi^{\frac{n}{2}}}{\Gamma(\frac{n+2}{2})}.$$

Consequently,

$$V_n(r) = \frac{\pi^{\frac{n}{2}}}{\Gamma(\frac{n+2}{2})}r^n,$$

or, what is the same,

$$V_n(r) = \frac{\pi^{\frac{n}{2}}}{\frac{n}{2}\Gamma(\frac{n}{2})}r^n. \tag{17.47}$$

Example 3 It is clear from geometric considerations that $dV_n(r) = S_{n-1}(r)\,dr$, where $S_{n-1}(r)$ is the $(n-1)$-dimensional surface area of the sphere bounding the n-dimensional ball of radius r in $\mathbb{R}^n$.

Thus $S_{n-1}(r) = \frac{dV_n}{dr}(r)$, and, taking account of (17.47), we obtain

$$S_{n-1}(r) = \frac{2\pi^{\frac{n}{2}}}{\Gamma(\frac{n}{2})}r^{n-1}.$$

17.3.5 Problems and Exercises

1. Show that

a) $B(1/2, 1/2) = \pi$;

b) $B(\alpha, 1 - \alpha) = \int_0^\infty \frac{x^{\alpha-1}}{1+x} dx$;

c) $\frac{\partial B}{\partial \alpha}(\alpha, \beta) = \int_0^1 x^{\alpha-1}(1-x)^{\beta-1} \ln x \, dx$;

d) $\int_0^{+\infty} \frac{x^p \, dx}{(a+bx^q)^r} = \frac{a^{-r}}{q}(\frac{a}{b})^{\frac{p+1}{q}} B(\frac{p+1}{q}, r - \frac{p+1}{q})$;

e) $\int_0^{+\infty} \frac{dx}{\sqrt[n]{1+x^n}} = \frac{\pi}{n \sin \frac{\pi}{n}}$;

f) $\int_0^{+\infty} \frac{dx}{1+x^3} = \frac{2\pi}{3\sqrt{3}}$;

g) $\int_0^{+\infty} \frac{x^{\alpha-1} \, dx}{1+x} = \frac{\pi}{\sin \pi\alpha}$ $(0 < \alpha < 1)$;

h) $\int_0^{+\infty} \frac{x^{\alpha-1} \ln^n x}{1+x} dx = \frac{d^n}{d\alpha^n}(\frac{\pi}{\sin \pi\alpha})$ $(0 < \alpha < 1)$;

i) the length of the curve defined in polar coordinates by the equation $r^n = a^n \cos n\varphi$, where $n \in \mathbb{N}$ and $a > 0$, is $a B(\frac{1}{2}, \frac{1}{2n})$.

2. Show that

a) $\Gamma(1) = \Gamma(2)$;

b) the derivative Γ' of Γ is zero at some point $x_0 \in \,]1, 2[$;

c) the function Γ' is monotonically increasing on the interval $]0, +\infty[$;

d) the function Γ is monotonically decreasing on $]0, x_0]$ and monotonically increasing on $[x_0, +\infty[$;

e) the integral $\int_0^1 (\ln \frac{1}{u})^{x-1} \ln \ln \frac{1}{u} \, du$ equals zero if $x = x_0$;

f) $\Gamma(\alpha) \sim \frac{1}{\alpha}$ as $\alpha \to +0$;

g) $\lim_{n\to\infty} \int_0^{+\infty} e^{-x^n} dx = 1$.

3. *Euler's formula* $E := \prod_{k=1}^{n-1} \Gamma(\frac{k}{n}) = \frac{(2\pi)^{\frac{n-1}{2}}}{\sqrt{n}}$.

a) Show that $E^2 = \prod_{k=1}^{n-1} \Gamma(\frac{k}{n})\Gamma(\frac{n-k}{n})$.

b) Verify that $E^2 = \frac{\pi^{n-1}}{\sin \frac{\pi}{n} \sin 2\frac{\pi}{n} \cdot \ldots \cdot \sin(n-1)\frac{\pi}{n}}$.

c) Starting from the identity $\frac{z^n-1}{z-1} = \prod_{k=1}^{n-1}(z - e^{i\frac{2k\pi}{n}})$, let z tend to 1 to obtain

$$n = \prod_{k=1}^{n-1}(1 - e^{i\frac{2k\pi}{n}}),$$

and from this relation derive the relation

$$n = 2^{n-1} \prod_{k=1}^{n-1} \sin \frac{k\pi}{n}.$$

d) Using this last equality, obtain Euler's formula.

4. *Legendre's formula* $\Gamma(\alpha)\Gamma(\alpha + \frac{1}{2}) = \frac{\sqrt{\pi}}{2^{2\alpha-1}}\Gamma(2\alpha)$.

a) Show that $B(\alpha, \alpha) = 2\int_0^{1/2}(\frac{1}{4} - (\frac{1}{2} - x)^2)^{\alpha-1}\,dx$.

b) By a change of variable in this last integral, prove that $B(\alpha, \alpha) = \frac{1}{2^{2\alpha-1}}B(\frac{1}{2}, \alpha)$.

c) Now obtain Legendre's formula.

5. Retaining the notation of Problem 5 of Sect. 17.1, show a route by which the second, more delicate part of the problem can be carried out using the Euler integrals.

a) Observe that $\tilde{k} = k$ for $k = \frac{1}{\sqrt{2}}$ and

$$\tilde{E} = E = \int_0^{\pi/2}\sqrt{1 - \frac{1}{2}\sin^2\varphi}\,d\varphi, \qquad \tilde{K} = K = \int_0^{\pi/2}\frac{d\varphi}{\sqrt{1 - \frac{1}{2}\sin^2\varphi}}.$$

b) After a suitable change of variable these integrals can be brought into a form from which it follows that for $k = 1/\sqrt{2}$

$$K = \frac{1}{2\sqrt{2}}B(1/4, 1/2) \quad \text{and} \quad 2E - K = \frac{1}{2\sqrt{2}}B(3/4, 1/2).$$

c) It now results that for $k = 1/\sqrt{2}$

$$E\tilde{K} + \tilde{E}K - K\tilde{K} = \pi/2.$$

6. *Raabe's*[2] *integral* $\int_0^1 \ln\Gamma(x)\,dx$.
Show that

a) $\int_0^1 \ln\Gamma(x)\,dx = \int_0^1 \ln\Gamma(1 - x)\,dx$.

b) $\int_0^1 \ln\Gamma(x)\,dx = \frac{1}{2}\ln\pi - \frac{1}{\pi}\int_0^{\pi/2}\ln\sin x\,dx$.

c) $\int_0^{\pi/2}\ln\sin x\,dx = \int_0^{\pi/2}\ln\sin 2x\,dx - \frac{\pi}{2}\ln 2$.

d) $\int_0^{\pi/2}\ln\sin x\,dx = -\frac{\pi}{2}\ln 2$.

e) $\int_0^1 \ln\Gamma(x)\,dx = \ln\sqrt{2\pi}$.

7. Using the equality

$$\frac{1}{x^s} = \frac{1}{\Gamma(s)}\int_0^{+\infty}y^{s-1}e^{-xy}\,dy$$

and justifying the reversal in the order of the corresponding integrations, verify that

a) $\int_0^{+\infty}\frac{\cos ax}{x^\alpha}\,dx = \frac{\pi a^{\alpha-1}}{2\Gamma(\alpha)\cos\frac{\pi\alpha}{2}}$ $(0 < \alpha < 1)$.

b) $\int_0^{+\infty}\frac{\sin bx}{x^\beta}\,dx = \frac{\pi b^{\beta-1}}{2\Gamma(\beta)\sin\frac{\pi\beta}{2}}$ $(0 < \beta < 2)$.

---

[2]J.L. Raabe (1801–1859) – Swiss mathematician and physicist.

c) Now obtain once again the value of the Dirichlet integral $\int_0^{+\infty} \frac{\sin x}{x}\,dx$ and the value of the Fresnel integrals $\int_0^{+\infty} \cos x^2\,dx$ and $\int_0^{+\infty} \sin x^2\,dx$.

8. Show that for $\alpha > 1$

$$\int_0^{+\infty} \frac{x^{\alpha-1}}{e^x - 1}\,dx = \Gamma(\alpha) \cdot \zeta(\alpha),$$

where $\zeta(a) = \sum_{n=1}^{\infty} \frac{1}{n^a}$ is the *Riemann zeta function*.

9. *Gauss' formula.* In Example 6 of Sect. 16.3 we exhibited the function

$$F(\alpha, \beta, \gamma, x) := 1 + \sum_{n=1}^{\infty} \frac{\alpha(\alpha+1)\cdots(\alpha+n-1)\beta(\beta+1)\cdots(\beta+n-1)}{n!\gamma(\gamma+1)\cdots(\gamma+n-1)} x^n,$$

which was introduced by Gauss and is the sum of this hypergeometric series. It turns out that the following formula of Gauss holds:

$$F(\alpha, \beta, \gamma, 1) = \frac{\Gamma(\gamma) \cdot \Gamma(\gamma-\alpha-\beta)}{\Gamma(\gamma-\alpha) \cdot \Gamma(\gamma-\beta)}.$$

a) By developing the function $(1-tx)^{-\beta}$ in a series, show that for $\alpha > 0$, $\gamma - \alpha > 0$, and $0 < x < 1$ the integral

$$P(x) = \int_0^1 t^{\alpha-1}(1-t)^{\gamma-\alpha-1}(1-tx)^{-\beta}\,dt$$

can be represented as

$$P(x) = \sum_{n=0}^{\infty} P_n \cdot x^n,$$

where $P_n = \frac{\beta(\beta+1)\cdots(\beta+n-1)}{n!} \cdot \frac{\Gamma(\alpha+n)\cdot\Gamma(\gamma-\alpha)}{\Gamma(\gamma+n)}$.

b) Show that

$$P_n = \frac{\Gamma(\alpha) \cdot \Gamma(\gamma-\alpha)}{\Gamma(\gamma)} \cdot \frac{\alpha(\alpha+1)\cdots(\alpha+n-1)\beta(\beta+1)\cdots(\beta+n-1)}{n!\gamma(\gamma+1)\cdots(\gamma+n-1)}.$$

c) Now prove that for $\alpha > 0$, $\gamma - \alpha > 0$, and $0 < x < 1$

$$P(x) = \frac{\Gamma(\alpha) \cdot \Gamma(\gamma-\alpha)}{\Gamma(\gamma)} \cdot F(\alpha, \beta, \gamma, x).$$

d) Under the additional condition $\gamma - \alpha - \beta > 0$ justify the possibility of passing to the limit as $x \to 1 - 0$ on both sides of the last equality, and show that

$$\frac{\Gamma(\alpha) \cdot \Gamma(\gamma-\alpha-\beta)}{\Gamma(\gamma-\beta)} = \frac{\Gamma(\alpha) \cdot \Gamma(\gamma-\alpha)}{\Gamma(\gamma)} F(\alpha, \beta, \gamma, 1),$$

from which Gauss' formula follows.

10. *Stirling's[3] formula.* Show that

a) $\ln \frac{1+x}{1-x} = 2x \sum_{m=0}^{\infty} \frac{x^{2m}}{2m+1}$ for $|x| < 1$;

b) $(n + \frac{1}{2}) \ln(1 + \frac{1}{n}) = 1 + \frac{1}{3} \frac{1}{(2n+1)^2} + \frac{1}{5} \frac{1}{(2n+1)^4} + \frac{1}{7} \frac{1}{(2n+1)^6} + \cdots$;

c) $1 < (n + \frac{1}{2}) \ln(1 + \frac{1}{n}) < 1 + \frac{1}{12n(n+1)}$ for $n \in \mathbb{N}$;

d) $1 < \dfrac{(1+\frac{1}{n})^{n+1/2}}{e} < \dfrac{e^{\frac{1}{12n}}}{e^{\frac{1}{12(n+1)}}}$;

e) $a_n = \frac{n! e^n}{n^{(n+1/2)}}$ is a monotonically decreasing sequence;

f) $b_n = a_n e^{-\frac{1}{12n}}$ is a monotonically increasing sequence;

g) $n! = c n^{n+1/2} e^{-n + \frac{\theta_n}{12n}}$, where $0 < \theta_n < 1$, and $c = \lim_{n \to \infty} a_n = \lim_{n \to \infty} b_n$;

h) the relation $\sin \pi x = \pi x \prod_{n=1}^{\infty} (1 - \frac{x^2}{n^2})$ with $x = 1/2$ implies Wallis' formula

$$\sqrt{\pi} = \lim_{n \to \infty} \frac{(n!)^2 2^{2n}}{(2n)!} \cdot \frac{1}{\sqrt{n}};$$

i) *Stirling's formula* holds:

$$n! = \sqrt{2\pi n} \left(\frac{n}{e}\right)^n e^{\frac{\theta_n}{12n}}, \quad 0 < \theta_n < 1;$$

j) $\Gamma(x + 1) \sim \sqrt{2\pi x} (\frac{x}{e})^x$ as $x \to +\infty$.

11. Show that $\Gamma(x) = \sum_{n=0}^{\infty} \frac{(-1)^n}{n+x} \cdot \frac{1}{n!} + \int_1^{\infty} t^{x-1} e^{-t} \, dt$. This relation makes it possible to define $\Gamma(z)$ for complex $z \in \mathbb{C}$ except at the points $0, -1, -2, \ldots$.

17.4 Convolution of Functions and Elementary Facts About Generalized Functions

17.4.1 Convolution in Physical Problems (Introductory Considerations)

A variety of devices and systems in the living and nonliving natural world carry out their functions responding to a stimulus f with an appropriate signal $\tilde{f}$. In other words, each such device or system is an operator A that transforms the incoming signal f into the outgoing signal $\tilde{f} = Af$. Naturally, each such operator has its own domain of perceivable signals (domain of definition) and its own form of response to them (range of values). A convenient mathematical model for a large class of actual processes and machines is a linear operator A that preserves translations.

---

[3]J. Stirling (1692–1770) – Scottish mathematician.

Definition 1 Let A be a linear operator acting on a vector space of real- or complex-valued functions defined on $\mathbb{R}$. We denote by T_{t_0} the *shift operator* or *translation operator* acting on the same space according to the rule

$$(T_{t_0}f)(t) := f(t - t_0).$$

The operator A is *translation-invariant* (or *preserves translations*) if

$$A(T_{t_0}f) = T_{t_0}(Af)$$

for every function f in the domain of definition of the operator A.

If t is time, the relation $A \circ T_{t_0} = T_{t_0} \circ A$ can be interpreted as the assumption that the properties of the device A are time-invariant: the reaction of the device to the signals $f(t)$ and $f(t - t_0)$ differ only in a shift by the amount t_0 in time, nothing more.

For every device A the following two fundamental problems arise: first, to predict the reaction $\tilde{f}$ of the device to an arbitrary input f; second, knowing the output $\tilde{f}$, to determine, if possible, the input signal f.

At this point we shall solve the first of these two problems heuristically in application to a translation-invariant linear operator A. It is a simple, but very important fact that in order to describe the response $\tilde{f}$ of such a device A to any input signal f, it suffices to know the response E of A to a pulse δ.

Definition 2 The response $E(t)$ of the device A to a unit pulse δ is called the *system function* of the device (in optics) or the *transient pulse function* of the device (in electrical engineering).

As a rule, we shall use the briefer term "system function".

Without going into detail just yet, we shall say that a pulse can be imitated, for example, by the function $\delta_\alpha(t)$ shown in Fig. 17.1, and this imitation is assumed to become closer as the duration α of the "pulse" gets shorter, preserving the relation $\alpha \cdot \frac{1}{\alpha} = 1$. Instead of step functions, one may imitate a pulse using smooth functions (Fig. 17.2) while preserving the natural conditions:

$$f_\alpha \geq 0, \qquad \int_{\mathbb{R}} f_\alpha(t)\, dt = 1, \qquad \int_{U(0)} f_\alpha(t)\, dt \to 1 \quad \text{as } \alpha \to 0,$$

where $U(0)$ is an arbitrary neighborhood of the point $t = 0$.

The response of the device A to an ideal unit pulse (denoted, following Dirac, by the letter δ) should be regarded as a function $E(t)$ to which the response of the device A to an input approximating δ tends as the imitation improves. Naturally a certain continuity of the operator A is assumed (not made precise as yet), that is, continuity of the change in the response $\tilde{f}$ of the device under a continuous change in the input f.

Fig. 17.1

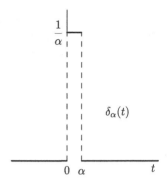

Fig. 17.2

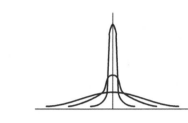

Fig. 17.3

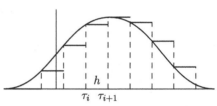

For example, if we take a sequence $\{\Delta_n(t)\}$ of step functions $\Delta_n(t) := \delta_{1/n}(t)$ (Fig. 17.1), then, setting $A\Delta_n =: E_n$, we obtain $A\delta := E = \lim_{n\to\infty} E_n = \lim_{n\to\infty} A\Delta_n$.

Let us now consider the input signal f in Fig. 17.3 and the piecewise constant function $l_h(t) = \sum_i f(\tau_i)\delta_h(t - \tau_i)h$. Since $l_h \to f$ as $h \to 0$, one must assume that

$$\tilde{l}_h = Al_h \to Af = \tilde{f} \quad \text{as } h \to 0.$$

But if the operator A is linear and preserves translates, then

$$\tilde{l}_h(t) = \sum_i f(\tau_i)E_h(t - \tau_i)h,$$

where $E_h = A\delta_h$. Thus, as $h \to 0$ we finally obtain

$$\tilde{f}(t) = \int_{\mathbb{R}} f(\tau)E(t - \tau)\,d\tau. \qquad (17.48)$$

Formula (17.48) solves the first of the two problems indicated above. It represents the response $\tilde{f}(t)$ of the device A in the form of a special integral depending on the parameter t. This integral is completely determined by the input signal $f(t)$ and the system function $E(t)$ of the device A. From the mathematical point of view the device A and the integral (17.48) are simply identical.

We note incidentally that the problem of determining the input signal from the output $\tilde{f}$ now reduces to solving the integral equation (17.48) for f.

Definition 3 The *convolution of the functions* $u : \mathbb{R} \to \mathbb{C}$ and $v : \mathbb{R} \to \mathbb{C}$ is the function $u * v : \mathbb{R} \to \mathbb{C}$ defined by the relation

$$(u * v)(x) := \int_{\mathbb{R}} u(y)v(x - y)\,dy, \qquad (17.49)$$

provided this improper integral exists for all $x \in \mathbb{R}$.

Thus formula (17.48) asserts that the response of a linear device A that preserves translates to an input given by the function f is the convolution $f * E$ of the function f and the system function E of the device A.

17.4.2 General Properties of Convolution

Now let us consider the basic properties of convolution from a mathematical point of view.

a. Sufficient Conditions for Existence

We first recall certain definitions and notation.

Let $f : G \to \mathbb{C}$ be a real- or complex-valued function defined on an open set $G \subset \mathbb{R}$.

The function f is *locally integrable on* G if every point $x \in G$ has a neighborhood $U(x) \subset G$ in which the function $f|_{U(x)}$ is integrable. In particular, if $G = \mathbb{R}$, the condition of local integrability of the function f is obviously equivalent to the relation $f|_{[a,b]} \in \mathcal{R}[a, b]$ for every closed interval $[a, b]$.

The *support of the function* f (denoted supp f) is the closure in G of the set $\{x \in G \mid f(x) \neq 0\}$.

A function f is *of compact support* (in G) if its support is a compact set.

The set of functions $f : G \to \mathbb{C}$ having continuous derivatives in G up to order m ($0 \leq m \leq \infty$) inclusive, is usually denoted $C^{(m)}(G)$ and the subset of it consisting of functions of compact support is denoted $C_0^{(m)}(G)$. In the case when $G = \mathbb{R}$, instead of $C^{(m)}(\mathbb{R})$ and $C_0^{(m)}(\mathbb{R})$ it is customary to use the abbreviation $C^{(m)}$ and $C_0^{(m)}$ respectively.

We now exhibit the most frequently encountered cases of convolution of functions, in which its existence can be established without difficulty.

Proposition 1 *Each of the conditions listed below is sufficient for the existence of the convolution $u * v$ of locally integrable functions $u : \mathbb{R} \to \mathbb{C}$ and $v : \mathbb{R} \to \mathbb{C}$.*

1) *The functions $|u|^2$ and $|v|^2$ are integrable on $\mathbb{R}$.*
2) *One of the functions $|u|$, $|v|$ is integrable on $\mathbb{R}$ and the other is bounded on $\mathbb{R}$.*
3) *One of the functions u and v is of compact support.*

Proof 1) By the Cauchy–Bunyakovskii inequality

$$\left(\int_{\mathbb{R}} |u(y)v(x-y)| \, dy \right)^2 \le \int_{\mathbb{R}} |u|^2(y) \, dy \int_{\mathbb{R}} |v|^2(x-y) \, dy,$$

from which it follows that the integral (17.49) exists, since

$$\int_{-\infty}^{+\infty} |v|^2(x-y) \, dy = \int_{-\infty}^{+\infty} |v|^2(y) \, dy.$$

2) If, for example, $|u|$ is integrable on $\mathbb{R}$ and $|v| \le M$ on $\mathbb{R}$, then

$$\int_{\mathbb{R}} |u(y)v(x-y)| \, dy \le M \int_{\mathbb{R}} |u|(y) \, dy < +\infty.$$

3) Suppose $\operatorname{supp} u \subset [a, b] \subset \mathbb{R}$. Then obviously

$$\int_{\mathbb{R}} u(y)v(x-y) \, dy = \int_{a}^{b} u(y)v(x-y) \, dy.$$

Since u and v are locally integrable, this last integral exists for every value of $x \in \mathbb{R}$.

The case when the function of compact support is v reduces to the one just considered by the change of variable $x - y = z$. $\qquad \square$

b. Symmetry

Proposition 2 *If the convolution $u * v$ exists, then the convolution $v * u$ also exists, and the following equality holds*:

$$u * v = v * u. \tag{17.50}$$

Proof Making the change of variable $x - y = z$ in (17.49), we obtain

$$u * v(x) := \int_{-\infty}^{+\infty} u(y)v(x-y) \, dy = \int_{-\infty}^{+\infty} v(z)u(x-z) \, dz =: v * u(x). \qquad \square$$

c. Translation Invariance

Suppose, as above, that T_{x_0} is the shift operator, that is, $(T_{x_0})f(x) = f(x - x_0)$.

Proposition 3 *If the convolution $u * v$ of the functions u and v exists, then the following equalities hold*:

$$T_{x_0}(u * v) = T_{x_0}u * v = u * T_{x_0}v. \tag{17.51}$$

Proof If we recall the physical meaning of formula (17.48), the first of these equalities becomes obvious, and the second can then be obtained from the symmetry of convolution. Nevertheless, let us give a formal verification of the first equality:

$$(T_{x_0})(u * v)(x) := (u * v)(x - x_0) :=$$

$$= \int_{-\infty}^{+\infty} u(y)v(x - x_0 - y)\,dy = \int_{-\infty}^{+\infty} u(y - x_0)v(x - y)\,dy =$$

$$= \int_{-\infty}^{+\infty} (T_{x_0}u)(y)v(x - y)\,dy =: \big((T_{x_0}u) * v\big)(x). \qquad \square$$

d. Differentiation of a Convolution

The convolution of functions is an integral depending on a parameter, and differentiation of it can be carried out in accordance with the general rules for differentiating such integrals, provided of course suitable hypotheses hold.

The conditions under which the convolution (17.49) of the functions u and v is continuously differentiable are demonstrably satisfied if, for example, u is continuous and v is a smooth function and one of the two is of compact support.

Proof Indeed, if we confine the variation of the parameter to any finite interval, then under these hypotheses the entire integral (17.49) reduces to the integral over a finite closed interval independent of x. Such an integral can be differentiated with respect to a parameter in accordance with the classical rule of Leibniz. $\qquad \square$

In general the following proposition holds:

Proposition 4 *If u is a locally integrable function and v is a $C_0^{(m)}$ function of compact support $(0 \le m \le +\infty)$, then $(u * v) \in C^{(m)}$, and*[4]

$$D^k(u * v) = u * \big(D^k v\big). \tag{17.52}$$

---

[4] Here D is differentiation, and, as usual $D^k v = v^{(k)}$.

Proof When u is a continuous function, the proposition follows immediately from what was just proved above. In its general form it can be obtained if we also keep in mind the observation made in Problem 6 of Sect. 17.1.   □

Remark 1 In view of the commutativity of convolution (formula (17.50)) Proposition 4 of course remains valid if u and v are interchanged, preserving the left-hand side of Eq. (17.52).

Formula (17.52) shows that convolution commutes with the differentiation operator, just as it commutes with translation (formula (17.51)). But while (17.51) is symmetric in u and v, one cannot in general interchange u and v in the right-hand side of (17.52), since u may fail to have the corresponding derivative. The fact that the convolution $u * v$, as one can see by (17.52), may still turn out to be a differentiable function, might suggest that the hypotheses of Proposition 4 are sufficient, but not necessary for differentiability of the convolution.

Example 1 Let f be a locally integrable function and δ_α the "step" function shown in Fig. 17.1. Then

$$(f * \delta_\alpha)(x) = \int_{-\infty}^{+\infty} f(y)\delta_\alpha(x - y)\,dy = \frac{1}{\alpha} \int_{x-\alpha}^{x} f(y)\,dy, \qquad (17.53)$$

and consequently if f is continuous at the points x and $x - \alpha$, then the convolution $f * \delta_\alpha$ is differentiable, due to the averaging (smoothing) property of the integral.

The conditions for differentiability of the convolution stated in Proposition 4 are, however, completely sufficient for practically all the cases one encounters in which formula (17.52) is applied. For that reason we shall not attempt to refine them any further, preferring to illustrate some beautiful new possibilities that open up as a result of the smoothing action of convolution just discovered.

17.4.3 Approximate Identities and the Weierstrass Approximation Theorem

We remark that the integral in (17.53) gives the average value of the function f on the interval $[x - \alpha, x]$, and therefore, if f is continuous at x, the relation $(f * \delta_\alpha)(x) \to f(x)$ obviously holds as $\alpha \to 0$. In accordance with the introductory considerations of Sect. 17.4.1 that gave a picture of the δ-function, we would like to write this last relation as the limiting equality

$$(f * \delta)(x) = f(x), \quad \text{if } f \text{ is continuous at } x. \qquad (17.54)$$

This equality shows that the δ-function can be interpreted as the identity (neutral) element with respect to convolution. Equality (17.54) can be regarded as making perfect sense if it is shown that every family of functions converging to the δ-function has the same property as the special family δ_α of (17.53).

Let us now pass to precise statements and introduce the following useful definition.

Definition 4 The family $\{\Delta_\alpha; \alpha \in A\}$ of functions $\Delta_\alpha : \mathbb{R} \to \mathbb{R}$ depending on the parameter $\alpha \in A$ forms an *approximate identity* over a base $\mathcal{B}$ in A if the following three conditions hold:

a) all the functions in the family are nonnegative ($\Delta_\alpha \geq 0$);
b) for every function Δ_α in the family, $\int_\mathbb{R} \Delta_\alpha(x)\,dx = 1$;
c) for every neighborhood U of $0 \in \mathbb{R}$, $\lim_\mathcal{B} \int_U \Delta_\alpha(x)\,dx = 1$.

Taking account of the first two conditions, we see that this last condition is equivalent to the relation $\lim_\mathcal{B} \int_{\mathbb{R}\setminus U} \Delta_\alpha(x)\,dx = 0$.

The original family of "step" functions δ_α considered in Example 1 of Sect. 17.4.1 is an approximate identity as $\alpha \to 0$. We shall now give other examples of approximate identities.

Example 2 Let $\varphi : \mathbb{R} \to \mathbb{R}$ be an arbitrary nonnegative function of compact support that is integrable over $\mathbb{R}$ and satisfies $\int_\mathbb{R} \varphi(x)\,dx = 1$. For $\alpha > 0$ we construct the functions $\Delta_\alpha(x) := \frac{1}{\alpha}\varphi(\frac{x}{\alpha})$. The family of these functions is obviously an approximate identity as $\alpha \to +0$ (see Fig. 17.2).

Example 3 Consider the sequence of functions

$$\Delta_n(x) = \begin{cases} \dfrac{(1-x^2)^n}{\int_{|x|<1}(1-x^2)^n\,dx} & \text{for } |x| \leq 1, \\ 0 & \text{for } |x| > 1. \end{cases}$$

To establish that this family is an approximate identity we need only verify that condition c) of Definition 4 holds in addition to a) and b). But for every $\varepsilon \in]0, 1]$ we have

$$0 \leq \int_\varepsilon^1 (1 - x^2)^n\,dx \leq \int_\varepsilon^1 (1 - \varepsilon^2)^n\,dx =$$

$$= (1 - \varepsilon^2)^n (1 - \varepsilon) \to 0, \quad \text{as } n \to \infty.$$

At the same time,

$$\int_0^1 (1 - x^2)^n\,dx > \int_0^1 (1 - x)^n\,dx = \frac{1}{n+1}.$$

Therefore condition c) holds.

Example 4 Let

$$\Delta_n(x) = \begin{cases} \cos^{2n}(x)/\int_{-\pi/2}^{\pi/2} \cos^{2n}(x)\,dx & \text{for } |x| \leq \pi/2, \\ 0 & \text{for } |x| > \pi/2. \end{cases}$$

As in Example 3, it remains only to verify condition c) here. We remark first of all that

$$\int_0^{\pi/2} \cos^{2n} x \, dx = \frac{1}{2} B\left(n + \frac{1}{2}, \frac{1}{2}\right) = \frac{1}{2} \frac{\Gamma(n + \frac{1}{2n})}{\Gamma(n)} \cdot \frac{\Gamma(\frac{1}{2})}{n} > \frac{\Gamma(\frac{1}{2})}{2n}.$$

On the other hand, for $\varepsilon \in \,]0, \pi/2[$

$$\int_\varepsilon^{\pi/2} \cos^{2n} x \, dx \leq \int_\varepsilon^{\pi/2} \cos^{2n} \varepsilon \, dx < \frac{\pi}{2} (\cos \varepsilon)^{2n}.$$

Combining the two inequalities just obtained, we conclude that for every $\varepsilon \in \,]0, \pi/2]$,

$$\int_\varepsilon^{\pi/2} \Delta_n(x) \, dx \to 0 \quad \text{as } n \to \infty,$$

from which it follows that condition c) of Definition 4 holds.

Definition 5 The function $f : G \to \mathbb{C}$ *is* uniformly continuous *on the set* $E \subset G$ *if for every* $\varepsilon > 0$ *there exists* $\rho > 0$ *such that* $|f(x) - f(y)| < \varepsilon$ *for every* $x \in E$ *and every* $y \in G$ *belonging to the* ρ-*neighborhood* $U_G^\rho(x)$ *of* x *in* G.

In particular, if $E = G$ we simply get back the definition of a function that is uniformly continuous on its entire domain of definition.

We now prove a fundamental proposition.

Proposition 5 *Let* $f : \mathbb{R} \to \mathbb{C}$ *be a bounded function and* $\{\Delta_\alpha; \alpha \in A\}$ *an approximate identity as* $\alpha \to \omega$. *If the convolution* $f * \Delta_\alpha$ *exists for every* $\alpha \in A$ *and the function* f *is uniformly continuous on the set* $E \subset \mathbb{R}$, *then*

$$(f * \Delta_\alpha)(x) \rightrightarrows f(x) \quad \text{on } E \text{ as } \alpha \to \omega.$$

Thus it is asserted that the family of functions $f * \Delta_\alpha$ converges uniformly to f on a set E on which it is uniformly continuous. In particular, if E consists of only one point, the condition of uniform continuity of f on E reduces to the condition that f be continuous at x, and we find that $(f * \Delta_\alpha)(x) \to f(x)$ as $\alpha \to \omega$. Previously this fact served as our motivation for writing relation (17.54).

Let us now prove Proposition 5.

Proof Suppose $|f(x)| \leq M$ on $\mathbb{R}$. Given a number $\varepsilon > 0$, we choose $\rho > 0$ in accordance with Definition 5 and denote the ρ-neighborhood of 0 in $\mathbb{R}$ by $U(0)$.

Taking account of the symmetry of convolution, we obtain the following two estimates, which hold simultaneously for all $x \in E$:

$$\left| (f * \Delta_\alpha)(x) - f(x) \right|$$

$$= \left| \int_{\mathbb{R}} f(x - y) \Delta_\alpha(y) \, dy - f(x) \right| =$$

$$= \left| \int_{\mathbb{R}} \left(f(x - y) - f(x) \right) \Delta_\alpha(y) \, dy \right| \leq$$

$$\leq \int_{U(0)} \left| f(x - y) - f(x) \right| \Delta_\alpha(y) \, dy + \int_{\mathbb{R} \setminus U(0)} \left| f(x - y) - f(x) \right| \Delta_\alpha(y) \, dy <$$

$$< \varepsilon \int_{U(0)} \Delta_\alpha(y) \, dy + 2M \int_{\mathbb{R} \setminus U(0)} \Delta_\alpha(y) \, dy \leq \varepsilon + 2M \int_{\mathbb{R} \setminus U(0)} \Delta_\alpha(y) \, dy.$$

As $\alpha \to \omega$, this last integral tends to zero, so that the inequality

$$\left| (f * \Delta_\alpha)(x) - f(x) \right| < 2\varepsilon$$

holds for all $x \in E$ from some point α_ε on. This completes the proof of Proposition 5. $\qquad\square$

Corollary 1 *Every continuous function of compact support on $\mathbb{R}$ can be uniformly approximated by infinitely differentiable functions.*

Proof Let verify that $C_0^{(\infty)}$ is everywhere dense in C_0 in this sense.
   We let, for example,

$$\varphi(x) = \begin{cases} k \cdot \exp(-\frac{1}{1-x^2}) & \text{for } |x| < 1, \\ 0 & \text{for } |x| \geq 1, \end{cases}$$

where k is chosen so that $\int_{\mathbb{R}} \varphi(x) \, dx = 1$.
   The function φ is of compact support and infinitely differentiable. In that case, the family of infinitely differentiable functions $\Delta_\alpha = \frac{1}{\alpha} \varphi(\frac{x}{\alpha})$, as observed in Example 2, is an approximate identity as $\alpha \to +0$. If $f \in C_0$, it is clear that $f * \Delta_\alpha \in C_0$. Moreover, by Proposition 4 we have $f * \Delta_\alpha \in C_0^\infty$. Finally, it follows from Proposition 5 that $f * \Delta_\alpha \rightrightarrows f$ on $\mathbb{R}$ as $\alpha \to +0$. $\qquad\square$

Remark 2 If the function $f \in C_0$ belongs to $C_0^{(m)}$, then for every value $n \in \{0, 1, \ldots, m\}$ we can guarantee that $(f * \Delta_\alpha)^{(n)} \rightrightarrows f^{(n)}$ on $\mathbb{R}$ as $\alpha \to +0$.

Proof Indeed, in this case $(f * \Delta_\alpha)^{(n)} = f^{(n)} * \Delta_\alpha$ (see Proposition 4 and Remark 1). All that now remains is to cite Corollary 1. $\qquad\square$

Corollary 2 (The Weierstrass approximation theorem) *Every continuous function on a closed interval can be uniformly approximated on that interval by an algebraic polynomial.*

Proof Since polynomials map to polynomials under a linear change of variable while the continuity and uniformity of the approximation of functions are preserved, it suffices to verify Corollary 2 on any convenient interval $[a, b] \subset \mathbb{R}$. For that reason we shall assume $0 < a < b < 1$. We continue the given function $f \in C[a, b]$ to a function F that is continuous on $\mathbb{R}$ by setting $F(x) = 0$ for $x \in \mathbb{R} \backslash]0, 1[$ and, for example, letting F be a linear function going from 0 to $f(a)$ and from $f(b)$ to 0 on the intervals $[0, a]$ and $[b, 1]$ respectively.

If we now take the approximate identity consisting of the functions Δ_n of Example 3, we can conclude from Proposition 5 that $F * \Delta_n \rightrightarrows f = F|_{[a,b]}$ on $[a, b]$ as $n \to \infty$. But for $x \in [a, b] \subset [0, 1]$ and $y \in [0, 1]$ we have $|x - y| \leq 1$, therefore

$$F * \Delta_n(x) := \int_{-\infty}^{\infty} F(y) \Delta_n(x - y) \, dy = \int_0^1 F(y) \Delta_n(x - y) \, dy =$$

$$= \int_0^1 F(y) p_n \cdot \left(1 - (x - y)^2\right)^n dy = \int_0^1 F(y) \left(\sum_{k=0}^{2n} a_k(y) x^k\right) dy =$$

$$= \sum_{k=0}^{2n} \left(\int_0^1 F(y) a_k(y) \, dy\right) x^k.$$

This last expression is a polynomial $P_{2n}(x)$ of degree $2n$ and we have shown that $P_{2n} \rightrightarrows f$ on $[a, b]$ as $n \to \infty$. □

Remark 3 By a slight extension of this reasoning one can show that Weierstrass theorem remains valid if the interval $[a, b]$ is replaced by an arbitrary compact subset of $\mathbb{R}$.

Remark 4 It is also not difficult to verify that for every open set G in $\mathbb{R}$ and every function $f \in C^{(m)}(G)$ there exists a sequence $\{P_k\}$ of polynomials such that $P_k^{(n)} \rightrightarrows f^{(n)}$ on every compact set $K \subset G$ for each $n \in \{0, 1, \ldots, m\}$ as $k \to \infty$.

If in addition the set G is bounded and $f \in C^{(m)}(\overline{G})$, then one can even get $P_k^{(n)} \rightrightarrows f^{(n)}$ on $\overline{G}$ as $k \to \infty$.

Remark 5 Just as the approximate identity of Example 3 was used in the proof of Corollary 2, one can use the sequence from Example 4 to prove that every 2π-periodic function on $\mathbb{R}$ can be uniformly approximated by trigonometric polynomials of the form

$$T_n(x) = \sum_{k=0}^{n} a_k \cos kx + b_k \sin kx.$$

We have used only approximate identities made up of functions of compact support above. However, it should be kept in mind that approximate identities of functions that are not of compact support play an important role in many cases. We shall give only two examples.

Example 5 The family of functions $\Delta_y(x) = \frac{1}{\pi} \cdot \frac{y}{x^2 + y^2}$ is an approximate identity on $\mathbb{R}$ as $y \to +0$, since $\Delta_y > 0$ for $y > 0$,

$$\int_{-\infty}^{\infty} \Delta_y(x) \, dx = \frac{1}{\pi} \arctan\left(\frac{x}{y}\right)\Bigg|_{x=-\infty}^{+\infty} = 1,$$

and for every $\rho > 0$ we have

$$\int_{-\rho}^{\rho} \Delta_y(x) \, dx = \frac{2}{\pi} \arctan \frac{\rho}{y} \to 1,$$

when $y \to +0$.

If $f : \mathbb{R} \to \mathbb{R}$ is a bounded continuous function, then the function

$$u(x, y) = \frac{1}{\pi} \int_{-\infty}^{\infty} \frac{f(\xi)y}{(x - \xi)^2 + y^2} \, d\xi, \tag{17.55}$$

which is the convolution $f * \Delta_y$, is defined for all $x \in \mathbb{R}$ and $y > 0$.

As one can easily verify using the Weierstrass M-test, the integral (17.55), which is called the *Poisson integral for the half-plane*, is a bounded infinitely differentiable function in the half-plane $\mathbb{R}_+^2 = \{(x, y) \in \mathbb{R}^2 \mid y > 0\}$. Differentiating it under the integral sign, we verify that for $y > 0$

$$\Delta u := \frac{\partial^2 u}{\partial x^2} + \frac{\partial^2 u}{\partial y^2} = f * \left(\frac{\partial^2}{\partial x^2} + \frac{\partial^2}{\partial y^2}\right) \Delta_y = 0,$$

that is, u is a harmonic function.

By Proposition 5 one can also guarantee that $u(x, y)$ converges to $f(x)$ as $y \to 0$. Thus, the integral (17.55) solves the problem of constructing a bounded function that is harmonic in the half-plane $\mathbb{R}_+^2$ and assumes prescribed boundary values f on $\partial \mathbb{R}_+^2$.

Example 6 The family of functions $\Delta_t = \frac{1}{2\sqrt{\pi t}} e^{-\frac{x^2}{4t}}$ is an approximate identity on $\mathbb{R}$ as $t \to +0$. Indeed, we certainly have $\Delta_t > 0$ and $\int_{-\infty}^{+\infty} \Delta_t(x) = 1$, since $\int_{-\infty}^{+\infty} e^{-v^2} \, dv = \sqrt{\pi}$ (the Euler–Poisson integral). Finally, for every $\rho > 0$ we have

$$\int_{-\rho}^{\rho} \frac{1}{2\sqrt{\pi t}} e^{-\frac{x^2}{4t}} \, dt = \frac{1}{\sqrt{\pi}} \int_{-\rho/2\sqrt{t}}^{\rho/2\sqrt{t}} e^{-v^2} \, dv \to 1, \quad \text{as } t \to +0.$$

If f is a continuous and, for example, bounded function on $\mathbb{R}$, then the function

$$u(x, t) = \frac{1}{2\sqrt{\pi t}} \int_{-\infty}^{+\infty} f(\xi) e^{-\frac{(x-\xi)^2}{4t}} \, d\xi, \tag{17.56}$$

which is the convolution $f * \Delta_t$, is obviously infinitely differentiable for $t > 0$.

By differentiating under the integral sign for $t > 0$, we find that

$$\frac{\partial u}{\partial t} - \frac{\partial^2 u}{\partial x^2} = f * \left(\frac{\partial}{\partial t} - \frac{\partial^2}{\partial x^2} \right) \Delta_t = 0,$$

that is, the function u satisfies the one-dimensional heat equation with the initial condition $u(x, 0) = f(x)$. This last equality should be interpreted as the limiting relation $u(x, t) \to f(x)$ as $t \to +0$, which follows from Proposition 5.

17.4.4 *Elementary Concepts Involving Distributions

a. Definition of Generalized Functions

In Sect. 17.4.1 of this section we derived the formula (17.48) on the heuristic level. This equation enabled us to determine the response of a linear transformation A to an input signal f given that we know the system function E of the device A. In determining the system function of a device we made essential use of a certain intuitive idea of a unit pulse action and the δ-function that describes it. It is clear, however, that the δ-function is really not a function in the classical sense of the term, since it must have the following properties, which contradict the classical point of view: $\delta(x) \geq 0$ on $\mathbb{R}$; $\delta(x) = 0$ for $x \neq 0$, $\int_{\mathbb{R}} \delta(x)\,dx = 1$.

The concepts connected with linear operators, convolution, the δ-function, and the system function of a device acquire a precise mathematical description in the so-called theory of generalized functions or the theory of distributions. We are now going to explain the basic principles and the elementary, but ever more widely used techniques of this theory.

Example 7 Consider a point mass m that can move along the axis and is attached to one end of an elastic spring whose other end is fixed at the origin; let k be the elastic constant of the spring. Suppose that a time-dependent force $f(t)$ begins to act on the point resting at the origin, moving it along the axis. By Newton's law,

$$m\ddot{x} + kx = f, \tag{17.57}$$

where $x(t)$ is the coordinate of the point (its displacement from its equilibrium position) at time t.

Under these conditions the function $x(t)$ is uniquely determined by the function f, and the solution $x(t)$ of the differential equation (17.57) is obviously a linear function of the right-hand side f. Thus we are dealing with the linear operator $f \xrightarrow{A} x$ inverse to the differential operator $x \xrightarrow{B} f$ (where $B = m\frac{d^2}{dt^2} + k$) that connects $x(t)$ and $f(t)$ by the relation $Bx = f$. Since the operator A obviously commutes with translations over time, it follows from (17.48) that in order to find the response $x(t)$ of this mechanical system to the function $f(t)$, it suffices to find its response to a unit pulse δ, that is, it suffices to know the so-called *fundamental solution* E of the equation

$$m\ddot{E} + kE = \delta. \tag{17.58}$$

Relation (17.58) would not raise any problems if δ actually denoted a function. However Eq. (17.58) is not yet clear. But being formally unclear is quite a different thing from being actually false. In the present case one need only explain the meaning of (17.58).

One route to such an explanation is already familiar to us: we can interpret δ as an approximate identity imitating the delta-function and consisting of classical functions $\Delta_\alpha(t)$; we interpret E as the limit to which the solution $E_\alpha(t)$ of the equation

$$m\ddot{E}_\alpha + kE_\alpha = \Delta_\alpha \tag{17.57'}$$

tends as the parameter α changes suitably.

A second approach to this problem, one that has significant advantages, is to make a fundamental enlargement of the idea of a function. It proceeds from the remark that in general objects of observation are characterized by their interaction with other ("test") objects. Thus we propose regarding a function not as a set of values at different points, but rather as an object that can act on other (test) objects in a certain manner. Let us try to make this statement, which as of now is too general, more specific.

Example 8 Let $f \in C(\mathbb{R}, \mathbb{R})$. As our test functions, we choose functions in C_0 (continuous functions of compact support on $\mathbb{R}$). A function f generates the following functional, which acts on C_0:

$$\langle f, \varphi \rangle := \int_{\mathbb{R}} f(x)\varphi(x)\,dx. \tag{17.59}$$

Using approximate identities consisting of functions of compact support, one can easily see that $\langle f, \varphi \rangle \equiv 0$ on C_0 if and only if $f(x) \equiv 0$ on $\mathbb{R}$.

Thus, each function $f \in C(\mathbb{R}, \mathbb{R})$ generates via (17.59) a linear functional $A_f : C_0 \to \mathbb{R}$ and, we emphasize, different functionals A_{f_1} and A_{f_2} correspond to different functions f_1 and f_2.

Hence formula (17.59) establishes an embedding (injective mapping) of the set of functions $C(\mathbb{R}, \mathbb{R})$ into the set $\mathcal{L}(C_0; \mathbb{R})$ of linear functionals on C_0, and consequently every function $f \in C(\mathbb{R}, \mathbb{R})$ can be interpreted as a certain functional $A_f \in \mathcal{L}(C_0; \mathbb{R})$.

If we consider the class of locally integrable functions on $\mathbb{R}$ instead of the set $C(\mathbb{R}, \mathbb{R})$ of continuous functions, we obtain by the same formula (17.59) a mapping of this set into the space $\mathcal{L}(C_0; \mathbb{R})$. Moreover $(\langle f, \varphi \rangle \equiv 0$ on $C_0) \Leftrightarrow (f(x) = 0$ at all points of continuity of f on $\mathbb{R}$, that is, $f(x) = 0$ almost everywhere on $\mathbb{R}$). Hence in this case we obtain an embedding of equivalence classes of functions into $\mathcal{L}(C_0; \mathbb{R})$ if each equivalence class contains locally integrable functions that differ only on a set of measure zero.

Thus, the locally integrable functions f on $\mathbb{R}$ (more precisely, equivalence classes of such functions) can be interpreted via (17.59) as linear functionals $A_f \in \mathcal{L}(C_0; \mathbb{R})$. The mapping $f \mapsto A_f = \langle f, \cdot \rangle$ provided by (17.59) of locally integrable functions into $\mathcal{L}(C_0; \mathbb{R})$ is not a mapping onto all of $\mathcal{L}(C_0; \mathbb{R})$. Therefore, interpreting functions as elements of $\mathcal{L}(C_0; \mathbb{R})$ (that is, as functionals) we obtain, besides the classical functions interpreted as functionals of the form (17.59), also new functions (functionals) that have no pre-image in the classical functions.

Example 9 The functional $\delta \in \mathcal{L}(C_0; \mathbb{R})$ is defined by the relation

$$\langle \delta, \varphi \rangle := \delta(\varphi) := \varphi(0), \tag{17.60}$$

which must hold for every function $\varphi \in C_0$.

We can verify (see Problem 7) that no locally integrable function f on $\mathbb{R}$ can represent the functional δ in the form (17.59).

Thus we have embedded the set of classical locally integrable functions into a larger set of linear functionals. These linear functionals are called generalized functions or distributions (a precise definition is given below). The widely used term "distribution" has its origin in physics.

Example 10 Suppose a unit mass (or unit charge) is distributed on $\mathbb{R}$. If this distribution is sufficiently regular, in the sense that it has, for example, a continuous or integrable density $\rho(x)$ on $\mathbb{R}$, the interaction of the mass M with other objects described by functions $\varphi_0 \in C_0^{(\infty)}$ can be defined as a functional

$$M(\varphi) = \int_{\mathbb{R}} \rho(x)\varphi(x)\,dx.$$

If the distribution is singular, for example, the whole mass M is concentrated at a single point, then by "smearing" the mass and interpreting the limiting point situation using an approximate identity made up of regular distributions, we find that the interaction of the mass M with the other objects mentioned above should be expressed by a formula

$$M(\varphi) = \varphi(0),$$

which shows that such a mass distribution on $\mathbb{R}$ should be identified with the δ-function (17.60) on R.

These preliminary considerations give some sense to the following general definition.

Definition 6 Let P be a vector space of functions, which will be called the space of *test functions* from now on, on which there is defined a notion of convergence.

The *space of generalized functions* or *distributions* on P is the vector space P' of continuous (real- or complex-valued) linear functionals on P. Here it is assumed that each element $f \in P$ generates a certain functional $A_f = \langle f, \cdot \rangle \in P'$ and that the mapping $f \mapsto A_f$ is a continuous embedding of P into P' if the convergence in P' is introduced as *weak ("pointwise") convergence of functionals*, that is,

$$P' \ni A_n \to A \in P' := \forall \varphi \in P \left(A_n(\varphi) \to A(\varphi) \right).$$

Let us make this definition more precise in the particular case when P is the vector space $C_0^{(\infty)}(G, \mathbb{C})$ of infinitely differentiable functions of compact support in G, where G is an arbitrary open subset of $\mathbb{R}$ (possibly $\mathbb{R}$ itself).

Definition 7 (The spaces $\mathcal{D}$ and $\mathcal{D}'$) We introduce *convergence in* $C_0^{(\infty)}(G, \mathbb{C})$ as follows: A sequence $\{\varphi_n\}$ of functions $\varphi_n \in C_0^{(\infty)}(G)$ converges to $\varphi \in C_0^{(\infty)}(G, \mathbb{C})$ if there exists a compact set $K \subset G$ that contains the supports of all the functions of the sequence $\{\varphi_n\}$ and $\varphi_n^{(m)} \rightrightarrows \varphi^{(m)}$ on K (and hence also on G) as $n \to \infty$ for all $m = 0, 1, 2, \ldots$.

The vector space obtained in this way with this convergence is usually denoted $\mathcal{D}(G)$, and when $G = \mathbb{R}$, simply $\mathcal{D}$.

We denote the space of generalized functions (distributions) corresponding to this space of basic (test) functions by $\mathcal{D}'(G)$ or $\mathcal{D}'$ respectively.

In this section and the one following we shall not consider any generalized functions other than the elements of $\mathcal{D}'(G)$ just introduced. For that reason we shall use the term distribution or generalized function to refer to elements of $\mathcal{D}'(G)$ without saying so explicitly.

Definition 8 A distribution $F \in \mathcal{D}'(G)$ *is regular* if it can be represented as

$$F(\varphi) = \int_G f(x)\varphi(x)\,\mathrm{d}x, \quad \varphi \in \mathcal{D}(G),$$

where f is a locally integrable function in G.

Nonregular distributions will be called *singular distributions* or *singular generalized functions*.

In accordance with this definition the δ-function of Example 9 is a singular generalized function.

The action of a generalized function (distribution) F on a test function φ, that is, the pairing of F and φ will be denoted, as before, by either of the equivalent expressions $F(\varphi)$ or $\langle F, \varphi \rangle$.

Before passing to the technical machinery connected with generalized functions, which was our motive for defining them, we note that the concept of a generalized function, like the majority of mathematical concepts, had a certain period of gestation, during which it developed implicitly in the work of a number of mathematicians.

Physicists, following Dirac, made active use of the δ-function as early as the late 1920s and early 1930s and operated with singular generalized functions without worrying about the absence of the necessary mathematical theory.

The idea of a generalized function was stated explicitly by S.L. Sobolev,[5] who laid the mathematical foundations of the theory of generalized functions in the mid-1930s. The current state of the machinery of the theory of distributions was largely the work of L. Schwartz.[6] What has just been said explains why, for example, the space $\mathcal{D}'$ of generalized functions is often referred to as the *Sobolev–Schwartz space of generalized functions.*

We shall now explain certain elements of the machinery of the theory of distributions. The development and extension of the use of this machinery continues even today, mainly in connection with the requirements of the theory of differential equations, the equations of mathematical physics, functional analysis, and their applications.

To simplify the notation we shall consider below only generalized functions in $\mathcal{D}'$, although all of their properties, as will be seen from their definitions and proofs, remain valid for distributions of any class $\mathcal{D}'(G)$, where G is an arbitrary open subset of $\mathbb{R}$.

Operations with distributions are defined by starting with the integral relations that are valid for classical functions, that is, for regular generalized functions.

b. Multiplication of a Distribution by a Function

If f is a locally integrable function on $\mathbb{R}$ and $g \in C^{(\infty)}$, then for any function $\varphi \in C_0^{(\infty)}$, on the one hand $g\varphi \in C_0^{(\infty)}$ and, on the other hand, we have the obvious equality

$$\int_{\mathbb{R}} (f \cdot g)(x)\varphi(x)\,\mathrm{d}x = \int_{\mathbb{R}} f(x)(g \cdot \varphi)(x)\,\mathrm{d}x$$

or, in other notation

$$\langle f \cdot g, \varphi \rangle = \langle f, g \cdot \varphi \rangle.$$

This relation, which is valid for regular generalized functions, provides the basis for the following definition of the distribution $F \cdot g$ obtained by *multiplying the distribution $F \in \mathcal{D}'$ by the function $g \in C^{(\infty)}$*:

$$\langle F \cdot g, \varphi \rangle := \langle F, g \cdot \varphi \rangle. \tag{17.61}$$

The right-hand side of Eq. (17.61) is defined, and thus defines the value of the functional $F \cdot g$ on any function $\varphi \in D$, that is, the functional $F \cdot g$ itself is defined.

---

[5]S.L. Sobolev (1908–1989) – one of the most prominent Soviet mathematicians.

[6]L. Schwartz (1915–2002) – well-known French mathematician. He was awarded the Fields medal, a prize for young mathematicians, at the International Congress of Mathematicians in 1950 for the above mentioned work.

Example 11 Let us see how the distribution $\delta \cdot g$ acts, where $g \in C^{(\infty)}$. In accordance with the definition (17.61) and the definition of δ, we obtain

$$\langle \delta \cdot g, \varphi \rangle := \langle \delta, g \cdot \varphi \rangle := (g \cdot \varphi)(0) = g(0) \cdot \varphi(0).$$

c. Differentiation of Generalized Functions

If $f \in C^{(1)}$ and $\varphi \in C_0^{(\infty)}$, integration by parts yields the equality

$$\int_{\mathbb{R}} f'(x)\varphi(x)\,\mathrm{d}x = -\int_{\mathbb{R}} f(x)\varphi'(x)\,\mathrm{d}x. \tag{17.62}$$

This equality is the point of departure for the following fundamental definition of *differentiation of a generalized function* $F \in \mathcal{D}'$:

$$\langle F', \varphi \rangle := -\langle F, \varphi' \rangle. \tag{17.63}$$

Example 12 If $f \in C^{(1)}$, the derivative of f in the classical sense equals its derivative in the distribution sense (provided, naturally, the classical function is identified with the regular generalized function corresponding to it). This follows from a comparison of relations (17.62) and (17.63), in which the right-hand sides are equal if the distribution F is generated by the function f.

Example 13 Take the *Heaviside*[7] *function*

$$H(x) = \begin{cases} 0 & \text{for } x < 0, \\ 1 & \text{for } x \geq 0, \end{cases}$$

sometimes called the *unit step*. Regarding it as a generalized function, let us find the derivative H' of this function, which is discontinuous in the classical sense.

From the definition of the regular generalized function H corresponding to the Heaviside function and relation (17.63) we find

$$\langle H', \varphi \rangle := -\langle H, \varphi' \rangle := -\int_{-\infty}^{+\infty} H(x)\varphi'(x)\,\mathrm{d}x = \int_0^{+\infty} \varphi'(x)\,\mathrm{d}x = \varphi(0),$$

since $\varphi \in C_0^{(\infty)}$. Thus $\langle H', \varphi \rangle = \langle \delta, \varphi \rangle$, for every function $\varphi \in C_0^{(\infty)}$. Hence $H' = \delta$.

Example 14 Let us compute $\langle \delta', \varphi \rangle$:

$$\langle \delta', \varphi \rangle := -\langle \delta, \varphi' \rangle = -\varphi'(0).$$

---

[7]O. Heaviside (1850–1925) – British physicist and engineer, who developed on the symbolic level the important mathematical machinery known as the *operational calculus*.

It is natural that in the theory of generalized functions, as in the theory of classical functions, the higher-order derivatives are defined by setting $F^{(n+1)} := (F^{(n)})'$.

Comparing the results of the last two examples, one can consequently write

$$\langle H'', \varphi \rangle = -\varphi'(0).$$

Example 15 Let us show that $\langle \delta^{(n)}, \varphi \rangle = (-1)^n \varphi^{(n)}(0)$.

Proof For $n = 0$ this is the definition of the δ-function.

We have seen in Example 14 that this equality holds for $n = 1$.

We now prove it by induction, assuming that it has been established for a fixed value $n \in \mathbb{N}$. Using definition (17.63), we find

$$\langle \delta^{(n+1)}, \varphi \rangle := \langle (\delta^{(n)})', \varphi \rangle := -\langle \delta^{(n)}, \varphi' \rangle =$$
$$= -(-1)^n (\varphi')^{(n)}(0) = (-1)^{n+1} \varphi^{(n+1)}(0). \qquad \square$$

Example 16 Suppose the function $f : \mathbb{R} \to \mathbb{C}$ is continuously differentiable for $x < 0$ and for $x > 0$, and suppose the one-sided limits $f(-0)$ and $f(+0)$ of the function exist at 0. We denote the quantity $f(+0) - f(-0)$, the saltus or jump of the function at 0, by $\int f(0)$, and by f' and $\{f'\}$ respectively the derivative of f in the distribution sense and the distribution defined by the function equal to the usual derivative of f for $x < 0$ and $x > 0$. At $x = 0$ this last function is not defined, but that is not important for the integral through which it defines the regular distribution $\{f'\}$.

In Example 12 we noted that if $f \in C^{(1)}$, then $f' = \{f'\}$. We shall show that in general this is not the case, but rather the following important formula holds:

$$f' = \{f'\} + \int f(0) \cdot \delta. \qquad (17.64)$$

Proof Indeed,

$$\langle f', \varphi \rangle = -\langle f, \varphi' \rangle = - \int_{-\infty}^{+\infty} f(x)\varphi'(x)\,dx =$$

$$= -\left(\int_{-\infty}^{0} + \int_{0}^{+\infty} \right) \left(f(x)\varphi'(x) \right) dx =$$

$$= -\left((f \cdot \varphi(x)) \big|_{x=-\infty}^{0} - \int_{-\infty}^{0} f'(x)\varphi(x)\,dx + (f \cdot \varphi)(x) \big|_{0}^{+\infty} - \right.$$
$$\left. - \int_{0}^{+\infty} f'(x)\varphi(x)\,dx \right) =$$

$$= \left(f(+0) - f(-0) \right)\varphi(0) + \int_{-\infty}^{+\infty} f'(x)\varphi(x)\,dx =$$

$$= \langle \int f(0) \cdot \delta, \varphi \rangle + \langle \{f'\}, \varphi \rangle. \qquad \square$$

If all derivatives up to order m of the function $f : \mathbb{R} \to \mathbb{C}$ exist on the intervals $x < 0$ and $x > 0$, and they are continuous and have one-sided limits at $x = 0$, then, repeating the reasoning used to derive (17.64), we obtain

$$f^{(m)} = \{ f^{(m)} \} + [f(0) \cdot \delta^{(m-1)} + [f'(0) \cdot \delta^{(m-2)} + \cdots$$

$$\cdots + [f^{(m-1)}(0) \cdot \delta. \tag{17.65}$$

We now exhibit some properties of the operation of differentiation of generalized functions.

Proposition 6 a) *Every generalized function $F \in \mathcal{D}'$ is infinitely differentiable.*
b) *The differentiation operation $D : \mathcal{D}' \to \mathcal{D}'$ is linear.*
c) *If $F \in \mathcal{D}'$ and $g \in C^{(\infty)}$, then $(F \cdot g) \in \mathcal{D}'$, and the Leibniz formula holds:*

$$(F \cdot g)^{(m)} = \sum_{k=0}^{m} \binom{m}{k} F^{(k)} \cdot g^{(m-k)}.$$

d) *The differentiation operation $D : \mathcal{D}' \to \mathcal{D}'$ is continuous.*
e) *If the series $\sum_{k=1}^{\infty} f_k(x) = S(x)$ formed from locally integrable functions $f_k : \mathbb{R} \to \mathbb{C}$ converges uniformly on each compact subset of $\mathbb{R}$, then it can be differentiated termwise any number of times in the sense of generalized functions, and the series so obtained will converge in $\mathcal{D}'$.*

Proof a) $\langle F^{(m)}, \varphi \rangle := -\langle F^{(m-1)}, \varphi' \rangle := (-1)^m \langle F, \varphi^{(m)} \rangle$.
b) Obvious.
c) Let us verify the formula for $m = 1$:

$$\langle (F \cdot g)', \varphi \rangle := -\langle Fg, \varphi' \rangle := -\langle F, g \cdot \varphi' \rangle = -\langle F, (g \cdot \varphi)' - g' \cdot \varphi \rangle =$$

$$= \langle F', g\varphi \rangle + \langle F, g' \cdot \varphi \rangle = \langle F' \cdot g, \varphi \rangle + \langle F \cdot g', \varphi \rangle = \langle F' \cdot g + F \cdot g', \varphi \rangle.$$

In the general case we can obtain the formula by induction.
d) Let $F_m \to F$ in $\mathcal{D}'$ as $m \to \infty$, that is, for every function $\varphi \in \mathcal{D} \langle F_m, \varphi \rangle \to \langle F, \varphi \rangle$ as $m \to \infty$. Then

$$\langle F'_m, \varphi \rangle := -\langle F_m, \varphi' \rangle \to -\langle F, \varphi' \rangle =: \langle F', \varphi \rangle.$$

e) Under these conditions the sum $S(x)$ of the series, being the uniform limit of locally integrable functions $S_m(x) = \sum_{k=1}^{m} f_k(x)$ on compact sets, is locally integrable. It remains to observe that for every function $\varphi \in \mathcal{D}$ (that is, of compact support and infinitely differentiable) we have the relation

$$\langle S_m, \varphi \rangle = \int_{\mathbb{R}} S_m(x)\varphi(x)\,dx \to \int_{\mathbb{R}} S(x)\varphi(x)\,dx = \langle S, \varphi \rangle.$$

We now conclude on the basis of what was proved in d) that $S'_m \to S'$ as $m \to \infty$. $\square$

We see that the operation of differentiation of generalized functions retains the most important properties of classical differentiation while acquiring a number of remarkable new properties that open up a great deal of freedom of operation, which did not exist in the classical case because of the presence of nondifferentiable functions there and the instability (lack of continuity) of classical differentiation under limiting processes.

d. Fundamental Solutions and Convolution

We began this subsection with intuitive ideas of the unit pulse and the system function of the device. In Example 7 we exhibited an elementary mechanical system that naturally generates a linear operator preserving time shifts. Studying it, we arrived at Eq. (17.58), which the system function E of that operator must satisfy.

We shall conclude this subsection by returning once again to these questions, but now with the goal of illustrating an adequate mathematical description in the language of generalized functions.

We begin by making sense of Eq. (17.58). On its right-hand side is the generalized function δ, so that relation (17.58) should be interpreted as equality of generalized functions. Since we know the operations of differentiating generalized functions and linear operations on distributions, it follows that the left-hand side of Eq. (17.58) is now also comprehensible, even if interpreted in the sense of generalized functions.

Let us now attempt to solve Eq. (17.58).

At times $t < 0$ the system was in a state of rest. At $t = 0$ the point received a unit pulse, thereby acquiring a velocity $v = v(0)$ such that $mv = 1$. For $t > 0$ there are no external forces acting on the system, and its law of motion $x = x(t)$ is subject to the usual differential equation

$$m\ddot{x} + kx = 0, \tag{17.66}$$

which are to be solved with the initial conditions $x(0) = 0$, $\dot{x}(0) = v = 1/m$.

Such a solution is unique and can be written out immediately:

$$x(t) = \frac{1}{\sqrt{km}} \sin \sqrt{\frac{k}{m}} t, \quad t \geq 0.$$

Since in the present case the system is at rest for $t < 0$, we can conclude that

$$E(t) = \frac{H(t)}{\sqrt{km}} \sin \sqrt{\frac{k}{m}} t, \quad t \in \mathbb{R}, \tag{17.67}$$

where H is the Heaviside function (see Example 13).

Let us now verify, using the rules for differentiating generalized functions and the results of the examples studied above, that the function $E(t)$ defined by Eq. (17.67) satisfies Eq. (17.58).

To simplify the writing we shall verify that the function

$$e(x) = H(x)\frac{\sin \omega x}{\omega} \tag{17.68}$$

satisfies (in the sense of distribution theory) the equation

$$\left(\frac{d^2}{dx^2} + \omega^2\right)e = \delta. \tag{17.69}$$

Indeed,

$$\left(\frac{d^2}{dx^2} + \omega\right)e = \frac{d^2}{dx^2}\left(H\frac{\sin \omega x}{\omega}\right) + \omega^2\left(H\frac{\sin \omega x}{\omega}\right) =$$

$$= H''\frac{\sin \omega x}{\omega} + 2H'\cos \omega x - \omega H(x)\sin \omega x +$$

$$+ \omega H(x)\sin \omega x = \delta'\frac{\sin \omega x}{\omega} + 2\delta \cos \omega x.$$

Further, for every function $\varphi \in \mathcal{D}$,

$$\left\langle \delta'\frac{\sin \omega x}{\omega} + 2\delta \cos \omega x, \varphi \right\rangle = \left\langle \delta', \frac{\sin \omega x}{\omega}\varphi \right\rangle + \langle \delta, 2(\cos \omega x)\varphi \rangle =$$

$$= -\left\langle \delta, \frac{d}{dx}\left(\frac{\sin \omega x}{\omega}\varphi\right) \right\rangle + 2\varphi(0) =$$

$$= -\left((\cos \omega x)\varphi(x) + \frac{\sin \omega x}{\omega}\varphi'(x)\right)\bigg|_{x=0} + 2\varphi(0) =$$

$$= \varphi(0) = \langle \delta, \varphi \rangle,$$

and it is thereby verified that the function (17.68) satisfies (17.69).

Finally, we introduce the following definition.

Definition 9 A *fundamental solution* or *Green's function* (*system function* or *influence function*) of the operator $A : \mathcal{D}' \to \mathcal{D}'$ is a generalized function $E \in \mathcal{D}'$ that is mapped by A to the function $\delta \in D'$, that is, $A(E) = \delta$.

Example 17 In accordance with this definition the function (17.68) is a fundamental solution for the operator $A = (\frac{d^2}{dx^2} + \omega^2)$, since it satisfies (17.69).

The function (17.67) satisfies Eq. (17.58), that is, it is a Green's function for the operator $A = (m\frac{d^2}{dt^2} + k)$. The fundamental role of the system function of a translation-invariant operator has already been discussed in Sect. 17.4.1, where formula (17.48) was obtained, on the basis of which one can now write the solution of

Eq. (17.57) corresponding to the initial conditions given in Example 7:

$$x(t) = (f * E)(t) = \int_{-\infty}^{+\infty} f(t - \tau) H(\tau) \frac{\sin\sqrt{\frac{k}{m}}\tau}{\sqrt{km}}\, d\tau, \qquad (17.70)$$

$$x(t) = \frac{1}{\sqrt{km}} \int_0^{+\infty} f(t - \tau) \sin\sqrt{\frac{k}{m}}\tau\, d\tau. \qquad (17.71)$$

When we take account of the important role of the convolution and the fundamental solution just illustrated, it becomes clear that it is desirable to define the convolution of generalized functions also. This is done in the theory of distributions, but we shall not take the time to do so. We note only that in the case of regular distributions the definition of the convolution of generalized functions is equivalent to the classical definition of the convolution of functions studied above.

17.4.5 Problems and Exercises

1. a) Verify that convolution is associative: $u * (v * w) = (u * v) * w$.

b) Suppose, as always, that $\Gamma(\alpha)$ is the Euler gamma function and $H(x)$ is the Heaviside function. We set

$$H_\lambda^\alpha(x) := H(x) \frac{x^{\alpha - 1}}{\Gamma(\alpha)} e^{\lambda x}, \quad \text{where } \alpha > 0, \text{ and } \lambda \in \mathbb{C}.$$

Show that $H_\lambda^\alpha * H_\lambda^\beta = H_\lambda^{\alpha + \beta}$.

c) Verify that the function $F = H(x) \frac{x^{n-1}}{(n-1)!} e^{\lambda x}$ is the nth convolution power of $f = H(x) e^{\lambda x}$, that is, $F = \underbrace{f * f * \cdots * f}_{n}$.

2. The function $G_\sigma(x) = \frac{1}{\sigma\sqrt{2\pi}} e^{-\frac{x^2}{2\sigma^2}}$, $\sigma > 0$, defines the probability density function for the Gaussian normal distribution.

a) Draw the graph of $G_\sigma(x)$ for different values of the parameter σ.

b) Verify that the mathematical expectation (mean value) of a random variable with the probability distribution G_σ is zero, that is, $\int_{\mathbb{R}} x G_\sigma(x)\, dx = 0$.

c) Verify that the standard deviation of x (the square root of the variance of x) is σ, that is $(\int_{\mathbb{R}} x^2 G_\sigma(x)\, dx)^{1/2} = \sigma$.

d) It is proved in probability theory that the probability density of the sum of two independent random variables is the convolution of the densities of the individual variables. Verify that $G_\alpha * G_\beta = c \sqrt{\alpha^2 + \beta^2}$.

e) Show that the sum of n independent identically distributed random variables (for example, n independent measurements of the same object), all distributed according to the normal law G_σ, is distributed according to the law $G_{\sigma\sqrt{n}}$. From this it follows in particular that the expected order of errors for the average of n such measurements when taken as the value of the measured quantity, equals $\sigma/\sqrt{n}$, where σ is the probable error of an individual measurement.

3. We recall that the function $A(x) = \sum_{n=0}^{\infty} \alpha_n x^n$ is called the *generating function of the sequence* $a_0, a_1, \ldots$.

Suppose given two sequences $\{a_k\}$ and $\{b_k\}$. If we assume that $a_k = b_k = 0$ for $k < 0$, then the convolution of the sequences $\{a_k\}$ and $\{b_k\}$ can be naturally defined as the sequence $\{c_k = \sum_m a_m b_{k-m}\}$. Show that the generating function of the convolution of two sequences equals the product of the generating functions of these sequences.

4. a) Verify that if the convolution $u * v$ is defined and one of the functions u and v is periodic with period T, then $u * v$ is also a function of period T.

b) Prove the Weierstrass theorem on approximation of a continuous periodic function by a trigonometric polynomial (see Remark 5).

c) Prove the strengthened versions of the Weierstrass approximation theorem given in Example 4.

5. a) Suppose the interior of the compact set $K \subset \mathbb{R}$ contains the closure $\overline{E}$ of the set E in Proposition 5. Show that in that case $\int_K f(y) \Delta_k(x - y)\, dy \rightrightarrows f(x)$ on E.

b) From the expansion $(1 - z)^{-1} = 1 + z + z^2 + \cdots$ deduce that $g(\rho, \theta) = \frac{1+\rho e^{i\theta}}{2(1-\rho e^{i\theta})} = \frac{1}{2} + \rho e^{i\theta} + \rho^2 e^{i2\theta} + \cdots$ for $0 \le \rho < 1$.

c) Verify that if $0 \le \rho < 1$ and

$$P_\rho(\theta) := \operatorname{Re} g(\rho, \theta) = \frac{1}{2} + \rho \cos\theta + \rho^2 \cos 2\theta + \cdots,$$

then the function $P_\rho(\theta)$ has the form

$$P_\rho(\theta) = \frac{1}{2} \frac{1 - \rho^2}{1 - 2\rho \cos\theta + \rho^2}$$

and is called the *Poisson kernel for the disk*.

d) Show that the family of functions $P_\rho(\theta)$ depending on the parameter ρ has the following set of properties: $P_\rho(\theta) \ge 0$, $\frac{1}{\pi} \int_0^{2\pi} P_\rho(\theta)\, d\theta = 1$, $\int_{\varepsilon > 0}^{2\pi - \varepsilon} P_\rho(\theta)\, d\theta \to 0$ as $\rho \to 1 - 0$.

e) Prove that if $f \in C[0, 2\pi]$, then the function

$$u(\rho, \theta) = \frac{1}{\pi} \int_0^{2\pi} P_\rho(\theta - t) f(t)\, dt$$

is a harmonic function in the disk $\rho < 1$ and $u(\rho, \theta) \rightrightarrows f(\theta)$ as $\rho \to 1 - 0$. Thus, the Poisson kernel makes it possible to construct a function harmonic in the disk having prescribed values on the boundary circle.

f) For locally integrable functions u and v that are periodic with the same period T, one can give an unambiguous definition of the convolution (convolution over a period) as follows:

$$\left(u \underset{T}{*} v\right)(x) := \int_a^{a+T} u(y)v(x-y)\,\mathrm{d}y.$$

The periodic functions on $\mathbb{R}$ can be interpreted as functions defined on the circle, so that this operation can naturally be regarded as the convolution of two functions defined on a circle.

Show that if $f(\theta)$ is a locally integrable 2π-periodic function on $\mathbb{R}$ (or, what is the same, f is a function on a circle), and the family $P_\rho(\theta)$ of functions depending on the parameter ρ has the properties of the Poisson kernel enumerated in d), then $(f \underset{2\pi}{*} P_\rho)(\theta) \to f(\theta)$ as $\rho \to 1-0$ at each point of continuity of f.

6. a) Suppose $\varphi(x) := a \exp(\frac{1}{|x|^2-1})$ for $|x| < 1$ and $\varphi(x) := 0$ for $|x| \geq 1$. Let the constant a be chosen so that $\int_{\mathbb{R}} \varphi(x)\,\mathrm{d}x = 1$. Verify that the family of functions $\varphi_\alpha(x) = \frac{1}{\alpha}\varphi(\frac{x}{\alpha})$ is an approximate identity as $\alpha \to +0$ consisting of functions in $C_0^{(\infty)}$ on $\mathbb{R}$.

b) For every interval $I \subset \mathbb{R}$ and every $\varepsilon > 0$ construct a function $e(x)$ of class $C_0^{(\infty)}$ such that $0 \leq e(x) \leq 1$ on $\mathbb{R}$, $e(x) = 1 \Leftrightarrow x \in I$, and finally, $\operatorname{supp} e \subset I_\varepsilon$, where I_ε is the ε-neighborhood (or the ε-inflation) of the set I in $\mathbb{R}$. (Verify that for a suitable value of $\alpha > 0$ one can take $e(x)$ to be $\chi_I * \varphi_\alpha$.)

c) Prove that for every $\varepsilon > 0$ there exists a countable set $\{e_k\}$ of functions $e_k \in C_0^{(\infty)}$ (an ε-partition of unity on $\mathbb{R}$) that possesses the following properties: $\forall k \in \mathbb{N}$, $\forall x \in \mathbb{R}$ $(0 \leq e_k(x) \leq 1)$; the diameter of the support $\operatorname{supp} e_k$ of every function in the family is at most $\varepsilon > 0$; every point $x \in \mathbb{R}$ belongs to only a finite number of the sets $\operatorname{supp} e_k$; $\sum_k e_k(x) \equiv 1$ on $\mathbb{R}$.

d) Show that for every open covering $\{U_\gamma, \gamma \in \Gamma\}$ of the open set $G \subset \mathbb{R}$ and every function $\varphi \in C^{(\infty)}(G)$ there exists a sequence $\{\varphi_k; k \in \mathbb{N}\}$ of functions $\varphi_k \in C_0^{(\infty)}$ that has the following properties: $\forall k \in \mathbb{N} \exists \gamma \in \Gamma \ (\operatorname{supp}\varphi_k \subset U_\gamma)$; every point $x \in G$ belongs to only a finite number of sets $\operatorname{supp}\varphi_k$; $\sum_k \varphi_k(x) = \varphi(x)$ on G.

e) Prove that the set of functions $C_0^{(\infty)}$ interpreted as generalized functions is everywhere dense in the corresponding set $C^{(\infty)}(G)$ of regular generalized functions.

f) Two generalized functions F_1 and F_2 in $\mathcal{D}'(G)$ are regarded as equal on an open set $U \subset G$ if $\langle F_1, \varphi \rangle = \langle F_2, \varphi \rangle$ for every function $\varphi \in \mathcal{D}(G)$ whose support is contained in U. Generalized functions F_1 and F_2 are regarded as locally equal at the point $x \in G$ if they are equal in some neighborhood $U(x) \subset G$ of that point. Prove that $(F_1 = F_2) \Leftrightarrow (F_1 = F_2$ locally at each point $x \in G)$.

7. a) Let $\varphi(x) := \exp(\frac{1}{|x|^2-1})$ for $|x| < 1$ and $\varphi(x) := 0$ for $|x| \geq 1$. Show that $\int_{\mathbb{R}} f(x)\varphi_\varepsilon(x)\,\mathrm{d}x \to 0$ as $\varepsilon \to +0$ for every function f that is locally integrable on $\mathbb{R}$, where $\varphi_\varepsilon(x) = \varphi(\frac{x}{\varepsilon})$.

b) Taking account of the preceding result and the fact that $\langle \delta, \varphi_\varepsilon \rangle = \varphi(0) \neq 0$, prove that the generalized function δ is not regular.

c) Show that there exists a sequence of regular generalized functions (even corresponding to functions of class $C_0^{(\infty)}$) that converges in $\mathcal{D}'$ to the generalized function δ. (In fact every generalized function is the limit of regular generalized functions corresponding to functions in $\mathcal{D} = C_0^{(\infty)}$. In this sense the regular generalized functions form an everywhere dense set in $\mathcal{D}'$, just as the rational numbers $\mathbb{Q}$ are everywhere dense in the real numbers $\mathbb{R}$.)

8. a) Compute the value $\langle F, \varphi \rangle$ of the generalized function $F \in \mathcal{D}'$ on the function $\varphi \in \mathcal{D}$ if $F = \sin x\delta$; $F = 2\cos x\delta$; $F = (1+x^2)\delta$.

b) Verify that the operation $F \to \psi F$ of multiplication by the function $\psi \in C^{(\infty)}$ is a continuous operation in $\mathcal{D}'$.

c) Verify that linear operations on generalized functions are continuous in $\mathcal{D}'$.

9. a) Show that if F is the regular distribution generated by the function $f(x) = \begin{cases} 0 \text{ for } x\le 0, \\ x \text{ for } x>0, \end{cases}$ then $F' = H$, where H is the distribution corresponding to the Heaviside function.

b) Compute the derivative of the distribution corresponding to the function $|x|$.

10. a) Verify that the following limiting passages in $\mathcal{D}'$ are correct:

$$\lim_{\alpha \to +0} \frac{\alpha}{x^2+\alpha^2} = \pi\delta; \qquad \lim_{\alpha \to +0} \frac{\alpha x}{\alpha^2+x^2} = \pi x\delta; \qquad \lim_{\alpha \to +0} \frac{x}{x^2+\alpha^2} = \ln|x|.$$

b) Show that if $f = f(x)$ is a locally integrable function on $\mathbb{R}$ and $f_\varepsilon = f(x+\varepsilon)$, then $f_\varepsilon \to f$ in $\mathcal{D}'$ as $\varepsilon \to 0$.

c) Prove that if $\{\Delta_\alpha\}$ is an approximate identity consisting of smooth functions as $\alpha \to 0$, then $F_\alpha = \int_{-\infty}^x \Delta_\alpha(t)\,dt \to H$ as $\alpha \to 0$, where H is the generalized function corresponding to the Heaviside function.

11. a) The symbol $\delta(x-a)$ usually denotes the δ-*function shifted to the point a*, that is, the generalized function acting on a function $\varphi \in \mathcal{D}$ according to the rule $\langle \delta(x-a), \varphi \rangle = \varphi(a)$. Show that the series $\sum_{k\in\mathbb{Z}} \delta(x-k)$ converges in $\mathcal{D}'$.

b) Find the derivative of the function $[x]$ (the integer part of x).

c) A 2π-periodic function on $\mathbb{R}$ is defined in the interval $]0, 2\pi]$ by the formula $f|_{]0,2\pi]}(x) = \frac{1}{2} - \frac{x}{2\pi}$. Show that $f' = -\frac{1}{2\pi} + \sum_{k\in\mathbb{Z}} \delta(x-2\pi k)$.

d) Verify that $\delta(x-\varepsilon) \to \delta(x)$ as $\varepsilon \to 0$.

e) As before, denoting the δ-function shifted to the point ε by $\delta(x-\varepsilon)$, show by direct computation that $\frac{1}{\varepsilon}(\delta(x-\varepsilon) - \delta(x)) \to -\delta'(x) = -\delta'$.

f) Starting from the preceding limiting passage, interpret $-\delta'$ as the distribution of charges corresponding to a dipole with electric moment $+1$ located at the point $x = 0$. Verify that $\langle -\delta', 1 \rangle = 0$ (the total charge of a dipole is zero) and that $\langle -\delta', x \rangle = 1$ (its moment is indeed 1).

g) An important property of the δ-function is its homogeneity: $\delta(\lambda x) = \lambda^{-1}\delta(x)$. Prove this equality.

12. a) For the generalized function F defined as $\langle F, \varphi \rangle = \int_0^\infty \sqrt{x}\varphi(x)\,dx$, verify the following equalities:

$$\langle F', \varphi \rangle = \frac{1}{2}\int_0^{+\infty} \frac{\varphi(x)}{\sqrt{x}}\,dx;$$

$$\langle F'', \varphi \rangle = -\frac{1}{4}\int_0^{+\infty} \frac{\varphi(x) - \varphi(0)}{x^{3/2}}\,dx;$$

$$\langle F''', \varphi \rangle = \frac{3}{8}\int_0^{+\infty} \frac{\varphi(x) - \varphi(0) - x\varphi'(0)}{x^{5/2}}\,dx;$$

$$\vdots$$

$$\langle F^{(n)}, \varphi \rangle = \frac{(-1)^{n-1}(2n-3)!!}{2^n} \times$$

$$\times \int_0^{+\infty} \frac{\varphi(x) - \varphi(0) - x\varphi'(0) - \cdots - \frac{x^{n-2}}{(n-2)!}\varphi^{(n-2)}(0)}{x^{\frac{2n+1}{2}}}\,dx.$$

b) Show that if $n - 1 < p < n$ and the generalized function x_+^{-p} is defined by the relation

$$\langle x_+^{-p}, \varphi \rangle := \int_0^{+\infty} \frac{\varphi(x) - \varphi(0) - x\varphi'(0) - \cdots - \frac{x^{n-2}}{(n-2)!}\varphi^{(n-2)}(0)}{x^p}\,dx.$$

Then its derivative is the function $-px_+^{-(p+1)}$ defined by the relation

$$\langle -px_+^{-(p+1)}, \varphi \rangle = -p\int_0^{+\infty} \frac{\varphi(x) - \varphi(0) - x\varphi'(0) - \cdots - \frac{x^{n-1}}{(n-1)!}\varphi^{(n-1)}(0)}{x^{p+1}}\,dx.$$

13. The generalized function defined by the equality

$$\langle F, \varphi \rangle := \mathrm{PV}\int_{-\infty}^{+\infty} \frac{\varphi(x)}{x}\,dx \left(:= \lim_{\varepsilon \to +0}\left(\int_{-\infty}^{-\varepsilon} + \int_{\varepsilon}^{+\infty}\right)\frac{\varphi(x)}{x}\,dx \right)$$

is denoted $\mathcal{P}\frac{1}{x}$. Show that

a) $\langle \mathcal{P}\frac{1}{x}, \varphi \rangle = \int_0^{+\infty} \frac{\varphi(x) - \varphi(-x)}{x}\,dx$.

b) $(\ln|x|)' = \mathcal{P}\frac{1}{x}$.

c) $\langle (\mathcal{P}\frac{1}{x})', x \rangle = \int_0^{+\infty} \frac{\varphi(x) + \varphi(-x) - 2\varphi(0)}{x^2}\,dx$.

d) $\frac{1}{x+i0} := \lim_{v \to +0}\frac{1}{x+iy} = -i\pi\delta + \mathcal{P}\frac{1}{x}$.

14. Some difficulties may arise with the definition of the multiplication of generalized functions: for example, the function $|x|^{-2/3}$ is absolutely (improperly) integrable on $\mathbb{R}$; it generates a corresponding generalized function $\int_{-\infty}^{+\infty} |x|^{-2/3}\varphi(x)\,dx$, but its square $|x|^{-4/3}$ is no longer an integrable function, even in the improper sense. The answers to the following questions show that it is theoretically impossible to define a natural associative and commutative operation of multiplication for any generalized functions.

a) Show that $f(x)\delta = f(0)\delta$ for every function $f \in C^{(\infty)}$.

b) Verify that $x\mathcal{P}\frac{1}{x} = 1$ in $\mathcal{D}'$.

c) If the operation of multiplication were extended to all pairs of generalized functions, it would at least not be associative and commutative. Otherwise,

$$0 = 0\mathcal{P}\frac{1}{x} = \left(x\delta(x)\right)\mathcal{P}\frac{1}{x} = \left(\delta(x)x\right)\mathcal{P}\frac{1}{x} = \delta(x)\left(x\mathcal{P}\frac{1}{x}\right) = \delta(x)1 = 1\delta(x) = \delta.$$

15. a) Show that a fundamental solution E for the linear operator $A : \mathcal{D}' \to \mathcal{D}'$ is in general ambiguously defined, up to any solution of the homogeneous equation $Af = 0$.

b) Consider the differential operator

$$P\left(x, \frac{d}{dx}\right) := \frac{d^n}{dx^n} + a_1(x)\frac{d^{n-1}}{dx^{n-1}} + \cdots + a_n(x).$$

Show that if $u_0 = u_0(x)$ is a solution of the equation $P(x, \frac{d}{dx})u_0 = 0$ that satisfies the initial conditions $u_0(0) = \cdots = u_0^{(n-2)}(0) = 0$ and $u_0^{(n-1)}(0) = 1$, then the function $E(x) = H(x)u_0(x)$ (where $H(X)$ is the Heaviside function) is a fundamental solution for the operator $P(x, \frac{d}{dx})$.

c) Use this method to find the fundamental solutions for the following operators:

$$\left(\frac{d}{dx} + a\right), \quad \left(\frac{d^2}{dx^2} + a^2\right), \quad \frac{d^m}{dx^m}, \quad \left(\frac{d}{dx} + a\right)^m, \quad m \in \mathbb{N}.$$

d) Using these results and the convolution, find solutions of the equations $\frac{d^m u}{dx^m} = f$, $(\frac{d}{dx} + a)^m = f$, where $f \in C(\mathbb{R}, \mathbb{R})$.

17.5 Multiple Integrals Depending on a Parameter

In the first two subsections of the present section we shall exhibit properties of proper and improper multiple integrals depending on a parameter. The total result of these subsections is that the basic properties of multiple integrals depending on a parameter do not differ essentially from the corresponding properties of one-dimensional integrals depending on a parameter studied above. In the third subsection we shall study the case of an improper integral whose singularity itself depends on a parameter, which is important in applications. Finally, in the fourth subsection we shall study the convolution of functions of several variables and some specifically multi-dimensional questions on generalized functions closely connected with integrals depending on a parameter and the classical integral formulas of analysis.

17.5.1 Proper Multiple Integrals Depending on a Parameter

Let X be a measurable subset of $\mathbb{R}^n$, for example, a bounded domain with smooth or piecewise-smooth boundary, and let Y be a subset of $\mathbb{R}^n$.

Consider the following integral depending on a parameter:

$$F(y) = \int_X f(x, y)\, dx, \tag{17.72}$$

where the function f is assumed to be defined on the set $X \times Y$ and integrable on X for each fixed value of $y \in Y$.

The following propositions hold.

Proposition 1 *If $X \times Y$ is a compact subset of $\mathbb{R}^{n+m}$ and $f \in C(X \times Y)$, then $F \in C(Y)$.*

Proposition 2 *If Y is a domain in $\mathbb{R}^m$, $f \in C(X \times Y)$, and $\frac{\partial f}{\partial y^i} \in C(X \times Y)$, then the function F is differentiable with respect to y^i in Y, where $y = (y^1, \ldots, y^i, \ldots, y^m)$ and*

$$\frac{\partial F}{\partial y^i}(y) = \int_X \frac{\partial f}{\partial y^i}(x, y)\, dx. \tag{17.73}$$

Proposition 3 *If X and Y are measurable compact subsets of $\mathbb{R}^n$ and $\mathbb{R}^m$ respectively, while $f \in C(X \times Y)$, then $F \in C(Y) \subset \mathcal{R}(Y)$, and*

$$\int_Y F(y)\, dy := \int_Y dy \int_X f(x, y)\, dx = \int_X dx \int_Y f(x, y)\, dy. \tag{17.74}$$

We note that the values of the function f here may lie in any normed vector space Z. The most important special cases occur when Z is $\mathbb{R}$, $\mathbb{C}$, $\mathbb{R}^n$, or $\mathbb{C}^n$. In these cases the verification of Propositions 1–3 obviously reduce to the case of their proof for $Z = \mathbb{R}$. But for $Z = \mathbb{R}$ the proofs of Propositions 1 and 2 are verbatim repetitions of the proof of the corresponding propositions for a one-dimensional integral (see Sect. 17.1), and Proposition 3 is a simple corollary of Proposition 1 and Fubini's theorem (Sect. 11.4).

17.5.2 Improper Multiple Integrals Depending on a Parameter

If the set $X \subset \mathbb{R}^n$ or the function $f(x, y)$ in the integral (17.72) is unbounded, it is understood as the limit of improper integrals over sets of a suitable exhaustion of X. In studying multiple improper integrals depending on a parameter, as a rule, one is interested in particular exhaustions like those that we studied in the one-dimensional case. In complete accord with the one-dimensional case, we remove the ε-neighborhood of the singularities,[8] find the integrals over the remaining parts X_ε of X and then find the limit of the values of the integrals over X_ε as $\varepsilon \to +0$.

If this limiting passage is uniform with respect to the parameter $y \in Y$, we say that the improper integral (17.72) converges uniformly on Y.

Example 1 The integral

$$F(\lambda) = \iint_{\mathbb{R}^2} e^{-\lambda(x^2+y^2)} \, dx \, dy$$

results from the limiting passage

$$\iint_{\mathbb{R}^2} e^{-\lambda(x^2+y^2)} \, dx \, dy := \lim_{\varepsilon \to +0} \iint_{x^2+y^2 \le 1/\varepsilon^2} e^{-\lambda(x^2+y^2)} \, dx \, dy$$

and, as one can easily verify using polar coordinates, it converges for $\lambda > 0$. Furthermore, it converges uniformly on the set $E_{\lambda_0} = \{\lambda \in \mathbb{R} \mid \lambda \ge \lambda_0 > 0\}$, since for $\lambda \in E_{\lambda_0}$,

$$0 < \iint_{x^2+y^2 \ge 1/\varepsilon^2} e^{-\lambda(x^2+y^2)} \, dx \, dy \le \iint_{x^2+y^2 \ge 1/\varepsilon^2} e^{-\lambda_0(x^2+y^2)} \, dx \, dy,$$

and this last integral tends to 0 as $\varepsilon \to 0$ (the original integral $F(\lambda)$ converges at $\lambda = \lambda_0 > 0$).

Example 2 Suppose, as always, that $B(a, r) = \{x \in \mathbb{R}^n \mid |x - a| < r\}$ is the ball of radius r with center at $a \in \mathbb{R}^n$, and let $y \in \mathbb{R}^n$. Consider the integral

$$F(y) = \int_{B(0,1)} \frac{|x - y|}{(1 - |x|)^\alpha} \, dx := \lim_{\varepsilon \to +0} \int_{B(0,1-\varepsilon)} \frac{|x - y|}{(1 - |x|)^\alpha} \, dx.$$

Passing to polar coordinates in $\mathbb{R}^n$, we verify that this integral converges only for $\alpha < 1$. If the value $\alpha < 1$ is fixed, the integral converges uniformly with respect to the parameter y on every compact set $Y \subset \mathbb{R}^n$, since $|x - y| \le M(Y) \in \mathbb{R}$ in that case.

&

We note that in these examples the set of singularities of the integral was independent of the parameter. Thus, if we adopt the concept given above of uniform convergence of an improper integral with a fixed set of singularities, it is clear that all the basic properties of such improper multiple integrals depending on a parameter can be obtained from the corresponding properties of proper multiple integrals and theorems on passage to the limit for families of functions depending on a parameter.

We shall not take the time to explain these facts again, which are theoretically already familiar to us, preferring instead to use the machinery we have developed

---

[8]That is, the points in every neighborhood of which the function f is unbounded. If the set X is also unbounded, we remove a neighborhood of infinity from it.

to study the following very important and frequently encountered situation in which the singularity of an improper integral (one-dimensional or multi-dimensional) itself depends on a parameter.

17.5.3 Improper Integrals with a Variable Singularity

Example 3 As is known, the potential of a unit charge located at the point $x \in \mathbb{R}^3$ is expressed by the formula $U(x, y) = \frac{1}{|x-y|}$, where y is a variable point of $\mathbb{R}^3$. If the charge is now distributed in a bounded region $X \subset \mathbb{R}^3$ with a bounded density $\mu(x)$ (equal to zero outside X), the potential of a charge distributed in this way can be written (by virtue of the additivity of potential) as

$$U(y) = \int_{\mathbb{R}^3} U(x, y)\mu(x)\,dx = \int_X \frac{\mu(x)\,dx}{|x - y|}. \tag{17.75}$$

The role of the parameter in this last integral is played by the variable point $y \in \mathbb{R}^3$. If the point y lies in the exterior of the set X, the integral (17.75) is a proper integral; but if $y \in \overline{X}$, then $|x - y| \to 0$ as $X \ni x \to y$, and y becomes a singularity of the integral. As y varies, this singularity thus moves.

Since $U(y) = \lim_{\varepsilon \to +0} U_\varepsilon(y)$, where

$$U_\varepsilon(y) = \int_{X \setminus B(y,\varepsilon)} \frac{\mu(x)}{|x - y|}\,dx,$$

it is natural to consider, as before, that the integral (17.75) with a variable singularity converges uniformly on the set Y if $U_\varepsilon(y) \rightrightarrows U(y)$ on Y as $\varepsilon \to +0$.

We have assumed that $|\mu(x)| \le M \in \mathbb{R}$ on X, and therefore

$$\left| \int_{X \cap B(y,\varepsilon)} \frac{\mu(x)\,dx}{|x - y|} \right| \le M \int_{B(y,\varepsilon)} \frac{dx}{|x - y|} = 2\pi M \varepsilon^2.$$

This estimate shows that $|U(y) - U_\varepsilon(y)| \le 2\pi M \varepsilon^2$ for every $y \in \mathbb{R}^3$, that is, the integral (17.75) converges uniformly on the set $Y = \mathbb{R}^3$.

In particular, if we verify that the function $U_\varepsilon(y)$ is continuous with respect to y, we will then be able to deduce from general considerations that the potential $U(y)$ is continuous. But the continuity of $U_\varepsilon(y)$ does not follow formally from Proposition 1 on the continuity of an improper integral depending on a parameter, since in the present case the domain of integration $X \setminus B(y, \varepsilon)$ changes when y changes. For that reason, we need to examine the question of the continuity of $U_\varepsilon(y)$ more closely.

We remark that for $|y - y_0| \le \varepsilon$,

$$U_\varepsilon(y) = \int_{X \setminus B(y_0, 2\varepsilon)} \frac{\mu(x)\,dx}{|x - y|} + \int_{(X \setminus B(y,\varepsilon)) \cap B(y_0, 2\varepsilon)} \frac{\mu(x)\,dx}{|x - y|}.$$

The first of these two integrals is continuous with respect to y assuming $|y - y_0| < \varepsilon$, being a proper integral with a fixed domain of integration. The absolute value of the second does not exceed

$$\int_{B(y_0, 2\varepsilon)} \frac{M \, dx}{|x - y|} = 8\pi M \varepsilon^2.$$

Hence the inequality $|U_\varepsilon(y) - U_\varepsilon(y_0)| < \varepsilon + 16\pi M \varepsilon^2$ holds for all values of y sufficiently close to y_0, which establishes that $U_\varepsilon(y)$ is continuous at the point $y_0 \in \mathbb{R}^3$.

Thus we have shown that the potential $U(y)$ is a continuous function in the whole space $\mathbb{R}^3$.

These examples provide the basis for adopting the following definition.

Definition 1 Suppose the integral (17.72) is an improper integral that converges for each $y \in Y$. Let X_ε be the portion of the set X obtained by removing from X the ε-neighborhood of the set of singularities of the integral,[9] and let $F_\varepsilon(y) = \int_{X_\varepsilon} f(x, y) \, dx$. We shall say that *the integral* (17.72) *converges uniformly on the set* Y if $F_\varepsilon(y) \rightrightarrows F(y)$ on Y as $\varepsilon \to +0$.

The following useful proposition is an immediate consequence of this definition and considerations similar to those illustrated in Example 3.

Proposition 4 *If the function* f *in the integral* (17.72) *admits the estimate* $|f(x, y)| \le \frac{M}{|x-y|^\alpha}$, *where* $M \in \mathbb{R}$, $x \in X \subset \mathbb{R}^n$, $y \in Y \subset \mathbb{R}^n$, *and* $\alpha < n$, *then the integral converges uniformly on* Y.

Example 4 In particular, we conclude on the basis of Proposition 4 that the integral

$$V_i(y) = \int_X \frac{\mu(x)(x^i - y^i)}{|x - y|^3} \, dx,$$

obtained by formal differentiation of the potential (17.75) with respect to the variable y^i ($i = 1, 2, 3$) converges uniformly on $Y = \mathbb{R}^3$, since $|\frac{\mu(x)(x^i - y^i)}{|x-y|^3}| \le \frac{M}{|x-y|^2}$.

As in Example 3, it follows from this that the function $V_i(y)$ is continuous on $\mathbb{R}^3$.

Let us now verify that the function $U(y)$ – the potential (17.75) – really does have a partial derivative $\frac{\partial U}{\partial y^i}$ and that $\frac{\partial U}{\partial y^i}(y) = V_i(y)$.

To do this it obviously suffices to verify that

$$\int_a^b V_i(y^1, y^2, y^3) \, dy^i = U(y^1, y^2, y^3)\big|_{y^i = a}^b.$$

---

[9]See the footnote on p. 473.

But in fact,

$$\int_a^b V_i(y)\,dy^i = \int_a^b dy^i \int_X \frac{\mu(x)(x^i - y^i)}{|x - y|^3}\,dx =$$

$$= \int_X \mu(x)\,dx \int_a^b \frac{(x^i - y^i)}{|x - y|^3}\,dy^i =$$

$$= \int_X \mu(x)\,dx \int_a^b \frac{\partial}{\partial y^i}\left(\frac{1}{|x - y|}\right)dy^i =$$

$$= \left(\int_X \frac{\mu(x)\,dx}{|x - y|}\right)\Bigg|_{y^i = a}^{b} = U(y)\Big|_{y^i = a}^{b}.$$

The only nontrivial point in this computation is the reversal of the order of integration. In general, in order to reverse the order of improper integrals, it suffices to have a multiple integral that converges absolutely with respect to the whole set of variables. This condition holds in the present case, so that the interchange is justified. Of course, it could also be justified directly due to the simplicity of the function involved.

Thus, we have shown that the potential $U(y)$ generated by a charge distributed in $\mathbb{R}^3$ with a bounded density is continuously differentiable in the entire space.

The techniques and reasoning used in Examples 3 and 4 enable us to study the following more general situation in a very similar way.

Let

$$F(y) = \int_X K\big(y - \varphi(x)\big)\psi(x, y)\,dx, \tag{17.76}$$

where X is a bounded measurable domain in $\mathbb{R}^n$, the parameter y ranges over the domain $Y \subset \mathbb{R}^m$, with $n \leq m$, $\varphi : X \to \mathbb{R}^m$ is a smooth mapping satisfying $\operatorname{rank}\varphi'(x) = n$, and $\|\varphi'(x)\| \geq c > 0$, that is, φ defines an n-dimensional parametrized surface, or, more precisely, an n-path in $\mathbb{R}^m$. Here $K \in C(\mathbb{R}^m \setminus 0, \mathbb{R})$, that is, the function $K(z)$ is continuous everywhere in $\mathbb{R}^m$ except at $z = 0$, near which it may be unbounded; and $\psi : X \times Y \to \mathbb{R}$ is a bounded continuous function. We shall assume that for each $y \in Y$ the integral (17.76) (which in general is an improper integral) exists.

In the integral (17.75) that we considered above, in particular, we had

$$n = m, \qquad \varphi(x) = x, \qquad \psi(x, y) = \mu(x), \qquad K(z) = |z|^{-1}.$$

It is not difficult to verify that under these restrictions on the function φ, Definition 1 of uniform convergence of the integral (17.76) means that for every $\alpha > 0$ one can choose $\varepsilon > 0$ such that

$$\left|\int_{|y - \varphi(x)| < \varepsilon} K\big(y - \varphi(x)\big)\psi(x, y)\,dx\right| < \alpha, \tag{17.77}$$

where the integral is taken over the set[10] $\{x \in X \mid |y - \varphi(x)| < \varepsilon\}$.

The following propositions hold for the integral (17.76).

Proposition 5 *If the integral* (17.76) *converges uniformly on Y under the hypotheses described above on the functions φ, ψ, and K, then $F \in C(Y, \mathbb{R})$.*

Proposition 6 *If it is known in addition that the function ψ in the integral* (17.76) *is independent of the parameter y (that is, $\psi(x, y) = \psi(x)$) and $K \in C^{(1)}(\mathbb{R}^m \backslash 0, \mathbb{R})$, then if the integral*

$$\int_X \frac{\partial K}{\partial y^i}(y - \varphi(x))\psi(x)\,dx$$

converges uniformly on the set $y \in Y$, one can say that the function F has a continuous partial derivative $\frac{\partial F}{\partial y^i}$, and

$$\frac{\partial F}{\partial y^i}(y) = \int_X \frac{\partial K}{\partial y^i}(y - \varphi(x))\psi(x)\,dx. \tag{17.78}$$

The proofs of these propositions, as stated, are completely analogous to those in Examples 3 and 4, and so we shall not take the time to give them.

We note only that the convergence of an improper integral (under an arbitrary exhaustion) implies its absolute convergence. In Examples 3 and 4 the hypothesis of absolute convergence was used in the estimates and in reversing the order of integration. As an illustration of the possible uses of Propositions 5 and 6, let us consider another example from potential theory.

Example 5 Suppose a charge is distributed on a smooth compact surface $S \subset \mathbb{R}^3$ with surface density $\nu(x)$. The potential of such a charge distribution is called a *single-layer potential* and is obviously represented by the surface integral

$$U(y) = \int_S \frac{\nu(x)\,d\sigma(x)}{|x - y|}. \tag{17.79}$$

Suppose ν is a bounded function. Then for $y \notin S$ this integral is proper, and the function $U(y)$ is infinitely differentiable outside S.

But if $y \in S$, the integral has an integrable singularity at the point y. The singularity is integrable because the surface S is smooth and differs by little from a piece of the plane $\mathbb{R}^2$ near the point $y \in S$; and we know that a singularity of type $1/r^\alpha$ is integrable in the plane if $\alpha < 2$. Using Proposition 5, we can turn this general consideration into a formal proof. If we represent S locally in a neighborhood V_y of

---

[10]Here we are assuming that the set X itself is bounded in $\mathbb{R}^n$. Otherwise one must supplement inequality (17.77) with the analogous inequality in which the integral is taken over the set $\{x \in X \mid |x| > 1/\varepsilon\}$.

the point $y \in S$ in the form $x = \varphi(t)$, where $t \in V_t \subset \mathbb{R}^2$ and $\operatorname{rank} \varphi' = 2$, then

$$\int_{V_y} \frac{v(x)\,d\sigma(x)}{|x - y|} = \int_{V_t} \frac{v(\varphi(t))}{|y - \varphi(t)|} \sqrt{\det\left\langle \frac{\partial\varphi}{\partial t^i}, \frac{\partial\varphi}{\partial t^j} \right\rangle}\,dt,$$

and, applying Proposition 2, we also verify that the integral (17.79) represents a function $U(y)$ that is continuous on the entire space $\mathbb{R}^3$.

Outside the support of the charge, as already noted, the three-dimensional potential (17.75) and the single-layer potential (17.79) are infinitely differentiable. Carrying out this differentiation under the integral sign, we verify in a unified manner that outside the support of the charge the potential, like the function $1/|x - y|$, satisfies Laplace's equation $\Delta U = 0$ in $\mathbb{R}^3$, that is, it is a harmonic function in this domain.

17.5.4 *Convolution, the Fundamental Solution, and Generalized Functions in the Multidimensional Case

a. Convolution in $\mathbb{R}^n$

Definition 2 The *convolution* $u * v$ of real- or complex-valued functions u and v defined in $\mathbb{R}^n$ is defined by the relation

$$(u * v)(x) := \int_{\mathbb{R}_n} u(y)v(x - y)\,dy. \tag{17.80}$$

Example 6 Comparing formulas (17.75) and (17.80), we can conclude, for example, that the potential U of a charge distributed in $\mathbb{R}^3$ with density $\mu(x)$ is the convolution $(\mu * E)$ of the function μ and the potential E of a unit charge located at the origin of $\mathbb{R}^3$.

Relation (17.80) is a direct generalization of the definition of convolution given in Sect. 17.4. For that reason, all the properties of the convolution considered in Sect. 17.4 for the case $n = 1$ and their proofs remain valid if $\mathbb{R}$ is replaced by $\mathbb{R}^n$.

An approximate identity in $\mathbb{R}^n$ is defined just as in $\mathbb{R}$ with $\mathbb{R}$ replaced by $\mathbb{R}^n$ and $U(0)$ understood to be a neighborhood of the point $0 \in \mathbb{R}^n$ in $\mathbb{R}^n$.

The concept of uniform continuity of a function $f : G \to \mathbb{C}$ on a set $E \subset G$, and with it the basic Proposition 5 of Sect. 17.4 on convergence of the convolution $f * \Delta_\alpha$ to f, also carry over in all its details to the multi-dimensional case.

We note only that in Example 3 and in the proof of Corollary 1 of Sect. 17.4 x must be replaced by $|x|$ in the definition of the functions $\Delta_n(x)$ and $\varphi(x)$. Only minor changes are needed in the approximate identity given in Example 4 of Sect. 17.4 for the proof of the Weierstrass theorem on approximation of periodic functions by trigonometric polynomials. In this case it is a question of approximating a function $f(x^1, \dots, x^n)$ that is continuous and periodic with periods $T_1, T_2, \dots, T_n$ respectively in the variables $x^1, x^2, \dots, x^n$.

The assertion amounts to the statement that for every $\varepsilon > 0$ one can exhibit a trigonometric polynomial in n variables with the respective periods $T_1, T_2, \ldots, T_n$ that approximates f on $\mathbb{R}^n$ within ε.

We confine ourselves to these remarks. An independent verification of the properties of the convolution (17.80) for $n \in \mathbb{N}$, which were proved for the case $n = 1$ in Sect. 17.4, will be an easy but useful exercise for the reader, helping to promote an adequate understanding of what was said in Sect. 17.4.

b. Generalized Functions of Several Variables

We now take up certain multi-dimensional aspects of the concepts connected with generalized functions, which were introduced in Sect. 17.4.

As before, let $C^{(\infty)}(G)$ and $C_0^{(\infty)}(G)$ denote respectively the sets of infinitely differentiable functions in the domain $G \subset \mathbb{R}^n$ and the set of infinitely differentiable functions of compact support in G. If $G = \mathbb{R}^n$, we shall use the respective abbreviations $C^{(\infty)}$ and $C_0^{(\infty)}$. Let $m := (m_1, \ldots, m_n)$ be a multi-index and

$$\varphi^{(m)} := \left(\frac{\partial}{\partial x^1} \right)^{m_1} \cdot \ldots \cdot \left(\frac{\partial}{\partial x^n} \right)^{m_n} \varphi.$$

In $C_0^{(\infty)}(G)$ we introduce *convergence of functions*. As in Definition 7 of Sect. 17.4, we consider that $\varphi_k \to \varphi$ in $C_0^{(\infty)}(G)$ as $k \to \infty$ if the supports of all the functions of the sequence $\{\varphi_k\}$ are contained in one compact subset of G and $\varphi_k^{(m)} \rightrightarrows \varphi^{(m)}$ on G for every multi-index m as $k \to \infty$, that is, the functions converge uniformly, and so do all of their partial derivatives.

Given this, we adopt the following definition.

Definition 3 The vector space $C_0^{(\infty)}(G)$ with this convergence is denoted $\mathcal{D}(G)$ (and simply $\mathcal{D}$ if $G = \mathbb{R}^n$) and is called the space of *fundamental* or *test functions*.

Continuous linear functionals on $\mathcal{D}(G)$ are called *generalized functions* or *distributions*. They form the *vector space of generalized functions*, denoted $\mathcal{D}'(G)$ (or $\mathcal{D}'$ when $G = \mathbb{R}^n$).

Convergence in $\mathcal{D}'(G)$, as in the one-dimensional case, is defined as weak (pointwise) convergence of functionals (see Definition 6 of Sect. 17.4).

The definition of a regular generalized function carries over verbatim to the multi-dimensional case.

The definition of the δ-function and the δ-function shifted to the point $x_0 \in G$ (denoted $\delta(x_0)$, or more often, but not always happily, $\delta(x - x_0)$) also remain the same.

Now let us consider some examples.

Example 7 Set

$$\Delta_t(x) := \frac{1}{(2a\sqrt{\pi t})^n} e^{-\frac{|x|^2}{4a^2 t}},$$

where $a > 0$, $t > 0$, $x \in \mathbb{R}^n$. We shall show that these functions, regarded as regular distributions in $\mathbb{R}^n$, converge to the δ-function on $\mathbb{R}^n$ as $t \to +0$.

For the proof it suffices to verify that the family of functions Δ_t is an approximate identity in $\mathbb{R}^n$ as $t \to +0$.

Using a change of variable, reduction of the multiple integral to an iterated integral, and the value of the Euler–Poisson integral, we find

$$\int_{\mathbb{R}^n} \Delta_t(x)\,dx = \frac{1}{(\sqrt{\pi})^n} \int_{\mathbb{R}^n} e^{-\left|\frac{x}{2a\sqrt{t}}\right|^2} d\left(\frac{x}{2a\sqrt{t}}\right) = \frac{1}{(\sqrt{\pi})^n}\left(\int_{-\infty}^{+\infty} e^{-u^2}\,du\right)^n = 1.$$

Next, for any fixed value of $r > 0$ we have

$$\int_{B(0,r)} \Delta_t(x)\,dx = \frac{1}{(\sqrt{\pi})^n} \int_{B(0,\frac{r}{2a\sqrt{t}})} e^{-|\xi|^2}\,d\xi \to 1,$$

as $t \to +0$.

Finally, taking account of the fact that $\Delta_t(x)$ is nonnegative, we conclude that these functions indeed constitute an approximate identity in $\mathbb{R}^n$.

Example 8 A generalization of the δ-function (corresponding, for example, to a unit charge located at the origin in $\mathbb{R}^n$) is the following generalized function δ_S (corresponding to a distribution of charge over a piecewise-smooth surface S with a distribution of unit surface density). The effect of δ_S on the function $\varphi \in \mathcal{D}$ is defined by the relation

$$\langle \delta_S, \varphi \rangle := \int_S \varphi(x)\,d\sigma.$$

Like the distribution δ, the distribution δ_S is not a regular generalized function.

Multiplication of a distribution by a function in $\mathcal{D}$ is defined in $\mathbb{R}^n$ just as in the one-dimensional case.

Example 9 If $\mu \in D$, then $\mu\delta_S$ is a generalized function acting according to the rule

$$\langle \mu\delta_S, \varphi \rangle = \int_S \varphi(x)\mu(x)\,d\sigma. \tag{17.81}$$

If the function $\mu(x)$ were defined only on the surface S, Eq. (17.81) could be regarded as the definition of the generalized function $\mu\delta_S$. By natural analogy, the generalized function introduced in this way is called a *single layer on the surface S with density μ*.

Differentiation of generalized functions in the multi-dimensional case is defined by the same principle as in the one-dimensional case, but has a few peculiarities.

If $F \in \mathcal{D}'(G)$ and $G \subset \mathbb{R}^n$, the generalized function $\frac{\partial F}{\partial x^i}$ is defined by the relation

$$\left\langle \frac{\partial F}{\partial x^i}, \varphi \right\rangle := -\left\langle F, \frac{\partial \varphi}{\partial x^i} \right\rangle.$$

It follows that

$$\left\langle F^{(m)}, \varphi \right\rangle = (-1)^{|m|} \left\langle F, \varphi^{(m)} \right\rangle, \tag{17.82}$$

where $m = (m_1, \ldots, m_k)$ is a multi-index and $|m| = \sum_{i=1}^{n} m_i$.

It is natural to verify the relation $\frac{\partial^2 F}{\partial x^i \partial x^j} = \frac{\partial^2 F}{\partial x^j \partial x^i}$. But that follows from the equality of the right-hand sides in the relations

$$\left\langle \frac{\partial^2 F}{\partial x^i \partial x^j}, \varphi \right\rangle = \left\langle F, \frac{\partial^2 \varphi}{\partial x^j \partial x^i} \right\rangle,$$

$$\left\langle \frac{\partial^2 F}{\partial x^j \partial x^i}, \varphi \right\rangle = \left\langle F, \frac{\partial^2 \varphi}{\partial x^i \partial x^j} \right\rangle,$$

which follows from the classical equality $\frac{\partial^2 \varphi}{\partial x^i \partial x^j} = \frac{\partial^2 \varphi}{\partial x^j \partial x^i}$, which holds for every function $\varphi \in \mathcal{D}$.

Example 10 Now consider an operator $D = \sum_m a_m D^m$, where $m = (m_1, \ldots, m_n)$ is a multi-index, $D^m = (\frac{\partial}{\partial x^1})^{m_1} \cdot \ldots \cdot (\frac{\partial}{\partial x^n})^{m_n}$, a_m are numerical coefficients, and the sum extends over a finite set of multi-indices. This is a differential operator.

The *transpose* or *adjoint* of D is the operator usually denoted $^t D$ or D^* and defined by the relation

$$\langle DF, \varphi \rangle =: \left\langle F, {}^t D\varphi \right\rangle,$$

which must hold for all $\varphi \in \mathcal{D}$ and $F \in \mathcal{D}'$. Starting from Eq. (17.82), we can now write the explicit formula

$$^t D = \sum_m (-1)^{|m|} a_m D^m$$

for the adjoint of the differential operator D.

In particular, if all the values of $|m|$ are even, the operator D is *self-adjoint*, that is, $^t D = D$.

It is clear that the operation of differentiation in $\mathcal{D}'(\mathbb{R}^n)$ preserves all the properties of differentiation in $\mathcal{D}'(\mathbb{R})$. However, let us consider the following important example, which is specific to the multi-dimensional case.

Example 11 Let S be a smooth $(n-1)$-dimensional submanifold of $\mathbb{R}^n$, that is, S is a smooth hypersurface. Assume that the function f defined on $\mathbb{R}^n \setminus S$ is infinitely differentiable and that all its partial derivatives have a limit at every point $x \in S$ under one-sided approach to x from either (local) side of the surface S.

The difference between these two limits will be the jump $\int \frac{\partial f}{\partial x^i}$ of the partial derivative under consideration at the point x corresponding to a particular direction of passage across the surface S at x. The sign of the jump changes if that direction is reversed. The jump can thus be regarded as a function defined on an oriented surface if, for example, we make the convention that the direction of passage is given by an orienting normal to the surface.

The function $\frac{\partial f}{\partial x^i}$ is defined, continuous, and locally bounded outside S, and by the assumptions just made f is locally ultimately bounded upon approach to the surface S itself. Since S is a submanifold of $\mathbb{R}^n$, no matter how we complete the definition of $\frac{\partial f}{\partial x^i}$ on S, we obtain a function with possible discontinuities on only S, and hence locally integrable in $\mathbb{R}^n$. But integrable functions that differ on a set of measure zero have equal integrals, and therefore, without worrying about the values on S, we may assume that $\frac{\partial f}{\partial x^i}$ generates some regular generalized function $\{\frac{\partial f}{\partial x^i}\}$ according to the rule

$$\left\langle \left\{ \frac{\partial f}{\partial x^i} \right\}, \varphi \right\rangle = \int_{\mathbb{R}^n} \left(\frac{\partial f}{\partial x^i} \cdot \varphi \right)(x)\, dx.$$

We shall now show that if f is regarded as a generalized function, then the following important formula holds in the sense of differentiation of generalized functions:

$$\frac{\partial f}{\partial x^i} = \left\{ \frac{\partial f}{\partial x^i} \right\} + (\int f)_S \cos \alpha_i \delta_S, \tag{17.83}$$

where the last term is understood in the sense of Eq. (17.81), $(\int f)_S$ is the jump of the function f at $x \in S$ corresponding to either of the two possible directions of the unit normal $\mathbf{n}$ to S at x, and $\cos \alpha_i$ is the projection of $\mathbf{n}$ onto the x^i-axis (that is, $\mathbf{n} = (\cos \alpha_1, \ldots, \cos \alpha_k)$).

Proof Formula (17.83) generalizes Eq. (17.64), which we use to derive it.

For definiteness we consider the case $i = 1$. Then

$$\left\langle \frac{\partial f}{\partial x^1}, \varphi \right\rangle := -\left\langle f, \frac{\partial \varphi}{\partial x^1} \right\rangle = -\int_{\mathbb{R}^n} \left(f \cdot \frac{\partial \varphi}{\partial x^1} \right)(x)\, dx =$$

$$= -\int \cdots \int_{x^2 \ldots x^n} dx^2 \cdots dx^n \int_{-\infty}^{+\infty} f \frac{\partial \varphi}{\partial x^1}\, dx^1 =$$

$$= \int \cdots \int_{x^2 \ldots x^n} dx^2 \cdots dx^n \left[(\int f)\varphi + \int_{-\infty}^{+\infty} \frac{\partial f}{\partial x^1} \varphi\, dx^1 \right] =$$

$$= \int_{\mathbb{R}_n} \frac{\partial f}{\partial x^1} \varphi\, dx + \int \cdots \int_{x^2 \ldots x^n} (\int f)\varphi\, dx^2 \cdots dx^n.$$

Here the jump $\int f$ of f is taken at the point $x = (x^1, x^2, \ldots, x^n) \in S$ as one passes through the surface at that point in the direction of the positive x_1-axis. The value of the function φ in computing the product $(\int f)\varphi$ is taken at the same point. Hence, this last integral can be written as a surface integral of first kind

$$\int_S (\int f)\varphi \cos\alpha_1 \, d\sigma,$$

where α_1 is the angle between the direction of the positive x_1-axis and the normal to S at x, direct so that in passing through x in the direction of that normal the function f has precisely the jump $\int f$. This means only that $\cos\alpha_1 \geq 0$. It remains only to remark that if we choose the other direction for the normal, the sign of the jump and the sign of the cosine would both reverse simultaneously; hence the product $(\int f)\cos\alpha_1$ does not change. $\qquad\square$

Remark 1 As can be seen from this proof, formula (17.83) holds once the jump $(\int f)_S$ of f is defined at each point $x \in S$, and a locally integrable partial derivative $\frac{\partial f}{\partial x^j}$ exists outside of S in $\mathbb{R}^n$, perhaps as an improper integral generating a regular generalized function $\{\frac{\partial f}{\partial\partial x^j}\}$.

Remark 2 At points $x \in S$ at which the direction of the x^1-axis is not transversal to S, that is, it is tangent to S, difficulties may arise in the definition of the jump $\int f$ in the given direction. But it can be seen from (17.83) that its last term is obtained from the integral

$$\int \cdots \int_{x^2 \cdots x^n} (\int f)\varphi \, dx^2 \cdots dx^n.$$

The projections of the set E on $x^2, \ldots, x^n$-hyperplane has $(n-1)$-dimensional measure zero and therefore has no effect on the value of the integral. Hence we can regard the form (17.83) as having meaning and being valid always if $(\int f)_S \cos\alpha_i$ is given the value 0 when $\cos\alpha_i = 0$.

Remark 3 Similar considerations make it possible to neglect sets of area zero; therefore one can regard formula (17.83) as proved for piecewise-smooth surfaces.

As our next example we shall show how the classical Gauss–Ostrogradskii formula can be obtained directly from the differential relation (17.83), and in a form that is maximally free of the extra analytic requirements that we informed the reader of previously.

Example 12 Let G be a finite domain in $\mathbb{R}^n$ bounded by a piecewise-smooth surface S. Let $\mathbf{A} = (A^1, \ldots, A^n)$ be a vector field that is continuous in $\overline{G}$ and such that the function $\operatorname{div}\mathbf{A} = \sum_{i=1}^n \frac{\partial A^i}{\partial x^i}$ is defined in G and integrable on G, possibly in the improper sense.

If we regard the field A as zero outside $\overline{G}$, then the jump of this field at each point x of the boundary S of the domain G when leaving G is $-\mathbf{A}(x)$. Assuming that $\mathbf{n}$ is a unit outward normal vector to S, applying formula (17.83) to each component A^i of the field $\mathbf{A}$ and summing these equalities, we arrive at the relation

$$\operatorname{div} \mathbf{A} = \{\operatorname{div} \mathbf{A}\} - (\mathbf{A} \cdot \mathbf{n}) \delta_S, \tag{17.84}$$

in which $\mathbf{A} \cdot \mathbf{n}$ is the inner product of the vectors A and $\mathbf{n}$ at the corresponding point $x \in S$.

Relation (17.84) is equality of generalized functions. Let us apply it to the function $\psi \in C_0^{(\infty)}$ equal to 1 on G (the existence and construction of such a function has been discussed more than once previously). Since for every function $\varphi \in \mathcal{D}$

$$\langle \operatorname{div} \mathbf{A}, \varphi \rangle = - \int_{\mathbb{R}^n} (\mathbf{A} \cdot \nabla \varphi) \, dx \tag{17.85}$$

(which follows immediately from the definition of the derivative of a generalized function), for the field $\mathbf{A}$ and the function ψ we obviously have $\langle \operatorname{div} \mathbf{A}, \psi \rangle = 0$. But, when we take account of Eq. (17.84) this gives the relation

$$0 = \langle \{\operatorname{div} \mathbf{A}\}, \psi \rangle - \langle (\mathbf{A} \cdot \mathbf{n}) \delta_S, \psi \rangle,$$

which in classical notation

$$0 = \int_G \operatorname{div} \mathbf{A} \, dx - \int_S (\mathbf{A} \cdot \mathbf{n}) \, d\sigma \tag{17.86}$$

is the same as the Gauss–Ostrogradskii formula.

Let us now consider several important examples connected with differentiation of generalized functions.

Example 13 We consider the vector field $\mathbf{A} = \frac{x}{|x|^3}$ defined in $\mathbb{R}^3 \setminus 0$ and show that in the space $\mathcal{D}'(\mathbb{R}^3)$ of generalized functions we have the equality

$$\operatorname{div} \frac{x}{|x|^3} = 4\pi \delta. \tag{17.87}$$

We remark first that for $x \neq 0$ we have $\operatorname{div} \frac{x}{|x|^3} = 0$ in the classical sense.

Now, using successively the definition of $\operatorname{div} \mathbf{A}$ in the form (17.85), the definition of an improper integral, the equality $\operatorname{div} \frac{x}{|x|^3} = 0$ for $x \neq 0$, the Gauss–Ostrogradskii formula (17.86), and the fact that φ has compact support, we obtain

$$\left\langle \operatorname{div} \frac{x}{|x|^3}, \varphi \right\rangle = - \int_{\mathbb{R}^3} \left(\frac{x}{|x|^3} \cdot \nabla \varphi(x) \right) dx =$$

$$= \lim_{\varepsilon \to +0} - \int_{\varepsilon < |x| < 1/\varepsilon} \left(\frac{x}{|x|^3} \cdot \nabla \varphi(x) \right) dx =$$

$$= \lim_{\varepsilon \to +0} - \int_{\varepsilon < |x| < 1/\varepsilon} \text{div}\left(\frac{x\varphi(x)}{|x|^3}\right) dx =$$

$$= \lim_{\varepsilon \to +0} - \int_{|x|=\varepsilon} \varphi(x)\frac{(x \cdot n)}{|x|^3} \, d\sigma = 4\pi\varphi(0) = \langle 4\pi\delta, \varphi \rangle.$$

For the operator $A : \mathcal{D}'(G) \to \mathcal{D}'(g)$, as before, we define a *fundamental solution* to be a generalized function $E \subset \mathcal{D}'(G)$ for which $A(E) = \delta$.

Example 14 We verify that the regular generalized function $E(x) = -\frac{1}{4\pi|x|}$ in $\mathcal{D}'(\mathbb{R}^3)$ is a fundamental solution of the Laplacian $\Delta = (\frac{\partial}{\partial x^1})^2 + (\frac{\partial}{\partial x^2})^2 + (\frac{\partial}{\partial x^3})^2$.

Indeed, $\Delta = \text{div}\,\text{grad}$, and $\text{grad}\,E(x) = \frac{x}{4\pi|x|^3}$ for $x \neq 0$, and therefore the equality $\text{div}\,\text{grad}\,E = \delta$ follows from relation (17.87).

As in Example 13, one can verify that for any $n \in \mathbb{N}$, $n \geq 2$, we have the following relation in $\mathbb{R}^n$:

$$\text{div}\,\frac{x}{|x|^n} = \sigma_n \delta, \qquad (17.87')$$

where $\sigma_n = \frac{2\pi^{n/2}}{\Gamma(n/2)}$ is the area of the unit sphere in $\mathbb{R}^n$.

Hence we can conclude upon taking account of the relation $\Delta = \text{div}\,\text{grad}$ that

$$\Delta \ln|X| = 2\pi\delta \quad \text{in } \mathbb{R}^2$$

and

$$\Delta \frac{1}{|x|^{n-2}} = -(n-2)\sigma_n\delta \quad \text{in } \mathbb{R}^n, \; n > 2.$$

Example 15 Let us verify that the function

$$E(x,t) = \frac{H(t)}{(2a\sqrt{\pi t})^n} e^{-\frac{|x|^2}{4a^2 t}},$$

where $x \in \mathbb{R}^n$, $t \in \mathbb{R}$, and H is the Heaviside function (that is, we set $E(x,t) = 0$ when $t < 0$) satisfies the equation

$$\left(\frac{\partial}{\partial t} - a^2\Delta\right)E = \delta.$$

Here Δ is the Laplacian with respect to x in $\mathbb{R}^n$, and $\delta = \delta(x,t)$ is the δ-function in $\mathbb{R}^n_x \times \mathbb{R}_t = \mathbb{R}^{n+1}$.

When $t > 0$, we have $E \in C^{(\infty)}(\mathbb{R}^{n+1})$ and by direct differentiation we verify that

$$\left(\frac{\partial}{\partial t} - a^2\Delta\right)E = 0 \quad \text{when } t > 0.$$

Taking this fact into account along with the result of Example 7, we obtain for any function $\varphi \in \mathcal{D}(\mathbb{R}^{n+1})$

$$\left\langle \left(\frac{\partial}{\partial t} - a^2 \Delta\right) E, \varphi \right\rangle =$$

$$= -\left\langle E, \left(\frac{\partial}{\partial t} + a^2 \Delta\right)\varphi \right\rangle =$$

$$= -\int_0^{+\infty} dt \int_{\mathbb{R}^n} E(x,t)\left(\frac{\partial \varphi}{\partial t} + a^2 \Delta\varphi\right) dx =$$

$$= -\lim_{\varepsilon \to +0} \int_\varepsilon^{+\infty} dt \int_{\mathbb{R}^n} E(x,t)\left(\frac{\partial \varphi}{\partial t} + a^2 \Delta\varphi\right) dx =$$

$$= \lim_{\varepsilon \to +0}\left[\int_{\mathbb{R}^n} E(x,\varepsilon)\varphi(x,0)\,dx + \int_\varepsilon^{+\infty} dt \int_{\mathbb{R}^n}\left(\frac{\partial E}{\partial t} - a^2 \Delta E\right)\varphi\,dx\right] =$$

$$= \lim_{\varepsilon \to +0}\left[\int_{\mathbb{R}^n} E(x,\varepsilon)\varphi(x,0)\,dx + \int_{\mathbb{R}^n} E(x,\varepsilon)\big(\varphi(x,\varepsilon) - \varphi(x,0)\big)\,dx\right] =$$

$$= \lim_{\varepsilon \to +0}\int_{\mathbb{R}^n} E(x,\varepsilon)\varphi(x,0)\,dx = \varphi(0,0) = \langle \delta, \varphi \rangle.$$

Example 16 Let us show that the function

$$E(x,t) = \frac{1}{2a} H\big(at - |x|\big),$$

where $a > 0$, $x \in \mathbb{R}_x^1$, $t \in \mathbb{R}_t^1$, and H is the Heaviside function, satisfies the equation

$$\left(\frac{\partial^2}{\partial t^2} - a^2 \frac{\partial^2}{\partial x^2}\right) E = \delta,$$

in which $\delta = \delta(x,t)$ is the δ-function in the space $\mathcal{D}'(\mathbb{R}_x^1 \times \mathbb{R}_t^1) = \mathcal{D}'(\mathbb{R}^2)$.

Let $\varphi \in \mathcal{D}(\mathbb{R}^2)$. Using the abbreviation $\Box_a := \frac{\partial^2}{\partial t^2} - a^2 \frac{\partial^2}{\partial x^2}$, we find

$$\langle \Box_a E, \varphi \rangle = \langle E, \Box_a \varphi \rangle = \int_{\mathbb{R}_x} dx \int_{\mathbb{R}_t} E(s,t)\Box_a \varphi(x,t)\,dt =$$

$$= \frac{1}{2a}\int_{-\infty}^{+\infty} dx \int_{\frac{|x|}{a}}^{+\infty} \frac{\partial^2 \varphi}{\partial t^2}\,dt - \frac{a}{2}\int_0^{+\infty} dt \int_{-at}^{at} \frac{\partial^2 \varphi}{\partial x^2}\,dx =$$

$$= -\frac{1}{2a}\int_{-\infty}^{+\infty} \frac{\partial \varphi}{\partial t}\left(x, \frac{|x|}{a}\right) dx - \frac{a}{2}\int_0^{+\infty}\left[\frac{\partial \varphi}{\partial x}(at,t) - \frac{\partial \varphi}{\partial x}(-at,t)\right] dt =$$

$$= -\frac{1}{2}\int_0^{+\infty} \frac{d\varphi}{dt}(at,t)\,dt - \frac{1}{2}\int_0^{+\infty} \frac{d\varphi}{dt}(-at,t)\,dt =$$

$$= \frac{1}{2}\varphi(0,0) + \frac{1}{2}\varphi(0,0) = \varphi(0,0) = \langle \delta, \varphi \rangle.$$

In Sect. 17.4 we have discussed in detail the role of the system function of the operator and the role of the convolution in the problem of determining the input u from the output $\widetilde{u}$ of a translation-invariant linear operator $Au = \widetilde{u}$. Everything that has been discussed on that score carries over to the multi-dimensional case without any changes. Hence, if we know the fundamental solution E of the operator A, that is, if $AE = \delta$, then one can present the solution u of the equation $Au = f$ as the convolution $u = f * E$.

Example 17 Using the function $E(x,t)$ of Example 16, one can thus present the solution

$$u(x,t) = \frac{1}{2a} \int_0^t d\tau \int_{x-a(t-\tau)}^{x+a(t-\tau)} f(\xi, \tau) \, d\xi$$

of the equation

$$\frac{\partial^2 u}{\partial t^2} - a^2 \frac{\partial^2 u}{\partial x^2} = f,$$

which is the convolution $f * E$ of the functions f and E and necessarily exists under the assumption, for example, that the function f is continuous. By direct differentiation of the resulting integral with respect to the parameters, one can easily verify that $u(x,t)$ is indeed a solution of the equation $\square_a u = f$.

Example 18 Similarly, on the basis of the result of Example 15 we find the solution

$$u(x,t) = \int_0^t d\tau \int_{\mathbb{R}^n} \frac{f(\xi, \tau)}{[2a\sqrt{\pi(t-\tau)}]^n} e^{-\frac{|x-\xi|^2}{4a^2(t-\tau)}} \, d\xi$$

of the equation $\frac{\partial u}{\partial t} - \Delta u = f$, for example, under the assumption that the function f is continuous and bounded, which guarantees the existence of the convolution $f * E$. We note that these assumptions are made only for example, and are far from obligatory. Thus, from the point of view of generalized functions one could pose the question of the solution of the equation $\frac{\partial u}{\partial t} - \Delta u = f$ taking as $f(x,t)$ the generalized function $\varphi(x) \cdot \delta(t)$, where $\varphi \in \mathcal{D}(\mathbb{R}^n)$ and $\delta \in \mathcal{D}'(\mathbb{R})$.

The formal substitution of such a function f under the integral sign leads to the relation

$$u(x,t) = \int_{\mathbb{R}^n} \frac{\varphi(\xi)}{[2a\sqrt{\pi t}]^n} e^{-\frac{|x-\xi|^2}{4a^2 t}} \, d\xi.$$

Applying the rule for differentiating an integral depending on a parameter one can verify that this function is a solution of the equation $\frac{\partial u}{\partial t} - a\Delta u = 0$ for $t > 0$. We note that $u(x,t) \to \varphi(x)$ as $t \to +0$. This follows from the result of Example 7, where it was established that the family of functions encountered here is an approximate identity.

Example 19 Finally, recalling the fundamental solution of the Laplace operator obtained in Example 14, we find the solution

$$u(x) = \int_{\mathbb{R}^n} \frac{f(\xi)\, d\xi}{|x - \xi|}$$

of the Poisson equation $\Delta u = -4\pi f$, which up to notation and relabeling is the same as the potential (17.75) for a charge distributed with density f, which we considered earlier.

If the function f is taken as $v(x)\delta_S$, where S is a piecewise smooth surface in $\mathbb{R}^3$, formal substitution into the integral leads to the function

$$u(x) = \int_S \frac{v(\xi)\, d\sigma(\xi)}{|x - \xi|},$$

which, as we know, is a single-layer potential; more precisely, the potential of a charge distributed over the surface $S \subset \mathbb{R}^3$ with surface density $v(x)$.

17.5.5 Problems and Exercises

1. a) Reasoning as in Example 3, where the continuity of the three-dimensional potential (17.75) was established, show that the single-layer potential (17.79) is continuous.

b) Verify the full proof of Propositions 4 and 5.

2. a) Show that for every set $M \subset \mathbb{R}^n$ and every $\varepsilon > 0$ one can construct a function f of class $C^{(\infty)}(\mathbb{R}^n, \mathbb{R})$ satisfying the following three conditions simultaneously: $\forall x \in \mathbb{R}^n$ $(0 \le f(x) \le 1)$; $\forall x \in M$ $(f(x) = 1)$; $\operatorname{supp} f \subset M_\varepsilon$, where M_ε is the ε-blowup (that is, the ε-neighborhood) of the set M.

b) Prove that for every closed set M in $\mathbb{R}^n$ there exists a nonnegative function $f \in C^{(\infty)}(\mathbb{R}^n, \mathbb{R})$ such that $(f(x) = 0) \Leftrightarrow (x \in M)$.

3. a) Solve Problems 6 and 7 of Sect. 17.4 in the context of a space $\mathbb{R}^n$ of arbitrary dimension.

b) Show that the generalized function δ_S (single layer) is not regular.

4. Using convolution, prove the following versions of the Weierstrass approximation theorem.

a) Any continuous function $f : I \to \mathbb{R}$ on a compact n-dimensional interval $I \subset \mathbb{R}^n$ can be uniformly approximated by an algebraic polynomial in n variables.

b) The preceding assertion remains valid even if I is replaced by an arbitrary compact set $K \subset \mathbb{R}$ and we assume that $f \in C(K, \mathbb{C})$.

c) For every open set $G \subset \mathbb{R}^n$ and every function $f \in C^{(m)}(G, \mathbb{R})$ there exists a sequence $\{P_k\}$ of algebraic polynomials in n variables such that $P_k^{(\alpha)} \rightrightarrows f^{(\alpha)}$ on each compact set $K \subset G$ as $k \to \infty$ for every multi-index $\alpha = (\alpha_1, \ldots, \alpha_n)$ such that $|\alpha| \le m$.

d) If G is a bounded open subset of $\mathbb{R}^n$ and $f \in C^{(\infty)}(\overline{G}, \mathbb{R})$, there exists a sequence $\{h\}$ of algebraic polynomials in n variables such that $P_k^{(\alpha)} \supset f^{(\alpha)}$ for every $\alpha = (\alpha_1, \ldots, \alpha_n)$ as $k \to \infty$.

e) Every periodic function $f \in C(\mathbb{R}^n, \mathbb{R})$ with periods $T_1, T_2, \ldots, T_n$ in the variables $x^1, \ldots, x^n$, can be uniformly approximated in $\mathbb{R}^n$ by trigonometric polynomials in n variables having the same periods $T_1, T_2, \ldots, T_n$ in the corresponding variables.

5. This problem contains further information on the averaging action of convolution.

a) Previously we obtained the integral Minkowski inequality

$$\left(\int_X \left| a(x) + b(x) \right|^p dx \right)^{1/p} \leq \left(\int_X |a|^p(x) dx \right)^{1/p} + \left(\int_X |b|^p(x) dx \right)^{1/p}$$

for $p \geq 1$ on the basis of this numerical Minkowski inequality.

The integral inequality in turn enables us to predict the following *generalized integral Minkowski inequality*:

$$\left(\int_X \left| \int_Y f(x, y) dy \right|^p dx \right)^{1/p} \leq \int_Y \left(\int_X |f|^p(x, y) dx \right)^{1/p} dy.$$

Prove this inequality, assuming that $p \geq 1$, that X and Y are measurable subsets (for example, intervals in $\mathbb{R}^m$ and $\mathbb{R}^n$ respectively), and that the right-hand side of the inequality is finite.

b) By applying the generalized Minkowski inequality to the convolution $f * g$, show that the relation $\|f * g\|_p \leq \|f\|_1 \cdot \|g\|_p$ holds for $p \geq 1$, where, as always, $\|u\|_p = (\int_{\mathbb{R}^n} |u|^p(x) dx)^{1/p}$.

c) Let $\varphi \in C_0^{(\infty)}(\mathbb{R}^n, \mathbb{R})$ with $0 \leq \varphi(x) \leq 1$ on $\mathbb{R}^n$ and $\int_{\mathbb{R}^n} \varphi(x) dx = 1$. Assume that $\varphi_\varepsilon(x) := \frac{1}{\varepsilon}\varphi(\frac{x}{\varepsilon})$ and $f_\varepsilon := f * \varphi_\varepsilon$ for $\varepsilon > 0$. Show that if $f \in \mathcal{R}_p(\mathbb{R}^n)$ (that is, if the integral $\int_{\mathbb{R}^n} |f|^p(x) dx$ exists), then $f_\varepsilon \in C^{(\infty)}(\mathbb{R}^n, \mathbb{R})$ and $\|f_\varepsilon\|_p \leq \|f\|_p$.

We note that the function f_ε is often called the *average of the function f with kernel φ_ε*.

d) Preserving the preceding notation, verify that the relation

$$\|f_\varepsilon - f\|_{p,I} \leq \sup_{|h|<\varepsilon} \|\tau_h f - f\|_{p,I},$$

holds on every interval $I \subset \mathbb{R}^n$, where $\|u\|_{p,I} = (\int_I |u|^p(x) dx)^{1/p}$ and $\tau_h f(x) = f(x - h)$.

e) Show that if $f \in \mathcal{R}_p(\mathbb{R}^n)$, then $\|\tau_h f - f\|_{p,I} \to 0$ as $h \to 0$.

f) Prove that $\|f_\varepsilon\|_p \leq \|f\|_p$ and $\|f_\varepsilon - f\|_p \to 0$ as $\varepsilon \to +0$ for every function $f \in \mathcal{R}_p(\mathbb{R}^n)$, $p \geq 1$.

g) Let $\mathcal{R}_p(G)$ be the normed vector space of functions that are absolutely integrable on the open set $G \subset \mathbb{R}^n$ with the norm $\| \ \|_{p,G}$. Show that the functions of class $C^{(\infty)}(G) \cap \mathcal{R}_p(G)$ form an everywhere-dense subset of $\mathcal{R}_p(G)$ and the same is true for the set $C_0^{(\infty)}(G) \cap \mathcal{R}_p(G)$.

h) The following proposition can be compared with the case $p = \infty$ in the preceding problem: *Every continuous function on G can be uniformly approximated on G by functions of class $C^{(\infty)}(G)$.*

i) If f is a T-periodic locally absolutely integrable function on $\mathbb{R}$, then, setting $\|f\|_{p,T} = (\int_a^{a+T} |f|^p(x)\, dx)^{1/p}$, we shall denote the vector space with this norm by $\mathcal{R}_p^T$. Prove that $\|f_\varepsilon - f\|_{p,T} \to 0$ as $\varepsilon \to +0$.

j) Using the fact that the convolution of two functions, one of which is periodic, is itself periodic, show that the smooth periodic functions of class $C^{(\infty)}$ are everywhere dense in $\mathcal{R}_p^T$.

6. a) Preserving the notation of Example 11 and using formula (17.83), verify that if $f \in C^{(1)}(\overline{\mathbb{R}^n \setminus S})$, then

$$\frac{\partial^2 f}{\partial x^i \partial x^j} = \left\{\frac{\partial^2 f}{\partial x^i \partial x^j}\right\} + \frac{\partial}{\partial x^j}\big((\int f)_S \cos\alpha_i \delta_S\big) + \left(\int \frac{\partial f}{\partial x^i}\right)_S \cos\alpha_j \delta_S.$$

b) Show that the sum $\sum_{i=1}^n (\int \frac{\partial f}{\partial x^i})_S \cos\alpha_i$ equals the jump $(\int \frac{\partial f}{\partial n_2})_S$ of the normal derivative of the function f at the corresponding point $x \in S$, this jump being independent of the direction of the normal and equal to the sum $(\frac{\partial f}{\partial n_1} + \frac{\partial f}{\partial n_2})(x)$ of the normal derivatives of f at the point x from the two sides of the surface S.

c) Verify the relation

$$\Delta f = \{\Delta f\} + \left(\int f \frac{\partial f}{\partial \mathbf{n}}\right)_S \delta_S + \frac{\partial}{\partial \mathbf{n}}\big((\int f)_S \delta_S\big),$$

where $\frac{\partial}{\partial \mathbf{n}}$ is the normal derivative, that is, $\langle \frac{\partial}{\partial \mathbf{n}} F, \varphi\rangle := -\langle F, \frac{\partial \varphi}{\partial \mathbf{n}}\rangle$, and $(\int f)_S$ is the jump of the function f at the point $x \in S$ in the direction of the normal $\mathbf{n}$.

d) Using the expression just obtained for Δf, prove the classical Green's formula

$$\int_G (f\Delta\varphi - \varphi\Delta f)\, dx = \int_S \left(f\frac{\partial \varphi}{\partial \mathbf{n}} - \varphi\frac{\partial f}{\partial \mathbf{n}}\right) d\sigma$$

under the assumption that G is a finite domain in $\mathbb{R}^n$ bounded by a piecewise-smooth surface S; f and φ belong to $C^{(1)}(G) \cap C^{(2)}(G)$, and the integral on the left-hand side exists, possibly as an improper integral.

e) Show that if the δ-function corresponds to a unit charge located at the origin 0 in $\mathbb{R}^n$, and the function $-\frac{\partial \delta}{\partial x^1}$ corresponds to a dipole with electric moment $+1$ located at 0 and oriented along the x^1-axis (see Problem 11e) of Sect. 17.4) and the function $v(x)\delta_S$ is the single layer corresponding to a charge distribution over the surface S with surface density $v(x)$, then the function $-\frac{\partial}{\partial \mathbf{n}}(v(x)\delta_S)$, called the *double layer*, corresponds to a distribution of dipoles over the surface S oriented by the normal $\mathbf{n}$ and having surface density moment $v(x)$.

f) Setting $\varphi = \frac{1}{|x-y|}$ in Green's formula and using the result of Example 14, show that every harmonic function f in the domain G in the class $C^{(1)}(\overline{G})$ can be represented as the sum of a single-layer and a double-layer potential located on the boundary S of G.

7. a) The function $\frac{1}{|x|}$ is the potential of the electric field intensity $\mathbf{A} = -\frac{x}{|x|^3}$ created in $\mathbb{R}^3$ by a unit charge located at the origin. We also know that

$$\operatorname{div}\left(-\frac{x}{|x|^3}\right) = 4\pi\delta, \qquad \operatorname{div}\left(-\frac{qx}{|x|^3}\right) = 4\pi q\delta, \qquad \operatorname{div grad}\left(\frac{q}{|x|}\right) = 4\pi\delta.$$

Starting from this, explain why it was necessary to assume that the function $U(x) = \int_{\mathbb{R}^3} \frac{\mu(\xi)\,d\xi}{|x-\xi|}$ must satisfy the equation $\Delta U = -4\pi\mu$. Verify that it does indeed satisfy the Poisson equation written here.

b) A physical corollary of the Gauss–Ostrogradskii formula, known in electromagnetic field theory as *Gauss' theorem* is that the flux across a closed surface S of the intensity of the electric field created by charges distributed in $\mathbb{R}^3$ equals Q/ε_0 (see pp. 279 and 280), where Q is the total charge in the region bounded by the surface S. Prove this theorem of Gauss.

8. Verify the following equalities, understood in the sense of the theory of generalized functions.

a) $\Delta E = \delta$, if

$$E(x) = \begin{cases} \frac{1}{2\pi}\ln|x| & \text{for } x \in \mathbb{R}^2, \\ -\frac{\Gamma(\frac{n}{2})}{2\pi^{n/2}(n-2)}|x|^{-n-2} & \text{for } x \in \mathbb{R}^n, n > 2. \end{cases}$$

b) $(\Delta + k^2)E = \delta$, if $E(x) = -\frac{e^{ik|x|}}{4\pi|x|}$ or if $E(x) = -\frac{e^{-ik|x|}}{4\pi|x|}$ and $x \in \mathbb{R}^3$.

c) $\Box_a E = \delta$, where $\Box_a = \frac{\partial^2}{\partial t^2} - a^2[(\frac{\partial}{\partial x^1})^2 + \cdots + (\frac{\partial}{\partial x^n})^2]$, and $E = \frac{H(at-|x|)}{2\pi a\sqrt{a^2 t^2 - |x|^2}}$

for $x \in \mathbb{R}^2$ or $E = \frac{H(t)}{4\pi a^2 t} \delta_{S_{at}} \equiv \frac{H(t)}{2\pi a}\delta(a^2 t^2 - |x|^2)$ for $x \in \mathbb{R}^3$, $t \in \mathbb{R}$. Here $H(t)$ is the Heaviside function, $S_{at} = \{x \in \mathbb{R}^3 \mid |x| = at\}$ is a sphere, and $a > 0$.

d) Using the preceding results, present the solution of the equation $Au = f$ for the corresponding differential operator A in the form of the convolution $f * E$ and verify, for example, assuming the function f continuous, that the integrals depending on a parameter that you have obtained indeed satisfy the equation $Au = f$.

9. *Differentiation of an integral over a liquid volume.*

Space is filled with a moving substance (a liquid). Let $v = v(t, x)$ and $\rho = \rho(t, x)$ be respectively the velocity of displacement and the density of the substance at time t at the point x. We observe the motion of a portion of the substance filling the domain Ω_0 at the initial moment of time.

a) Express the mass of the substance filling the domain Ω_t obtained from Ω_0 at time t and write the law of conservation of mass.

b) By differentiating the integral $F(t) = \int_{\Omega_t} f(t, x)\, d\omega$ with variable domain of integration Ω_t (the volume of liquid), show that $F'(t) = \int_{\Omega_t} \frac{\partial f}{\partial t}\, d\omega + \int_{\partial\Omega_t} f\langle v, n\rangle\, d\sigma$, where $\Omega_t, \partial\Omega_t, d\omega, d\sigma, n, v, \langle,\rangle$ are respectively the domain, its boundary, the element of volume, the element of area, the unit outward normal, the flow velocity at time t at corresponding points, and the inner product.

c) Show that $F'(t)$ in problem b) can be represented in the form $F'(t) = \int_{\Omega_t} (\frac{\partial f}{\partial t} + \operatorname{div}(f v))\, d\omega$.

d) Comparing the results of problems a), b), and c), obtain the equation of continuity $\frac{\partial \rho}{\partial t} + \operatorname{div}(\rho v) = 0$. (In this connection, see also Sect. 14.4.2.)

e) Let $|\Omega_t|$ be the volume of the domain Ω_t. Show that $\frac{d|\Omega_t|}{dt} = \int_{\Omega_t} \operatorname{div} v\, d\omega$.

f) Show that the velocity field v of the flow of an incompressible liquid is divergence-free ($\operatorname{div} v = 0$) and that this condition is the mathematical expression of the incompressibility (conservation of volume) of any portion of the evolving medium.

g) The phase velocity field $(\dot{p}, \dot{q})$ of a Hamiltonian system of classical mechanics satisfies the Hamilton equations $\dot{p} = -\frac{\partial H}{\partial q}$, $\dot{q} = \frac{\partial H}{\partial p}$, where $H = H(p, q)$ is the Hamiltonian of the system. Following Liouville, show that a Hamiltonian flow preserves the phase volume. Verify also that the Hamiltonian H (energy) is constant along the streamlines (trajectories).

Chapter 18
Fourier Series and the Fourier Transform

18.1 Basic General Concepts Connected with Fourier Series

18.1.1 Orthogonal Systems of Functions

a. Expansion of a Vector in a Vector Space

During this course of analysis we have mentioned several times that certain classes of functions form vector spaces in relation to the standard arithmetic operations. Such, for example, are the basic classes of analysis, which consist of smooth, continuous, or integrable real-, complex-, or vector-valued functions on a domain $X \subset \mathbb{R}^n$.

From the point of view of algebra the equality

$$f = \alpha_1 f_1 + \cdots + \alpha_n f_n,$$

where $f, f_1, \ldots, f_n$ are functions of the given class and α_i are coefficients from $\mathbb{R}$ or $\mathbb{C}$, simply means that the vector f is a linear combination of the vectors $f_1, \ldots, f_n$ of the vector space under consideration.

In analysis, as a rule, it is necessary to consider "infinite linear combinations" – series of functions of the form

$$f = \sum_{k=1}^{\infty} \alpha_k f_k. \tag{18.1}$$

The definition of the sum of the series requires that some topology (in particular, a metric) be defined in the vector space in question, making it possible to judge whether the difference $f - S_n$ tends to zero or not, where $S_n = \sum_{k=1}^{n} \alpha_k f_k$.

The main device used in classical analysis to introduce a metric on a vector space is to define some norm of a vector or inner product of vectors in that space. Section 10.1 was devoted to a discussion of these concepts.

We are now going to consider only spaces endowed with an inner product (which, as before, we shall denote $\langle , \rangle$). In such spaces one can speak of orthogonal vectors,

© Springer-Verlag Berlin Heidelberg 2016
V.A. Zorich, *Mathematical Analysis II*, Universitext,
DOI 10.1007/978-3-662-48993-2_10

orthogonal systems of vectors, and orthogonal bases, just as in the case of three-dimensional Euclidean space familiar from analytic geometry.

Definition 1 The vectors x and y in a vector space endowed with an inner product $\langle , \rangle$ are *orthogonal* (with respect to that inner product) if $\langle x, y \rangle = 0$.

Definition 2 The system of vectors $\{x_k; k \in K\}$ is *orthogonal* if the vectors in it corresponding to different values of the index k are pairwise orthogonal.

Definition 3 The system of vectors $\{e_k; k \in K\}$ is *orthonormalized* (or *orthonormal*) if $\langle e_i, e_j \rangle = \delta_{i,j}$ for every pair of indices $i, j \in K$, where $\delta_{i,j}$ is the Kronecker symbol, that is, $\delta_{i,j} = \begin{cases} 1, & \text{if } i=j, \\ 0, & \text{if } i \neq j. \end{cases}$

Definition 4 A finite system of vectors $x_1, \ldots, x_n$ is *linearly independent* if the equality $\alpha_1 x_1 + \alpha_2 x_2 + \cdots + \alpha_n x_n = 0$ is possible only when $\alpha_1 = \alpha_2 = \cdots = \alpha_n = 0$ (in the first equality 0 is the zero vector and in the second it is the zero of the coefficient field).

An arbitrary system of vectors of a vector space is a *system of linearly independent vectors* if every finite subsystem of it is linearly independent.

The main question that will interest us now is the question of expanding a vector in a given system of linearly independent vectors.

Having in mind later applications to spaces of functions (which may be infinite-dimensional as well) we must reckon with the fact that such an expansion may, in particular, lead to a series of the type (18.1). That is precisely where analysis enters into the study of the fundamental and essentially algebraic question we have posed.

As is known from analytic geometry, expansions in orthogonal and orthonormal systems have many technical advantages over expansions in arbitrary linearly independent systems. (The coefficients of the expansion are easy to compute; it is easy to compute the inner product of two vectors from their coefficients in an orthonormal basis, and so on.)

It is for that reason that we shall be mainly interested in expansions in orthonormal systems. In function spaces these will be *expansions in orthogonal systems of functions* or Fourier[1] *series*, to the study of which this chapter is devoted.

---

[1] J.-B.J. Fourier (1768–1830) – French mathematician. His most important work *Théorie analytique de la chaleur* (1822) contained the heat equation derived by Fourier and the method of separation of variables (the Fourier method) of solving it (see p. 510). The key element in the Fourier method is the expansion of a function in a trigonometric (Fourier) series. Many outstanding mathematicians later undertook the study of the possibility of such a representation. This, in particular, led to the creation of the theory of functions of a real variable and set theory, and it helped to promote the development of the very concept of a function.

b. Examples of Orthogonal Systems of Functions

Extending Example 12 of Sect. 10.1, we introduce an inner product

$$\langle f, g \rangle := \int_X (f \cdot \overline{g})(x)\, dx \tag{18.2}$$

on the vector space $\mathcal{R}_2(X, \mathbb{C})$ consisting of functions on the set $X \subset \mathbb{R}^n$ that are locally square-integrable (as proper or improper integrals).

Since $|f \cdot \overline{g}| \le \frac{1}{2}(|f|^2 + |g|^2)$, the integral in (18.2) converges and hence defines $\langle f, g \rangle$ unambiguously.

If we are discussing real-valued functions, relation (18.2) in the real space $\mathcal{R}_2(x, \mathbb{R})$ reduces to the equality

$$\langle f, g \rangle := \int_X (f \cdot g)(x)\, dx. \tag{18.3}$$

Relying on properties of the integral, one can easily verify that all the axioms for an inner product listed in Sect. 10.1 are satisfied in this case, provided we identify two functions that differ only on a set of n-dimensional measure zero. Throughout the following, in the text portion of the section, inner products of functions will be understood in the sense of Eqs. (18.2) and (18.3).

Example 1 We recall that for integers m and n

$$\int_{-\pi}^{\pi} e^{imx} \cdot e^{-inx}\, dx = \begin{cases} 0, & \text{if } m \ne n, \\ 2\pi, & \text{if } m = n; \end{cases} \tag{18.4}$$

$$\int_{-\pi}^{\pi} \cos mx \cos nx\, dx = \begin{cases} 0, & \text{if } m \ne n, \\ \pi, & \text{if } m = n \ne 0, \\ 2\pi, & \text{if } m = n = 0; \end{cases} \tag{18.5}$$

$$\int_{-\pi}^{\pi} \cos mx \sin nx\, dx = 0; \tag{18.6}$$

$$\int_{-\pi}^{\pi} \sin mx \sin nx\, dx = \begin{cases} 0, & \text{if } m \ne n, \\ \pi, & \text{if } m = n \ne 0. \end{cases} \tag{18.7}$$

These relations show that $\{e^{inx}; n \in \mathbb{Z}\}$ is an orthogonal system of vectors in the space $\mathcal{R}_2([-\pi, \pi], \mathbb{C})$ relative to the inner product (18.2) and the *trigonometric system* $\{1, \cos nx, \sin nx; n \in \mathbb{N}\}$ is orthogonal in $\mathcal{R}_2([-\pi, \pi], \mathbb{R})$. If we regard the trigonometric system as a set of vectors in $\mathcal{R}_2([-\pi, \pi], \mathbb{C})$, that is, if we allow linear combinations of them to have complex coefficients, then by Euler's formulas $e^{inx} = \cos nx + i \sin nx$, $\cos nx = \frac{1}{2}(e^{inx} + e^{-inx})$, $\sin nx = \frac{1}{2i}(e^{inx} - e^{-inx})$, we see that these two systems can be expressed linearly in terms of each other, that is, they are algebraically equivalent. For that reason the exponential system $\{e^{inx}; n \in \mathbb{Z}\}$ is also called the trigonometric system or more precisely the *trigonometric system in complex notation*.

Relations (18.4)–(18.7) show that these systems are orthogonal, but not normalized, while the systems $\{\frac{1}{\sqrt{2\pi}}e^{inx}; n \in \mathbb{Z}\}$ and

$$\left\{ \frac{1}{\sqrt{2\pi}}, \frac{1}{\sqrt{\pi}}\cos nx, \frac{1}{\sqrt{\pi}}\sin nx; n \in \mathbb{N} \right\}$$

are orthonormal.

If the closed interval $[-\pi, \pi]$ is replaced by an arbitrary closed interval $[-l, l] \subset \mathbb{R}$, then by a change of variable one can obtain the analogous systems $\{e^{i\frac{\pi}{l}nx}; n \in \mathbb{Z}\}$ and $\{1, \cos\frac{\pi}{l}nx, \sin\frac{\pi}{l}nx; n \in \mathbb{N}\}$, which are orthogonal in the spaces $\mathcal{R}_2([-l, l], \mathbb{C})$ and $\mathcal{R}_2([-l, l], \mathbb{R})$ and also the corresponding orthonormal systems

$$\left\{ \frac{1}{\sqrt{2l}}e^{i\frac{\pi}{l}nx}; n \in \mathbb{Z} \right\} \quad \text{and} \quad \left\{ \frac{1}{\sqrt{2l}}, \frac{1}{\sqrt{l}}\cos\frac{\pi}{l}nx, \frac{1}{\sqrt{l}}\sin\frac{\pi}{l}nx; n \in \mathbb{N} \right\}.$$

Example 2 Let I_x be an interval in $\mathbb{R}^m$ and I_y an interval in $\mathbb{R}^n$, and let $\{f_i(x)\}$ be an orthogonal system of functions in $\mathcal{R}_2(I_x, \mathbb{R})$ and $\{g_j(y)\}$ an orthogonal system of functions in $\mathcal{R}_2(I_y, \mathbb{R})$. Then, as follows from Fubini's theorem, the system of functions $\{u_{ij}(x, y) := f_j(x)g_j(y)\}$ is orthogonal in $\mathcal{R}_2(I_x \times I_y, \mathbb{R})$.

Example 3 We remark that for $\alpha \neq \beta$

$$\int_0^l \sin\alpha x \sin\beta x \, dx = \frac{1}{2}\left(\frac{\sin(\alpha - \beta)l}{\alpha - \beta} - \frac{\sin(\alpha + \beta)l}{\alpha + \beta} \right) =$$

$$= \cos\alpha l \cos\beta l \cdot \frac{\beta\tan\alpha l - \alpha\tan\beta l}{\alpha^2 - \beta^2}.$$

Hence, if α and β are such that $\frac{\tan\alpha l}{\alpha} = \frac{\tan\beta l}{\beta}$, the original integral equals zero. Consequently, if $\xi_1 < \xi_2 < \cdots < \xi_n < \cdots$ is a sequence of roots of the equation $\tan\xi l = c\xi$, where c is an arbitrary constant, then the system of functions $\{\sin(\xi_n x); n \in \mathbb{N}\}$ is orthogonal on the interval $[0, l]$. In particular, for $c = 0$, we obtain the familiar system $\{\sin(\frac{\pi}{l}nx); n \in \mathbb{N}\}$.

Example 4 Consider the equation

$$\left(\frac{d^2}{dx^2} + q(x) \right)u(x) = \lambda u(x),$$

where $q \in C^{(\infty)}([a, b], \mathbb{R})$ and λ is a numerical coefficient. Let us assume that the functions $u_1, u_2, \ldots$ are of class $C^{(2)}([a, b], \mathbb{R})$ and vanish at the endpoints of the closed interval $[a, b]$ and that each of them satisfies the given equation with particular values $\lambda_1, \lambda_2, \ldots$ of the coefficient λ. We shall show that if $\lambda_i \neq \lambda_j$, then the functions u_i and u_j are orthogonal on $[a, b]$.

Indeed, integrating by parts, we find that

$$\int_a^b \left[\left(\frac{d^2}{dx^2} + q(x) \right) u_i(x) \right] u_j(x) \, dx = \int_a^b u_i(x) \left[\left(\frac{d^2}{dx^2} + q(x) \right) u_j(x) \right] dx.$$

According to the equation, we obtain from this the relation

$$\lambda_i \langle u_i, u_j \rangle = \lambda_j \langle u_i, u_j \rangle;$$

and, since $\lambda_i \neq \lambda_j$, we now conclude that $\langle u_i, u_j \rangle = 0$.

In particular, if $q(x) \equiv 0$ on $[a, b]$ and $[a, b] = [0, \pi]$ we again find that the system $\{\sin nx; n \in \mathbb{N}\}$ is orthogonal on $[0, \pi]$.

Further examples, including examples of orthogonal systems of importance in mathematical physics, will be found in the problems at the end of this section.

c. Orthogonalization

It is well-known that in a finite-dimensional Euclidean space, starting with a linearly independent system of vectors, there is a canonical way of constructing an orthogonal and even orthonormal system of vectors equivalent to the given system, using the Gram[2]–Schmidt[3] orthogonalization process. By the same method one can obviously orthonormalize any linearly independent system of vectors $\psi_1, \psi_2, \ldots$ in any vector space having an inner product.

We recall that the orthogonalization process leading to the orthonormal system $\varphi_1, \varphi_2, \ldots$ is described by the following relations:

$$\varphi_1 = \frac{\psi_1}{\|\psi_1\|}, \qquad \varphi_2 = \frac{\psi_2 - \langle \psi_2, \varphi_1 \rangle \varphi_1}{\|\psi_2 - \langle \psi_2, \varphi_1 \rangle \varphi_1\|},$$

$$\varphi_n = \frac{\psi_n - \sum_{k=1}^{n-1} \langle \psi_n, \varphi_k \rangle \varphi_k}{\|\psi_n - \sum_{k=1}^{n-1} \langle \psi_n, \varphi_k \rangle \varphi_k\|}.$$

Example 5 The process of orthogonalizing the linearly independent system $\{1, x, x^2, \ldots\}$ in $\mathcal{R}_2([-1, 1], \mathbb{R})$ leads to the system of orthogonal polynomials known as the *Legendre polynomials*. We note that the name *Legendre polynomials* is often given not to the orthonormal system, but to a system of polynomials proportional

---

[2]J.P. Gram (1850–1916) – Danish mathematician who continued the research of P.L. Chebyshev and exhibited the connection between orthogonal series expansions and the problem of best least-squares approximation (see Fourier series below). It was in these investigations that the orthogonalization process and the famous Gram matrix arose (see p. 187 and the system (18.18) on p. 504).

[3]E. Schmidt (1876–1959) – German mathematician who studied the geometry of Hilbert space in connection with integral equations and described it in the language of Euclidean geometry.

to these polynomials. The proportionality factor can be chosen from various considerations, for example, requiring the leading coefficient to be 1 or requiring the polynomial to have the value 1 at $x = 1$. The orthogonality of the system is unaffected by these requirements, but in general orthonormality is lost.

We have already encountered the standard Legendre polynomials, which are defined by Rodrigues' formula

$$P_n(x) = \frac{1}{n!2^n} \frac{d^n(x^2-1)^n}{dx^n}.$$

For these polynomials $P_n(1) = 1$. Let us write out the first few Legendre polynomials, normalized by requiring the leading coefficient to be 1:

$$\widetilde{P}_0(x) \equiv 1, \qquad \widetilde{P}_1(x) = x, \qquad \widetilde{P}_2(x) = x^2 - \frac{1}{3}, \qquad \widetilde{P}_3(x) = x^3 - \frac{3}{5}x.$$

The orthonormalized Legendre polynomials have the form

$$\hat{P}_n(x) = \sqrt{\frac{2n+1}{2}} P_n(x),$$

where $n = 0, 1, 2, \ldots$.

One can verify by direct computation that these polynomials are orthogonal on the closed interval $[-1, 1]$. Taking Rodrigues' formula as the definition of the polynomial $P_n(x)$, let us verify that the system of Legendre polynomials $\{P_n(x)\}$ is orthogonal on the closed interval $[-1, 1]$. To do this, it suffices to verify that $P_n(x)$ is orthogonal to $1, x, \ldots, x^{n-1}$, since all polynomials P_k of degree $k < n$ are linear combinations of these.

Integrating by parts for $k < n$, we indeed find that

$$\int_{-1}^{1} x^k P_n(x)\, dx = \frac{1}{k!2^k} \int_{-1}^{1} \frac{d^{k+1}x^k}{dx^{k+1}} \cdot \frac{d^{n-k-1}(x^2-1)^n}{dx^{n-k-1}}\, dx = 0.$$

A certain picture of the origin of orthogonal systems of functions in analysis will be given in the last subsection of this section and in the problems at the end of the section. At present we shall return to the fundamental general problems connected with the expansion of a vector in terms of vectors of a given system of vectors in a vector space with inner product.

d. Continuity of the Inner Product and the Pythagorean Theorem

We shall have to work not only with finite sums of vectors but also with infinite sums (series). In this connection we note that the inner product is a continuous function, enabling us to extend the ordinary algebraic properties of the inner product to the case of series.

Let X be a vector space with an inner product $\langle , \rangle$ and the norm it induces $\|x\| := \sqrt{\langle x, x, \rangle}$ (see Sect. 10.1). Convergence of a series $\sum_{i=1}^{\infty} x_i = x$ of vectors $x_i \in X$ to the vector $x \in X$ will be understood in the sense of convergence in this norm.

Lemma 1 (Continuity of the inner product) *Let* $\langle , \rangle : X \to \mathbb{C}$ *be an inner product in the complex vector space X. Then*

a) *the function* $(x, y) \mapsto \langle x, y \rangle$ *is continuous jointly in the two variables;*
b) *if* $x = \sum_{i=1}^{\infty} x_i$, *then* $\langle x, y \rangle = \sum_{i=1}^{\infty} \langle x_i, y \rangle$;
c) *if* $e_1, e_2, \ldots$, *is an orthonormal system of vectors in X and* $x = \sum_{i=1}^{\infty} x^i e_i$ *and* $y = \sum_{i=1}^{\infty} y^i e_i$, *then* $\langle x, y \rangle = \sum_{i=1}^{\infty} x^i \bar{y}^i$.

Proof Assertion a) follows from the Cauchy–Bunyakovskii inequality (see Sect. 10.1):

$$\left| \langle x - x_0, y - y_0 \rangle \right|^2 \le \|x - x_0\|^2 \cdot \|y - y_0\|^2.$$

Assertion b) follows from a), since

$$\langle x, y \rangle = \sum_{i=1}^{n} \langle x_i, y \rangle + \left\langle \sum_{i=n+1}^{\infty} x_i, y \right\rangle,$$

and $\sum_{i=n+1}^{\infty} x_i \to 0$ as $n \to \infty$.

Assertion c) follows by repeated application of b), taking account of the relation $\langle x, y \rangle = \overline{\langle y, x \rangle}$. $\qquad\square$

The following result is an immediate consequence of the lemma.

Theorem 1 (Pythagoras[4])

a) *If* $\{x_i\}$ *is a system of mutually orthogonal vectors and* $x = \sum_i x_i$, *then* $\|x\|^2 = \sum_i \|x_i\|^2$.
b) *If* $\{e_i\}$ *is an orthonormalized system of vectors and* $x = \sum_i x^i e_i$, *then* $\|x\|^2 = \sum_i |x^i|^2$.

18.1.2 Fourier Coefficients and Fourier Series

a. Definition of the Fourier Coefficients and the Fourier Series

Let $\{e_i\}$ be an orthonormal system and $\{l_i\}$ an orthogonal system of vectors in a space X with inner product $\langle , \rangle$.

---

[4]Pythagoras of Samos (conjectured to be 580–500 BCE) – famous ancient Greek mathematician and idealist philosopher, founder of the Pythagorean school, which, in particular, made the discovery that the side and diagonal of a square are incommensurable, a discovery that disturbed the ancients. The classical Pythagorean theorem itself was known in a number of countries long before Pythagoras (possibly without proof, to be sure).

Suppose that $x = \sum_i x^i l_i$. The coefficients x^i in this expansion of the vector x can be found directly:

$$x^i = \frac{\langle x, l_i \rangle}{\langle l_i, l_i \rangle}.$$

If $l_i = e_i$, the expression becomes even simpler:

$$x^i = \langle x, e_i \rangle.$$

We remark that the formulas for x^i make sense and are completely determined if the vector x itself and the orthogonal system $\{l_i\}$ (or $\{e_i\}$) are given. The equality $x = \sum_i x^i l_i$ (or $x = \sum_i x^i e_i$) is no longer needed to compute x^i from these formulas.

Definition 5 The numbers $\{\frac{\langle x, l_i \rangle}{\langle l_i, l_i \rangle}\}$ are the *Fourier coefficients of the vector* $x \in X$ *in the orthogonal system* $\{l_i\}$.

If the system $\{e_i\}$ is orthonormal, the Fourier coefficients have the form $\{\langle x, e_i \rangle\}$.

From the geometric point of view the ith Fourier coefficient $\langle x, e_i \rangle$ of the vector $x \in X$ is the projection of that vector in the direction of the unit vector e_i. In the familiar case of three-dimensional Euclidean space E^3 with a given orthonormal frame e_1, e_2, e_3 the Fourier coefficients $x^i = \langle x, e_i \rangle$, $i = 1, 2, 3$, are the coordinates of the vector x in the basis e_1, e_2, e_3 appearing in the expansion $x = x^1 e_1 + x^2 e_2 + x^3 e_3$.

If we were given only the two vectors e_1 and e_2 instead of all three e_1, e_2, e_3, the expansion $x = x^1 e_1 + x^2 e_2$ in this system would certainly not be valid for all vectors $x \in E^3$. Nevertheless, the Fourier coefficients $x^i = \langle x, e_i \rangle$, $i = 1, 2$, would be defined in this case and the vector $x_e = x^1 e_1 + x^2 e_2$ would be the orthogonal projection of the vector x onto the plane L of the vectors e_1 and e_2. Among all the vectors in that plane, the vector x_e is distinguished by being closest to x in the sense that $\|x - y\| \geq \|x - x_e\|$ for any vector $y \in L$. This is the remarkable extremal property of the Fourier coefficients, to which we shall return below in the general situation.

Definition 6 If X is a vector space with inner product $\langle, \rangle$ and $l_1, l_2, \ldots, l_n, \ldots$ is an orthogonal system of nonzero vectors in X, then for each vector $x \in X$ one can form the series

$$x \sim \sum_{k=1}^{\infty} \frac{\langle x, l_k \rangle}{\langle l_k, l_k \rangle} l_k. \tag{18.8}$$

This series is the *Fourier series* of x in the orthogonal system $\{l_k\}$.

If the system $\{l_k\}$ is finite, the Fourier series reduces to its finite sum.

In the case of an orthonormal system $\{e_k\}$ the Fourier series of a vector $x \in X$ has a particularly simple expression:

$$x \sim \sum_{k=1}^{\infty} \langle x, e_k \rangle e_k. \tag{18.8'}$$

Example 6 Let $X = \mathcal{R}_2([-\pi, \pi], \mathbb{R})$. Consider the orthogonal system

$$\{1, \cos kx, \sin kx; k \in \mathbb{N}\}$$

of Example 1. To the function $f \in \mathcal{R}_2([-\pi, \pi], \mathbb{R})$ there corresponds a Fourier series

$$f \sim \frac{a_0(f)}{2} + \sum_{k=1}^{\infty} a_k(f) \cos kx + b_k(f) \sin kx$$

in this system. The coefficient $\frac{1}{2}$ is included in the zeroth term so as to give a unified appearance to the following formulas, which follow from the definition of the Fourier coefficients:

$$a_k(f) = \frac{1}{\pi} \int_{-\pi}^{\pi} f(x) \cos kx \, dx, \quad k = 0, 1, 2, \dots \tag{18.9}$$

$$b_k(f) = \frac{1}{\pi} \int_{-\pi}^{\pi} f(x) \sin kx \, dx, \quad k = 1, 2, \dots \tag{18.10}$$

Let us set $f(x) = x$. Then $a_k = 0$, $k = 0, 1, 2, \dots$, and $b_k = (-1)^{k+1} \frac{2}{k}$, $k = 1, 2, \dots$. Hence in this case we obtain

$$f(x) = x \sim \sum_{k=1}^{\infty} (-1)^{k+1} \frac{2}{k} \sin kx.$$

Example 7 Let us consider the orthogonal system $\{e^{ikx}; k \in \mathbb{Z}\}$ of Example 1 in the space $\mathcal{R}_2([-\pi, \pi], \mathbb{C})$. Let $f \in \mathcal{R}_2([-\pi, \pi], \mathbb{C})$. According to Definition 5 and relations (18.4), the Fourier coefficients $\{c_k(f)\}$ of f in the system $\{e^{ikx}\}$ are expressed by the following formula:

$$c_k(f) = \frac{1}{2\pi} \int_{-\pi}^{\pi} f(x) e^{-ikx} \, dx \left(= \frac{\langle f(x), e^{ikx} \rangle}{\langle e^{ikx}, e^{ikx} \rangle} \right). \tag{18.11}$$

Comparing Eqs. (18.9), (18.10), and (18.11) and taking account of Euler's formula $e^{i\varphi} = \cos \varphi + i \sin \varphi$, we obtain the following relations between the Fourier coefficients of a given function in the trigonometric systems written in real and complex forms:

$$c_k = \begin{cases} \frac{1}{2}(a_k - ib_k), & \text{if } k \geq 0, \\ \frac{1}{2}(a_{-k} + ib_{-k}), & \text{if } k < 0. \end{cases} \tag{18.12}$$

In order that the case $k = 0$ not be an exception in formulas (18.9) and (18.12), it is customary to use a_0 to denote not the Fourier coefficient itself, but rather its double, as was done above.

b. Basic General Properties of Fourier Coefficients and Series

The following geometric observation is key in this section.

Lemma (Orthogonal complement) *Let* $\{l_k\}$ *be a finite or countable system of nonzero pairwise orthogonal vectors in* X, *and suppose the Fourier series of* $x \in X$ *in the system* $\{l_k\}$ *converges to* $x_l \in X$.
  Then in the representation $x = x_l + h$ *the vector* h *is orthogonal to* x_l; *moreover,* h *is orthogonal to the entire linear subspace generated by the system of vectors* $\{l_k\}$, *and also to its closure in* X.

Proof Taking account of the properties of the inner product, we see that it suffices to verify that $\langle h, l_m \rangle = 0$ for every $l_m \in \{l_k\}$.
  We are given that

$$h = x - x_l = x - \sum_k \frac{\langle x, l_k \rangle}{\langle l_k, l_k \rangle} l_k.$$

Hence

$$\langle h, l_m \rangle = \langle x, l_m \rangle = \sum_k \frac{\langle x, l_k \rangle}{\langle l_k, l_k \rangle} \langle l_k, l_m \rangle = \langle x, l_m \rangle - \frac{\langle x, l_m \rangle}{\langle l_m, l_m \rangle} \langle l_m, l_m \rangle = 0. \qquad \square$$

Geometrically this lemma is transparent, and we have already essentially pointed it out when we considered a system of two orthogonal vectors in three-dimensional Euclidean space in Sect. 18.1.2a.
  On the basis of this lemma we can draw a number of important general conclusions on the properties of Fourier coefficients and Fourier series.

Bessel's Inequality

Taking account of the orthogonality of the vectors x_l and h in the decomposition $x = x_l + h$, we find by the Pythagorean theorem that $\|x\|^2 = \|x_l\|^2 + \|h\|^2 \geq \|x_l\|^2$ (the hypotenuse is never smaller than the leg). This relation, written in terms of Fourier coefficients, is called Bessel's inequality.
  Let us write it out. By the Pythagorean theorem

$$\|x_l\|^2 = \sum_k \left| \frac{\langle x, l_k \rangle}{\langle l_k, l_k \rangle} \right|^2 \langle l_k, l_k \rangle. \tag{18.13}$$

Hence

$$\sum_k \frac{|\langle x, l_k \rangle|^2}{\langle l_k l_k \rangle} \leq \|x\|^2. \tag{18.14}$$

This is *Bessel's inequality*. It has a particularly simply appearance for an orthonormal system of vectors $\{e_k\}$:

$$\sum_k |\langle x, e_k \rangle|^2 \leq \|x\|^2. \tag{18.15}$$

In terms of the Fourier coefficients α_k themselves Bessel's inequality (18.14) can be written as $\sum_k |\alpha_k|^2 \|l_k\|^2 \leq \|x\|^2$, which in the case of an orthonormal system reduces to $\sum_k |\alpha_k|^2 \leq \|x\|^2$.

We have included the absolute value sign in the Fourier coefficient, since we are allowing complex vectors spaces X. In this case the Fourier coefficient may assume complex values.

We note that in deriving Bessel's inequality we made use of the assumption that the vector x_l exists and that Eq. (18.13) holds. But if the system $\{l_k\}$ is finite, there is no doubt that the vector x_l does exist (that is, that the Fourier series converges in X). Hence inequality (18.14) holds for every finite subsystem of $\{l_k\}$, and then it must hold for the whole system as well.

Example 8 For the trigonometric system (see formulas (18.9) and (18.10)) Bessel's inequality has the form

$$\frac{|a_0(f)|^2}{2} + \sum_{k=1}^{\infty} |a_k(f)|^2 + |b_k(f)|^2 \leq \frac{1}{\pi} \int_{-\pi}^{\pi} |f|^2(x)\,dx. \tag{18.16}$$

For the system $\{e^{ikx}; k \in \mathbb{Z}\}$ (see formula (18.11)) Bessel's inequality can be written in a particularly elegant form:

$$\sum_{-\infty}^{+\infty} |c_k(f)|^2 \leq \frac{1}{2\pi} \int_{-\pi}^{\pi} |f|^2(x)\,dx. \tag{18.17}$$

Convergence of Fourier Series in a Complete Space

Suppose $\sum_k x^k e_k = \sum_k \langle x, e_k \rangle e_k$ is the Fourier series of the vector $x \in X$ in the orthonormal system $\{e_k\}$. By Bessel's inequality (18.15) the series $\sum_k |x^k|^2$ converges. By the Pythagorean theorem

$$\left\| x^m e_m + \cdots + x^n e_n \right\|^2 = |x^m|^2 + \cdots + |x^n|^2.$$

By the Cauchy convergence criterion for a series, the right-hand side of this equality becomes less than any $\varepsilon > 0$ for all sufficiently large values of m and $n > m$. Hence we then have

$$\left\| x^m e_m + \cdots + x^n e_n \right\| < \sqrt{\varepsilon}.$$

Consequently, the Fourier series $\sum_k x^k e_k$ satisfies the hypotheses of the Cauchy convergence criterion for series and therefore converges provided the original space X is complete in the metric induced by the norm $\|x\| = \sqrt{\langle x, x \rangle}$.

To simplify the writing we have carried out the reasoning for a Fourier series in an orthonormal system. But everything can be repeated for a Fourier series in any orthogonal system.

The Extremal Property of the Fourier Coefficients

We shall show that if the Fourier series $\sum_k x^k e_k = \sum_k \frac{\langle x, e_k \rangle}{\langle e_k, e_k \rangle} e_k$ of the vector $x \in X$ in the orthonormal system $\{e_k\}$ converges to a vector $x_l \in X$, then the vector x_l is precisely the one that gives the best approximation of x among all vectors $y = \sum_{k=1}^{\infty} \alpha_k e_k$ of the space L spanned by $\{e_k\}$, that is, for every $y \in L$,

$$\|x - x_l\| \le \|x - y\|,$$

and equality holds only for $y = x_l$.

Indeed, by the orthogonal complement lemma and the Pythagorean theorem,

$$\|x - y\|^2 = \left\|(x - x_l) + (x_l - y)\right\|^2 = \left\|h + (x_l - y)\right\|^2 =$$
$$= \|h\|^2 + \|x_l - y\|^2 \ge \|h\|^2 = \|x - x_l\|^2.$$

Example 9 Digressing slightly from our main purpose, which is the study of expansions in orthogonal systems, let us assume that we have an arbitrary system of linearly independent vectors $x_1, \ldots, x_n$ in X and are seeking the best approximation of a given vector $x \in X$ by linear combinations $\sum_{k=1}^{n} \alpha_k x_k$ of vectors of the system. Since we can use the orthogonalization process to construct an orthonormal system $e_1, \ldots, e_n$ that generates the same space L that is generated by the vectors $x_1, \ldots, x_n$, we can conclude from the extremal property of the Fourier coefficients that there exists a unique vector $x_l \in L$ such that $\|x - x_l\| = \inf_{y \in L} \|x - y\|$. Since the vector $h = x - x_l$ is orthogonal to the space L, from the equality $x_l + h = x$ we obtain the system of equations

$$\begin{cases} \langle x_1, x_1 \rangle \alpha_1 + \cdots + \langle x_n, x_1 \rangle \alpha_n = \langle x, x_1 \rangle, \\ \vdots \\ \langle x_1, x_n \rangle \alpha_1 + \cdots + \langle x_n, x_n \rangle \alpha_n = \langle x, x_n \rangle \end{cases} \tag{18.18}$$

for the coefficients $\alpha_1, \ldots, \alpha_n$ of the expansion $x_l = \sum_{k=1}^{n} \alpha_k x_k$ of the unknown vector x_l in terms of the vectors of the system $x_1, \ldots, x_n$. The existence and uniqueness of the solution of this system follow from the existence and uniqueness of the vector x_l. In particular, it follows from this by Cramer's theorem that the determinant of this system is nonzero. In other words, we have shown as a by-product that the Gram determinant of a system of linearly independent vectors is nonzero.

This approximation problem and the system of Eqs. (18.18) corresponding to it arise, as we have already noted, for example, in processing experimental data by Gauss' least-squares method. (See also Problem 1.)

c. Complete Orthogonal Systems and Parseval's Equality

Definition 7 The system $\{x_\alpha; \alpha \in A\}$ of vectors of a normed space X is *complete with respect to the set* $E \subset X$ (or *complete in* E) if every vector $x \in E$ can be approximated with arbitrary accuracy in the sense of the norm of X by finite linear combinations of vectors of the system.

If we denote by $L\{x_\alpha\}$ the linear span in X of the vectors of the system (that is, the set of all finite linear combinations of vectors of the system), Definition 7 can be restated as follows:

The system $\{x_\alpha\}$ is *complete with respect to the set* $E \subset X$ if E is contained in the closure $\overline{L}\{x_\alpha\}$ of the linear span of the vectors of the system.

Example 10 If $X = E^3$ and e_1, e_2, e_3 is a basis in E^3, then the system $\{e_1, e_2, e_2\}$ is complete in X. The system $\{e_1, e_2\}$ is not complete in X, but it is complete relative to the set $L\{e_1, e_2\}$ or any subset E of it.

Example 11 Let us regard the sequence of functions $1, x, x^2, \ldots$ as a system of vectors $\{x^k; k = 0, 1, 2, \ldots\}$ in the space $\mathcal{R}_2([a, b], \mathbb{R})$ or $\mathcal{R}_2([a, b], \mathbb{C})$. If $C[a, b]$ is a subspace of the continuous functions, then this system is complete with respect to the set $C[a, b]$.

Proof Indeed, for any function $f \in C[a, b]$ and for every number $\varepsilon > 0$, the Weierstrass approximation theorem implies that there exists an algebraic polynomial $P(x)$ such that $\max_{x \in [a,b]} |f(x) - P(x)| < \varepsilon$. But then

$$\|f - P\| := \sqrt{\int_a^b |f - P|^2(x)\, \mathrm{d}x} < \varepsilon\sqrt{b - a}$$

and hence one can approximate the function f in the sense of the norm of the space $\mathcal{R}_2([a, b])$ with arbitrary accuracy. $\square$

We note that, in contrast to the situation in Example 9, in the present case not every continuous function on the closed interval $[a, b]$ is a finite linear combination of the functions of this system; rather, such a function can only be approximated by such linear combinations. Thus $C[a, b] \subset \overline{L}\{x^n\}$ in the sense of the norm of the space $\mathcal{R}_2[a, b]$.

Example 12 If we remove one function, for example the function 1, from the system $\{1, \cos kx, \sin kx; k \in \mathbb{N}\}$, the remaining system $\{\cos kx, \sin kx; k \in \mathbb{N}\}$ is no longer complete in $\mathcal{R}_2([-\pi, \pi], \mathbb{C})$ or $\mathcal{R}_2([-\pi, \pi], \mathbb{R})$.

Proof Indeed, by the extremal property of the Fourier coefficients the best approximation of the function $f(x) \equiv 1$ among all the finite linear combinations

$$T_n(x) = \sum_{k=1}^{n} (a_k \cos kx + b_k \sin kx)$$

of any length n is given by the trigonometric polynomial $T_n(x)$ in which a_k and b_k are the Fourier coefficients of the function 1 with respect to the orthogonal system $\{\cos kx, \sin kx; k \in \mathbb{N}\}$. But by relations (18.5), such a polynomial of best approximation must be zero. Hence we always have

$$\|1 - T_n\| \geq \|1\| = \sqrt{\int_{-\pi}^{\pi} 1 \, dx} = \sqrt{2\pi} > 0,$$

and it is impossible to approximate 1 more closely than $\sqrt{2\pi}$ by linear combinations of functions of this system.   □

Theorem (Completeness conditions for an orthogonal system) *Let X be a vector space with inner product $\langle\ ,\ \rangle$, and $l_1, l_2, \ldots, l_n, \ldots$ a finite or countable system of nonzero pairwise orthogonal vectors in X. Then the following conditions are equivalent:*

a) *the system $\{l_k\}$ is complete with respect to the set*[5] *$E \subset X$;*
b) *for every vector $x \in E \subset X$ the following (Fourier series) expansion holds:*

$$x = \sum_k \frac{\langle x, l_k \rangle}{\langle l_k, l_k \rangle} l_k; \tag{18.19}$$

c) *for every vector $x \in E \subset X$ Parseval's*[6] *equality holds:*

$$\|x\|^2 = \sum_k \frac{|\langle x, l_k \rangle|^2}{\langle l_k, l_k \rangle}. \tag{18.20}$$

Equations (18.19) and (18.20) have a particularly simple form in the case of an orthonormal system $\{e_k\}$. In that case

$$x = \sum_k \langle x, e_k \rangle e_k \tag{18.19'}$$

and

$$\|x\| = \sum_k |\langle x, e_k \rangle|^2. \tag{18.20'}$$

---

[5]The set E may, in particular, consist of a single vector that is of interest for one reason or another.
[6]M.A. Parseval (1755–1836) – French mathematician who discovered this relation for the trigonometric system in 1799.

Thus the important Parseval equality (18.20) or (18.20′) is the Pythagorean theorem written in terms of the Fourier coefficients.

Let us now prove this theorem.

Proof a) ⇒ b) by virtue of the extremal property of Fourier coefficients;

b) ⇒ c) by the Pythagorean theorem;

c) ⇒ a) since by the lemma on the orthogonal complement (see Sect. b) above) the Pythagorean theorem implies

$$\left\| x - \sum_{k=1}^{n} \frac{\langle x, l_k \rangle}{\langle l_k, l_k \rangle} l_k \right\|^2 = \|x\|^2 - \left\| \sum_{k=1}^{n} \frac{\langle x, l_k \rangle}{\langle l_k, l_k \rangle} l_k \right\|^2 = \|x\|^2 - \sum_{k=1}^{n} \frac{|\langle x, l_k \rangle|^2}{\langle l_k, l_k \rangle}. \qquad \square$$

Remark We note that Parseval's equality implies the following simple necessary condition for completeness of an orthogonal system with respect to a set $E \subset X$: E does not contain a nonzero vector orthogonal to all the vectors in the system.

As a useful supplement to this theorem and the remark just made, we prove the following general proposition.

Proposition *Let X be a vector space with an inner product and $x_1, x_2, \ldots$ a system of linearly independent vectors in X. In order for the system $\{x_k\}$ to be complete in X,*

a) *a necessary condition is that there be no nonzero vector in X orthogonal to all the vectors in the system;*

b) *if X is a complete (Hilbert) space, it suffices that X contain no nonzero vector orthogonal to all the vectors in the system.*

Proof a) If the vector h is orthogonal to all the vectors in the system $\{x_k\}$, we conclude by the Pythagorean theorem that no linear combination of vectors in the system can differ from h by less than $\|h\|$. Hence, if the system is complete, then $\|h\| = 0$.

b) By the orthogonalization process we can obtain an orthonormal system $\{e_k\}$ whose linear span $L\{e_k\}$ is the same as the linear span $L\{x_k\}$ of the original system.

We now take an arbitrary vector $x \in X$. Since the space X is complete, the Fourier series of x in the system $\{e_k\}$ converges to a vector $x_e \in X$. By the lemma on the orthogonal complement, the vector $h = x - x_e$ is orthogonal to the space $L\{e_k\} = L\{x_k\}$. By hypothesis $h = 0$, so that $x = x_e$, and the Fourier series converges to the vector x itself. Thus the vector x can be approximated arbitrarily closely by finite linear combinations of vectors of the system $\{e_k\}$ and hence also by finite linear combinations of the vectors of the system $\{x_k\}$. $\qquad \square$

The hypothesis of completeness in part b) of this proposition is essential, as the following example shows.

Fig. 18.1

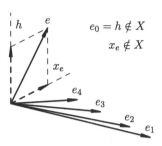

Example 13 Consider the space l_2 (see Sect. 10.1) of real sequences $a = (a^1, a^2, \ldots)$
for which $\sum_{j=1}^{\infty}(a^j)^2 < \infty$. We define the inner product of the vectors $a = (a^1, a^2, \ldots)$ and $b = (b^1, b^2, \ldots)$ in l_2 in the standard way: $\langle a, b \rangle := \sum_{j=1}^{\infty} a^j b^j$.

Now consider the orthonormal system $e_k = (\underbrace{0, \ldots, 0}_{k}, 1, 0, 0, \ldots)$, $k = 1, 2, \ldots$.

The vector $e_0 = (1, 0, 0, \ldots)$ does not belong to this system. We now add to the
system $\{e_k; k \in \mathbb{N}\}$ the vector $e = (1, 1/2, 1/2^2, 1/2^3, \ldots)$ and consider the linear
span $L\{e, e_1, e_2, \ldots\}$ of these vectors. We can regard this linear span as a vector
space X (a subspace of l_2) with the inner product from l_2.

We note that the vector $e_0 = (1, 0, 0, \ldots)$ obviously cannot be obtained as a finite
linear combination of vectors in the system $e, e_1, e_2, \ldots$, and therefore it does not
belong to X. At the same time, it can be approximated as closely as desired in l_2 by
such linear combinations, since $e - \sum_{k=1}^{n} \frac{1}{2^k} e_k = (1, 0, \ldots, 0, \frac{1}{2^{n+1}}, \frac{1}{2^{n+2}}, \ldots)$.

Hence we have established simultaneously that X is not closed in l_2 (and therefore X, in contrast to l_2, is not a complete metric space) and that the closure of X in
l_2 coincides with l_2, since the system $e_0, e_1, e_2, \ldots$ generates the entire space l_2.

We now observe that in $X = L\{e, e_1, e_2, \ldots\}$ there is no nonzero vector orthogonal to all the vectors $e_1, e_2, \ldots$.

Indeed, let $x \in X$, that is, $x = \alpha e + \sum_{k=1}^{n} \alpha_k e_k$, and let $\langle x, e_k \rangle = 0$, $k = 1, 2, \ldots$.
Then $\langle x, e_{n+1} \rangle = \frac{\alpha}{2^{n+1}} = 0$, that is, $\alpha = 0$. But then $\alpha_k = \langle x, e_k \rangle = 0$, $k = 1, \ldots, n$.

Hence we have constructed the required example: the orthogonal system
$e_1, e_2, \ldots$ is not complete in X, sine it is not complete in the closure of X, which
coincides with l_2.

This example is of course typically infinite-dimensional. Figure 18.1 represents
an attempt to illustrate what is going on.

We note that in the infinite-dimensional case (which is so characteristic of analysis) the possibility of approximating a vector arbitrarily closely by linear combinations of vectors of a system and the possibility of expanding the vector in a series of
vectors of the system are in general different properties of the system.

A discussion of this problem and the concluding Example 14 will clarify the
particular role of orthogonal systems and Fourier series for which these properties
hold or do not hold simultaneously (as the theorem proved above shows).

Definition 8 The system $x_1, x_2, \ldots, x_n, \ldots$ of vectors of a normed vector space X
is a *basis of* X if every finite subsystem of it consists of linearly independent vectors

and every vector $x \in X$ can be represented as $x = \sum_k \alpha_k x_k$, where α_k are coefficients from the scalar field of X and the convergence (when the sum is infinite) is understood in the sense of the norm on X.

How is the completeness of a system of vectors related to the property of being a basis?

In a finite-dimensional space X completeness of a system of vectors in X, as follows from considerations of compactness and continuity, is obviously equivalent to being a basis in X. In the infinite-dimensional case that is in general not so.

Example 14 Consider the set $C([-1, 1], \mathbb{R})$ of real-valued functions that are continuous on $[-1, 1]$ as a vector space over the field $\mathbb{R}$ with the standard inner product defined by (18.3). We denote this space by $C_2([-1, 1], \mathbb{R})$ and consider the system of linearly independent vectors $1, x, x^2, \ldots$ in it.

This system is complete in $C_2([-1, 1], \mathbb{R})$ (see Example 11), but is not a basis.

Proof We first show that if the series $\sum_{k=0}^{\infty} \alpha_k x^k$ converges in $C_2([-1, 1], \mathbb{R})$, that is, in the mean-square sense on $[-1, 1]$, then, regarded as a power series, it converges pointwise on the open interval $]-1, 1[$.

Indeed, by the necessary condition for convergence of a series, we have $\|\alpha_k x^k\| \to 0$ as $k \to \infty$. But

$$\|\alpha_k x^k\|^2 = \int_{-1}^{1} (\alpha_k x^k)^2 \, dx = \alpha_k^2 \frac{2}{2k+1}.$$

Hence $|\alpha_k| < \sqrt{2k+1}$ for all sufficiently large values of k. In that case the power series $\sum_{k=0}^{\infty} \alpha_k x^k$ definitely converges on the interval $]-1, 1[$.

We now denote the sum of this power series on $]-1, 1[$ by φ. We remark that on every closed interval $[a, b] \subset \,]-1, 1[$ the power series converges uniformly to $\varphi|_{[a,b]}$. Consequently it also converges in the sense of mean-square deviation.

It now follows that if a continuous function f is the sum of this series in the sense of convergence in $C_2([-1, 1], \mathbb{R})$, then f and φ are equal on $]-1, 1[$. But the function φ is infinitely differentiable. Hence if we take any function in $C_2([-1, 1], \mathbb{R})$ that is not infinitely differentiable on $]-1, 1[$ it cannot be expanded in a series in the system $\{x^k; k = 0, 1, \ldots\}$. $\qquad\square$

Thus, if we take, for example, the function $x = |x|$ and the sequence of numbers $\{\varepsilon_n = \frac{1}{n}; n \in \mathbb{N}\}$, we can construct a sequence $\{P_n(x); n \in \mathbb{N}\}$ of finite linear combinations $P_n(x) = \alpha_0 + \alpha_1 x + \cdots + \alpha_n x^n$ of elements of the system $\{x^k; k \in \mathbb{N}\}$ such that $\|f - P_n\| < \frac{1}{n}$, that is, $P_n \to f$ as $n \to \infty$. If necessary, one could assume that in each such linear combination $P_n(x)$ the coefficients can be assumed to have been chosen in the unique best-possible way (see Example 9). Nevertheless, the expansion $f = \sum_{k=0}^{\infty} \alpha_k x^k$ will not arise since in passing from $P_n(x)$ to $P_{n+1}(x)$, not only the coefficient α_{n+1} changes, but also possibly the coefficients $\alpha_0, \ldots, \alpha_n$.

If the system is orthogonal, this does not happen ($\alpha_0, \ldots, \alpha_n$ do not change) because of the extremal property of Fourier coefficients.

For example, one could pass from the system of monomials $\{x^k\}$ to the orthogonal system of Legendre polynomials and expand $f(x) = |x|$ in a Fourier series in that system.

18.1.3 *An Important Source of Orthogonal Systems of Functions in Analysis

We now give an idea as to how various orthogonal systems of functions and Fourier series in those systems arise in specific problems.

Example 15 (The Fourier method) Let us regard the closed interval $[0, l]$ as the equilibrium position of a homogeneous elastic string fastened at the endpoints of this interval, but otherwise free and capable of making small transverse oscillations about this equilibrium position. Let $u(x, t)$ be a function that describes these oscillations, that is, at each fixed instant of time $t = t_0$ the graph of the function $u(x, t_0)$ over the closed interval $0 \leq x \leq l$ gives the shape of the string at time t_0. This in particular, means that $u(0, t) = u(l, t) = 0$ at every instant t, since the ends of the string are clamped.

It is known (see for example Sect. 14.4) that the function $u(x, t)$ satisfies the equation

$$\frac{\partial^2 u}{\partial t^2} = a^2 \frac{\partial^2 u}{\partial x^2}, \tag{18.21}$$

where the positive coefficient a depends on the density and elastic constant of the string.

Equation (18.21) alone is of course insufficient to determine the function $u(x, t)$. From experiment we know that the motion $u(x, t)$ is uniquely determined if, for example, we prescribe the position $u(x, 0) = \varphi(x)$ of the string at some time $t = 0$ (which we shall call the initial instant) and the velocity $\frac{\partial u}{\partial t}(x, 0) = \psi(x)$ of the points of the string at that time. Thus, if we stretch the string into the shape $\varphi(x)$ and let it go, then $\psi(x) \equiv 0$.

Hence the problem of free oscillations of the string[7] that is fixed at the ends of the closed interval $[0, l]$ has been reduced to finding a solution $u(x, t)$ of Eq. (18.21) together with the boundary conditions

$$u(0, t) = u(l, t) = 0 \tag{18.22}$$

---

[7]We note that the foundations of the mathematical investigation of the oscillations of a string were laid by Brook Taylor.

and the initial conditions

$$u(x, 0) = \varphi(x), \qquad \frac{\partial u}{\partial t}(x, 0) = \psi(x). \tag{18.23}$$

To solve such problems there exists a very natural procedure called the method of *separation of variables* or the *Fourier method* in mathematics. It consists of the following. The solution $u(x, t)$ is sought in the form of a series $\sum_{n=1}^{\infty} X_n(x) T_n(t)$ whose terms $X(x) T(t)$ are solutions of an equation of special form (with variables separated) and satisfy the boundary conditions. In the present case, as we see, this is equivalent to expanding the oscillations $u(x, t)$ into a sum of simple harmonic oscillations (more precisely a sum of standing waves).

Indeed, if the function $X(x) T(t)$ satisfies Eq. (18.21), then $X(x) T''(t) = a^2 X''(x) T(t)$, that is,

$$\frac{T''(t)}{a^2 T(t)} = \frac{X''(x)}{X(x)}. \tag{18.24}$$

In Eq. (18.24) the independent variables x and t are on opposite sides of the equation (they have been separated), and therefore both sides actually represent the same constant λ. If we also take into account the boundary conditions $X(0) T(t) = X(l) T(t) = 0$ that the solution of stationary type must satisfy, we see that finding such a solution reduces to solving simultaneously the two equations

$$T''(t) = \lambda a^2 T(t), \tag{18.25}$$

$$X''(x) = \lambda X(x) \tag{18.26}$$

under the condition that $X(0) = X(l) = 0$.

It is easy to write the general solution of each of these equations individually:

$$T(t) = A \cos \sqrt{\lambda} at + B \sin \sqrt{\lambda} at, \tag{18.27}$$

$$X(x) = C \cos \sqrt{\lambda} x + D \sin \sqrt{\lambda} x. \tag{18.28}$$

If we attempt to satisfy the conditions $X(0) = X(l) = 0$, we find that for $\lambda \neq 0$ we must have $C = 0$, and, rejecting the trivial solution $D = 0$, we find that $\sin \sqrt{\lambda} l = 0$, from which we find $\sqrt{\lambda} = \pm n\pi / l$, $n \in N$.

Thus it turns out that the number λ in Eqs. (18.25) and (18.26) can be chosen only among a certain special series of numbers (the so-called *eigenvalues of the problem*), $\lambda_n = (n\pi / l)^2$, where $n \in N$. Substituting these values of λ into the expressions (18.27) and (18.28), we obtain a series of special solutions

$$u_n(x, t) = \sin n \frac{\pi}{l} x \left(A_n \cos n \frac{\pi a}{l} t + B_n \sin n \frac{\pi a}{l} t \right), \tag{18.29}$$

satisfying the boundary conditions $u_n(0, t) = u_n(l, t) = 0$ (and describing a standing wave of the form $\Phi(x) \cdot \sin(\omega t + \theta)$, in which each point $x \in [0, l]$ undergoes simple harmonic oscillations with its own amplitude $\Phi(x)$ but the same frequency ω for all points).

The quantities $\omega_n = n\frac{\pi a}{l}$, $n \in \mathbb{N}$, are called, for natural reasons, the *natural frequencies of the string*, and its simplest harmonic oscillations (18.29) are called the natural oscillations of the string. The oscillation $u_1(x, t)$ with smallest natural frequency is often called the *fundamental tone of the string* and the other natural frequencies $u_2(x, t), u_3(x, t), \dots$ are called *overtones* (it is the overtones that form the sound quality, called the *timbre*, characteristic of each particular musical instrument).

We now wish to represent the oscillation $u(x, t)$ we are seeking as a sum $\sum_{n=1}^{\infty} u_n(x, t)$ of the natural oscillations of the string. The boundary conditions (18.22) are automatically satisfied in this case, and we need worry only about the initial conditions (18.23), which mean that

$$\varphi(x) = \sum_{n=1}^{\infty} A_n \sin n\frac{\pi}{l}x \tag{18.30}$$

and

$$\psi(x) = \sum_{n=1}^{\infty} n\frac{\pi a}{l} B_n \sin n\frac{\pi}{l}x. \tag{18.31}$$

Thus the problem has been reduced to finding the coefficients A_n and B_n, which up to now have been free, or, what is the same, to expanding the functions φ and ψ in Fourier series in the system $\{\sin n\frac{\pi}{l}x; n \in \mathbb{N}\}$, which is orthogonal on the interval $[0, l]$.

It is useful to remark that the functions $\{\sin n\frac{\pi}{l}x; n \in \mathbb{N}\}$, which arose from Eq. (18.26) can be regarded as eigenvectors of the linear operator $A = \frac{d^2}{dx^2}$ corresponding to the eigenvalues $\lambda_n = n\frac{\pi}{l}$, which in turn arose from the condition that the operator A acts on the space of functions in $C^{(2)}[0, l]$ that vanish at the endpoints of the closed interval $[0, l]$. Hence Eqs. (18.30) and (18.31) can be interpreted as expansions in eigenvectors of this linear operator.

The linear operators connected with particular problems are one of the main sources of orthogonal systems of functions in analysis.

We now recall another fact known from algebra, which reveals the reason why such systems are orthogonal.

Let Z be a vector space with inner product $\langle , \rangle$, and let E be a subspace (possibly equal to Z itself) that is dense in Z. A linear operator $A : E \to Z$ is *symmetric* if $\langle Ax, y \rangle = \langle x, Ay \rangle$ for every pair of vectors $x, y \in E$. Then: *eigenvectors of a symmetric operator corresponding to different eigenvalues are orthogonal.*

Proof Indeed, if $Au = \alpha u$ and $Av = \beta v$, and $\alpha \neq \beta$, then

$$\alpha \langle u, v \rangle = \langle Au, v \rangle = \langle u, Av \rangle = \beta \langle u, v \rangle,$$

from which it follows that $\langle u, v \rangle = 0$.   □

It is now useful to look at Example 3 from this point of view. There we were essentially considering the eigenfunctions of the operator $A = (\frac{d^2}{dx^2} + q(x))$ operating on the space of functions in $C^{(2)}[a, b]$ that vanish at the endpoints of the closed interval $[a, b]$. Through integration by parts one can verify that this operator is symmetric on this space (with respect to the standard inner product (18.4)), so that the result of Example 4 is a particular manifestation of this algebraic fact.

In particular, when $q(x) \equiv 0$ the operator A becomes $\frac{d^2}{dx^2}$, which for $[a, b] = [0, l]$ occurred in the last example (Example 15).

We note also that in this example the question reduced to expanding the functions φ and ψ (see relations (18.30) and (18.31)) in a series of eigenfunctions of the operator $A = \frac{d^2}{dx^2}$. Here of course the question arises whether it is theoretically possible to form such an expansion, and this question is equivalent, as we now understand, to the question of the completeness of the system of eigenfunctions for the operator in question in the given space of functions.

The completeness of the trigonometric system (and certain other particular systems of orthogonal functions) in $\mathcal{R}_2[-\pi, \pi]$ seems to have been stated explicitly for the first time by Lyapunov.[8] The completeness of the trigonometric system in particular was implicitly present in the work of Dirichlet devoted to studying the convergence of trigonometric series. Parseval's equality, which is equivalent to completeness for the trigonometric system, as already noted, was discovered by Parseval at the turn of the nineteenth century. In its general form, the question of completeness of orthogonal systems and their application in the problems of mathematical physics were one of the main subjects of the research of Steklov,[9] who introduced the very concept of completeness (closedness) of an orthogonal system into mathematics. In studying completeness problems, by the way, he made active use of the method of integral averaging (smoothing) of a function (see Sects. 17.4 and 17.5), which for that reason is often called the *Steklov averaging method*.

18.1.4 Problems and Exercises

1. *The method of least squares.* The dependence $y = f(x_1, \ldots, x_n)$ of the quantity y on the quantities $x_1, \ldots, x_n$ is studied experimentally. As a result of m $(\geq n)$

---

[8] A.M. Lyapunov (1857–1918) – Russian mathematician and specialist in mechanics, a brilliant representative of the Chebyshev school, creator of the theory of stability of motion. He successfully studied various areas of mathematics and mechanics.

[9] V.A. Steklov (1864–1926) – Russian/Soviet mathematician, a representative of the Petersburg mathematical school founded by Chebyshev and founder of the school of mathematical physics in the USSR. The Mathematical Institute of the Russian Academy of Sciences bears his name.

experiments, a table was obtained

x_1	x_2	$\cdots$	x_n	y
a_1^1	a_2^1	$\cdots$	a_n^1	b^1
$\vdots$	$\vdots$	$\ddots$	$\vdots$	$\vdots$
a_1^m	a_2^m	$\cdots$	a_n^m	b^m

each of whose rows contains a set $(a_1^i, a_2^i, \ldots, a_n^i)$ of values of the parameters $x_1, x_2, \ldots, x_n$ and the value b^i of the quantity y corresponding to them, measured by some device with a certain precision. From these experimental data we would like to obtain an empirical formula of the form $y = \sum_{i=1}^n \alpha_i x_i$ convenient for computation. The coefficients $\alpha_1, \alpha_2, \ldots, \alpha_n$ of the required linear function are to be chosen so as to minimize the quantity $\sqrt{\sum_{k=1}^m (b^k - \sum_{k=1}^n \alpha_i a_i^k)^2}$, which is the mean-square deviation of the data obtained using the empirical formula from the results obtained in the experiments.

Interpret this problem as the problem of best approximation of the vector $(b^1, \ldots, b^m)$ be linear combinations of the vectors $(a_i^1, \ldots, a_i^m)$, $i = 1, \ldots, n$ and show that the question reduces to solving a system of linear equations of the same type as Eq. (18.18).

2. a) Let $C[a, b]$ be the vector space of functions that are continuous on the closed interval $[a, b]$ with the metric of uniform convergence and $C_2[a, b]$ the same vector space but with the metric of mean-square deviation on that closed interval (that is, $d(f, g) = \sqrt{\int_a^b |f - g|^2(x)\,dx}$. Show that if functions converge in $C[a, b]$, they also converge in $C_2[a, b]$, but not conversely, and that the space $C_2[a, b]$ is not complete, in contrast to $C[a, b]$.

b) Explain why the system of functions $\{1, x, x^2, \ldots\}$ is linearly independent and complete in $C_2[a, b]$, but is not a basis in that space.

c) Explain why the Legendre polynomials are a complete orthogonal system and also a basis in $C_2[-1, 1]$.

d) Find the first four terms of the Fourier expansion of the function $\sin \pi x$ on the interval $[-1, 1]$ in the system of Legendre polynomials.

e) Show that the square of the norm $\| P_n \|$ in $C_2[-1, 1]$ of the nth Legendre polynomial is

$$\frac{2}{2n+1} \quad \left(= (-1)^n \frac{(n+1)(n+2)\cdots 2n}{n!2^{2n}} \int_{-1}^1 (x^2 - 1)^n \, dx \right).$$

f) Prove that among all polynomials of given degree n with leading coefficient 1, the Legendre polynomial $\widetilde{P}_n(x)$ is the one closest to zero on the interval $[-1, 1]$.

g) Explain why the equality

$$\int_{-1}^1 |f|^2(x)\,dx = \sum_{n=0}^{\infty} \left(n + \frac{1}{2} \right) \left| \int_{-1}^1 f(x) P_n(x)\,dx \right|^2,$$

where $\{P_0, P_1, \ldots\}$ is the system of Legendre polynomials, necessarily holds for every function $f \in C_2([-1, 1], \mathbb{C})$.

3. a) Show that if the system $\{x_1, x_2, \ldots\}$ of vectors is complete in the space X and X is an everywhere-dense subset of Y, then $\{x_1, x_2, \ldots\}$ is also complete in Y.

b) Prove that the vector space $C[a, b]$ of functions that are continuous on the closed interval $[a, b]$ is everywhere dense in the space $\mathcal{R}_2[a, b]$. (It was asserted in Problem 5g of Sect. 17.5 that this is true even for infinitely differentiable functions of compact support on $[a, b]$.)

c) Using the Weierstrass approximation theorem, prove that the trigonometric system $\{1, \cos kx, \sin kx; k \in \mathbb{N}\}$ is complete in $\mathcal{R}_2[-\pi, \pi]$.

d) Show that the systems $\{1, x, x^2, \ldots\}$ and $\{1, \cos kx, \sin kx; k \in \mathbb{N}\}$ are both complete in $\mathcal{R}_2[-\pi, \pi]$, but the first is not a basis in this space and the second is.

e) Explain why Parseval's equality

$$\frac{1}{\pi} \int_{-\pi}^{\pi} |f|^2(x) \, dx = \frac{|a_0|^2}{2} + \sum_{k=1}^{\infty} |a_k|^2 + |b_k|^2$$

holds, where the numbers a_k and b_k are defined by (18.9) and (18.10).

f) Using the result of Example 8, now show that $\sum_{n=1}^{\infty} \frac{1}{n^2} = \frac{\pi^2}{6}$.

4. Orthogonality with a weight function.

a) Let $p_0, p_1, \ldots, p_n$ be continuous functions that are positive in the domain D. Verify that the formula

$$\langle f, g \rangle = \sum_{k=0}^{n} \int_D p_k(x) f^{(k)}(x) \overline{g}^{(k)}(x) \, dx$$

defines an inner product in $C^{(n)}(D, \mathbb{C})$.

b) Show that when functions that differ only on sets of measure zero are identified, the inner product

$$\langle f, g \rangle = \int_D p(x) f(x) \overline{g}(x) \, dx,$$

involving a positive continuous function p can be introduced in the space $\mathcal{R}(D, \mathbb{C})$.

The function p here is called a *weight function*, and if $\langle f, g \rangle = 0$, we say that *the functions f and g are orthogonal with weight p*.

c) Let $\varphi : D \to G$ be a diffeomorphism of the domain $D \subset \mathbb{R}^n$ onto the domain $G \subset \mathbb{R}^n$, and let $\{u_k(y); k \in \mathbb{N}\}$ be a system of functions in G that is orthogonal with respect to the standard inner product (18.2) or (18.3). Construct a system of functions that are orthogonal in D with weight $p(x) = |\det \varphi'(x)|$ and also a system of functions that are orthogonal in D in the sense of the standard inner product.

d) Show that the system of functions $\{e_{m,n}(x, y) = e^{i(mx+ny)}; m, n \in \mathbb{N}\}$ is orthogonal on the square $I = \{(x, y) \in \mathbb{R}^2 \mid |x| \le \pi \wedge |y| \le \pi\}$.

e) Construct a system of functions orthogonal on the two-dimensional torus $T^2 \subset \mathbb{R}^2$ defined by the parametric equations given in Example 4 of Sect. 12.1. The inner product of functions f and g on the torus is understood as the surface integral $\int_{T^2} f\bar{g}\,d\sigma$.

5. a) It is known from algebra (and we have also proved it in the course of discussing constrained extremal problems) that every symmetric operator $A : E^n \to E^n$ on n-dimensional Euclidean space E^n has nonzero eigenvectors. In the infinite-dimensional case this is generally not so.

Show that the linear operator $f(x) \to xf(x)$ of multiplication by the independent variable is symmetric in $C_2([a, b], \mathbb{R})$, but has no nonzero eigenvectors.

b) A *Sturm*[10]*–Liouville problem* that often arises in the equations of mathematical physics is to find a nonzero solution of an equation $u''(x) + [q(x) + \lambda p(x)]u(x) = 0$ on the interval $[a, b]$ satisfying certain boundary conditions, for example $u(a) = u(b) = 0$.

Here it is assumed that the functions $p(x)$ and $q(x)$ are known and continuous on the interval $[a, b]$ in question and that $p(x) > 0$ on $[a, b]$.

We have encountered such a problem in Example 15, where it was necessary to solve Eq. (18.26) under the condition $X(0) = X(l) = 0$. In this case we had $q(x) \equiv 0$, $p(x) \equiv 1$, and $[a, b] = [0, l]$. We have verified that a Sturm–Liouville problem may in general turn out to be solvable only for certain special values of the parameter λ, which are therefore called the *eigenvalues* of the corresponding *Sturm–Liouville problem*.

Show that if the functions f and g are solutions of a Sturm–Liouville problem corresponding to eigenvalues $\lambda_f \neq \lambda_g$, then the equality $\frac{d}{dx}(g'f - f'g) = (\lambda_f - \lambda_g)pfg$ holds on $[a, b]$ and the functions f and g are orthogonal on $[a, b]$ with weight p.

c) It is known (see Sect. 14.4) that the small oscillations of an inhomogeneous string fastened at the ends of the closed interval $[a, b]$ are described by the equation $(pu'_x)'_x = \rho u''_{tt}$, where $u = u(x, t)$ is the function that gives the shape of the string at each time t, $\rho = \rho(x)$ is the linear density, and $p = p(x)$ is the elastic constant at the point $x \in [a, b]$. The clamping conditions mean that $u(a, t) = u(b, t) = 0$.

Show that if we seek the solution of this equation in the form $X(x)T(t)$, the question reduces to a system $T'' = \lambda T$, $(pX')' = \lambda \rho X$, in which λ is the same number in both equations.

Thus a Sturm–Liouville problem arises for the function $X(x)$ on the closed interval $[a, b]$, which is solvable only for particular values of the parameter λ (the eigenvalues). (Assuming that $p(x) > 0$ on $[a, b]$ and that $p \in C^{(1)}[a, b]$ we can obviously bring the equation $(pX')' = \lambda \rho X$ into a form in which the first derivative is missing by the change of variable $x = \int_a^x \frac{d\xi}{p(\xi)}$.)

---

[10] J.Ch.F. Sturm (1803–1855) – French mathematician (and, as it happens, an honorary foreign member of the Petersburg Academy of Sciences); his main work was in the solution of boundary-value problems for the equations of mathematical physics.

d) Verify that the operator $S(u) = (p(x)u'(x))' - q(x)u(x)$ on the space of functions in $C^{(2)}[a, b]$ that satisfy the condition $u(a) = u(b) = 0$ is symmetric on this space. (That is, $\langle Su, v \rangle = \langle u, Sv \rangle$, where $\langle , \rangle$ is the standard inner product of real-valued functions.) Verify also that the eigenfunctions of the operator S corresponding to different eigenvalues are orthogonal.

e) Show that the solutions X_1 and X_2 of the equation $(pX')' = \lambda \rho X$ corresponding to different values λ_1 and λ_2 of the parameter λ and vanishing at the endpoints of the closed interval $[a, b]$ are orthogonal on $[a, b]$ with weight $\rho(x)$.

6. *The Legendre polynomials as eigenfunctions.*

a) Using the expression of the Legendre polynomials $P_n(x)$ given in Example 5 and the equality $(x^2 - 1)^n = (x - 1)^n(x + 1)^n$, show that $P_n(1) = 1$.

b) By differentiating the identity $(x^2 - 1)\frac{d}{dx}(x^2 - 1)^n = 2nx(x^2 - 1)^n$, show that $P_n(x)$ satisfies the equation

$$\left(x^2 - 1\right) \cdot P_n''(x) + 2x \cdot P_n'(x) - n(n + 1)P_n(x) = 0.$$

c) Verify that the operator

$$A := \left(x^2 - 1\right)\frac{d^2}{dx^2} + 2x\frac{d}{dx} = \frac{d}{dx}\left[(x^2 - 1)\frac{d}{dx}\right]$$

is symmetric in the space $C^{(2)}[-1, 1] \subset \mathcal{R}_2[-1, 1]$. Then, starting from the relation $A(P_n) = n(n + 1)P_n$, explain why the Legendre polynomials are orthogonal.

d) Using the completeness of the system $\{1, x, x^2, \ldots\}$ in $C^{(2)}[-1, 1]$, show that the dimension of the eigenspace of the operator A corresponding to the eigenvalue $n(n + 1)$ cannot be larger than 1.

e) Prove that the operator $A = \frac{d}{dx}[(x^2 - 1)\frac{d}{dx}]$ cannot have eigenfunctions in the space $C^{(2)}[-1, 1]$ except those in the system $\{P_0(x), P_1(x), \ldots\}$ of Legendre polynomials, nor any eigenvalues different from the number $\{n(n + 1); n = 0, 1, 2, \ldots\}$.

7. *Spherical functions.*

a) In solving various problems in $\mathbb{R}^3$ (for example, problems of potential theory connected with Laplace's equation $\Delta u = 0$) the solutions are sought in the form of a series of solutions of a special form. Such solutions are taken to be homogeneous polynomials $S_n(x, y, z)$ of degree n satisfying the equation $\Delta u = 0$. Such polynomials are called *harmonic polynomials*. In spherical coordinates (r, φ, θ) a harmonic polynomial $S_n(x, y, z)$ obviously has the form $r^n Y_n(\theta, \varphi)$. The functions $Y_n(\theta, \varphi)$ that arise in this way, depending only on the coordinates $0 \le \theta \le \pi$ and $0 \le \varphi \le 2\pi$ on the sphere, are called *spherical functions*. (They are trigonometric polynomials in two variables with $2n + 1$ free coefficients in Y_n, this number coming from the condition $\Delta S_n = 0$.)

Using Green's formula, show that for $m \ne n$ the functions Y_m and Y_n are orthogonal on the unit sphere in $\mathbb{R}^3$ (in the sense of the inner product $\langle Y_m, Y_n \rangle = \int\int Y_m \cdot Y_n \, d\sigma$, where the surface integral extends over the sphere $r = 1$).

b) Starting from the Legendre polynomials, one can also introduce the polyno-
mials $P_{n,m} = (1 - x^2)^{m/2} \frac{d^m P_n}{dx^m}(x)$, $m = 1, 2, \ldots, n$, and consider the functions

$$P_n(\cos\theta), \quad P_{n,m}(\cos\theta)\cos m\varphi, \quad P_{n,m}(\sin\theta)\sin m\varphi. \qquad (*)$$

It turns out that every spherical function $Y_n(\theta, \varphi)$ with index n is a linear combination of these functions. Accepting this result and taking account of the orthogonality of the trigonometric system, show that the functions of the system (*) form
an orthogonal basis in the $(2n + 1)$-dimensional space of spherical functions of a
given index n.

8. *The Hermite polynomials.* In the study of the equation of a linear oscillator in
quantum mechanics it becomes necessary to consider functions of class $C^{(2)}(\mathbb{R})$
with the inner product $\langle f, g \rangle = \int_{-\infty}^{+\infty} f\overline{g}\,dx$ in $C^{(2)}(\mathbb{R}) \subset (\mathbb{R}, \mathbb{C})$, and also the special functions $H_n(x) = (-1)^n e^{x^2} \frac{d^n}{dx^n}(e^{-x^2})$, $n = 0, 1, 2, \ldots$.

a) Show that $H_0(x) = 1$, $H_1(x) = 2x$, $H_2(x) = 4x^2 - 2$.
b) Prove that $H_n(x)$ is a polynomial of degree n. The system of functions
$\{H_0(x), H_1(x), \ldots\}$ is called the system of *Hermite polynomials*.
c) Verify that the function $H_n(x)$ satisfies the equation $H_n''(x) - 2xH_n'(x) + 2nH_n(x) = 0$.
d) The functions $\psi_n(x) = e^{-x^2} H_n(x)$ are called the *Hermite functions*. Show
that $\psi_n''(x) + (2n + 1 - x^2)\psi_n(x) = 0$, and that $\psi_n(x) \to 0$ as $x \to \infty$.
e) Verify that $\int_{-\infty}^{+\infty} \psi_n\psi_m\,dx = 0$ if $m \neq n$.
f) Show that the Hermite polynomials are orthogonal on $\mathbb{R}$ with weight e^{-x^2}.

9. *The Chebyshev–Laguerre*[11] *polynomials* $\{L_n(x); n = 0, 1, 2, \ldots\}$ can be defined
by the formula $L_n(x) := e^x \frac{d(x^n e^{-x})}{dx^n}$.
    Verify that

a) $L_n(x)$ is a polynomial of degree n;
b) the function $L_n(x)$ satisfies the equation

$$xL_n''(x) + (1 - x)L_n'(x) + nL_n(x) = 0;$$

c) the system $\{L_n; n = 0, 1, 2, \ldots\}$ of Chebyshev–Laguerre polynomials is orthogonal with weight e^{-x} on the half-line $[0, +\infty[$.

10. *The Chebyshev polynomials* $\{T_0(x) \equiv 1, T_n(x) = 2^{1-n}\cos n(\arccos x); n \in \mathbb{N}\}$
for $|x| < 1$ can be defined by the formula

$$T_n(x) = \frac{(-2)^n n!}{(2n)!}\sqrt{1 - x^2}\frac{d^n}{dx^n}(1 - x^2)^{n-\frac{1}{2}}.$$

Show that:

---

[11] E.N. Laguerre (1834–1886) – French mathematician.

a) $T_n(x)$ is a polynomial of degree n;

b) $T_n(x)$ satisfies the equation

$$\left(1 - x^2\right)T_n''(x) - xT_n'(x) + n^2 T_n(x) = 0;$$

c) the system $\{T_n; n = 0, 1, 2, \ldots\}$ of Chebyshev polynomials is orthogonal with weight $p(x) = \dfrac{1}{\sqrt{1-x^2}}$ on the interval $]-1, 1[$.

11. a) In probability theory and theory of functions one encounters the following *system of Rademacher*[12] *functions*: $\{\psi_n(x) = \varphi(2^n x); n = 0, 1, 2, \ldots\}$, where $\varphi(t) = \operatorname{sgn}(\sin 2\pi t)$. Verify that this is an orthonormal system on the closed interval $[0, 1]$.

b) The *system of Haar*[13] *functions* $\{\chi_{n,k}(x)\}$, where $n = 0, 1, 2, \ldots$ and $k = 1, 2, 2^2, \ldots$ is defined by the relations

$$\chi_{n,k}(x) = \begin{cases} 1, & \text{if } \frac{2k-2}{2^{n+1}} < x < \frac{2k-1}{2^{n+1}}, \\ -1, & \text{if } \frac{2k-1}{2^{n+1}} < x < \frac{2k}{2^{n+1}}, \\ 0 & \text{at all other points of } [0, 1]. \end{cases}$$

Verify that Haar system is orthogonal on the closed interval $[0, 1]$.

12. a) Show that every n-dimensional vector space with an inner product is isometrically isomorphic to the arithmetic Euclidean space $\mathbb{R}^n$ of the same dimensions.

b) We recall that a metric space is called separable if it contains a countable everywhere-dense subset. Prove that if a vector space with inner product is separable as a metric space with the metric induced by the inner product, then it has a countable orthonormal basis.

c) Let X be a separable Hilbert space (that is, X is a separable and complete metric space with the metric induced by the inner product in X). Taking an orthonormal basis $\{e_i; i \in \mathbb{N}\}$ in X, we construct the mapping $X \ni x \mapsto (c_1, c_2, \ldots)$, where $c_i = \langle x, e_i \rangle$ are the Fourier coefficients of the expansion of the vector in the basis $\{e_i\}$. Show that this mapping is a bijective, linear, and isometric mapping of X onto the space l_2 considered in Example 14.

d) Using Fig. 18.1, exhibit the basic idea of the construction of Example 14, and explain why it comes about precisely because the space in question is infinite-dimensional.

e) Explain how to construct an analogous example in the space of functions $C[a, b] \subset \mathcal{R}_2[a, b]$.

---

[12]H.A. Rademacher (1892–1969) – German mathematician (American after 1936).

[13]A. Haar (1885–1933) – Hungarian mathematician.

18.2 Trigonometric Fourier Series

18.2.1 Basic Types of Convergence of Classical Fourier Series

a. Trigonometric Series and Trigonometric Fourier Series

A classical trigonometric series is a series of the form[14]

$$\frac{a_0}{2} + \sum_{k=1}^{\infty} a_k \cos kx + b_k \sin kx, \qquad (18.32)$$

obtained on the basis of the trigonometric system $\{1, \cos kx, \sin kx; k \in \mathbb{N}\}$. The coefficients $\{a_0, a_k, b_k; k \in \mathbb{N}\}$ here are real or complex numbers. The partial sums of the trigonometric series (18.32) are the *trigonometric polynomials*

$$T_n(x) = \frac{a_0}{2} + \sum_{k=1}^{n} a_k \cos kx + b_k \sin kx \qquad (18.33)$$

corresponding to degree n.

If the series (18.32) converges pointwise on $\mathbb{R}$, its sum $f(x)$ is obviously a function of period 2π on $\mathbb{R}$. It is completely determined by its restriction to any closed interval of length 2π.

Conversely, given a function of period 2π on $\mathbb{R}$ (oscillations, a signal, and the like) that we wish to expand into a sum of certain canonical periodic functions, the first claimants for such canonical status are the simplest functions of period 2π, namely $\{1, \cos kx, \sin kx; k \in \mathbb{N}\}$, which are simple harmonic oscillations of entire frequencies.

Suppose we have succeeded in representing a continuous function as the sum

$$f(x) = \frac{a_0}{2} + \sum_{k=1}^{\infty} a_k \cos kx + b_k \sin kx \qquad (18.34)$$

of a trigonometric series that converges uniformly to it. Then the coefficients of the expansion (18.34) can be easily and uniquely found.

Multiplying Eq. (18.34) successively by each function of the system

$$\{1, \cos ks, \sin kx; k \in \mathbb{N}\},$$

using the fact that termwise integration is possible in the resulting uniformly convergent series, and taking account of the relations

$$\int_{-\pi}^{\pi} 1^2 \, dx = 2\pi,$$

---

[14]Writing the constant term in the form $a_0/2$, which is convenient for Fourier series, is not obligatory here.

$$\int_{-\pi}^{\pi} \cos mx \cos nx \, dx = \int_{-\pi}^{\pi} \sin mx \sin nx \, dx = 0 \quad \text{for } m \neq n, m, n \in \mathbb{N},$$

$$\int_{-\pi}^{\pi} \cos^2 nx \, dx = \int_{-\pi}^{\pi} \sin^2 nx \, dx = \pi, \quad n \in \mathbb{N},$$

we find the coefficients

$$a_k = a_k(f) = \frac{1}{\pi} \int_{-\pi}^{\pi} f(x) \cos kx \, dx, \quad k = 0, 1, \ldots, \tag{18.35}$$

$$b_k = b_k(f) = \frac{1}{\pi} \int_{-\pi}^{\pi} f(x) \sin kx \, dx, \quad k = 1, 2, \ldots \tag{18.36}$$

of the expansion (18.34) of the function f in a trigonometric series.

We have arrived at the same coefficients that we would have had if we had regarded (18.34) as the expansion of the vector $f \in \mathcal{R}_2[-\pi, \pi]$ in the orthogonal system $\{1, \cos kx, \sin kx; k \in \mathbb{N}\}$. This is not surprising, since the uniform convergence of the series (18.34) of course implies convergence in the mean on the closed interval $[-\pi, \pi]$, and then the coefficients of (18.34) must be the Fourier coefficients of the function f in the given orthogonal system (see Sect. 18.1).

Definition 1 If the integrals (18.35) and (18.36) have meaning for a function f, then the trigonometric series

$$f \sim \frac{a_0(f)}{2} + \sum_{k=1}^{\infty} a_k(f) \cos kx + b_k(f) \sin kx \tag{18.37}$$

assigned to f is called the *trigonometric Fourier series of f*.

Since there will be no Fourier series in this section except trigonometric Fourier series, we shall occasionally allow ourselves to drop the word "trigonometric" and speak of just "the Fourier series of f".

In the main we shall be dealing with functions of class $\mathcal{R}([-\pi, \pi], \mathbb{C})$, or, slightly more generally, with functions whose squared absolute values are integrable (possibly in the improper sense) on the open interval $]-\pi, \pi[$. We retain our previous notation $\mathcal{R}_2[-\pi, \pi]$ to denote the vector space of such functions with the standard inner product

$$\langle f, g \rangle = \int_{-\pi}^{\pi} f \overline{g} \, dx. \tag{18.38}$$

Bessel's inequality

$$\frac{|a_0(f)|^2}{2} + \sum_{k=1}^{\infty} |a_k(f)|^2 + |b_k(f)|^2 \leq \frac{1}{\pi} \int_{-\pi}^{\pi} |f|^2(x) \, dx, \tag{18.39}$$

which holds for every function $f \in \mathcal{R}_2([-\pi, \pi], \mathbb{C})$, shows that by no means every trigonometric series (18.32) can be the Fourier series of some function $f \in \mathcal{R}_2[-\pi, \pi]$.

Example 1 The trigonometric series

$$\sum_{k=1}^{\infty} \frac{\sin kx}{\sqrt{k}},$$

as we already know (see Example 7 of Sect. 16.2) converges on $\mathbb{R}$, but is not the Fourier series of any function $f \in \mathcal{R}_2[-\pi, \pi]$, since the series $\sum_{k=1}^{\infty}(\frac{1}{\sqrt{k}})^2$ diverges.

Thus, arbitrary trigonometric series (18.32) will not be studied here, only Fourier series (18.37) of functions in $\mathcal{R}_2[-\pi, \pi]$, and functions that are absolutely integrable on $]-\pi, \pi[$.

b. Mean Convergence of a Trigonometric Fourier Series

Let

$$S_n(x) = \frac{a_0(f)}{2} + \sum_{k=1}^{n} a_k(f) \cos kx + b_k(f) \sin kx, \qquad (18.40)$$

be the nth partial sum of the Fourier series of the function $f \in \mathcal{R}_2[-\pi, \pi]$. The deviation of S_n from f can be measured both in the natural metric of the space $\mathcal{R}_2[-\pi, \pi]$ induced by the inner product (18.38), that is, in the sense of the *mean-square deviation*

$$\|f - S_n\| = \sqrt{\int_{-\pi}^{\pi} |f - S_n|^2(x)\, dx} \qquad (18.41)$$

of S_n from f on the interval $[-\pi, \pi]$, and in the sense of pointwise convergence on that interval.

The first of these two kinds of convergence was studied for an arbitrary series in Sect. 18.1. Making the results obtained there specific in the context of a trigonometric Fourier series involves first of all noting that the trigonometric system $\{1, \cos kx, \sin kx; k \in \mathbb{N}\}$ is complete in $\mathcal{R}_2[-\pi, \pi]$. (This has already been noted in Sect. 18.1 and will be proved independently in Sect. 18.2.4 of the present section.)

Hence, the fundamental theorem of Sect. 18.1 enables us to say in the present case that the following theorem is true.

Theorem 1 (Mean convergence of a trigonometric Fourier series) *The Fourier series* (18.37) *of any function* $f \in \mathcal{R}_2([-\pi, \pi], \mathbb{C})$ *converges to the function in the* mean (18.41), *that is,*

$$f(x) \underset{\mathcal{R}_2}{=} \frac{a_0(f)}{2} + \sum_{k=1}^{\infty} a_k(f) \cos kx + b_k(f) \sin kx,$$

and Parseval's equality holds:

$$\frac{1}{\pi} \int_{-\pi}^{\pi} |f|^2(x) \, dx = \frac{|a_0(f)|^2}{2} + \sum_{k=1}^{\infty} |a_k(f)|^2 + |b_k(f)|^2. \tag{18.42}$$

We shall often use the more compact complex notation for trigonometric polynomials and trigonometric series, based on the Euler formulas $e^{ix} = \cos x + i \sin x$, $\cos x = \frac{1}{2}(e^{ix} + e^{-ix})$, $\sin x = \frac{1}{2i}(e^{ix} - e^{-ix})$. Using them, we can write the partial sum (18.40) of the Fourier series as

$$S_n(x) = \sum_{k=-n}^{n} c_k e^{ikx}, \tag{18.40'}$$

and the series (18.37) itself as

$$f \sim \sum_{-\infty}^{+\infty} c_k e^{ikx}, \tag{18.37'}$$

where

$$c_k = \begin{cases} \frac{1}{2}(a_k - ib_k), & \text{if } k > 0, \\ \frac{1}{2}a_0, & \text{if } k = 0, \\ \frac{1}{2}(a_{-k} + ib_{-k}), & \text{if } k < 0, \end{cases} \tag{18.43}$$

that is,

$$c_k = c_k(f) = \frac{1}{2\pi} \int_{-\pi}^{\pi} f(x) e^{-ikx} \, dx, \quad k \in \mathbb{Z}, \tag{18.44}$$

and hence the numbers c_k are simply the Fourier coefficients of f in the system $\{e^{ikx}; k \in \mathbb{Z}\}$.

We call attention to the fact that summation of the Fourier series (18.37') is understood in the sense of the convergence of the sums (18.40').

In complex notation Theorem 1 means that for every function $f \in \mathcal{R}_2([-\pi, \pi], \mathbb{C})$

$$f(x) \underset{\mathcal{R}_2}{=} \sum_{-\infty}^{\infty} c_k(f) e^{ikx}$$

and

$$\frac{1}{2\pi} \|f\|^2 = \sum_{-\infty}^{\infty} |c_k(f)|^2. \tag{18.45}$$

c. Pointwise Convergence of a Trigonometric Fourier Series

Theorem 1 gives a complete solution to the problem of mean convergence of a Fourier series (18.37), that is, convergence in the norm of the space $\mathcal{R}_2[-\pi, \pi]$. The remainder of this section will be mainly devoted to studying the conditions for and the nature of pointwise convergence of a trigonometric series. We shall consider only the simplest aspects of this problem. The study of pointwise convergence of a trigonometric series, as a rule, is such a delicate matter that, despite the traditional central position occupied by Fourier series after the work of Euler, Fourier, and Riemann, there is still no intrinsic description of the class of functions that can be represented by trigonometric series converging to them at every point (the *Riemann problem*). Until recently it was not even known whether the Fourier series of a continuous function must converge to it almost everywhere (it was known that convergence need not occur at every point). Previously A.N. Kolmogorov[15] had even given an example of a function $f \in L[-\pi, \pi]$ whose Fourier series diverged everywhere (where $L[-\pi, \pi]$ is the space of Lebesgue-integrable functions on the interval $[-\pi, \pi]$, obtainable as the metric completion of the space $\mathcal{R}[-\pi, \pi]$), and D.E. Men'shov[16] constructed a trigonometric series (18.32) with coefficients not all zero that nevertheless converged to zero almost everywhere (*Men'shov's null-series*). The problem posed by N.N. Luzin[17] (*Luzin's problem*) of determining whether the Fourier series of every function $f \in L_2[-\pi, \pi]$ converges almost everywhere (where $L_2[-\pi, \pi]$ is the metric completion of $\mathcal{R}_2[-\pi, \pi]$) was answered in the affirmative only in 1966 by L. Carleson.[18] It follows in particular from Carleson's result that the Fourier series of every function $f \in \mathcal{R}_2[-\pi, \pi]$ (for example a continuous function) must converge at almost all points of the closed interval $[-\pi, \pi]$.

18.2.2 Investigation of Pointwise Convergence of a Trigonometric Fourier Series

a. Integral Representation of the Partial Sum of a Fourier Series

Let us now turn our attention to the partial sum (18.40) of the Fourier series (18.37) and, using the complex notation (18.40′) for the expression (18.44) for the Fourier

---

[15]A.N. Kolmogorov (1903–1987) – outstanding Soviet scholar, who worked in probability theory, mathematical statistics, theory of functions, functional analysis, topology, logic, differential equations, and applied mathematics.

[16]D.E. Men'shov (1892–1988) – Soviet mathematician, one of the greatest specialists in the theory of functions of a real variable.

[17]N.N. Luzin (1883–1950) – Russian/Soviet mathematician, one of the greatest specialists in the theory of functions of a real variable, founder of the large Moscow mathematical school ("Lusitania").

[18]L. Carleson (b. 1928) – outstanding Swedish mathematician whose main works *are* in various areas of modern analysis.

coefficients, we make the following transformations:

$$S_n(x) = \sum_{k=-n}^{n} \left(\frac{1}{2\pi} \int_{-\pi}^{\pi} f(t)e^{-ikt}\, dt \right) e^{ikx} =$$

$$= \frac{1}{2\pi} \int_{-\pi}^{\pi} f(t) \left(\sum_{k=-n}^{n} e^{ik(x-t)} \right) dt. \tag{18.46}$$

But

$$D_n(u) := \sum_{k=-n}^{n} e^{iku} = \frac{e^{i(n+1)u} - e^{-inu}}{e^{iu} - 1} = \frac{e^{i(n+\frac{1}{2})u} - e^{-i(n+\frac{1}{2})u}}{e^{i\frac{1}{2}u} - e^{-i\frac{1}{2}u}}, \tag{18.47}$$

and, as can be seen from the very definition, $D_n(u) = (2n+1)$ if $e^{iu} = 1$.
Hence

$$D_n(u) = \frac{\sin(n + \frac{1}{2})u}{\sin \frac{1}{2} u}, \tag{18.48}$$

where the ratio is regarded as $2n+1$ when the denominator of the fraction becomes zero.

Continuing the computation (18.46), we now have

$$S_n(x) = \frac{1}{2\pi} \int_{-\pi}^{\pi} f(t) D_n(x - t)\, dt. \tag{18.49}$$

We have thus represented $S_n(x)$ as the convolution of the function f with the function (18.48), which is called the *Dirichlet kernel*.

As can be seen from the original definition (18.47) of the function $D_n(u)$, the Dirichlet kernel is of period 2π and even, and, in addition

$$\frac{1}{2\pi} \int_{-\pi}^{\pi} D_n(u)\, du = \frac{1}{\pi} \int_{0}^{\pi} D_n(u)\, du = 1. \tag{18.50}$$

Assuming the function f is of period 2π on $\mathbb{R}$ or is extended from $[-\pi, \pi]$ to $\mathbb{R}$ so as to have period 2π, and making a change of variable in (18.49), we obtain

$$S_n(x) = \frac{1}{2\pi} \int_{-\pi}^{\pi} f(x-t) D_n(t)\, dt = \frac{1}{2\pi} \int_{-\pi}^{\pi} f(x-t) \frac{\sin(n+\frac{1}{2})t}{\sin \frac{1}{2} t}\, dt. \tag{18.51}$$

In carrying out the change of variable here, we used the fact that the integral of a periodic function is the same over every interval whose length equals a period.

Taking account of the fact that $D_n(t)$ is an even function, we can rewrite Eq. (18.51) as

$$
S_n(x) = \frac{1}{2\pi} \int_0^\pi \big(f(x-t) + f(x+t) \big) D_n(t) \, dt =
$$

$$
= \frac{1}{2\pi} \int_0^\pi \big(f(x-t) + f(x+t) \big) \frac{\sin(n+\frac{1}{2})t}{\sin\frac{1}{2}t} \, dt. \tag{18.52}
$$

The Riemann–Lebesgue Lemma and the Localization Principle

The representation (18.52) for the partial sum of a trigonometric Fourier series, together with an observation of Riemann stated below, forms the basis for studying the pointwise convergence of a trigonometric Fourier series.

Lemma 1 (Riemann–Lebesgue) *If a locally integrable function* $f :]\omega_1, \omega_2[\to \mathbb{R}$ *is absolutely integrable (perhaps in the improper sense) on an open interval* $]\omega_1, \omega_2[$, *then*

$$
\int_{\omega_1}^{\omega_2} f(x) e^{i\lambda x} \, dx \to 0 \quad as \ \lambda \to \infty, \ \lambda \in \mathbb{R}. \tag{18.53}
$$

Proof If $]\omega_1, \omega_2[$ is a finite interval and $f(x) \equiv 1$, then Eq. (18.53) can be verified by direct integration and passage to the limit.

We shall reduce the general case to this simplest one.

Fixing an arbitrary $\varepsilon > 0$, we first choose an interval $[a, b] \subset]\omega_1, \omega_2[$ such that for every $\lambda \in \mathbb{R}$

$$
\left| \int_{\omega_1}^{\omega_2} f(x) e^{i\lambda x} \, dx - \int_a^b f(x) e^{i\lambda x} \, dx \right| < \varepsilon. \tag{18.54}
$$

In view of the estimates

$$
\left| \int_{\omega_1}^{\omega_2} f(x) e^{i\lambda x} \, dx - \int_a^b f(x) e^{i\lambda x} \, dx \right| \le
$$

$$
\le \int_{\omega_1}^a \left| f(x) e^{i\lambda x} \right| dx + \int_b^{\omega_2} \left| f(x) e^{i\lambda x} \right| dx = \int_{\omega_1}^a |f|(x) \, dx + \int_b^{\omega_2} |f|(x) \, dx
$$

and the absolute integrability of f on $]\omega_1, \omega_2[$, there does of course exist such a closed interval $[a, b]$.

Since $f \in \mathcal{R}([a, b], \mathbb{R})$ (more precisely $f|_{[a,b]} \in \mathcal{R}([a, b])$), there exists a lower Darboux sum $\sum_{j=1}^n m_j \Delta x_j$, where $m_j = \inf_{x \in [x_{j-1}, x_j]} f(x)$, such that

$$
0 < \int_a^b f(x) \, dx - \sum_{j=1}^n m_j \Delta x_j < \varepsilon.
$$

Now introducing the piecewise constant function $g(x) = m_j$ for $x \in [x_{j-1}, x_j]$, $j = 1, \dots, n$, we find that $g(x) \leq f(x)$ on $[a, b]$ and

$$0 \leq \left| \int_a^b f(x) e^{i\lambda x}\, dx - \int_a^b g(x) e^{i\lambda x}\, dx \right| \leq$$

$$\leq \int_a^b |f(x) - g(x)| |e^{i\lambda x}|\, dx = \int_a^b \left(f(x) - g(x) \right) dx < \varepsilon, \qquad (18.55)$$

but

$$\int_a^b g(x) e^{i\lambda x}\, dx = \sum_{j=1}^n \int_{x_{j-1}}^{x_j} m_j e^{i\lambda x}\, dx =$$

$$= \frac{1}{i\lambda} \sum_{j=1}^n (m_j e^{i\lambda x}) \big|_{x_{j-1}}^{x_j} \to 0 \quad \text{as } \lambda \to \infty, \ \lambda \in \mathbb{R}. \qquad (18.56)$$

Comparing relations (18.53)–(18.56), we find what was asserted. □

Remark 1 Separating the real and imaginary parts in (18.53), we find that

$$\int_{\omega_1}^{\omega_2} f(x) \cos \lambda x\, dx \to 0 \quad \text{and} \quad \int_{\omega_1}^{\omega_2} f(x) \sin \lambda x\, dx \to 0 \qquad (18.57)$$

as $\lambda \to \infty$, $\lambda \in \mathbb{R}$. If the function f in the preceding integrals had been complex, then, separating the real and imaginary parts in them, we would have found that relations (18.57), and consequently (18.53), would actually be valid for complex-valued functions $f :]\omega_1, \omega_2[\to \mathbb{C}$.

Remark 2 If it is known that $f \in \mathcal{R}_2[-\pi, \pi]$, then by Bessel's inequality (18.39) we can conclude immediately that

$$\int_{-\pi}^{\pi} f(x) \cos nx\, dx \to 0 \quad \text{and} \quad \int_{-\pi}^{\pi} f(x) \sin nx\, dx \to 0$$

as $n \to \infty$, $n \in \mathbb{N}$. Theoretically, we could have gotten by with just this discrete version of the Riemann–Lebesgue lemma for the elementary investigations of the classical Fourier series that will be carried out here.

Returning now to the integral representation (18.52) of the partial sum of a Fourier series, we remark that if the function f satisfies the hypotheses of the Riemann–Lebesgue lemma, then, since $\sin \frac{1}{2} t \geq \sin \frac{1}{2} \delta > 0$ for $0 < \delta \leq t \leq \pi$, we can use (18.57) to write

$$S_n(x) = \frac{1}{2\pi} \int_0^\delta \left(f(x - t) + f(x + t) \right) \frac{\sin(n + \frac{1}{2}) t}{\sin \frac{1}{2} t}\, dt + o(1) \quad \text{as } n \to \infty. \ (18.58)$$

The important conclusion one can deduce from (18.58) is that the convergence or divergence of a Fourier series at a point is completely determined by the behavior of the function in an arbitrarily small neighborhood of the point.

We state this principle as the following proposition.

Theorem 2 (Localization principle) *Let f and g be real- or complex-valued locally integrable functions on* $]-\pi, \pi[$ *and absolutely integrable on the whole interval (possibly in the improper sense).*

If the functions f and g are equal in any (arbitrarily small) neighborhood of the point $x_0 \in]-\pi, \pi[$, *then their Fourier series*

$$f(x) \sim \sum_{-\infty}^{+\infty} c_k(f)e^{ikx}, \qquad g(x) \sim \sum_{-\infty}^{+\infty} c_k(g)e^{ikx}$$

either both converge or both diverge at x_0. *When they converge, their limits are equal.*[19]

Remark 3 As can be seen from the reasoning used in deriving Eqs. (18.52) and (18.58), the point x_0 in the localization principle may also be an endpoint of the closed interval $[-\pi, \pi]$, but then (and this is essential!) in order for the periodic extensions of the functions f and g from the closed interval $[-\pi, \pi]$ to $\mathbb{R}$ to be equal on a neighborhood of x_0 it is necessary (and sufficient) that the original functions be equal on a neighborhood of both endpoints of the closed interval $[-\pi, \pi]$.

c. Sufficient Conditions for a Fourier Series to Converge at a Point

Definition 2 A function $f : \dot{U} \to \mathbb{C}$ defined on a deleted neighborhood of a point $x \in \mathbb{R}$ satisfies the *Dini conditions* at x if

a) both one-sided limits

$$f(x_-) = \lim_{t \to +0} f(x - t), \qquad f(x_+) = \lim_{t \to +0} f(x + t)$$

exist at x;

b) the integral

$$\int_{+0} \frac{(f(x - t) - f(x_-)) + (f(x + t) - f(x_+))}{t} \, dt$$

converges absolutely.[20]

---

[19] Although the limit need not be $f(x_0) = g(x_0)$.

[20] What is meant is that the integral $\int_0^\varepsilon$ converges absolutely for some value $\varepsilon > 0$.

Example 2 If f is a continuous function in $U(x)$ satisfying the Hölder condition

$$\left| f(x+t) - f(x) \right| \leq M|t|^{\alpha}, \quad 0 < \alpha \leq 1,$$

then, since the estimate

$$\left| \frac{f(x+t) - f(x)}{t} \right| \leq \frac{M}{|t|^{1-\alpha}}$$

now holds, the function f satisfies the Dini conditions at x.

It is also clear that if a continuous function f defined in a deleted neighborhood $\overset{\circ}{U}(x)$ of x has one-sided limits $f(x_-)$ and $f(x_+)$ and satisfies one-sided Hölder conditions

$$\left| f(x+t) - f(x_+) \right| \leq Mt^{\alpha},$$
$$\left| f(x-t) - f(x_-) \right| \leq Mt^{\alpha},$$

where $t > 0$, $0 < \alpha \leq 1$, and M is a positive constant, then f will satisfy the Dini conditions for the same reason as above.

Definition 3 We shall call a real- or complex-valued function f *piecewise-continuous on the closed interval* $[a, b]$ if there is a finite set of points $a = x_0 < x_1 < \cdots < x_n = b$ in this interval such that f is defined on each interval $]x_{j-1}, x_j[$, $j = 1, \ldots, n$, and has one-sided limits on approach to its endpoints.

Definition 4 A function having a piecewise-continuous derivative on a closed interval is *piecewise continuously differentiable* on that interval.

Example 3 If a function is piecewise continuously differentiable on a closed interval, then it satisfies the Hölder conditions with exponent $\alpha = 1$ at every point of the interval, as follows from Lagrange's finite-increment (mean-value) theorem. Hence, by Example 1, such a function satisfies Dini's conditions at every point of the interval. At the endpoints of the interval, of course only the corresponding one-sided pair of Dini's conditions needs to be verified.

Example 4 The function $f(x) = \operatorname{sgn} x$ satisfies Dini's conditions at every point $x \in \mathbb{R}$, even at zero.

Theorem 3 (Sufficient conditions for convergence of a Fourier series at a point) *Let $f : \mathbb{R} \to \mathbb{C}$ be a function of period 2π that is absolutely integrable on the closed interval $[-\pi, \pi]$. If f satisfies the Dini conditions at a point $x \in \mathbb{R}$, then its Fourier series converges at that point, and*

$$\sum_{-\infty}^{+\infty} c_k(f)e^{ikx} = \frac{f(x_-) + f(x_+)}{2}. \tag{18.59}$$

Proof By relations (18.52) and (18.50)

$$S_n(x) - \frac{f(x_-) + f(x_+)}{2} =$$

$$= \frac{1}{\pi} \int_0^\pi \frac{(f(x-t) - f(x_-)) + (f(x+t) - f(x_+))}{2 \sin \frac{1}{2}t} \sin\left(n + \frac{1}{2}\right) t \, dt.$$

Since $2 \sin \frac{1}{2}t \sim t$ as $t \to +0$, by the Dini conditions and the Riemann–Lebesgue lemma we see that this last integral tends to zero as $n \to \infty$.   □

Remark 4 In connection with the theorem just proved and the localization principle, we note that changing the value of the function at a point has no influence on the Fourier coefficients or the Fourier series or the partial sums of the Fourier series. Therefore the convergence and the sum of such a series at a point is determined not by the particular value of the function at the point, but by the integral mean of its values in an arbitrarily small neighborhood of the point. It is this fact that is reflected in Theorem 3.

Example 5 In Example 6 of Sect. 18.1 we found the Fourier series

$$x \sim \sum_{k=1}^\infty 2\frac{(-1)^{k+1}}{k} \sin kx \qquad (18.60)$$

for the function $f(x) = x$ on the closed interval $[-\pi, \pi]$. Extending the function $f(x)$ periodically from the interval $]-\pi, \pi[$ to the whole real line, we may assume that the series (18.60) is the Fourier series of this extended function. Then, on the basis of Theorem 3 we find that

$$\sum_{k=1}^\infty 2\frac{(-1)^{k+1}}{k} \sin kx = \begin{cases} x, & \text{if } |x| < \pi, \\ 0, & \text{if } |x| = \pi. \end{cases}$$

In particular, for $x = \frac{\pi}{2}$ it follows from this relation that

$$\sum_{n=0}^\infty \frac{(-1)^n}{2n+1} = \frac{\pi}{4}.$$

Example 6 Let $\alpha \in \mathbb{R}$ and $|\alpha| < 1$. Consider the 2π-periodic function $f(x)$ defined on the closed interval $[-\pi, \pi]$ by the formula $f(x) = \cos \alpha x$.

By formulas (18.35) and (18.36) we find its Fourier coefficients

$$a_n(f) = \frac{1}{\pi} \int_{-\pi}^\pi \cos \alpha x \cos nx \, dx = \frac{(-1)^n \sin \pi \alpha}{\pi} \cdot \frac{2\alpha}{\alpha^2 - n^2},$$

$$b_n(f) = \frac{1}{\pi} \int_{-\pi}^\pi \cos \alpha x \sin nx \, dx = 0.$$

By Theorem 3 the following equality holds at each point $x \in [-\pi, \pi]$:

$$\cos \alpha x = \frac{2\alpha \sin \pi \alpha}{\pi} \left(\frac{1}{2\alpha^2} + \sum_{n=1}^{\infty} \frac{(-1)^n}{\alpha^2 - n^2} \cos nx \right).$$

When $x = \pi$, this relation implies

$$\cot \pi \alpha - \frac{1}{\pi \alpha} = \frac{2\alpha}{\pi} \sum_{n=1}^{\infty} \frac{1}{\alpha^2 - n^2}. \tag{18.61}$$

If $|\alpha| \leq \alpha_0 < 1$, then $|\frac{1}{\alpha^2 - n^2}| \leq \frac{1}{n^2 - \alpha_0^2}$, and hence the series on the right-hand side of Eq. (18.61) converges uniformly with respect to α on every closed interval $|\alpha| \leq \alpha_0 < 1$. Hence it is legitimate to integrate it termwise, that is,

$$\int_0^{\pi} \left(\cot \pi \alpha - \frac{1}{\pi \alpha} \right) d\alpha = \sum_{n=1}^{\infty} \int_0^{\pi} \frac{2\alpha \, d\alpha}{\alpha^2 - n^2},$$

and

$$\ln \frac{\sin \pi \alpha}{\pi \alpha} \Big|_0^{\pi} = \sum_{n=1}^{\infty} \ln |\alpha^2 - n^2| \Big|_0^{\pi},$$

yielding

$$\ln \frac{\sin \pi x}{\pi x} = \sum_{n=1}^{\infty} \ln \left(1 - \frac{x^2}{n^2} \right),$$

and finally,

$$\frac{\sin \pi x}{\pi x} = \prod_{n=1}^{\infty} \left(1 - \frac{x^2}{n^2} \right) \quad \text{when } |x| < 1. \tag{18.62}$$

We have thus proved relation (18.62), which we mentioned earlier when deriving the complement formula for Euler's function $\Gamma(x)$ (Sect. 17.3).

d. Fejér's Theorem

Let us now consider the sequence of functions

$$\sigma_n(x) := \frac{S_0(x) + \cdots + S_n(x)}{n + 1},$$

that are the arithmetic means of the corresponding partial sums $S_0(x), \ldots, S_n(x)$ of the trigonometric Fourier series (18.37) of a function $f : \mathbb{R} \to \mathbb{C}$ of period 2π.

From the integral representation (18.51) of the partial sum of the Fourier series we have

$$\sigma_n(x) = \frac{1}{2\pi} \int_{-\pi}^{\pi} f(x-t)\mathcal{F}_n(t)\,dt,$$

where

$$\mathcal{F}_n(t) = \frac{1}{n+1}\big(D_0(t) + \cdots + D_n(t)\big).$$

Recalling the explicit form (18.48) of the Dirichlet kernel and taking account of the relation

$$\sum_{k=0}^{n} \sin\Big(k + \frac{1}{2}\Big)t = \frac{1}{2}\Big(\sin\frac{1}{2}t\Big)^{-1} \sum_{k=0}^{n}(\cos kt - \cos(k+1)t) = \frac{\sin^2(\frac{n+1}{2})t}{\sin\frac{1}{2}t},$$

we find

$$\mathcal{F}_n(t) = \frac{\sin^2\frac{n+1}{2}t}{(n+1)\sin^2\frac{1}{2}t}.$$

The function $\mathcal{F}_n$ is called the *Fejér kernel*, more precisely the *nth Fejér kernel*.[21]

Taking account of the original definition (18.47) of the Dirichlet kernel D_n, one can conclude that the Fejér kernel is a smooth function of period 2π whose value equals $(n+1)$ where the denominator of this last fraction equals zero.

The properties of the Fejér and Dirichlet kernels are similar in many ways, but in contrast to the Dirichlet kernel, the Fejér kernel is also nonnegative, so that we have the following lemma.

Lemma 2 *The sequence of functions*

$$\Delta_n(x) = \begin{cases} \frac{1}{2\pi}\mathcal{F}_n(x), & if\ |x| \le \pi, \\ 0, & if\ |x| > \pi \end{cases}$$

is an approximate identity on $\mathbb{R}$.

Proof The nonnnegativity of $\Delta_n(x)$ is clear.

Equality (18.50) enables us to conclude that

$$\int_{-\infty}^{\infty} \Delta_n(x)\,dx = \int_{-\pi}^{\pi} \Delta_n(x)\,dx = \frac{1}{2\pi}\int_{-\pi}^{\pi}\mathcal{F}_n(x)\,dx =$$

$$= \frac{1}{2\pi(n+1)}\sum_{k=0}^{n}\int_{-\pi}^{\pi} D_k(x)\,dx = 1.$$

---

[21]L. Fejér (1880–1956) – well-known Hungarian mathematician.

Finally, for every $\delta > 0$

$$0 \le \int_{-\infty}^{-\delta} \Delta_n(x)\,\mathrm{d}x = \int_{\delta}^{+\infty} \Delta_n(x)\,\mathrm{d}x = \int_{\delta}^{\pi} \Delta_n(x)\,\mathrm{d}x \le$$

$$\le \frac{1}{2\pi(n+1)} \int_{\delta}^{\pi} \frac{\mathrm{d}x}{\sin^2 \frac{1}{2}x} \to 0$$

as $n \to +\infty$. $\qquad\square$

Theorem 4 (Fejér) *Let* $f : \mathbb{R} \to \mathbb{C}$ *be a function of period* 2π *that is absolutely integrable on the closed interval* $[-\pi, \pi]$. *If*

a) *f is uniformly continuous on the set* $E \subset \mathbb{R}$, *then*

$$\sigma_n(x) \rightrightarrows f(x) \quad \text{on } E \text{ as } n \to \infty;$$

b) $f \in C(\mathbb{R}, \mathbb{C})$, *then*

$$\sigma_n(x) \rightrightarrows f(x) \quad \text{on } \mathbb{R} \text{ as } n \to \infty;$$

c) *f is continuous at the point* $x \in \mathbb{R}$, *then*

$$\sigma_n(x) \to f(x) \quad \text{as } n \to \infty.$$

Proof Statements b) and c) are special cases of a).

Statement a) itself is a special case of the general Proposition 5 of Sect. 17.4 on the convergence of a convolution, since

$$\sigma_n(x) = \frac{1}{2\pi} \int_{-\pi}^{\pi} f(x-t)\mathcal{F}_n(t)\,\mathrm{d}t = (f * \Delta_n)(x). \qquad\square$$

Corollary 1 (Weierstrass' theorem on approximation by trigonometric polynomials) *If a function* $f : [-\pi, \pi] \to \mathbb{C}$ *is continuous on the closed interval* $[-\pi, \pi]$ *and* $f(-\pi) = f(\pi)$, *then this function can be approximated uniformly on the closed interval* $[-\pi, \pi]$ *with arbitrary precision by trigonometric polynomials.*

Proof Extending f as a function of period 2π, we obtain a continuous 2π-periodic function on $\mathbb{R}$, to which the trigonometric polynomials $\sigma_n(x)$ converge uniformly by Fejér's theorem. $\qquad\square$

Corollary 2 *If f is continuous at x, its Fourier series either diverges at x or converges to $f(x)$.*

Proof Only the case of convergence requires formal verification. If the sequence $S_n(x)$ has a limit as $n \to \infty$, then the sequence $\sigma_n(x) = \frac{S_0(x)+\cdots+S_n(x)}{n+1}$ has that same limit. But by Fejér's theorem $\sigma_n(x) \to f(x)$ as $n \to \infty$, and hence $S_n(x) \to f(x)$ also whenever the limit $S_n(x)$ exists as $n \to \infty$. $\qquad\square$

Remark 5 We note that the Fourier series of a continuous function really can diverge at some points.

18.2.3 Smoothness of a Function and the Rate of Decrease of the Fourier Coefficients

a. An Estimate of the Fourier Coefficients of a Smooth Function

We begin with a simple, yet important and useful lemma.

Lemma 3 (Differentiation of a Fourier series) *If a continuous function* $f \in C([-\pi, \pi], \mathbb{C})$ *assuming equal values at the endpoints of the closed interval* $[-\pi, \pi]$ *is piecewise continuously differentiable on* $[-\pi, \pi]$*, then the Fourier series of its derivative*

$$f' \sim \sum_{-\infty}^{\infty} c_k(f') e^{ikx}$$

can be obtained by differentiating formally the Fourier series

$$f \sim \sum_{-\infty}^{\infty} c_k(f) e^{ikx}$$

of the function itself, that is,

$$c_k(f') = ikc_k(f), \quad k \in \mathbb{Z}. \tag{18.63}$$

Proof Starting from the definition of the Fourier coefficients (18.44), we find through integration by parts that

$$c_k(f') = \frac{1}{2\pi} \int_{-\pi}^{\pi} f'(x) e^{-ikx} \, dx = \frac{1}{2\pi} f(x) e^{-ikx} \Big|_{-\pi}^{\pi} + \frac{ik}{2\pi} \int_{-\pi}^{\pi} f(x) e^{-ikx} \, dx =$$

$$= ikc_k(f),$$

since $f(\pi)e^{-ik\pi} - f(-\pi)e^{ik\pi} = 0$. $\square$

Proposition 1 (Connection between smoothness of a function and the rate of decrease of its Fourier coefficients) *Let* $f \in C^{(m-1)}([-\pi, \pi], \mathbb{C})$ *and* $f^{(j)}(-\pi) = f^{(j)}(\pi)$, $j = 0, 1, \ldots, m - 1$. *If the function* f *has a piecewise-continuous derivative* $f^{(m)}$ *of order* m *on the closed interval* $[-\pi, \pi]$*, then*

$$c_k(f^{(m)}) = (ik)^m c_k(f), \quad k \in \mathbb{Z}, \tag{18.64}$$

and

$$\left|c_k(f)\right| = \frac{\gamma_k}{|k|^m} = o\left(\frac{1}{k^m}\right) \quad \text{as } k \to \infty, \ k \in \mathbb{Z}; \tag{18.65}$$

moreover, $\sum_{-\infty}^{\infty} \gamma_k^2 < \infty$.

Proof Relation (18.64) follows from an *m*-fold application of Eq. (18.63):

$$c_k\left(f^{(m)}\right) = (ik)c_k\left(f^{(m-1)}\right) = \cdots = (ik)^m c_k(f).$$

Setting $\gamma_k = |c_k(f^{(m)})|$ and using Bessel's inequality

$$\sum_{-\infty}^{\infty} \left|c_k\left(f^{(m)}\right)\right|^2 \le \frac{1}{2\pi} \int_{-\pi}^{\pi} \left|f^{(m)}\right|^2(x)\,\mathrm{d}x,$$

we obtain (18.65) from (18.64). $\qquad\square$

Remark 6 In the proposition just proved, as in Lemma 3, instead of assuming the conditions $f^{(j)}(-\pi) = f^{(j)}(\pi)$, we could have assumed that f is a function of period 2π on the entire line.

Remark 7 If a trigonometric Fourier series were written in the form (18.37), rather than in the complex form (18.37′), it would be necessary to replace the simple relations (18.64) by noticeably more cumbersome equalities, whose meaning, however, would be the same: under these hypotheses a Fourier series can be differentiated termwise whichever from it is written in, (18.37) or (18.37′). As for the estimates of the Fourier coefficients, $a_k(f)$ and $b_k(f)$ of (18.37), since $a_k(f) = c_k(f) + c_{-k}(f)$ and $b_k(f) = i(c_k(f) - c_{-k}(f))$, (see formulas (18.43)) it follows from (18.65) that if a function f satisfies the hypotheses of the proposition, then

$$\left|a_k(f)\right| = \frac{\alpha_k}{k^m}, \qquad \left|b_k(f)\right| = \frac{\beta_k}{k^m}, \quad k \in \mathbb{N}, \tag{18.64′}$$

where $\sum_{k=1}^{\infty} \alpha_k^2 < \infty$ and $\sum_{k=1}^{\infty} \beta_k^2 < \infty$, and we can assume that $\alpha_k = \beta_k = \gamma_k + \gamma_{-k}$.

b. Smoothness of a Function and the Rate of Convergence of Its Fourier Series

Theorem 5 *If the function $f : [-\pi, \pi] \to \mathbb{C}$ is such that*

a) $f \in C^{(m-1)}[-\pi, \pi]$, $m \in \mathbb{N}$,
b) $f^{(j)}(-\pi) = f^{(j)}(\pi)$, $j = 0, 1, \dots, m-1$,
c) *f has a piecewise continuous mth derivative $f^{(m)}$ on $[-\pi, \pi]$, $m \ge 1$, then the Fourier series of f converges absolutely and uniformly on $[-\pi, \pi]$ to f, and*

the deviation of the nth partial sum $S_n(x)$ of the Fourier series from $f(x)$ has the following estimate on the entire interval:

$$\left| f(x) - S_n(x) \right| \le \frac{\varepsilon_n}{n^{m-1/2}},$$

where $\{\varepsilon_n\}$ is a sequence of positive numbers tending to zero.

Proof We write the partial sum (18.40) of the Fourier series in the compact notation (18.40′):

$$S_n(x) = \sum_{-n}^{n} c_k(f) e^{ikx}.$$

According to the assumptions on the function f and Proposition 1 we have $|c_k(f)| = \gamma_k / |k|^m$, and $\sum \gamma_k / |k|^m < \infty$: since $0 \le \gamma_k / |k|^m \le \frac{1}{2}(\gamma_k^2 + 1/k^{2m})$ and $m \ge 1$, we have $\sum \gamma_k / |k|^m < \infty$. Hence the sequence $S_n(x)$ converges uniformly on $[-\pi, \pi]$ (by the Weierstrass M-test for series and the Cauchy criterion for sequences).

By Theorem 3 the limit $S(x)$ of $S_n(x)$ equals $f(x)$, since the function f satisfies the Dini conditions at each point of the closed interval $[-\pi, \pi]$ (see Example 3). And, since $f(-\pi) = f(\pi)$, the function f can be extended to $\mathbb{R}$ as a periodic function with the Dini conditions holding at each point $x \in \mathbb{R}$.

Now, using relation (18.63), we can proceed to obtain an estimate:

$$\left| f(x) - S_n(x) \right| = \left| S(x) - S_n(x) \right| = \left| \sum_{\pm k = n+1}^{\infty} c_k(f) e^{ikx} \right| \le$$

$$\le \sum_{\pm k = n+1}^{\infty} \left| c_k(f) \right| = \sum_{\pm k = n+1}^{\infty} \gamma_k / |k|^m \le$$

$$\le \left(\sum_{\pm k = n+1}^{\infty} \gamma_k^2 \right)^{1/2} \left(\sum_{\pm k = n+1}^{\infty} 1/k^{2m} \right)^{1/2}.$$

The first factor on the right-hand side of the Cauchy–Bunyakovskii inequality here tends to zero as $n \to \infty$, since $\sum_{-\infty}^{\infty} \gamma_k^2 < \infty$.

Next (see Fig. 18.2)

$$\sum_{k=n+1}^{\infty} 1/k^{2m} \le \int_n^{\infty} \frac{dx}{x^{2m}} = \frac{1}{2m-1} \cdot \frac{1}{n^{2m-1}}.$$

We thus obtain the assertion of Theorem 5. □

In connection with these results we now make a number of useful remarks.

Fig. 18.2

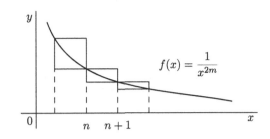

$$f(x) = \frac{1}{x^{2m}}$$

Fig. 18.3

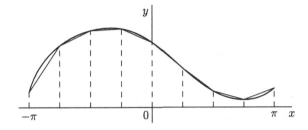

Remark 8 One can now easily obtain again the Weierstrass approximation theorem stated in Corollary 1 from Theorem 5 (and Theorem 3, of which essential use was made in the proof of Theorem 5), independently of Fejér's theorem.

Proof It suffices to prove this result for real-valued functions. Using the uniform continuity of f on $[-\pi, \pi]$, we approximate f on this closed interval uniformly within $\varepsilon/2$ by a piecewise-linear continuous function $\varphi(x)$ assuming the same values as f at the endpoints, that is, $\varphi(-\pi) = \varphi(\pi) = f(\pi)$ (Fig. 18.3). By Theorem 5 the Fourier series of φ converges to φ uniformly on the closed interval $[-\pi, \pi]$. Taking a partial sum of this series that differs from $\varphi(x)$ by less than $\varepsilon/2$, we obtain a trigonometric polynomial that approximates the original function f within ε on the whole interval $[-\pi, \pi]$. □

Remark 9 Let us assume that we have succeeded in representing a function f having a jump singularity as the sum $f = \varphi + \psi$ of a certain smooth function ψ and a simple function φ having the same singularity as f (Fig. 18.4a, c, b). Then the Fourier series of f is the sum of the Fourier series of ψ, which converges rapidly and uniformly by Theorem 5, and the Fourier series of the function φ. The latter can be regarded as known, if we take the standard function φ (shown in the figure as $\varphi(x) = -\pi - x$ for $-\pi < x < 0$ and $\varphi(x) = \pi - x$ for $0 < x < \pi$).

This observation can be used both in theoretical and computational problems connected with series (it is Krylov's[22] *method of separating singularities* and im-

---

[22] A.N. Krylov (1863–1945) – Russian/Soviet specialist in mechanics and mathematics, who made a large contribution to computational mathematics, especially in methods of computing the elements of ships.

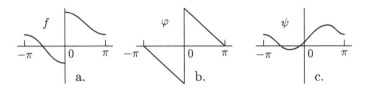

Fig. 18.4

proving the convergence of series) and in the theory of trigonometric Fourier series itself (see for example the *Gibbs*[23] *phenomenon*, described below in Problem 11).

Remark 10 (Integration of a Fourier series) By Theorem 5 we can state and prove the following complement to Lemma 3 on differentiation of a Fourier series.

Proposition 2 *If the function* $f : [-\pi, \pi] \to \mathbb{C}$ *is piecewise continuous, then after integration the correspondence* $f(x) \sim \sum_{-\infty}^{\infty} c_k(f) e^{ikx}$ *becomes the equality*

$$\int_0^x f(t) \, dt = c_0(f)x + \sum_{-\infty}^{\infty}{}' \frac{c_k(f)}{ik} \left(e^{ikx} - 1 \right),$$

where the prime indicates that the term with index $k = 0$ *is omitted from the sum; the summation is the limit of the symmetric partial sums* $\sum_{-n}^{n}$, *and the series converges uniformly on the closed interval* $[-\pi, \pi]$.

Proof Consider the auxiliary function

$$F(x) = \int_0^x f(t) \, dt - c_0(f)x$$

on the interval $[-\pi, \pi]$. Obviously $F \in C[-\pi, \pi]$. Also $F(-\pi) = F(\pi)$, since

$$F(\pi) - F(-\pi) = \int_{-\pi}^{\pi} f(t) \, dt - 2\pi c_0(f) = 0,$$

as follows from the definition of $c_0(f)$. Since the derivative $F'(x) = f(x) - c_0(f)$ of the function F is piecewise continuous, its Fourier series $\sum_{-\infty}^{\infty} c_k(F) e^{ikx}$ converges uniformly to F on the interval $[-\pi, \pi]$ by Theorem 5. By Lemma 3 we have $c_k(F) = \frac{c_k(F')}{ik}$ for $k \neq 0$. But $c_k(F') = c_k(F)$ if $k \neq 0$. Now writing the equality $F(x) = \sum_{-\infty}^{\infty} c_k(F) e^{ikx}$ in terms of the function f and noting that $F(0) = 0$, we obtain the assertion of the proposition.   □

---

[23] J.W. Gibbs (1839–1903) – American physicist and mathematician, one of the founders of thermodynamics and statistical mechanics.

18.2.4 Completeness of the Trigonometric System

a. The Completeness Theorem

In conclusion we return once again from pointwise convergence of the Fourier series to its mean convergence (18.41). More precisely, using the facts we have accumulated on the nature of pointwise convergence of the Fourier series, we now give a proof of the completeness of the trigonometric system $\{1, \cos kx, \sin kx; k \in \mathbb{N}\}$ in $\mathcal{R}_2([-\pi, \pi], \mathbb{R})$ independent of the proof already given in the problems. In doing so, as in Sect. 18.2.1, we take $\mathcal{R}_2([-\pi, \pi], \mathbb{R})$ or $\mathcal{R}_2([-\pi, \pi], \mathbb{C})$ to mean the vector space of real- or complex-valued functions that are locally integrable on $]-\pi, \pi[$ and whose squared absolute values are integrable on $]-\pi, \pi[$ (possibly in the improper sense). This vector space is assumed to be endowed with the standard inner product (18.38) generating the norm in terms of which convergence is mean convergence (18.41).

The theorem we are about to prove asserts simply that the system of trigonometric functions is complete in $\mathcal{R}_2([-\pi, \pi], \mathbb{C})$. But we shall state the theorem in such a way that the statement itself will contain the key to the proof. It is based on the obvious fact that the property of completeness is transitive: if A approximates B and B approximates C, then A approximates C.

Theorem 6 (Completeness of the trigonometric system) *Every function $f \in \mathcal{R}_2[-\pi, \pi]$ can be approximated arbitrarily closely in mean*

a) *by functions of compact support in $]-\pi, \pi[$ that are Riemann integrable over the closed interval $[-\pi, \pi]$;*

b) *by piecewise-constant functions of compact support on the closed interval $[-\pi, \pi]$;*

c) *by piecewise-linear continuous functions of compact support on the closed interval $[-\pi, \pi]$;*

d) *by trigonometric polynomials.*

Proof Since it obviously suffices to prove the theorem for real-valued functions, we confine ourselves to this case.

a) It follows from the definition of the improper integral that

$$\int_{-\pi}^{\pi} f^2(x)\,dx = \lim_{\delta \to +0} \int_{-\pi+\delta}^{\pi-\delta} f^2(x)\,dx.$$

Hence, for every $\varepsilon > 0$ there exists $\delta > 0$ such that the function

$$f_\delta(x) = \begin{cases} f(x), & \text{if } |x| < \pi - \delta, \\ 0 & \text{if } \pi - \delta \leq |x| \leq \pi, \end{cases}$$

will differ in mean from f on $[-\pi, \pi]$ by less than ε, since

$$\int_{-\pi}^{\pi} (f - f_\delta)^2(x)\,dx = \int_{-\pi}^{-\pi+\delta} f^2(x)\,dx + \int_{\pi-\delta}^{\pi} f^2(x)\,dx.$$

b) It suffices to verify that every function of the form f_δ can be approximated in $\mathcal{R}_2([-\pi, \pi], \mathbb{R})$ by piecewise-constant functions of compact support in $[-\pi, \pi]$. But the function f_δ is Riemann-integrable on $[-\pi + \delta, \pi - \delta]$. Hence it is bounded there by a constant M and moreover there exists a partition $-\pi + \delta = x_0 < x_1 < \cdots < x_n = \pi - \delta$ of this closed interval such that the corresponding lower Darboux sum $\sum_{i=1}^{n} m_i \Delta x_i$ of the function f_δ differs from the integral of f_δ over $[-\pi + \delta, \pi - \delta]$ by less than $\varepsilon > 0$.

Now setting

$$g(x) = \begin{cases} m_i, & \text{if } x \in \,]x_{i-1}, x_i[,\, i = 1, \ldots, n, \\ 0, & \text{at all other points of } [-\pi, \pi], \end{cases}$$

we obtain

$$\int_{-\pi}^{\pi} (f_\delta - g)^2(x)\,dx \le \int_{-\pi}^{\pi} |f_\delta - g||f_\delta + g|(x)\,dx \le$$

$$\le 2M \int_{-\pi+\delta}^{\pi-\delta} (f_\delta - g)(x)\,dx \le 2M\varepsilon,$$

and hence f_δ really can be approximated arbitrarily closely in the mean on $[-\pi, \pi]$ by piecewise-constant functions on the interval that vanish in a neighborhood of the endpoints of the interval.

c) It now suffices to learn how to approximate the functions in b) in mean. Let g be such a function. All of its points of discontinuity $x_1, \ldots, x_n$ lie in the open interval $]-\pi, \pi[$. There are only finitely many of them, so that for every $\varepsilon > 0$ one can choose $\delta > 0$ so small that the δ-neighborhoods of the points $x_1, \ldots, x_n$ are disjoint and contained strictly inside the interval $]-\pi, \pi[$, and $2\delta n M < \varepsilon$, where $M = \sup_{|x| \le \pi} |g(x)|$. Now replacing the function g on $[x_i - \delta, x_i + \delta]$, $i = 1, \ldots, n$, by the linear interpolation between the values $g(x_i - \delta)$ and $g(x_i + \delta)$ that g assumes at the endpoints of this interval, we obtain a piecewise linear continuous function g_δ that is of compact support in $[-\pi, \pi]$. By construction $|g_\delta(x)| \le M$ on $[-\pi, \pi]$, so that

$$\int_{-\pi}^{\pi} (g - g_\delta)^2(x)\,dx \le 2M \int_{-\pi}^{\pi} |g - g_\delta|(x)\,dx =$$

$$= 2M \sum_{i=1}^{n} \int_{x_i-\delta}^{x_i+\delta} |g - g_\delta|(x)\,dx \le 2M \cdot (2M \cdot 2\delta) \cdot n <$$

$$< 4M\varepsilon,$$

and the possibility of the approximation is now proved.

d) It remains only to show that one can approximate any function of class c) in mean on $[-\pi, \pi]$ by a trigonometric polynomial. But for every $\varepsilon > 0$ and every function of type g_δ, Theorem 5 enables us to find a trigonometric polynomial T_n that approximates g_δ uniformly within ε on the closed interval $[-\pi, \pi]$. Hence $\int_{-\pi}^{\pi} (g_\delta - T_n)^2 \, dx < 2\pi\varepsilon^2$, and the possibility of an arbitrarily precise approximation in mean by trigonometric polynomials on $[-\pi, \pi]$ for any function of class c) is now established.

By the triangle inequality in $\mathcal{R}_2[-\pi, \pi]$ we now conclude that all of Theorem 6 on the completeness of these classes in $\mathcal{R}_2[-\pi, \pi]$ is also proved. $\square$

b. The Inner Product and Parseval's Equality

Now that the completeness of the trigonometric system in $\mathcal{R}_2([-\pi, \pi], \mathbb{C})$ has been proved, we can use Theorem 1 to assert that

$$f = \frac{a_0(f)}{2} + \sum_{k=1}^{\infty} a_k(f) \cos kx + b_k(f) \sin kx \qquad (18.66)$$

for every function $f \in \mathcal{R}_2([-\pi, \pi], \mathbb{C})$, or, in complex notation,

$$f = \sum_{-\infty}^{\infty} c_k(f) e^{ikx} \qquad (18.67)$$

where the convergence is understood as convergence in the norm of $\mathcal{R}_2[-\pi, \pi]$, that is, as mean convergence, and the limiting passage in (18.67) is the limit of sums of the form $S_n(x) = \sum_{-n}^{n} c_k(f) e^{ikx}$ as $n \to \infty$.

If we rewrite Eqs. (18.66) and (18.67) as

$$\frac{1}{\sqrt{\pi}} f = \frac{a_0(f)}{\sqrt{2}} \frac{1}{\sqrt{2\pi}} + \sum_{k=1}^{\infty} a_k(f) \frac{\cos kx}{\sqrt{\pi}} + b_k(f) \frac{\sin kx}{\sqrt{\pi}}, \qquad (18.66')$$

$$\frac{1}{\sqrt{2\pi}} f = \sum_{-\infty}^{\infty} c_k(f) \frac{e^{ikx}}{\sqrt{2\pi}}, \qquad (18.67')$$

then the right-hand sides contain series in the orthonormal systems

$$\left\{ \frac{1}{\sqrt{2\pi}}, \frac{1}{\sqrt{\pi}} \cos kx, \frac{1}{\sqrt{\pi}} \sin kx; k \in \mathbb{N} \right\}$$

and $\{ \frac{1}{\sqrt{2\pi}} e^{ikx}; k \in \mathbb{Z} \}$. Hence by the general rule for computing the inner product of vectors from their coordinates in an orthonormal basis, we can assert that the equality

$$\frac{1}{\pi} \langle f, g \rangle = \frac{a_0(f)\bar{a}_0(g)}{2} + \sum_{k=1}^{\infty} a_k(f)\bar{a}_k(g) + b_k(f)\bar{b}_k(g) \qquad (18.68)$$

holds for functions f and g in $\mathcal{R}_2([-\pi, \pi], \mathbb{C})$, or, in other notation,

$$\frac{1}{2\pi} \langle f, g \rangle = \sum_{-\infty}^{\infty} c_k(f) \overline{c}_k(g), \tag{18.69}$$

where, as always,

$$\langle f, g \rangle = \int_{-\pi}^{\pi} f(x) \overline{g}(x) \, dx.$$

In particular, if $f = g$, we obtain the classical Parseval equality from (18.68) and (18.69) in two equivalent forms:

$$\frac{1}{\pi} \|f\|^2 = \frac{|a_0(f)|^2}{2} + \sum_{k=1}^{\infty} |a_k(f)|^2 + |b_k(f)|^2, \tag{18.70}$$

$$\frac{1}{2\pi} \|f\|^2 = \sum_{-\infty}^{\infty} |c_k(f)|^2. \tag{18.71}$$

We have already noted that from the geometric point of view Parseval's equality can be regarded as an infinite-dimensional version of the Pythagorean theorem.

Parseval's relation provides the basis for the following useful proposition.

Proposition 3 (Uniqueness of Fourier series) *Let f and g be two functions in $\mathcal{R}_2[-\pi, \pi]$. Then*

a) *if the trigonometric series*

$$\frac{a_0}{2} + \sum_{k=1}^{\infty} a_k \cos kx + b_k \sin kx \quad \left(= \sum_{-\infty}^{\infty} c_k e^{ikx} \right)$$

converges in mean to f on the interval $[-\pi, \pi]$ it is the Fourier series of f;

b) *if the functions f and g have the same Fourier series, they are equal almost everywhere on $[\pi, \pi]$, that is, $f = g$ in $\mathcal{R}_2[-\pi, \pi]$.*

Proof Assertion a) is actually a special case of the general fact that the expansion of a vector in an orthogonal system is unique. The inner product, as we know (see Lemma 1b) shows immediately that the coefficients of such an expansion are the Fourier coefficients and no others.

Assertion b) can be obtained from Parseval's equality taking account of the completeness of the trigonometric system in $\mathcal{R}_2([-\pi, \pi], \mathbb{C})$, which was just proved.

Since the difference $(f - g)$ has a zero Fourier series, it follows from Parseval's equality that $\|f - g\|_{\mathcal{R}_2} = 0$. Hence the functions f and g are equal at all points of continuity, that is, almost everywhere.   □

Remark 11 When studying Taylor series $\sum_{n=0}^{\infty} \frac{f^{(n)}(a)}{n!}(x-a)^n$ we noted previously that different functions of class $C^{(\infty)}(\mathbb{R}, \mathbb{R})$ can have the same Taylor series (at some points $a \in \mathbb{R}$). This contrast with the uniqueness theorem just proved for the Fourier series should not be taken too seriously, since every uniqueness theorem is a relative one in the sense that it involves a particular space and a particular type of convergence.

For example, in the space of analytic functions (that is, functions that can be represented as power series $\sum_{n=0}^{\infty} a_n(z-z_0)^n$ converging to them pointwise), two different functions have distinct Taylor series about every point.

If, in turn, in studying trigonometric series we abandon the space $\mathcal{R}_2[-\pi, \pi]$ and study pointwise convergence of a trigonometric series, then, as already noted (p. 524) one can construct a trigonometric series not all of whose coefficients are zero, which nevertheless converges to zero almost everywhere. According to Proposition 3 such a null-series of course does not converge to zero in the mean-square sense.

In conclusion, we illustrate the use of the properties of trigonometric Fourier series obtained here by studying the following derivation, due to Hurwitz,[24] of the classical isoperimetric inequality in the two-dimensional case. In order to avoid cumbersome expressions and accidental technical difficulties, we shall use complex notation.

Example 7 Between the volume V of a domain in the Euclidean space E^n, $n \geq 2$, and the $(n-1)$-dimensional surface area F of the hypersurface that bounds it, the following relation holds:

$$n^n v_n V^{n-1} \leq F^n, \tag{18.72}$$

called the *isoperimetric inequality*. Here v_n is the volume of the n-dimensional unit ball in E^n. Equality in the isoperimetric inequality (18.72) holds only for the ball.

The name "isoperimetric" comes from the classical problem of finding the closed plane curve of a given length L that encloses the largest area S. In this case inequality (18.72) means that

$$4\pi S \leq L^2. \tag{18.73}$$

It is this inequality that we shall now prove, assuming that the curve in question is smooth and is defined parametrically as $x = \varphi(s)$, $y = \psi(s)$, where s is arc length along the curve and φ and ψ belong to $C^{(1)}[0, L]$. The condition that the curve be closed means that $\varphi(0) = \varphi(L)$ and $\psi(0) = \psi(L)$.

We now pass from the parameter s to the parameter $t = 2\pi \frac{s}{L} - \pi$, which ranges from $-\pi$ to π, and we shall assume that our curve is defined parametrically as

$$x = x(t), \quad y = y(t), \quad -\pi \leq t \leq \pi, \tag{18.74}$$

---

[24] A. Hurwitz (1859–1919) – German mathematician, a student of F. Klein.

with

$$x(-\pi) = x(\pi), \qquad y(-\pi) = y(\pi). \tag{18.75}$$

We write (18.74) as a single complex-valued function

$$z = z(t), \quad -\pi \le t \le \pi, \tag{18.74'}$$

where $z(t) = x(t) + iy(t)$ and by (18.75) $z(-\pi) = z(\pi)$.

We remark that

$$\left|z'(t)\right|^2 = \left(x'(t)\right)^2 + \left(y'(t)\right)^2 = \left(\frac{ds}{dt}\right)^2,$$

and hence under our choice of parameter

$$\left|z'(t)\right|^2 = \frac{L^2}{4\pi^2}. \tag{18.76}$$

Next, taking into account the relations $\bar{z}z' = (x - iy)(x' + iy') = (xx' + yy') + i(xy' - x'y)$, and using Eqs. (18.75), we write the formula for the area of the region bounded by the closed curve (18.74):

$$S = \frac{1}{2}\int_{-\pi}^{\pi}(xy' - yx')(t)\,dt = \frac{1}{2i}\int_{-\pi}^{\pi}z'(t)\bar{z}(t)\,dt. \tag{18.77}$$

We now write the Fourier series expansion of the function (18.74'):

$$z(t) = \sum_{-\infty}^{\infty}c_k e^{ikt}.$$

Then

$$z'(t) \sim \sum_{-\infty}^{\infty}ikc_k e^{ikt}.$$

Equalities (18.76) and (18.77) mean in particular that

$$\frac{1}{2\pi}\|z'\|^2 = \frac{1}{2\pi}\int_{-\pi}^{\pi}\left|z'(t)\right|^2\,dt = \frac{L^2}{4\pi^2},$$

and

$$\frac{1}{2\pi}\langle z', z\rangle = \frac{1}{2\pi}\int_{-\pi}^{\pi}z'(t)\bar{z}(t)\,dt = \frac{i}{\pi}S.$$

In terms of Fourier coefficients, as follows from Eqs. (18.69) and (18.71), these relations assume the form

$$L^2 = 4\pi^2\sum_{-\infty}^{\infty}|kc_k|^2,$$

$$S = \pi \sum_{-\infty}^{\infty} k c_k \bar{c}_k.$$

Thus,

$$L^2 - 4\pi S = 4\pi^2 \sum_{-\infty}^{\infty} (k^2 - k)|c_k|^2.$$

The right-hand side of this equality is obviously nonnegative and vanishes only if $c_k = 0$ for all $k \in \mathbb{Z}$ except $k = 0$ and $k = 1$.

Thus, inequality (18.73) is proved, and at the same time we have found the equation

$$z(t) = c_0 + c_1 e^{it}, \quad -\pi \le t \le \pi,$$

of the curve for which it becomes equality. This is the complex form of the parametric equation of a circle with center at c_0 in the complex plane and radius $|c_1|$.

18.2.5 Problems and Exercises

1. a) Show that

$$\sum_{n=1}^{\infty} \frac{\sin nx}{n} = \frac{\pi - x}{2} \quad \text{for } 0 < x < 2\pi,$$

and find the sum of the series at all other points of $\mathbb{R}$.

Using the preceding expansion and the rules for operating with trigonometric Fourier series, now show that the following equalities are true.

b) $\sum_{k=1}^{\infty} \frac{\sin 2kx}{2k} = \frac{\pi}{4} - \frac{x}{2}$ for $0 < x < \pi$.

c) $\sum_{k=1}^{\infty} \frac{\sin(2k-1)x}{2k-1} = \frac{\pi}{4}$ for $0 < x < \pi$.

d) $\sum_{n=1}^{\infty} \frac{(-1)^{n-1}}{n} \sin nx = \frac{x}{2}$ for $|x| < \pi$.

e) $x^2 = \frac{\pi}{3} + 4 \sum_{n=1}^{\infty} \frac{(-1)^n}{n^2} \cos nx$ for $|x| < \pi$.

f) $x = \frac{\pi}{2} - \frac{4}{\pi} \sum_{k=1}^{\infty} \frac{\cos(2k-1)x}{(2k-1)^2}$ for $0 \le x \le \pi$.

g) $\frac{3x^2 - 6\pi x + 2\pi^2}{12} = \sum_{n=1}^{\infty} \frac{\cos nx}{n^2}$ for $0 \le x \le \pi$.

h) Sketch the graph of the sums of the trigonometric series here over the entire real line $\mathbb{R}$. Using the results obtained, find the sums of the following numerical series:

$$\sum_{n=0}^{\infty} \frac{(-1)^n}{2n + 1}, \quad \sum_{n=1}^{\infty} \frac{1}{n^2}, \quad \sum_{n=1}^{\infty} \frac{(-1)^n}{n^2}.$$

2. Show that:

a) if $f : [-\pi, \pi] \to \mathbb{C}$ is an odd (resp. even) function, then its Fourier coefficients have the following property: $a_k(f) = 0$ (resp. $b_k(f) = 0$) for $k = 0, 1, 2, \ldots$;

b) if $f : \mathbb{R} \to \mathbb{C}$ has period $2\pi/m$ then its Fourier coefficients $c_k(f)$ can be nonzero only when k is a multiple of m;

c) if $f : [-\pi, \pi] \to \mathbb{R}$ is real-valued, then $c_k(f) = \bar{c}_{-k}(f)$ for all $k \in \mathbb{N}$;

d) $|a_k(f)| \le 2 \sup_{|x|<\pi} |f(x)|$, $|b_k(f)| \le 2 \sup_{|x|<\pi} |f(x)|$, $|c_k(f)| \le \sup_{|x|<\pi} |f(x)|$.

3. a) Show that each of the systems $\{\cos kx; k = 0, 1, \ldots\}$ and $\{\sin kx; k \in \mathbb{N}\}$ is complete in the space $\mathcal{R}_2[a, a + \pi]$ for any value of $a \in \mathbb{R}$.

b) Expand the function $f(x) = x$ in the interval $[0, \pi]$ with respect to each of these two systems.

c) Draw the graphs of the sums of the series just found over the entire real line.

d) Exhibit the trigonometric Fourier series of the function $f(x) = |x|$ on the closed interval $[-\pi, \pi]$ and determine whether it converges uniformly to this function on the entire closed interval $[-\pi, \pi]$.

4. The Fourier series $\sum_{-\infty}^{\infty} c_k(f)e^{ikt}$ of a function f can be regarded as a special case of a power series $\sum_{-\infty}^{\infty} c_k z^k$ $(= \sum_{-\infty}^{-1} c_k z^k + \sum_{0}^{\infty} c_k z^k)$, in which z is restricted to the unit circle in the complex plane, that is, $z = e^{it}$.

Show that if the Fourier coefficients $c_k(f)$ of the function $f : [-\pi, \pi] \to \mathbb{C}$ vanish so rapidly that $\underline{\lim}_{k \to -\infty} |c_k(f)|^{1/k} = c_- > 1$ and $\overline{\lim}_{k \to +\infty} |c_k(f)|^{1/k} = c_+ < 1$, then

a) the function f can be regarded as the image of the unit circle under a function represented in the annulus $c_-^{-1} < |z| < c_+^{-1}$ by the series $\sum_{-\infty}^{\infty} c_k z^k$;

b) for $z = x + iy$ and $\ln \frac{1}{c_-} < y < \ln \frac{1}{c_+}$ the series $\sum_{-\infty}^{\infty} c_k(f)e^{ikx}$ converges absolutely (and, in particular, its sum is independent of the order of summation of the terms);

c) in any strip of the complex plane defined by the conditions $a \le \operatorname{Im} z \le b$, where $\ln \frac{1}{c_-} < a < b < \ln \frac{1}{c_+}$, the series $\sum_{-\infty}^{\infty} c_k(f)e^{ikx}$ converges absolutely and uniformly;

d) using the expansion $e^z = 1 + \frac{z}{1!} + \frac{z^2}{2!} + \cdots$ and Euler's formula $e^{ix} = \cos x + i \sin x$, show that

$$1 + \frac{\cos x}{1!} + \cdots + \frac{\cos nx}{n!} + \cdots = e^{\cos x} \cos(\sin x),$$

$$\frac{\sin x}{1!} + \cdots + \frac{\sin nx}{n!} + \cdots = e^{\cos x} \sin(\sin x);$$

e) using the expansions $\cos z = 1 - \frac{z^2}{2!} + \frac{z^4}{4!} - \cdots$ and $\sin z = z - \frac{z^3}{3!} + \frac{z^5}{5!} - \cdots$, verify that

$$\sum_{n=0}^{\infty} (-1)^n \frac{\cos(2n + 1)x}{(2n + 1)!} = \sin(\cos x)\cosh(\sin x),$$

$$\sum_{n=0}^{\infty} (-1)^n \frac{\sin(2n + 1)x}{(2n + 1)!} = \sin(\cos x)\sinh(\sin x),$$

$$\sum_{n=0}^{\infty}(-1)^n \frac{\cos 2nx}{(2n)!} = \cos(\cos x)\cosh(\sin x),$$

$$\sum_{n=0}^{\infty}(-1)^n \frac{\sin 2nx}{(2n!)} = \cos(\cos x)\sinh(\sin x).$$

5. Verify that

a) the systems $\{1, \cos k\frac{2\pi}{T}x, \sin k\frac{2\pi}{T}x; k \in \mathbb{N}\}$ and $\{e^{ik\frac{2\pi}{T}x}; k \in \mathbb{Z}\}$ are orthogonal in the space $\mathcal{R}_2([a, a+T], \mathbb{C})$ for every $a \in \mathbb{R}$;

b) the Fourier coefficients $a_k(f)$, $b_k(f)$, and $c_k(f)$ of a T-periodic function f in these systems are the same whether the Fourier expansion is done on the interval $[-\frac{T}{2}, \frac{T}{2}]$ or any other closed interval of the form $[a, a+T]$;

c) if $c_k(f)$ and $c_k(g)$ are the Fourier coefficients of T-periodic functions f and g, then

$$\frac{1}{T}\int_a^{a+T} f(x)\overline{g}(x)\,dx = \sum_{-\infty}^{\infty} c_k(f)\overline{c}_k(g);$$

d) the Fourier coefficients $c_k(h)$ normalized by the factor $\frac{1}{T}$ of the "convolution"

$$h(x) = \frac{1}{T}\int_0^T f(x-t)g(t)\,dt$$

of T-periodic smooth functions f and g and the coefficients $c_k(f)$ and $c_k(g)$ of the functions themselves are related by $c_k(h) = c_k(f)c_k(g)$, $k \in \mathbb{Z}$.

6. Prove that if α is incommensurable with π, then

a) $\lim_{N\to\infty} \frac{1}{N}\sum_{n=1}^{N} e^{ik(x+n\alpha)} = \frac{1}{2\pi}\int_{-\pi}^{\pi} e^{ikt}\,dt$;

b) for every continuous 2π-periodic function $f : \mathbb{R} \to \mathbb{C}$

$$\lim_{N\to\infty} \frac{1}{N}\sum_{n=1}^{N} f(x+n\alpha) = \frac{1}{2\pi}\int_{-\pi}^{\pi} f(t)\,dt.$$

7. Prove the following propositions.

a) If the function $f : \mathbb{R} \to \mathbb{C}$ is absolutely integrable on $\mathbb{R}$, then

$$\left|\int_{-\infty}^{\infty} f(x)e^{i\lambda x}\,dx\right| \le \int_{-\infty}^{\infty}\left|f\left(x+\frac{\pi}{\lambda}\right) - f(x)\right|dx.$$

b) If the functions $f : \mathbb{R} \to \mathbb{C}$ and $g : \mathbb{R} \to \mathbb{C}$ are absolutely integrable on $\mathbb{R}$ and g is bounded in absolute value on $\mathbb{R}$, then

$$\int_{-\infty}^{\infty} f(x+t)g(t)e^{i\lambda t}\,dt =: \varphi_\lambda(x) \rightrightarrows 0 \quad \text{on } \mathbb{R} \text{ as } \lambda \to \infty.$$

c) If $f : \mathbb{R} \to \mathbb{C}$ is a 2π-periodic function that is absolutely integrable over a period, then the remainder $S_n(x) - f(x)$ of its trigonometric Fourier series can be represented as

$$S_n(x) - f(x) = \frac{1}{\pi} \int_0^{\pi} \left(\Delta^2 f\right)(x, t) D_n(t) \, dt,$$

where D_n is the nth Dirichlet kernel, and $(\Delta^2 f)(x, t) = f(x + t) - 2f(x) + f(x - t)$.

d) For every $\delta \in \,]0, \pi[$ the formula for the remainder just obtained can be brought into the form

$$S_n(x) - f(x) = \frac{1}{\pi} \int_0^{\delta} \frac{\sin nt}{t} \left(\Delta^2 f\right)(x, t) \, dt + o(1),$$

where $o(1)$ trends to zero as $n \to \infty$, and uniformly on each closed interval $[a, b]$ on which f is bounded.

e) If the function $f : [-\pi, \pi] \to \mathbb{C}$ satisfies the Hölder condition $|f(x_1) - f(x_2)| \leq M |x_1 - x_2|^{\alpha}$ on $[-\pi, \pi]$ (where M and α are positive numbers) and in addition $f(-\pi) = f(\pi)$, then the Fourier series of f converges to it uniformly on the entire interval.

8. a) Prove that if $f : \mathbb{R} \to \mathbb{R}$ is a 2π-periodic function having a piecewise continuous derivative $f^{(m)}$ of order m ($m \in \mathbb{N}$), then f can be represented as

$$f(x) = \frac{\alpha_0}{2} + \frac{1}{\pi} \int_{-\pi}^{\pi} B_m(t - x) f^{(m)}(t) \, dt,$$

where $B_m(u) = \sum_{n=1}^{\infty} \frac{\cos(ku + \frac{m\pi}{2})}{k^m}$, $m \in \mathbb{N}$.

b) Using the Fourier series expansion obtained in Problem 1 for the function $\frac{\pi - x}{2}$ on the interval $[0, 2\pi]$, prove that $B_1(u)$ is a polynomial of degree 1 and $B_m(u)$ is a polynomial of degree m on the interval $[0, 2\pi]$. These polynomials are called the *Bernoulli polynomials*.

c) Verify that $\int_0^{2\pi} B_m(u) \, du = 0$ for every $m \in \mathbb{N}$.

9. a) Let $x_m = \frac{2\pi m}{2n+1}$, $m = 0, 1, \ldots, 2n$. Verify that

$$\frac{2}{2n + 1} \sum_{m=0}^{2n} \cos k x_m \cos l x_m = \delta_{kl},$$

$$\frac{2}{2n + 1} \sum_{m=0}^{2n} \sin k x_m \sin l x_m = \delta_{kl},$$

$$\sum_{m=0}^{2n} \sin k x_m \cos l x_m = 0,$$

where k and l are nonnegative integers, $\delta_{kl} = 0$ for $k \neq l$, and $\delta_{kl} = 1$ for $k = l$.

b) Let $f : \mathbb{R} \to \mathbb{R}$ be a 2π-periodic function that is absolutely integrable over a period. Let us partition the closed interval $[0, 2\pi]$ into $2n + 1$ equal parts by the points $x_m = \frac{2\pi m}{2n+1}$, $m = 0, 1, \ldots, 2n$. Let us compute the integrals

$$a_k(f) = \frac{1}{\pi} \int_0^{2\pi} f(x) \cos kx \, dx, \qquad b_k(f) = \frac{1}{\pi} \int_0^{2\pi} f(x) \sin kx \, dx$$

approximately using the rectangular method corresponding to this partition of the interval $[0, 2\pi]$. We obtain the quantities

$$\tilde{a}_k(f) = \frac{2}{2n + 1} \sum_{m=0}^{2n} f(x_m) \cos kx_m,$$

$$\tilde{b}_k(f) = \frac{2}{2n + 1} \sum_{m=0}^{2n} f(x_m) \sin kx_m,$$

which we place in the nth partial sum $S_n(f, x)$ of the Fourier series of f instead of the respective coefficients $a_k(f)$ and $b_k(f)$.

Prove that when this is done the result is a trigonometric polynomial $\widetilde{S}_n(f, x)$ of order n that interpolates the function f at the nodes x_m, $m = 0, 1, \ldots, 2n$, that is, at these points $f(x_m) = \widetilde{S}_n(x, m)$.

10. a) Suppose the function $f : [a, b] \to \mathbb{R}$ is continuous and piecewise differentiable, and suppose that the square of its derivative f' is integrable over the interval $]a, b[$. Using Parseval's equality, prove the following:

a) if $[a, b] = [0, \pi]$, then either of the conditions $f(0) = f(\pi) = 0$ or $\int_0^\pi f(x) \, dx = 0$ implies *Steklov's inequality*

$$\int_0^\pi f^2(x) \, dx \le \int_0^\pi (f')^2(x) \, dx,$$

in which equality is possible only for $f(x) = a \cos x$;

b) if $[a, b] = [-\pi, \pi]$ and the conditions $f(-\pi) = f(\pi)$ and $\int_{-\pi}^\pi f(x) \, dx = 0$ both hold, then *Wirtinger's inequality* holds:

$$\int_{-\pi}^\pi f^2(x) \, dx \le \int_{-\pi}^\pi (f')^2(x) \, dx,$$

where equality is possible only if $f(x) = a \cos x + b \sin x$.

11. The *Gibbs phenomenon* is the behavioral property of the partial sums of a trigonometric Fourier series described below, first observed by Wilbraham (1848) and later (1898) rediscovered by Gibbs (*Mathematical Encyclopedia*, Vol. 1, Moscow, 1977).

a) Show that

$$\operatorname{sgn} x = \frac{4}{\pi} \sum_{k=1}^{\infty} \frac{\sin(2k-1)x}{2k-1} \quad \text{for } |x| < \pi.$$

b) Verify that the function $S_n(x) = \frac{4}{\pi} \sum_{k=1}^{n} \frac{\sin(2k-1)x}{2k-1}$ has a maximum at $x = \frac{\pi}{2n}$ and that as $n \to \infty$

$$S_n\left(\frac{\pi}{2n}\right) = \frac{2}{\pi} \sum_{k=1}^{n} \frac{\sin(2k-1)\frac{\pi}{2n}}{(2k-1)\frac{\pi}{2n}} \cdot \frac{\pi}{n} \to \frac{2}{\pi} \int_0^{\pi} \frac{\sin x}{x}\, dx \approx 1.179.$$

Thus the oscillation of $S_n(x)$ near $x = 0$ as $n \to \infty$ exceeds the jump of the function $\operatorname{sgn} x$ itself at that point by approximately 18 % (the jump of $S_n(x)$ "due to inertia").

c) Describe the limit of the graphs of the functions $S_n(x)$ in problem b).

Now suppose that $S_n(f, x)$ is the nth partial sum of the trigonometric Fourier series of a function f and suppose that $S_n(f, x) \to f(x)$ in a deleted neighborhood $0 < |x - \xi| < \delta$ of the point ξ as $n \to \infty$ and that f has one-sided limits $f(\xi_-)$ and $f(\xi_+)$ at ξ. For definiteness we shall assume that $f(\xi_-) \le f(\xi_+)$.

We say that *Gibbs' phenomenon occurs for the sums* $S_n(f, x)$ at the point ξ if $\underline{\lim}_{n \to \infty} S_n(f, x) < f(\xi_-) \le f(\xi_+) < \overline{\lim}_{n \to \infty} S_n(f, x)$.

d) Using Remark 9 show that Gibbs' phenomenon occurs at the point ξ for every function of the form $\varphi(x) + c \operatorname{sgn}(x - \xi)$, where $c \ne 0$, $|\xi| < \pi$, and $\varphi \in C^{(1)}[-\pi, \pi]$.

12. *Multiple trigonometric Fourier series.*

a) Verify that the system of functions $\frac{1}{(2\pi)^{n/2}} e^{ikx}$, where $k = (k_1, \ldots, k_n)$, $x = (x_1, \ldots, x_n)$, $kx = k_1 x_1 + \cdots + k_n x_n$, and $k_1, \ldots, k_n \in \mathbb{Z}$, is orthonormal on any n-dimensional cube $I = \{x \in \mathbb{R}^n \mid a_j \le x_j \le a_j + 2\pi, j = 1, 2, \ldots, n\}$.

b) To a function f that is integrable over I we assign the sum $f \sim \sum_{-\infty}^{\infty} c_k(f) e^{ikx}$, which is called the *Fourier series of f in the system* $\{\frac{1}{(2\pi)^{n/2}} e^{ikx}\}$, if $c_k(f) = \frac{1}{(2\pi)^{n/2}} \int_I f(x) e^{-ikx}\, dx$. The numbers $c_k(f)$ are called the *Fourier coefficients of f in the system* $\{\frac{1}{(2\pi)^{n/2}} e^{ikx}\}$.

In the multidimensional case the Fourier series is often summed via the partial sums

$$S_N(x) = \sum_{|k| \le N} c_k(f) e^{ikx},$$

where $|k| \le N$ means that $N = (N_1, \ldots, N_k)$ and $|k_j| \le N_j$, $j = 1, \ldots, n$.

Show that for every function $f(x) = f(x_1, \ldots, x_n)$ that is 2π-periodic in each variable

$$S_n(x) = \frac{1}{\pi^n} \int_I \prod_{j=1}^{n} D_{N_j}(t_j - x_j) f(t) \, dt =$$

$$= \frac{1}{\pi^n} \int_{-\pi}^{\pi} \cdots \int_{-\pi}^{\pi} f(t - x) \prod_{j=1}^{n} D_{N_j}(t_j) \, dt_1 \cdots dt_n,$$

where $D_{N_j}(u)$ is the N_jth one-dimensional Dirichlet kernel.

c) Show that the *Fejér sum*

$$\sigma_N(x) := \frac{1}{N+1} \sum_{k=0}^{N} S_k(x) = \frac{1}{(N_1+1) \cdot \ldots \cdot (N_n+1)} \sum_{k_1=0}^{N_1} \cdots \sum_{k_n=0}^{N_n} S_{k_1 \ldots k_n}(x)$$

of a function $f(x) = f(x_1, \ldots, x_n)$ that is 2π-periodic in each variable can be represented as

$$\sigma_N(x) = \frac{1}{\pi^n} \int_I f(t - x) \Phi_N(t) \, dt,$$

where $\Phi_N(u) = \prod_{j=1}^{n} \mathcal{F}_{N_j}(u_j)$ and $\mathcal{F}_{N_j}$ is the N_jth one-dimensional Fejér kernel.

d) Now extend Fejér's theorem to the n-dimensional case.

e) Show that if a function f that is 2π-periodic in each variable is absolutely integrable over a period I, possibly in the improper sense, then $\int_I |f(x + u) - f(x)| \, dx \to 0$ as $u \to 0$ and $\int_I |f - \sigma_N|(x) \, dx \to 0$ as $N \to \infty$.

f) Prove that two functions f and g that are absolutely integrable over the cube I can have equal Fourier series (that is, $c_k(f) = c_k g$ for every multi-index k) only if $f(x) = g(x)$ almost everywhere on I. This is a strengthening of Proposition 3 on uniqueness of Fourier series.

g) Verify that the original orthonormal system $\{\frac{1}{(2\pi)^{n/2}} e^{ikx}\}$ is complete in $\mathcal{R}_2(I)$, so that the Fourier series of every function $f \in \mathcal{R}_2(I)$ converges to f in the mean on I.

h) Let f be a function in $C^{(\infty)}(\mathbb{R}^n)$ of period 2π in each variable. Verify that $c_k(f^{(\alpha)}) = i^{|\alpha|} k^\alpha c_k(f)$, where $\alpha = (\alpha_1, \ldots, \alpha_n)$, $k = (k_1, \ldots, k_n)$, $|\alpha| = |\alpha_1| + \cdots + |\alpha_n|$, $k^\alpha = k_1^{\alpha_1} \cdot \ldots \cdot k_n^{\alpha_n}$, and α_j are nonnegative integers.

i) Let f be a function of class $C^{(mn)}(\mathbb{R}^n)$ of period 2π in each variable. Show that if the estimate

$$\frac{1}{(2\pi)^n} \int_I |f^{(\alpha)}|^2(x) \, dx \leq M^2$$

holds for each multi-index $\alpha = (\alpha_1, \ldots, \alpha_n)$ such that α_j is 0 or m (for every $j = 1, \ldots, n$), then

$$|f(x) - S_n(x)| \leq \frac{CM}{N^{m-\frac{1}{2}}},$$

where $\underline{N} = \min\{N_1, \ldots, N_n\}$ and C is a constant depending on m but not on N or $x \in I$.

j) Notice that if a sequence of continuous functions converges in mean on the interval I to a function f and simultaneously converges uniformly to φ, then $f(x) = \varphi(x)$ on I.

Using this observation, prove that if a function $f : \mathbb{R}^n \to \mathbb{C}$ of period 2π in each variable belongs to $C^{(1)}(\mathbb{R}^n, \mathbb{C})$, then the trigonometric Fourier series of f converges to f uniformly on the entire space $\mathbb{R}^n$.

13. *Fourier series of generalized functions.* Every 2π-periodic function $f : \mathbb{R} \to \mathbb{C}$ can be regarded as a function $f(s)$ of a point on the unit circle Γ (the point is fixed by the value of the arc-length parameter s, $0 \le s \le 2\pi$).

Preserving the notation of Sect. 17.4, we consider the space $\mathcal{D}(\Gamma)$ on Γ consisting of functions in $C^{(\infty)}(\Gamma)$ and the space $\mathcal{D}'(\Gamma)$ of generalized functions, that is, continuous linear functionals on $\mathcal{D}(\Gamma)$. The value of the functional $F \in \mathcal{D}'(\Gamma)$ on the function $\varphi \in \mathcal{D}(\Gamma)$ will be denoted $F(\varphi)$, so as to avoid the symbol $\langle F, \varphi \rangle$ used in the present chapter to denote the Hermitian inner product (18.38).

Each function f that is integrable on Γ can be regarded as an element of $\mathcal{D}'(\Gamma)$ (a regular generalized function) acting on the function $\varphi \in \mathcal{D}(\Gamma)$ according to the formula

$$f(\varphi) = \int_0^{2\pi} f(s)\varphi(s)\,\mathrm{d}s.$$

Convergence of a sequence $\{F_n\}$ of generalized functions in $\mathcal{D}'(\Gamma)$ to a generalized function $F \in \mathcal{D}'(\Gamma)$, as usual, means that for every function $\varphi \in \mathcal{D}(\Gamma)$

$$\lim_{n \to \infty} F_n(\varphi) = F(\varphi).$$

a) Using the fact that by Theorem 5 the relation $\varphi(s) = \sum_{-\infty}^{\infty} c_k(\varphi)\mathrm{e}^{ikx}$ holds for every function $\varphi \in C^{(\infty)}(\Gamma)$, and, in particular, $\varphi(0) = \sum_{-\infty}^{\infty} c_k(\varphi)$, show that in the sense of convergence in the space of generalized functions $\mathcal{D}'(\Gamma)$ we have

$$\sum_{k=-n}^{n} \frac{1}{2\pi}\mathrm{e}^{iks} \to \delta \quad \text{as } n \to \infty.$$

Here δ is the element of $\mathcal{D}'(\Gamma)$ whose effect on the function $\varphi \in \mathcal{D}(\Gamma)$ is defined by $\delta(\varphi) = \varphi(0)$.

b) If $f \in \mathcal{R}(\Gamma)$, the Fourier coefficients of the function f in the system $\{\mathrm{e}^{iks}\}$ defined in the standard manner, can be written as

$$c_k(f) = \frac{1}{2\pi} \int_0^{2\pi} f(s)\mathrm{e}^{-ikx}\,\mathrm{d}x = \frac{1}{2\pi}f\left(\mathrm{e}^{-iks}\right).$$

By analogy we now define the Fourier coefficients $c_k(F)$ of any generalized function $F \in \mathcal{D}'(\Gamma)$ by the formula $c_k(F) = \frac{1}{2\pi}F(\mathrm{e}^{-iks})$, which makes sense because $\mathrm{e}^{-iks} \in \mathcal{D}(\Gamma)$.

Thus to every generalized function $F \in \mathcal{D}'(\Gamma)$ we assign the Fourier series

$$F \sim \sum_{-\infty}^{\infty} c_k(F) e^{iks}.$$

Show that $\delta \sim \sum_{-\infty}^{\infty} \frac{1}{2\pi} e^{iks}$.

c) Prove the following fact, which is remarkable for its simplicity and the freedom of action that it provides: the Fourier series of every generalized function $F \in \mathcal{D}'(\Gamma)$ converges to F (in the sense of convergence in the space $\mathcal{D}'(\Gamma)$).

d) Show that the Fourier series of a function $F \in \mathcal{D}'(\Gamma)$ (like the function F itself, and like every convergent series of generalized functions) can be differentiated termwise any number of times.

e) Starting from the equality $\delta = \sum_{-\infty}^{\infty} \frac{1}{2\pi} e^{iks}$, find the Fourier series of δ'.

f) Let us now return from the circle Γ to the line $\mathbb{R}$ and study the functions e^{iks} as regular generalized functions in $\mathcal{D}'(\mathbb{R})$ (that is, as continuous linear functionals on the space $\mathcal{D}(\mathbb{R})$ of functions in the class $C_0^{(\infty)}(\mathbb{R})$ of infinitely differentiable functions of compact support in $\mathbb{R}$).

Every locally integrable function f can be regarded as an element of $\mathcal{D}'(\mathbb{R})$ (a *regular generalized function in* $\mathcal{D}'(\mathbb{R})$) whose effect on the function $\varphi \in C_0^{(\infty)}(\mathbb{R}, \mathbb{C})$ is given by the rule $f(\varphi) = \int_{\mathbb{R}} f(x)\varphi(x)\, dx$. Convergence in $\mathcal{D}'(\mathbb{R})$ is defined in the standard way:

$$\left(\lim_{n \to \infty} F_n = F \right) := \forall \varphi \in \mathcal{D}(\mathbb{R}) \left(\lim_{n \to \infty} F_n(\varphi) = F(\varphi) \right).$$

Show that the equality

$$\frac{1}{2\pi} \sum_{-\infty}^{\infty} e^{ikx} = \sum_{-\infty}^{\infty} \delta(x - 2\pi k)$$

holds in the sense of convergence in $\mathcal{D}'(\mathbb{R})$. In both sides of this equality a limiting passage is assumed as $n \to \infty$ over symmetric partial sums $\sum_{-n}^{n}$, and $\delta(x - x_0)$, as always, denotes the δ-function of $\mathcal{D}'(\mathbb{R})$ shifted to the point x_0, that is, $\delta(x - x_0)(\varphi) = \varphi(x_0)$.

18.3 The Fourier Transform

18.3.1 Representation of a Function by Means of a Fourier Integral

a. The Spectrum and Harmonic Analysis of a Function

Let $f(t)$ be a T-periodic function, for example a periodic signal with frequency $\frac{1}{T}$ as a function of time. We shall assume that the function f is absolutely integrable over a period. Expanding f in a Fourier series (when f is sufficiently regular, as we know, the Fourier series converges to f) and transforming that series,

$$f(t) = \frac{a_0(f)}{2} + \sum_{k=1}^{\infty} a_k(f) \cos k\omega_0 t + b_k(f) \sin k\omega_0 t =$$

$$= \sum_{-\infty}^{\infty} c_k(f) e^{ik\omega_0 t} = c_0 + 2 \sum_{k=1}^{\infty} |c_k| \cos(k\omega_0 t + \arg c_k), \qquad (18.78)$$

we obtain a representation of f as a sum of a constant term $\frac{a_0}{2} = c_0$ – the *mean value of f over a period* – and *sinusoidal components* with frequencies $\nu_0 = \frac{1}{T}$ (the *fundamental frequency*), $2\nu_0$ (the *second harmonic frequency*), and so on. In general the kth *harmonic component* $2|c_k| \cos(k\frac{2\pi}{T}t + \arg c_k)$ of the signal has *frequency* $k\nu_0 = \frac{k}{T}$, *cyclic frequency* $k\omega_0 = 2\pi k\nu_0 = \frac{2\pi}{T}k$, *amplitude* $2|c_k| = \sqrt{a_k^2 + b_k^2}$, and *phase* $\arg c_k = -\arctan \frac{b_k}{a_k}$.

The expansion of a periodic function (signal) into a sum of simple harmonic oscillations is called the *harmonic* analysis of f. The numbers $\{c_k(f); k \in \mathbb{Z}\}$ or $\{a_0(f), a_k(f), b_k(f); k \in \mathbb{N}\}$ are called the *spectrum of the function* (signal) f. A periodic function thus has a *discrete spectrum*.

Let us now set out (on a heuristic level) what happens to the expansion (18.78) when the period T of the signal increases without bound.

Simplifying the notation by writing $l = \frac{T}{2}$ and $\alpha_k = k\frac{\pi}{l}$, we rewrite the expansion

$$f(t) = \sum_{-\infty}^{\infty} c_k e^{ik\frac{\pi}{l}t}$$

as follows:

$$f(t) = \sum_{-\infty}^{\infty} \left(c_k \frac{l}{\pi} \right) e^{ik\frac{\pi}{l}t} \frac{\pi}{l}, \qquad (18.79)$$

where

$$c_k = \frac{1}{2l} \int_{-l}^{l} f(t) e^{-i\alpha_k t} \, dt$$

and hence

$$c_k \frac{l}{\pi} = \frac{1}{2\pi} \int_{-l}^{l} f(t) e^{-i\alpha_k t} \, dt.$$

Assuming that in the limit as $l \to +\infty$ we arrive at an arbitrary function f that is absolutely integrable over $\mathbb{R}$, we introduce the auxiliary function

$$c(\alpha) = \frac{1}{2\pi} \int_{-\infty}^{\infty} f(t) e^{-i\alpha t} \, dt, \qquad (18.80)$$

whose values at points $\alpha = \alpha_k$ differ only slightly from the quantities $c_k \frac{l}{\pi}$ in formula (18.79). In that case

$$f(t) \approx \sum_{-\infty}^{\infty} c(\alpha_k)e^{i\alpha_k t} \frac{\pi}{l}, \tag{18.81}$$

where $\alpha_k = k\frac{\pi}{l}$ and $\alpha_{k+1} - \alpha_k = \frac{\pi}{l}$. This last integral resembles a Riemann sum, and as the partition is refined, which occurs as $l \to \infty$, we obtain

$$f(t) = \int_{-\infty}^{\infty} c(\alpha)e^{i\alpha t}\, d\alpha. \tag{18.82}$$

Thus, following Fourier, we have arrived at the expansion of the function f into a continuous linear combination of harmonics of variable frequency and phase.

The integral (18.82) will be called the Fourier integral below. It is the continuous equivalent of a Fourier series. The function $c(\alpha)$ in it is the analog of the Fourier coefficient, and will be called the Fourier transform of the function f (defined on the entire line $\mathbb{R}$). Formula (18.80) for the Fourier transform is completely equivalent to the formula for the Fourier coefficients. It is natural to regard the function $c(\alpha)$ as the *spectrum of the function* (signal) f. In contrast to the case of a periodic signal f considered above and the discrete spectrum (of Fourier coefficients) corresponding to it, the spectrum $c(\alpha)$ of an arbitrary signal may be nonzero on whole intervals and even on the entire line (*continuous spectrum*).

Example 1 Let us find the function having the following spectrum of compact support:

$$c(\alpha) = \begin{cases} h, & \text{if } |\alpha| \le a, \\ 0, & \text{if } |\alpha| > a. \end{cases} \tag{18.83}$$

Proof By formula (18.82) we find, for $t \ne 0$

$$f(t) = \int_{-a}^{a} he^{i\alpha t}\, d\alpha = h\frac{e^{iat} - e^{-iat}}{it} = 2h\frac{\sin at}{t}, \tag{18.84}$$

and when $t = 0$, we obtain $f(0) = 2ha$, which equals the limit of $2h\frac{\sin at}{t}$ as $t \to 0$.   □

The representation of a function in the form (18.82) is called its *Fourier integral representation*. We shall discuss below the conditions under which such a representation is possible. Right now, we consider another example.

Example 2 Let P be a device having the following properties: it is a linear signal transform, that is, $P(\sum_j a_j f_j) = \sum_j a_j P(f_j)$, and it preserves the periodicity of a signal, that is, $P(e^{i\omega t}) = p(\omega)e^{i\omega t}$, where the coefficient $p(\omega)$ depends on the frequency ω of the periodic signal $e^{i\omega t}$.

We use the compact complex notation, although of course everything could be rewritten in terms of $\cos \omega t$ and $\sin \omega t$.

The function $p(\omega) =: R(\omega)e^{i\varphi(\omega)}$ is called the *spectral characteristic of the device P*. Its absolute value $R(\omega)$ is usually called the *frequency characteristic* and its argument $\varphi(\omega)$ the *phase characteristic* of the device P. A signal $e^{i\omega t}$, after passing through the device, emerges transformed into the signal $R(\omega)e^{i(\omega t+\varphi(\omega))}$, its amplitude changed as a result of the factor $R(\omega)$ and its phase shifted due to the presence of the term $\varphi(\omega)$.

Let us assume that we know the spectral characteristic $p(\omega)$ of the device P and the signal $f(t)$ that enters the device; we ask how to find the signal $x(t) = P(f)(t)$ that emerges from the device.

Representing the signal $f(t)$ as the Fourier integral (18.82) and using the linearity of the device and the integral, we find

$$x(t) = P(f)(t) = \int_{-\infty}^{\infty} c(\omega)p(\omega)e^{i\omega t}\, d\omega.$$

In particular, if

$$p(\omega) = \begin{cases} 1 & \text{for } |\omega| \le \Omega, \\ 0 & \text{for } |\omega| > \Omega, \end{cases} \tag{18.85}$$

then

$$x(t) = \int_{-\Omega}^{\Omega} c(\omega)e^{i\omega t}\, d\omega$$

and, as one can see from the spectral characteristics of the device,

$$P(e^{i\omega t}) = \begin{cases} e^{i\omega t} & \text{for } |\omega| \le \Omega, \\ 0 & \text{for } |\omega| > \Omega. \end{cases}$$

A device P with the spectral characteristic (18.85) transmits (filters) frequencies not greater than Ω without distortion and truncates all of the signal involved with higher frequencies (larger than Ω). For that reason, such a device is called an *ideal low-frequency filter* (with *upper frequency limit* Ω) in radio technology.

Let us now turn to the mathematical side of the matter and to a more careful study of the concepts that arise.

b. Definition of the Fourier Transform and the Fourier Integral

In accordance with formulas (18.80) and (18.82) we make the following definition.

Definition 1 The function

$$\mathcal{F}[f](\xi) := \frac{1}{2\pi} \int_{-\infty}^{\infty} f(x)e^{-i\xi x}\, dx \tag{18.86}$$

is the *Fourier transform* of the function $f : \mathbb{R} \to \mathbb{C}$.

The integral here is understood in the sense of the principal value

$$\int_{-\infty}^{\infty} f(x)e^{-i\xi x}\,dx := \lim_{A\to+\infty} \int_{-A}^{A} f(x)e^{-i\xi x}\,dx,$$

and we assume that it exists.

If $f : \mathbb{R} \to \mathbb{C}$ is absolutely integrable on $\mathbb{R}$, then, since $|f(x)e^{-ix\xi}| = |f(x)|$ for $x, \xi \in \mathbb{R}$, the Fourier transform (18.86) is defined, and the integral (18.86) converges absolutely and uniformly with respect to ξ on the entire line $\mathbb{R}$.

Definition 2 If $c(\xi) = \mathcal{F}[f](\xi)$ is the Fourier transform of $f : \mathbb{R} \to \mathbb{C}$, then the integral assigned to f,

$$f(x) \sim \int_{-\infty}^{\infty} c(\xi)e^{ix\xi}\,d\xi, \tag{18.87}$$

understood as a principal value, is called the *Fourier integral* of f.

The Fourier coefficients and the Fourier series of a periodic function are thus the discrete analog of the Fourier transform and the Fourier integral respectively.

Definition 3 The following integrals, understood as principal values,

$$\mathcal{F}_c[f](\xi) := \frac{1}{\pi} \int_{-\infty}^{\infty} f(x)\cos \xi x\,dx, \tag{18.88}$$

$$\mathcal{F}_s[f](\xi) := \frac{1}{\pi} \int_{-\infty}^{\infty} f(x)\sin \xi x\,dx, \tag{18.89}$$

are called respectively the *Fourier cosine transform* and the *Fourier sine transform* of the function f.

Setting $c(\xi) = \mathcal{F}[f](\xi)$, $a(\xi) = \mathcal{F}_c[f](\xi)$, and $b(\xi) = \mathcal{F}_s[f](\xi)$, we obtain the relation that is already partly familiar to us from Fourier series

$$c(\xi) = \frac{1}{2}\big(a(\xi) - ib(\xi)\big). \tag{18.90}$$

As can be seen from relations (18.88) and (18.89),

$$a(-\xi) = a(\xi), \qquad b(-\xi) = -b(\xi). \tag{18.91}$$

Formulas (18.90) and (18.91) show that Fourier transforms are completely determined on the entire real line $\mathbb{R}$ if they are known for nonnegative values of the argument.

From the physical point of view this is a completely natural fact – the spectrum of a signal needs to be known for frequencies $\omega \geq 0$; the negative frequencies α in (18.80) and (18.82) – result from the form in which they are written. Indeed,

$$\int_{-A}^{A} c(\xi) e^{ix\xi}\, d\xi = \left(\int_{-A}^{0} + \int_{0}^{A} \right) c(\xi) e^{ix\xi}\, d\xi = \int_{0}^{A} \left(c(\xi) e^{ix\xi} + c(-\xi) e^{ix\xi} \right) d\xi =$$

$$= \int_{0}^{A} \left(a(\xi) \cos x\xi + b(\xi) \sin x\xi \right) d\xi,$$

and hence the Fourier integral (18.87) can be represented as

$$\int_{0}^{\infty} \left(a(\xi) \cos x\xi + b(\xi) \sin x\xi \right) d\xi, \tag{18.87'}$$

which is in complete agreement with the classical form of a Fourier series. If the function f is real-valued, it follows from formulas (18.90) and (18.91) that

$$c(-\xi) = \overline{c(\xi)}, \tag{18.92}$$

since in this case $a(\xi)$ and $b(\xi)$ are real-valued functions on $\mathbb{R}$, as one can see from their definitions (18.88) and (18.89). On the other hand, under the assumption $\overline{f}(x) = f(x)$, Eq. (18.92) can be obtained immediately from the definition (18.86) of the Fourier transform, if we take into account that the conjugation sign can be moved under the integral sign. This last observation allows us to conclude that

$$\mathcal{F}[\overline{f}](-\xi) = \overline{\mathcal{F}[f](\xi)} \tag{18.93}$$

for every function $f : \mathbb{R} \to \mathbb{C}$.

It is also useful to note that if f is a real-valued even function, that is, $\overline{f(x)} = f(x) = f(-x)$, then

$$\overline{\mathcal{F}_c[f](\xi)} = \mathcal{F}_c[f](\xi), \qquad \mathcal{F}_s[f](\xi) \equiv 0,$$
$$\overline{\mathcal{F}[f](\xi)} = \mathcal{F}[f](\xi) = \mathcal{F}[f](-\xi); \tag{18.94}$$

and if f is a real-valued odd function, that is, $\overline{f(x)} = f(x) = -f(-x)$, then

$$\mathcal{F}_c[f](\xi) \equiv 0, \qquad \overline{\mathcal{F}_s[f](\xi)} = \mathcal{F}_s[f](\xi),$$
$$\overline{\mathcal{F}[f](\xi)} = -\mathcal{F}[f](\xi) = \mathcal{F}[f](-\xi); \tag{18.95}$$

and if f is a purely imaginary function, that is, $\overline{f(x)} = -f(x)$, then

$$\mathcal{F}[\overline{f}](-\xi) = -\overline{\mathcal{F}[f](\xi)}. \tag{18.96}$$

We remark that if f is a real-valued function, its Fourier integral (18.87') can also be written as

$$\int_0^\infty \sqrt{a^2(\xi) + b^2(\xi)} \cos\big(x\xi + \varphi(\xi)\big)\, d\xi = 2 \int_0^\infty |c(\xi)| \cos\big(x\xi + \varphi(\xi)\big)\, d\xi,$$

where $\varphi(\xi) = -\arctan \frac{b(\xi)}{a(\xi)} = \arg c(\xi)$.

Example 3 Let us find the Fourier transform of $f(t) = \frac{\sin at}{t}$ (assuming $f(0) = a \in \mathbb{R}$).

$$\mathcal{F}[f](\alpha) = \lim_{A \to +\infty} \frac{1}{2\pi} \int_{-A}^{A} \frac{\sin at}{t} e^{-i\alpha t}\, dt =$$

$$= \lim_{A \to +\infty} \frac{1}{2\pi} \int_{-A}^{A} \frac{\sin at \cos \alpha t}{t}\, dt = \frac{2}{2\pi} \int_0^{+\infty} \frac{\sin at \cos \alpha t}{t}\, dt =$$

$$= \frac{1}{2\pi} \int_0^{+\infty} \left(\frac{\sin(a + \alpha)t}{t} + \frac{\sin(a - \alpha)t}{t} \right) dt =$$

$$= \frac{1}{2\pi} \big(\operatorname{sgn}(a + \alpha) + \operatorname{sgn}(a - \alpha)\big) \int_0^\infty \frac{\sin u}{u}\, du = \begin{cases} \frac{1}{2} \operatorname{sgn} a, & \text{if } |\alpha| \le |a|, \\ 0, & \text{if } |\alpha| > |a|, \end{cases}$$

since we know the value of the Dirichlet integral

$$\int_0^\infty \frac{\sin u}{u}\, du = \frac{\pi}{2}. \tag{18.97}$$

Hence if we assume $a \ge 0$ and take the function $f(t) = 2h\frac{\sin at}{t}$ of Eq. (18.84), we find, as we should have expected, that the Fourier transform is the spectrum of this function exhibited in relations (18.83).

The function f in Example 3 is not absolutely integrable on $\mathbb{R}$, and its Fourier transform has discontinuities. That the Fourier transform of an absolutely integrable function has no discontinuities is attested by the following lemma.

Lemma 1 *If the function $f : \mathbb{R} \to \mathbb{C}$ is locally integrable and absolutely integrable on $\mathbb{R}$, then*

a) *its Fourier transform $\mathcal{F}[f](\xi)$ is defined for every value $\xi \in \mathbb{R}$;*
b) $\mathcal{F}[f] \in C(\mathbb{R}, \mathbb{C})$;
c) $\sup_\xi |\mathcal{F}[g](\xi)| \le \frac{1}{2\pi} \int_{-\infty}^\infty |f(x)|\, dx$;
d) $\mathcal{F}[f](\xi) \to 0$ *as* $\xi \to \infty$.

Proof We have already noted that $|f(x)e^{ix\xi}| \le |f(x)|$, from which it follows that the integral (18.86) converges absolutely and uniformly with respect to $\xi \in \mathbb{R}$. This fact simultaneously proves parts a) and c).

Part d) follows from the Riemann–Lebesgue lemma (see Sect. 18.2).

For a fixed finite $A \geq 0$, the estimate

$$\left| \int_{-A}^{A} f(x) \left(e^{-ix(\xi + h)} - e^{-ix\xi} \right) dx \right| \leq \sup_{|x| \leq A} \left| e^{-ixh} - 1 \right| \int_{-A}^{A} |f(x)| dx$$

establishes that the integral

$$\frac{1}{2\pi} \int_{-A}^{A} f(x) e^{-ix\xi} dx,$$

is continuous with respect to ξ; and the uniform convergence of this integral as $A \to +\infty$ enables us to conclude that $\mathcal{F}[f] \in C(\mathbb{R}, \mathbb{C})$.   □

Example 4 Let us find the Fourier transform of the function $f(t) = e^{-t^2/2}$:

$$\mathcal{F}[f](\alpha) = \int_{-\infty}^{+\infty} e^{-t^2/2} e^{-i\alpha t} dt = \int_{-\infty}^{+\infty} e^{-t^2/2} \cos \alpha t \, dt.$$

Differentiating this last integral with respect to the parameter α and then integrating by parts, we find that

$$\frac{d\mathcal{F}[f]}{d\alpha}(\alpha) + \alpha \mathcal{F}[f](\alpha) = 0,$$

or

$$\frac{d}{d\alpha} \ln \mathcal{F}[f](\alpha) = -\alpha.$$

It follows that $\mathcal{F}[f](\alpha) = ce^{-\alpha^2/2}$, where c is a constant which, using the Euler–Poisson integral (see Example 17 of Sect. 17.2) we find from the relation

$$c = \mathcal{F}[f](0) = \int_{-\infty}^{+\infty} e^{-t^2/2} dt = \sqrt{2\pi}.$$

Thus we have found that $\mathcal{F}[f](\alpha) = \sqrt{2\pi} e^{-\alpha^2/2}$, and simultaneously shown that $\mathcal{F}_c[f](\alpha) = \sqrt{2\pi} e^{-\alpha^2/2}$ and $\mathcal{F}_s[f](\alpha) \equiv 0$.

c. Normalization of the Fourier Transform

We obtained the Fourier transform (18.80) and the Fourier integral (18.82) as the natural continuous analogs of the Fourier coefficients $c_k = \frac{1}{2\pi} \int_{-\pi}^{\pi} f(x) e^{-ikx} dx$ and the Fourier series $\sum_{-\infty}^{\infty} c_k e^{ikx}$ of a periodic function f in the trigonometric system $\{e^{ikx}; k \in \mathbb{Z}\}$. This system is not orthonormal, and only the ease of writing a trigonometric Fourier series in it has caused it to be used traditionally instead of the more natural orthonormal system $\{\frac{1}{\sqrt{2\pi}} e^{ikx}; k \in \mathbb{Z}\}$. In this normalized system the Fourier series has the form $\sum_{-\infty}^{\infty} \hat{c}_k \frac{1}{\sqrt{2\pi}} e^{ikx}$, and the Fourier coefficients are defined by the formulas $\hat{c}_k = \frac{1}{\sqrt{2\pi}} \int_{-\pi}^{\pi} f(x) e^{-ikx} \, dx$.

The continuous analogs of such natural Fourier coefficients and such a Fourier series would be the Fourier transform

$$\hat{f}(\xi) := \frac{1}{\sqrt{2\pi}} \int_{-\infty}^{\infty} f(x) e^{-ix\xi} \, dx \qquad (18.98)$$

and the Fourier integral

$$f(x) = \frac{1}{\sqrt{2\pi}} \int_{-\infty}^{\infty} \hat{f}(\xi) e^{ix\xi} \, d\xi, \qquad (18.99)$$

which differ from those considered above only in the normalizing coefficient.

In the symmetric formulas (18.98) and (18.99) the Fourier "coefficient" and the Fourier "series" practically coalesce, and so in the future we shall essentially be interested only in the properties of the integral transform (18.98), calling it the *normalized Fourier transform* or, where no confusion can arise, simply the *Fourier transform* of the function f.

In general the name *integral operator* or *integral transform* is customarily given to an operator A that acts on a function f according to a rule

$$A(f)(y) = \int_X K(x, y) f(x) \, dx,$$

where $K(x, y)$ is a given function called the *kernel of the integral operator*, and $X \subset \mathbb{R}^n$ is the set over which the integration extends and on which the integrands are assumed to be defined. Since y is a free parameter in some set Y, it follows that $A(f)$ is a function on Y.

In mathematics there are many important integral transforms, and among them the Fourier transform occupies one of the most key positions. The reasons for this situation go very deep and involve the remarkable properties of the transformation (18.98), which we shall to some extent describe and illustrate in action in the remaining part of this section.

Thus, we shall study the normalized Fourier transform (18.98).

Along with the notation $\hat{f}$ for the normalized Fourier transform, we introduce the notation

$$\tilde{f}(\xi) := \frac{1}{\sqrt{2\pi}} \int_{-\infty}^{\infty} f(x) e^{i\xi x} \, dx, \qquad (18.100)$$

that is, $\tilde{f}(\xi) = \hat{f}(-\xi)$.

Formulas (18.98) and (18.99) say that

$$\tilde{\hat{f}} = \hat{\tilde{f}} = f, \qquad (18.101)$$

that is, the integral transforms (18.98) and (18.99) are mutually inverse to each other. Hence if (18.98) is the Fourier transform, then it is natural to call the integral operator (18.100) the *inverse Fourier transform*.

We shall discuss in detail below certain remarkable properties of the Fourier transform and justify them. For example

$$\widehat{f^{(n)}}(\xi) = (i\xi)^n \hat{f}(\xi),$$
$$\widehat{f * g} = \sqrt{2\pi}\,\hat{f}\cdot\hat{g},$$
$$\|\hat{f}\| = \|f\|.$$

That is, the Fourier transform maps the operator of differentiation into the operator of multiplication by the independent variable; the Fourier transform of the convolution of functions amounts to multiplying the transforms; the Fourier transform preserves the norm (Parseval's equality), and is therefore an isometry of the corresponding function space.

But we shall begin with the inversion formula (18.101).

For another convenient normalization of the Fourier transform see Problem 10 below.

d. Sufficient Conditions for a Function to be Representable as a Fourier Integral

We shall now prove a theorem that is completely analogous in both form and content to the theorem on convergence of a trigonometric Fourier series at a point. To preserve the familiar appearance of our earlier formulas and transformations to the maximum extent, we shall use the nonnormalized Fourier transform $c(\xi)$ in the present part of this subsection, together with its rather cumbersome but sometimes convenient notation $\mathcal{F}[f](\xi)$. Afterwards, when studying the integral Fourier transform as such, we shall as a rule work with the normalized Fourier transform $\hat{f}$ of the function f.

Theorem 1 (Convergence of the Fourier integral at a point) *Let $f : \mathbb{R} \to \mathbb{C}$ be an absolutely integrable function that is piecewise continuous on each finite closed interval of the real axis $\mathbb{R}$.*

If the function f satisfies the Dini conditions at a point $x \in \mathbb{R}$, then its Fourier integral (18.82), (18.87), (18.87′), (18.99) *converges at that point to the value $\frac{1}{2}(f(x_-) + f(x_+))$, equal to half the sum of the left and right-hand limits of the function at that point.*

Proof By Lemma 1 the Fourier transform $c(\xi) = \mathcal{F}[f](\xi)$ of the function f is continuous on $\mathbb{R}$ and hence integrable on every interval $[-A, A]$. Just as we transformed the partial sum of the Fourier series, we now carry out the following transformations of the partial Fourier integral:

$$S_A(x) = \int_{-A}^{A} c(\xi)e^{ix\xi}\,d\xi = \int_{-A}^{A}\left(\frac{1}{2\pi}\int_{-\infty}^{\infty} f(t)e^{-it\xi}\,dt\right)e^{ix\xi}\,d\xi =$$

$$= \frac{1}{2\pi} \int_{-\infty}^{\infty} f(t) \left(\int_{-A}^{A} e^{i(x-t)\xi} \, d\xi \right) dt =$$

$$= \frac{1}{2\pi} \int_{-\infty}^{\infty} f(t) \frac{e^{i(x-t)A} - e^{-i(x-t)A}}{i(x-t)} \, dt =$$

$$= \frac{1}{\pi} \int_{-\infty}^{\infty} f(t) \frac{\sin(x-t)A}{x-t} \, dt = \frac{1}{\pi} \int_{-\infty}^{\infty} f(x+u) \frac{\sin Au}{u} \, du =$$

$$= \frac{1}{\pi} \int_{0}^{\infty} \left(f(x-u) + f(x+u) \right) \frac{\sin Au}{u} \, du.$$

The change in the order of integration at the second equality from the beginning of the computation is legal. In fact, in view of the piecewise continuity of f, for every finite $B > 0$ we have the equality

$$\int_{-A}^{A} \left(\frac{1}{2\pi} \int_{-B}^{B} f(t) e^{-it\xi} \, dt \right) e^{ix\xi} \, d\xi = \frac{1}{2\pi} \int_{-B}^{B} f(t) \left(\int_{-A}^{A} e^{i(x-t)\xi} \, d\xi \right) dt,$$

from which as $B \to +\infty$, taking account of the uniform convergence of the integral $\int_{-B}^{B} f(x) e^{-it\xi} \, dt$ with respect to ξ, we obtain the equality we need.

We now use the value of the Dirichlet integral (18.97) and complete our transformation:

$$S_A(x) - \frac{f(x_-) + f(x_+)}{2} =$$

$$= \frac{1}{\pi} \int_{0}^{+\infty} \frac{(f(x-u) - f(x_-)) + (f(x+u) - f(x_+))}{u} \sin Au \, du.$$

The resulting integral tends to zero as $A \to \infty$. We shall explain this and thereby finish the proof of the theorem.

We represent this integral as the sum of the integrals over the interval $]0, 1]$ and over the interval $[1, +\infty[$. The first of these two integrals tends to zero as $A \to +\infty$ in view of the Dini conditions and the Riemann–Lebesgue lemma. The second integral is the sum of four integrals corresponding to the four terms $f(x-u)$, $f(x+u)$, $f(x_-)$ and $f(x_+)$. The Riemann–Lebesgue lemma applies to the first two of these four integrals, and the last two can be brought into the following form, up to a constant factor:

$$\int_{1}^{+\infty} \frac{\sin Au}{u} \, du = \int_{A}^{+\infty} \frac{\sin v}{v} \, dv.$$

But as $A \to +\infty$ this last integral tends to zero, since the Dirichlet integral (18.97) converges.   □

Remark 1 In the proof of Theorem 1 we have actually studied the convergence of the integral as a principal value. But if we compare the notations (18.87) and (18.87′), it becomes obvious that it is precisely this interpretation of the integral that corresponds to convergence of the integral (18.87′).

From this theorem we obtain in particular

Corollary 1 *Let $f : \mathbb{R} \to \mathbb{C}$ be a continuous absolutely integrable function.*

If the function f is differentiable at each point $x \in \mathbb{R}$ or has finite one-sided derivatives or satisfies a Hölder condition, then it is represented by its Fourier integral.

Hence for functions of these classes both equalities (18.80) and (18.82) or (18.98) and (18.99) hold, and we have thus proved the inversion formula for the Fourier transform for such functions.

Let us consider several examples.

Example 5 Assume that the signal $v(t) = P(f)(t)$ emerging from the device P considered in Example 2 is known, and we wish to find the input signal $f(t)$ entering the device P.

In Example 2 we have shown that f and v are connected by the relation

$$v(t) = \int_{-\infty}^{\infty} c(\omega) p(\omega) e^{i\omega t} \, d\omega,$$

where $c(\omega) = \mathcal{F}[f](\omega)$ is the spectrum of the signal F (the nonnormalized Fourier transform of the function f) and p is the spectral characteristic of the device P. Assuming all these functions are sufficiently regular, from the theorem just proved we conclude that then

$$c(\omega) p(\omega) = \mathcal{F}[v](\omega).$$

From this we find $c(\omega) = \mathcal{F}[f](\omega)$. Knowing $c(\omega)$, we find the signal f using the Fourier integral (18.87).

Example 6 Let $a > 0$ and

$$f(x) = \begin{cases} e^{-ax} & \text{for } x > 0, \\ 0 & \text{for } x \leq 0. \end{cases}$$

Then

$$\mathcal{F}[f](\xi) = \frac{1}{2\pi} \int_{0}^{+\infty} e^{-ax} e^{-i\xi x} \, dx = \frac{1}{2\pi} \frac{1}{a + i\xi}.$$

In discussing the definition of the Fourier transform, we have already noted a number of its obvious properties in Part b of the present subsection. We note further that if $f_-(x) := f(-x)$, then $\mathcal{F}[f_-](\xi) = \mathcal{F}[f](-\xi)$. This is an elementary change of variable in the integral.

We now take the function $e^{-a|x|} = f(x) + f(-x) =: \varphi(x)$.

Then

$$\mathcal{F}[\varphi](\xi) = \mathcal{F}[f](\xi) + \mathcal{F}[f](-\xi) = \frac{1}{\pi} \frac{a}{a^2 + \xi^2}.$$

If we now take the function $\psi(x) = f(x) - f(-x)$, which is an odd extension of the function e^{-ax}, $x > 0$, to the entire real line, then

$$\mathcal{F}[\psi](\xi) = \mathcal{F}[f](\xi) - \mathcal{F}[f](-\xi) = -\frac{i}{\pi}\frac{\xi}{a^2 + \xi^2}.$$

Using Theorem 1, or more precisely the corollary to it, we find that

$$\frac{1}{2\pi}\int_{-\infty}^{\infty}\frac{e^{ix\xi}}{a + i\xi}\,d\xi = \begin{cases} e^{-ax}, & \text{if } x > 0, \\ \frac{1}{2}, & \text{if } x = 0, \\ 0, & \text{if } x < 0; \end{cases}$$

$$\frac{1}{\pi}\int_{-\infty}^{+\infty}\frac{ae^{ix\xi}}{a^2 + \xi^2}\,d\xi = e^{-a|x|};$$

$$\frac{i}{\pi}\int_{-\infty}^{+\infty}\frac{\xi e^{ix\xi}}{a^2 + \xi^2}\,d\xi = \begin{cases} e^{-ax}, & \text{if } x > 0, \\ 0, & \text{if } x = 0, \\ -e^{ax}, & \text{if } x < 0. \end{cases}$$

All the integrals here are understood in the sense of the principal value, although the second one, in view of its absolute convergence, can also be understood in the sense of an ordinary improper integral.

Separating the real and imaginary parts in these last two integrals, we find the Laplace integrals we have encountered earlier

$$\int_0^{+\infty}\frac{\cos x\xi}{a^2 + \xi^2}\,d\xi = \frac{\pi}{2a}e^{-a|x|},$$

$$\int_0^{+\infty}\frac{\sin x\xi}{a^2 + \xi^2}\,d\xi = \frac{\pi}{2}e^{-a|x|}\operatorname{sgn} x.$$

Example 7 On the basis of Example 4 it is easy to find (by an elementary change of variable) that if

$$f(x) = e^{-a^2 x^2}, \quad \text{then} \quad \hat{f}(\xi) = \frac{1}{\sqrt{2}a}e^{-\frac{\xi^2}{4a^2}}.$$

It is very instructive to trace the simultaneous evolution of the graphs of the functions f and $\hat{f}$ as the parameter a varies from $1/\sqrt{2}$ to 0. The more "concentrated" one of the functions is, the more "smeared" the other is. This circumstance is closely connected with the Heisenberg uncertainty principle in quantum mechanics. (In this connection see Problems 6 and 7.)

Remark 2 In completing the discussion of the question of the possibility of representing a function by a Fourier integral, we note that, as Examples 1 and 3 show, the conditions on f stated in Theorem 1 and its corollary are sufficient but not necessary for such a representation to be possible.

18.3.2 The Connection of the Differential and Asymptotic Properties of a Function and Its Fourier Transform

a. Smoothness of a Function and the Rate of Decrease of Its Fourier Transform

It follows from the Riemann–Lebesgue lemma that the Fourier transform of any absolutely integrable function on $\mathbb{R}$ tends to zero at infinity. This has already been noted in Lemma 1 proved above. We now show that, like the Fourier coefficients, the smoother the function, the faster its Fourier transform tends to zero. The dual fact is that the faster a function tends to zero, the smoother its Fourier transform.

We begin with the following auxiliary proposition.

Lemma 2 *Let $f : \mathbb{R} \to \mathbb{C}$ be a continuous function having a locally piecewise continuous derivative f' on $\mathbb{R}$. Given this,*

a) *if the function f' is integrable on $\mathbb{R}$, then $f(x)$ has a limit both as $x \to -\infty$ and as $\to +\infty$;*

b) *if the functions f and f' are integrable on $\mathbb{R}$, then $f(x) \to 0$ as $\to \infty$.*

Proof Under these restrictions on the functions f and f' the Newton–Leibniz formula holds

$$f(x) = f(0) + \int_0^x f'(t)\,dt.$$

In conditions a) the right-hand side of this equality has a limit both as $x \to +\infty$ and as $x \to -\infty$.

If a function f having a limit at infinity is integrable on $\mathbb{R}$, then both of these limits must obviously be zero. $\square$

We now prove

Proposition 1 (Connection between the smoothness of a function and the rate of decrease of its Fourier transform) *If $f \in C^{(k)}(\mathbb{R}, \mathbb{C})$ $(k = 0, 1, \ldots)$ and all the functions $f, f', \ldots, f^{(k)}$ are absolutely integrable on $\mathbb{R}$, then*

a) *for every $n \in \{0, 1, \ldots, k\}$*

$$\widehat{f^{(n)}}(\xi) = (i\xi)^n \hat{f}(\xi), \tag{18.102}$$

b) $\hat{f}(\xi) = o\left(\frac{1}{\xi^k}\right)$ *as $\xi \to 0$.*

Proof If $k = 0$, then a) holds trivially and b) follows from the Riemann–Lebesgue lemma.

Let $k > 0$. By Lemma 2 the functions $f, f', \ldots, f^{(k-1)}$ tend to zero as $x \to \infty$. Taking this into account, we integrate by parts,

$$\widehat{f^{(k)}}(\xi) := \frac{1}{\sqrt{2\pi}} \int_{-\infty}^{\infty} f^{(k)}(x)e^{-i\xi x}\,dx =$$

$$= \frac{1}{\sqrt{2\pi}} \left(f^{(k-1)}(x)e^{-i\xi x}\Big|_{x=-\infty}^{+\infty} + (i\xi) \int_{-\infty}^{\infty} f^{(k-1)}(x)e^{-i\xi x}\,dx \right) =$$

$$= \cdots = \frac{(i\xi)^k}{\sqrt{2\pi}} \int_{-\infty}^{\infty} f(x)e^{-i\xi x}\,dx = (i\xi)^k \hat{f}(\xi).$$

Thus Eq. (18.102) is established. This is a very important relation, and we shall return to it.

We have shown that $\hat{f}(\xi) = (i\xi)^{-k}\widehat{f^{(k)}}(\xi)$, but by the Riemann–Lebesgue lemma $\widehat{f^{(k)}}(\xi) \to 0$ as $\xi \to 0$ and hence b) is also proved.   □

b. The Rate of Decrease of a Function and the Smoothness of Its Fourier Transform

In view of the nearly complete identity of the direct and inverse Fourier transforms the following proposition, dual to Proposition 1, holds.

Proposition 2 (The connection between the rate of decrease of a function and the smoothness of its Fourier transform) *If a locally integrable function $f : \mathbb{R} \to \mathbb{C}$ is such that the function $x^k f(x)$ is absolutely integrable on $\mathbb{R}$, then*

a) *the Fourier transform of f belongs to $C^{(k)}(\mathbb{R}, \mathbb{C})$.*
b) *the following equality holds:*

$$\hat{f}^{(k)}(\xi) = (-i)^k \widehat{x^k f(x)}(\xi). \tag{18.103}$$

Proof For $k = 0$ relation (18.103) holds trivially, and the continuity of $\hat{f}(\xi)$ has already been proved in Lemma 1. If $k > 0$, then for $n < k$ we have the estimate $|x^n f(x)| \leq |x^k f(x)|$ at infinity, from which it follows that $x^n f(x)$ is absolutely integrable. But $|x^n f(x)e^{-i\xi x}| = |x^n f(x)|$, which enables us to invoke the uniform convergence of these integrals with respect to the parameter ξ and successively differentiate them under the integral sign:

$$\hat{f}(\xi) = \frac{1}{\sqrt{2\pi}} \int_{-\infty}^{\infty} f(x)e^{-i\xi x}\,dx,$$

$$\hat{f}'(\xi) = \frac{-i}{\sqrt{2\pi}} \int_{-\infty}^{\infty} xf(x)e^{-i\xi x}\,dx,$$

$$\vdots$$

$$\hat{f}^{(k)}(\xi) = \frac{(-i)^k}{\sqrt{2\pi}} \int_{-\infty}^{\infty} x^k f(x) e^{-i\xi x} \, dx.$$

By Lemma 1 this last integral is continuous on the entire real line. Hence indeed $\hat{f} \in C^{(k)}(\mathbb{R}, \mathbb{C})$. □

c. The Space of Rapidly Decreasing Functions

Definition 4 We denote the set of functions $f \in C^{(\infty)}(\mathbb{R}, \mathbb{C})$ satisfying the condition

$$\sup_{x \in \mathbb{R}} \left| x^\beta f^{(\alpha)}(x) \right| < \infty$$

for all nonnegative integers α and β by $S(\mathbb{R}, \mathbb{C})$ or more briefly by S. Such functions are called *rapidly decreasing functions* (as $x \to \infty$).

The set of rapidly decreasing functions obviously forms a vector space under the standard operations of addition of functions and multiplication of a function by a complex number.

Example 8 The function e^{-x^2} or, for example, all functions of compact support in $C_0^{(\infty)}(\mathbb{R}, \mathbb{C})$ belong to S.

Lemma 3 *The restriction of the Fourier transform to S is a vector-space automorphism of S.*

Proof We first show that $(f \in S) \Rightarrow (\hat{f} \in S)$.

To do this we first remark that by Proposition 2a we have $\hat{f} \in C^{(\infty)}(\mathbb{R}, \mathbb{C})$.

We then remark that the operation of multiplication by x^α ($\alpha \geq 0$) and the operation D of differentiation do not lead outside the class of rapidly decreasing functions. Hence, for any nonnegative integers α and β the relation $f \in S$ implies that the function $D^\beta(x^\alpha f(x))$ belongs to the space S. Its Fourier transform tends to zero at infinity by the Riemann–Lebesgue lemma. But by formulas (18.102) and (18.103)

$$D^\beta\big(\widehat{x^\alpha f(x)}\big)(\xi) = i^{\alpha+\beta} \xi^\beta \hat{f}^{(\alpha)}(\xi),$$

and we have shown that $\xi^\beta \hat{f}^{(\alpha)}(\xi) \to 0$ as $\xi \to \infty$, that is, $\hat{f} \in S$.

We now show that $\hat{S} = S$, that is, that the Fourier transform maps S onto the whole space S.

We recall that the direct and inverse Fourier transforms are connected by the simple relation $\hat{f}(\xi) = \tilde{f}(-\xi)$. Reversing the sign of the argument of the function obviously is an operation that maps the set S into itself. Hence the inverse Fourier transform also maps S into itself.

Finally, if f is an arbitrary function in S, then by what has been proved $\varphi = \tilde{f} \in S$ and by the inversion formula (18.101) we find that $f = \hat{\varphi}$.

The linearity of the Fourier transform is obvious, so that Lemma 3 is now completely proved. □

18.3.3 The Main Structural Properties of the Fourier Transform

a. Definitions, Notation, Examples

We have made a rather detailed study above of the Fourier transform of a function $f : \mathbb{R} \to \mathbb{C}$ defined on the real line. In particular, we have clarified the connection that exists between the regularity properties of a function and the corresponding properties of its Fourier transform. Now that this question has been theoretically answered, we shall study the Fourier transform only of sufficiently regular functions so as to exhibit the fundamental technical properties of the Fourier transform in concentrated form and without technical complications. In compensation we shall consider not only one-dimensional but also the multi-dimensional Fourier transform and derive its basic properties practically independently of what was discussed above.

Those wishing to confine themselves to the one-dimensional case may assume that $n = 1$ below.

Definition 5 Suppose $f : \mathbb{R}^n \to \mathbb{C}$ is a locally integrable function on $\mathbb{R}^n$. The function

$$\hat{f}(\xi) := \frac{1}{(2\pi)^{n/2}} \int_{\mathbb{R}^n} f(x) e^{-i(\xi, x)} \, dx \qquad (18.104)$$

is called the *Fourier transform of the function* f.

Here we mean that $x = (x_1, \ldots, x_n)$, $\xi = (\xi_1, \ldots, \xi_n)$, $(\xi, x) = \xi_1 x_1 + \cdots + \xi_n x_n$, and the integral is regarded as convergent in the following sense of principal value:

$$\int_{\mathbb{R}^n} \varphi(x_1, \ldots, x_n) \, dx_1 \cdots dx_n := \lim_{A \to +\infty} \int_{-A}^{A} \cdots \int_{-A}^{A} \varphi(x_1, \ldots, x_n) \, dx_1 \cdots dx_n.$$

In this case the multidimensional Fourier transform (18.104) can be regarded as n one-dimensional Fourier transforms carried out with respect to each of the variables $x_1, \ldots, x_n$.

Then, when the function f is absolutely integrable, the question of the sense in which the integral (18.104) is to be understood does not arise at all.

Let $\alpha = (\alpha_1, \ldots, \alpha_n)$ and $\beta = (\beta_1, \ldots, \beta_n)$ be multi-indices consisting of nonnegative integers α_j, β_j, $j = 1, \ldots, n$, and suppose, as always, that D^α denotes the differentiation operator $\frac{\partial^{|\alpha|}}{\partial x_1^{\alpha_1} \cdots \partial x_n^{\alpha_n}}$ of order $|\alpha| := \alpha_1 + \cdots + \alpha_n$ and $x^\beta := x_1^{\beta_1} \cdot \ldots \cdot x_n^{\beta_n}$.

Definition 6 We denote the set of functions $f \in C^{(\infty)}(\mathbb{R}^n, \mathbb{C})$ satisfying the condition

$$\sup_{x \in \mathbb{R}^n} \left| x^\beta D^\alpha f(x) \right| < \infty$$

for all nonnegative multi-indices α and β by the symbol $S(\mathbb{R}^n, \mathbb{C})$, or by S where no confusion can arise. Such functions are said to be *rapidly decreasing* (as $x \to \infty$).

The set S with the algebraic operations of addition of functions and multiplication of a function by a complex number is obviously a vector space.

Example 9 The function $e^{-|x|^2}$, where $|x|^2 = x_1^2 + \cdots + x_n^2$, and all the functions in $C_0^{(\infty)}(\mathbb{R}^n, \mathbb{C})$ of compact support belong to S.

If $f \in S$, then integral in relation (18.104) obviously converges absolutely and uniformly with respect to ξ on the entire space $\mathbb{R}^n$. Moreover, if $f \in S$, then by standard rules this integral can be differentiated as many times as desired with respect to any of the variables $\xi_1, \ldots, \xi_n$. Thus if $f \in S$, then $\hat{f} \in C^{(\infty)}(\mathbb{R}, \mathbb{C})$.

Example 10 Let us find the Fourier transform of the function $\exp(-|x|^2/2)$. When integrating rapidly decreasing functions one can obviously use Fubini's theorem and if necessary change the order of improper integrations without difficulty.

In the present case, using Fubini's theorem and Example 4, we find

$$\frac{1}{(2\pi)^{n/2}} \int_{\mathbb{R}^n} e^{-|x|^2/2} \cdot e^{-i(\xi, x)} \, dx =$$

$$= \prod_{j=1}^n \frac{1}{\sqrt{2\pi}} \int_{-\infty}^\infty e^{-x_j^2/2} e^{-i\xi_j x_j} \, dx_j = \prod_{j=1}^n e^{-\xi_j^2/2} = e^{-|\xi|^2/2}.$$

We now state and prove the basic structural properties of the Fourier transform, assuming, so as to avoid technical complications, that the Fourier transform is being applied to functions of class S. This is approximately the same as learning to operate (compute) with rational numbers rather than the entire space $\mathbb{R}$ all at once. The process of completion is of the same type. On this account, see Problem 5.

b. Linearity

The linearity of the Fourier transform is obvious; it follows from the linearity of the integral.

c. The Relation Between Differentiation and the Fourier Transform

The following formulas hold

$$\widehat{D^\alpha f}(\xi) = i^{|\alpha|} \xi^\alpha \hat{f}(\xi), \tag{18.105}$$

$$\left(\widehat{x^\alpha f(x)}\right)(\xi) = i^{|\alpha|} D^\alpha \hat{f}(\xi). \tag{18.106}$$

Proof The first of these can be obtained, like formula (18.102), via integration by parts (of course, with a preliminary use of Fubini's theorem in the case of a space $\mathbb{R}^n$ of dimension $n > 1$).

Formula (18.106) generalizes relation (18.103) and is obtained by direct differentiation of (18.104) with respect to the parameters $\xi_1, \ldots, \xi_n$. □

Remark 3 In view of the obvious estimate

$$\left|\hat{f}(\xi)\right| \le \frac{1}{(2\pi)^{n/2}} \int_{\mathbb{R}^n} \left|f(x)\right| dx < +\infty,$$

it follows from (18.105) that $\hat{f}(\xi) \to 0$ as $\xi \to \infty$ for every function $f \in S$, since $D^\alpha f \in S$.

Next, the simultaneous use of formulas (18.105) and (18.106) enables us to write that

$$D^\beta\left(\widehat{x^\alpha f(x)}\right)(\xi) = (i)^{|\alpha|+|\beta|} \xi^\beta D^\alpha \hat{f}(\xi),$$

from which it follows that if $f \in S$, then for any nonnegative multi-indices α and β we have $\xi^\beta D^\alpha \hat{f}(\xi) \to 0$ when $\xi \to \infty$ in $\mathbb{R}^n$. Thus we have shown that

$$(f \in S) \Rightarrow (\hat{f} \in S).$$

d. The Inversion Formula

Definition 7 The operator defined (together with its notation) by the equality

$$\tilde{f}(\xi) := \frac{1}{(2\pi)^{n/2}} \int_{\mathbb{R}^n} f(x) e^{i(\xi,x)} dx, \tag{18.107}$$

is called the *inverse Fourier transform*.

The following *Fourier inversion formula* holds:

$$\tilde{\hat{f}} = \hat{\tilde{f}} = f, \tag{18.108}$$

or in the form of the Fourier integral:

$$f(x) = \frac{1}{(2\pi)^{n/2}} \int_{\mathbb{R}^n} \hat{f}(\xi) e^{i(x,\xi)} d\xi. \tag{18.109}$$

Using Fubini's theorem one can immediately obtain formula (18.108) from the corresponding formula (18.101) for the one-dimensional Fourier transform, but, as promised, we shall give a brief independent proof of the formula.

Proof We first show that

$$\int_{\mathbb{R}^n} g(\xi)\hat{f}(\xi)e^{i(x,\xi)}\,d\xi = \int_{\mathbb{R}^n} \hat{g}(\xi)f(x+y)\,dy \qquad (18.110)$$

for any functions $f, g \in S(\mathbb{R}, \mathbb{C})$. Both integrals are defined, since $f, g \in S$ and so by Remark 3 we also have $\hat{f}, \hat{g} \in S$.

Let us transform the integral on the left-hand side of the equality to be proved:

$$\int_{\mathbb{R}^n} g(\xi)\hat{f}(\xi)e^{i(x,\xi)}\,d\xi =$$

$$= \int_{\mathbb{R}^n} g(\xi)\left(\frac{1}{(2\pi)^{n/2}}\int_{\mathbb{R}^n} f(y)e^{-i(\xi,y)}\,dy\right)e^{i(x,\xi)}\,d\xi =$$

$$= \frac{1}{(2\pi)^{n/2}}\int_{\mathbb{R}^n}\left(\int_{\mathbb{R}^n} g(\xi)e^{-i(\xi,y-x)}\,d\xi\right)f(y)\,dy =$$

$$= \int_{\mathbb{R}^n} \hat{g}(y-x)f(y)\,dy = \int_{\mathbb{R}^n} \hat{g}(y)f(x+y)\,dy.$$

There is no doubt as to the legitimacy of the reversal in the order of integration, since f and g are rapidly decreasing functions. Thus (18.110) is now verified.

We now remark that for every $\varepsilon > 0$

$$\frac{1}{(2\pi)^{n/2}}\int_{\mathbb{R}^n} g(\varepsilon\xi)e^{i(y,\xi)}\,d\xi = \frac{1}{(2\pi)^{n/2}\varepsilon^n}\int_{\mathbb{R}^n} g(u)e^{-i(y,u/\varepsilon)}\,du = \varepsilon^{-n}\hat{g}(y/\varepsilon),$$

so that, by Eq. (18.110)

$$\int_{\mathbb{R}^n} g(\varepsilon\xi)\hat{f}(\xi)e^{i(x,\xi)}\,d\xi = \int_{\mathbb{R}^n} \varepsilon^{-n}\hat{g}(y/\varepsilon)f(x+y)\,dy = \int_{\mathbb{R}^n} \hat{g}(u)f(x+\varepsilon u)\,du.$$

Taking account of the absolute and uniform convergence with respect to ε of the extreme integrals in the last chain of equalities, we find, as $\varepsilon \to 0$,

$$g(0)\int_{\mathbb{R}^n} \hat{f}(\xi)e^{i(x,\xi)}\,d\xi = f(x)\int_{\mathbb{R}^n} \hat{g}(u)\,du.$$

Here we set $g(x) = e^{-|x|^2/2}$. In Example 10 we saw that $\hat{g}(u) = e^{-|u|^2/2}$. Recalling the Euler–Poisson integral $\int_{-\infty}^{\infty} e^{-x^2}\,dx = \sqrt{\pi}$ and using Fubini's theorem, we conclude that $\int_{\mathbb{R}^n} e^{-|u|^2/2}\,du = (2\pi)^{n/2}$, and as a result, we obtain Eq. (18.109). $\square$

Remark 4 In contrast to the single equality (18.109), which means that $\tilde{\hat{f}} = f$, relations (18.108) also contain the equality $\hat{\tilde{f}}$. But this relation follows immediately from the one proved, since $\tilde{f}(\xi) = \hat{f}(-\xi)$ and $\widehat{f(-x)} = \widehat{f}(x)$.

Remark 5 We have already seen (see Remark 3) that if $f \in S$, then $\hat{f} \in S$, and hence $\tilde{f} \in S$ also, that is, $\hat{S} \subset S$ and $\tilde{S} \subset S$. We now conclude from the relations $\hat{\tilde{f}} = \tilde{\hat{f}} = f$ that $\tilde{S} = \hat{S} = S$.

e. Parseval's Equality

This is the name given to the relation

$$\langle f, g \rangle = \langle \hat{f}, \hat{g} \rangle, \tag{18.111}$$

which in expanded form means that

$$\int_{\mathbb{R}^n} f(x)\overline{g}(x)\,dx = \int_{\mathbb{R}^n} \hat{f}(\xi)\overline{\hat{g}}(\xi)\,d\xi. \tag{18.111'}$$

It follows in particular from (18.111) that

$$\|f\|^2 = \langle f, f \rangle = \langle \hat{f}, \hat{f} \rangle = \|\hat{f}\|^2. \tag{18.112}$$

From the geometric point of view, Eq. (18.111) means that the Fourier transform preserves the inner product between functions (vectors of the space S), and hence is an isometry of S.

The name "Parseval's equality" is also sometimes given to the relation

$$\int_{\mathbb{R}^n} \hat{f}(\xi)g(\xi)\,d\xi = \int_{\mathbb{R}^n} f(x)\hat{g}(x)\,dx, \tag{18.113}$$

which is obtained from (18.110) by setting $x = 0$. The main Parseval equality (18.111) is obtained from (18.113) by replacing g with $\overline{\hat{g}}$ and using the fact that $(\overline{\hat{\hat{g}}}) = \overline{g}$, since $\overline{\tilde{g}} = \tilde{\overline{g}}$ and $\hat{\tilde{g}} = \tilde{\hat{g}} = g$.

f. The Fourier Transform and Convolution

The following important relations hold

$$\widehat{(f * g)} = (2\pi)^{n/2}\hat{f} \cdot \hat{g}, \tag{18.114}$$

$$\widehat{(f \cdot g)} = (2\pi)^{-n/2}\hat{f} * \hat{g} \tag{18.115}$$

(sometimes called *Borel's formulas*), which connect the operations of convolution and multiplication of functions through the Fourier transform.

Let us prove these formulas:

Proof

$$\widehat{(f * g)}(\xi) = \frac{1}{(2\pi)^{n/2}} \int_{\mathbb{R}^n} (f * g)(x)e^{-i(\xi,x)}\,dx =$$

$$= \frac{1}{(2\pi^{n/2})} \int_{\mathbb{R}^n} \left(\int_{\mathbb{R}^n} f(x-y)g(y)\,dy \right) e^{-i(\xi,x)}\,dx =$$

$$= \frac{1}{(2\pi)^{n/2}} \int_{\mathbb{R}^n} g(y)e^{-i(\xi,y)} \left(\int_{\mathbb{R}^n} f(x-y)e^{-i(\xi,x-y)}\,dx \right) dy =$$

$$= \frac{1}{(2\pi)^{n/2}} \int_{\mathbb{R}^n} g(y)e^{-i(\xi,y)} \left(\int_{\mathbb{R}^n} f(u)e^{-i(\xi,u)}\,du \right) dy =$$

$$= \int_{\mathbb{R}^n} g(y)e^{-i(\xi,y)} \hat{f}(\xi)\,dy = (2\pi)^{n/2} \hat{f}(\xi)\hat{g}(\xi).$$

The legitimacy of the change in order of integration is not in doubt, given that $f, g \in S$.

Formula (18.115) can be obtained by a similar computation if we use the inversion formula (18.109). However, Eq. (18.115) can be derived from relation (18.114) already proved if we recall that $\hat{\tilde{f}} = \tilde{\hat{f}} = f$, $\tilde{\overline{f}} = \overline{\tilde{f}}$, $\hat{\overline{f}} = \overline{\tilde{f}}$, and that $\overline{u \cdot v} = \overline{u} \cdot \overline{v}$, $\overline{u * v} = \overline{u} * \overline{v}$. □

Remark 6 If we set $\tilde{f}$ and $\tilde{g}$ in place of f and g in formulas (18.114) and (18.115) and apply the inverse Fourier transform to both sides of the resulting equalities, we arrive at the relations

$$\widetilde{f \cdot g} = (2\pi)^{-n/2}(\tilde{f} * \tilde{g}), \tag{18.114'}$$

$$\widetilde{f * g} = (2\pi)^{n/2}(\tilde{f} \cdot \tilde{g}). \tag{18.115'}$$

18.3.4 Examples of Applications

Let us now illustrate the Fourier transform (and some of the machinery of Fourier series) in action.

a. The Wave Equation

The successful use of the Fourier transform in the equations of mathematical physics is bound up (in its mathematical aspect) primarily with the fact that the Fourier transform replaces the operation of differentiation with the algebraic operation of multiplication.

For example, suppose we are seeking a function $u : \mathbb{R} \to \mathbb{R}$ satisfying the equation

$$a_0 u^{(n)}(x) + a_1 u^{(n-1)}(x) + \cdots + a_n u(x) = f(x),$$

where $a_0, \ldots, a_n$ are constant coefficients and f is a known function. Applying the Fourier transform to both sides of this equation (assuming that the functions u and

f are sufficiently regular), by relation (18.105) we obtain the algebraic equation

$$\big(a_0(i\xi)^n + a_1(i\xi)^{n-1} + \cdots + a_n\big)\hat{u}(\xi) = \hat{f}(\xi)$$

for $\hat{u}$. After finding $\hat{u}(\xi) = \frac{\hat{f}(\xi)}{P(i\xi)}$ from the equation, we obtain $u(x)$ by applying the inverse Fourier transform.

We now apply this idea to the search for a function $u = u(x,t)$ satisfying the one-dimensional wave equation

$$\frac{\partial^2 u}{\partial t^2} = a^2 \frac{\partial^2 u}{\partial x^2} \quad (a > 0)$$

and the initial conditions

$$u(x,0) = f(x), \qquad \frac{\partial u}{\partial t}(x,0) = g(x)$$

in $\mathbb{R} \times \mathbb{R}$.

Here and in the next example we shall not take the time to justify the intermediate computations because, as a rule, it is easier simply to find the required function and verify directly that it solves the problem posed than to justify and overcome all the technical difficulties that arise along the way. As it happens, generalized functions, which have already been mentioned, play an essential role in the theoretical struggle with these difficulties.

Thus, regarding t as a parameter, we carry out a Fourier transform on x on both sides of the equation. Then, assuming on the one hand that differentiation with respect to the parameter under the integral sign is permitted and using formula (18.105) on the other hand, we obtain

$$\hat{u}''_{tt}(\xi,t) = -a^2\xi^2 \hat{u}(\xi,t),$$

from which we find

$$\hat{u}(\xi,t) = A(\xi)\cos a\xi t + B(\xi)\sin a\xi t.$$

By the initial conditions, we have

$$\hat{u}(\xi,0) = \hat{f}(\xi) = A(\xi),$$
$$\hat{u}'_t(\xi,0) = \widehat{(u'_t)}(\xi,0) = \hat{g}(\xi) = a\xi B(\xi).$$

Thus,

$$\hat{u}(\xi,t) = \hat{f}(\xi)\cos a\xi t + \frac{\hat{g}(\xi)}{a\xi}\sin a\xi t =$$
$$= \frac{1}{2}\hat{f}(\xi)\big(e^{ia\xi t} + e^{-ia\xi t}\big) + \frac{1}{2}\frac{\hat{g}(\xi)}{ia\xi}\big(e^{ia\xi t} - e^{-ia\xi t}\big).$$

Multiplying this equality by $\frac{1}{\sqrt{2\pi}}e^{ix\xi}$ and integrating with respect to ξ – in short, taking the inverse Fourier transform – and using formula (18.105) we obtain immediately

$$u(x,t) = \frac{1}{2}\big(f(x-at) + f(x+at)\big) + \frac{1}{2}\int_0^t \big(g(x-a\tau) + g(x+a\tau)\big)\,d\tau.$$

b. The Heat Equation

Another element of the machinery of Fourier transforms (specifically, formulas (18.114′) and (18.115′)) which remained in the background in the preceding example, manifests itself quite clearly when we seek a function $u = u(x,t)$, $x \in \mathbb{R}^n$, $t \geq 0$, that satisfies the heat equation

$$\frac{\partial u}{\partial t} = a^2 \Delta u \quad (a > 0)$$

and the initial condition $u(x,0) = f(x)$ on all of $\mathbb{R}^n$.

Here, as always $\Delta = \frac{\partial^2}{\partial x_1^2} + \cdots + \frac{\partial^2}{\partial x_n^2}$.

Carrying out a Fourier transform with respect to the variable $x \in \mathbb{R}^n$, (assuming that this is possible to do) we find by (18.105) the ordinary equation

$$\frac{\partial \hat{u}}{\partial t}(\xi,t) = a^2(i)^2\big(\xi_1^2 + \cdots + \xi_n^2\big)\hat{u}(\xi,t),$$

from which it follows that

$$\hat{u}(\xi,t) = c(\xi)e^{-a^2|\xi|^2 t},$$

where $|\xi|^2 = \xi_1^2 + \cdots + \xi_n^2$. Taking into account the relation $\hat{u}(\xi,0) = \hat{f}(\xi)$, we find

$$\hat{u}(\xi,t) = \hat{f}(\xi) \cdot e^{-a^2|\xi|^2 t}.$$

Now applying the inverse Fourier transform, taking account of (18.114′), we obtain

$$u(x,t) = (2\pi)^{-n/2}\int_{\mathbb{R}_n} f(y)E_0(y-x,t)\,dy,$$

where $E_0(x,t)$ is the function whose Fourier transform with respect to x is $e^{-a^2|\xi|^2 t}$. The inverse Fourier transform with respect to ξ of the function $e^{-a^2|\xi|^2 t}$ is essentially already known to us from Example 10. Making an obvious change of variable, we find

$$E_0(x,t) = \frac{1}{(2\pi)^{n/2}}\left(\frac{\sqrt{\pi}}{a\sqrt{t}}\right)^n e^{-\frac{|x|^2}{4a^2 t}}.$$

Setting $E(x, t) = (2\pi)^{-n/2} E_0(x, t)$, we find the fundamental solution

$$E(x, t) = (2a\sqrt{\pi t})^{-n} e^{-\frac{|x|^2}{4a^2 t}} \quad (t > 0),$$

of the heat equation, which was already familiar to us (see Example 15 of Sect. 17.4), and the formula

$$u(x, t) = (f * E)(x, t)$$

for the solution satisfying the initial condition $u(x, 0) = f(x)$.

c. The Poisson Summation Formula

This is the name given to the following relation

$$\sqrt{2\pi} \sum_{n=-\infty}^{\infty} \varphi(2\pi n) = \sum_{n=-\infty}^{\infty} \hat{\varphi}(n) \tag{18.116}$$

between a function $\varphi : \mathbb{R} \to \mathbb{C}$ (assume $\varphi \in S$) and its Fourier transform $\hat{\varphi}$. Formula (18.116) is obtained by setting $x = 0$ in the equality

$$\sqrt{2\pi} \sum_{n=-\infty}^{\infty} \varphi(x + 2\pi n) = \sum_{n=-\infty}^{\infty} \hat{\varphi}(n) e^{inx}, \tag{18.117}$$

which we shall prove assuming that φ is a rapidly decreasing function.

Proof Since φ and $\hat{\varphi}$ both belong to S, the series on both sides of (18.117) converge absolutely (and so they can be summed in any order), and uniformly with respect to x on the entire line $\mathbb{R}$. Moreover, since the derivatives of a rapidly decreasing function are themselves in class S, we can conclude that the function $f(x) = \sum_{n=-\infty}^{\infty} \varphi(x + 2\pi n)$ belongs to $C^{(\infty)}(\mathbb{R}, \mathbb{C})$. The function f is obviously of period 2π. Let $\{\hat{c}_k(f)\}$ be its Fourier coefficients in the orthonormal system $\{\frac{1}{\sqrt{2\pi}} e^{ikx}; k \in \mathbb{Z}\}$, then

$$\hat{c}_k(f) := \frac{1}{\sqrt{2\pi}} \int_0^{2\pi} f(x) e^{-ikx} \, dx = \sum_{n=-\infty}^{\infty} \frac{1}{\sqrt{2\pi}} \int_0^{2\pi} \varphi(x + 2\pi n) e^{-ikx} \, dx =$$

$$= \sum_{n=-\infty}^{\infty} \frac{1}{\sqrt{2\pi}} \int_{2\pi n}^{2\pi(n+1)} \varphi(x) e^{-ikx} \, dx = \frac{1}{\sqrt{2\pi}} \int_{-\infty}^{\infty} \varphi(x) e^{-ikx} \, dx =: \hat{\varphi}(k).$$

But f is a smooth 2π-periodic function and so its Fourier series converges to it at every point $x \in \mathbb{R}$. Hence, at every point $x \in \mathbb{R}$ we have the relation

$$\sum_{n=-\infty}^{\infty} \varphi(x + 2\pi n) = f(x) = \sum_{n=-\infty}^{\infty} \hat{c}_n(f) \frac{e^{inx}}{\sqrt{2\pi}} = \frac{1}{\sqrt{2\pi}} \sum_{n=-\infty}^{\infty} \hat{\varphi}(n) e^{inx}. \qquad \square$$

Remark 7 As can be seen from the proof, relations (18.116) and (18.117) by no means hold only for functions of class S. But if φ does happen to belong to S, then Eq. (18.117) can be differentiated termwise with respect to x any number of times, yielding as a corollary new relations between $\varphi, \varphi', \ldots,$ and $\hat{\varphi}$.

d. Kotel'nikov's Theorem (Whittaker–Shannon Sampling Theorem)[25]

This example, based like the preceding one on a beautiful combination of the Fourier series and the Fourier integral, has a direct relation to the theory of information transmission in a communication channel. To keep it from appearing artificial, we recall that because of the limited capabilities of our sense organs, we are able to perceive signals only in a certain range of frequencies. For example, the ear "hears" in the range from 20 Hz to 20 kHz. Thus, no matter what the signals are, we, like a filter (see Sect. 18.3.1) cut out only a bounded part of their spectra and perceive them as band-limited signals (having a bounded spectrum).

For that reason, we shall assume from the outset that the transmitted or received signal $f(t)$ (where t is time, $-\infty < t < \infty$) is band-limited, the spectrum being nonzero only for frequencies whose magnitudes do not exceed a certain critical value $a > 0$. Thus $\hat{f}(\omega) \equiv 0$ for $|\omega| > a$, and so for a band-limited function the representation

$$f(t) = \frac{1}{\sqrt{2\pi}} \int_{-\infty}^{\infty} \hat{f}(\omega) e^{i\omega t} \, d\omega$$

reduces to the integral over just the interval $[-a, a]$:

$$f(t) = \frac{1}{\sqrt{2\pi}} \int_{-a}^{a} \hat{f}(\omega) e^{i\omega t} \, d\omega. \tag{18.118}$$

On the closed interval $[-a, a]$ we expand the function $\hat{f}(\omega)$ in a Fourier series

$$\hat{f}(\omega) = \sum_{-\infty}^{\infty} c_k(\hat{f}) e^{i \frac{\pi \omega}{a} k} \tag{18.119}$$

in the system $\{e^{i \frac{\pi \omega}{a} k}; k \in \mathbb{Z}\}$ which is orthogonal and complete in that interval. Taking account of formula (18.118), we find the following simple expression for the coefficients $c_k(\hat{f})$ of this series:

$$c_k(\hat{f}) := \frac{1}{2a} \int_{-a}^{a} \hat{f}(\omega) e^{-i \frac{\pi \omega}{a} k} \, d\omega - a = \frac{\sqrt{2\pi}}{2a} f\left(-\frac{\pi}{a} k\right). \tag{18.120}$$

---

[25]V.A. Kotel'nikov (b. 1908) – Soviet scholar, a well-known specialist in the theory of radio communication.
J.M. Whittaker (1905–1984) – British mathematician who worked mainly in complex analysis.
C.E. Shannon (1916–2001) – American mathematician and engineer, one of the founders of information theory and inventor of the term "bit" as an abbreviation of "binary digit".

Substituting the series (18.119) into the integral (18.118), taking account of relations (18.120), we find

$$f(t) = \frac{1}{\sqrt{2\pi}} \int_{-a}^{a} \left(\frac{\sqrt{2\pi}}{2a} \sum_{k=-\infty}^{\infty} f\left(\frac{\pi}{a}k\right) e^{i\omega t - i\frac{\pi k}{a}\omega} \right) d\omega =$$

$$= \frac{1}{2a} \sum_{k=-\infty}^{\infty} f\left(\frac{\pi}{a}k\right) \int_{-a}^{a} e^{i\omega(t - \frac{\pi}{a}k)} d\omega.$$

Calculating these elementary integrals, we arrive at *Kotel'nikov's formula*

$$f(t) = \sum_{k=-\infty}^{\infty} f\left(\frac{\pi}{a}k\right) \frac{\sin a(t - \frac{\pi}{a}k)}{a(t - \frac{\pi}{a}k)}. \tag{18.121}$$

Formula (18.121) shows that, in order to reconstruct a message described by a band-limited function $f(t)$ whose spectrum is concentrated in the frequency range $|\omega| \le a$, it suffices to transmit over the channel only the values $f(k\Delta)$ (called *marker values*) of the function at equal time intervals $\Delta = \pi/a$.

This proposition, together with formula (18.121) is due to V.A. Kotel'nikov and is called *Kotel'nikov's theorem* or the *sampling theorem*.

Remark 8 The interpolation formula (18.121) itself was known in mathematics before Kotel'nikov's 1933 paper, but this paper was the first to point out the fundamental significance of the expansion (18.121) for the theory of transmission of continuous messages over a communication channel. The idea of the derivation of formula (18.121) given above is also due to Kotel'nikov. In the general case this question was later studied by the outstanding American engineer and mathematician Claude Shannon, whose work in 1948 provided the fundamentals the information theory.

Remark 9 In reality the transmission and receiving time of a communication is also limited, so that instead of the entire series (18.121) we take one of its partial sums $\sum_{-N}^{N}$. Special research has been devoted to estimating the errors that thereby arise.

Remark 10 If we assume that the amount of information transmitted over the communication channel is proportional to the amount of reference values, then accordingly to formula (18.121) the communication channel capacity is proportional to its bandwidth frequency.

18.3.5 Problems and Exercises

1. a) Write out the proof of relations (18.93)–(18.96) in detail.

b) Regarding the Fourier transform as a mapping $f \mapsto \hat{f}$, show that it has the following frequently used properties:

$$f(at) \mapsto \frac{1}{a}\hat{f}\left(\frac{\omega}{a}\right)$$

(the change of scale rule);

$$f(t-t_0) \mapsto \hat{f}(\omega)e^{-i\omega t_0}$$

(time shift of the input signal – the Fourier pre-image – or the *translation theorem*)

$$\left[f(t+t_0) \pm f(t-t_0)\right] \mapsto \begin{cases} \hat{f}(\omega)2\cos\omega t_0, \\ \hat{f}(\omega)2\sin\omega t_0; \end{cases}$$

$$f(t)e^{\pm i\omega_0 t} \mapsto \hat{f}(\omega \pm \omega_0)$$

(frequency shift of the Fourier transform);

$$f(t)\cos\omega_0 t \mapsto \frac{1}{2}\left[\hat{f}(\omega - \omega_0) + \hat{f}(\omega + \omega_0)\right],$$

$$f(t)\sin\omega_0 t \mapsto \frac{1}{2}\left[\hat{f}(\omega - \omega_0) - \hat{f}(\omega + \omega_0)\right]$$

(amplitude modulation of a harmonic signal);

$$f(t)\sin^2\frac{\omega_0 t}{2} \mapsto \frac{1}{4}\left[2\hat{f}(\omega) - \hat{f}(\omega - \omega_0) - \hat{f}(\omega + \omega_0)\right].$$

c) Find the Fourier transforms (or, as we say, the *Fourier images*) of the following functions:

$$\Pi_A(t) = \begin{cases} \frac{1}{2A} & \text{for } |t| \le A, \\ 0 & \text{for } |t| > A \end{cases}$$

(the *rectangular pulse*);

$$\Pi_A(t)\cos\omega_0 t$$

(a harmonic signal modulated by a rectangular pulse);

$$\Pi_A(t+2A) + \Pi_A(t-2A)$$

(two rectangular pulses of the same polarity);

$$\Pi_A(t-A) - \Pi_A(t+A)$$

(two rectangular pulses of opposite polarity);

$$\Lambda_A(t) = \begin{cases} \frac{1}{A}\left(1 - \frac{|t|}{A}\right) & \text{for } |t| \le A, \\ 0 & \text{for } |t| > A \end{cases}$$

(a triangular pulse);

$$\cos at^2 \quad \text{and} \quad \sin at^2 \quad (a > 0);$$

$$|t|^{-\frac{1}{2}} \quad \text{and} \quad |t|^{-\frac{1}{2}}e^{-a|t|} \quad (a > 0).$$

d) Find the Fourier pre-images of the following functions:

$$\operatorname{sinc}\frac{\omega A}{\pi}, \quad 2i\frac{\sin^2 \omega A}{\omega A}, \quad 2\operatorname{sinc}^2\frac{\omega A}{\pi},$$

where $\operatorname{sinc}\frac{x}{\pi} := \frac{\sin x}{x}$ is the *sample function (cardinal sine)*.

e) Using the preceding results, find the values of the following integrals, which we have already encountered:

$$\int_{-\infty}^{\infty}\frac{\sin x}{x}\,dx, \quad \int_{-\infty}^{\infty}\frac{\sin^2 x}{x^2}\,dx, \quad \int_{-\infty}^{\infty}\cos x^2\,dx, \quad \int_{-\infty}^{\infty}\sin x^2\,dx.$$

f) Verify that the Fourier integral of a function $f(t)$ can be written in any of the following forms:

$$f(t) \sim \int_{-\infty}^{\infty} \hat{f}(\omega)e^{it\omega}\,d\omega = \frac{1}{2\pi}\int_{-\infty}^{\infty}d\omega\int_{-\infty}^{\infty}f(x)e^{-i\omega(x-t)}\,dx =$$

$$= \frac{1}{\pi}\int_{0}^{\infty}d\omega\int_{-\infty}^{\infty}f(x)\cos 2\omega(x-t)\,dx.$$

2. Let $f = f(x, y)$ be a solution of the two-dimensional Laplace equation $\frac{\partial^2 f}{\partial x^2} + \frac{\partial^2 f}{\partial y^2} = 0$ in the half-plane $y \geq 0$ satisfying the conditions $f(x, 0) = g(x)$ and $f(x, y) \to 0$ as $y \to +\infty$ for every $x \in \mathbb{R}$.

a) Verify that the Fourier transform $\hat{f}(\xi, y)$ of f on the variable x has the form $\hat{g}(\xi)e^{-y|\xi|}$.

b) Find the Fourier pre-image of the function $e^{-y|\xi|}$ on the variable ξ.

c) Now obtain the representation of the function f as a Poisson integral

$$f(x, y) = \frac{1}{\pi}\int_{-\infty}^{\infty}\frac{y}{(x-\xi)^2 + y^2}g(\xi)\,d\xi,$$

which we have met in Example 5 of Sect. 17.4.

3. We recall that the nth *moment* of the function $f : \mathbb{R} \to \mathbb{C}$ is the quantity $M_n(f) = \int_{-\infty}^{\infty}x^n f(x)\,dx$. In particular, if f is the density of a probability distribution, that is, $f(x) \geq 0$ and $\int_{-\infty}^{\infty}f(x)\,dx = 1$, then $x_0 = M_1(f)$ is the mathematical expectation of a random variable x with the distribution f and the variance $\sigma^2 := \int_{-\infty}^{\infty}(x - x_0)^2 f(x)\,dx$ of this random variable can be represented as $\sigma^2 = M_2(f) - M_1^2(f)$.

Consider the Fourier transform

$$\hat{f}(\xi) = \int_{-\infty}^{\infty} f(x)e^{-i\xi x}\,dx$$

of the function f. By expanding $e^{-i\xi x}$ in a series, show that

a) $\hat{f}(\xi) = \sum_{n=0}^{\infty} \frac{(-i)^n M_n(f)}{n!}\xi^n$ if, for example, $f \in S$.

b) $M_n(f) = (i)^n \hat{f}^{(n)}(0), n = 0, 1, \ldots$.

c) Now let f be real-valued, and let $\hat{f}(\xi) = A(\xi)e^{i\varphi(\xi)}$, where $A(\xi)$ is the absolute value of $\hat{f}(\xi)$ and $\varphi(\xi)$ is its argument; then $A(\xi) = A(-\xi)$ and $\varphi(-\xi) = -\varphi(\xi)$. To normalize the problem, assume that $\int_{-\infty}^{\infty} f(x)\,dx = 1$. Verify that in that case

$$\hat{f}(\xi) = 1 + i\varphi'(0)\xi + \frac{A''(0) - (\varphi'(0))^2}{2}\xi^2 + o(\xi^2) \quad (\xi \to 0)$$

and

$$x_0 := M_1(f) = -\varphi'(0), \quad \text{and} \quad \sigma^2 = M_2(f) - M_1^2(f) = -A''(0).$$

4. a) Verify that the function $e^{-a|x|}$ $(a > 0)$, like all its derivatives, which are defined for $x \neq 0$, decreases at infinity faster than any negative power of $|x|$ and yet this function does not belong to the class S.

b) Verify that the Fourier transform of this function is infinitely differentiable on $\mathbb{R}$, but does not belong to S (and all because $e^{-a|x|}$ is not differentiable at $x = 0$).

5. a) Show that the functions of class S are dense in the space $\mathcal{R}_2(\mathbb{R}^n, \mathbb{C})$ of functions $f : \mathbb{R}^n \to \mathbb{C}$ whose squares are absolutely integrable, endowed with the inner product $\langle f, g \rangle = \int_{\mathbb{R}^n} (f \cdot \overline{g})(x)\,dx$ and the norm it generates $\|f\| = (\int_{\mathbb{R}^n} |f|^2(x)\,dx)^{1/2}$ and the metric $d(f, g) = \|f - g\|$.

b) Now let us regard S as a metric space (S, d) with this metric d (convergence in the mean-square sense on $\mathbb{R}^n$). Let $L_2(\mathbb{R}^n, \mathbb{C})$ or, more briefly, L_2, denote the completion of the metric space (S, d) (see Sect. 9.5). Each element $f \in L_2$ is determined by a sequence $\{\varphi_k\}$ of functions $\varphi_k \in S$ that is a Cauchy sequence in the sense of the metric d.

Show that in that case the sequence $\{\hat{\varphi}\}$ of Fourier images of the functions φ_k is also a Cauchy sequence in S and hence defines a certain element $\hat{f} \in L_2$, which it is natural to call the Fourier transform of $f \in L_2$.

c) Show that a vector-space structure and an inner product can be introduced in a natural way on L_2, and in these structures the Fourier transform $L_2 \widehat{\to} L_2$ turns out to be a linear isometry of L_2 onto itself.

d) Using the example of the function $f(x) = \frac{1}{\sqrt{1+x^2}}$, one can see that if $f \in \mathcal{R}_2(\mathbb{R}, \mathbb{C})$ we do not necessarily have $f \in \mathcal{R}(\mathbb{R}, \mathbb{C})$. Nevertheless, if $f \in \mathcal{R}_2(\mathbb{R}, \mathbb{C})$, then, since f is locally integrable, one can consider the function

$$\hat{f}_A(\xi) = \frac{1}{\sqrt{2\pi}} \int_{-A}^{A} f(x)e^{-i\xi x}\,dx.$$

Verify that $\hat{f}_A \in C(\mathbb{R}, \mathbb{C})$ and $\hat{f}_A \in \mathcal{R}_2(\mathbb{R}, \mathbb{C})$.

e) Prove that $\hat{f}_A$ converges in L_2 to some element $\hat{f} \in L_2$ and $\|\hat{f}_A\| \to \|\hat{f}\| = \|f\|$ as $A \to +\infty$ (this is Plancherel's theorem[26]).

6. *The uncertainty principle.* Let $\varphi(x)$ and $\psi(p)$ be functions of class S (or elements of the space L_2 of Problem 5), with $\psi = \hat{\varphi}$ and $\int_{-\infty}^{\infty} |\varphi|^2(x)\,dx = \int_{-\infty}^{\infty} |\psi|^2(p)\,dp = 1$. In this case the functions $|\varphi|^2$ and $|\psi|^2$ can be regarded as probability densities for random variables x and p respectively.

a) Show that by a shift in the argument of φ (a special choice of the point from which the argument is measured) one can obtain a new function φ such that $M_1(|\varphi|) = \int_{-\infty}^{\infty} x|\varphi|^2(x)\,dx = 0$ without changing the value of $\|\hat{\varphi}\|$, and then, without changing the relation $M_1(|\varphi|) = 0$ one can, by a similar shift in the argument of ψ arrange that $M_1(|\psi|) = \int_{-\infty}^{\infty} p|\psi|^2(p)\,dp = 0$.

b) For real values of the parameter α consider the quantity

$$\int_{-\infty}^{\infty} \left| \alpha x \varphi(x) + \varphi'(x) \right|^2 dx \geq 0$$

and, using Parseval's equality and the formula $\hat{\varphi}'(p) = ip\hat{\varphi}(p)$, show that $\alpha^2 M_2(|\varphi|) - \alpha + M_2(|\psi|) \geq 0$. (For the definitions of M_1 and M_2 see Problem 3.)

c) Obtain from this the relation

$$M_2(|\varphi|) \cdot M_2(|\psi|) \geq 1/4.$$

This relation shows that the more "concentrated" the function φ itself is, the more "smeared" its Fourier transform, and vice versa (see Examples 1 and 7 and Problem 7b).

In quantum mechanics this relation, called the *uncertainty principle*, assumes a specific physical meaning. For example, it is impossible to measure precisely both the coordinate of a quantum particle and its momentum. This fundamental fact (called *Heisenberg's*[27] *uncertainty principle*), is mathematically the same as the relation between $M_2(|\varphi|)$ and $M_2(|\psi|)$ found above.

The next three problems give an elementary picture of the Fourier transform of generalized functions.

7. a) Using Example 1, find the spectrum of the signal expressed by the functions

$$\Delta_\alpha(t) = \begin{cases} \frac{1}{2\alpha} & \text{for } |t| \leq \alpha, \\ 0 & \text{for } |t| > \alpha. \end{cases}$$

b) Examine the variation of the function $\Delta_\alpha(t)$ and its spectrum as $\alpha \to +0$ and tell what, in your opinion, should be regarded as the spectrum of a unit pulse, expressed by the δ-function.

---

[26]M. Plancherel (1885–1967) – Swiss mathematician.

[27]W. Heisenberg (1901–1976) – German physicist, one of the founders of quantum mechanics.

c) Using Example 2, now find the signal $\varphi(t)$ emerging from an ideal low-frequency filter (with upper frequency limit a) in response to a unit pulse $\delta(t)$.

d) Using the result just obtained, now explain the physical meaning of the terms in the Kotel'nikov series (18.121) and propose a theoretical scheme for transmitting a band-limited signal $f(t)$, based on Kotel'nikov's formula (18.121).

8. *The space of L. Schwartz.* Verify that

a) If $\varphi \in S$ and P is a polynomial, then $(P \cdot \varphi) \in S$.

b) If $\varphi \in S$, then $D^\alpha \varphi \in S$ and $D^\beta(P D^\alpha \varphi) \in S$, where α and β are nonnegative multi-indices and P is a polynomial.

c) We introduce the following notion of convergence in S. A sequence $\{\varphi_k\}$ of functions $\varphi_k \in S$ converges to zero if for all nonnegative multi-indices α and β the sequence of functions $\{x^\beta D^\alpha \varphi(x)\}$ converges uniformly to zero on $\mathbb{R}^n$. The relation $\varphi_k \to \varphi \in S$ will mean that $(\varphi - \varphi_k) \to 0$ in S.

The vector space S of rapidly decreasing functions with this convergence is called the *Schwartz space*.

Show that if $\varphi_k \to \varphi$ in S, then $\hat{\varphi}_k \to \hat{\varphi}$ in S as $k \to \infty$. Thus the Fourier transform is a continuous linear operator on the Schwartz space.

9. *The space S' of tempered distributions.* The continuous linear functionals defined on the space S of rapidly decreasing functions are called *tempered distributions*. The vector space of such functionals (the conjugate of S) is denoted S'. The value of the functional $F \in S'$ on a function $\varphi \in S$ will be denoted $F(\varphi)$.

a) Let $P : \mathbb{R}^n \to \mathbb{C}$ be a polynomial in n variables and $f : \mathbb{R}^n \to \mathbb{C}$ a locally integrable function admitting the estimate $|f(x)| \le |P(x)|$ at infinity (that is, it may increase as $x \to \infty$, but only moderately: not faster than power growth). Show that f can then be regarded as a (regular) element of S' if we set

$$f(\varphi) = \int_{\mathbb{R}^n} f(x)\varphi(x)\,\mathrm{d}x \quad (\varphi \in S).$$

b) Multiplication of a tempered distribution $F \in S'$ by an ordinary function $f : \mathbb{R}^n \to \mathbb{C}$ is defined, as always, by the relation $(fF)(\varphi) := F(f\varphi)$. Verify that for tempered distributions multiplication is well defined, not only by functions $f \in S$, but also by polynomials $P : \mathbb{R}^n \to \mathbb{C}$.

c) Differentiation of tempered distributions $F \in S'$ is defined in the traditional way: $(D^\alpha F)(\varphi) := (-1)^{|\alpha|} F(D^\alpha \varphi)$.

Show that this is correctly defined, that is, if $F \in S'$, then $D^\alpha F \in S'$ for every nonnegative integer multi-index $\alpha = (\alpha_1, \dots, \alpha_n)$.

d) If f and φ are sufficiently regular functions (for example, functions in S), then, as relation (18.113) shows, the following equality holds:

$$\hat{f}(\varphi) = \int_{\mathbb{R}^n} \hat{f}(x)\varphi(x)\,\mathrm{d}x = \int_{\mathbb{R}^n} f(x)\hat{\varphi}(x)\,\mathrm{d}x = f(\hat{\varphi}).$$

This equality (Parseval's equality) is made the basis of the definition of the Fourier transform $\hat{F}$ of a tempered distribution $F \in S'$. By definition we set $\hat{F}(\varphi) := F(\hat{\varphi})$.

Due to the invariance of S under the Fourier transform, this definition is correct for every element $F \in S'$.

Show that it is not correct for generalized functions in $\mathcal{D}'(\mathbb{R}^n)$ mapping the space $\mathcal{D}(\mathbb{R}^n)$ of smooth functions of compact support. This fact explains the role of the Schwartz space S in the theory of the Fourier transform and its application to generalized functions.

e) In Problem 7 we acquired a preliminary idea of the Fourier transform of the δ-function. The Fourier transform of the δ-function could have been sought directly from the definition of the Fourier transform of a regular function. In that case we would have found that

$$\hat{\delta}(\xi) = \frac{1}{(2\pi)^{n/2}} \int_{\mathbb{R}^n} \delta(x) e^{-i(\xi, x)} \, dx = \frac{1}{(2\pi)^{n/2}}.$$

Now show that when we seek the Fourier transform of the tempered distribution $\delta \in S'(\mathbb{R}^n)$ correctly, that is, starting from the equality $\hat{\delta}(\varphi) = \delta(\hat{\varphi})$, the result (still the same) is that $\delta(\hat{\varphi}) = \hat{\varphi}(0) = \frac{1}{(2\pi)^{n/2}}$. (One can renormalize the Fourier transform so that this constant equals 1; see Problem 10.)

f) Convergence in S', as always in generalized functions, is understood in the following sense: $(F_n \to F)$ in S' as $n \to \infty := (\forall \varphi \in S \ (F_n(\varphi) \to F(\varphi)$ as $n \to \infty))$.

Verify the Fourier inversion formula (the Fourier integral formula) for the δ-function:

$$\delta(x) = \lim_{A \to +\infty} \frac{1}{(2\pi)^{n/2}} \int_{-A}^{A} \cdots \int_{-A}^{A} \hat{\delta}(\xi) e^{i(x, \xi)} \, d\xi.$$

g) Let $\delta(x - x_0)$, as usual, denote the shift of the δ-function to the point x_0, that is, $\delta(x - x_0)(\varphi) = \varphi(x_0)$. Verify that the series

$$\sum_{n=-\infty}^{\infty} \delta(x - n) \quad \left(= \lim_{N \to \infty} \sum_{-N}^{N} \delta(x - n) \right)$$

converges in $S'(\mathbb{R}^n)$. (Here $\delta \in S'(\mathbb{R}^n)$ and $n \in \mathbb{Z}$.)

h) Using the possibility of differentiating a convergent series of generalized functions termwise and taking account of the equality from Problem 13f of Sect. 18.2, show that if $F = \sum_{n=-\infty}^{\infty} \delta(x - n)$, then

$$\hat{F} = \sqrt{2\pi} \sum_{n=-\infty}^{\infty} \delta(x - 2\pi n).$$

i) Using the relation $\hat{F}(\varphi) = F(\hat{\varphi})$, obtain the Poisson summation formula from the preceding result.

j) Prove the following relation (the θ-formula)

$$\sum_{n=-\infty}^{\infty} e^{-tn^2} = \sqrt{\frac{\pi}{t}} \sum_{n=-\infty}^{\infty} e^{-\frac{\pi^2}{t}n^2} \quad (t > 0),$$

which plays an important role in the theory of elliptic functions and the theory of heat conduction.

10. If the Fourier transform $\check{\mathcal{F}}[f]$ of a function $f : \mathbb{R} \to \mathbb{C}$ is defined by the formulas

$$\check{f}(v) := \check{\mathcal{F}}[f](v) := \int_{-\infty}^{\infty} f(t)e^{-2\pi i v t}\, dt,$$

many of the formulas relating to the Fourier transform become particularly simple and elegant.

a) Verify that $\hat{f}(u) = \frac{1}{\sqrt{2\pi}} \check{f}(\frac{u}{2\pi})$.

b) Show that $\check{\mathcal{F}}[\check{\mathcal{F}}[f]](t) = f(-t)$, that is,

$$f(t) = \int_{-\infty}^{\infty} \check{f}(v)e^{2\pi i v t}\, dv.$$

This is the most natural form of the expansion of f in harmonics of different frequencies v, and $\check{f}(v)$ in this expansion is the *frequency spectrum of* f.

c) Verify that $\check{\delta} = 1$ and $\check{1} = \delta$.

d) Verify that the Poisson summation formula (18.116) now assumes the particularly elegant form

$$\sum_{n=-\infty}^{\infty} \varphi(n) = \sum_{n=-\infty}^{\infty} \check{\varphi}(n).$$

Chapter 19
Asymptotic Expansions

The majority of phenomena that we have to deal with can be characterized mathematically by a certain set of numerical parameters having rather complicated interrelations. However the description of a phenomenon as a rule becomes significantly simpler if it is known that some of these parameters or some combination of them is very large or, contrariwise, very small.

Example 1 In describing relative motions occurring with speeds v that are much smaller than the speed of light ($|v| \ll c$) we may use, instead of the Lorentz transformations (Example 3 of Sect. 1.3)

$$x' = \frac{x - vt}{\sqrt{1 - (\frac{v}{c})^2}}, \qquad t' = \frac{t - (\frac{v}{c^2})x}{\sqrt{1 - (\frac{v}{c})^2}},$$

the Galilean transformation

$$x' = x - vt, \qquad t' = t,$$

since $v/c \approx 0$.

Example 2 The period

$$T = 4\sqrt{\frac{l}{g}} \int_0^{\pi/2} \frac{\mathrm{d}\theta}{\sqrt{1 - k^2 \sin^2 \theta}}$$

of oscillations of a pendulum is connected with the maximal angle of deviation φ_0 from its equilibrium position via the parameter $k^2 = \sin^2 \frac{\varphi_0}{2}$ (see Sect. 6.4). If the oscillations are small, that is, $\varphi_0 \approx 0$, we obtain the simple formula

$$T \approx 2\pi \sqrt{\frac{l}{g}}$$

for the period of such oscillations.

© Springer-Verlag Berlin Heidelberg 2016
V.A. Zorich, *Mathematical Analysis II*, Universitext,
DOI 10.1007/978-3-662-48993-2_11

Example 3 Suppose a restoring force acting on a particle m is returning it to its equilibrium position and that the force is proportional to the displacement (a spring with spring constant k, for example). Suppose also that the resisting force of the medium is proportional to the square of the velocity (with coefficient of proportionality α). The equation of motion in that case has the following form (see Sect. 5.6):

$$m\ddot{x} + \alpha\dot{x}^2 + kx = 0.$$

If the medium "rarefies", then $\alpha \to 0$ and one may assume that the motion is approximated by the motion described by the equation

$$m\ddot{x} + kx = 0$$

(harmonic oscillations with frequency $\sqrt{\frac{k}{m}}$), and if the medium "condenses", then $\alpha \to \infty$, and, dividing by α, we find in the limit the equation $\dot{x}^2 = 0$, that is, $x(t) \equiv$ const.

Example 4 If $\pi(x)$ is the number of primes not larger than $x \in \mathbb{R}$, then, as is known (see Sect. 3.2), for large x the quantity $\pi(x)$ can be found with small relative error by the formula

$$\pi(x) \approx \frac{x}{\ln x}.$$

Example 5 It would be difficult to find more trivial, yet nevertheless important relations than

$$\sin x \approx x \quad \text{or} \quad \ln(1+x) \approx x,$$

in which the relative error becomes smaller as x approaches 0 (see Sect. 5.3). These relations can be made more precise if desired, namely

$$\sin x \approx x - \frac{1}{3!}x^3, \qquad \ln(1+x) \approx x - \frac{1}{2}x^2,$$

by adjoining one or more of the following terms obtained from the Taylor series.

Thus the problem is to find a clear, convenient, and essentially correct description of a phenomenon being studied using the specifics of the situation that arises when some parameter (or combination of parameters) that characterizes the phenomenon is small (tends to zero) or, contrariwise, large (tends to infinity).

Hence, we are once again essentially discussing the theory of limits.

Problems of this type are called *asymptotic problems*. They arise, as one can see, in practically all areas of mathematics and natural science.

The solution of an asymptotic problem usually consists of the following stages: passing to the limit and finding the (main term of the) asymptotics, that is, a convenient simplified description of the phenomenon; estimating the error that arises in using the asymptotic formula so found, and determining its range of applicability;

then sharpening the main term of the asymptotics, analogous to the process of adjoining the next term in Taylor's formula (but far from being equally algorithmic in every case).

The methods of solving asymptotic problems (called *asymptotic methods*) are usually closely connected with the specifics of a problem. Among the few rather general and at the same time elementary asymptotic methods one finds Taylor's formula, one of the most important relations in differential calculus.

The present chapter should give the reader a beginning picture of the elementary asymptotic methods of analysis.

In the first section we shall introduce the general concepts and definitions relating to elementary asymptotic methods; in the second we shall use them in discussing Laplace's method of constructing the asymptotic expansion of Laplace transforms. This method, which was discovered by Laplace in his research on the limit theorems of probability theory, is an important component of the saddle-point method later developed by Riemann, usually discussed in a course of complex analysis. Further information on various asymptotic methods of analysis can be found in the specialized books cited in the bibliography. These books also contain an extensive bibliography on this circle of questions.

19.1 Asymptotic Formulas and Asymptotic Series

19.1.1 Basic Definitions

a. Asymptotic Estimates and Asymptotic Equalities

For the sake of completeness we begin with some recollections and clarifications.

Definition 1 Let $f : X \to Y$ and $g : X \to Y$ be real- or complex- or in general vector-valued functions defined on a set X and let $\mathcal{B}$ be a base in X. Then the relations

$$f = O(g) \quad \text{or} \quad f(x) = O\big(g(x)\big) \quad x \in X$$
$$f = O(g) \quad \text{or} \quad f(x) = O\big(g(x)\big) \quad \text{over the base } \mathcal{B}$$
$$f = o(g) \quad \text{or} \quad f(x) = o\big(g(x)\big) \quad \text{over the base } \mathcal{B}$$

mean by definition that in the equality $|f(x)| = \alpha(x)|g(x)|$, the real-valued function $\alpha(x)$ is respectively bounded on X, ultimately bounded over the base $\mathcal{B}$, and infinitesimal over the base $\mathcal{B}$.

These relations are usually called *asymptotic estimates* (of f).
The relation

$$f \sim g \quad \text{or} \quad f(x) \sim g(x) \quad \text{over the base } \mathcal{B},$$

which by definition means that $f(x) = g(x) + o(g(x))$ over the base $\mathcal{B}$, is usually called *asymptotic equivalence* or *asymptotic equality*[1] of the functions *over the base $\mathcal{B}$*.

Asymptotic estimates and asymptotic equalities unite in the term *asymptotic formulas*.

Wherever it is not important to indicate the argument of a function the abbreviated notations $f = o(g)$, $f = O(g)$, or $f \sim g$ are used, and we shall make systematic use of this abbreviation.

If $f = O(g)$ and simultaneously $g = O(f)$, we write $f \asymp g$ and say that f and g are quantities *of the same order* over the given base.

In what we are going to be doing below, $Y = \mathbb{C}$ or $Y = \mathbb{R}$, $X \subset \mathbb{C}$ or $X \subset \mathbb{R}$; $\mathcal{B}$ as a rule is one of the bases $X \ni x \to 0$ or $X \ni x \to \infty$. Using this notation one can write in particular that

$$\cos x = O(1), \quad x \in \mathbb{R},$$

$$\cos z \neq O(1), \quad z \in \mathbb{C},$$

$$\ln e^z = 1 + z + o(z) \quad \text{as } z \to 0,\ z \in \mathbb{C},$$

$$(1 + x)^\alpha = 1 + \alpha x + o(x) \quad \text{as } x \to 0,\ x \in \mathbb{R},$$

$$\pi(x) = \frac{x}{\ln x} + o\left(\frac{x}{\ln x}\right) \quad \text{as } x \to +\infty,\ x \in \mathbb{R}.$$

Remark 1 In regard to asymptotic equalities it is useful to note that they are only limiting relations whose use is permitted for computational purposes, but only after some additional work is done to find an estimate of the remainder. We have already mentioned this when discussing Taylor's formula. In addition, one must keep in mind that asymptotic equivalence in general makes it possible to compute with small relative error, but not small absolute error. Thus, for example, as $x \to +\infty$, the difference $\pi(x) - \frac{x}{\ln x}$ does not tend to zero, since $\pi(x)$ jumps by 1 at each prime integer value of x. At the same time, the relative error in replacing $\pi(x)$ by $\frac{x}{\ln x}$ tends to zero:

$$\frac{o\left(\frac{x}{\ln x}\right)}{\left(\frac{x}{\ln x}\right)} \to 0 \quad \text{as } x \to +\infty.$$

This circumstance, as we shall see below, leads to asymptotic series that have computational importance when one considers the relative error but not the absolute error; for that reason these series are often divergent, in contrast to classical series, for which the absolute value of the difference between the function being approximated and the nth partial sum of the series tends to zero as $n \to +\infty$.

Let us consider some examples of ways of obtaining asymptotic formulas.

---

[1]It is also useful to keep in mind the symbol $\simeq$ often used to denote asymptotic equivalence.

Example 6 The labor involved in computing the values of $n!$ or $\ln n!$ increase as $n \in \mathbb{N}$ increases. We shall use the fact that n is large, however, and obtain under that assumption a convenient asymptotic formula for computing $\ln n!$ approximately.

It follows from the obvious relations

$$\int_1^n \ln x \, dx = \sum_{k=2}^n \int_{k-1}^k \ln x \, dx < \sum_{k=1}^n \ln k < \sum_{k=2}^n \int_k^{k+1} \ln x \, dx = \int_2^{n+1} \ln x \, dx$$

that

$$0 < \ln n! - \int_1^n \ln x \, dx < \int_1^2 \ln x \, dx + \int_n^{n+1} \ln x \, dx < \ln 2(n+1).$$

But

$$\int_1^n \ln x \, dx = n(\ln n - 1) + 1 = n \ln n - (n-1),$$

and therefore as $n \to \infty$

$$\ln n! = \int_1^n \ln x \, dx + O\left(\ln 2(n+1)\right) =$$

$$= n \ln n - (n-1) + O(\ln n) = n \ln n + O(n).$$

Since $O(n) = o(n \ln n)$ when $n \to +\infty$, the relative error of the formula $\ln n! \approx n \ln n$ tends to zero as $n \to +\infty$.

Example 7 We shall show that as $x \to +\infty$ the function

$$f_n(x) = \int_1^x \frac{e^t}{t^n} \, dt \quad (n \in \mathbb{R})$$

is asymptotically equivalent to the function $g_n(x) = x^{-n} e^x$. Since $g_n(x) \to +\infty$ as $x \to +\infty$, applying L'Hôpital's rule we find

$$\lim_{x \to +\infty} \frac{f_n(x)}{g_n(x)} = \lim_{x \to +\infty} \frac{f_n'(x)}{g_n'(x)} = \lim_{x \to \infty} \frac{x^{-n} e^x}{x^{-n} e^x - n x^{-n-1} e^x} = 1.$$

Example 8 Let us find the asymptotic behavior of the function

$$f(x) = \int_1^x \frac{e^t}{t} \, dt$$

more precisely. It differs from the exponential integral

$$\text{Ei}(x) = \int_{-\infty}^x \frac{e^t}{t} \, dt$$

only by a constant term.

Integrating by parts, we obtain

$$f(x) = \frac{e^t}{t}\Big|_1^x + \int_1^x \frac{e^t}{t^2}\,dt = \left(\frac{e^t}{t} + \frac{e^t}{t^2}\right)\Big|_1^x + \int_1^x \frac{2e^t}{t^3}\,dt =$$

$$= \left(\frac{e^t}{t} + \frac{1!e^t}{t^2} + \frac{2!e^t}{t^3}\right)\Big|_1^x + \int_1^x \frac{3!e^t}{t^4}\,dt =$$

$$= e^t\left(\frac{0!}{t} + \frac{1!}{t^2} + \frac{2!}{t^3} + \cdots + \frac{(n-1)}{t^n}\right)\Big|_1^x + \int_1^x \frac{n!e^t}{t^{n+1}}\,dt.$$

This last integral, as shown in Example 7, is $O(x^{-(n+1)}e^x)$ as $x \to +\infty$. Including in the term $O(x^{-(n+1)}e^x)$ the constant $-e\sum_{k=1}^n (k-1)!$ obtained when $t = 1$ is substituted, we find that

$$f(x) = e^x \sum_{k=1}^n \frac{(k-1)!}{x^k} + O\left(\frac{e^x}{x^{n+1}}\right) \qquad \text{as } x \to +\infty.$$

The error $O(\frac{e^x}{x^{n+1}})$ in the approximate equality

$$f(x) \approx \sum_{k=1}^n \frac{(k-1)!}{x^k}e^x$$

is asymptotically infinitesimal compared with each term of the sum, including the last. As the same time, as $x \to +\infty$ each successive term of the sum is infinitesimal compared with its predecessor; therefore it is natural to write the continually sharper sequence of such formulas as a series generated by f:

$$f(x) \simeq e^x \sum_{k=1}^\infty \frac{(k-1)!}{x^k}.$$

We note that this series obviously diverges for every value of $x \in \mathbb{R}$, so that we cannot write

$$f(x) = e^x \sum_{k=1}^\infty \frac{(k-1)!}{x^k}.$$

Thus we are dealing here with a new and clearly useful *asymptotic* interpretation of a series connected, in contrast with the classical case, with the relative rather than the absolute error of approximation of the function. The partial sums of such a series, in contrast to the classical case, are used not so much to approximate the values of the function at specific points as to describe their collective behavior under the limiting passage in question (which in the present example occurs as $x \to +\infty$).

b. Asymptotic Sequences and Asymptotic Series

Definition 2 A sequence of asymptotic formulas

$$f(x) = \psi_0(x) + o(\psi_0(x)),$$
$$f(x) = \psi_0(x) + \psi_1(x) + o(\psi_1(x)),$$
$$\vdots$$
$$f(x) = \psi_0(x) + \psi_1(x) + \cdots + \psi_n(x) + o(\psi_n(x)),$$
$$\vdots$$

that are valid over a base $\mathcal{B}$ in the set X where the functions are defined, is written as the relation

$$f(x) \simeq \psi_0(x) + \psi_1(x) + \cdots + \psi_n(x) + \cdots$$

or, more briefly, as $f(x) \simeq \sum_{k=0}^{\infty} \psi_k(x)$. It is called an *asymptotic expansion of f in the given base $\mathcal{B}$.*

It is clear from this definition that in asymptotic expansions we always have

$$o(\psi_n(x)) = \psi_{n+1}(x) + o(\psi_{n+1}(x)) \quad \text{over the base } \mathcal{B},$$

and hence for any $n = 0, 1, 2, \ldots$ we have

$$\psi_{n+1}(x) = o(\psi_n(x)) \quad \text{over the base } \mathcal{B},$$

that is, each successive term of the expansion contributes its correction, which is asymptotically more precise in comparison with its predecessor.

Asymptotic expansions usually arise in the form of a linear combination

$$c_0\varphi_0(x) + c_1\varphi_1(x) + \cdots + c_n\varphi_n(x) + \cdots$$

of functions of some sequence $\{\varphi_n(x)\}$ that is convenient for the specific problem.

Definition 3 Let X be a set with a base $\mathcal{B}$ defined on it. The sequence $\{\varphi_n(x)\}$ of functions defined on X is called an *asymptotic sequence over the base $\mathcal{B}$* if $\varphi_{n+1}(x) = o(\varphi_n(x))$ over the base $\mathcal{B}$ (for any two adjacent terms φ_n and φ_{n+1} of the sequence) and if none of the functions $\varphi_n \in \{\varphi_n(x)\}$ is identically zero on any element of $\mathcal{B}$.

Remark 2 The condition that $(\varphi_n|_B)(x) \not\equiv 0$ on the elements B of the base $\mathcal{B}$ is natural, since otherwise all the functions $\varphi_{n+1}, \varphi_{n+2}, \ldots$ would also be zero on B and the system $\{\varphi_n\}$ would be trivial in respect to its asymptotics.

Example 9 The following sequences are obviously asymptotic:

a) $1, x, x^2, \ldots, x^n, \ldots$ as $x \to 0$;

b) $1, \frac{1}{x}, \frac{1}{x^2}, \ldots, \frac{1}{x^n}, \ldots$ as $x \to \infty$;

c) $x^{p_1}, x^{p_2}, \ldots, x^{p_n}, \ldots$

    in the base $x \to 0$ if $p_1 < p_2 < \cdots < p_n < \cdots$,

    in the base $x \to \infty$ if $p_1 > p_2 > \cdots > p_n > \cdots$;

d) the sequence $\{g(x)\varphi_n(x)\}$ obtained from an asymptotic sequence through multiplication of all its terms by the same function.

Definition 4 If $\{\varphi_n\}$ is an asymptotic sequence over the base $\mathcal{B}$, then an asymptotic expansion of the form

$$f(x) \simeq c_0\varphi_0(x) + c_1\varphi_1(x) + \cdots + c_n\varphi_n(x) + \cdots$$

is called an *asymptotic expansion* or *asymptotic series of the function f with respect to the asymptotic sequence* $\{\varphi_n\}$ *over the base* $\mathcal{B}$.

Remark 3 The concept of an asymptotic series (in the context of power series) was stated by Poincaré (1886), who made vigorous use of asymptotic expansions in his work on celestial mechanics. But asymptotic series themselves, like some of the methods of obtaining them, had been encountered earlier. In regard to the possible generalization of the concept of an asymptotic expansion in the sense of Poincaré (which we have discussed in Definitions 2–4) see Problem 5 at the end of this section.

19.1.2 General Facts About Asymptotic Series

a. Uniqueness of an Asymptotic Expansion

When we speak of the asymptotic behavior of a function over a base $\mathcal{B}$, we are interested only in the nature of the limiting behavior of the function, so that if two generally different functions f and g are equal on some element of the base $\mathcal{B}$, they have the same asymptotic behavior over $\mathcal{B}$ and should be considered equal in the asymptotic sense.

Moreover, if we fix in advance some asymptotic sequence $\{\varphi_n\}$ in terms of which it is desirable to carry out an asymptotic expansion, we must reckon with the limited possibilities of any such system of functions $\{\varphi_n\}$. To be specific, there will be functions that are infinitesimal with respect to every term φ_n of the given asymptotic system.

Example 10 Let $\varphi_n(x) = \frac{1}{x^n}$, $n = 0, 1, \ldots$; then $\mathrm{e}^{-x} = o(\varphi_n(x))$ as $x \to +\infty$.

For that reason it is natural to adopt the following definitions.

Definition 5 If $\{\varphi_n(x)\}$ is an asymptotic sequence over the base $\mathcal{B}$, a function f such that $f(x) = o(\varphi_n(x))$ over $\mathcal{B}$ for each $n = 0, 1, \ldots$ is called an *asymptotic zero with respect to* $\{\varphi_n(x)\}$.

Definition 6 Functions f and g are *asymptotically equal over the base $\mathcal{B}$ with respect to a sequence of functions* $\{\varphi_n\}$ that is asymptotic over $\mathcal{B}$ if the difference $f - g$ is an asymptotic zero with respect to $\{\varphi_n\}$.

Proposition 1 (Uniqueness of an asymptotic expansion) *Let $\{\varphi_n\}$ be an asymptotic sequence of functions over a base $\mathcal{B}$.*

a) *If a function f admits an asymptotic expansion with respect to the sequence $\{\varphi_n\}$ over $\mathcal{B}$, then that expansion is unique.*

b) *If the functions f and g admit an asymptotic expansion in the system $\{\varphi_n\}$, then these expansions are the same if and only if the functions f and g are asymptotically equal over $\mathcal{B}$ with respect to $\{\varphi_n\}$.*

Proof a) Suppose the function φ is not identically zero on any element of $\mathcal{B}$.

We shall show that if $f(x) = o(\varphi(x))$ over $\mathcal{B}$, and at the same time $f(x) = c\varphi(x) + o(\varphi(x))$ over $\mathcal{B}$, then $c = 0$.

Indeed, $|f(x)| \geq |c\varphi(x)| - |o(\varphi(x))| = |c||\varphi(x)| - o(|\varphi(x)|)$ over $\mathcal{B}$, and so if $|c| > 0$, there exists $B_1 \in \mathcal{B}$ at each point of which $|f(x)| \geq \frac{|c|}{2}|\varphi(x)|$. But if $f(x) = o(\varphi(x))$ over $\mathcal{B}$, then there exists $B_2 \in \mathcal{B}$ at each point of which $|f(x)| \leq \frac{|c|}{3}|\varphi(x)|$. Hence at each point $x \in B_1 \cap B_2$ we would have to have $\frac{|c|}{2}|\varphi(x)| \leq \frac{|c|}{3}|\varphi(x)|$ or, assuming $|c| \neq 0, 3|\varphi(x)| \leq 2|\varphi(x)|$. But this is impossible if $\varphi(x) \neq 0$ at even one point of $B_1 \cap B_2$.

Now let us consider the asymptotic expansion of a function f with respect to the sequence $\{\varphi_n\}$.

Let $f(x) = c_0\varphi_0(x) + o(\varphi_0(x))$ and $f(x) = \tilde{c}_0\varphi(x) + o(\varphi_0(x))$ over $\mathcal{B}$. Subtracting the second equality from the first, we find that $0 = (c_0 - \tilde{c}_0)\varphi_0(x) + o(\varphi_0(x))$ over $\mathcal{B}$. But $0 = o(\varphi_0(x))$ over $\mathcal{B}$ and so, by what has been proved, $c_0 - \tilde{c}_0 = 0$.

If we have proved that $c_0 = \tilde{c}_0, \ldots, c_{n-1} = \tilde{c}_{n-1}$ for two expansions of the function f in the system $\{\varphi_n\}$, then by the equalities

$$f(x) = c_0\varphi_0(x) + \cdots + c_{n-1}\varphi_{n-1}(x) + c_n\varphi_n(x) + o(\varphi_n(x)),$$
$$f(x) = c_0\varphi_0(x) + \cdots + c_{n-1}\varphi_{n-1}(x) + \tilde{c}_n\varphi_n(x) + o(\varphi_n(x))$$

we find in the same way that $c_n = \tilde{c}_n$.

By induction we now conclude that a) is true.

b) If $f(x) = c_0\varphi_0(x) + \cdots + c_n\varphi_n(x) + o(\varphi_n(x))$ and $g(x) = c_0\varphi_0(x) + \cdots + c_n\varphi_n(x) + o(\varphi_n(x))$ over $\mathcal{B}$, then $f(x) - g(x) = o(\varphi_n(x))$ over $\mathcal{B}$ for each $n = 0, 1, \ldots$, and hence the functions f and g are asymptotically equal with respect to the sequence $\{\varphi_n(x)\}$.

The converse follows from a), since an asymptotic zero, which we take to be the difference $f - g$, can have only the zero asymptotic expansion. $\qquad\square$

Remark 4 We have discussed the question of uniqueness of an asymptotic expansion. We emphasize, however, that an asymptotic expansion of a function with respect to a preassigned asymptotic sequence is by no means always possible. Two functions f and g in general need not always be connected by one of the asymptotic relations $f = O(g)$, $f = o(g)$ or $f \sim g$ over a base $\mathcal{B}$.

The very general asymptotic Taylor formula, for example, exhibits a specific class of functions (having derivatives of order up to n at $x = 0$), each of which admits the asymptotic representation

$$f(x) = f(0) + \frac{1}{1!}f'(0)x + \cdots + \frac{1}{n!}f^{(n)}(0)x^n + o\left(x^n\right)$$

as $x \to 0$. But even the function $x^{1/2}$ cannot be expanded asymptotically in the system $1, x, x^2, \ldots$. Thus one must not identify an asymptotic sequence and an asymptotic expansion with any canonical base and the expansion of any asymptotic in it. There are many more possible types of asymptotic behavior than can be described by any fixed asymptotic sequence, so that the description of the asymptotic behavior of a function is not so much an expansion in terms of a preassigned asymptotic system as it is the search for such a system. One cannot, for example, when computing the indefinite integral of an elementary function, require in advance that the result be a composition of certain elementary functions, because it may not be an elementary function at all. The search for asymptotic formulas, like the computation of indefinite integrals, is of interest only to the extent that the result is simpler and more accessible to investigation than the original expression.

b. Admissible Operations with Asymptotic Formulas

The elementary arithmetic properties of the symbols o and O (such properties as $o(g) + o(g) = o(g)$, $o(g) + O(g) = O(g) + O(g) = O(g)$, and the like) have been studied along with the theory of limits (Proposition 4 of Sect. 3.2). The following obvious proposition follows from these properties and the definition of an asymptotic expansion.

Proposition 2 (Linearity of asymptotic expansions) *If the functions f and g admit asymptotic expansions $f \simeq \sum_{n=0}^{\infty} a_n\varphi_n$ and $g \simeq \sum_{n=0}^{\infty} b_n\varphi_n$ with respect to the asymptotic sequence $\{\varphi_n\}$ over the base $\mathcal{B}$, then a linear combination of them $\alpha f + \beta g$ admits such an expansion, and $(\alpha f + \beta g) \simeq \sum_{n=0}^{\infty}(\alpha a_n + \beta b_n)\varphi_n$.*

Further properties of asymptotic expansions and asymptotic formulas in general will involve more and more specialized cases.

Proposition 3 (Integration of asymptotic equalities) *Let f be a continuous function on the interval $I = [a, \omega[$ (or $I =]\omega, a]$).*

a) *If the function $g(x)$ is continuous and nonnegative on I and the integral $\int_a^\omega g(x)\,dx$ diverges, then the relations*

$$f(x) = O\big(g(x)\big), \qquad f(x) = o\big(g(x)\big), \qquad f(x) \sim g(x) \quad \text{as } I \ni x \to \omega$$

imply respectively that

$$F(x) = O\big(G(x)\big), \qquad F(x) = o\big(G(x)\big), \quad \text{and} \quad F(x) \sim G(x),$$

where

$$F(x) = \int_a^x f(t)\,dt \quad \text{and} \quad G(x) = \int_a^x g(t)\,dt.$$

b) *If the functions $\varphi_n(x)$, $n = 0, 1, \ldots$, which are continuous and positive on $I = [a, \omega[$ form an asymptotic sequence as $I \ni x \to \omega$ and the integrals $\Phi_n(x) = \int_x^\omega \varphi_n(t)\,dt$ converge for $x \in I$ then the functions $\Phi_n(x)$, $n = 0, 1, \ldots$ also form an asymptotic sequence over the base $I \ni x \to \omega$.*

c) *If the integral $\mathcal{F}(x) = \int_x^\omega f(x)\,dx$ converges and f has the asymptotic expansion $f(x) \simeq \sum_{n=0}^\infty c_n \varphi_n(x)$ as $I \ni x \to \omega$ with respect to the asymptotic sequence $\{\varphi_n(x)\}$ of b), then $\mathcal{F}(x)$ has the asymptotic expansion $\mathcal{F}(x) \simeq \sum_{n=0}^\infty c_n \Phi_n(x)$.*

Proof a) If $f(x) = O(g(x))$ as $I \ni x \to \omega$, there exists $x_0 \in I$ and a constant M such that $|f(x)| \leq Mg(x)$ for $x \in [x_0, \omega[$. It follows that for $x \in [x_0, \omega[$, we have $|\int_a^x f(t)\,dt| \leq |\int_a^{x_0} f(t)\,dt| + M \int_{x_0}^x g(t)\,dt = O(\int_a^x g(t)\,dt)$.

To prove the other two relations one can use L'Hôpital's rule (as in Example 7), taking account of the relation $G(x) = \int_a^x g(t)\,dt \to \infty$ as $I \ni x \to \omega$. As a result, we find

$$\lim_{I \ni x \to \omega} \frac{F(x)}{G(x)} = \lim_{I \ni x \to \omega} \frac{F'(x)}{G'(x)} = \lim_{I \ni x \to \omega} \frac{f(x)}{g(x)}.$$

b) Since $\Phi_n(x) \to 0$ as $I \ni x \to \omega$ $(n = 0, 1, \ldots)$ applying L'Hôpital's rule again, we find that

$$\lim_{I \ni x \to \omega} \frac{\Phi_{n+1}(x)}{\Phi_n(x)} = \lim_{I \ni x \to \omega} \frac{\Phi'_{n+1}(x)}{\Phi'_n(x)} = \lim_{I \ni x \to \omega} \frac{\varphi_{n+1}(x)}{\varphi_n(x)} = 0.$$

c) The function $r_n(x)$ in the relation

$$f(x) = c_0 \varphi_0(x) + c_1 \varphi_1(x) + \cdots + c_n \varphi_n(x) + r_n(x),$$

being the difference of continuous functions on I, is itself continuous on I, and we obviously have $R_n(x) = \int_x^\omega r_n(t)\,dt \to 0$ as $I \ni x \to \omega$. But $r_n(x) = o(\varphi_n(x))$ as $I \ni x \to \omega$ and $\Phi_n(x) \to 0$ as $I \ni x \to \omega$. Therefore, again by L'Hôpital's rule, it follows that in the equality

$$\mathcal{F}(x) = c_0 \Phi_0(x) + c_1 \Phi_1(x) + \cdots + c_n \Phi_n(x) + R_n(x)$$

the quantity $R_n(x)$ is $o(\Phi_n(x))$ as $I \ni x \to \omega$. $\square$

Remark 5 Differentiation of asymptotic equalities and asymptotic series is generally not legitimate.

Example 11 The function $f(x) = e^{-x} \sin(e^x)$ is continuously differentiable on $\mathbb{R}$ and is an asymptotic zero with respect to the asymptotic sequence $\{\frac{1}{x^n}\}$ as $x \to +\infty$. The derivatives of the functions $\frac{1}{x^n}$, up to a constant factor, again have the form $\frac{1}{x^k}$. However the function $f'(x) = -e^x \sin(e^x) + \cos(e^x)$ not only fails to be an asymptotic zero; it doesn't even have an asymptotic expansion with respect to the sequence $\{\frac{1}{x^n}\}$ as $x \to +\infty$.

19.1.3 Asymptotic Power Series

In conclusion, let us examine asymptotic power series in some detail, since they are encountered relatively often, although in a rather generalized form, as was the case in Example 8.

We shall study expansions with respect to the sequence $\{x^n; n = 0, 1, \ldots\}$, which is asymptotic as $x \to 0$ and with respect to $\{\frac{1}{x^n}; n = 0, 1, \ldots\}$, which is asymptotic as $x \to \infty$. Since these are both the same object up to the change of variable $x = \frac{1}{u}$, we state the next proposition only for expansions with respect to the first sequence and then note the specifics of certain of the formulations given in the case of expansions with respect to the second sequence.

Proposition 4 *Let 0 be a limit point of E and let*

$$f(x) \simeq a_0 + a_1 x + a_2 x^2 + \cdots,$$
$$g(x) \simeq b_0 + b_1 x + b_2 x^2 + \cdots \qquad as\ E \ni x \to 0.$$

Then as $E \ni x \to 0$,

a) $(\alpha f + \beta g) \simeq \sum_{n=0}^{\infty} (\alpha a_n + \beta b_n) x^n$;

b) $(f \cdot g)(x) \simeq \sum_{n=0}^{\infty} c_n x^n$, *where* $c_n = a_0 b_n + a_1 b_{n-1} + \cdots + a_n b_0$, $n = 0, 1, \ldots$;

c) *if $b_0 \neq 0$, then* $(\frac{f}{g})(x) \simeq \sum_{n=0}^{\infty} d_n x^n$, *where the coefficients d_n can be found from the recurrence relations*

$$a_o = b_0 d_0, \qquad a_1 = b_0 d_1 + b_1 d_0, \qquad \ldots, \qquad a_n = \sum_{k=0}^{n} b_k d_{n-k}, \qquad \ldots;$$

d) *if E is a deleted neighborhood or one-sided neighborhood of 0 and f is continuous on E, then*

$$\int_0^x f(t)\,dt \simeq a_0 x + \frac{a_1}{2} x^2 + \cdots + \frac{a_{n-1}}{n} x^n + \cdots;$$

e) *if in addition to the assumptions of d) we also have* $f \in C^{(1)}(E)$ *and*

$$f'(x) \simeq a_0' + a_1'x + \cdots,$$

then $a_n' = (n+1)a_{n+1}$, $n = 0, 1, \ldots$.

Proof a) This is a special case of Proposition 2.

b) Using the properties of $o(\ )$ (see Proposition 4 of Sect. 3.2), we find that

$$(f \cdot g)(x) =$$
$$= f(x) \cdot g(x) =$$
$$= \left(a_0 + a_1x + \cdots + a_nx^n + o(x^n)\right)\left(b_0 + b_1x + \cdots + b_nx^n + o(x^n)\right) =$$
$$= (a_0b_0) + (a_0b_1 + a_1b_0)x + \cdots + (a_0b_n + a_1b_{n-1} + \cdots + a_nb_0)x^n + o(x^n)$$

as $E \ni x \to 0$.

c) If $b_0 \neq 0$, then $g(x) \neq 0$ for x close to zero, and therefore we can consider the ratio $\frac{f(x)}{g(x)} = h(x)$. Let us verify that if the coefficients $d_0, \ldots, d_n$ in the representation $h(x) = d_0 + d_1x + \cdots + d_nx^n + r_n(x)$ have been chosen in accordance with c), then $r_n(x) = o(x^n)$ as $E \ni x \to 0$. From the identity $f(x) = g(x)h(x)$, we find that

$$a_0 + a_1x + \cdots + a_nx^n + o(x^n) =$$
$$= \left(b_0 + b_1x + \cdots + b_nx^n + o(x^n)\right)\left(d_0 + d_1x + \cdots + d_nx^n + r_n(x)\right) =$$
$$= (b_0d_0) + (b_0d_1 + b_1d_0)x + \cdots + (b_0d_n + b_1d_{n-1} + \cdots + b_nd_0)x^n +$$
$$+ b_0r_n(x) + o(r_n(x)) + o(x^n),$$

from which it follows that $o(x^n) = b_0r_n(x) + o(r_n(x)) + o(x^n)$, or $r_n(x) = o(x^n)$ as $E \ni x \to 0$, since $b_0 \neq 0$.

d) This follows from part c) of Proposition 3 if we set $\omega = 0$ there and recall that $-\int_x^0 f(t)\,dt = \int_0^x f(t)\,dt$.

e) Since the function $f'(x)$ is continuous on $]0, x]$ (or $[x, 0[$) and bounded (it tends to a_0' as $x \to 0$), the integral $\int_0^x f'(t)\,dt$ exists. Obviously $f(x) = a_0 + \int_0^x f'(t)\,dt$, since $f(x) \to a_0$ as $x \to 0$. Substituting the asymptotic expansion of $f'(x)$ into this equality and using what was proved in d), we find that

$$f(x) \simeq a_0 + a_0'x + \frac{a_1'}{2}x^2 + \cdots + \frac{a_{n-1}'}{n}x^n + \cdots.$$

It now follows from the uniqueness of asymptotic expansions (Proposition 1) that $a_n' = (n+1)a_n$, $n = 0, 1, \ldots$. $\qquad\square$

Corollary 1 *If U is a neighborhood (or one-sided neighborhood) of infinity in $\mathbb{R}$ and the function f is continuous in U and has the asymptotic expansion*

$$f(x) \simeq a_0 + \frac{a_1}{x} + \frac{a_2}{x^2} + \cdots + \frac{a_n}{x^n} + \cdots \qquad as\ U \ni x \to \infty,$$

then the integral

$$\mathcal{F}(x) = \int_x^\infty \left(f(t) - a_0 - \frac{a_1}{t} \right) dt$$

over an interval contained in U converges and has the following asymptotic expansion:

$$\mathcal{F}(x) \simeq \frac{a_2}{x} + \frac{a_3}{2x^2} + \cdots + \frac{a_n}{nx^n} + \cdots \quad \text{as } U \ni x \to \infty.$$

Proof The convergence of the integral is obvious, since

$$f(t) - a_0 - \frac{a_1}{t} \sim \frac{a_2}{t^2} \quad \text{as } U \ni t \to \infty.$$

It remains only to integrate the asymptotic expansion

$$f(t) - a_0 - \frac{a_1}{t} \simeq \frac{a_2}{t^2} + \frac{a_3}{t^3} + \cdots + \frac{a_n}{t^n} \cdots \quad \text{as } U \ni t \to \infty,$$

citing, for example, Proposition 3d). $\square$

Corollary 2 *If in addition to the hypotheses of Corollary 1 it is known that $f \in C^{(1)}(U)$ and f' admits the asymptotic expansion*

$$f'(x) \simeq a_0' + \frac{a_1'}{x} + \frac{a_2'}{x^2} + \cdots + \frac{a'}{n} + \cdots \quad \text{as } U \ni x \to \infty,$$

then this expansion can be obtained by formally differentiating the expansion of the function f, and

$$a_n' = -(n-1)a_{n-1}, \quad n = 2, 3, \ldots \quad \text{and} \quad a_0' = a_1' = 0.$$

Proof Since $f'(x) = a_0' + \frac{a_1'}{x} + O(1/x^2)$ as $U \ni x \to \infty$, we have

$$f(x) = f(x_0) + \int_{x_0}^x f'(t)\, dt = a_0' x + a_1' \ln x + O(1)$$

as $U \ni x \to \infty$; and since $f(x) \simeq a_0 + \frac{a_1}{x} + \frac{a_2}{x^2} + \cdots$ and the sequence x, $\ln x$, 1, $\frac{1}{x}$, $\frac{1}{x^2}, \ldots$ is an asymptotic sequence as $U \ni x \to \infty$, Proposition 1 enables us to conclude that $a_0' = a_1' = 0$. Now, integrating the expansion $f'(x) \simeq \frac{a_2'}{x^2} + \frac{a_3'}{x^3} + \cdots$, by Corollary 1 we obtain the expansion of $f(x)$, and by the uniqueness of the expansion we arrive at the relations $a_n' = -(n-1)a_{n-1}$ for $n = 2, 3, \ldots$. $\square$

19.1.4 Problems and Exercises

1. a) Let $h(z) = \sum_{n=0}^\infty a_n z^{-n}$ for $|z| > R$, $z \in \mathbb{C}$. Show that then $h(z) \simeq \sum_{n=0}^\infty a_n z^{-n}$ as $\mathbb{C} \ni z \to \infty$.

b) Assuming that the required solution $y(x)$ of the equation $y'(x) + y^2(x) = \sin\frac{1}{x^2}$ has an asymptotic expansion $y(x) \simeq \sum_{n=0}^{\infty} c_n x^{-n}$ as $x \to \infty$, find the first three terms of this expansion.

c) Prove that if $f(z) = \sum_{n=0}^{\infty} a_n z^n$ for $|z| < r$, $z \in \mathbb{C}$, and $g(z) \simeq b_1 z + b_2 z^2 + \cdots$ as $\mathbb{C} \ni z \to 0$, then the function $f \circ g$ is defined in some deleted neighborhood of $0 \in \mathbb{C}$ and $(f \circ g)(z) \simeq c_0 + c_1 z + c_2 z^2 + \cdots$ as $\mathbb{C} \ni z \to 0$, where the coefficients $c_0, c_1, \ldots$ can be obtained by substituting the series in the series, just as for convergent power series.

2. Show the following.

a) If f is a continuous, positive, monotonic function for $x \geq 0$, then

$$\sum_{k=0}^{n} f(k) = \int_0^n f(x)\,dx + O\big(f(n)\big) + O(1) \quad \text{as } n \to \infty;$$

b) $\sum_{k=1}^{n} \frac{1}{k} = \ln n + c + o(1)$ as $n \to \infty$;

c) $\sum_{k=1}^{n} k^\alpha (\ln k)^\beta \sim \frac{n^{\alpha+1}(\ln n)^\beta}{\alpha+1}$ as $n \to \infty$ for $\alpha > -1$.

3. Through integration by parts find the asymptotic expansions of the following functions as $x \to +\infty$:

a) $\Gamma_s(x) = \int_x^{+\infty} t^{s-1} e^{-t}\,dt$ – the incomplete gamma function;

b) $\mathrm{erf}(x) = \frac{1}{\sqrt{\pi}} \int_{-x}^{x} e^{-t^2}\,dt$ – the probability error function (we recall that $\int_{-\infty}^{\infty} e^{-x^2}\,dx = \sqrt{\pi}$ is the Euler–Poisson integral);

c) $F(x) = \int_x^{+\infty} \frac{e^{it}}{t^\alpha}\,dt$ if $\alpha > 0$.

4. Using the result of the preceding problem, find the asymptotic expansions of the following functions as $x \to +\infty$:

a) $\mathrm{Si}(x) = \int_0^x \frac{\sin t}{t}\,dt$ – the sine integral (we recall that $\int_0^\infty \frac{\sin x}{x}\,dx = \frac{\pi}{2}$ is the Dirichlet integral).

b) $C(x) = \int_0^x \cos\frac{\pi}{2} t^2\,dt$, $S(x) = \int_0^x \sin\frac{\pi}{2} t^2\,dt$ – the Fresnel integrals (we recall that $\int_0^{+\infty} \cos x^2\,dx = \int_0^\infty \sin x^2\,dx = \frac{1}{2}\sqrt{\frac{\pi}{2}}$).

5. The following generalization of the concept of an expansion in an asymptotic sequence $\{\varphi_n(x)\}$ introduced by Poincaré and studied above is due to Erdélyi.[2]

Let X be a set, $\mathcal{B}$ a base in X, $\{\varphi_n(x)\}$ an asymptotic sequence of functions on X. If the functions $f(x), \psi_0(x), \psi_1(x), \psi_2(x), \ldots$ are such that the equality

$$f(x) = \sum_{k=0}^{n} \psi_k(x) + o\big(\varphi_n(x)\big) \quad \text{over the base } \mathcal{B}$$

---

[2] A. Erdélyi (1908–1977) – Hungarian/British mathematician.

holds for every $n = 0, 1, \ldots$, we write

$$f(x) \simeq \sum_{n=0}^{\infty} \psi_n(x), \quad \{\varphi_n(x)\} \quad \text{over the base } \mathcal{B},$$

and we say that we have the *asymptotic expansion of the function f over the base $\mathcal{B}$ in the sense of Erdélyi.*

a) Please note that in Problem 4 you obtained the asymptotic expansion in the sense of Erdélyi if you assume $\varphi_n(x) = x^{-n}$, $n = 0, 1, \ldots$.

b) Show that asymptotic expansions in the sense of Erdélyi do not have the property of uniqueness (the functions ψ_n can be changed).

c) Show that if a set X, a base $\mathcal{B}$ in X, a function f on X, and sequences $\{\mu_n(x)\}$ and $\{\varphi_n(x)\}$, the second of which is asymptotic over the base $\mathcal{B}$, are given, then the expansion

$$f(x) \simeq \sum_{n=0}^{\infty} \alpha_n \mu_n(x), \quad \{\varphi_n(x)\} \quad \text{over the base } \mathcal{B},$$

where a_n are numerical coefficients, is either impossible or unique.

6. *Uniform asymptotic estimates.* Let X be a set and $\mathcal{B}_X$ a base in X, and let $f(x, y)$ and $g(x, y)$ be (vector-valued) functions defined on X and depending on the parameter $y \in Y$. Set $|f(x, y)| = \alpha(x, y)|g(x, y)|$. We say that the asymptotic relations

$$f(x, y) = o\big(g(x, y)\big), \qquad f(x, y) = O\big(g(x, y)\big), \qquad f(x, y) \sim g(x, y)$$

are *uniform with respect to the parameter y on the set Y* if (respectively) $\alpha(x, y) \rightrightarrows 0$ on Y over the base $\mathcal{B}_X$; $\alpha(x, y)$ is ultimately bounded over the base $\mathcal{B}_x$ uniformly with respect to $y \in Y$; and finally $f = \alpha \cdot g + o(g)$, where $\alpha(x, y) \rightrightarrows 1$ on Y over the base $\mathcal{B}_x$.

Show that if we introduce the base $\mathcal{B} = \{\mathcal{B}_x \times Y\}$ in $X \times Y$ whose elements are the direct products of the elements B_x of the base $\mathcal{B}_X$ and the set Y, then these definitions are equivalent respectively to the following:

$$f(x, y) = o\big(g(x, y)\big), \qquad f(x, y) = O\big(g(x, y)\big), \qquad f(x, y) \sim g(x, y)$$

over the base $\mathcal{B}$.

7. *Uniform asymptotic expansions.* The asymptotic expansion

$$f(x, y) \simeq \sum_{n=0}^{\infty} a_n(y)\varphi_n(x) \quad \text{over the base } \mathcal{B}_X$$

is *uniform with respect to the parameter y on Y* if the estimate $r_n(x, y) = o(\varphi_n(x))$ over the base $\mathcal{B}_X$ in X holds uniformly on Y in the equalities

$$f(x, y) = \sum_{k=0}^{n} a_k(y)\varphi_k(x) + r_n(x, y), \quad n = 0, 1, \ldots.$$

a) Let Y be a (bounded) measurable set in $\mathbb{R}^n$, and suppose that for each fixed $x \in X$ the functions $f(x, y), a_0(y), a_1(y), \ldots$ are integrable over Y. Show that if the asymptotic expansion $f(x, y) \simeq \sum_{n=0}^{\infty} a_n(y)\varphi_n(x)$ over the base $\mathcal{B}_X$ is uniform with respect to the parameter $y \in Y$, then the following asymptotic expansion also holds

$$\int_Y f(x, y)\, dy \simeq \sum_{n=0}^{\infty} \left(\int_Y a_n(y)\, dy \right) \varphi_n(x) \quad \text{over the base } \mathcal{B}_X.$$

b) Let $Y = [c, d] \subset \mathbb{R}$. Assume that the function $f(x, y)$ is continuously differentiable with respect to y on the closed interval Y for each fixed $x \in X$ and for some $y_0 \in Y$ admits the asymptotic expansion

$$f(x, y_0) \simeq \sum_{n=0}^{\infty} a_n(y_0)\varphi_n(x) \quad \text{over the base } \mathcal{B}_X.$$

Prove that if the asymptotic expansion

$$\frac{\partial f}{\partial y}(x, y) \simeq \sum_{n=0}^{\infty} \alpha_n(y)\varphi_n(x) \quad \text{over the base } \mathcal{B}_X$$

holds uniformly with respect to $y \in Y$ with coefficients $\alpha_n(y)$ that are continuous in y, $n = 0, 1, \ldots$, then the original function $f(x, y)$ has an asymptotic expansion $f(x, y) \simeq \sum_{n=0}^{\infty} a_n(y)\varphi_n(x)$ over the base $\mathcal{B}_X$ that is uniform with respect to $y \in Y$, its coefficients $a_n(y)$, $n = 0, 1, \ldots$ are smooth functions of y on the interval Y and $\frac{da_n}{dy}(y) = \alpha_n(y)$.

8. Let $p(x)$ be a smooth function that is positive on the closed interval $c \le x \le d$.

a) Solve the equation $\frac{\partial^2 u}{\partial x^2}(x, \lambda) = \lambda^2 p(x)u(x, \lambda)$ in the case when $p(x) \equiv 1$ on $[c, d]$.

b) Let $0 < m \le p(x) \le M < +\infty$ on $[c, d]$ and let $u(c, \lambda) = 1$, $\frac{\partial u}{\partial x}(c, \lambda) = 0$. Estimate the quantity $u(x, \lambda)$ from above and below for $x \in [c, d]$.

c) Assuming that $\ln u(x, \lambda) \simeq \sum_{n=0}^{\infty} c_n(x)\lambda^{1-n}$ as $\lambda \to +\infty$, where $c_0(x)$, $c_1(x), \ldots$ are smooth functions and, using the fact that $(\frac{u'}{u})' = \frac{u''}{u} - (\frac{u'}{u})^2$, show that $c_0'^2(x) = p(x)$ and $(c''_{n-1} + \sum_{k=0}^{n} c_k' \cdot c_{n-k}')(x) = 0$.

19.2 The Asymptotics of Integrals (Laplace's Method)

19.2.1 The Idea of Laplace's Method

In this subsection we shall discuss Laplace's method – one of the few reasonably general methods of constructing the asymptotics of an integral depending on a parameter. We confine our attention to integrals of the form

$$F(\lambda) = \int_a^b f(x) e^{\lambda S(x)} \, dx, \qquad (19.1)$$

where $S(x)$ is a real-valued function and λ is a parameter. Such integrals are usually called *Laplace integrals*.

Example 1 The *Laplace transform*

$$L(f)(\xi) = \int_0^{+\infty} f(x) e^{-\xi x} \, dx$$

is a special case of a Laplace integral.

Example 2 Laplace himself applied his method to integrals of the form $\int_a^b f(x) \varphi^n(x) \, dx$, where $n \in \mathbb{N}$ and $\varphi(x) > 0$ on $]a, b[$. Such an integral is also a special case of a general Laplace integral (19.1), since $\varphi^n(x) = \exp(n \ln \varphi(x))$.

We shall be interested in the asymptotics of the integral (19.1) for large values of the parameter λ, more precisely as $\lambda \to +\infty$, $\lambda \in \mathbb{R}$.

So as not to become distracted with secondary issues when describing the basic idea of Laplace's method, we shall assume that $[a, b] = I$ is a finite closed interval in the integral (19.1), that the functions $f(x)$ and $S(x)$ are smooth on I, and that $S(x)$ has a unique, strict maximum $S(x_0)$ at the point $x_0 \in I$. Then the function $\exp(\lambda S(x))$ also has a strict maximum at x_0, which rises higher above the other values of this function on the interval I as the value of the parameter λ increases. As a result, if $f(x) \not\equiv 0$ in a neighborhood of x_0, the entire integral (19.1) can be replaced by the integral over an arbitrarily small neighborhood of x_0, thereby admitting a relative error that tends to zero as $\lambda \to +\infty$. This observation is called the *localization principle*. Reversing the historical sequence of events, one might say that this localization principle for Laplace integrals resembles the principal of local action of approximate identities and the δ-function.

Now that the integral is being taken over only a small neighborhood of x_0, the functions $f(x)$ and $S(x)$ can be replaced by the main terms of their Taylor expansions as $I \ni x \to x_0$.

It remains to find the asymptotics of the resulting canonical integral, which can be done without any particular difficulty.

It is in the sequential execution of these steps that the essence of Laplace's method of finding the asymptotics of an integral is to be found.

Example 3 Let $x_0 = a$, $S'(a) \neq 0$, and $f(a) \neq 0$, which happens, for example, when the function $S(x)$ is monotonically decreasing on $[a, b]$. Under these conditions $f(x) = f(a) + o(1)$ and $S(x) = S(a) + (x - a)S'(a) + o(1)$, as $I \ni x \to a$. Carrying out the idea of Laplace's method, for a small $\varepsilon > 0$ and $\lambda \to +\infty$, we find that

$$F(\lambda) \sim \int_a^{a+\varepsilon} f(x)e^{\lambda S(x)}\, dx \sim$$

$$\sim f(a)e^{\lambda S(a)} \int_0^{\varepsilon} e^{\lambda t\, S'(a)}\, dt = -\frac{f(a)e^{\lambda S(a)}}{\lambda S'(a)}\left(1 - e^{\lambda S'(a)\varepsilon}\right).$$

Since $S'(a) < 0$, it follows that in the case in question

$$F(\lambda) \sim -\frac{f(a)e^{\lambda S(a)}}{\lambda S'(a)} \qquad \text{as } \lambda \to +\infty. \tag{19.2}$$

Example 4 Let $a < x_0 < b$. Then $S'(x_0) = 0$, and we assume that $S''(x_0) \neq 0$, that is, $S''(x_0) < 0$, since x_0 is a maximum.

Using the expansions $f(x) = f(x_0) + o(x - x_0)$ and $S(x) = S(x_0) + \frac{1}{2}S''(x_0)(x - x_0)^2 + o((x - x_0)^2)$, which hold as $x \to x_0$, we find that for small $\varepsilon > 0$ and $\lambda \to +\infty$

$$F(\lambda) \sim \int_{x_0-\varepsilon}^{x_0+\varepsilon} f(x)e^{\lambda S(x)}\, dx \sim f(x_0)e^{\lambda S(x_0)} \int_{-\varepsilon}^{\varepsilon} e^{\frac{1}{2}\lambda S''(x_0)t^2}\, dt.$$

Making the change of variable $\frac{1}{2}\lambda S''(x_0)t^2 = -u^2$ (since $S''(x_0) < 0$), we obtain

$$\int_{-\varepsilon}^{\varepsilon} e^{\frac{1}{2}\lambda S''(x_0)t^2}\, dt = \sqrt{-\frac{2}{\lambda S''(x_0)}} \int_{-\varphi(\lambda,\varepsilon)}^{\varphi(\lambda,\varepsilon)} e^{-u^2}\, du,$$

where $\varphi(\lambda, \varepsilon) = \sqrt{-\frac{\lambda S''(x_0)}{2}}\varepsilon \to +\infty$ as $\lambda \to +\infty$.

Taking account of the equality

$$\int_{-\infty}^{+\infty} e^{-u^2}\, du = \sqrt{\pi},$$

we now find the principal term of the asymptotics of the Laplace integral in this case:

$$F(\lambda) \sim \sqrt{-\frac{2\pi}{\lambda S''(x_0)}} f(x_0)e^{\lambda S(x_0)} \qquad \text{as } \lambda \to +\infty. \tag{19.3}$$

Example 5 If $x_0 = a$, but $S'(x_0) = 0$ and $S''(x_0) < 0$, then, reasoning as in Example 4, we find this time that

$$F(\lambda) \sim \int_a^{a+\varepsilon} f(x)e^{\lambda S(x)}\, dx \sim f(x_0)e^{\lambda S(x_0)} \int_0^{\varepsilon} e^{\frac{1}{2}\lambda S''(x_0)t^2}\, dt,$$

and so

$$F(\lambda) \sim \frac{1}{2}\sqrt{-\frac{2\pi}{\lambda S''(x_0)}} f(x_0)e^{\lambda S(x_0)} \qquad \text{as } \lambda \to +\infty. \tag{19.4}$$

We have now obtained on a heuristic level the three very useful formulas (19.2)–(19.4) involving the asymptotics of the Laplace integral (19.1).

It is clear from these considerations that Laplace's method can be used successfully in the study of the asymptotics of any integral

$$\int_X f(x, \lambda)\, dx \quad \text{as } \lambda \to +\infty \tag{19.5}$$

provided (a) the localization principle holds for the integral, that is, the integral can be replaced by one equivalent to it as $\lambda \to +\infty$ extending over arbitrarily small neighborhoods of the distinguished points, and (b) the integrand in the localized integral can be replaced by a simpler one for which the asymptotics is on the one hand the same as that of the integral being investigated and on the other hand easy to find.

If, for example, the function $S(x)$ in the integral (19.1) has several local maxima $x_0, x_1, \ldots, x_n$ on the closed interval $[a, b]$, then, using the additivity of the integral, we replace it with small relative error by the sum of similar integrals taken over neighborhoods $U(x_j)$ of the maxima $x_0, x_1, \ldots, x_n$ so small that each contains only one such point. The asymptotic behavior of the integral

$$\int_{U(x_j)} f(x) e^{\lambda S(x)}\, dx \quad \text{as } \lambda \to +\infty,$$

as already mentioned, is independent of the size of the neighborhood $U(x_j)$ itself, and hence the asymptotic expansion of this integral as $\lambda \to +\infty$ is denoted $F(\lambda, x_j)$ and called the *contribution of the point x_j to the asymptotics of the integral* (19.1).

In its general formulation the localization principle thus means that the asymptotic behavior of the integral (19.5) is obtained as the sum $\sum_j F(\lambda, x_j)$ of the contributions of all the points of the integrand that are critical in some respect.

For the integral (19.1) these points are the maxima of the function $S(x)$, and, as one can see from formulas (19.2)–(19.4), the main contribution comes entirely from the local maximum points at which the absolute maximum of $S(x)$ on $[a, b]$ is attained.

In the following subsections of this section we shall develop the general considerations stated here and then consider some useful applications of Laplace's method. For many applications what we have already discussed is sufficient. It will also be shown below how to obtain not only the main term of the asymptotics, but also the entire asymptotic series.

19.2.2 The Localization Principle for a Laplace Integral

Lemma 1 (Exponential estimate) *Let $M = \sup_{a < x < b} S(x) < \infty$, and suppose that for some value $\lambda_0 > 0$ the integral* (19.1) *converges absolutely. Then it converges* *absolutely for every* $\lambda \geq \lambda_0$ *and the following estimate holds for such values of* λ:

$$|F(\lambda)| \leq \int_a^b |f(x)e^{\lambda S(x)}| \, dx \leq Ae^{\lambda M}, \qquad (19.6)$$

where $A \in \mathbb{R}$.

Proof Indeed, for $\lambda \geq \lambda_0$,

$$|F(\lambda)| = \left| \int_a^b f(x)e^{\lambda S(x)} \, dx \right| = \left| \int_a^b f(x)e^{\lambda_0 S(x)} e^{(\lambda - \lambda_0)S(x)} dx \right| \leq$$

$$\leq e^{(\lambda - \lambda_0)M} \int_a^b |f(x)e^{\lambda_0 S(x)}| \, dx = \left(e^{-\lambda_0 M} \int_a^b |f(x)e^{\lambda_0 S(x)}| \, dx \right) e^{\lambda M}. \quad \square$$

Lemma 2 (Estimate of the contribution of a maximum point) *Suppose the integral* (19.1) *converges absolutely for some value* $\lambda = \lambda_0$, *and suppose that in the interior or on the boundary of the interval* I *there is a point* x_0 *at which* $S(x_0) = \sup_{a < x < b} S(x) = M$. *If* $f(x)$ *and* $S(x)$ *are continuous at* x_0 *and* $f(x_0) \neq 0$, *then for every* $\varepsilon > 0$ *and every sufficiently small neighborhood* $U_I(x_0)$ *of* x_0 *in* I *we have the estimate*

$$\left| \int_{U_I(x_0)} f(x)e^{\lambda S(x)} \, dx \right| \geq Be^{\lambda(S(x_0)-\varepsilon)} \qquad (19.7)$$

with a constant $B > 0$, *valid for* $\lambda \geq \max\{\lambda_0, 0\}$.

Proof For a fixed $\varepsilon > 0$ let us take any neighborhood $U_I(x_0)$ inside which $|f(x)| \geq \frac{1}{2}|f(x_0)|$ and $S(x_0) - \varepsilon \leq S(x) \leq S(x_0)$. Assuming that f is real-valued, we can now conclude that f is of constant sign inside $U_I(x)$. This enables us to write for $\lambda \geq \max\{\lambda_0, 0\}$

$$\left| \int_{U_I(x_0)} f(x)e^{\lambda S(x)} \, dx \right| = \int_{U_I(x_0)} |f(x)| e^{\lambda S(x)} \, dx \geq$$

$$\geq \int_{U_I(x_0)} \frac{1}{2} |f(x_0)| e^{\lambda(S(x_0)-\varepsilon)} dx = Be^{\lambda(S(x_0)-\varepsilon)}. \quad \square$$

Proposition 1 (Localization principle) *Suppose the integral* (19.1) *converges absolutely for a value* $\lambda = \lambda_0$, *and suppose that inside or on the boundary of the interval* I *of integration the function* $S(x)$ *has a unique point* x_0 *of absolute maximum, that is, outside every neighborhood* $U(x_0)$ *of the point* x_0 *we have*

$$\sup_{I \setminus U(x_0)} S(x) < S(x_0).$$

If the functions $f(x)$ *and* $S(x)$ *are continuous at* x_0 *and* $f(x_0) \neq 0$, *then*

$$F(\lambda) = F_{U_I(x_0)}(\lambda)\big(1 + O(\lambda^{-\infty})\big) \quad as \ \lambda \to +\infty, \qquad (19.8)$$

where $U_I(x_0)$ is an arbitrary neighborhood of x_0 in I,

$$F_{U_I(x_0)}(\lambda) := \int_{U_I(x_0)} f(x)e^{\lambda S(x)}\,dx,$$

and $O(\lambda^{-\infty})$ denotes a function that is $o(\lambda^{-n})$ as $\lambda \to +\infty$ for every $n \in \mathbb{N}$.

Proof It follows from Lemma 2 that if the neighborhood $U_I(x_0)$ is sufficiently small, then the following inequality holds ultimately as $\lambda \to +\infty$ for every $\varepsilon > 0$

$$\left|F_{U_I(x_0)}(\lambda)\right| > e^{\lambda(S(x_0)-\varepsilon)}. \tag{19.9}$$

At the same time, by Lemma 1 for every neighborhood $U(x_0)$ of the point x_0 we have the estimate

$$\int_{I\setminus U(x_0)} |f(x)|e^{\lambda S(x)}\,dx \le Ae^{\lambda \mu} \quad \text{as } \lambda \to +\infty, \tag{19.10}$$

where $A > 0$ and $\mu = \sup_{x\in I\setminus U(x_0)} S(x) < S(x_0)$.

Comparing this estimate with inequality (19.9), it is easy to conclude that inequality (19.9) holds ultimately as $\lambda \to +\infty$ for every neighborhood $U_I(x_0)$ of x_0.

It now remains only to write

$$F(\lambda) = F_I(\lambda) = F_{U_I(x_0)}(\lambda) + F_{I\setminus U(x_0)}(\lambda),$$

and, citing estimates (19.9) and (19.10), conclude that (19.8) holds.   □

Thus it is now established that with a relative error of the order $O(\lambda^{-\infty})$ as $\lambda \to +\infty$ when estimating the asymptotic behavior of the Laplace integral, one can replace it by the integral over an arbitrarily small neighborhood $U_I(x_0)$ of the point x_0 where the absolute maximum of $S(x)$ occurs on the interval I of integration.

19.2.3 Canonical Integrals and Their Asymptotics

Lemma 3 (Canonical form of the function in the neighborhood of a critical point) *If the real-valued function $S(x)$ has smoothness $C^{(n+k)}$ in a neighborhood (or one-sided neighborhood) of a point $x_0 \in \mathbb{R}$, and*

$$S'(x_0) = \cdots = S^{(n-1)}(x_0) = 0, \qquad S^{(n)}(x_0) \neq 0,$$

and $k \in \mathbb{N}$ or $k = \infty$, then there exist neighborhoods (or one-sided neighborhoods) I_x of x_0 and I_y of 0 in $\mathbb{R}$ and a diffeomorphism $\varphi \in C^{(k)}(I_y, I_x)$ such that

$$S\big(\varphi(y)\big) = S(x_0) + sy^n, \quad \text{when } y \in I_y \text{ and } s = \operatorname{sgn} S^{(n)}(x_0).$$

Here

$$\varphi(0) = x_0 \quad and \quad \varphi'(0) = \left(\frac{n!}{|S^{(n)}(x_0)|} \right)^{1/n}.$$

Proof Using Taylor's formula with the integral form of the remainder,

$$S(x) = S(x_0) + \frac{(x-x_0)^n}{(n-1)!} \int_0^1 S^{(n)}\big(x_0 + t(x-x_0)\big)(1-t)^{n-1}\, dt,$$

we represent the difference $S(x) - S(x_0)$ in the form

$$S(x) - S(x_0) = (x-x_0)^n r(x),$$

where the function

$$r(x) = \frac{1}{(n-1)!} \int_0^1 S^{(n)}\big(x_0 + t(x-x_0)\big)(1-t)^{n-1}\, dt,$$

by virtue of the theorem on differentiation of an integral with respect to the parameter x, belongs to class $C^{(k)}$, and $r(x_0) = \frac{1}{n!} S^{(n)}(x_0) \neq 0$. Hence the function $y = \psi(x) = (x-x_0)\sqrt[n]{|r(x)|}$ also belongs to $C^{(k)}$ in some neighborhood (or one-sided neighborhood) I_x of x_0 and is even monotonic, since

$$\psi'(x_0) = \sqrt[n]{|r(x_0)|} = \left(\frac{|S^{(n)}(x_0)|}{n!} \right)^{1/n} \neq 0.$$

In this case the function ψ on I_x has an inverse $\psi^{-1} = \varphi$ defined on the interval $I_y = \psi(I_x)$ containing the point $0 = \psi(x_0)$. Here $\varphi \in C^{(k)}(I_y, I_x)$.

Further, $\varphi'(0) = (\psi'(x_0))^{-1} = (\frac{n!}{|S^{(n)}(x_0)|})^{1/n}$. Finally, by construction $S(\varphi(y)) = S(x_0) + sy^n$, where $s = \operatorname{sgn} r(x_0) = \operatorname{sgn} S^{(n)}(x_0)$.   □

Remark 1 The cases $n = 1$ or $n = 2$ and $k = 1$ or $k = \infty$ are usually the ones of most interest.

Proposition 2 (Reduction) *Suppose the interval of integration $I = [a, b]$ in the integral (19.1) is finite and the following conditions hold:*

a) $f, S \in C(I, \mathbb{R})$;
b) $\max_{x \in I} S(x)$ *is attained only at the one point $x_0 \in I$;*
c) $S \in C^{(n)}(U_I(x_0), \mathbb{R})$ *in some neighborhood $U_I(x_0)$ of x_0 (inside the interval I);*
d) $S^{(n)}(x_0) \neq 0$ *and if $1 < n$, then $S^{(1)}(x_0) = \cdots = S^{(n-1)}(x_0) = 0$.*

Then as $\lambda \to +\infty$ the integral (19.1) can be replaced by an integral of the form

$$R(\lambda) = e^{\lambda S(x_0)} \int_{I_y} r(y) e^{-\lambda y^n}\, dy$$

with a relative error defined by the localization principle (19.8), *where* $I_y = [-\varepsilon, \varepsilon]$ *or* $I_y = [0, \varepsilon]$, ε *is an arbitrarily small positive number, and the function r has the same degree of smoothness* I_y *that f has in a neighborhood of* x_0.

Proof Using the localization principle, we replace the integral (19.1) with the integral over a neighborhood $I_x = U_I(x_0)$ of x_0 in which the hypotheses of Lemma 3 hold. Making the change of variable $x = \varphi(y)$, we obtain

$$\int_{I_x} f(x) e^{\lambda S(x)} \, dx = \left(\int_{I_y} f(\varphi(y)) \varphi'(y) e^{-\lambda y^n} \, dy \right) e^{\lambda S(x_0)}. \tag{19.11}$$

The negative sign in the exponent $(-\lambda y^n)$ comes from the fact that by hypothesis $x_0 = \varphi(0)$ is a maximum.   □

The asymptotic behavior of the canonical integrals to which the Laplace integral (19.1) reduces in the main cases is given by the following lemma.

Lemma 4 (Watson[3]) *Let* $\alpha > 0$, $\beta > 0$, $0 < a \leq \infty$, *and* $f \in C([0, a], \mathbb{R})$. *Then with respect to the asymptotics of the integral*

$$W(\lambda) = \int_0^a x^{\beta-1} f(x) e^{-\lambda x^\alpha} \, dx \tag{19.12}$$

as $\lambda \to +\infty$, *the following assertions hold*:

a) *The main term of the asymptotics of* (19.12) *has the form*

$$W(\lambda) = \frac{1}{\alpha} f(0) \Gamma(\beta/\alpha) \lambda^{-\frac{\beta}{\alpha}} + O(\lambda^{-\frac{\beta+1}{\alpha}}), \tag{19.13}$$

if it is known that $f(x) = f(0) + O(x)$ *as* $\to 0$.
   b) *If* $f(x) = a_0 + a_1 x + \cdots + a_n x^n + O(x^{n+1})$ *as* $x \to 0$, *then*

$$W(\lambda) = \frac{1}{\alpha} \sum_{k=0}^n a_k \Gamma\left(\frac{k+\beta}{\alpha}\right) \lambda^{-\frac{k+\beta}{\alpha}} + O(\lambda^{-\frac{n+\beta+1}{\alpha}}). \tag{19.14}$$

c) *If f is infinitely differentiable at* $x = 0$, *then the following asymptotic expansion holds*:

$$W(\lambda) \simeq \frac{1}{\alpha} \sum_{k=0}^\infty \frac{f^{(k)}(0)}{k!} \Gamma\left(\frac{k+\beta}{\alpha}\right) \lambda^{-\frac{k+\beta}{\alpha}}, \tag{19.15}$$

which can be differentiated any number of times with respect to λ.

---

[3]G.H. Watson (1886–1965) – British mathematician.

Proof We represent the integral (19.12) as a sum of integrals over the interval $]0, \varepsilon]$ and $[\varepsilon, a[$, where ε is an arbitrarily small positive number.

By Lemma 1

$$\left| \int_\varepsilon^a x^{\beta-1} f(x) e^{-\lambda x^\alpha} \, dx \right| \le A e^{-\lambda \varepsilon^\alpha} = O\left(\lambda^{-\infty}\right) \quad \text{as } \lambda \to +\infty,$$

and therefore

$$W(\lambda) = \int_0^\varepsilon x^{\beta-1} f(x) e^{-\lambda x^\alpha} \, dx + O\left(\lambda^{-\infty}\right) \quad \text{as } \lambda \to +\infty.$$

In case b) we have $f(x) = \sum_{k=0}^n a_k x^k + r_n(x)$, where $r_n \in C[0, \varepsilon]$ and $|r_n(x)| \le C x^{n+1}$ on the interval $[0, \varepsilon]$. Hence

$$W(\lambda) = \sum_{k=0}^n a_k \int_0^\varepsilon x^{k+\beta-1} e^{-\lambda x^\alpha} \, dx + c(\lambda) \int_0^\varepsilon x^{n+\beta} e^{-\lambda x^\alpha} \, dx + o\left(\lambda^{-\infty}\right),$$

where $c(\lambda)$ is bounded as $\lambda \to +\infty$.

By Lemma 1, as $\lambda \to +\infty$,

$$\int_0^\varepsilon x^{k+\beta-1} e^{-\lambda x^\alpha} \, dx = \int_0^{+\infty} x^{k+\beta-1} e^{-\lambda x^\alpha} \, dx + O\left(\lambda^{-\infty}\right).$$

But

$$\int_0^{+\infty} x^{k+\beta-1} e^{-\lambda x^\alpha} \, dx = \frac{1}{\alpha} \Gamma\left(\frac{k+\beta}{\alpha}\right) \lambda^{-\frac{k+\beta}{\alpha}},$$

from which formula (19.14) and the special case of it, formula (19.13), now follow.

The expansion (19.15) now follows from (19.14) and Taylor's formula.

The possibility of differentiating (19.15) with respect to λ follows from the fact that the derivative of the integral (19.12) with respect to the parameter λ is an integral of the same type as (19.12) and for $W'(\lambda)$ one can use formula (19.15) to present explicitly an asymptotic expansion as $\lambda \to +\infty$ that is the same as the one obtained by formal differentiation of the original expansion (19.15). □

Example 6 Consider the Laplace transform

$$F(\lambda) = \int_0^{+\infty} f(x) e^{-\lambda x} \, dx,$$

which we have already encountered in Example 1. If this integral converges absolutely for some value $\lambda = \lambda_0$ and the function f is infinitely differentiable at $x = 0$, then by formula (19.15) we find that

$$F(\lambda) \simeq \sum_{k=0}^\infty f^{(k)}(0) \lambda^{-(k+1)} \quad \text{as } \lambda \to +\infty.$$

19.2.4 The Principal Term of the Asymptotics of a Laplace Integral

Theorem 1 (A typical principal term of the asymptotics) *Suppose the interval of integration $I = [a, b]$ in the integral (19.1) is finite, $f, S \in C(I, \mathbb{R})$, and $\max_{x \in I} S(x)$ is attained only at one point $x_0 \in I$.*

Suppose it is also known that $f(x_0) \neq 0$, $f(x) = f(x_0) + O(x - x_0)$ for $I \ni x \to x_0$, and the function S belongs to $C^{(k)}$ in a neighborhood of x_0.

The following statements hold.

a) *If $x_0 = a$, $k = 2$, and $S'(x_0) \neq 0$ (that is, $S'(x_0) < 0$), then*

$$F(\lambda) = \frac{f(x_0)}{-S'(x_0)} e^{\lambda S(x_0)} \lambda^{-1} \left[1 + O(\lambda^{-1}) \right] \quad as \; \lambda \to +\infty; \qquad (19.2')$$

b) *if $a < x_0 < b$, $k = 3$, and $S''(x_0) \neq 0$ (that is, $S''(x_0) < 0$), then*

$$F(\lambda) = \sqrt{\frac{2\pi}{-S''(x_0)}} f(x_0) e^{\lambda S(x_0)} \lambda^{-1/2} \left[1 + O(\lambda^{-1/2}) \right] \quad as \; \lambda \to +\infty; \quad (19.3')$$

c) *if $x_0 = a$, $k = 3$, $S'(a) = 0$, and $S''(a) \neq 0$ (that is, $S''(a) < 0$), then*

$$F(\lambda) = \sqrt{\frac{\pi}{-2S''(x_0)}} f(x_0) e^{\lambda S(x_0)} \lambda^{-1/2} \left[1 + O(\lambda^{-1/2}) \right] \quad as \; \lambda \to +\infty. \quad (19.4')$$

Proof Using the localization principle and making the change of variable $x = \varphi(y)$ shown in Lemma 3, according to the reduction in Proposition 2, we arrive at the following relations:

a) $F(\lambda) = e^{\lambda S(x_0)} \left(\displaystyle\int_0^\varepsilon (f \circ \varphi)(y) \varphi'(y) e^{-\lambda y} \, dy + O(\lambda^{-\infty}) \right);$

b) $F(\lambda) = e^{\lambda S(x_0)} \left(\displaystyle\int_{-\varepsilon}^\varepsilon (f \circ \varphi)(y) \varphi'(y) e^{-\lambda y^2} \, dy + O(\lambda^{-\infty}) \right) =$

$\qquad = e^{\lambda S(x_0)} \left(\displaystyle\int_0^\varepsilon \big((f \circ \varphi)(y) \varphi'(y) + (f \circ \varphi)(-y) \varphi'(-y) \big) e^{-\lambda y^2} \, dy + \right.$

$\qquad \qquad \left. + O(\lambda^{-\infty}) \right);$

c) $F(\lambda) = e^{\lambda S(x_0)} \left(\displaystyle\int_0^\varepsilon (f \circ \varphi)(y) \varphi'(y) e^{-\lambda y^2} \, dy + O(\lambda^{-\infty}) \right).$

Under the requirements stated above, the function $(f \circ \varphi)\varphi'$ satisfies all the hypotheses of Watson's lemma. It now remains only to apply Watson's lemma (formula (19.14) for $n = 0$) and to recall the expressions for $\varphi(0)$ and $\varphi'(0)$ indicated in Lemma 3. $\qquad \square$

Thus we have justified formula (19.2)–(19.4) together with the remarkably simple, clear and effective recipe that led us to these formulas in Sect. 19.1.

Now let us consider some examples of the application of this theorem.

Example 7 The asymptotics of the gamma function. The function

$$\Gamma(\lambda+1) = \int_0^{+\infty} t^\lambda e^{-t}\, dt \quad (\lambda > -1)$$

can be represented as a Laplace integral

$$\Gamma(\lambda+1) = \int_0^{+\infty} e^{-t} e^{\lambda \ln t}\, dt,$$

and if for $\lambda > 0$ we make the change of variable $t = \lambda x$, we arrive at the integral

$$\Gamma(\lambda+1) = \lambda^{\lambda+1} \int_0^{+\infty} e^{-\lambda(x-\ln x)}\, dx,$$

which can be studied using the methods of the theorem.

The function $S(x) = \ln x - x$ has a unique maximum $x = 1$ on the interval $]0, +\infty[$, and $S''(1) = -1$. By the localization principle (Proposition 1) and assertion b) of Theorem 1, we conclude that

$$\Gamma(\lambda+1) = \sqrt{2\pi\lambda}\left(\frac{\lambda}{e}\right)^\lambda \left[1 + O\left(\lambda^{-1/2}\right)\right] \quad \text{as } \lambda \to +\infty.$$

In particular, recalling that $\Gamma(n+1) = n!$ for $n \in \mathbb{N}$, we obtain the classical *Stirling's formula*[4]

$$n! = \sqrt{2\pi n}(n/e)^n\left[1 + O\left(n^{-1/2}\right)\right] \quad \text{as } n \to \infty,\ n \in \mathbb{N}.$$

Example 8 The asymptotics of the Bessel function

$$I_n(x) = \frac{1}{\pi}\int_0^\pi e^{x\cos\theta}\cos n\theta\, d\theta,$$

where $n \in \mathbb{N}$. Here $f(\theta) = \cos n\theta$, $S(\theta) = \cos\theta$, $\max_{0 \le x \le \pi} S(\theta) = S(0) = 1$, $S'(0) = 0$, and $S''(0) = -1$, so that by assertion c) of Theorem 1

$$I_n(x) = \frac{e^x}{\sqrt{2\pi x}}\left[1 + O\left(x^{-1/2}\right)\right] \quad \text{as } x \to +\infty.$$

Example 9 Let $f \in C^{(1)}([a, b], \mathbb{R})$, $S \in C^{(2)}([a, b], \mathbb{R})$, with $S(x) > 0$ on $[a, b]$, and $\max_{a \le x \le b} S(x)$ is attained only at the one point $x_0 \in [a, b]$. If $f(x_0) \ne 0$, $S'(x_0) = 0$, and $S''(x_0) \ne 0$, then, rewriting the integral

$$\mathcal{F}(\lambda) = \int_a^b f(x)\left[S(x)\right]^\lambda dx$$

---

[4] See also Problem 10 of Sect. 7.3.

in the form of a Laplace integral

$$\mathcal{F}(\lambda) = \int_a^b f(x) e^{\lambda \ln S(x)} \, dx,$$

on the basis of assertions b) and c) of Theorem 1, we find that as $\lambda \to +\infty$

$$\mathcal{F}(\lambda) = \varepsilon f(x_0) \sqrt{\frac{2\pi}{-S''(x_0)}} \left[S(x_0) \right]^{\lambda+1/2} \lambda^{-1/2} \left[1 + O\left(\lambda^{-1/2} \right) \right],$$

where $\varepsilon = 1$ if $a < x_0 < b$ and $\varepsilon = 1/2$ if $x_0 = a$ or $x_0 = b$.

Example 10 The *asymptotics of the Legendre polynomials*

$$P_n(x) = \frac{1}{\pi} \int_0^\pi \left(x + \sqrt{x^2 - 1} \cos\theta \right)^n \, d\theta$$

in the domain $x > 1$ as $n \to \infty$, $n \in \mathbb{N}$, can be obtained as a special case of the preceding example when $f \equiv 1$,

$$S(\theta) = x + \sqrt{x^2 - 1} \cos\theta, \qquad \max_{0 \le \theta \le \pi} S(\theta) = S(0) = x + \sqrt{x^2 - 1},$$

$$S'(0) = 0, \qquad S''(0) = -\sqrt{x^2 - 1}.$$

Thus,

$$P_n(x) = \frac{(x + \sqrt{x^2 - 1})^{n+1/2}}{\sqrt{2\pi n} \sqrt[4]{x^2 - 1}} \left[1 + O\left(n^{-1/2} \right) \right] \quad \text{as } n \to +\infty, \; n \in \mathbb{N}.$$

19.2.5 *Asymptotic Expansions of Laplace Integrals*

Theorem 1 gives only the principal terms of the characteristic asymptotics of a Laplace integral (19.1) and even that under the condition that $f(x_0) \neq 0$. On the whole this is of course a typical situation, and for that reason Theorem 1 is undoubtedly a valuable result. However, Watson's lemma shows that the asymptotics of a Laplace integral can sometimes be brought to an asymptotic expansion. Such a possibility is especially important when $f(x_0) = 0$ and Theorem 1 gives no result.

It is naturally impossible to get rid of the hypothesis $f(x_0) \neq 0$ completely, without replacing it with anything, while remaining within the limits of Laplace's method: after all, if $f(x) \equiv 0$ in a neighborhood of a maximum x_0 of the function $S(x)$ or if $f(x)$ tends to zero very rapidly as $x \to x_0$, then x_0 may not be responsible for the asymptotics of the integral. Now that we have arrived at a certain type of asymptotic sequence $\{ e^{\lambda c} \lambda^{-p_k} \}$ $(p_0 < p_1 < \cdots)$ as $\lambda \to +\infty$ as a result of the considerations we have studied, we can speak of an asymptotic zero in relation to such a sequence and, without assuming that $f(x_0) \neq 0$, we can state the localization principle as follows: *Up to an asymptotic zero with respect to the asymptotic sequence* $\{e^{\lambda S(x_0)}\lambda^{-p_k}\}$ ($p_0 < p_1 < \cdots$) *the asymptotic behavior of the Laplace integral* (19.1) *as* $\lambda \to +\infty$ *equals the asymptotics of the portion of this integral taken over an arbitrarily small neighborhood of the point* x_0, *provided that this point is the unique maximum of the function* $S(x)$ *on the interval of integration.*

However, we shall not go back and re-examine these questions in order to sharpen them. Rather, assuming f and S belong to $C^{(\infty)}$, we shall give a derivation of the corresponding asymptotic expansions using Lemma 1 on the exponential estimate, Lemma 3 on the change of variable, and Watson's lemma (Lemma 4).

Theorem 2 (Asymptotic expansion) *Let* $I = [a, b]$ *be a finite interval,* $f, S \in C(I, \mathbb{R})$, *and assume* $\max_{x \in I} S(x)$ *is attained only at the point* $x_0 \in I$ *and* $f, S \in C^{(\infty)}(U_I(x_0), \mathbb{R})$ *in some neighborhood* $U_I(x_0)$ *of* x_0. *Then in relation to the asymptotics of the integral* (19.1) *the following assertions hold.*

a) *If* $x_0 = a$, $S^{(m)}(a) \neq 0$, $S^{(j)}(a) = 0$ *for* $1 \leq j < m$, *then*

$$F(\lambda) \simeq \lambda^{-1/m} e^{\lambda S(a)} \sum_{k=0}^{\infty} a_k \lambda^{-k/m} \quad \text{as } \lambda \to +\infty, \qquad (19.16)$$

where

$$a_k = \frac{(-1)^{k+1} m^k}{k!} \Gamma\left(\frac{k+1}{m}\right) \left(h(x, a)\frac{d}{dx}\right)^k (f(x)h(x, a))\Big|_{x=a},$$

$$h(x, a) = (S(a) - S(x))^{1-1/m} / S'(x).$$

b) *If* $a < x_0 < b$, $S^{(2m)}(x_0) \neq 0$, *and* $S^{(j)}(x_0) = 0$ *for* $1 \leq j < 2m$, *then*

$$F(\lambda) \simeq \lambda^{-1/2m} e^{\lambda S(x_0)} \sum_{k=0}^{\infty} c_k \lambda^{-k/m} \quad \text{as } \lambda \to +\infty, \qquad (19.17)$$

where

$$c_k = 2\frac{(-1)^{2k+1}(2m)^{2k}}{(2k)!} \Gamma\left(\frac{2k+1}{2m}\right) \left(h(x, x_0)\frac{d}{dx}\right)^{2k} (f(x)h(x, x_0))\Big|_{x=x_0},$$

$$h(x, x_0) = (S(x_0) - S(x))^{1-\frac{1}{2m}} / S'(x).$$

c) *If* $f^{(n)}(x_0) \neq 0$ *and* $f(x) \sim \frac{1}{n!} f^{(n)}(x_0)(x - x_0)^n$ *as* $x \to x_0$, *then the main term of the asymptotics in the cases a) and b) respectively has the form*

$$F(\lambda) = \frac{1}{m} \lambda^{-\frac{n+1}{m}} e^{\lambda S(a)} \Gamma\left(\frac{n+1}{m}\right) \left(\frac{m!}{|S^m(a)|}\right)^{\frac{n+1}{m}} \times$$

$$\times \left[\frac{1}{n!} f^{(n)}(a) + O\left(\lambda^{-\frac{n+1}{m}}\right)\right], \qquad (19.18)$$

$$F(\lambda) = \frac{1}{m}\lambda^{-\frac{n+1}{2m}} e^{\lambda S(x_0)} \Gamma\left(\frac{n+1}{2m}\right)\left(\frac{(2m)!}{|S^{2m}(x_0)|}\right)^{\frac{n+1}{2m}} \times$$

$$\times \left[\frac{1}{n!} f^{(n)}(x_0) + O\left(\lambda^{-\frac{n+1}{2m}}\right)\right]. \tag{19.19}$$

d) *The expansions* (19.16) *and* (19.17) *can be differentiated with respect to* λ *any number of times.*

Proof It follows from Lemma 1 that under these hypotheses the integral (19.1) can be replaced by an integral over an arbitrarily small neighborhood of x_0 up to a quantity of the form $e^{\lambda S(x_0)} O(\lambda^{-\infty})$ as $\lambda \to \infty$.

Making the change of variable $x = \varphi(y)$ from Lemma 3 in such a neighborhood, we bring the last integral into the form

$$e^{-\lambda S(x_0)} \int_{I_y} (f \circ \varphi)(y)\varphi'(y)e^{-\lambda y^\alpha}\, dy, \tag{19.20}$$

where $I_y = [0, \varepsilon]$, $\alpha = m$ if $x_0 = a$, and $I_y = [-\varepsilon, \varepsilon]$, $\alpha = 2m$ if $a < x_0 < b$.

The neighborhood in which the change of variable $x = \varphi(y)$ took place can be assumed so small that both functions f and S are infinitely differentiable in it. Then the resulting integrand $(f \circ \varphi)(y)\varphi'(y)$ in the integral (19.20) can also be assumed infinitely differentiable.

If $I_y = [0, \varepsilon]$, that is, when $x_0 = a$, Watson's lemma is immediately applicable to the integral (19.20) and the existence of the expansion (19.16) is thereby proved.

If $I_y = [-\varepsilon, \varepsilon]$, that is, in the case $a < x_0 < b$, we bring the integral (19.20) into the form

$$e^{\lambda S(x_0)} \int_0^\varepsilon \left[(f \circ \varphi)(y)\varphi'(y) + (f \circ \varphi)(-y)\varphi'(-y)\right]e^{-\lambda y^{2m}}\, dy, \tag{19.21}$$

and, once again applying Watson's lemma, we obtain the expansion (19.17).

The possibility of differentiating the expansions (19.16) and (19.17) follows from the fact that under our assumptions the integral (19.1) can be differentiated with respect to λ to yield a new integral satisfying the hypotheses of the theorem. We write out the expansions (19.16) and (19.17) for it, and we can verify immediately that these expansions really are the same as those obtained by formal differentiation of the expansions (19.16) and (19.17) for the original integrals.

We now take up the formulas for the coefficients a_k and c_k. By Watson's lemma $a_k = \frac{1}{k!m}\frac{d^k\Phi}{dy^k}(0)\Gamma(\frac{k+1}{m})$, where $\Phi(y) = (f \circ \varphi)(y)\varphi'(y)$.

However, taking account of the relations

$$S(\varphi(y)) - S(a) = -y^m,$$

$$S'(x)\varphi'(y) = -my^{m-1},$$

$$\varphi'(y) = -m(S(a) - S(x))^{1-\frac{1}{m}}/S'(x),$$

$$\frac{\mathrm{d}}{\mathrm{d}y} = \varphi'(y)\frac{\mathrm{d}}{\mathrm{d}x},$$

$$\Phi(y) = f(x)\varphi'(y),$$

we obtain

$$\frac{\mathrm{d}^k \Phi}{\mathrm{d}y^k}(0) = (-m)^{k+1}\left(h(x,a)\frac{\mathrm{d}}{\mathrm{d}x}\right)^k \left(f(x)h(x,a)\right)\big|_{x=a},$$

where $h(x,a) = (S(a) - S(x))^{1-\frac{1}{m}}/S'(x)$.

Formulas for the coefficients c_k can be obtained similarly by applying Watson's lemma to the integral (19.21).

Setting $\psi(y) = f(\varphi(y))\varphi'(y) + f(\varphi(-y))\varphi'(-y)$, we can write, as $\lambda \to +\infty$,

$$\int_0^\varepsilon \psi(y)e^{-\lambda y^{2m}}\,\mathrm{d}y \simeq \frac{1}{2m}\sum_{n=0}^\infty \frac{\psi^{(n)}(0)}{n!}\Gamma\left(\frac{n+1}{2m}\right)\lambda^{-\frac{n+1}{2m}}.$$

But, $\psi^{(2k+1)}(0) = 0$ since $\psi(y)$ is an even function; therefore this last asymptotic expansion can be rewritten as

$$\int_0^\varepsilon \psi(y)e^{-\lambda y^{2m}}\,\mathrm{d}y \simeq \frac{1}{2m}\sum_{k=0}^\infty \frac{\psi^{(2k)}(0)}{(2k)!}\Gamma\left(\frac{2k+1}{2m}\right)\lambda^{-\frac{2k+1}{2m}}.$$

It remains only to note that $\psi^{(2k)}(0) = 2\Phi^{(2k)}(0)$, where $\Phi(y) = f(\varphi(y))\varphi'(y)$. The formula for c_k can now be obtained from the already established formula for a_k by replacing k with $2k$ and doubling the result of the substitution.

To obtain the principal terms (19.18) and (19.19) in the asymptotic expansions (19.16) and (19.17) under the condition $f(x) = \frac{1}{n!}f^{(n)}(x_0)(x - x_0)^n + O((x - x_0)^{n+1})$ indicated in c), where $f^{(n)}(x_0) \neq 0$, it suffices to recall that $x = \varphi(y)$, $x_0 = \varphi(0)$, $x - x_0 = \varphi'(0)y + O(y^2)$, that is,

$$(f \circ \varphi)(y) = y^n\left(\frac{f^{(n)}(x_0)}{n!}(\varphi'(0))^n + O(y)\right)$$

and

$$(f \circ \varphi)(y)\varphi'(y) = y^n\left(\frac{f^{(n)}(x_0)}{n!}(\varphi'(0))^{n+1} + O(y)\right)$$

as $y \to 0$, since $\varphi'(0) = (\frac{m!}{|S^{(m)}(a)|})^{1/m} \neq 0$ if $x_0 = a$ and $\varphi'(0) = (\frac{(2m)!}{|S^{(2m)}(x_0)|})^{1/2m} \neq 0$ if $a < x_0 < b$.

It now remains only to substitute these expressions into the integrals (19.20) and (19.21) respectively and use formula (19.13) from Watson's lemma. $\square$

Remark 2 We again get formula (19.2′) from formula (19.18) when $n = 0$ and $m = 1$.

Similarly, when $n = 0$ and $m = 1$ formula (19.19) yields (19.3′) again.
Finally, Eq. (19.4′) comes from Eq. (19.18) with $n = 0$ and $m = 2$.
All of this, of course, assumes the hypotheses of Theorem 2.

Remark 3 Theorem 2 applies to the case where the function $S(x)$ has a unique maximum on the interval $I = [a, b]$. If there are several such points $x_1, \ldots, x_n$, the integral (19.1) is partitioned into a sum of such integrals, each of whose asymptotics is described by Theorem 2. That is, in this case the asymptotic behavior is obtained as the sum $\sum_{j=1}^{n} F(\lambda, x_j)$ of the contributions of these maximum points.

It is easy to see that when this happens, some or even all of the terms may cancel one another.

Example 11 If $S \in C^{(\infty)}(\mathbb{R}, \mathbb{R})$ and $S(x) \to -\infty$ as $x \to \infty$, then

$$F(\lambda) = \int_{-\infty}^{\infty} S'(x) e^{\lambda S(x)} \, dx \equiv 0 \quad \text{for } \lambda > 0.$$

Hence, in this case such an interference of the contributions must necessarily occur. From the formal point of view this example may seem unconvincing, since previously we had been considering the case of a finite interval of integration. However, those doubts are removed by the following important remark.

Remark 4 To simplify what were already very cumbersome statements in Theorems 1 and 2 we assumed that the interval of integration I was finite and that the integral (19.1) was a proper integral. In fact, however, if the inequality $\sup_{I \setminus U(x_0)} S(x) < S(x_0)$ holds outside an interval $U(x_0)$ of the maximum point $x_0 \in I$, then Lemma 1 enables us to conclude that the integrals over intervals strictly outside of $U(x_0)$ are exponentially small in comparison with $e^{\lambda S(x_0)}$ as $\lambda \to +\infty$ (naturally, under the assumption that the integral (19.1) converges absolutely for at least one value $\lambda = \lambda_0$).

Thus both Theorem 1 and Theorem 2 are also applicable to improper integrals if the conditions just mentioned are met.

Remark 5 Due to their cumbersome nature, the formulas for the coefficients obtained in Theorem 2 can normally be used only for obtaining the first few terms of the asymptotics needed in specific computations. It is extremely rare that one can obtain the general form of the asymptotic expansion of even a simpler function than appears in Theorem 2 from these formulas for the coefficients a_k and c_k. Nevertheless, such situations do arise. To clarify the formulas themselves, let us consider the following examples.

Example 12 The asymptotic behavior of the function

$$\text{Erf}(x) = \int_{x}^{+\infty} e^{-u^2} \, du$$

as $x \to +\infty$ is easy to obtain through integration by parts:

$$\mathrm{Erf}(x) = \frac{e^{-x^2}}{2x} - \frac{1}{2} \int_x^{+\infty} u^{-2} e^{-u^2} \, du = \frac{e^{-x^2}}{2x} - \frac{3 e^{-x^2}}{2^2 x^3} + \int_x^{+\infty} u^{-4} e^{-u^2} \, du = \cdots ,$$

from which, after obvious estimates, it follows that

$$\mathrm{Erf}(x) \simeq \frac{e^{-x^2}}{2x} \sum_{k=0}^{\infty} \frac{(-1)^k (2k-1)!!}{2^k} x^{-2k} \quad \text{as } x \to +\infty. \tag{19.22}$$

Let us now obtain this expansion from Theorem 2.

By the change of variable $u = xt$ we arrive at the representation

$$\mathrm{Erf}(x) = x \int_1^{+\infty} e^{-x^2 t^2} \, dt.$$

Setting $\lambda = x^2$ here and denoting the variable of integration, as in Theorem 2, by x, we reduce the problem to finding the asymptotic behavior of the integral

$$F(\lambda) = \int_1^\infty e^{-\lambda x^2} \, dx, \tag{19.23}$$

since $\mathrm{Erf}(x) = x F(x^2)$.

When Remark 4 is taken into account, the integral (19.23) satisfies the hypotheses of Theorem 2: $S(x) = -x^2$, $S'(x) = -2x < 0$ for $1 \le x < +\infty$, $S'(1) = -2$, $S(1) = -1$.

Thus, $x_0 = a = 1$, $m = 1$, $f(x) \equiv 1$, $h(x, a) = \frac{1}{-2x}$, $h(x, a)\frac{d}{dx} = \frac{1}{-2x}\frac{d}{dx}$.

Hence,

$$\left(\frac{1}{-2x} \frac{d}{dx} \right)^0 \left(-\frac{1}{2x} \right) = -\frac{1}{2x} = \left(-\frac{1}{2} \right) x^{-1},$$

$$\left(\frac{1}{-2x} \frac{d}{dx} \right)^1 \left(-\frac{1}{2x} \right) = -\frac{1}{2x} \frac{d}{dx} \left(-\frac{1}{2x} \right) = \left(-\frac{1}{2} \right)^2 (-1) x^{-3},$$

$$\left(\frac{1}{-2x} \frac{d}{dx} \right)^2 \left(-\frac{1}{2x} \right) = \left(-\frac{1}{2x} \frac{d}{dx} \right)^1 \left(\left(-\frac{1}{2} \right)^2 (-1) x^{-3} \right) =$$

$$= \left(-\frac{1}{2} \right)^3 (-1)(-3) x^{-5},$$

$$\vdots$$

$$\left(\frac{1}{-2x} \frac{d}{dx} \right)^k \left(-\frac{1}{2x} \right) = -\frac{(2k-1)!!}{2^{k+1}} x^{-(2k+1)}.$$

Setting $x = 1$, we find that

$$a_k = \frac{(-1)^{k+1}}{k!} \Gamma(k+1) \left(-\frac{(2k-1)!!}{2^{k+1}} \right) = (-1)^k \frac{(2k-1)!!}{2^{k+1}}.$$

Now writing out the asymptotic expansion (19.16) for the integral (19.23) taking account of the relations $\mathrm{Erf}(x) = x F(x^2)$, we obtain the expansion (19.22) for the function $\mathrm{Erf}(x)$ as $x \to +\infty$.

Example 13 In Example 7, starting from the representation

$$\Gamma(\lambda+1) = \lambda^{\lambda+1} \int_0^{+\infty} e^{-\lambda(x-\ln x)} \, dx, \tag{19.24}$$

we obtained the principal term of the asymptotics of the function $\Gamma(\lambda+1)$ as $\lambda \to +\infty$. Let us now sharpen the formula obtained earlier, using Theorem 2b).

To simplify the notation a bit, let us replace x by $x + 1$ in the integral (19.24). We then find that

$$\Gamma(\lambda+1) = \lambda^{\lambda+1} e^{-\lambda} \int_{-1}^{+\infty} e^{\lambda(\ln(1+x)-x)} \, dx$$

and the question reduces to studying the asymptotics of the integral

$$F(\lambda) = \int_{-1}^{+\infty} e^{\lambda(\ln(1+x)-x)} \, dx \tag{19.25}$$

as $\lambda \to +\infty$. Here $S(x) = \ln(1+x) - x$, $S'(x) = \frac{1}{1+x} - 1$, $S'(0) = 0$, that is, $x_0 = 0$, $S''(x) = -\frac{1}{(1+x)^2}$, $S''(0) = -1 \neq 0$. That is, taking account of Remark 4, we see that the hypotheses b) of Theorem 2 are satisfied, where we must also set $f(x) \equiv 1$ and $m = 1$, since $S''(0) \neq 0$.

In this case the function $h(x, x_0) = h(x)$ has the following form:

$$h(x) = -\frac{1+x}{x} \left(x - \ln(1+x) \right)^{1/2}.$$

If we wish to find the first two terms of the asymptotics, we need to compute the following at $x = 0$:

$$\left(h(x) \frac{d}{dx} \right)^0 \left(h(x) \right) = h(x),$$

$$\left(h(x) \frac{d}{dx} \right)^1 \left(h(x) \right) = h(x) \frac{dh}{dx}(x),$$

$$\left(h(x)\frac{d}{dx}\right)^2(h(x)) = \left(h(x)\frac{d}{dx}\right)\left(h(x)\frac{dh}{dx}(x)\right) =$$

$$= h(x)\left[\left(\frac{dh}{dx}\right)^2(x) + h(x)\frac{d^2h}{dx^2}(x)\right].$$

This computation, as one can see, is easily done if we find the values $h(0)$, $h'(0)$, $h''(0)$, which in turn can be obtained from the Taylor expansion of $h(x)$, $x \geq 0$ in a neighborhood of 0:

$$h(x) = -\frac{1+x}{x}\left[x - \left(x - \frac{1}{2}x^2 + \frac{1}{3}x^3 - \frac{1}{4}x^2 + O(x^5)\right)\right]^{1/2} =$$

$$= -\frac{1+x}{x}\left[\frac{1}{2}x^2 - \frac{1}{3}x^3 + \frac{1}{4}x^4 + O(x^5)\right]^{1/2} =$$

$$= -\frac{1+x}{\sqrt{2}}\left[1 - \frac{2}{3}x + \frac{2}{4}x^2 + O(x^3)\right]^{1/2} =$$

$$= -\frac{1+x}{\sqrt{2}}\left(1 - \frac{1}{3}x + \frac{7}{36}x^2 + O(x^3)\right) =$$

$$= -\frac{1}{\sqrt{2}} - \frac{\sqrt{2}}{3}x + \frac{5}{36\sqrt{2}}x^2 + O(x^3).$$

Thus, $h(0) = -\frac{1}{\sqrt{2}}$, $h'(0) = -\frac{\sqrt{2}}{3}$, $h''(0) = \frac{5}{18\sqrt{2}}$,

$$\left(h(x)\frac{d}{dx}\right)^0(h(x))\Big|_{x=0} = -\frac{1}{\sqrt{2}},$$

$$\left(h(x)\frac{d}{dx}\right)^1(h(x))\Big|_{x=0} = -\frac{1}{3},$$

$$\left(h(x)\frac{d}{dx}\right)^2(h(x))\Big|_{x=0} = -\frac{1}{12\sqrt{2}},$$

$$c_0 = -2\Gamma\left(\frac{1}{2}\right)\left(-\frac{1}{\sqrt{2}}\right) = \sqrt{2\pi},$$

$$c_1 = -2\frac{2^2}{2!}\Gamma\left(\frac{3}{2}\right)\left(-\frac{1}{12\sqrt{2}}\right) = 4 \cdot \frac{1}{2}\Gamma\left(\frac{1}{2}\right)\frac{1}{12\sqrt{2}} = \frac{\sqrt{2\pi}}{12}.$$

Hence, as $\lambda \to \infty$,

$$F(\lambda) = \sqrt{2\pi}\lambda^{-1/2}\left(1 + \frac{1}{12}\lambda^{-1} + O(\lambda^{-2})\right),$$

that is, as $\lambda \to +\infty$,

$$\Gamma(\lambda+1) = \sqrt{2\pi\lambda}\left(\frac{\lambda}{e}\right)^\lambda \left(1 + \frac{1}{12}\lambda^{-1} + O(\lambda^{-2})\right). \qquad (19.26)$$

It is useful to keep in mind that the asymptotic expansions (19.16) and (19.17) can also be found by following the proof of Theorem 2 without invoking the expressions for the coefficients shown in the statement of Theorem 2.

As an example, we once again obtain the asymptotics of the integral (19.25), but in a slightly different way.

Using the localization principle and making a change of variable $x = \varphi(y)$ in a neighborhood of zero such that $0 = \varphi(0)$, $S(\varphi(y)) = \ln(1 + \varphi(y)) - \varphi(y) = -y^2$, we reduce the problem to studying the asymptotics of the integral

$$\int_{-\varepsilon}^{\varepsilon} \varphi'(y) e^{-\lambda y^2}\, dy = \int_0^{\varepsilon} \psi(y) e^{-\lambda y^2}\, dy,$$

where $\psi(y) = \varphi'(y) + \varphi'(-y)$. The asymptotic expansion of this last integral can be obtained from Watson's lemma

$$\int_0^{\varepsilon} \psi(y) e^{-\lambda y^2}\, dy \simeq \frac{1}{2}\sum_{k=0}^{\infty} \frac{\psi^{(k)}(0)}{k!}\Gamma\left(\frac{k+1}{2}\right)\lambda^{-(k+1)/2} \qquad \text{as } \lambda \to +\infty,$$

which by the relations $\psi^{(2k+1)}(0) = 0$, $\psi^{(2k)}(0) = 2\varphi^{(2k+1)}(0)$ yields the asymptotic series

$$\sum_{k=0}^{\infty} \frac{\varphi^{(2k+1)}(0)}{(2k)!}\Gamma\left(k+\frac{1}{2}\right)\lambda^{-(k+1/2)} = \lambda^{-1/2}\Gamma\left(\frac{1}{2}\right)\sum_{k=0}^{\infty} \frac{\varphi^{(2k+1)}(0)}{k!2^{2k}}\lambda^{-k}.$$

Thus for the integral (19.25) we obtain the following asymptotic expansion

$$F(\lambda) \simeq \lambda^{-1/2}\sqrt{\pi}\sum_{k=0}^{\infty} \frac{\varphi^{(2k+1)}(0)}{k!2^{2k}}\lambda^{-k}, \qquad (19.27)$$

where $x = \varphi(y)$ is a smooth function such that $x - \ln(1+x) = y^2$ in a neighborhood of zero (for both x and y).

If we wish to know the first two terms of the asymptotics, we must put the specific values $\varphi'(0)$ and $\varphi^{(3)}(0)$ into formula (19.27).

It may be of some use to illustrate the following device for computing these values, which can be used generally to obtain the Taylor expansion of an inverse function from the expansion of the direct function.

Assuming $x > 0$ and $y > 0$, from the relation

$$x - \ln(1+x) = y^2$$

we obtain successively

$$\frac{1}{2}x^2\left(1-\frac{2}{3}x+\frac{1}{2}x^2+O(x^3)\right)=y^2,$$

$$x=\sqrt{2}y\left(1-\frac{2}{3}x+\frac{1}{2}x^2+O(x^3)\right)^{-1/2}=$$

$$=\sqrt{2}y\left(1+\frac{1}{3}x-\frac{1}{12}x^2+O(x^3)\right)=$$

$$=\sqrt{2}y+\frac{\sqrt{2}}{3}yx-\frac{\sqrt{2}}{12}yx^2+O(yx^3).$$

But $x\sim\sqrt{2}y$ as $y\to 0$ ($x\to 0$), and therefore, using the representation of x already found, one can continue this computation and find that as $y\to 0$

$$x=\sqrt{2}y+\frac{\sqrt{2}}{3}y\left(\sqrt{2}y+\frac{\sqrt{2}}{3}yx+O(y^3)\right)-\frac{\sqrt{2}}{12}y(\sqrt{2}y)^2+O(y^4)=$$

$$=\sqrt{2}y+\frac{2}{3}y^2+\frac{2}{9}y^2x-\frac{\sqrt{2}}{6}y^3+O(y^4)=$$

$$=\sqrt{2}y+\frac{2}{3}y^2+\frac{2}{9}y^2(\sqrt{2}y)-\frac{\sqrt{2}}{6}y^3+O(y^4)=$$

$$=\sqrt{2}y+\frac{2}{3}y^2+\frac{\sqrt{2}}{18}y^3+O(y^4).$$

Thus for the quantities $\varphi'(0)$ and $\varphi^{(3)}(0)$ of interest to us we find the following values: $\varphi'(0)=\sqrt{2}$, $\varphi^{(3)}(0)=\frac{\sqrt{2}}{3}$.

Substituting them into formula (19.27), we find that

$$F(\lambda)=\lambda^{-1/2}\sqrt{2\pi}\left(1+\frac{1}{12}\lambda^{-1}+O(\lambda^{-2})\right)\quad\text{as }\lambda\to+\infty,$$

from which we again obtain formula (19.26).

In conclusion we shall make two more remarks on the problems discussed in this section.

Remark 6 (Laplace's method in the multidimensional case) We note that Laplace's method can also be successfully applied in studying the asymptotics of multiple Laplace integrals

$$F(\lambda)=\int_X f(x)e^{\lambda S(x)}\,dx,$$

in which $x\in\mathbb{R}^n$, X is a domain in $\mathbb{R}^n$, and f and S are real-valued functions in X.

Lemma 1 on the exponential estimate holds for such integrals, and by this lemma the study of the asymptotics of such an integral reduces to studying the asymptotics of a part of it

$$\int_{U(x_0)} f(x)e^{\lambda S(x)}\,dx,$$

taken over a neighborhood of a maximum point x_0 of the function $S(x)$.

If this is a nondegenerate maximum, that is, $S''(x_0) \neq 0$, then by Morse's lemma (see Sect. 8.6 of Part 1) there exists a change of variable $x = \varphi(y)$ such that $S(x_0) - S(\varphi(y)) = |y|^2$, where $|y|^2 = (y^1)^2 + \cdots + (y^n)^2$. Thus the question reduces to the canonical integral

$$\int_I (f \circ \varphi)(y)\det\varphi'(y)e^{-\lambda|y|^2}\,dy,$$

which in the case of smooth functions f and S can be studied by applying Fubini's theorem and using Watson's lemma proved above (see Problems 8–11 in this connection).

Remark 7 (The stationary phase method) In a wider interpretation, Laplace's method, as we have already noted, consists of the following:

1^0 a certain localization principle (Lemma 1 on the exponential estimate),

2^0 a method of locally reducing an integral to canonical form (Morse's lemma), and

3^0 a description of the asymptotics of canonical integrals (Watson's lemma).

We have met the idea of localization previously in our study of approximate identities, and also in studying Fourier series and the Fourier transform (the Riemann–Lebesgue lemma, smoothness of a function and the rate at which its Fourier transform decreases, convergence of Fourier series and integrals).

Integrals of the form

$$\widetilde{F}(\lambda) = \int_X f(x)e^{i\lambda S(x)}\,dx,$$

where $x \in \mathbb{R}^n$, called *Fourier integrals*, occupy an important place in mathematics and its applications. A Fourier integral differs from a Laplace integral only in the modest factor i in the exponent. This leads, however, to the relation $|e^{i\lambda S(x)}| = 1$ when λ and $S(x)$ are real, and hence the idea of a dominant maximum is not applicable to the study of the asymptotics of a Fourier integral.

Let $X = [a, b] \subset \mathbb{R}^1$, $f \in C_0^{(\infty)}([a, b], \mathbb{R})$, (that is, f is of compact support on $[a, b]$), $S \in C^{(\infty)}([a, b], \mathbb{R})$ and $S'(x) \neq 0$ on $[a, b]$.

Integrating by parts and using the Riemann–Lebesgue lemma (see Problem 12), we find that

$$\int_a^b f(x)e^{i\lambda S(x)}\,dx = \frac{1}{i\lambda}\int_a^b \frac{f(x)}{S'(x)}\,de^{i\lambda S(x)} =$$

$$= -\frac{1}{i\lambda} \int_a^b \frac{\mathrm{d}}{\mathrm{d}x}\left(\frac{f}{S'}\right)(x) e^{i\lambda S(x)} \, \mathrm{d}x =$$

$$= \frac{1}{\lambda} \int_a^b f_1(x) e^{i\lambda S(x)} \, \mathrm{d}x = \cdots = \frac{1}{\lambda^n} \int_a^b f_n(x) e^{i\lambda S(x)} \, \mathrm{d}x =$$

$$= o\left(\lambda^{-n}\right) \quad \text{as } \lambda \to \infty.$$

Thus if $S'(0) \neq 0$ on the closed interval $[a, b]$, then because of the constantly increasing frequencies of oscillation of the function $e^{i\lambda S(x)}$ as $\lambda \to \infty$, the Fourier integral over the closed interval $[a, b]$ turns out to be a quantity of type $O(\lambda^{-\infty})$.

The function $S(x)$ in the Fourier integral is called the *phase function*. Thus the Fourier integral has its own localization principle called the *stationary phase principle*. According to this principle, the asymptotic behavior of the Fourier integral as $\lambda \to \infty$ (when $f \in C_0^{(\infty)}$) is the same as the asymptotics of the Fourier integral taken over a neighborhood $U(x_0)$ of a stationary point x_0 of the phase function (that is, a point x_0 at which $S'(x_0) = 0$) up to a quantity $O(\lambda^{-\infty})$.

After this, by a change of variable the question reduces to the canonical integral

$$E(\lambda) = \int_0^\varepsilon f(x) e^{i\lambda x^\alpha} \, \mathrm{d}x$$

whose asymptotic behavior is described by a special lemma of Erdélyi, which plays the same role for the Fourier integral that Watson's lemma plays for the Laplace integral.

This scheme for investigating the asymptotics of a Fourier integral is called the *stationary phase method*.

The nature of the localization principle in the stationary phase method is completely different from its nature in the case of the Laplace integral, but the general scheme of Laplace's method, as one can see, remains applicable even here.

Certain details relating to the stationary phase method will be found in Problems 12–17.

19.2.6 Problems and Exercises

Laplace's Method in the One-Dimensional Case

1. a) For $\alpha > 0$ the function $h(x) = e^{-\lambda x^\alpha}$ attains its maximum when $x = 0$. Here $h(x)$ is a quantity of order 1 in a δ-neighborhood of $x = 0$ of size $\delta = O(\lambda^{-1/\alpha})$.

Using Lemma 1, show that if $0 < \delta < 1$, then the integral

$$W(\lambda) = \int_{c(\lambda, \delta)}^a x^{\beta-1} f(x) e^{-\lambda x^\alpha} \, \mathrm{d}x,$$

where $c(\lambda, \delta) = \lambda^{\frac{\delta-1}{\alpha}}$ has order $O(e^{-A\lambda^\delta})$ as $\lambda \to +\infty$, A being a positive constant.

b) Prove that if the function f is continuous at $x = 0$, then

$$W(\lambda) = \alpha^{-1}\Gamma(\beta/\alpha)\big[f(0) + o(1)\big]\lambda^{-\beta/\alpha} \quad \text{as } \lambda \to +\infty.$$

c) In Theorem 1a), the hypothesis $f(x) = f(x_0) + O(x - x_0)$ can be weakened and replaced by the condition that f be continuous at x_0. Show that when this is done the same principal term of the asymptotics is obtained, but in general not Eq. (19.2') itself, in which $O(x - x_0)$ is now replaced by $o(1)$.

2. a) The Bernoulli numbers B_{2k} are defined by the relations

$$\frac{1}{t} - \frac{1}{1-e^{-t}} = -\frac{1}{2} - \sum_{k=1}^{\infty} \frac{B_{2k}}{(2k)!}t^{2k-1}, \quad |t| < 2\pi.$$

It is known that

$$\left(\frac{\Gamma'}{\Gamma}\right)(x) = \ln x + \int_0^{\infty}\left(\frac{1}{t} - \frac{1}{1-e^{-t}}\right)e^{-tx}\,dt.$$

Show that

$$\left(\frac{\Gamma'}{\Gamma}\right)(x) \simeq \ln x - \frac{1}{2x} - \sum_{k=0}^{\infty}\frac{B_{2k}}{2k}x^{-2k} \quad \text{as } x \to +\infty.$$

b) Prove that as $x \to +\infty$

$$\ln \Gamma(x) \simeq \left(x - \frac{1}{2}\right)\ln x - x + \frac{1}{2}\ln 2\pi + \sum_{k=1}^{\infty}\frac{B_{2k}}{2k(2k-1)}x^{-2k+1}.$$

This asymptotic expansion is called *Stirling's series*.

c) Using Stirling's series, obtain the first two terms of the asymptotics of $\Gamma(x+1)$ as $x \to +\infty$ and compare your result with what was obtained in Example 13.

d) Following the method of Example 13 and independently of it using Stirling's series, show that

$$\Gamma(x+1) = \sqrt{2\pi x}\left(\frac{x}{e}\right)^x\left(1 + \frac{1}{12x} + \frac{1}{288x^2} + O\left(\frac{1}{x^3}\right)\right) \quad \text{as } x \to +\infty.$$

3. a) Let $f \in C([0,a], \mathbb{R})$, $S \in C^{(1)}([0,a], \mathbb{R})$, $S(x) > 0$ on $[0,a]$, and suppose $S(x)$ attains its maximum at $x = 0$, with $S'(0) \neq 0$. Show that if $f(0) \neq 0$, then

$$I(\lambda) := \int_0^a f(x)S^\lambda(x)\,dx \sim -\frac{f(0)}{\lambda S'(0)}S^{\lambda+1}(0) \quad \text{as } \lambda \to +\infty.$$

b) Obtain the asymptotic expansion

$$I(\lambda) \simeq S^{\lambda+1}(0) \sum_{k=0}^{\infty} a_k \lambda^{-(k+1)} \quad \text{as } \lambda \to +\infty,$$

if it is known in addition that $f, S \in C^{(\infty)}([0, a], \mathbb{R})$.

4. a) Show that

$$\int_0^{\pi/2} \sin^n t \, dt = \sqrt{\frac{\pi}{2n}} \left(1 + O\left(n^{-1}\right)\right) \quad \text{as } n \to +\infty.$$

b) Express this integral in terms of Eulerian integrals and show that for $n \in \mathbb{N}$ it equals $\frac{(2n-1)!!}{(2n)!!} \cdot \frac{\pi}{2}$.

c) Obtain Wallis' formula $\pi = \lim_{n\to\infty} \frac{1}{n} \left(\frac{(2n)!!}{(2n-1)!!}\right)^2$.

d) Find the second term in the asymptotic expansion of the original integral as $n \to +\infty$.

5. a) Show that $\int_{-1}^{1} (1 - x^2)^n \, dx \sim \sqrt{\frac{\pi}{n}}$ as $n \to +\infty$.

b) Find the next term in the asymptotics of this integral.

6. Show that if $\alpha > 0$, then as $x \to +\infty$

$$\int_0^{+\infty} t^{-\alpha t} t^x \, dt \sim \sqrt{\frac{2\pi}{e^\alpha}} x^{\frac{1}{2\alpha}} \exp\left(\frac{\alpha}{e} x^{\frac{1}{\alpha}}\right).$$

7. a) Find the principal term of the asymptotics of the integral

$$\int_0^{+\infty} (1+t)^n e^{-nt} \, dt \quad \text{as } n \to +\infty.$$

b) Using this result and the identity $k! n^{-k} = \int_0^{+\infty} e^{-nt} t^k \, dt$, show that

$$\sum_{k=0}^{n} c_n^k k! n^{-k} = \sqrt{\frac{\pi n}{2}} \left(1 + O\left(n^{-1}\right)\right) \quad \text{as } n \to +\infty.$$

Laplace's Method in the Multidimensional Case

8. *The exponential estimate lemma.* Let $M = \sup_{x \in D} S(x)$, and suppose that for some $\lambda = \lambda_0$ the integral

$$F(\lambda) = \int_{D \subset \mathbb{R}^n} f(x) e^{\lambda S(x)} \, dx \tag{*}$$

converges absolutely. Show that it then converges absolutely for $\lambda \geq \lambda_0$ and

$$|f(\lambda)| \leq \int_D |f(x) e^{\lambda S(x)}| \, dx \leq A e^{\lambda M} \quad (\lambda \geq \lambda_0),$$

where A is a positive constant.

9. *Morse's lemma.* Let x_0 be a nondegenerate critical point of the function $S(x)$, $x \in \mathbb{R}^n$, defined and belonging to class $C^{(\infty)}$ in a neighborhood of x_0. Then there exist neighborhoods U and V of $x = x_0$ and $y = 0$ and a diffeomorphism $\varphi : V \to U$ of class $C^{(\infty)}(V, U)$ such that

$$S\big(\varphi(y)\big) = S(x_0) + \frac{1}{2}\sum_{j=1}^{n} v_j \big(y^j\big)^2,$$

where $\det \varphi'(0) = 1$, $v_1, \ldots, v_n$ are the eigenvalues of the matrix $S''_{xx}(x_0)$, and $y^1, \ldots, y^n$ are the coordinates of $y \in \mathbb{R}^n$.

Prove this slightly more specific form of Morse's lemma starting from Morse's lemma itself, which is discussed in Sect. 8.6 of Part 1.

10. *Asymptotics of a canonical integral.*

a) Let $t = (t_1, \ldots, t_n)$, $V = \{t \in \mathbb{R}^n \mid |t_j| \leq \delta, j = 1, 2, \ldots, n\}$, and $a \in C^{(\infty)}(V, \mathbb{R})$. Consider the function

$$F_1\big(\lambda, t'\big) = \int_{-\delta}^{\delta} a(t_1, \ldots, t_n) e^{-\frac{\lambda v_1}{2} t_1^2}\, dt_1,$$

where $t' = (t_2, \ldots, t_n)$ and $v_1 > 0$. Show that $F_1(\lambda, t') \simeq \sum_{k=0}^{\infty} a_k(t')\lambda^{-(k+\frac{1}{2})}$ as $\lambda \to +\infty$. This expansion is uniform in $t' \in V' = \{t' \in \mathbb{R}^{n-1} \mid |t^j| \leq \delta, j = 2, \ldots, n\}$ and $a_k \in C^{(\infty)}(V', \mathbb{R})$ for every $k = 0, 1, \ldots$.

b) Multiplying $F_1(\lambda, t')$ by $e^{-\frac{\lambda v_2}{2} t_2^2}$ and justifying the termwise integration of the corresponding asymptotic expansion, obtain the asymptotic expansion of the function

$$F_2\big(\lambda, t''\big) = \int_{-\delta}^{\delta} F_1\big(\lambda, t'\big) e^{-\frac{\lambda v_2}{2} t_2^2}\, dt_2 \quad \text{as } \lambda \to +\infty,$$

where $t'' = (t_3, \ldots, t_n)$, $v_2 > 0$.

c) Prove that for the function

$$A(\lambda) = \int_{-\delta}^{\delta} \cdots \int_{-\delta}^{\delta} a(t_1, \ldots, t_n) e^{-\frac{\lambda}{2}\sum_{j=1}^{n} v_j t_j^2}\, dt_1 \cdots dt_n,$$

where $v_j > 0$, $j = 1, \ldots, n$, the following asymptotic expansion holds:

$$A(\lambda) \simeq \lambda^{-n/2} \sum_{k=0}^{\infty} a_k \lambda^{-k} \quad \text{as } \lambda \to +\infty,$$

where $\alpha_0 = \sqrt{\dfrac{(2\pi)^n}{v_1 \cdots v_n}}\, a(0)$.

11. *The asymptotics of the Laplace integral in the multidimensional case.*

a) Let D be a closed bounded domain in $\mathbb{R}^n$, $f, S \in C(D, \mathbb{R})$, and suppose $\max_{x \in D} S(x)$ is attained only at one interior point x_0 of D. Let f and S be $C^{(\infty)}$ in some neighborhood of x_0 and $\det S''(x_0) \neq 0$.

Prove that if the integral (*) converges absolutely for some value $\lambda = \lambda_0$, then

$$F(\lambda) \simeq e^{\lambda S(x_0)} \lambda^{-n/2} \sum_{k=0}^{\infty} a_k \lambda^{-k} \quad \text{as } \lambda \to +\infty,$$

and this expansion can be differentiated with respect to λ any number of times, and its principal term has the form

$$F(\lambda) = e^{\lambda S(x_0)} \lambda^{-n/2} \sqrt{\frac{(2\pi)^n}{|\det S''(x_0)|}} \left(f(x_0) + O\left(\lambda^{-1}\right) \right).$$

b) Verify that if instead of the relation $f, S \in C^{(\infty)}$ all we know is that $f \in C$ and $S \in C^{(3)}$ in a neighborhood of x_0, then the principal term of the asymptotics as $\lambda \to +\infty$ remains the same with $O(\lambda^{-1})$ replaced by $o(1)$ as $\lambda \to +\infty$.

The Stationary Phase Method in the One-Dimensional Case

12. *Generalization of the Riemann–Lebesgue lemma.*

a) Prove the following generalization of the Riemann–Lebesgue lemma.

Let $S \in C^{(1)}([a, b], \mathbb{R})$ and $S'(x) \neq 0$ on $[a, b] =: I$. Then for every function f that is absolutely integrable on the interval I the following relation holds:

$$\widetilde{F}(\lambda) = \int_a^b f(x) e^{i\lambda S(x)} \, dx \to 0 \quad \text{as } \lambda \to \infty, \ \lambda \in \mathbb{R}.$$

b) Verify that if it is known in addition that $f \in C^{(n+1)}(I, \mathbb{R})$ and $S \in C^{(n+2)}(I, \mathbb{R})$, then as $\lambda \to \infty$

$$\widetilde{F}(\lambda) = \sum_{k=0}^{n} (i\lambda)^{-(k+1)} \left(\frac{1}{S'(x)} \frac{d}{dx} \right)^k \frac{f(x)}{S'(x)} \bigg|_a^b + o\left(\lambda^{-(n+1)}\right).$$

c) Write out the principal term of the asymptotics of the function $\widetilde{F}(\lambda)$ as $\lambda \to \infty$, $\lambda \in \mathbb{R}$.

d) Show that if $S \in C^{(\infty)}(I, \mathbb{R})$ and $f|_{[a,c]} \in C^{(2)}[a, c]$, $f|_{[c,b]} \in C^{(2)}[c, b]$, but $f \notin C^{(2)}[a, b]$, then the function $\widetilde{F}(\lambda)$ is not necessarily $o(\lambda^{-1})$ as $\lambda \to \infty$.

e) Prove that when $f, S \in C^{(\infty)}(I, \mathbb{R})$, the function $\widetilde{F}(\lambda)$ admits an asymptotic series expansion as $\lambda \to \infty$.

f) Find asymptotic expansions as $\lambda \to \infty$, $\lambda \in \mathbb{R}$, for the following integrals: $\int_0^\varepsilon (1 + x)^{-\alpha} \psi_j(x, \lambda) \, dx$, $j = 1, 2, 3$, if $\alpha > 0$ and $\psi_1 = e^{i\lambda x}$, $\psi_2 = \cos \lambda x$, and $\psi_3 = \sin \lambda x$.

13. *The localization principle.*

a) Let $I = [a, b] \subset \mathbb{R}$, $f \in C_0^{(\infty)}(I, \mathbb{R})$, $S \in C^{(\infty)}(I, \mathbb{R})$, and $S'(x) \neq 0$ on I. Prove that in this case

$$\widetilde{F}(\lambda) := \int_a^b f(x) e^{i\lambda S(x)} \, dx = O\left(|\lambda|^{-\infty}\right) \quad \text{as } \lambda \to \infty.$$

b) Suppose $f \in C_0^{(\infty)}(I, \mathbb{R})$, $S \in C^{(\infty)}(I, \mathbb{R})$, and $x_1, \ldots, x_m$ is a finite set of stationary points of $S(x)$ outside which $S'(x) \neq 0$ on I. We denote by $\widetilde{F}(\lambda, x_j)$ the integral of the function $f(x) e^{i\lambda S(x)}$ over a neighborhood $U(x_j)$ of the point x_j, $j = 1, \ldots, m$, not containing any other critical points in its closure. Prove that

$$\widetilde{F}(\lambda) = \sum_{j=1}^m \widetilde{F}(\lambda, x_j) + O\left(|\lambda|^{-\infty}\right) \quad \text{as } \lambda \to \infty.$$

14. *Asymptotics of the Fourier integral in the one-dimensional case.*

a) In a reasonably general situation finding the asymptotics of a one-dimensional Fourier integral can be reduced through the localization principle to describing the asymptotics of the canonical integral

$$E(\lambda) = \int_0^a x^{\beta-1} f(x) e^{i\lambda x^\alpha} \, dx,$$

for which the following lemma holds.

Erdélyi's lemma *Let* $\alpha \geq 1$, $\beta > 0$, $f \in C^{(\infty)}([0, a], \mathbb{R})$ *and* $f^{(k)}(a) = 0$, $k = 0, 1, 2, \ldots$. *Then*

$$E(\lambda) \simeq \sum_{k=0}^\infty a_k \lambda^{-\frac{k+\beta}{\alpha}} \quad \text{as } \lambda \to +\infty,$$

where

$$a_k = \frac{1}{\alpha} \Gamma\left(\frac{k+\beta}{\alpha}\right) e^{i\frac{\pi}{2}\frac{k+\beta}{\alpha}} \frac{f^{(k)}(0)}{k!},$$

and this expansion can be differentiated any number of times with respect to λ.

Using Erdélyi's lemma, prove the following assertion.

Let $I = [x_0 - \delta, x_0 + \delta]$ be a finite closed interval, let $f, S \in C^{(\infty)}(I, \mathbb{R})$ with $f \in C_0(I, \mathbb{R})$, and let S have a unique stationary point x_0 on I, where $S'(x_0) = 0$ but $S''(x_0) \neq 0$. Then as $\lambda \to +\infty$

$$\widetilde{F}(\lambda, x_0) := \int_{x_0-\delta}^{x_0+\delta} f(x) e^{i\lambda S(x)} \, dx \simeq e^{i\frac{\pi}{4} S''(x_0)} e^{i\lambda S(x_0)} \lambda^{-\frac{1}{2}} \sum_{k=0}^\infty a_k \lambda^{-k}$$

and the principal term of the asymptotics has the form

$$\widetilde{F}(\lambda, x_0) = \sqrt{\frac{2\pi}{\lambda |S''(x_0)|}} e^{i(\frac{\pi}{4} \operatorname{sgn} S''(x_0) + \lambda S(x_0))} \left(f(x_0) + O(\lambda^{-1})\right).$$

b) Consider the Bessel function of integer order $n \geq 0$:

$$J_n(x) = \frac{1}{\pi} \int_0^\pi \cos(x \sin \varphi - n\varphi) \, d\varphi.$$

Show that

$$J_n(x) = \sqrt{\frac{2}{\pi x}} \left[\cos\left(x - \frac{n\pi}{2} - \frac{\pi}{4}\right) + O(x^{-1}) \right] \quad \text{as } x \to +\infty.$$

The Stationary Phase Method in the Multidimensional Case

15. *The localization principle.*

a) Prove the following assertion. Let D be a domain in $\mathbb{R}^n$, $f \in C_0^{(\infty)}(D, \mathbb{R})$, $S \in C^{(\infty)}(D, \mathbb{R})$, $\operatorname{grad} S(x) \neq 0$ for $x \in \operatorname{supp} f$, and

$$\widetilde{F}(\lambda) := \int_D f(x) e^{i\lambda S(x)} \, dx. \tag{**}$$

Then for every $k \in \mathbb{N}$ there exists a positive constant $A(k)$ such that the estimate $|\widetilde{F}(\lambda)| \leq A(k)\lambda^{-k}$ holds for $\lambda \geq 1$, and hence $\widetilde{F}(\lambda) = O(\lambda^{-\infty})$ as $\lambda \to +\infty$.

b) Suppose as before that $f \in C_0^{(\infty)}(D, \mathbb{R})$, $S \in C^{(\infty)}(D, \mathbb{R})$, but S has a finite number of critical points $x_1, \ldots, x_m$ outside which $\operatorname{grad} S(x) \neq 0$. We denote by $\widetilde{F}(\lambda, x_j)$ the integral of the function $f(x)e^{i\lambda S(x)}$ over a neighborhood $U(x_j)$ of x_j whose closure contains no critical points except x_j. Prove that

$$\widetilde{F}(\lambda) = \sum_{j=1}^m \widetilde{F}(\lambda, x_j) + O(\lambda^{-\infty}) \quad \text{as } \lambda \to +\infty.$$

16. *Reduction to a canonical integral.* If x_0 is a nondegenerate critical point of the function $S \in C^{(\infty)}(D, \mathbb{R})$ defined in a domain $D \subset \mathbb{R}^n$, then by Morse's lemma (see Problem 9) there exists a local change of variable $x = \varphi(y)$ such that $x_0 = \varphi(0)$, $S(\varphi(y)) = S(x_0) + \frac{1}{2}\sum_{j=1}^n \varepsilon_j (y^j)^2$, where $\varepsilon_j = \pm 1$, $y = (y^1, \ldots, y^n)$, and $\det \varphi'(y) > 0$.

Using the localization principle (Problem 15), now show that if $f \in C_0^{(\infty)}(D, \mathbb{R})$, $S \in C^{(\infty)}(D, \mathbb{R})$, and S has at most a finite number of critical points in D, all non-degenerate, then the study of the asymptotics of the integral (**) reduces to studying the asymptotics of the special integral

$$\Phi(\lambda) := \int_{-\delta}^{\delta} \cdots \int_{-\delta}^{\delta} \psi(y^1, \ldots, y^n) e^{\frac{i\lambda}{2} \sum_{j=1}^n \varepsilon_j (y^j)^2} \, dy^1 \cdots dy^n.$$

17. *Asymptotics of a Fourier integral in the multidimensional case.* Using Erdélyi's lemma (Problem 14a)) and the scenario described in Problem 10, prove that if D is a domain in $\mathbb{R}^n$, $f, S \in C^{(\infty)}(D, \mathbb{R})$, supp f is a compact subset of D, and x_0 is the only critical point of S in D and is nondegenerate, then for the integral (**) the following asymptotic expansion holds as $\lambda \to +\infty$:

$$\widetilde{F}(\lambda) \simeq \lambda^{-n/2} e^{i\lambda S(x_0)} \sum_{k=0}^{\infty} a_k \lambda^{-k},$$

which can be differentiated any number of times with respect to λ.

The main term of the asymptotics has the form

$$\widetilde{F}(\lambda) = \left(\frac{2\pi}{\lambda}\right)^{n/2} \exp\left[i\lambda S(x_0) + \frac{i\pi}{4} \operatorname{sgn} S''(x_0)\right] \times$$

$$\times \left|\det S''(x_0)\right|^{-1/2} \left[f(x_0) + O\left(\lambda^{-1}\right)\right] \quad \text{as } \lambda \to +\infty.$$

Here $S''(x)$ is the symmetric and by hypothesis nonsingular matrix of second derivatives of the function S at x_0 (the *Hessian*), and sgn $S''(x_0)$ is the signature of this matrix (or the quadratic form corresponding to it), that is, the difference $\nu_+ - \nu_-$ between the number of positive and negative eigenvalues of the matrix $S''(x_0)$.

Topics and Questions for Midterm Examinations

1 Series and Integrals Depending on a Parameter

1. The Cauchy criterion for convergence of a series. The comparison theorem and the basic sufficient conditions for convergence (majorant, integral, Abel–Dirichlet). The series $\zeta(s) = \sum_{n=1}^{\infty} n^{-s}$.

2. Uniform convergence of families and series of functions. The Cauchy criterion and the basic sufficient conditions for uniform convergence of a series of functions (M-test, Abel–Dirichlet).

3. Sufficient conditions for two limiting passages to commute. Continuity, integration, and differentiation and passage to the limit.

4. The region of convergence and the nature of convergence of a power series. The Cauchy–Hadamard formula. Abel's (second) theorem. Taylor expansions of the basic elementary functions. Euler's formula. Differentiation and integration of a power series.

5. Improper integrals. The Cauchy criterion and the basic sufficient conditions for convergence (M-test, Abel–Dirichlet).

6. Uniform convergence of an improper integral depending on a parameter. The Cauchy criterion and the basic sufficient conditions for uniform convergence (majorant, Abel–Dirichlet).

7. Continuity, differentiation, and integration of a proper integral depending on a parameter.

8. Continuity, differentiation, and integration of an improper integral depending on a parameter. The Dirichlet integral.

9. The Eulerian integrals. Domains of definition, differential properties, reduction formulas, various representations, interconnections. The Poisson integral.

10. Approximate identities. The theorem on convergence of the convolution. The classical Weierstrass theorem on uniform approximation of a continuous function by an algebraic polynomial.

© Springer-Verlag Berlin Heidelberg 2016
V.A. Zorich, *Mathematical Analysis II*, Universitext,
DOI 10.1007/978-3-662-48993-2

2 Problems Recommended as Midterm Questions

Problem 1 P is a polynomial. Compute $(e^{t\frac{d}{dx}})P(x)$.

Problem 2 Verify that the vector-valued function $e^{tA}x_0$ is a solution of the Cauchy problem $\dot{x} = Ax$, $x(0) = x_0$. (Here $\dot{x} = Ax$ is a system of equations defined by the matrix A.)

Problem 3 Find up to order $o(1/n^3)$ the asymptotics of the positive roots $\lambda_1 < \lambda_2 < \cdots < \lambda_n < \cdots$ of the equation $\sin x + 1/x = 0$ as $n \to \infty$.

Problem 4 a) Show that $\ln 2 = 1 - 1/2 + 1/3 - \cdots$. How many terms of this series must be taken to determine $\ln 2$ within 10^{-3}?

b) Verify that $\frac{1}{2}\ln\frac{1+t}{1-t} = t + \frac{1}{3}t^3 + \frac{1}{5}t^5 + \cdots$. Using this expansion it becomes convenient to compute $\ln x$ by setting $x = \frac{1+t}{1-t}$.

c) Setting $t = 1/3$ in b), obtain the equality

$$\frac{1}{2}\ln 2 = \frac{1}{3} + \frac{1}{3}\left(\frac{1}{3}\right)^3 + \frac{1}{5}\left(\frac{1}{3}\right)^5 + \cdots.$$

How many terms of this series must one take to find $\ln 2$ within 10^{-3}? Compare this with the result of a).

This is one of the methods of improving convergence.

Problem 5 Verify that in the sense of Abel summation

a) $1 - 1 + 1 \cdots = \frac{1}{2}$.

b) $\sum_{k=1}^{\infty} \sin k\varphi = \frac{1}{2}\cdot\frac{1}{2}\varphi$, $\varphi \neq 2\pi n$, $n \in \mathbb{Z}$.

c) $\frac{1}{2} + \sum_{k=1}^{\infty} \cos k\varphi = 0$, $\varphi \neq 2\pi n$, $n \in \mathbb{Z}$.

Problem 6 Prove Hadamard's lemma:

a) If $f \in C^{(1)}(U(x_0))$, then $f(x) = f(x_0) + \varphi(x)(x - x_0)$, where $\varphi \in C(U(x_0))$ and $\varphi(x_0) = f'(x_0)$.

b) If $f \in C^{(n)}(U(x_0))$, then

$$f(x) = f(x_0) + \frac{1}{1!}f'(x_0)(x - x_0) + \cdots +$$

$$+ \frac{1}{(n-1)!}f^{(n-1)}(x_0)(x - x_0)^{n-1} + \varphi(x)(x - x_0)^n,$$

where $\varphi \in C(U(x_0))$ and $\varphi(x_0) = \frac{1}{n!}f^{(n)}(x_0)$.

c) What do these relations look like in coordinate form, when $x = (x^1, \ldots, x^n)$, that is, when f is a function of n variables?

Problem 7 a) Verify that the function

$$J_0(x) = \frac{1}{\pi} \int_0^1 \frac{\cos xt}{\sqrt{1-t^2}} \, dt$$

satisfies Bessel's equation $y'' + \frac{1}{x}y' + y = 0$.
  b) Try to solve this equation using power series.
  c) Find the power-series expansion of the function $J_0(x)$.

Problem 8 Verify that the following asymptotic expansions hold

  a) $\Gamma(\alpha, x) := \int_x^{+\infty} t^{\alpha-1} e^{-t} \, dt \simeq e^{-\alpha} \sum_{k=1}^{\infty} \frac{\Gamma(\alpha)}{\Gamma(\alpha-k+1)} x^{\alpha-k}$,
  b) $\mathrm{Erf}(x) := \int_x^{+\infty} e^{-t^2} \, dt \simeq \frac{1}{2}\sqrt{\pi} e^{-x^2} \sum_{k=1}^{\infty} \frac{1}{\Gamma(3/2-k)x^{2k-1}}$

as $x \to +\infty$.

Problem 9 a) Following Euler, obtain the result that the series $1 - 1!x + 2!x^2 - 3!x^3 + \cdots$ is connected with the function

$$S(x) := \int_0^{+\infty} \frac{e^{-t}}{1+xt} \, dt.$$

  b) Does this series converge?
  c) Does it give the asymptotic expansion of $S(x)$ as $x \to 0$?

Problem 10 a) A linear device A whose characteristics are constant over time responds to a signal $\delta(t)$ in the form of a δ-function by giving out the signal (function) $E(t)$. What will the response of this device be to an input signal $f(t)$, $-\infty < t < +\infty$?
  b) Can the input signal f always be recovered uniquely from the transformed signal $\widehat{f} := Af$?

3 Integral Calculus (Several Variables)

1. Riemann integral on an n-dimensional interval. Lebesgue criterion for existence of the integral.
2. Darboux criterion for existence of the integral of a real-valued function on an n-dimensional interval.
3. Integral over a set. Jordan measure (content) of a set and its geometric meaning. Lebesgue criterion for existence of the integral over a Jordan-measurable set. Linearity and additivity of the integral.
4. Estimates of the integral.
5. Reduction of a multiple integral to an iterated integral. Fubini's theorem and its most important corollaries.

6. Formula for change of variables in a multiple integral. Invariance of measure and the integral.

7. Improper multiple integrals: basic definitions, majorant criterion for convergence, canonical integrals. Computation of the Euler–Poisson integral.

8. Surfaces of dimension k in $\mathbb{R}^n$ and basic methods of defining them. Abstract k-dimensional manifolds. Boundary of a k-dimensional manifold as a $(k-1)$-dimensional manifold without boundary.

9. Orientable and nonorientable manifolds. Methods of defining the orientation of an abstract manifold and a (hyper)surface in $\mathbb{R}^n$.

Orientability of the boundary of an orientable manifold. Orientation induced on the boundary from the manifold.

10. Tangent vectors and the tangent space to a manifold at a point. Interpretation of a tangent vector as a differential operator.

11. Differential forms in a region $D \subset \mathbb{R}^n$. Examples: differential of a function, work form, flux form. Coordinate expression of a differential form. Exterior derivative operator.

12. Mapping of objects and the adjoint mapping of functions on these objects. Transformation of points and vectors of tangent spaces at these points under a smooth mapping. Transfer of functions and differential forms under a smooth mapping. A recipe for carrying out the transfer of forms in coordinate form.

13. Commutation of transfer of differential forms with exterior multiplication and differentiation. Differential forms on a manifold. Invariance (unambiguous nature) of operations on differential forms.

14. A scheme for computing work and flux. Integral of a k-form over a k-dimensional smooth oriented surface, taking account of orientation. Independence of the integral of the choice of parametrization. General definition of the integral of a differential k-form over a k-dimensional compact oriented manifold.

15. Green's formula on a square, its derivation, interpretation, and expression in the language of integrals of the corresponding differential forms. The general Stokes formula. Reduction to a k-dimensional interval and proof for a k-dimensional interval. The classical integral formulas of analysis as particular versions of the general Stokes formula.

16. The volume element on $\mathbb{R}^n$ and on a surface. Dependence of the volume element on orientation. The integral of first kind and its independence of orientation. Area and mass of a material surface as an integral of first kind. Expression of the volume element of a k-dimensional surface $S^k \subset \mathbb{R}^n$ in local parameters and the expression of the volume element of a hypersurface $S^{n-1} \subset \mathbb{R}^n$ in Cartesian coordinates of the ambient space.

17. Basic differential operators of field theory (grad, curl, div) and their connection with the exterior derivative operator d in oriented Euclidean space $\mathbb{R}^3$.

18. Expression of work and flux of a field as integrals of first kind. The basic integral formulas of field theory in $\mathbb{R}^3$ as the vector expression of the classical integral formulas of analysis.

19. A potential field and its potential. Exact and closed forms. A necessary differential condition for a form to be exact and for a vector field to be a potential field. Its sufficiency in a simply connected domain. Integral criterion for exactness of 1-forms and vector fields.

20. Local exactness of a closed form (the Poincaré lemma). Global analysis. Homology and cohomology. De Rham's theorem (statement).

21. Examples of the application of the Stokes (Gauss–Ostrogradskii) formula: derivation of the basic equations of the mechanics of continuous media. Physical meaning of the gradient, curl, and divergence.

22. Hamilton's nabla operator and work with it. The gradient, curl, and divergence in triorthogonal curvilinear coordinates.

4 Problems Recommended for Studying the Midterm Topics

The numbers followed by closing parentheses below refer to the topics 1–22 just listed. The closing parentheses dashes are followed by section numbers (for example 13.4 means Sect. 4 of Chap. 13), which in turn are separated by a dash from the numbers of the problems from the section related to the topic from the list above.

1) 11.1—2,3; 2) 11.1—4; 3) 11.2—1,3,4; 4) 11.3—1,2,3,4; 5) 11.4—6,7 and 13.2—6; 6) 11.5—9 and 12.5—5,6; 7) 11.6—1,5,7; 8) 12.1—2,3 and 12.4—1,4; 9) 12.2—1,2,3,4 and 12.5—11; 10) 15.3—1,2; 11) 12.5—9 and 15.3—3; 12) 15.3—4; 13) 12.5—8,10; 14) 13.1—3,4,5,9; 15) 13.1—1,10,13,14; 16) 12.4—10 and 13.2—5; 17) 14.1—1,2; 18) 14.2—1,2,3,4,8; 19) 14.3—7,13,14; 20) 14.3—11,12; 21) 13.3—1 and 14.1—8; 22) 14.1—4,5,6.

Examination Topics

1 Series and Integrals Depending on a Parameter

1. Cauchy criterion for convergence of a series. Comparison theorem and the basic sufficient conditions for convergence (majorant, integral, Abel–Dirichlet). The series $\zeta(s) = \sum_{n=1}^{\infty} n^{-s}$.

2. Uniform convergence of families and series of functions. Cauchy criterion and the basic sufficient conditions for uniform convergence of a series of functions (M-test, Abel–Dirichlet).

3. Sufficient conditions for commutativity of two limiting passages. Continuity, integration, and differentiation and passage to the limit.

4. Region of convergence and the nature of convergence of a power series. Cauchy–Hadamard formula. Abel's (second) theorem. Taylor expansions of the basic elementary functions. Euler's formula. Differentiation and integration of a power series.

5. Improper integrals. Cauchy criterion and the basic sufficient conditions for convergence (majorant, Abel–Dirichlet).

6. Uniform convergence of an improper integral depending on a parameter. Cauchy criterion and the basic sufficient conditions for uniform convergence (M-test, Abel–Dirichlet).

7. Continuity, differentiation, and integration of a proper integral depending on a parameter.

8. Continuity, differentiation, and integration of an improper integral depending on a parameter. Dirichlet integral.

9. Eulerian integrals. Domains of definition, differential properties, reduction formulas, various representations, interconnections. Poisson integral.

10. Approximate identities. Theorem on convergence of the convolution. Classical Weierstrass theorem on uniform approximation of a continuous function by an algebraic polynomial.

11. Vector spaces with an inner product. Continuity of the inner product and algebraic properties connected with it. Orthogonal and orthonormal systems of vectors.

© Springer-Verlag Berlin Heidelberg 2016
V.A. Zorich, *Mathematical Analysis II*, Universitext,
DOI 10.1007/978-3-662-48993-2

Pythagorean theorem. Fourier coefficients and Fourier series. Examples of inner products and orthogonal systems in spaces of functions.

12. Orthogonal complement. Extremal property of Fourier coefficients. Bessel's inequality and convergence of the Fourier series. Conditions for completeness of an orthonormal system. Method of least squares.

13. Classical (trigonometric) Fourier series in real and complex form. Riemann–Lebesgue lemma. Localization principle and convergence of a Fourier series at a point. Example: expansion of $\cos(\alpha x)$ in a Fourier series and the expansion of $\sin(\pi x)/\pi x$ in an infinite product.

14. Smoothness of a function, rate of decrease of its Fourier coefficients, and rate of convergence of its Fourier series.

15. Completeness of the trigonometric system and mean convergence of a trigonometric Fourier series.

16. Fourier transform and the Fourier integral (the inversion formula). Example: computation of $\widehat{f}$ for $f(x) := \exp(-a^2 x^2)$.

17. Fourier transform and the derivative operator. Smoothness of a function and the rate of decrease of its Fourier transform. Parseval's equality. The Fourier transform as an isometry of the space of rapidly decreasing functions.

18. Fourier transform and convolution. Solution of the one-dimensional heat equation.

19. Recovery of a transmitted signal from the spectral function of a device and the signal received. Sampling theorem (Kotel'nikov–Shannon formula).

20. Asymptotic sequences and asymptotic series. Example: asymptotic expansion of $\mathrm{Ei}(x)$. Difference between convergent and asymptotic series. Asymptotic Laplace integral (principal term). Stirling's formula.

2 Integral Calculus (Several Variables)

1. Riemann integral on an n-dimensional interval. Lebesgue criterion for existence of the integral.

2. Darboux criterion for the existence of the integral of a real-valued function on an n-dimensional interval.

3. Integral over a set. Jordan measure (content) of a set and its geometric meaning. Lebesgue criterion for existence of the integral over a Jordan-measurable set. Linearity and additivity of the integral.

4. Estimates of the integral.

5. Reduction of a multiple integral to an iterated integral. Fubini's theorem and its most important corollaries.

6. Formula for change of variables in a multiple integral. Invariance of measure and the integral.

7. Improper multiple integrals: basic definitions, the majorant criterion for convergence, canonical integrals. Computation of the Euler–Poisson integral.

8. Surfaces of dimension k in $\mathbb{R}^n$ and the basic methods of defining them. Abstract k-dimensional manifolds. Boundary of a k-dimensional manifold as a $(k-1)$-dimensional manifold without boundary.

9. Orientable and nonorientable manifolds. Methods of defining the orientation of an abstract manifold and a (hyper)surface in $\mathbb{R}^n$.

Orientability of the boundary of an orientable manifold. Orientation on the boundary induced from the manifold.

10. Tangent vectors and the tangent space to a manifold at a point. Interpretation of a tangent vector as a differential operator.

11. Differential forms in a region $D \subset \mathbb{R}^n$. Examples: differential of a function, work form, flux form. Coordinate expression of a differential form. Exterior derivative operator.

12. Mapping of objects and the adjoint mapping of functions on these objects. Transformation of points and vectors of tangent spaces at these points under a smooth mapping. Transfer of functions and differential forms under a smooth mapping. A recipe for carrying out the transfer of forms in coordinate form.

13. Commutation of the transfer of differential forms with exterior multiplication and differentiation. Differential forms on a manifold. Invariance (unambiguous nature) of operations on differential forms.

14. A scheme for computing work and flux. Integral of a k-form over a k-dimensional smooth oriented surface. Taking account of orientation. Independence of the integral of the choice of parametrization. General definition of the integral of a differential k-form over a k-dimensional compact oriented manifold.

15. Green's formula on a square, its derivation, interpretation, and expression in the language of integrals of the corresponding differential forms. General Stokes formula. Reduction to a k-dimensional interval and proof for a k-dimensional interval. Classical integral formulas of analysis as particular versions of the general Stokes formula.

16. Volume element on $\mathbb{R}^n$ and on a surface. Dependence of volume element on orientation. The integral of first kind and its independence of orientation. Area and mass of a material surface as an integral of first kind. Expression of volume element of a k-dimensional surface $S^k \subset \mathbb{R}^n$ in local parameters and expression of volume element of a hypersurface $S^{n-1} \subset \mathbb{R}^n$ in Cartesian coordinates of the ambient space.

17. Basic differential operators of field theory (grad, curl, div) and their connection with the exterior derivative operator d in oriented Euclidean space $\mathbb{R}^3$.

18. Expression of work and flux of a field as integrals of first kind. Basic integral formulas of field theory in $\mathbb{R}^3$ as the vector expression of the classical integral formulas of analysis.

19. A potential field and its potential. Exact and closed forms. A necessary differential condition for a form to be exact and for a vector field to be a potential field. Its sufficiency in a simply connected domain. Integral criterion for exactness of 1-forms and vector fields.

20. Local exactness of a closed form (Poincaré's lemma). Global analysis. Homology and cohomology. De Rham theorem (formulation).

21. Examples of the application of the Stokes (Gauss–Ostrogradskii) formula: derivation of the basic equations of the mechanics of continuous media. Physical meaning of the gradient, curl, and divergence.

22. Hamilton's Nabla operator, and computation of work with it. Gradient, curl and divergence in a 3-dimensional orthogonal system of curvilinear coordinates.

Examination Problems
(Series and Integrals Depending on a Parameter)

1. We shall consider a sequence of real-valued functions $\{f_n\}$ defined on the interval $[0, 1]$, for example.

   a) What types of convergence for a sequence of functions do you know?

   b) Provide the definition of each of them.

   c) What are the relations between them? (Prove the relation or give an explanatory example when there is no such relation.)

2. Let f be a periodic function with period 2π. Suppose it is identically zero on the interval $]-\pi, 0[$ and $f(x) = 2x$ on the interval $[0, \pi]$. Calculate the sum S of the standard trigonometric Fourier series of this function.

3. a) We know the expansion in power series of the function $(1 + x)^{-1}$ (geometric progression). Obtain from it the expansion in a power series of the function $\ln(1+x)$ and justify your steps.

   b) What is the radius of convergence of the obtained series?

   c) Does this series converge at $x = 1$, and if so, is its sum equal to $\ln 2$? Why?

4. a) It is known that the spectral function (characteristic function) p of a linear device (operator) A is everywhere nonzero. How can we find the transmitted signal f if we know the function p and the received signal $g = Af$.

   b) Let the function p be defined by $p(\omega) \equiv 1$ for $|\omega| \leq 10$ and $p(\omega) \equiv 0$ for $|\omega| > 10$. Suppose that we know the spectrum $\widehat{g}$ (Fourier transform) of the received signal g and that it is exactly $\widehat{g}(\omega) \equiv 1$ for $|\omega| \leq 1$ and $\widehat{g}(\omega) \equiv 0$ for $|\omega| > 1$. Finally, suppose that it is also known that the input signal f does not contain some other frequencies apart from the frequencies transmitted by the device A (i.e., beyond the frequencies $|\omega| \leq 10$). Find the input signal f.

5. Using Euler's Γ function and Laplace's method, obtain the very useful asymptotic Stirling's formula $n! \sim \sqrt{2\pi n}(\frac{n}{e})^n$.

© Springer-Verlag Berlin Heidelberg 2016
V.A. Zorich, *Mathematical Analysis II*, Universitext,
DOI 10.1007/978-3-662-48993-2

Intermediate Problems
(Integral Calculus of Several Variables)

.

1. Compute the values of the following forms ω in $\mathbb{R}^n$ on the given set of vectors.

a) $\omega = x^2 \, dx^1$ applied to the vector $\xi = (1, 2, 3) \in T\mathbb{R}^3_{(1,2,3)}$;

b) $\omega = dx^1 \wedge dx^3 + x^1 \, dx^2 \wedge dx^4$ applied to the ordered pair of vectors $(\xi_1, \xi_2) \in T\mathbb{R}^4_{(1,0,0,0)}$. (Set $\xi_1 = (\xi_1^1, \ldots, \xi_1^4)$, $\xi_2 = (\xi_2^1, \ldots, \xi_2^4)$.)

2. Let $f^1, \ldots, f^n$ be smooth functions with argument $x = (x^1, \ldots, x^n) \in \mathbb{R}^n$. Express the form $df^1 \wedge \cdots \wedge df^n$ in terms of the forms $dx^1, \ldots, dx^n$.

3. Let F be a vector field of a force acting on a domain $D \subset \mathbb{R}^3$. By the action of this vector field an object was transferred along a smooth path $\gamma \subset D$ from the point $a \in D$ to the point $b \in D$. Calculate the work done by the vector field in this process.

a) Write the formula for the calculation of this work as an integral of the first type and as an integral of the second type (i.e., in terms of ds and dx, dy, dz, respectively).

b) Prove that in the case of the gravitational vector field, this work does not depend on the path and that it is equal to ...?

4. Consider the following problem about the flux of a vector field.

a) One has the vector field V (for instance, the vector field velocity of some current) on the domain $D \in \mathbb{R}^3$. Write a formula for the calculation of the flux of the vector field V through the oriented surface $S = S_+^2 \subset D$ as an integral of the first type and as an integral of the second type (i.e., in terms of $d\sigma$ and $dy \wedge dz$, $dz \wedge dx$, $dx \wedge dy$ respectively).

b) Consider a convex polyhedral domain $D \subset \mathbb{R}^3$. On each of its faces is constructed a vector pointing toward the exterior normal direction with magnitude equal to the area of the corresponding side. Physics states that the sum of these vectors is equal to zero (otherwise, we could build a perpetual motion device). Mathematics agrees. Prove this fact.

c) Deduce Archimedes's law by a direct computation (calculate the buoyancy force acting on a submerged body in a bathtub completely filled with water, for example, as the resulting pressure on the surface of the body).

© Springer-Verlag Berlin Heidelberg 2016
V.A. Zorich, *Mathematical Analysis II*, Universitext,
DOI 10.1007/978-3-662-48993-2

Appendix A
Series as a Tool
(Introductory Lecture)

When a geological deposit is discovered, it is explored and then exploited. In mathematics, it is also like that. Axiomatics and useful formalisms arise as the result of solving concrete questions and problems. They do not fall down from the sky, as it seems to inexperienced students when everything starts with axioms.

This course is largely dedicated to series, i.e., basically limits of sequences. We shall give at least an initial idea of how and where this tool works, in order to convince ourselves that the study of this remarkably effective machinery, namely the theory of series, does not reduce to the abstract study of the convergence of series (the existence of a limit).

A.1 Getting Ready

A.1.1 The Small Bug on the Rubber Rope

(Problem proposed by the academician L.B. Okun to the academician A.D. Sakharov.)[1]

Problem 1 You hold one end of a 1 km long rubber rope. A small bug crawls toward you from the other end, which is fixed, with a speed of 1 cm/s. As soon as it crawls one centimeter, you stretch the rubber rope another kilometer every time. Does the insect ever reach your hand? And if it does, how long will it take?

---

[1]Martin Gardner in his book *Time Travel and Other Mathematical Bewilderments* (New York: W. H. Freeman & Company, 1987, English, p. 295) writes, "This delightful problem, which has the flavor of a Zeno paradox, was devised by Denys Wilquin of New Caledonia. It appeared first in December 1972 in Pierre Berloquin's lively puzzle column in the French monthly *Science et Vie*."

© Springer-Verlag Berlin Heidelberg 2016
V.A. Zorich, *Mathematical Analysis II*, Universitext,
DOI 10.1007/978-3-662-48993-2

A.1.2 Integral and Estimation of Sums

After some thinking, it may occur to you that the following sum might be useful in finding the answer $S_n = 1 + \frac{1}{2} + \frac{1}{3} + \cdots + \frac{1}{n}$.

Problem 2 Recall the integral, and show that $S_n - 1 < \int_1^n \frac{1}{x} \, dx < S_{n-1}$.

A.1.3 From Monkeys to Doctors of Science Altogether in 10^6 Years

Littlewood in his famous book *Littlewood's Miscellany*, speaking about large numbers, wrote that 10^6 years is the time needed to convert monkeys into doctors of science.[2]

Problem 3 Would the little bug arrive in time for the thesis defense or at least before the end of the universe?

A.2 The Exponential Function

A.2.1 Power Series Expansion of the Functions exp, sin, cos

According to Taylor's formula with remainder in Lagrange's form, one has

$$e^x = 1 + \frac{1}{1!}x + \frac{1}{2!}x^2 + \cdots + \frac{1}{n!}x^n + r_n(x),$$

where $r_n(x) = \frac{1}{(n+1)!} e^{\xi} \cdot x^{n+1}$ and $|\xi| < |x|$;

$$\cos x = 1 - \frac{1}{2!}x^2 + \frac{1}{4!}x^4 - \cdots + \frac{(-1)^n}{2n!}x^{2n} + r_{2n}(x),$$

where $r_{2n}(x) = \frac{1}{(2n+1)!} \cos(\xi + \frac{\pi}{2}(2n+1))x^{2n+1}$ and $|\xi| < |x|$;

$$\sin x = x - \frac{1}{3!}x^3 + \frac{1}{5!}x^5 - \cdots + \frac{(-1)^n}{(2n+1)!}x^{2n+1} + r_{2n+1}(x),$$

where $r_{2n+1}(x) = \frac{1}{(2n+2)!} \sin(\xi + \frac{\pi}{2}(2n+2))x^{2n+2}$ and $|\xi| < |x|$. Since for every fixed value $x \in \mathbb{R}$, the remainder in each of the above formulas clearly tends to zero

---

[2] John E. Littlewood, *Littlewood's Miscellany*. Cambridge: Cambridge University Press, 1986, English, p. 212.

as $n \to \infty$, we can write

$$e^x = 1 + \frac{1}{1!}x + \frac{1}{2!}x^2 + \frac{1}{3!}x^3 + \frac{1}{4!}x^4 + \frac{1}{5!}x^5 + \cdots + \frac{1}{n!}x^n + \cdots,$$

$$\cos x = 1 - \frac{1}{2!}x^2 + \frac{1}{4!}x^4 - \cdots + \frac{(-1)^n}{2n!}x^{2n} + \cdots,$$

$$\sin x = x - \frac{1}{3!}x^3 + \frac{1}{5!}x^5 - \cdots + \frac{(-1)^n}{(2n+1)!}x^{2n+1} + \cdots.$$

A.2.2 Exit to the Complex Domain and Euler's Formula

We substitute x for the complex number ix in the right-hand side of the first of these equalities. Then, after some simple arithmetic manipulations, we obtain Euler's outstanding relationship

$$e^{ix} = \cos x + i \sin x.$$

Setting $x = \pi$, we find that $e^{i\pi} + 1 = 0$. This is the famous equation connecting the fundamental constants of mathematics: e from analysis, i from algebra, π from geometry, 1 from arithmetic, and 0 from logic.

We defined the function exp for purely imaginary values of the argument and obtained Euler's formula $e^{ix} = \cos x + i \sin x$, from which, clearly, it also follows that

$$\cos x = \frac{1}{2}\left(e^{ix} + e^{-ix}\right) \quad \text{and} \quad \sin x = \frac{1}{2i}\left(e^{ix} - e^{-ix}\right).$$

A.2.3 The Exponential Function as a Limit

We know that $(1 + \frac{x}{n})^n \to e^x$ as $n \to \infty$ for $x \in \mathbb{R}$. It is natural to assume that $e^z = \lim_{n \to \infty}(1 + \frac{z}{n})^n$, where now $z = x + iy$ is an arbitrary complex number. A computation of this limit gives $e^z = e^x(\cos y + i \sin y)$.

Problem 4 Verify this and obtain a formula for $\cos z$ and $\sin z$.

A.2.4 Multiplication of Series and the Basic Property of the Exponential Function

The expression $e^z = e^x(\cos y + i \sin y)$ for e^{x+iy} can be naturally obtained from the relation $e^{x+iy} = e^x e^{iy}$ if it is valid for complex values of the argument of the function exp.

We shall prove this by direct multiplication. Let u and v be complex numbers. Setting $e^u := \sum_{k=0}^{\infty} \frac{1}{k!} u^k$ and $e^v := \sum_{m=0}^{\infty} \frac{1}{m!} v^m$ we find that

$$e^u \cdot e^v = \left(\sum_{k=0}^{\infty} \frac{1}{k!} u^k \right) \cdot \left(\sum_{m=0}^{\infty} \frac{1}{m!} v^m \right) = \sum_{k=0}^{\infty} \sum_{m=0}^{\infty} \frac{1}{k!} \frac{1}{m!} u^k v^m =$$

$$= \sum_{n=0}^{\infty} \sum_{n=k+m} \frac{1}{k!} \frac{1}{m!} u^k v^m = \sum_{n=0}^{\infty} \frac{1}{n!} (u+v)^n = e^{u+v}.$$

We used here the fact that $\sum_{n=k+m} \frac{n!}{k! m!} u^k v^m = (u+v)^n$, provided that $uv = vu$.

A.2.5 Exponential of a Matrix and the Role of Commutativity

What happens if in the expression

$$e^A = 1 + \frac{1}{1!} A + \frac{1}{2!} A^2 + \cdots + \frac{1}{n!} A^n + \cdots ,$$

we consider A a square matrix, and 1 is the identity matrix of the same size? For example, if A is the identity matrix, then it is easy to check that e^A turns out to be a diagonal matrix, with elements e on the main diagonal.

Problem 5 a) Calculate $\exp A$ for the following matrices A:

$$\begin{pmatrix} 0 & 0 \\ 0 & 0 \end{pmatrix}, \quad \begin{pmatrix} 1 & 0 \\ 0 & -1 \end{pmatrix}, \quad \begin{pmatrix} 0 & 1 \\ -1 & 0 \end{pmatrix}, \quad \begin{pmatrix} 0 & 0 \\ 1 & 0 \end{pmatrix}, \quad \begin{pmatrix} 0 & 1 \\ 0 & 0 \end{pmatrix}, \quad \begin{pmatrix} 0 & 1 & 0 \\ 0 & 0 & 1 \\ 0 & 0 & 0 \end{pmatrix}.$$

b) Let A_1 and A_2 be the last two matrices of order two. Find e^{A_1}, e^{A_2} and check that $e^{A_1} \cdot e^{A_2} \neq e^{A_1 + A_2}$. What is going on here?

c) Show that $e^{tA} = I + tA + o(t)$, for $t \to 0$.

d) Check that $\det(I + tA) = 1 + t \cdot (\operatorname{tr} A) + o(t)$, where $\operatorname{tr} A$ is the trace of the square matrix A.

e) Prove the important relationship $\det e^A = e^{\operatorname{tr} A}$.

A.2.6 Exponential of Operators and Taylor's Formula

Let $P(x)$ be a polynomial and $A = \frac{d}{dx}$ the differentiation operator. Then $(AP)(x) = \frac{dP}{dx}(x) = P'(x)$.

Problem 6 a) Check that the relation $\exp(t\frac{d}{dx})P(x) = P(x+t)$ is what you know as Taylor's formula.

b) By the way, how many terms of the series e^x do you have to consider in order to obtain a polynomial that allows you to calculate e^x on the interval $[-3, 5]$ with an accuracy up to 10^{-2}?

A.3 Newton's Binomial

A.3.1 Expansion in Power Series of the Function $(1+x)^\alpha$

Newton knew the validity, for every natural number α, of the formula for the binomial expansion

$$(1+x)^\alpha = 1 + \frac{\alpha}{1!}x + \frac{\alpha(\alpha-1)}{2!}x^2 + \cdots + \frac{\alpha(\alpha-1)\cdots(\alpha-n+1)}{2!}x^2 + \cdots,$$

and then he remarked that this formula remains valid for arbitrary α, but the number of terms in the sum might be infinite.

For instance, $(1+x)^{-1} = 1 - x + x^2 - x^3 + \cdots$ if $|x| < 1$.

A.3.2 Integration of a Series and Expansion of $\ln(1+x)$

By integrating the last series over the interval $[0, x]$, we find that

$$\ln(1+x) = x - \frac{1}{2}x^2 + \frac{1}{3}x^3 - \cdots \quad \text{for } |x| < 1.$$

A.3.3 Expansion of the Functions $(1+x^2)^{-1}$ and $\arctan x$

Analogously, we write the expansion $(1+x^2)^{-1} = 1 - x^2 + x^4 - x^6 + \cdots$, we integrate its terms over the interval $[0, x]$, and we obtain

$$\arctan x = x - \frac{1}{3}x^3 + \frac{1}{5}x^5 - \cdots.$$

If we set $x = 1$, this expansion seems to imply that $\frac{\pi}{4} = 1 - \frac{1}{3} + \frac{1}{5} - \frac{1}{7} + \cdots$.

Perhaps this is true (and certainly it is), but we have the feeling that we are already going beyond the limits of what is permitted. The following example will only reinforce our concerns.

A.3.4 Expansion of $(1 + x)^{-1}$ and Computing Curiosities

For $x = 1$, the expansion $(1 + x)^{-1} = 1 - x + x^2 - x^3 + \cdots$ leads to the equality
$\frac{1}{2} = 1 - 1 + 1 - 1 + \cdots$.

By grouping terms, we can obtain $\frac{1}{2} = (1 - 1) + (1 - 1) + \cdots = 0$ and we can
obtain $\frac{1}{2} = 1 + (-1 + 1) + (-1 + 1) + \cdots = 1$.

After this, it is necessary to question almost everything that we have done so
successfully and nonchalantly by multiplying the infinite sums (series), rearranging
and grouping their terms, and integrating them. All this must obviously be clarified.
We shall do it soon, but before that, we mention yet another area where series are
commonly used.

A.4 Solution of Differential Equations

A.4.1 Method of Undetermined Coefficients

Consider the simplest equation $\ddot{x} + x = 0$ of harmonic oscillations. We shall look for
the solution as a series $x(t) = a_0 + a_1 t + a_2 t^2 + \cdots$. Substituting the series into the
equation, grouping the terms with equal powers of t, and equating the coefficients
with the same powers in t on both sides of the equation, we obtain an infinite system
of equations:

$$2a_2 + a_0 = 0, \qquad 2 \cdot 3 a_3 + a_1 = 0, \qquad 3 \cdot 4 a_4 + a_2 = 0, \quad \ldots.$$

If the initial conditions $x(0) = x_0$ and $x'(0) = v_0$ are given, then from the series
$x(t) = a_0 + a_1 t + a_2 t^2 + \cdots$, and $x'(t) = a_1 + 2a_2 t + \cdots$, we find that $a_0 = x_0$ and
$a_1 = v_0$. If we know a_0 and a_1, we can find successively and uniquely the remaining
coefficients of the expansion.

For example, if $x(0) = 0$ and $x'(0) = 1$, then

$$x(t) = t - \frac{1}{3!} t^3 + \frac{1}{5!} t^5 - \cdots = \sin t,$$

and if $x(0) = 1$ and $x'(0) = 0$, then

$$x(t) = 1 - \frac{1}{2!} t^2 + \frac{1}{4!} t^4 - \cdots = \cos t.$$

A.4.2 Use of the Exponential Function

What happens if the solution that we are looking for has the form $x(t) = e^{\lambda t}$? Then
$\ddot{x} + x = e^{\lambda t}(\lambda^2 - 1) = 0$, and therefore $\lambda^2 + 1 = 0$, i.e., $\lambda = i$ or $\lambda = -i$. But what
are these strange complex oscillations $x(t) = e^{it}$, $x(t) = e^{-it}$, and $x(t) = c_1 e^{it} + c_2 e^{-it}$?

Problem 7 Analyze the situation and solve the problem, for example, if $x(0) = 0$ and $x'(0) = 1$ or if $x(0) = 1$ and $x'(0) = 0$. Recall Euler's formula and compare your results with those obtained above.

A.5 The General Idea About Approximation and Expansion

A.5.1 The Meaning of a Positional Number System. Irrational Numbers

Recall the usual representation of the number $\pi = 3.1415926\ldots$ or in general a decimal expansion $a_0.a_1a_2a_3\ldots$: this is the sum $a_0 10^0 + a_1 10^{-1} + a_2 10^{-2} + a_3 10^{-3} + \cdots$.

We know that a finite expansion corresponds to a rational number, and the representation of an irrational number requires an infinite number of decimal digits, and therefore requires the study of an infinite number of terms and infinite sums, i.e., series.

If we truncate a series at some point, we get a rational number. We usually work such numbers. What happened here? We have simplified the object, allowing some error. This means that we are approximating a complex object (an irrational number in this case) through some other objects (the rational numbers here), while allowing some error, which we call the degree of precision of the approximation. An improvement in the precision leads to the complication of the object that we use as an approximation. A compromise has to be found depending on the concrete circumstances.

A.5.2 Expansion of a Vector in a Basis and Some Analogies with Series

In linear algebra and in geometry, we decompose vectors in terms of a basis. For mathematical analysis, the traditional representation

$$f(x) = f(0) + \frac{1}{1!}f'(0)x + \frac{1}{2!}f''(0)x^2 + \cdots$$

actually means the same thing if we consider that the basis is the set of functions $e_n = x^n$. This is the Taylor series of the function f at the point $x_0 = 0$.

Analogously, if some periodic signal or process $f(t)$ is subjected to spectral analysis, then one is interested in its decomposition $f(t) = \sum_{n=0}^{\infty} a_n \cos nt + b_n \sin nt$ into the simplest harmonic oscillations. Such series are called classical (or trigonometric) Fourier series.

What is new in this situation, in comparison with that in linear algebra, is that we consider here an infinite sum, which is understood as the limit of finite sums.

Thus in the space of our objects one must define the concept of proximity between the objects, in addition to the structure of a linear space, allowing one to be able to consider the limit of the sequence of the objects themselves or their sum.

A.5.3 Distance

The proximity between objects is determined by the presence of a particular concept, the concept of neighborhood of an object (neighborhood of a point in the space). This is the same as specifying a topology in the space. In topological spaces it is possible to speak about limits and continuity.

If in a space, a distance between objects, i.e., the points of the space, is somehow introduced, then the neighborhoods of a point are automatically defined, and even more specifically, the δ-neighborhoods of a point.

The distance between points of the same space can be measured in different ways. For example, the distance between two continuous functions over an interval can be measured by the maximum of the absolute value of the difference between the values of the functions on this interval (uniform metric), and it is also possible to measure it by the integral of the absolute value of the difference of the functions over this interval (integral metric). The choice of the metric is dictated by the problem under consideration.

Appendix B
Change of Variables in Multiple Integrals (Deduction and First Discussion of the Change of Variables Formula)[1]

B.1 Formulation of the Problem and a Heuristic Derivation of the Change of Variables Formula

By studying the integral in the one-dimensional case, at some moment we obtained an important change of variables formula for such an integral. Our task now is to find a change of variables formula in the general case. We formulate the problem more precisely.

Let D_x be a set in $\mathbb{R}^n$, f an integrable function on D_x, and $\varphi : D_t \to D_x$ a mapping $t \mapsto \varphi(t)$ from the set $D_t \subset \mathbb{R}^n$ to D_x. The question is, what is the law, assuming that we know f and φ, that allows us to find a function ψ on D_t such that we have the equality

$$\int_{D_x} f(x)\,dx = \int_{D_t} \psi(t)\,dt,$$

which reduces the computation of an integral over D_x to an integral over D_t?

We suppose first that D_t is an n-dimensional interval $I \subset \mathbb{R}^n$ and $\varphi : I \to D_x$ is a diffeomorphic mapping from I onto D_x. To every partition of the interval I into subintervals $I_1, I_2, \ldots, I_k$ corresponds a partition of D_x into subsets $\varphi(I_i)$, $i = 1, \ldots, k$. If all these sets are measurable and intersect pairwise only on sets of measure zero, then by the additivity of the integral,

$$\int_{D_x} f(x)\,dx = \sum_{i=1}^{k} \int_{\varphi(I_i)} f(x)\,dx. \tag{B.1}$$

If the function f is continuous on D_x, then the mean value theorem implies

$$\int_{\varphi(I_i)} f(x)\,dx = f(\xi_i)\mu\big(\varphi(I_i)\big),$$

---

[1] Fragment of a lecture with an alternative and independent proof of the change of variables formula.

© Springer-Verlag Berlin Heidelberg 2016
V.A. Zorich, *Mathematical Analysis II*, Universitext,
DOI 10.1007/978-3-662-48993-2

where $\xi_i \in \varphi(I_i)$. Since $f(\xi_i) = f(\varphi(\tau_i))$, with $\tau_i = \varphi^{-1}(\xi_i)$, then it remains for us to link $\mu(\varphi(I_i))$ with $\mu(I_i) = |I_i|$.

If φ is a linear transform, then $\varphi(I_i)$ is a parallelepiped, whose volume we know from analytical geometry and algebra and is equal to $|\det \varphi'|\mu(I_i)$. But a diffeomorphism is locally almost a linear map. Therefore, if the size of the intervals I_i is sufficiently small, then it can be assumed, with a small relative error, that $\mu(\varphi(I_i)) \approx |\det \varphi'(\tau_i)|\mu(I_i)$ (it is possible to prove that with a proper choice of the point $\tau_i \in I_i$, one has the exact equality). In this way,

$$\sum_{i=1}^{k} \int_{\varphi(I_i)} f(x)\, dx \approx \sum_{i=1}^{k} f\big(\varphi(\tau_i)\big)\big|\det \varphi'(\tau_i)\big| \cdot |I_i|. \tag{B.2}$$

However, on the right-hand side of this approximate equality there is the integral sum of the function $f(\varphi(t))|\det \varphi'(t)|$ over the interval I, corresponding to the partition P of this interval with marked points τ. In the limit $\lambda(P) \to 0$, from equations (B.1) and (B.2) we get

$$\int_{D_x} f(x)\, dx = \int_{D_t} f\big(\varphi(t)\big)\big|\det \varphi'(t)\big|\, dt. \tag{B.3}$$

This is the required formula together with its explanation. Note that it is possible to justify rigorously each step of this deduction, which led us to the formula. Strictly speaking, we need to prove only the validity of the last passage to the limit, that the integral on the right-hand side of (B.3) exists, and also to explain the approximation $\mu(\varphi(I_i)) \approx |\det \varphi'(\tau_i)| \cdot |I_i|$.

Let us do it.

B.2   Some Properties of Smooth Mappings and Diffeomorphisms

a) Recall that a smooth mapping φ from a closed and bounded interval $I \subset \mathbb{R}^n$ (or from any other convex compact subset) is a Lipschitz function. This follows from the mean value theorem and the boundedness of φ' (because of the continuity) over a compact set

$$\big|\varphi(t_2) - \varphi(t_1)\big| \leq \sum_{\tau \in [t_1, t_2]} \big\|\varphi'(\tau)\big\| \cdot |t_2 - t_1| \leq L|t_2 - t_1|. \tag{B.4}$$

b) Thus, the distance between the images of the points under the mapping φ cannot exceed L times the distance between the points.

For instance, if some subset $E \subset I$ has diameter d, then the diameter of its image $\varphi(E)$ is not more than Ld, and the set $\varphi(E)$ can be covered with (n-dimensional) cubes with edges of size Ld and volume $(Ld)^n$.

Thus if E is a cube with edges of size δ and volume δ^n, then its image is covered by a standard coordinate cube of volume $(L\sqrt{n}\delta)^n$.

c) It follows from this that the image under smooth mappings of 0-measure sets have also measure 0 (in the sense of n-dimensional objects). [After all, in the definition of a set of measure zero, it is possible to consider coverings by cubes, instead of a covering with general n-dimensional intervals, i.e., "rectangular parallelepipeds", as we can easily see.]

If a smooth mapping $\varphi : D_t \to D_x$ has also an inverse smooth mapping $\varphi^{-1} :$ $D_x \to D_t$, i.e., if φ is a diffeomorphism, then it is clear that the pre-image of a set with measure zero also has measure zero.

d) Since under a diffeomorphism, the Jacobian of the mapping $\det \varphi'$ is everywhere different from zero, and the mapping itself is bijective, then (due to the inverse function theorem) the interior points of any set under such a mapping are transformed into the interior points of the image of this set, and the boundary points are transformed into the boundary points of the image.

Recall the definition of an admissible (Jordan-measurable) set, as a bounded set whose boundary set has measure zero; thus we can conclude that under diffeomorphisms, the image of a measurable set is again a measurable set.

(This is also true for any smooth mapping. However, for diffeomorphisms it is even true that the pre-image of a measurable set is also a measurable set.)

e) This latter in particular means that if $\varphi : D_t \to D_x$ is a diffeomorphism, then from the existence of the integral on the left-hand side of formula (B.3) there follows (based on Lebesgue's criterion) the existence of the integral on the right-hand side.

B.3 Relation Between the Measures of the Image and the Pre-image Under Diffeomorphisms

We shall show that if $\varphi : I \to \varphi(I)$ is a diffeomorphism, then

$$\mu\big(\varphi(I)\big) = \int_I \det \varphi'(t)\, dt, \tag{B.5}$$

under the assumption that the integrand $\det \varphi'$ is positive.

Hence, by the mean value theorem, in particular, we find that there is a point $\tau \in I$ such that

$$\mu\big(\varphi(I)\big) = \det \varphi'(\tau)|I|. \tag{B.6}$$

Formula (B.5) is actually a particular case of (B.3), when $f \equiv 1$.

For linear mappings, this formula is already known, although perhaps without discussing those details related to the fact that it is valid (for linear maps) not only for simple parallelepipeds but for all measurable sets. Let us clarify this. We know that a linear map is the composite of elementary linear mappings, which, up to a possible permutation of a pair of coordinates, are reduced to a change in only one of these coordinates: multiplying or adding a number of any one of the coordinates to another one. Fubini's theorem allows us to determine that in the first case, the volume of any measurable set is multiplied by the same factor that multiplies the

B Change of Variables in Multiple Integrals

coordinate (more precisely, its absolute value if we consider nonoriented volume). In the second case, although the face changes, its volume remains the same, since the corresponding one-dimensional section only moves, keeping its linear measure. Finally, a permutation of a pair of coordinates changes the orientation of the spatial frame (the determinant of such a linear transformation is -1), but it does not change the nonoriented volume of the face. (In the language of Fubini's theorem, this is just a change in the order of two integrations.)

It now remains to recall that the determinant of the composition of linear mappings is the product of the determinants of the factors.

Thus, considering that for linear and affine mappings the formula (B.5) is already established, we prove it for an arbitrary diffeomorphism with positive Jacobian.

a) We use again the finite-increment theorem, but now to estimate the possible deviation of the mapping $\varphi : I \to \varphi(I)$ from the affine mapping $t \mapsto A(t) = \varphi(a) + \varphi'(a)(t - a)$, where t is a variable, and a is a fixed point in the interval I. The mapping $A : I \to A(I)$ is simply the linear part of the Taylor expansion of the function φ at the point $a \in I$.

If we apply the finite-increment (mean value) theorem to the function $t \to \varphi(t) - \varphi'(a)(t - a)$, we obtain

$$\left|\varphi(t) - \varphi(a) - \varphi'(a)(t - a)\right| \leq \sup_{\tau \in [a,t]} \left\|\varphi'(\tau) - \varphi'(a)\right\| \cdot |t - a|. \tag{B.7}$$

Given the uniform continuity of the continuous function φ' on the compact set I, from equation (B.7) we conclude that there is a nonnegative function $\delta \to \varepsilon(\delta)$, tending to zero as $\delta \to +0$, such that for any two points $t, a \in I \subset \mathbb{R}^n$,

$$|t - a| \leq \sqrt{n}\delta \Longrightarrow \left|\varphi(t) - A(t)\right| = \left|\varphi(t) - \varphi(a) - \varphi'(a)(t - a)\right| \leq \varepsilon(\delta)\delta. \tag{B.8}$$

b) Now we go back to the proof of formula (B.5). First we shall carry out a small technical simplification: we shall assume that the lengths of the edges of the parallelepiped I are commensurable and that therefore, they can be divided into equal cubes $\{I\}$ with arbitrarily small (as necessary) edges $\delta_i = \delta$ and volume $\delta_i^n = \delta^n$, i.e., $I = \bigcup_i I_i$ and $|I| = \sum_i |I_i| = \sum_i \delta_i^n$.

In every cube I_i, we fix a point a_i, we build the corresponding affine mapping $A_i(t) = \varphi(a_i) - \varphi'(a_i)(t - a_i)$, we consider the image $A_i(\partial I_i)$ of the cube's I_i boundary ∂I_i under the mapping A_i, and we consider the $\varepsilon(\delta)\delta$-neighborhood of this image, which we denote by Δ_i. By (B.8), the image $\varphi(\partial I_i)$ of the boundary ∂I_i of the cube I_i lies in Δ_i under the diffeomorphism φ. Thus, one has the following inclusions and inequalities:

$$A_i(I_i) \setminus \Delta_i \subset \varphi(I_i) \subset A_i(I_i) \cup \Delta_i,$$

$$\left|A_i(I_i)\right| - |\Delta_i| \leq \left|\varphi(I_i)\right| \leq \left|A_i(I_i)\right| + |\Delta_i|.$$

When we take the sum over all indices, we have

$$\sum_i \left|A_i(I_i)\right| - \sum_i |\Delta_i| \leq \left|\varphi(I)\right| = \sum_i \left|\varphi(I_i)\right| \leq \sum_i \left|A_i(I_i)\right| + \sum_i |\Delta_i|. \tag{B.9}$$

As $\delta \to +0$,

$$\sum_i |A_i(I_i)| = \sum_i \det \varphi'(a_i)|I_i| \to \int_I \det \varphi'(t)\,dt.$$

Therefore, to prove formula (B.5) in our case, it remains to verify that $\sum_i |\Delta_i| \to 0$ if $\delta \to +0$.

c) We estimate from above the volume $|\Delta_i|$, based on the estimates (B.4) and (B.8). According to (B.4), the edges of the parallelepiped $A_i(I_i)$ have length not greater than $L\delta$, where $\delta = \delta_i$ is the length of the edge of a cube I_i. Thus the $(n-1)$-dimensional "area" of any of the $2n$ faces of the parallelepiped $A_i(I_i)$ is not greater than $(L\delta)^{n-1}$. We take an $\varepsilon(\delta)\delta$-neighborhood of such a face. Its volume is estimated with the value $(2+2)\varepsilon(\delta)\delta(L\delta)^{n-1}$, where the second 2 appearing in the formula is the absorption contribution of the rounded parts of this neighborhood, occurring near the boundary of the face. In this way, $|\Delta_i| < 2n \cdot 4L^{n-1}\varepsilon(\delta)\delta^n$; therefore,

$$\sum_i |\Delta_i| < 8nL^{n-1} \sum_i \varepsilon(\delta)\delta_i^n = 8nL^{n-1}\varepsilon(\delta)|I|,$$

and we see that $\sum_i |\Delta_i| \to 0$ for $\delta \to +0$.

d) The estimated values for $|\Delta_i|$ show at the same time that no matter how arbitrarily small the reduction of the edges of the original interval I becomes, which one might need in order to obtain their commensurability, in the limit this does not affect the result.

B.4 Some Examples, Remarks, and Generalizations

Thus formula (B.3) for the case $D_t = I$ and a continuous function f is already proved. We shall consider and discuss some examples. These will show at the same time that in fact, we have already proved formula (B.3) not only for the case $D_t = I$ and not only for a continuous function f.

a) *Negligible sets.* As it is used in practice, replacing variables or the use of a coordinate transformation formula sometimes has several special features (for example, somewhere there might be a violation of mutual uniqueness, vanishing of the Jacobian, or lack of differentiability). Typically, these special features occur on sets of measure zero, and are therefore relatively easy to overcome.

For example, if you need to go from an integral over a circle to an integral over a rectangle, we often make the change of variables

$$x = r\cos\varphi, \qquad y = r\sin\varphi. \tag{B.10}$$

These are the well-known formulas for the transition from polar coordinates to Cartesian coordinates in the plane. The rectangle $I = \{(r, \varphi) \in \mathbb{R}^2 \mid 0 \le r \le R, 0 \le \varphi \le 2\pi\}$ under this mapping is transformed into the circle $K = \{(x, y) \in \mathbb{R}^2 \mid$

$x^2 + y^2 \le R^2\}$. This mapping is smooth, but it is not a diffeomorphism: the whole
side of the rectangle I on which $r = 0$ is transformed under this mapping into the
point $(0,0)$; the images of the points $(r, 0)$ and $(r, 2\pi)$ coincide. However, if we consider, for example, the sets $I \setminus \partial I$ and $K \setminus E$, where E is the union of the boundary
∂K of the circle K and the radius going to the point $(0, R)$, then the restriction of
the mapping (B.10) to the domain $I \setminus \partial I$ is a diffeomorphism with the set $K \setminus E$.
Therefore, if instead of the rectangle I, we take a slightly smaller rectangle I_δ lying
strictly in the interior of I, then we can apply formula (B.10) to this rectangle I_δ
and its image K_δ. And then, exhausting the rectangle I with such rectangles I_δ and
noticing that their images exhaust the circle K, that $|I_\delta| \to |I|$ and $|K_\delta| \to |K|$, in
the limit we obtain formula (B.3) applied to the original pair K, I.

This applies, of course, to the general polar (spherical) coordinates system in $\mathbb{R}^n$.
We shall now develop these observations.

b) *Exhaustions and limit transitions.* We define an *exhaustion* of a set $E \subset \mathbb{R}^n$
to be a sequence of measurable sets $\{E_n\}$ such that $E_n \subset E_{n+1} \subset E$ for every $n \in \mathbb{N}$
and $\bigcup_{n=1}^{\infty} E_n = E$.

Lemma 1 *If $\{E_n\}$ is an exhaustion of a measurable set E, then*

a) $\lim_{n\to\infty} \mu(E_n) = \mu(E)$;
b) *for every function $f \in \mathcal{R}(E)$, one has $f|_{E_n} \in \mathcal{R}(E_n)$ and*

$$\lim_{n\to\infty} \int_{E_n} f(x)\,dx = \int_E f(x)\,dx.$$

Proof a) Since $E_n \subset E_{n+1} \subset E$, then $\mu(E_n) \le \mu(E_{n+1}) \le \mu(E)$ and
$\lim_{n\to\infty} \mu(E_n) \le \mu(E)$. For proving the equality in a), we shall show that the inequality $\lim_{n\to\infty} \mu(E_n) \ge \mu(E)$ also holds.

The boundary ∂E of the set E is compact and has measure zero. Therefore, it can
be covered with a finite number of open intervals such that the sum of their volumes
is less than ε for a given $\varepsilon > 0$. Let Δ be the union of these open intervals. Then the
set $O = E \cup \Delta$ is open in $\mathbb{R}^m$; by construction, O contains the closure $\overline{E}$ of the set
E; and $\mu(O) \le \mu(E) + \mu(\Delta) < \mu(E) + \varepsilon$.

For every set E_n of the exhaustion $\{E_n\}$ we repeat the construction above with
$\varepsilon_n = \varepsilon/2^n$. We obtain then a sequence of open sets $O_n = E_n \cup \Delta_n$ such that $E_n \subset$
O_n, $\mu(O_n) \le (E_n) + \mu(\Delta_n) < \mu(E_n) + \varepsilon_n$ and $\bigcup_{n=1}^{\infty} O_n \supset \bigcup_{n=1}^{\infty} E_n \supset E$.

The system of open sets $\Delta, O_1, O_2, \dots$ is an open cover of the compact set $\overline{E}$.
Let $\Delta, O_1, O_2, \dots, O_k$ be a finite open subcover of the compact set $\overline{E}$. Since
$E_1 \subset E_2 \subset \cdots \subset E_k$, the sets $\Delta, \Delta_1, \dots, \Delta_k, E_k$ are also a cover of $\overline{E}$, and then

$$\mu(E) \le \mu(\overline{E}) + \mu(\Delta) + \mu(\Delta_1) + \cdots + \mu(\Delta_k) < \mu(E_k) + 2\varepsilon.$$

It follows from this that $\mu(E) \le \lim_{n\to\infty} \mu(E_n)$.

b) The fact that $f|_{E_n} \in \mathcal{R}(E_n)$ is known to us, and it follows from Lebesgue's
criterion for the existence of the integral over a measurable set. By the hypothe-
sis $f \in \mathcal{R}(E)$, there exists a constant M such that $|f(x)| \le M$ over E. From the

additivity of the integral and the general estimates for the integral, we get

$$\left| \int_E f(x)\,dx - \int_{E_n} f(x)\,dx \right| = \left| \int_{E\setminus E_n} f(x)\,dx \right| \le M\mu(E \setminus E_n).$$

Hence, taking into account what we proved in a), we conclude that assertion b) holds. $\qquad\square$

The additivity of the integral and the possibility of exhausting the domain of integration with the domains where the change of variables formula works (i.e., it is directly applicable) allow us to apply the formula to the original domain. In general, the idea of exhaustion lies at the heart of many constructions in analysis. In particular, it is fundamental in the definition of improper integrals.

Appendix C
Multidimensional Geometry and Functions of a Very Large Number of Variables (Concentration of Measures and Laws of Large Numbers)

C.1 An Observation

Almost the entire volume of a multidimensional body is concentrated in a small neighborhood of the boundary of the body.

Problem 1 a) Check this in the examples of the cube and the ball. Show that if we remove the shell with thickness 1 cm from a 1000-dimensional watermelon with 1 meter radius, then there remains less than a thousandth of the original watermelon.

b) If we project the sphere $S^{n-1}(r) \subset \mathbb{R}^n$ orthogonally onto a hyperplane passing through the center of the sphere, then we obtain a ball (double covered) with the same dimension $n-1$ and the same radius r. Considering what we obtain above, notice (still on a qualitative level), that almost all the area of the sphere $S^{n-1}(r)$ for $n \gg 1$ is concentrated in a small neighborhood of the equator, the intersection of the sphere with the former hyperplane.

C.2 Sphere and Random Vectors

Problem 2 a) The sphere $S^{n-1}(r)$ with radius r and center at the origin of the n-dimensional Euclidean space $\mathbb{R}^n$ is projected orthogonally onto a coordinate axis. We get the interval $[-r, r]$. We fix another interval $[a, b] \subset [-r, r]$. Let $S[a, b]$ be the area of the part $S^{n-1}_{[a,b]}(r)$ of the sphere $S^{n-1}(r)$ that is projected onto the interval $[a, b]$. Find the quotient $\frac{S[a,b]}{S[-r,r]}$, i.e., the probability $\mathrm{Pr}_n[a, b]$ that a randomly chosen point on the sphere will be on the layer $S^{n-1}_{[a,b]}(r)$ over the interval $[a, b]$, considering that the points are uniformly distributed over the sphere.
Answer:

$$\mathrm{Pr}_n[a, b] = \frac{\int_a^b (1 - (x/r)^2)^{\frac{n-3}{2}}\, \mathrm{d}x}{\int_{-r}^r (1 - (x/r)^2)^{\frac{n-3}{2}}\, \mathrm{d}x}.$$

© Springer-Verlag Berlin Heidelberg 2016
V.A. Zorich, *Mathematical Analysis II*, Universitext,
DOI 10.1007/978-3-662-48993-2

b) Let $\delta \in (0, 1)$ and $[a, b] = [\delta r, r]$. Show that as $n \to \infty$,

$$\Pr_n[\delta r, r] \sim \frac{1}{\delta\sqrt{2\pi n}} e^{-\frac{1}{2}\delta^2 n}.$$

Hint: You can use Laplace's method for obtaining asymptotics of the integral over a large parameter.

c) The result obtained in b) implies that the vast majority of the area of a multidimensional sphere is concentrated in a small neighborhood of the equatorial plane, in the layer $S^{n-1}_{[-\delta r, \delta r]}(r)$ over the interval $[-\delta r, \delta r]$.

Deduce from this that if we take independently and randomly a pair of vectors in $\mathbb{R}^n$, then for $n \gg 1$, it is very likely that they will be almost orthogonal, i.e., their scalar product will be close to zero. Estimate the probability that the scalar product is greater than $\varepsilon > 0$ and calculate its variance for $n \gg 1$.

d) Prove, based on the result proved in a), that for $r = \sigma\sqrt{n}$ and $n \to \infty$, one has

$$\Pr_n[a, b] \to \frac{1}{\sqrt{2\pi}\sigma} \int_a^b e^{-\frac{x^2}{2\sigma^2}} \, dx.$$

e) Considering the result obtained in b), prove now Gauss's law on the distribution of measurement errors and Maxwell's laws on the distribution of gas molecules according to speed and energy (considering in the first case that the observations are independent and their mean square stabilizes as the number of observations increases, and in the second case considering that the gas is homogeneous and that the total energy of the molecules in a portion of the gas is proportional to the number of molecules in this portion).

C.3 Multidimensional Sphere, Law of Large Numbers, and Central Limit Theorem

By solving this problem, you will discover the following fact, important in many aspects and manifested in many areas (for example, in statistical physics).

Let S^m be the unit sphere in the Euclidean space $\mathbb{R}^{m+1}$ with a very large dimension $m + 1$. Suppose also that we are given a sufficiently regular real-valued function on the sphere (for example, from a fixed Lipschitz class). We take randomly and independently two points and calculate the value of the function at these points. With a high probability, the values will almost coincide and they will be close to a certain number M_f.

(This, still hypothetical, number M_f is called the *median value of the function* or *function median*. It is also called the *average value of the function in the sense of Lévy*.[1] The motivation for these terms will soon be clear, together with a precise definition of the number M_f.)

---

[1] P. Lévy (1886–1971) – famous French mathematician, student of J. Hadamard.

We introduce some notation and conventions. We define the distance between two points on the sphere $S^m \subset \mathbb{R}^{m+1}$, understood in terms of its geodesic metric ρ. We denote by A_δ a δ-neighborhood in S^m of the set $A \subset S^m$. We replace the standard mass of the sphere with a uniformly distributed probability measure μ, i.e., $\mu(S^m) = 1$.

We have the following assertion proved by Paul Lévy, commonly called *Lévy's isoperimetric inequality*.

For every $0 < a < 1$ *and* $\delta > 0$, *there exists* $\min\{\mu(A_\delta) \mid A \subset S^m, \mu(A) = a\}$, *and it is attained on the spherical cap* A^0 *with measure* a.

Here $A^0 = B(r)$, where $B(r) = B(x_0, r) = \{x \in S^m \mid \rho(x_0, x) < r\}$ and $\mu(B(r)) = a$.

Problem 3 a) For $a = 1/2$, i.e., when A^0 is a hemisphere, obtain the following result:

If the subset $A \subset S^{n+1}$ *is such that* $\mu(A) \geq 1/2$, *then* $\mu(A_\delta) \geq 1 - \sqrt{\pi/8}e^{-\delta^2 n/2}$.
(If $n \to \infty$, we can change here $\sqrt{\pi/8}$ for $1/2$.)

b) We denote by M_f the number such that

$$\mu\{x \in S^n \mid f(x) \leq M_f\} \geq 1/2 \quad \text{and} \quad \mu\{x \in S^n \mid f(x) \geq M_f\} \geq 1/2.$$

It is called the *median* or *average value in the sense of Lévy of the function* $f : S^n \to \mathbb{R}$. (If the M_f-level of the function f on the sphere has measure zero, then the measure of each of these two sets mentioned above will be equal to exactly half of the μ-area of the sphere S^m.)

Obtain the following lemma due to Lévy:

If $f \in C(S^{n+1})$ *and* $A = \{x \in S^{n+1} \mid f(x) = M_f\}$, *then* $\mu(A_\delta) \geq 1 - \sqrt{\pi/2} \times e^{-\delta^2 n/2}$.

c) Let $\omega_f(\delta) = \sup\{|f(x) - f(y)| \mid \rho(x, y) \leq \delta\}$ be the *modulus of continuity of the function* f.

The values of the function f on the set A_δ are close to M_f. More precisely, if $\omega_f(\delta) \leq \varepsilon$, then $|f(x) - M_f| \leq \varepsilon$ on A_δ. Thus Lévy's lemma shows that "good" functions are actually almost constant in almost their entire domain of definition S^m when the dimension m is very large.

Considering that $f \in \mathrm{Lip}(S^{n-1}, \mathbb{R})$ and L is the Lipschitz constant of the function f, estimate the probability $\Pr\{|f(x) - M_f| > \varepsilon\}$ and the dispersion value $|f(x) - M_f|$ for $n \gg 1$.

d) Obtain, as above, estimates in the case that the function f is not defined on the unit sphere but in the sphere $S^{n-1}(r)$ with radius r.

e) If f is a smooth function, then we can clearly take the maximum modulus of its gradient as the Lipschitz constant L. For example, the linear function $S_n = \frac{1}{n}(x_1 + \cdots + x_n)$ has $L = L_n = \frac{1}{\sqrt{n}}$. Suppose that we have a sequence of Lipschitz functions $f_n \in \mathrm{Lip}(S^{n-1}(r_n), \mathbb{R})$, for which $L_n = O(\frac{1}{\sqrt{n}})$ and $r_n = \sqrt{n}$.

Estimate $\Pr\{|f_n(x) - M_{f_n}| > \varepsilon\}$ and the dispersion value $|f_n(x) - M_{f_n}|$ for $n \gg 1$.

In particular, for $f_n = S_n$ deduce the standard law of large numbers.

666C Multidimensional Geometry and Functions

f) Let $f_n = x_1 + \cdots + x_n$. The levels of this function are hyperplanes in $\mathbb{R}^n$ orthogonal to the vector $(1, \ldots, 1)$. The same can be said about the linear function $\Sigma_n = \frac{1}{\sqrt{n}}(x_1 + \cdots + x_n)$, with the only difference that under the movement from the origin in the direction of $(1, \ldots, 1)$, its values coincide with the distances to the origin. For this reason, its values are distributed on the sphere $S^{n-1}(r_n)$ exactly as they are on each of the coordinates.

Using this discussion and the result of Problem 2.d), setting $r_n = \sigma\sqrt{n}$, obtain your own version of the central limit theorem.

C.4 Multidimensional Intervals (Multidimensional Cubes)

Problem 4 a) Let I be the standard unit interval $[0, 1]$ of the real line $\mathbb{R}$, and I^n the standard n-dimensional interval in $\mathbb{R}^n$, usually called the n-dimensional unit cube. This is a unit of volume in $\mathbb{R}^n$, but its diameter $\sqrt{n}$ for $n \gg 1$ is extremely huge. Thus, even Lipschitz functions on I^n with Lipschitz constant L can have values spread within $L\sqrt{n}$.

Yet here, as in the above case of a sphere, there is a phenomenon of asymptotic stabilization (concentration) of values of such functions in the limit $n \to \infty$.

Now, try to find the proper formulations of the problem and study the phenomenon, up to the level of your ability (then check Sect. C.5 of this appendix).

b) Suppose we have n independent random variables x_i, taking values in the unit interval $[0, 1]$ and having distribution probabilities $p_i(x)$, which are uniformly separated from zero (in particular, all $p_i(x)$ may coincide). Then as n grows, the large majority of the random points $(x_1, \ldots, x_n) \in I^n$ will lie in close proximity to the border of the cube.

Explain this, and considering the result in a), obtain your own general law of large numbers.

c) Show with an example that if the probability density of the random variables in b) is concentrated in the vertices of the cube as point masses, then the asymptotic stabilization of values for Lipschitz functions in the limit $n \to \infty$ may not occur.

d) We noted above that although the volume of the cube I^n in $\mathbb{R}^n$ is equal to 1, its diameter $\sqrt{n}$ increases for $n \gg 1$, which creates difficulties. However, we have the following useful compensating observation: if each of two subsets A and B of the cube I^n has measure greater than an arbitrarily small fixed positive number ε, then the distance between A and B is bounded from above by a constant depending only on ε (and not depending on n). Prove this, and use this result if you need it.

e) Calculate the volume of the unit ball in $\mathbb{R}^n$ and show that the radius of the ball with volume one increases as $\sqrt{n/(2\pi e)}$ as $n \to \infty$. Go back to Sects. C.1 and C.2 and convince yourself again that the normal distribution and the laws related to it are closely linked in the geometric aspect with a simple multidimensional object, namely with the ball of unit volume.

C.5 Gaussian Measures and Their Concentration

Problem 5 a) We mentioned in Sect. C.2 of this appendix the isoperimetric inequality on the sphere, in connection with the discussion of the observed stabilization of values (constancy) of regular functions on the multidimensional sphere. The same problem about minimizing the measure of a δ-blowup of a set is important, and for the same reason it is also interesting in relation to other spaces that serve as natural domains for the relevant functions.

For example, in the case of the Gaussian probability measures defined by the normal probability distribution in the standard Euclidean space $\mathbb{R}^n$, the answer is also known (obtained by Borel). In this case, the extreme domain (with the fixed initial value of the Gaussian measure and a δ-blowup, understood in the sense of the Euclidean metric) turns out to be a half-space.

In particular, if we take the half-space with Gaussian measure $\frac{1}{2}$ and we directly calculate the value of the Gaussian measure of the complement in its Euclidean δ-blowup, then considering Borel's isoperimetric inequality, we can deduce that for any set A having a Gaussian measure $\frac{1}{2}$ in the space $\mathbb{R}^n$, the measure of its δ-blowup can be estimated from below with $\mu(A_\delta) \geq 1 - I_\delta$, where I_δ is the integral of the density $(2\pi)^{\frac{n}{2}} \exp(-\frac{|x|^2}{2})$ of the Gaussian measure of the half-space, given with Euclidean distance δ from the origin.

An estimate from above of the integral I_δ, for example, allows us to claim that $\mu(A_\delta) \geq 1 - 2\exp(-\frac{\delta^2}{2})$. Prove this.

b) This is a rough estimate, but it shows the rapid growth of $\mu(A_\delta)$, with an increase of δ, whatever the initial set A of measure $\frac{1}{2}$ is.

It is very interesting to notice (and considering the possible transition to infinite-dimensional spaces, even quite useful) that the last estimate does not depend on the dimension of the space. It may seem that this absence of the dimension is a great loss and weakness in the estimates within the context of concentration measures discussed and in the stabilization of values of functions of several variables. In fact, this estimate even contains the principle of the concentration of a measure on the unit sphere of large dimension, discussed above.

It is enough to prove (prove it) that the main part of the Gaussian probability measure of the Euclidean space $\mathbb{R}^n$ for $n \gg 1$ is concentrated in the vicinity of the unit Euclidean sphere of radius $\sqrt{n}$. This means that at the intersection of this neighborhood with the half-space, which is distant from the origin, the proportion of this measure is exponentially small. Therefore, the main part of the measure is in this neighborhood of the sphere of radius $\sqrt{n}$, which falls in the layer between two close parallel hyperplanes, symmetric with respect to the origin. If now we move through a homothety from the sphere of radius $\sqrt{n}$ to the unit sphere, then we obtain the principle of concentration of measure on the unit sphere, which we have already discussed (do the necessary calculations). In the statement of this principle, the dimension of the space occurs explicitly. This dimension was also present in the Gaussian case, but it was hidden in the size $\sqrt{n}$ of the sphere, and the main part of the measure of the whole space is concentrated in a neighborhood of this sphere.

C.6 A Little Bit More About the Multidimensional Cube

In the Euclidean space $\mathbb{R}^n$ we consider the n-dimensional unit interval ("cube")

$$I^n := \left\{ x = (x^1, \ldots, x^n) \in \mathbb{R}^n \mid |x^i| \leq \frac{1}{2}, i = 1, 2, \ldots, n \right\}.$$

Its volume is equal to one, although the diameter is $\sqrt{n}$. (Recall that the Euclidean ball of volume one in $\mathbb{R}^n$ has radius of order $\sqrt{n}$, as mentioned above.) We shall consider the standard probability measure uniformly distributed on the cube I^n.

Let $a = (a^1, \ldots, a^n)$ be a unit vector, and $x = (x^1, \ldots, x^n)$ an arbitrary point in the cube I^n.

The following inequality holds (probability estimate of Bernstein type):

$$\Pr_n \left\{ \left| \sum_{i=1}^{n} a^i x^i \right| \geq t \right\} \leq 2 \exp(-6t^2).$$

If we interpret the sum $\sum_{i=1}^{n} a^i x^i$ as a scalar product $\langle a, x \rangle$, we notice that this can be large (of order $\sqrt{n}$) if the vector a is not directed along any edge of the cube, but along the main diagonal, mixing all coordinate directions equally. If we take $a = (\frac{1}{\sqrt{n}}, \ldots, \frac{1}{\sqrt{n}})$ in the previous estimate, we deduce that the volume of the n-dimensional cube I^n concentrates, as n increases, in a small neighborhood of the hyperplane passing through the origin and orthogonal to the vector $(\frac{1}{\sqrt{n}}, \ldots, \frac{1}{\sqrt{n}})$.

In particular, if we consider a billiard in such a cube as a dynamical system (gas) composed with noninteracting particles, then for $n \gg 1$, the large majority of particle trajectories will go in a direction nearly perpendicular to the fixed vector $(\frac{1}{\sqrt{n}}, \ldots, \frac{1}{\sqrt{n}})$, and they are a large part of the time in a neighborhood of the above hyperplane.

C.7 The Coding of a Signal in a Channel with Noise

We point out in conclusion another area where the functions with a very large number of variables also appear naturally and where the principle of concentration of a measure is shown and also used substantially.

We are already used to the digital (discrete) coding and transmission of signals (music, images, messages, information) on a communication channel. In this form, a message can be thought of as a vector $x = (x^1, \ldots, x^n)$ in the space $\mathbb{R}^n$ with a very large dimension. The transmission of such messages requires an energy E, which is proportional to $\|x\|^2 = |x^1|^2 + \cdots + |x^n|^2$ (like the total kinetic energy of the gas molecules, discussed above). If T is the duration of the transmitted message x, then $P = E/T$ is the average power required to transfer one character (a coordinate of the vector x). If Δ is the average time required to transfer a single coordinate of the vector x, then $T = n\Delta$ and $E = nP\Delta$.

The transmitting and receiving devices are aligned in a such a way that the transmitter transforms (encodes) the original message to be transmitted in the form of the vector x. It sends it over the communication channel, and the receiver, knowing the code, decrypts x, transforming it into the form of the original message.

If we need to transmit M messages $A_1, \ldots, A_M$ of length n, then it is enough to fix n points in the ball of radius $\sqrt{E}$, agreeing on this selection with the receiving end of the communication channel. If in the communication channel there is no interference, then having received the vector from the agreed set, the receiver decodes it correctly into the corresponding message A.

If in the channel we do have interference (which is often the case), then because of the interference, a random vector $\xi = (\xi^1, \ldots, \xi^n)$ shifts the transmitted vector a, and the vector $a + \xi$ arrives at the receiver, and this vector must be properly decoded.

If the points $a_1, \ldots, a_n$ were chosen in such a way that the balls of radius $\|\xi\|$ with these points as center do not intersect, then an unambiguous deciphering is still possible. But if we want to meet this requirement, then we cannot take just any points $a_1, \ldots, a_M$, and there is a problem of dense packing of spheres. This is a difficult problem, whose solution in the present situation can be avoided, as was shown by Shannon, given that here the dimension n of the space $\mathbb{R}^n$ is huge.

We shall allow ourselves sometimes to make mistakes while interpreting the received message. However, we require the probability of error to be arbitrarily small (less than any fixed positive number).

Shannon showed that even in the presence of random noise (white noise) in the communication channel with limited capacities, by choosing a long enough code (i.e., for a large value of n), it is possible to achieve velocities of transmission close to the velocities of transmission of information in channels without noise, with an arbitrarily small probability of error.

The geometric idea of Shannon's theorem is directly related to the characteristics discussed above of the distribution measures (volumes) of domains in a space with large dimension. Let us explain this.

Suppose that two identical balls in the space $\mathbb{R}^n$ intersect. If the received signal lies in this intersection, then it is possible to have errors in the interpretation of the message sent by the source. But if the probability of falling into such an area is considered proportional to the relative volume of the region, then it is natural to compare the volume of the intersection of the balls with the volume of a ball. We carry out the proper estimations. If the centers of two balls of radius 1 are separated by the distance ε ($0 < \varepsilon < 2$), then the intersection of these balls is contained in a ball of radius $\sqrt{1 - (\varepsilon/2)^2}$ with center in the middle of the segment connecting the centers of the original balls. Hence, the ratio between the volume of the intersection of the two balls and their own original volume does not exceed $(1 - (\varepsilon/2)^2)^{n/2}$. It is clear now that for every fixed ε, this value can be made arbitrarily small by choosing a sufficiently large value of n.

Appendix D
Operators of Field Theory in Curvilinear Coordinates

Introduction

Almost any book with mathematical problems and even any textbook of mathematical analysis states something like the following. "Children, remember":

We call the *gradient* of a function $U(u, x, z)$ the vector

$$\operatorname{grad} U := \left(\frac{\partial U}{\partial x}, \frac{\partial U}{\partial y}, \frac{\partial U}{\partial z} \right).$$

The *curl* of a vector field $A = (P, Q, R)(x, y, z)$ is the vector

$$\operatorname{curl} A := \left(\frac{\partial R}{\partial y} - \frac{\partial Q}{\partial z}, \frac{\partial P}{\partial z} - \frac{\partial R}{\partial x}, \frac{\partial Q}{\partial x} - \frac{\partial P}{\partial y} \right).$$

The *divergence* of a vector field $B = (P, Q, R)(x, y, z)$ is the function

$$\operatorname{div} B := \frac{\partial P}{\partial x} + \frac{\partial Q}{\partial y} + \frac{\partial R}{\partial z}.$$

The fact that this is true only in Cartesian coordinates is not usually discussed, as well as what should be done if the coordinate system is different. This is understandable, since the very formulation of this problem already requires some suitable definition of these objects.

D.1 Reminders of Algebra and Geometry

D.1.1 Bilinear Forms and Their Coordinate Representation

a. Scalar Product and General Linear Forms

We shall consider a vector space with a scalar product $\langle, \rangle$. We can still consider that $\langle, \rangle$ denotes an arbitrary bilinear form on an n-dimensional vector space X. If we

© Springer-Verlag Berlin Heidelberg 2016
V.A. Zorich, *Mathematical Analysis II*, Universitext,
DOI 10.1007/978-3-662-48993-2

choose a basis of the space $\xi_1, \ldots, \xi_n$, then the objects of the space (in particular, vectors and forms) will have a coordinate representation. We recall the coordinate representation of the bilinear form $\langle , \rangle$.

If we take two vectors $x = x^i \xi_i$, $y = y^j \xi_j$ and their decomposition in terms of the basis, then we have $\langle x, y \rangle = \langle x^i \xi_i, y^j \xi_j \rangle = \langle \xi_i, \xi_j \rangle x^i y^j = g_{ij} x^i y^j$. As usual, summation over repeated indices is understood. Thus if a basis of the space is given, the choice of values $\langle \xi_i, \xi_j \rangle = g_{ij}$ completely defines the bilinear form.

If the form is a scalar product, then a basis is orthogonal if $g_{ij} = 0$ for $i \neq j$. It is assumed here that the form is nondegenerate, of course.

b. Nondegeneracy of Bilinear Forms

A bilinear form is called nondegenerate if once we fix a value in one of its arguments, then the bilinear form is identically zero with respect to the other argument if and only if the fixed value is zero (the zero vector).

The nondegeneracy of the form is equivalent to the fact that the determinant of the matrix (g_{ij}) is different from zero. Indeed, if the fixed vector $x = x^i \xi_i$ is such that $\langle x, y \rangle \equiv 0$ with respect to y, then $\langle \xi_i, \xi_j \rangle x^i = 0$ and $g_{ij} x^i = 0$ for every value $j \in \{i, \ldots, n\}$. This homogeneous system of equations has a unique solution (zero) if and only if the determinant of the matrix (g_{ij}) of the system is nonzero.

D.1.2 Correspondence Between Forms and Vectors

a. 1-Forms in the Presence of 2-Forms and Their Correspondence with Vectors

If one has a 2-form $\langle , \rangle$, then each vector A can be associated with a 1-form, namely the linear form $\langle A, x \rangle$. If the 2-form is nondegenerate, then the correspondence is one-to-one. Indeed, if we are given such a linear function $a(x) = a_j x^j$ (where $a_j = a(\xi_j)$) and we want to represent it in the form $\langle A, x \rangle$, where $A = \xi_i A^i$, then in the coordinates of the vector A we have the system of equations $a(\xi) = \langle \xi_i, \xi_j \rangle A^i$, $j = 1, \ldots, n$, which is uniquely solvable if the determinant of the matrix (g_{ij}) is different from zero.

Thus, the coordinates of the vector $A = A^i \xi_i$ and the coefficients of the 1-form a in the same basis $\{\xi_i\}$ are linked by the mutually inverse relations

$$a_j = g_{ij} A^i, \qquad A^i = g^{ij} a_j.$$

b. Correspondence Between a Vector and an $(n-1)$-Form

Similarly, if one has a nondegenerate n-form Ω, each vector B can be associated with an $(n-1)$-form, namely the form $\Omega(B, \ldots)$.

We shall deal below with vector fields A, B and carry out this described method on the tangent space, for example in relation to the form of work $\omega_A^1 = \langle A, \cdot \rangle$ and the form flux $\omega_B^{n-1} = \Omega^n(B, \dots)$, in the presence of the inner product $\langle, \rangle$ and the volume form Ω^n, respectively.

D.1.3 Curvilinear Coordinates and Metric

a. Curvilinear Coordinates, Metric, and Volume Form

Suppose that in an n-dimensional surface (manifold) we have a metric, which in local coordinates $(t^1, \dots, t^n)$ (in the local charts) is given by the form $g_{ij}(t)\,dt^i\,dt^j$, determined by the scalar product $\langle, \rangle(t)$, with the corresponding parameter t of the tangent plane (tangent space) to the surface.

For example, if the surface (or curve) is given in a parametric form, it is embedded into the Euclidean space, and then the scalar product in the tangent planes (spaces) to the surface is naturally induced from that in the ambient space.

We even know how to find the area of such a surface (n-measure), i.e., it is necessary to integrate the volume form

$$\Omega = \sqrt{\det g_{ij}(t)}\,dt^1 \wedge \cdots \wedge dt^n.$$

b. Orthogonal Systems of Curvilinear Coordinates and Unit Vectors

Recall that a system of curvilinear coordinates $(t^1, \dots, t^n)$ is called orthogonal if $g_{ij} \equiv 0$ for $i \neq j$.

The length element in an orthogonal system of curvilinear coordinates is written in a particularly simple form:

$$ds^2 = g_{11}(t)\left(dt^1\right)^2 + \cdots + g_{nn}(t)\left(dt^n\right)^2.$$

It is often rewritten in the more compact notation

$$ds^2 = E_1(t)\left(dt^1\right)^2 + \cdots + E_n(t)\left(dt^n\right)^2.$$

The vectors $\xi_1 = (1, 0, \dots, 0), \dots, \xi_n = (0, \dots, 0, 1)$ of the coordinate directions form a basis of the tangent space, corresponding to the value of the parameter t. But the norm (length) of these vectors is, in general, not equal to one. We have always, independent of whether the system of coordinates is orthogonal, $\langle \xi_i, \xi_i \rangle(t) = g_{ii}(t)$, i.e., $\|\xi_i\| = \sqrt{g_{ii}(t)}$, $i \in \{1, \dots, n\}$.

Thus, the unit vectors $(e_1, \dots, e_n)$ (vectors of length one) of the coordinate directions have the following coordinate representation:

$$e_1 = \left(\frac{1}{\sqrt{g_{11}}}, 0, \dots, 0\right), \quad \dots, \quad e_n = \left(0, \dots, 0, \frac{1}{\sqrt{g_{nn}}}\right).$$

In particular, if the system of curvilinear coordinates is orthogonal, then the following system of vectors of coordinate directions will be an orthonormal basis in the corresponding tangent space:

$$e_1 = \left(\frac{1}{\sqrt{E_1}}, 0, \ldots, 0 \right), \quad \ldots, \quad e_n = \left(0, \ldots, 0, \frac{1}{\sqrt{E_n}} \right).$$

c. Cartesian, Cylindrical, and Spherical Coordinates

As examples of orthogonal coordinate systems we have the standard Cartesian, cylindrical, and spherical coordinates in $\mathbb{R}^3$.

Problem 1 Write down the metric $g_{ij}(t)\, dt^i\, dt^j$ in each of these coordinate systems and find an orthonormal basis (e_1, e_2, e_3).

Answer In Cartesian coordinates (x, y, z), cylindrical coordinates (r, φ, z), and spherical coordinates (R, φ, θ) of the Euclidean space $\mathbb{R}^3$, the quadratic form $g_{ij}(t)\, dt^i\, dt^j$ has the following form:

$$ds^2 = dx^2 + dy^2 + dz^2 =$$
$$= dr^2 + r^2\, d\varphi^2 + dz^2 =$$
$$= dR^2 + R^2 \cos^2\theta\, d\varphi^2 + R^2\, d\theta^2.$$

In Cartesian, cylindrical, and spherical coordinates, the triples of unit vectors of coordinate directions are the following, respectively:

$$e_x = (1, 0, 0), \qquad e_y = (0, 1, 0), \qquad e_z = (0, 0, 1);$$

$$e_r = (1, 0, 0), \qquad e_\varphi = \left(0, \frac{1}{r}, 0 \right), \qquad e_z = (0, 0, 1);$$

$$e_R = (1, 0, 0), \qquad e_\varphi = \left(0, \frac{1}{R \cos\theta}, 0 \right), \qquad e_\theta = \left(0, 0, \frac{1}{R} \right).$$

D.2   Operators grad, curl, div in Curvilinear Coordinates

D.2.1   Differential Forms and Operators grad, curl, div

The differential dU of a function U is a 1-form. When one has a scalar product $\langle,\rangle$, as we know, to the 1-form dU corresponds a vector A such that $dU = \langle A, \cdot \rangle$. This vector is called the *gradient of the function* U and is denoted by $\operatorname{grad} U$.

Thus, $dU = \langle \operatorname{grad} U, \cdot \rangle$.

Suppose that in the Euclidean space $\mathbb{R}^3$ (or in any three-dimensional Riemannian manifold) we have the 1-form $\omega_A^1 = \langle A, \cdot \rangle$ corresponding to the field A. The differential $d\omega_A^1$ of this form is a 2-form ω_B^2, corresponding, in the presence of a volume form Ω^3, to some vector field B (i.e., $\omega_B^2 = \Omega^3(B, \cdot, \cdot)$). Then the field B is called the *curl of the vector field* A, and is denoted by curl A.

Thus, $d\omega_A^1 = \omega_{\text{curl }A}^2$.

If one has a volume form Ω^n on an n-dimensional surface (for example on $\mathbb{R}^n$), then there is defined an $(n-1)$-form for the flux of a vector field B, namely the form $\omega_B^{n-1} = \Omega^n(B, \cdot, \cdot)$. The differential $d\omega_B^{n-1}$ of this $(n-1)$-form is an n-form, which therefore has the type $\rho\Omega^n$. The proportionality factor, the function ρ, is called the *divergence of the vector field* B and is denoted by div B.

Thus, $d\omega_B^{n-1} = (\text{div } B)\Omega^n$.

D.2.2 Gradient of a Function and Its Coordinate Representation

a. Coordinate Representation for the Correspondence Between a Vector and a 1-Form

In Sect. D.1.2, we derived a relation between the coefficients of a 1-form $\omega_A^1 = \langle A, \cdot \rangle$ and the coordinates of the vector $A = A^i \xi_i$. If we take the unit vectors e_i instead of the vectors ξ_i, and since $\xi_i = \sqrt{g_{ii}} e_i$, then the coordinates of the vector $A = A_e^i e_i$ in the basis $\{e_i\}$ and its former coordinates are related through the equation $A_e^i = A^i \sqrt{g_{ii}}$ for $i \in \{1, \dots, n\}$.

Hence all new related formulas have the form

$$a_j = g_{ij} \frac{A_e^i}{\sqrt{g_{ii}}}, \qquad \frac{A_e^i}{\sqrt{g_{ii}}} = g^{ij} a_j.$$

These formulas allow us to write, in terms of the vector $A = A_e^i e_i$, the corresponding form $\omega_A^1 = \langle A, \cdot \rangle = a_j \, dt^j$ and conversely, to write the vector $A = A_e^i e_i$ in terms of the 1-form $\omega^1 = a_j \, dt^j$.

Problem 2 Write down in Cartesian, cylindrical, and spherical coordinates of the Euclidean space $\mathbb{R}^3$ the explicit form of the 1-form $\omega_A^1 = \langle A, \cdot \rangle$, corresponding to the vector $A = A^i e_i$.

Answer The 1-form ω_A^1 has the following form, in Cartesian coordinates (x, y, z), cylindrical coordinates (r, φ, z), and spherical coordinates (R, φ, θ) of the Euclidean space $\mathbb{R}^3$, respectively:

$$\omega_A^1 = A_x \, dx + A_y \, dy + A_z \, dz =$$
$$= A_r \, dr + A_\varphi r \, d\varphi + A_z \, dz =$$
$$= A_R \, dR + A_\varphi R \cos \varphi \, d\varphi + A_\theta R \, d\theta.$$

b. Differential of a Function and the Gradient

We shall apply the general formula relating the vector A and the form ω_A^1 in the case of the form $dU = \langle \operatorname{grad} U, \cdot \rangle$, in order to find the decomposition $\operatorname{grad} U = A_e^i e_i$. Since $dU = \frac{\partial U}{\partial t^j} dt^j$, i.e., $a_j = \frac{\partial U}{\partial t^j}$, then we have $A_e^i = g^{ij} \sqrt{g_{ii}} \frac{\partial U}{\partial t^j}$.

In the case of an orthogonal system of curvilinear coordinates, the matrix (g_{ij}) is diagonal, as well as its inverse matrix (g^{ij}). Moreover, $g^{ii} = 1/g_{ii}$. Hence in this case,

$$\operatorname{grad} U = \frac{1}{\sqrt{g_{11}}} \frac{\partial U}{\partial t^1} e_1 + \cdots + \frac{1}{\sqrt{g_{nn}}} \frac{\partial U}{\partial t^n} e_n.$$

c. Gradient in Cartesian, Cylindrical, and Spherical Coordinates

Problem 3 Write down the vector $\operatorname{grad} U = A_e^i e_i$ in Cartesian, cylindrical, and spherical coordinates of the Euclidean space $\mathbb{R}^3$.

Answer The vector $\operatorname{grad} U$ has the following form in Cartesian (x, y, z), cylindrical (r, θ, z), and spherical (R, φ, θ) coordinates of the Euclidean space $\mathbb{R}^3$, respectively:

$$\operatorname{grad} U = \frac{\partial U}{\partial x} e_x + \frac{\partial U}{\partial y} e_y + \frac{\partial U}{\partial z} e_z =$$

$$= \frac{\partial U}{\partial r} e_r + \frac{1}{r} \frac{\partial U}{\partial \varphi} e_\varphi + \frac{\partial U}{\partial z} e_z =$$

$$= \frac{\partial U}{\partial R} e_R + \frac{1}{R \cos \theta} \frac{\partial U}{\partial \varphi} e_\varphi + \frac{1}{R^2} \frac{\partial U}{\partial \theta} e_\theta.$$

D.2.3 Divergence and Its Coordinate Representation

a. Coordinate Representation for the Correspondence Between a Vector and an $(n-1)$-Form

We know that if there exists a nondegenerate n-form Ω^n in an n-dimensional vector space, then one can establish a one-to-one correspondence between a vector B and the $(n-1)$-form $\omega_B^{n-1} = \Omega^n(B, \ldots)$. We wish to write down an explicit formula relating the coordinates of the vector $B = B^i \xi_i$ and the coefficients of the form $\omega_B^{n-1} = b_i x^1 \wedge \cdots \widehat{x^i} \wedge \cdots \wedge x^n$, considering that both objects are expressed in terms of the one basis $\{\xi_i\}$ of the space. Here, x^i is a linear function as usual, whose action is given by assigning the i-coordinate of a vector, i.e., $x^i(v) := v^i$; the symbol $\widehat{x^i}$ means that the corresponding factor is omitted. The n-form Ω^n in the n-dimensional vector space is $x^1 \wedge \cdots \wedge x^n$ or proportional to this standard volume form, equal to one on the set of the basis vectors $(\xi_1, \ldots, \xi_n)$.

In general, the value of the form $\Omega^1 = x^1 \wedge \cdots \wedge x^n$ on any vector set $(v_1, \ldots, v_n)$ is equal to the determinant of the matrix (v_i^j) consisting of the coordinates of these vectors. Hence if we consider the rule for the expansion of the determinant on a row, we can write

$$\Omega^n(B, \ldots) = \sum_{i=1}^{n} (-1)^{i-1} B^i x^1 \wedge \cdots \wedge \widehat{x^i} \wedge \cdots \wedge x^n.$$

However, $\omega_B^{n-1} = \Omega^n(B, \ldots)$; thus

$$\sum_{i=1}^{n} b_i x^1 \wedge \cdots \wedge \widehat{x^i} \wedge \cdots \wedge x^n = \sum_{i=1}^{n} (-1)^{i-1} B^i x^1 \wedge \cdots \wedge \widehat{x^i} \wedge \cdots \wedge x^n.$$

Therefore, $b_i = (-1)^{i-1} B^i$ for every $i \in \{1, \ldots, n\}$. If instead we had the form $c\omega^n = cx^1 \wedge \cdots \wedge x^n$, then we would have the equation $b_i = (-1)^{i-1} cB^i$ for every $i \in \{1, \ldots, n\}$.

Recall also that if there is an inner product $\langle, \rangle$ and a fixed basis $\{\xi_i\}$ in a vector space, then there is also a natural volume form $\sqrt{\det g_{ij}} x^1 \wedge \cdots \wedge x^n$ defined, as well as the scalar product itself, in terms of the values $g_{ij} = \langle \xi_i, \xi_j \rangle$.

Finally, recall that in this case, the unit vectors (with respect to the norm) are not in general the vectors $\{\xi_i\}$, but the vectors $e_i = \xi_i / \sqrt{g_{ii}}$. Since $\xi_i = \sqrt{g_{ii}} e_i$, the original decomposition of the vector $B = B^i \xi_i$ in the basis $\{e_i\}$ becomes $B = B_e^i e_i$, where $B_e^i = \sqrt{g_{ii}} B^i$.

Therefore, if one has a scalar product on the space, then there is a natural volume form $\Omega_g^n = \sqrt{\det g_{ij}} x^1 \wedge \cdots \wedge x^n$, and if $\omega_B^{n-1} = \Omega_g^n(B, \ldots)$, then the coefficients of the form $\omega_B^{n-1} = b_i x^1 \wedge \cdots \wedge \widehat{x^i} \wedge \cdots \wedge x^n$ and the coordinates of the vector B in the decomposition $B = B_e^i e_i$ in terms of the basis of unit vectors $e_i = \xi_i / \sqrt{g_{ii}}$ are related by the equations

$$b_i = (-1)^{i-1} \sqrt{\det g_{ij}} \frac{B_e^i}{\sqrt{g_{ii}}}.$$

In an orthogonal basis, $\det g_{ij} = g_{11} \cdots g_{nn}$. In this case,

$$b_i = (-1)^{i-1} \sqrt{g_{11} \cdots \widehat{g_{ii}} \cdots g_{nn}} B_e^i.$$

All of the above remains valid when it is applied to the case of the vector field $B(t)$ and the differential form $\omega_B^{n-1} = \Omega_g^n(B, \ldots)$ of the field generated by the volume form.

Thus if $\Omega_g^n = \sqrt{\det g_{ij}}(t) \, dt^1 \wedge \cdots \wedge dt^n$,

$$\omega_B^{n-1} = b_i(t) \, dt^1 \wedge \cdots \wedge \widehat{dt^i} \wedge \cdots dt^n,$$

and $B(t) = B_e^i(t)e_i(t)$ is the decomposition in terms of the unit vectors of the curvilinear coordinates $(t^1, \ldots, t^n)$, then

$$b_i = (-1)^{i-1} \frac{\sqrt{\det g_{ij}}}{\sqrt{g_{ii}}} B_e^i, \qquad B_e^i = (-1)^{i-1} \frac{\sqrt{g_{ii}}}{\sqrt{\det g_{ij}}} b_i.$$

If the system of curvilinear coordinates is orthogonal, we come back to the relation $b_i = (-1)^{i-1} \sqrt{g_{11} \cdots \widehat{g_{ii}} \cdots g_{nn}} B_e^i$.

In particular, for a 3-dimensional orthogonal system of curvilinear coordinates (t^1, t^2, t^3), using the same notation $E_i = g_{ii}$ mentioned at the beginning, it is possible to write the following coordinate representation of the form ω_B^2 corresponding to the vector $B = B_e^1 e_1 + B_e^2 e_2 + B_e^3 e_3$:

$$\omega_B^2 = B_e^1 \sqrt{E_2 E_3} \, dt^2 \wedge dt^3 + B_e^2 \sqrt{E_3 E_1} \, dt^3 \wedge dt^1 + B_e^3 \sqrt{E_1 E_2} \, dt^1 \wedge dt^2 =$$

$$= \sqrt{E_1 E_2 E_3} \left(\frac{B_e^1}{\sqrt{E_1}} \, dt^2 \wedge dt^3 + \frac{B_e^2}{\sqrt{E_2}} \, dt^3 \wedge dt^1 + \frac{B_e^3}{\sqrt{E_3}} \, dt^1 \wedge dt^2 \right).$$

(Bear in mind that in the 3-dimensional case, the 2-form ω^2 is not usually written as $b_1 \, dt^2 \wedge dt^3 + a_2 \, dt^1 \wedge dt^3 + b_3 \, dt^1 \wedge dt^2$, but as $a_1 \, dt^2 \wedge dt^3 + a_2 \, dt^3 \wedge dt^1 + a_3 \, dt^1 \wedge dt^2$; for example, $P \, dy \wedge dz + Q \, dz \wedge dx + R \, dx \wedge dy$.)

Problem 4 Specify the explicit form of the 2-form $\omega_B^2 = \Omega_g^3(B, \ldots)$ corresponding to the vector field $B = B_e^i e_i$ in Cartesian, cylindrical, and spherical coordinates of the Euclidean space $\mathbb{R}^3$.

Answer The form ω_B^2 has the following form in Cartesian (x, y, z), cylindrical (r, θ, z), and spherical (R, φ, θ) coordinates of the Euclidean space $\mathbb{R}^3$:

$$\omega_B^2 = B_x \, dy \wedge dz + B_y \, dz \wedge dx + B_z \, dx \wedge dy =$$

$$= B_r r \, d\varphi \wedge dz + B_\varphi \, dz \wedge dr + B_z r \, dr \wedge d\varphi =$$

$$= B_R R^2 \cos\theta \, d\varphi \wedge d\theta + B_\varphi R \, d\theta \wedge dR + B_\theta R \cos\theta \, dR \wedge d\varphi.$$

b. The Differential Form of a Flux and the Divergence of the Velocity Field

The form $\omega_B^{n-1} = \Omega_g^n(B, \ldots)$ is often called a form of a flux, since when B is the flux velocity field (at least for $n = 3$), one has to integrate exactly this form to find the outflow (flux) through a surface.

The differential of the form of a flux ω_B^{n-1} is an n-form, proportional to the volume form. The coefficients of proportionality are called the divergence field B, as we know. Thus $d\omega_B^{n-1} = \text{div } B \cdot \Omega_g^n$.

We want to study the field $B = B_e^i e_i$ itself and find its divergence div B. We already know how to find the form of a flux ω_B^{n-1} from the field $B = B_e^i e_i$. We shall find it, compute its differential, and obtain an n-form, proportional to the volume form, whose coefficients of proportionality are the divergence of the field B.

Let us show this. We write the $(n-1)$-form ω_B^{n-1} in the following form:

$$\omega_B^{n-1} = b_1(t)\, dt^1 \wedge \cdots \wedge \widehat{dt^i} \wedge \cdots \wedge dt^n.$$

We compute its differential

$$d\omega_B^{n-1} = \left(\sum_{n=1}^{n} \frac{\partial b_i}{\partial t^i} (-1)^{i-1} \right) dt^1 \wedge \cdots \wedge dt^n.$$

We express the coefficients b_i of the form ω_B^{n-1} through the coordinates B_e^i of the vector $B = B_e^i e_i$:

$$d\omega_B^{n-1} = \left(\sum_{n=1}^{n} \frac{\partial}{\partial t^i} \left(\frac{\sqrt{\det g_{ij}}}{\sqrt{g_{ii}}} B_e^i \right) \right) dt^1 \wedge \cdots \wedge dt^n.$$

We compare this form with the volume form

$$\Omega_g^n = \sqrt{\det g_{ij}}(t)\, dt^1 \wedge \cdots \wedge dt^n,$$

and we obtain

$$\operatorname{div} B = \frac{1}{\sqrt{\det g_{ij}}} \left(\sum_{n=1}^{n} \frac{\partial}{\partial t^i} \left(\frac{\sqrt{\det g_{ij}}}{\sqrt{g_{ii}}} B_e^i \right) \right).$$

In an orthogonal system of curvilinear coordinates, this formula takes the form

$$\operatorname{div} B = \frac{1}{\sqrt{g_{11} \cdots g_{nn}}} \left(\sum_{n=1}^{n} \frac{\partial}{\partial t^i} \left(\frac{\sqrt{g_{11} \cdots g_{nn}}}{\sqrt{g_{ii}}} B_e^i \right) \right).$$

c. Divergence in Cartesian, Cylindrical, and Spherical Coordinates

Problem 5 Write down formulas to calculate the divergence of a vector field $B = B_e^i e_i$ in Cartesian, cylindrical, and spherical coordinates of the Euclidean space $\mathbb{R}^3$.

Answer In Cartesian coordinates (x, y, z), cylindrical coordinates (r, φ, z), and spherical coordinates (R, φ, θ) of the Euclidean space $\mathbb{R}^3$, the divergence $\operatorname{div} B$ of the vector field $B = B_e^i e_i$ can be calculated according to the formula

$$\text{div } B = \frac{\partial B_x}{\partial x} + \frac{\partial B_y}{\partial y} + \frac{\partial B_z}{\partial z} =$$

$$= \frac{1}{r}\left(\frac{\partial r B_r}{\partial r} + \frac{\partial B_\varphi}{\partial \varphi}\right) + \frac{\partial B_z}{\partial z} =$$

$$= \frac{1}{R^2 \cos\varphi}\left(\frac{\partial R^2 \cos\theta B_R}{\partial R} + \frac{\partial R B_\varphi}{\partial \varphi} + \frac{\partial R \cos\theta B_\theta}{\partial \theta}\right).$$

D.2.4 Curl of a Vector Field and Its Coordinate Representation

a. Correspondence Between a Vector Field A and the Vector Field $B = \text{curl } A$

We shall now consider the special 3-dimensional case. We shall assume, as before, that we are given a metric $g_{ij}(t)\,dt^i\,dt^j$ in the curvilinear coordinates (t^1, t^2, t^3), generating at the same time the volume form $\Omega_g^3 = \sqrt{\det g_{ij}}(t)\,dt^1 \wedge dt^2 \wedge dt^3$.

In this case the vector field $A = A_e^i e_i$ corresponds to the 1-form ω_A^1, and the differential $d\omega_A^1$ of this form, as a 2-form ($(n-1)$-form), corresponds to a vector field $B = B_e^i e_i$ such that $d\omega_A^1 = \omega_B^2$. This vector field B is called, as we know, the curl of the original field A and is denoted by $\text{curl } A$.

b. The Coordinate Representation of the Correspondence Between Vector Fields A and $B = \text{curl } A$

We wish to learn how to calculate the coordinates of the field $B = \text{curl } A$ in terms of the coordinates of the vector field A. According to the procedure described above, from the vector field $A = A_e^i e_i$ we build its corresponding 1-form $\omega_A^1 = \langle A, \cdot\rangle$:

$$\omega_A^1 = a_i\,dt^i = \frac{g_{ij}}{\sqrt{g_{jj}}} A_e^j\,dt^i.$$

We take its differential

$$d\omega_A^1 = \frac{\partial}{\partial t^k}\left(\frac{g_{ij}}{\sqrt{g_{jj}}} A_e^j\right) dt^k \wedge dt^i =$$

$$= \left(\frac{\partial}{\partial t^2}\left(\frac{g_{3j}}{\sqrt{g_{jj}}} A_e^j\right) - \frac{\partial}{\partial t^3}\left(\frac{g_{2j}}{\sqrt{g_{jj}}} A_e^j\right)\right) dt^2 \wedge dt^3 +$$

$$+ \left(\frac{\partial}{\partial t^3}\left(\frac{g_{1j}}{\sqrt{g_{jj}}} A_e^j\right) - \frac{\partial}{\partial t^1}\left(\frac{g_{3j}}{\sqrt{g_{jj}}} A_e^j\right)\right) dt^3 \wedge dt^1 +$$

$$+ \left(\frac{\partial}{\partial t^1}\left(\frac{g_{2j}}{\sqrt{g_{jj}}} A_e^j\right) - \frac{\partial}{\partial t^2}\left(\frac{g_{1j}}{\sqrt{g_{jj}}} A_e^j\right)\right) dt^1 \wedge dt^2,$$

considering this form a form of type ω_B^2. By comparing the coefficients, we have $\omega_B^2 = d\omega_A^1 = b_1 \, dt^2 \wedge b_2 \, dt^3 \wedge dt^1 + b_3 \, dt^1 \wedge dt^2$. We obtain the coordinates $B_e^i = \frac{\sqrt{g_{ii}}}{\sqrt{\det(g_{ij})}} b_i$ of the vector $B = \text{curl } A$.

In the case of a 3-dimensional orthogonal system of curvilinear coordinates (t^1, t^2, t^3), the formula simplifies. In this case,

$$d\omega_A^1 = \frac{\partial}{\partial t^k} \left(\sqrt{g_{ii}} A_e^i \right) dt^k \wedge dt^i =$$

$$= \left(\frac{\partial}{\partial t^2} \left(\sqrt{g_{33}} A_e^3 \right) - \frac{\partial}{\partial t^3} \left(\sqrt{g_{22}} A_e^2 \right) \right) dt^2 \wedge dt^3 +$$

$$+ \left(\frac{\partial}{\partial t^3} \left(\sqrt{g_{11}} A_e^1 \right) - \frac{\partial}{\partial t^1} \left(\sqrt{g_{33}} A_e^3 \right) \right) dt^3 \wedge dt^1 +$$

$$+ \left(\frac{\partial}{\partial t^1} \left(\sqrt{g_{22}} A_e^2 \right) - \frac{\partial}{\partial t^2} \left(\sqrt{g_{11}} A_e^1 \right) \right) dt^1 \wedge dt^2,$$

and using the notation $E_i = g_{ii}$, it is possible to write the coordinates of the vector $\text{curl } A = B = B_e^1 e_1 + B_e^2 e_2 + B_e^3 e_3$:

$$B_e^1 = \frac{1}{\sqrt{E_2 E_3}} \left(\frac{\partial A_e^3 \sqrt{E_3}}{\partial t^2} - \frac{\partial A_e^2 \sqrt{E_2}}{\partial t^3} \right),$$

$$B_e^2 = \frac{1}{\sqrt{E_3 E_1}} \left(\frac{\partial A_e^1 \sqrt{E_1}}{\partial t^3} - \frac{\partial A_e^3 \sqrt{E_3}}{\partial t^1} \right),$$

$$B_e^3 = \frac{1}{\sqrt{E_1 E_2}} \left(\frac{\partial A_e^2 \sqrt{E_2}}{\partial t^1} - \frac{\partial A_e^1 \sqrt{E_1}}{\partial t^2} \right),$$

which means that

$$\text{curl } A = \frac{1}{\sqrt{E_1 E_2 E_3}} \begin{vmatrix} \sqrt{E_1} e_1 & \sqrt{E_2} e_2 & \sqrt{E_3} e_3 \\ \partial_1 & \partial_2 & \partial_3 \\ \sqrt{E_1} A_e^1 & \sqrt{E_2} A_e^2 & \sqrt{E_3} A_e^3 \end{vmatrix}.$$

c. Curl in Cartesian, Cylindrical, and Spherical Coordinates

Problem 6 Write down the formula to calculate the curl of a vector field $A = A_e^1 e_1 + A_e^2 e_2 + A_e^3 e_3$ in Cartesian, cylindrical, and spherical coordinates of the Euclidean space $\mathbb{R}^3$.

Answer In Cartesian (x, y, z), cylindrical (r, φ, z), and spherical (R, φ, θ) coordinates of the Euclidean space, the curl $(\text{curl } A)$ of the vector field $A = A_e^1 e_1 + A_e^2 e_2 +$

$A_3^3 e_3$ is calculated according to the formula

$$\operatorname{curl} A = \left(\frac{\partial A_z}{\partial y} - \frac{\partial A_y}{\partial z} \right) e_x + \left(\frac{\partial A_x}{\partial z} - \frac{\partial A_z}{\partial x} \right) e_y + \left(\frac{\partial A_y}{\partial x} - \frac{\partial A_x}{\partial y} \right) e_z =$$

$$= \frac{1}{r} \left(\frac{\partial A_z}{\partial \varphi} - \frac{\partial r A_\varphi}{\partial z} \right) e_r + \left(\frac{\partial A_r}{\partial z} - \frac{\partial A_z}{\partial r} \right) e_\varphi + \frac{1}{r} \left(\frac{\partial r A_\varphi}{\partial r} - \frac{\partial A_r}{\partial \varphi} \right) e_z =$$

$$= \frac{1}{R \cos \theta} \left(\frac{\partial A_\theta}{\partial \varphi} - \frac{\partial A_\varphi \cos \theta}{\partial \theta} \right) e_R + \frac{1}{R} \left(\frac{\partial A_R}{\partial \theta} - \frac{\partial R A_\theta}{\partial R} \right) e_\varphi +$$

$$+ \frac{1}{R} \left(\frac{\partial R A_\varphi}{\partial R} - \frac{1}{\cos \theta} \frac{\partial A_R}{\partial \varphi} \right) e_\theta .$$

Appendix E
Modern Formula of Newton–Leibniz and the Unity of Mathematics (Final Survey)

E.1 Reminders

E.1.1 Differential, Differential Form, and the General Stokes's Formula

a. What Happened and Was the Reason That Brought Us to This Kind of Life

We already began the ascent to the modern Newton–Leibniz formula at the very beginning of this course of mathematical analysis, when we defined the differential $df(x)$ of a function $f : X \to Y$ at the point x. By analyzing this concept gradually in detail, we found that it is a linear function operating on a linear vector space $T_x X$ of displacements from the point under consideration with values in the space $T_y Y$ of displacements from the point $y = f(x)$. The spaces $T_x X$ and $T_y Y$ are called *tangent spaces* to X and Y at the corresponding points. The differential itself is also called the *tangent mapping* or *total derivative* with respect to the original mapping (function) $f : X \to Y$ at the point x.

Once one has become acquainted with the concept of tangent line or tangent plane to a surface, one understands the origin and the geometric meaning of this terminology.

Passing to functions of several variables and mappings of multidimensional objects, we left the definition of the differential unchanged, but every time, we explicitly deciphered the coordinate representation of the differential. In this way, the notion of the Jacobian matrix of a mapping appeared.

We know that the differential of a function $f : \mathbb{R}^n \to \mathbb{R}$ has the form

$$df(x) = \frac{\partial f}{\partial x^1} dx^1 + \cdots + \frac{\partial f}{\partial x^n} dx^n,$$

i.e., it is a linear combination of differentials of simple functions, the coordinate functions, and the value of the differential $df(x)(\xi)$ at the vector $\xi \in T_x \mathbb{R}^n$ coincides with the value of the derivative $D_\xi f(x)$ of the function on this vector, and

© Springer-Verlag Berlin Heidelberg 2016
V.A. Zorich, *Mathematical Analysis II*, Universitext,
DOI 10.1007/978-3-662-48993-2

since $dx^i(\xi) = \xi^i$, one has

$$df(x)(\xi) = \frac{\partial f}{\partial x^1}\xi^1 + \cdots + \frac{\partial f}{\partial x^n}\xi^n.$$

If you are acquainted with the linear algebra of linear, multilinear, and skew-symmetric forms and the operation of their external product, you could, by applying this to differentials, write a differential form of the type

$$\omega^k(x) = a_{i_1 \ldots i_k}(x)\, dx^{i_1} \wedge \cdots \wedge dx^{i_k},$$

realizing that this is a skew-symmetric k-form on the tangent space whose value on the set of vectors $(\xi_1, \ldots, \xi_k)$ can be calculated if the value of $dx^{i_1} \wedge \cdots \wedge dx^{i_k}(\xi_1, \ldots, \xi_k)$ is known. Lastly, this is equal to the determinant of the matrix

$$\begin{pmatrix} \xi_1^{i_1} & \cdots & \xi_1^{i_k} \\ \vdots & \ddots & \vdots \\ \xi_k^{i_1} & \cdots & \xi_k^{i_k} \end{pmatrix},$$

as we know from algebra (given that $dx^i(\xi) = \xi^i$).

Recall that we were led to differential forms by the change of variables formula for a multiple integral. For a one-dimensional integral, the form $f(x)\,dx$, standing under the integral sign, dictated the correct change of variable formula $f(\varphi(t))\,d\varphi(t)$. We were concerned, as Euler was, about the fact that this was not the case for higher-dimensional integrals. We wanted to correct this deficiency and at the same time understand what we are actually integrating, since the result should not depend on the choice of the system of coordinates.

Analyzing this problem, we also had to figure out a number of concepts, not only in algebra but also in geometry. We understood what a k-dimensional surface is, curvilinear coordinates, local charts, local maps and atlas, what the orientation of a surface is, and how it is specified, what the border of a surface and the induced orientation on the border are, and finally what all of this looks like in the general case of manifolds of dimension k.

We had to analyze what occurs with our objects and operations under a change of coordinate system. We also had to figure out the direction in which points, vectors, and functions on those objects are transferred, in particular forms under smooth mappings, and how exactly to implement the corresponding transfer in the coordinates. At the same time, we convinced ourselves that the operation of differentiation on forms is indeed invariant with respect to the choice of coordinate system. The differentiation of forms, in the coordinate representation, is realized in the most simple and natural way,

$$d\omega^k(x) = da_{i_1 \ldots i_k}(x)\, dx^{i_1} \wedge \cdots \wedge dx^{i_k},$$

which it is often taken, for this reason, as the original definition of this operation.

Appealing to some suggestions from physics (computation of work, flux), we realized that we integrate differential forms not only because they solve the original problem about the change of variables formula in multiple integrals, but also they lead to the following far-reaching generalization of the classical Newton–Leibniz formula:

$$\int_{M_+^k} \mathrm{d}\omega^{k-1} = \int_{\partial M_+^k} \omega^{k-1}.$$

This formula, frequently called the *general Stokes's formula*, rightfully should be called the Newton–Leibniz–Gauss–Ostrogradskii–Green–Maxwell–Cartan–Poincaré formula.

b. The Problem of Primitives Yesterday and Today

One of the very first questions in classical mathematical analysis is the question about the inversion of the operation of differentiation, more precisely, the question of whether every function f (for example, continuous) is the derivative of some other function, and if so, how to find the antiderivative or primitive F of the given function. In the language of forms, this question is whether a 1-form $f(x)\,\mathrm{d}x$ is the differential $\mathrm{d}F$ of some 0-form, i.e., a function F.

We gave a positive answer to this question, considering everything over a numerical interval. We did not even consider any other situation. If you ask yourself the same question, for example, for a function identically equal to one on the circle or for an appropriate form $\mathrm{d}\varphi$, you will immediately realize that the answer is negative. There is no differentiable function on the circle whose derivative everywhere is equal to one.

This is one of the manifestations of a relation between a question of global analysis and the topology of the domain, where the question is posed and solved.

A significant part of the following text is devoted to a deeper, although not complete, discussion of this relation.

Generalizing the classical situation, we shall ask the following question: *Given a differential k-form ω^k, we look for a $(k-1)$-form ω^{k-1} such that $\omega^k = \mathrm{d}\omega^{k-1}$.*

c. Closed and Exact Differential Forms

Differential forms ω^k having a primitive (i.e., being the differential of some form ω^{k-1}: $\omega^k = \mathrm{d}\omega^{k-1}$) are called *exact forms*.

We shall easily prove that an obvious necessary differential condition for the exactness of a form ω^k is the equality $\mathrm{d}\omega^k = 0$, due to the fact that the external redifferentiation of any differential form is identically zero.

If the differential of a form is equal to zero, the form is called *closed*.

Thus, closedness is a necessary condition for the exactness of a form.

Previously, we considered in all details and interpretations the case of 1-forms. We also convinced ourselves that although closedness is a necessary condition for exactness, this condition is not sufficient, and it is significantly associated with the topology of the domain in which the problem is posed.

In physics, potential vector fields play an important role. If we have a scalar product $\langle , \rangle$ (or a nondegenerate bilinear form) in some space, then there arises a correspondence between linear functions (forms) and vector fields, defined by the equality $\omega_A^1(x)(\xi) = \langle A(x), \xi \rangle$. Incidentally, when we want to calculate the work that should be done by a vector field along a path γ, then we just integrate the form ω_A^1, called a *work form*. The remarkable characteristic of potential vector fields is that the work on those fields depends only on the beginning and the end of the path of transition and is equal to the difference between the values of the potential generating this field. In particular, the work on a closed contour (a cycle) with such a vector field is zero.

In the language of vector fields, the differential characteristic of a potential vector field is, as we know, that they have no rotation (their curl vanishes). We also know that irrotational vector fields are not always potential vector fields, and it depends on the topology of the domain on which they act. In a simply connected domain, this necessary characteristic is also sufficient. For example, in a three-dimensional ball or a ball with deleted center, or in a cut-out ball, every irrotational field is a potential field; in the two-dimensional disk this is also the case, but in the disk with the center deleted, it is no longer the case. (Recall the typical example: in writing the form $\mathrm{d}\varphi$ in Cartesian coordinates (x, y), we considered the vector field $(-y, x)/(x^2 + y^2)$ corresponding to it.)

Along with the necessary differential condition of exactness of a form, which "feels" the form locally, we had an integral criterion for exactness of 1-forms, consisting in the fact that the integral of a form over any cycle (closed path) lying in the considered domain is always equal to zero.

This *integral criterion for the exactness of forms* remains true with respect to forms of any degree, with the proper understanding of what the cycle of the corresponding dimension should be.

This is one of de Rham's theorems, which has as a consequence a much older theorem, also called *Poincaré's lemma*, asserting that in the space $\mathbb{R}^n$, in a ball, or on any other domain homeomorphic to it, every closed form is exact.

E.1.2 Manifolds, Chains, and the Boundary Operator

a. Cycles and Boundaries

In the previous Stokes's formula we have geometric objects (curves, surfaces, manifolds, and their boundary, i.e., the border), on which we integrate the corresponding differential forms.

Similar to the operator d of differentiation, we have the operator ∂, which maps surfaces to their boundary. The boundary ∂M^k of a manifold M^k is also a manifold, but with one dimension fewer. Moreover, the variety ∂M^k no longer has a boundary, i.e., the reapplication of the operator ∂ always gives the empty set. In this sense, the operators d and ∂ are similar. But if the operator d increases the dimension of the object by 1, the operator ∂ reduces the dimension by 1.

The concepts of closedness and exactness in forms correspond here to the concepts of cycles and boundaries.

A compact surface, a manifold M^k (later we shall say also chain) of dimension k, is called *a cycle of dimension k* if $\partial M = \emptyset$, i.e., M does not have any boundary points.

Thus, the sphere of dimension k is a cycle of dimension k.

A surface, manifold M^k (a chain), is called a *boundary* if it has a "primitive" in the sense that there is a surface or manifold M^{k+1} (chain) such that $\partial M^{k+1} = M^k$.

It is clear that if the surface or manifold is the boundary of some other compact manifold, then it must be a cycle. However, the situation here is similar to that of forms, where the conditions are necessary but in general not sufficient to ensure that in the domain where this cycle lies, there is also a manifold such that the cycle is the boundary of that manifold.

Take, for example, a circular ring, or annulus, in the plane. Then every circle containing the hole is a cycle, but it is not the boundary of a manifold lying on the annulus. But if instead of an annulus we consider a disk, then the situation is radically different.

Let us consider the boundary of the annulus, and we shall recall the following fact. The operator ∂ acting on boundaries is not a simple set-theoretic transformation. On an atlas of the surface or manifold, this operator gives an atlas of the boundary, which is called the *induced atlas of the boundary*. If the original atlas consists of compatible charts, then under this operator, the induced atlas will also have this property. Thus if the manifold is orientable, then its boundary possesses an orientation, which is called the *induced orientation* or *agreed* or *compatible orientation of the boundary*.

If the annulus G that we just discussed is oriented with the standard left frame of the Cartesian coordinates in the plane, then its boundary, consisting of two circles γ_1, γ_2, will be oriented such that the outer circle γ_2 goes in the positive direction (counterclockwise) and the inner circle is negatively oriented (clockwise). The integral in such a boundary is reduced to the difference between the integrals over γ_1 and γ_2. It is useful to write that as $\partial G = \gamma_{2+} - \gamma_{1+}$.

For example, if you need to calculate the work that is accomplished by five turns along the path γ_{2+}, then three along the path γ_{1+}, and finally two along γ_{2-}, then you have to integrate over the chain $5\gamma_{2+} + 3\gamma_{1+} + 2\gamma_{2-} = 5\gamma_{2+} + 3\gamma_{1+} - 2\gamma_{2+} = 3\gamma_{2+} + 3\gamma_{1+}$. The integration over such chain corresponds, of course, to a linear combination of the integrals over γ_{1+} and γ_{2+}.

This discussion illustrates why it is useful to consider linear combinations of geometric objects. These are called *chains*. We have explained here only where the concept of chains comes from, what are they in general, and where and why they are useful. We are not going into general and formal definitions, since we do not need them here in the more general form, and they can be found in the book. Analogously, just as in analysis, when we are forced to go from the usual ordinary functions to generalized functions, in geometry one goes from the simplest objects like cubes and chains of cubes to their generalizations like singular cubes and chains of singular cubes. Moreover, we then do the next extension and invent the concept of flux, which combines differential forms, generalized functions, and manifolds.

b. Homological Cycles

We shall see below that it is sometimes possible to calculate the integral of a form over a cycle by going to some other cycle, sometimes significantly simpler, which is in some way associated with the original cycle. This is a remarkable, important, and useful fact, which is used in different areas of mathematics and its applications.

In order to understand the relation between cycles, we have to consider the following fact: their difference must be the boundary of an object lying on the domain we are considering. We say that such cycles are *homologous* in this domain.

For example, two closed oriented paths γ_{1+}, γ_{2+} on a domain D or on a manifold M are homologous if we can find an orientable surface $S_+^2 \subset D$ ($S_+^2 \subset M$) such that $\partial S_+^2 = \gamma_{2+} - \gamma_{1+}$.

Thus, the circles γ_{1+}, γ_{2+} considered above are homologous in the annulus G_+.

Since the operator ∂ acts on boundaries and is extended by linearity over chains, it is possible to determine the homology of chains.

For instance, the chains γ_{1+} and $2\gamma_{2+}$ are not homologous on the annulus G_+.

We shall discuss the role and applications of the concept of homology of cycles in the context of the integration of differential forms.

E.2 Pairing

E.2.1 The Integral as a Bilinear Function and General Stokes's Formula

a. The Integral of an Exact Form over a Cycle and of a Closed Form over a Boundary

We introduce first some useful notation.

Let $\Omega(M)$ denote the whole set of differential forms on a manifold (or surface) M, and let $\Omega^k(M)$ denote the subset of forms of order k (i.e., k-forms), $Z^k(M)$ its subset of closed k-forms, and $B^k(M)$ its subset of exact k-forms.

Analogously, let $C(M)$ be the set of chains on a manifold (or surface) M, and let $C_k(M)$ be the subset of chains of dimension k (k-chains), $Z_k(M)$ the subset of cycles (k-cycles), and $B_k(M)$ its subset of boundary cycles (k-boundaries).

Thus, $\Omega(M) \supset \Omega^k(M) \supset Z^k(M) \supset B^k(M)$ and $C(M) \supset C_k(M) \supset Z_k(M) \supset B_k(M)$.

As long as we do not change the manifold M on which we wish to calculate something, in order to simplify the notation we shall remove the symbol M whenever it does not lead to confusion, that is present in the just-discussed notation.

Now we shall make a concluding remark.

Consider the integral of an exact form $b^k \in B^k$ over the cycle $z_k \in Z_k$ and of a closed form $z^k \in Z^k$ over a boundary $b_k \in B_k$. Employing Stokes's formula, we find that

$$\int_{z_k} b^k = \int_{z_k} d\omega^{k-1} = \int_{\partial z_k} \omega^{k-1} = \int_{\emptyset} \omega^{k-1} = 0$$

and

$$\int_{b_k} z^k = \int_{\partial c_{k+1}} z^k = \int_{c_{k+1}} dz^k = \int_{c_{k+1}} 0 = 0.$$

b. Integral of a Closed Form over a Cycle and Its Invariance Under Certain Changes of the Form and the Cycle

The remark that we just made leads to the following important and very useful conclusion.

We shall consider now the integral of a closed form z^k over a cycle z_k. Given that the addition of an exact form b^k to a closed form z^k gives again a closed form (since $d(z^k + b^k) = dz^k + db^k = 0$), and the addition of a boundary cycle b_k to a cycle z_k gives again a cycle (since $\partial(b_k + z_k) = \partial b_k + \partial z_k = 0$), recalling the remark we just made, we can now write the following chain of equalities:

$$\int_{z_k} z^k = \int_{z_k} (z^k + b^k) = \int_{z_k + b_k} (z^k + b^k) = \int_{[z_k]} [z^k].$$

Here $[z^k]$ means the class of forms that differ from the original form z^k modulo an exact form, and $[z_k]$ is the class of cycles differing from the original one up to a boundary cycle.

Thus by calculating the integral of a closed form z^k over a cycle z_k, we can afford to choose, without changing the value of the integral, any cycle from the class $[z_k]$ and any form from the class $[z^k]$.

E.2.2 Equivalence Relations (Homology and Cohomology)

a. Toward Uniformity in Terminology: Cycles and Cocycles, Boundaries and Coboundaries

Along with the unification of notation, it is convenient to agree on the following standardization of terminology. Since the elements of the sets Z_k and B_k are called *cycles* and *boundaries*, respectively, we shall call the elements of Z^k and B^k *cocycles* and *coboundaries*, respectively.

Thus a cocycle is a closed differential form, and a coboundary is an exact differential form.

b. Homology and Cohomology

A class $[z_k]$, or more precisely a class $[z_k](M)$, is called a *homology class* of the cycle z_k on the manifold (or surface) M.

A class $[z^k]$, or more precisely a class $[z^k](M)$, is called a *cohomology class* of the cocycle z^k on the manifold (or surface) M.

The operator ∂ taking boundary chains is called a *boundary operator*, and the operator d acting on differential forms is called a *coboundary operator*.

Two cycles are *homologous on the manifold* (or surface) M if their difference is the boundary of a chain lying on M.

Two cocycles are *cohomologous on the manifold* (or surface) M if their difference is a coboundary on M (i.e., two closed forms are cohomologous on the manifold if their difference is an exact form on the manifold).

E.2.3 Pairing of Homology and Cohomology Classes

a. The Integral as a Bilinear Function

The integral $\int_{c_k} \omega^k$ of a k-form over a chain on some manifold M can be considered a pairing $\langle \omega^k, c_k \rangle$ of objects from two vector spaces, namely the linear space of k-forms Ω^k and the linear space of k-chains C_k.

We can conclude, knowing the properties of the integral, that the operation $\langle \omega^k, c_k \rangle$ is bilinear.

b. Nondegeneracy of the Bilinear Form of Pairing (de Rham Theorem)

When we considered the above pairing between cycles and cocycles, we obtained an important result, which can be stated now in the following form:

$$\langle z^k, z_k \rangle = \langle [z^k], [z_k] \rangle.$$

Recalling the definition of the cohomology and homology classes $[z^k]$, $[z_k]$, we can say that they are elements of the quotient space $H^k := Z^k / B^k$ and $H_k := Z_k / B_k$, respectively.

The vector spaces H^k and H_k, whose complete notation is $H^k(M)$ and $H_k(M)$, are called the *space of k-dimensional cohomology of the manifold M* and the *space of k-dimensional homology of the manifold M*, respectively.

Thus, the integral actually also pairs cohomology and homology classes. The pairing $\langle [z^k], [z_k] \rangle$ is clearly linear and is *nondegenerate*, as was shown by de Rham.

(Recall that a bilinear form $\langle, \rangle$ is called nondegenerate if once we fix one of the arguments with a nonzero value, the form is not identically zero with respect to the other argument.)

c. Integral Criterion for the Exactness of a Closed Form

De Rham's theorem that we just mentioned implies the following criterion of exactness of a closed form: *A closed form $z^k = \omega^k$ on a manifold (surface, domain) M is exact on M if and only if the integral of this form over every k-dimensional cycle lying on M is equal to zero.*

Indeed, if $\langle z^k, z_k \rangle = 0$ for every cycle z_k lying on M, then according to de Rham's theorem, $[z^k] = 0$ in $H^k = Z^k/B^k$. This means that $z^k \in B^k$.

We have examined in detail all aspects for the case of 1-forms, and we also proved this criterion in this case. We have now established this criterion in general.

In particular, you can now say by looking at a manifold or domain where there is an irrotational vector field or a divergence-free vector field whether the vector field is a potential, or it has a vector potential (i.e., it is the curl of some vector field), respectively.

We can also use de Rham's theorem on the second argument, of course. For example, if we know that on some manifold all the closed k-forms are exact, we can say that on this manifold every k-cycle is a boundary cycle (homologous to zero). Thus, we have a conclusion about the topology of the manifold.

E.2.4 Another Interpretation of Homology and Cohomology

a. Duality of Operators d and ∂

In the notation of the pairing $\langle \omega^k, c_k \rangle$, Stokes's formula has the form

$$\langle d\omega^{k-1}, c_k \rangle = \langle \omega^{k-1}, \partial c_k \rangle,$$

showing the duality between the operators d and ∂.

b. The Operators d and ∂ as Mappings

In some cases, it is useful to write the full notation of the operators d and ∂, for example, in the notation of the following sequences of linear mappings:

$$\cdots \xrightarrow{d_{k-2}} \Omega^{k-1} \xrightarrow{d_{k-1}} \Omega^k \xrightarrow{d_k} \Omega^{k+1} \xrightarrow{d_{k+1}} \cdots,$$

$$\cdots \xleftarrow{\partial_{k-1}} C_{k-1} \xleftarrow{\partial_k} C_k \xleftarrow{\partial_{k+1}} C_{k+1} \xleftarrow{\partial_{k+2}} \cdots.$$

Using the standard notations Ker and Im for the kernel and the image of a linear mapping, we can write, for example, that

$$Z^k = \operatorname{Ker} d_k, \qquad Z_k = \operatorname{Ker} \partial_k, \qquad B^k = \operatorname{Im} d_{k-1}, \qquad B_k = \operatorname{Im} \partial_{k+1},$$

and thus

$$H^k = \operatorname{Ker} d_k / \operatorname{Im} d_{k-1} \quad \text{and} \quad H_k = \operatorname{Ker} \partial_k / \operatorname{Im} \partial_{k+1}.$$

E.2.5 Remarks

A few words as a conclusion. I repeat that this is just an overview, an overview of the principles that does not go into details. The details are covered in the textbook, and numerous developments are given in the specialized literature, which is easier to read with an initial idea of the subject, of course.

In physics and mechanics, we often speak in the language of vector fields. However, you now know how to translate problems in the language of vector fields into the language of differential forms, and conversely you know how to relate standard operators like grad, curl, div with the operator d of the exterior differentiation of forms.

In continuum mechanics, the Hamiltonian operator ∇ is used. Some techniques that are used with it are presented in the text. There you will also find the answer to the question of how to represent and calculate the operators grad, curl, div in curvilinear coordinates.

All of this, including Stokes's formula, has numerous applications. For example, look at the deduction of Euler's equation in continuum mechanics, or write down Maxwell's equations for an electromagnetic field. I shall not mention the internal mathematical applications in analysis, especially complex analysis, geometry, algebraic topology...

References

1 Classic Works

1.1 Primary Sources

Newton, I.:

- a. (1687): Philosophiæ Naturalis Principia Mathematica. Jussu Societatis Regiæ ac typis Josephi Streati, London. English translation from the 3rd edition (1726): University of California Press, Berkeley, CA (1999).
- b. (1967–1981): The Mathematical Papers of Isaac Newton, D.T. Whiteside, ed., Cambridge University Press.

Leibniz, G.W. (1971): Mathematische Schriften. C.I. Gerhardt, ed., G. Olms, Hildesheim.

1.2 Major Comprehensive Expository Works

Euler, L.

- a. (1748): Introductio in Analysin Infinitorum. M.M. Bousquet, Lausanne. English translation: Springer-Verlag, Berlin – Heidelberg – New York (1988–1990).
- b. (1755): Institutiones Calculi Differentialis. Impensis Academiæ Imperialis Scientiarum, Petropoli. English translation: Springer, Berlin – Heidelberg – New York (2000).
- c. (1768–1770): Institutionum Calculi Integralis. Impensis Academiæ Imperialis Scientiarum, Petropoli.

Cauchy, A.-L.

- a. (1989): Analyse Algébrique. Jacques Gabay, Sceaux.
- b. (1840–1844): Leçons de Calcul Différential et de Calcul Intégral. Bachelier, Paris.

© Springer-Verlag Berlin Heidelberg 2016
V.A. Zorich, *Mathematical Analysis II*, Universitext,
DOI 10.1007/978-3-662-48993-2

1.3 Classical Courses of Analysis from the First Half of the Twentieth Century

Courant, R. (1988): Differential and Integral Calculus. Translated from the German. Vol. 1 reprint of the second edition 1937. Vol. 2 reprint of the 1936 original. Wiley Classics Library. A Wiley-Interscience Publication. John Wiley & Sons, Inc., New York.

de la Vallée Poussin, Ch.-J. (1954, 1957): Cours d'Analyse Infinitésimale. (Tome 1 11 éd., Tome 2 9 éd., revue et augmentée avec la collaboration de Fernand Simonart.) Librairie universitaire, Louvain. English translation of an earlier edition: Dover Publications, New York (1946).

Goursat, É. (1992): Cours d'Analyse Mathématiques. (Vol. 1 reprint of the 4th ed. 1924, Vol. 2 reprint of the 4th ed. 1925) Les Grands Classiques Gauthier-Villars. Jacques Gabay, Sceaux. English translation: Dover Publ. Inc., New York (1959).

2 Textbooks[1]

Apostol, T.M. (1974): Mathematical Analysis. 2nd ed. World Student Series Edition. Addison-Wesley Publishing Co., Reading, Mass. – London – Don Mills, Ont.

Courant, R., John F. (1999): Introduction to Calculus and Analysis. Vol. I. Reprint of the 1989 edition. Classics in Mathematics. Springer-Verlag, Berlin.

Courant, R., John F. (1989): Introduction to Calculus and Analysis. Vol. II. With the assistance of Albert A. Blank and Alan Solomon. Reprint of the 1974 edition. Springer-Verlag, New York.

Nikolskii, S.M. (1990): A Course of Mathematical Analysis. Vols. 1, 2. Nauka, Moscow. English translation of an earlier addition: Mir, Moscow (1985).

Rudin, W. (1976): Principals of Mathematical Analysis. McGraw-Hill, New York.

Rudin, W. (1987): Real and Complex Analysis. 3rd ed., McGraw-Hill, New York.

Spivak, M. (1965): Calculus on Manifolds: A Modern Approach to the Classical Theorems of Advanced Calculus. W. A. Benjamin, New York.

Whittaker, E.T., Watson, J.N. (1979): A Course of Modern Analysis. AMS Press, New York.

3 Classroom Materials

Biler, P., Witkowski, A. (1990): Problems in Mathematical Analysis. Monographs and Textbooks in Pure and Applied Mathematics, 132. Marcel Dekker, New York.

---

[1] For the convenience of the Western reader the bibliography of the English edition has substantially been revised.

Demidovich, B.P. (1990): A Collection of Problems and Exercises in Mathematical Analysis. Nauka, Moscow. English translation of an earlier edition: Gordon and Breach, New York – London – Paris (1969).

Gelbaum, B. (1982): Problems in Analysis. Problem Books in Mathematics. Springer-Verlag, New York – Berlin.

Gelbaum, B., Olmsted, J. (1964): Counterexamples in Analysis. Holden-Day, San Francisco.

Makarov, B.M., Goluzina, M.G., Lodkin, A.A., Podkorytov, A.N. (1992): Selected Problems in Real Analysis. Nauka, Moscow. English translation: Translations of Mathematical Monographs, 107, American Mathematical Society, Providence (1992).

Pólya, G., Szegő, G. (1970/1971): Aufgaben und Lehrsätze aus der Analysis. Springer-Verlag, Berlin – Heidelberg – New York. English translation: Springer-Verlag, Berlin – Heidelberg – New York (1972–1976).

4 Further Reading

Arnol'd, V.I.

  - a. (1989a): Huygens and Barrow, Newton and Hooke: Pioneers in Mathematical Analysis and Catastrophe Theory, from Evolvents to Quasicrystals. Nauka, Moscow. English translation: Birkhäuser, Boston (1990).
  - b. (1989b): Mathematical Methods of Classical Mechanics. Nauka, Moscow. English translation: Springer-Verlag, Berlin – Heidelberg – New York (1997).

Avez, A. (1986): Differential Calculus. Translated from the French. A Wiley-Interscience Publication. John Wiley & Sons, Ltd., Chichester.

Bourbaki, N. (1969): Éléments d'Histoire des Mathématiques. 2e édition revue, corrigée, augmentée. Hermann, Paris. English translation: Springer-Verlag, Berlin – Heidelberg – New York (1994).

Cartan, H. (1977): Cours de Calcul Différentiel. Hermann, Paris. English translation of an earlier edition: Differential Calculus. Exercised by C. Buttin, F. Riedeau and J.L. Verley. Houghton Mifflin Co., Boston, Mass. (1971).

de Bruijn, N.G. (1958): Asymptotic Methods in Analysis. North-Holland, Amsterdam.

Dieudonné, J. (1969): Foundations of Modern Analysis. Enlarged and corrected printing. Academic Press, New York.

Dubrovin, B.A., Novikov, S.P., Fomenko, A.T. (1986): Modern Geometry – Methods and Applications. Nauka, Moscow. English translation: Springer-Verlag, Berlin – Heidelberg – New York (1992).

Einstein, A. (1982): Ideas and Opinions. Three Rivers Press, New York. Contains translations of the papers "Principles of Research" (original German title: "Motive des Forschens"), pp. 224–227, and "Physics and Reality," pp. 290–323.

Evgrafov, M.A. (1979): Asymptotic Estimates and Entire Functions. 3rd ed. Nauka, Moscow. English translation from the first Russian edition: Gordon & Breach, New York (1961).

Fedoryuk, M.V. (1977): The Saddle-Point Method. Nauka, Moscow (Russian).

Feynman, R., Leighton, R., Sands, M. (1963–1965): The Feynman Lectures on Physics, Vol. 1: Modern Natural Science. The Laws of Mechanics. Addison-Wesley, Reading, Mass.

Gel'fand, I.M. (1998): Lectures on Linear Algebra. Dobrosvet, Moscow. English translation of an earlier edition: Dover, New York (1989).

Halmos, P. (1974): Finite-Dimensional Vector Spaces. Springer-Verlag, Berlin – Heidelberg – New York.

Jost, J. (2003): Postmodern Analysis. 2nd ed. Universitext. Springer, Berlin.

Klein, F. (1926): Vorlesungen über die Entwicklung der Mathematik im 19 Jahrhundert. Springer-Verlag, Berlin.

Kolmogorov, A.N., Fomin, S.V. (1989): Elements of the Theory of Functions and Functional Analysis. 6th ed., revised, Nauka, Moscow. English translation of an earlier edition: Graylock Press, Rochester, New York (1957).

Kostrikin, A.I., Manin, Yu.I. (1986): Linear Algebra and Geometry. Nauka, Moscow. English translation: Gordon and Breach, New York (1989).

Landau, L.D., Lifshits, E.M. (1988): Field Theory. 7th ed., revised, Nauka, Moscow. English translation of an earlier edition: Pergamon Press, Oxford – New York (1975).

Lax, P.D., Burstein S.Z., Lax A. (1972): Calculus with Applications and Computing. Vol. I. Notes based on a course given at New York University. Courant Institute of Mathematical Sciences, New York University, New York.

Manin, Yu. I. (1979): Mathematics and Physics. Znanie, Moscow. English translation: Birkhäuser, Boston (1979).

Milnor, J. (1963): Morse Theory. Princeton University Press.

Narasimhan, R. (1968): Analysis on Real and Complex Manifolds. Masson, Paris.

Olver, F.W.J. (1997): Asymptotics and Special Functions. Reprint. AKP Classics. A K Peters, Wellesley, MA.

Pham, F. (1992): Géometrie et calcul différentiel sur les variétés. [Course, studies and exercises for Masters in mathematics] Inter Editions, Paris.

Poincaré, H. (1982): The Foundations of Science. Authorized translation by George Bruce Halstead, and an introduction by Josiah Royce, preface to the UPA edition by L. Pearce Williams. University Press of America, Washington, DC.

Pontryagin, L.S. (1974): Ordinary Differential Equations. Nauka, Moscow. English translation of an earlier edition: Addison-Wesley, Reading, Mass. (1962).

Shilov, G.E.

  – a. (1996): Elementary Real and Complex Analysis. Revised English edition translated from Russian and edited by Richard A. Silverman. Corrected reprint of the 1973 English edition. Dover Publications Inc., Mineola, NY.

  – b. (1969): Mathematical Analysis. Functions of one variable. Pts. 1 and 2. Nauka, Moscow (Russian).

– c. (1972): Mathematical Analysis. Functions of several variables (3 pts.). Nauka, Moscow (Russian).

Schwartz, L. (1998): Analyse. Hermann, Paris.

Weyl, H. (1926): Die heutige Erkenntnislage in der Mathematik. Weltkreis-Verlag, Erlangen. Russian translations of eighteen of Weyl's essays, with an essay on Weyl: Mathematical Thought. Nauka, Moscow (1989).

Zel'dovich, Ya.B., Myshkis, A.D. (1967): Elements of Applied Mathematics. Nauka, Moscow. English translation: Mir, Moscow (1976).

Zorich, V.A. (2011): Mathematical Analysis of Problems in the Natural Sciences. Springer, Heidelberg.

Index of Basic Notation

Logical symbols

$\Longrightarrow$ logical consequence (implication)

$\Longleftrightarrow$ logical equivalence

$:=\,\Big\}$ equality by definition; colon
$=:$ on the side of the object defined

Sets

$\overline{E}$ closure of the set E

∂E boundary of the set E

$\overset{\circ}{E} := E \backslash \partial E$ interior of the set E

$B(x, r)$ ball of radius r with center at x

$S(x, r)$ sphere of radius r with center at x

Spaces

(X, d) metric space with metric d

(X, τ) topological space with system τ of open sets

$\mathbb{R}^n (\mathbb{C}^n)$ n-dimensional real (complex) space

$\mathbb{R}^1 = \mathbb{R}$ $(\mathbb{C}^1 = \mathbb{C})$ set of real (complex) numbers

$x = (x^1, \ldots, x^n)$ coordinate expression of a point of n-dimensional space

$C(X, Y)$ set (space) of continuous functions on X with values in Y

$C[a, b]$ abbreviation for $C([a, b], \mathbb{R})$ or $C([a, b], \mathbb{C})$

$C^{(k)}(X, Y)$ set of mappings from X into Y that are k times continuously differentiable

$C^{(k)}[a, b]$ abbreviation for $C^{(k)}([a, b], \mathbb{R})$ or $C^{(k)}([a, b], \mathbb{C})$

$C_p[a, b]$ space $C[a, b]$ endowed with norm $\|f\|_p$

$C_2[a, b]$ space $C[a, b]$ with Hermitian inner product $\langle f, g \rangle$ of functions or mean-square deviation norm

$\mathcal{R}(E)$ set (space) of functions that are Riemann integrable over the set E

$\mathcal{R}[a, b]$ space $\mathcal{R}(E)$ when $E = [a, b]$

© Springer-Verlag Berlin Heidelberg 2016
V.A. Zorich, *Mathematical Analysis II*, Universitext,
DOI 10.1007/978-3-662-48993-2

$\widetilde{\mathcal{R}}(E)$ space of classes of Riemann integrable functions on E that are equal almost everywhere on E

$\widetilde{\mathcal{R}}_p(E)(\mathcal{R}_p(E))$ space $\widetilde{\mathcal{R}}(E)$ endowed with norm $\|f\|_p$

$\widetilde{\mathcal{R}}_2(E)(\mathcal{R}_2(E))$ space $\widetilde{\mathcal{R}}(E)$ endowed with Hermitian inner product $\langle f, g\rangle$ or mean-square deviation norm

$\mathcal{R}_p[a, b], \mathcal{R}_2[a, b]$ spaces $\mathcal{R}_p(E)$ and $\mathcal{R}_2(E)$ when $E = [a, b]$

$\mathcal{L}(X; Y), (\mathcal{L}(X_1, \ldots, X_n; Y))$ space of linear (n-linear) mappings from X (from $(X_1 \times \cdots \times X_n)$) into Y

TM_p or $TM(p), T_pM, T_p(M)$ tangent space to the surface (manifold) M at the point $p \in M$

S Schwartz space of rapidly decreasing functions

$\mathcal{D}(G)$ space of fundamental functions of compact support in the domain G

$\mathcal{D}'(G)$ space of generalized functions on the domain G

$\mathcal{D}$ an abbreviation for $\mathcal{D}(G)$ when $G = \mathbb{R}^n$

$\mathcal{D}'$ an abbreviation for $\mathcal{D}'(G)$ when $G = \mathbb{R}^n$

Metrics, norms, inner products

$d(x_1, x_2)$ distance between points x_1 and x_2 in the metric space (X, d)

$|x|, \|x\|$ absolute value (norm) of a vector $x \in X$ in a normed vector space

$\|A\|$ norm of the linear (multilinear) operator A

$\|f\|_p := (\int_E |f|^p(x)\,\mathrm{d}x)^{1/p}$, $p \geq 1$ integral norm of the function f

$\|f\|_2$ mean-square deviation norm ($\|f\|_p$ when $p = 2$)

$\langle \mathbf{a}, \mathbf{b}\rangle$ Hermitian inner product of the vectors $\mathbf{a}$ and $\mathbf{b}$

$\langle f, g\rangle := \int_E(f \cdot \overline{g})(x)\,\mathrm{d}x$ Hermitian inner product of the functions f and g

$\mathbf{a} \cdot \mathbf{b}$ inner product of $\mathbf{a}$ and $\mathbf{b}$ in $\mathbb{R}^3$

$\mathbf{a} \times \mathbf{b}$ vector (cross) product of vectors $\mathbf{a}$ and $\mathbf{b}$ in $\mathbb{R}^3$

$(\mathbf{a}, \mathbf{b}, \mathbf{c})$ scalar triple product of vectors $\mathbf{a}, \mathbf{b}, \mathbf{c}$ in $\mathbb{R}^3$

Functions

$g \circ f$ composition of functions f and g

f^{-1} inverse of the function f

$f(x)$ value of the function f at the points x; a function of x

$f(x^1, \ldots, x^n)$ value of the function f at the point $x = (x^1, \ldots, x^n) \in X$ in the n-dimensional space X; a function depending on n variables $x^1, \ldots, x^n$

supp f support of the function f

$\int f(x)$ jump of the function f at the point x

$\{f_t : t \in T\}$ a family of functions depending on the parameter $t \in T$

$\{f_n; n \in \mathbb{N}\}$ or $\{f_n\}$ a sequence of functions

$f_t \xrightarrow[\mathcal{B}]{} f$ on E convergence of the family of functions $\{f_t; t \in T\}$ to the function f on the set E over the base $\mathcal{B}$ in T

$f_t \underset{\mathcal{B}}{\rightrightarrows} f$ on E uniform convergence of the family of functions $\{f_t; t \in T\}$ to the function f on the set E over the base B in T

$\left.\begin{array}{l} f = o(g) \text{ over } \mathcal{B} \\ f = O(g) \text{ over } \mathcal{B} \\ f \sim g \text{ or } f \simeq g \text{ over } \mathcal{B} \end{array}\right\}$ asymptotic formulas (the symbols of comparative asymptotic behavior of the functions f and g over the base $\mathcal{B}$)

$f(x) \simeq \sum_{n=1}^{\infty} \varphi_n(x)$ over $\mathcal{B}$ expansion in an asymptotic series

$\mathcal{D}(x)$ Dirichlet function

$\exp(A)$ exponential of a linear operator A

$B(\alpha, \beta)$ Euler beta function

$\Gamma(\alpha)$ Euler gamma function

χ_E characteristic function of the set E

Differential calculus

$f'(x), f_x(x), df(x), Df(x)$ tangent mapping to f (differential of f) at the point x

$\frac{\partial f}{\partial x^i}, \partial_i f(x), D_i f(x)$ partial derivative (partial differential) of a function f depending on variables $x^1, \ldots, x^n$ at the point $x = (x^1, \ldots, x^n)$ with respect to
the variable x^i

$D_{\mathbf{v}} f(x)$ derivative of the function f with respect to the vector $\mathbf{v}$ at the point x

∇ Hamilton's nabla operator

grad f gradient of the function f

div $\mathbf{A}$ divergence of the vector field $\mathbf{A}$

curl $\mathbf{B}$ curl of the vector field $\mathbf{B}$

Integral calculus

$\mu(E)$ measure of the set E

$\left.\begin{array}{l}\int_E f(x)\,dx \\ \int_E f(x^1, \ldots, x^n)\,dx^1 \cdots dx^n \\ \int \cdots \int_E f(x^1, \ldots, x^n)\,dx^1 \cdots dx^n\end{array}\right\}$ integral of the function f over the set $E \subset \mathbb{R}^n$

$\int_Y dy \int_X f(x, y)\,dx$ iterated integral

$\left.\begin{array}{l}\int_\gamma P\,dx + Q\,dy + R\,dz \\ \int_\gamma \mathbf{F} \cdot d\mathbf{s}, \int_\gamma \langle \mathbf{F}, d\mathbf{s}\rangle\end{array}\right\}$ curvilinear integral (of second kind) or the work of the field $\mathbf{F} = (P, Q, R)$ along the pathγ

$\int_\gamma f\,ds$ curvilinear integral (of first kind) of the function f along the curve γ

$\left.\begin{array}{l}\iint_S P\,dy \wedge dz + Q\,dz \wedge dx + R\,dx \wedge dy \\ \iint_S \mathbf{F} \cdot d\boldsymbol{\sigma}, \quad \iint_X \langle \mathbf{F}, d\boldsymbol{\sigma}\rangle\end{array}\right\}$ integral (of second kind) over the surface S in $\mathbb{R}^3$; flux of the field $\mathbf{F} = (P, Q, R)$ across the surface S

$\iiint_S f\,d\sigma$ surface integral (of first kind) of f over the surface S

Differential forms

$\omega\ (\omega^p)$ a differential form (of degree p)

$\omega^p \wedge \omega^q$ exterior product of forms ω^p and ω^q

$d\omega$ (exterior) derivative of the form ω

$\int_M \omega$ integral of the form ω over the surface (manifold) M

$\omega_{\mathbf{F}}^1 := \langle \mathbf{F}, \cdot \rangle$ work form

$\omega_{\mathbf{V}}^2 := (\mathbf{V}, \cdot, \cdot)$ flux form

Subject Index

Name Index

A

Abel, N., 376, 378, 383, 418
Alexander, J., 164
Ampére, A., 235
Archimedes ('Aρχιμήδης), 245
Arzelà, C., 396, 402
Ascoli, G., 396, 402

B

Banach, S., 35
Bernoulli, D., 310
Bernoulli, Jacob, 548, 626
Bernoulli, Johann, 93
Bessel, F., 390, 408, 411, 430, 502
Biot, J., 237
Borel, É., 573
Brouwer, L., 180, 243, 322
Brunn, H., 123
Bunyakovskii, V., 46, 448, 499

C

Cantor, G., 23, 33, 113
Carleson, L., 524
Carnot, S., 227
Cartan, É., 251
Cauchy, A., 21, 30, 46, 157, 288, 310, 369, 374, 375, 377, 419, 448, 499
Cavalieri, B., 134
Cesàro, E., 393
Chebyshev, P., 3, 518
Clapeyron, E., 223
Clausius, R., 227
Coulomb, Ch., 280, 289

D

D'Alembert, J., 307
Darboux, G., 116–118, 122, 131, 394

De Rham, G., 360
Dini, U., 384, 426, 528, 563
Dirac, P., 238, 281, 459
Dirichlet, P., 287, 364, 376, 378, 379, 418, 431, 432, 443, 513, 525, 551, 563, 601

E

Earnshaw, S., 286
Erdélyi, A., 601, 625
Euler, L., 92, 307, 429, 433, 436, 438, 441, 524, 560, 572, 601

F

Faraday, M., 235, 238
Federer, H., 237
Fejér, L., 531
Feynman, R., 262
Fourier, J., 493–586, 624, 625
Fréchet, M., 11
Frenet, J., 73
Fresnel, A., 431, 443, 601
Fubini, G., 129, 572

G

Galilei, Galileo, 587
Gauss, C., 238, 243, 250, 278, 279, 285, 391, 436, 443, 466, 483, 491
Gibbs, J., 538, 549
Gram, J., 187, 497, 504
Grassmann, H., 319
Green, G., 238, 285, 465, 491

H

Haar, A., 519
Hadamard, J., 375, 412
Hamilton, W., 262, 265
Hardy, G., 393

© Springer-Verlag Berlin Heidelberg 2016
V.A. Zorich, *Mathematical Analysis II*, Universitext,
DOI 10.1007/978-3-662-48993-2

Printed in the United States
By Bookmasters